Metabolic Engineering

PRINCIPLES AND METHODOLOGIES

Metabolic Engineering

PRINCIPLES AND METHODOLOGIES

Gregory N. Stephanopoulos

Department of Chemical Engineering
Massachusetts Institute of Technology
Cambridge, Massachusetts

Aristos A. Aristidou

Department of Chemical Engineering
Massachusetts Institute of Technology
Cambridge, Massachusetts

Jens Nielsen

Department of Biotechnology
Technical University of Denmark
Lyngby, Denmark

ACADEMIC PRESS

San Diego   London   Boston   New York   Sydney   Tokyo   Toronto

Copyright © 1998 by ACADEMIC PRESS

All Rights Reserved.
No part of this publication may be reproduced or transmitted in any form or by any
means, electronic or mechanical, including photocopy, recording, or any information
storage and retrieval system, without permission in writing from the publisher.

Academic Press
a division of Harcourt Brace & Company
525 B Street, Suite 1900, San Diego, California 92101-4495, USA
http://www.apnet.com

Academic Press
24-28 Oval Road, London NW1 7DX, UK
http://www.hbuk.co.uk/ap/

Library of Congress Catalog Card Number: 98-84372

International Standard Book Number: 0-12-666260-6

PRINTED IN THE UNITED STATES OF AMERICA
98 99 00 01 02 03 MM 9 8 7 6 5 4 3 2 1

To Our Families

CONTENTS

PREFACE

Metabolic engineering is about the analysis and modification of metabolic pathways. The field emerged during the past decade, and powered by techniques from applied molecular biology and reaction engineering, it is becoming a focal point of research activity in biological and biochemical engineering, cell physiology, and applied microbiology. Although the notion of pathway manipulation had been discussed before, the vision of metabolic engineering as defining a discipline in its own right was first suggested by Bailey in 1991. It was embraced soon thereafter by both engineers and life scientists who saw in it the opportunity to capture the potential of sequence and other information generated from genomics research.

We first attempted to convey the excitement and basic concepts of metabolic engineering to our students in a course that was taught at MIT in 1993. The experiment was repeated again in 1995 and 1997, at which time a definite syllabus and tentative set of notes had emerged as a result of these offerings. A similar development occurred at the Technical University of Denmark (DTU), where metabolic engineering has been a central topic in biochemical engineering courses at both the undergraduate and the graduate level. In 1996, a standard one-semester course on metabolic engineering was offered for the first time. The growing interest in metabolic engineering and requests to share the course material led us to the decision to write this book. In so doing, we have tried to formulate a framework of quantitative biochemistry for the analysis of pathways of enzymatic reactions. In this sense, the book reflects a shift of focus from equipment toward single cells, as it concentrates on the elucidation and manipulation of their biochemical functions. As such, this text can support a graduate or advanced undergraduate course on metabolic engineering to complement current offerings in biochemical engineering.

The book manuscript was used to teach courses on metabolic engineering at MIT and DTU, as well as a summer course at MIT. The material can be covered in a single semester with no prerequisites, although some prior exposure to an introductory biochemistry course is helpful. Assigned readings from biochemistry texts during the first quarter of the semester can complement the first part of the book. Problem sets aiding the understanding of basic concepts will be posted periodically at the web site listed below. Although the focus of this book is on metabolism, the concepts of pathway analysis are broad and as such generally applicable to other types of reaction sequences, including those involved in protein expression and post-translational modification or in signal transduction pathways.

Writing a book on a subject that is still in its formative stage is a challenge that carries with it increased responsibility. For this reason, we set as our goal to define core *principles* central to pathway design and analysis, complemented with specific *methods* derived from recent research. We expect these methods to evolve further and hope that this book will play a role in catalyzing such activity. Software implementing the various methods can be found at the book web site, http://www.cpb.dtu.dk/cpb/metabol.htm, with hyperlinks to other sites where public domain software is available. To facilitate broad interdisciplinary participation, only codes with a minimum of service and user-friendliness have been selected. Furthermore, the mathematical complexity of the book has been kept to an absolute minimum, and background material has been provided wherever possible to assist the less mathematically inclined. We are aware of the challenges of this task and the difficulties in satisfying all segments of the readership spectrum. We encourage readers to continue their review of the book undeterred by any temporary difficulties.

We are indebted to many individuals for their direct contributions or indirect input in planning and executing this project. First, we thank our students for their boundless energy and refreshing creativity, in particular Maria Klapa, for a thorough review of metabolic flux analysis, and Troy Simpson, whose research provided the basis for complex pathway analysis. Also, we thank Martin Bastian Pedersen for drafting many of the figures, and Christian Müller, Susanne Sloth Larsen, Birgitte Karsbøl, and Kristen Nielsen for their help in finalizing the manuscript. We thank our colleagues, in particular Tony Sinskey, for his enthusiasm about the unlimited possibilities of metabolic engineering, and Sue Harrison and Eduardo Agosin for their most constructive comments. Finally, we thank our collaborators and friends, in particular, Barry Buckland, Bernhard Palsson, John Villadsen, Maish Yarmush, and D. Ramkrishna. Their vision and unwavering support when it mattered meant a lot.

Gregory N. Stephanopoulos
Aristos A. Aristidou
Jens Nielsen

LIST OF SYMBOLS

Below is a list of the symbols most frequently used throughout this text. The unit specified is the most typically applied unit, in some cases the symbols may have another unit.

a_{cell} Specific surface area of the cells (m^2 $(g\,DW)^{-1}$)

a   Row vector containing weights of individual variables on the objective function in eq. (8.26)

A_i Affinity of the ith reaction ($kJ\,mole^{-1}$)

c Concentration ($mmoles\,L^{-1}$)

c_i Concentration of the ith compound ($mmoles\,L^{-1}$)

c_i^f Concentration of the ith compound in the feed to the bioreactor ($mmoles\,L^{-1}$)

$C_i^{J_j}$ Flux control coefficient for the ith enzyme on the jth steady state flux J_j

$*C_i^{J_j}$ Group flux control coefficient for the ith group on the steady state flux J_j

$C_i^{X_j}$ Concentration control coefficient for the ith enzyme on the jth metabolite concentration

$*C_i^{X_j}$ Group concentration control coefficient for the ith group on the jth metabolite concentration

C^J Matrix containing flux control coefficients

C^X Matrix containing concentration control coefficients

d_{mem} Thickness of the cytoplasmic membrane (m)

D Dilution rate (h^{-1})

D_{mem} Diffusion coefficient for membrane diffusion ($m^2\,s^{-1}$)

D_i^J Deviation index given by eq. (11.84)

E_i Activity (or concentration) of the ith enzyme

E   Elemental composition matrix or matrix containing elasticity coefficients

E_c Elemental composition matrix for non-measured compounds
E_m Elemental composition matrix for measured compounds
f Flux amplification factor given by eq. (11.87)
f_{ij} The ratio of flux j to flux i (given by eq. (11.50))
F Volumetric flow rate into the bioreactor (L h^{-1})
F_{out} Volumetric flow rate out of the bioreactor (L h^{-1})
F Variance-covariance matrix
g_{ij} Stoichiometric coefficient for the ith intracellular metabolite in the jth reaction
G Gibbs function (kJ mole^{-1})
ΔG Gibbs free energy change (kJ mole^{-1})
$\Delta G^{0,}$ Gibbs free energy change with all reactants and products at their standard states (kJ mole^{-1})
G Matrix containing the stoichiometric coefficients for the intracellular metabolites
G_c Matrix containing the stoichiometric coefficients for intracellular metabolites in reactions for which fluxes are not measured
G_m Matrix containing the stoichiometric coefficients for intracellular metabolites in reactions for which fluxes are measured
G_{ex} Stoichiometric matrix for a metabolic model containing the stoichiometry for all reactions both in a forward and in a reverse direction
h Test function given by eq. (4.29)
$h_{s,i}$ Carbon content in the ith substrate (C-moles mole^{-1})
$h_{p,i}$ Carbon content in the ith metabolic product (C-moles mole^{-1})
H Enthalpy function (kJ mole^{-1})
H_j Thermodynamic function given by eq. (14.29)
I Identity matrix, $i.e.$, matrix with all diagonal elements being 1 and all other elements being 0
j Flow ratio given by eq. (14.46)
J_i Steady state flux through the ith pathway branch (mmoles (g DW h)$^{-1}$)
J Vector of steady state fluxes (mmoles (g DW h)$^{-1}$)
J_{dep} Vector of dependent fluxes (mmoles (g DW h)$^{-1}$)
J_{in} Vector of independent fluxes (mmoles (g DW h)$^{-1}$)
k Rate constant (h^{-1})
K The number of intracellular metabolites considered in the analysis
K_{eq} Equilibrium constant
K_{par} Partitioning coefficient between the lipid membrane and the medium (dimensionless)

K_m Michaelis-Menten constant (or saturation constant) (mmoles L^{-1})

K_i Inhibition constant (mmoles L^{-1})

$\mathbf{K}$ Kernel matrix which fulfills eq. (12.7)

L_{ij} Phenomenological coefficients

m_{ATP} ATP requirements for maintenance metabolism (mmoles ATP $(g\,DW\,h)^{-1}$)

M The number of metabolic products considered in the analysis

N The number of substrates considered in the analysis

p Parameters that influence reaction rate (used in eq. (11.5))

P Permeability coefficient (m s^{-1})

P_i The ith metabolic product

$\mathbf{P}$ Variance-covariance matrix of residuals (given by eq. (4.24)) or matrix containing the parameter elasticity coefficients

q The degree of coupling

Q The number of macromolecular pools considered in the analysis

Q_{heat} Heat production associated with biomass growth (kJ (C-moles biomass)$^{-1}$)

r Specific rate (mmoles $(g\,DW\,h)^{-1}$)

r_{ATP} Rate of ATP production (mmoles $(g\,DW\,h)^{-1}$)

r_i Activity amplification factor given by eq. (11.86)

$r_{macro,i}$ Specific rate of formation of the ith macromolecular pool $(g\,(g\,DW\,h)^{-1})$

$r_{met,i}$ Specific rate of formation of the ith intracellular metabolite (mmoles $(g\,DW\,h)^{-1}$)

r_p Specific product formation rate (mmoles $(g\,DW\,h)^{-1}$)

r_s Specific substrate uptake rate (mmoles $(g\,DW\,h)^{-1}$)

r_{tran} Specific rate of transport across the cytoplasmic membrane (mmoles $(g\,DW\,h)^{-1}$)

$\mathbf{r}_c$ Vector of non-measured specific rates (mmoles $(g\,DW\,h)^{-1}$)

$\mathbf{r}_m$ Vector of measured specific rates (mmoles $(g\,DW\,h)^{-1}$)

$\mathbf{r}_{macro}$ Vector containing the specific rates of macromolecular formation $(g\,(g\,DW\,h)^{-1})$

$\mathbf{r}_{met}$ Vector containing the specific rates of intracellular metabolite formation (mmoles $(g\,DW\,h)^{-1}$)

$\mathbf{r}_p$ Vector containing the specific rates of metabolic product formation (mmoles $(g\,DW\,h)^{-1}$)

$\mathbf{r}_s$ Vector containing the specific rates of substrate uptake (mmoles $(g\,DW\,h)^{-1}$)

R Gas constant (= 0.008314 kJ (K-mole)$^{-1}$)

R_i Activity amplification parameter given by eq. (13.40)

$R_{X_i}^{J_j}$ Response coefficient given by eq. (11.7)

$\mathbf{R}$ Redundancy matrix given by eq. (4.17)

$\mathbf{R}_r$ Reduced redundancy matrix containing independent rows of $\mathbf{R}$

S Entropy function $(kJ \, (K \, mole)^{-1})$

S_i The ith substrate

T Temperature (K)

$\mathbf{T}$ Matrix containing stoichiometric coefficients as specified by eq. (8.12)

v_j Specific rate of the jth relation $(mmoles \, (g \, DW \, h)^{-1})$

$*v^i$ Overall specific rate (or activity) of the ith reaction group $(mmoles \, (g \, DW \, h)^{-1})$

v_{max} Maximum specific rate of an enzyme catalyzed reaction $(mmoles \, h^{-1})$

$\mathbf{v}$ Vector of reaction rates (or intracellular steady state fluxes) $(mmoles \, (g \, DW \, h)^{-1})$

$\mathbf{v}_c$ Vector of non-measured reaction rates $(mmoles \, (g \, DW \, h)^{-1})$

$\mathbf{v}_m$ Vector of measured reaction rates $(mmoles \, (g \, DW \, h)^{-1})$

V Volume of the bioreactor (L)

x Biomass concentration $(g \, L^{-1})$

$X_{macro, i}$ Concentration of ith macromolecular pool (units) $(g \, (g \, DW)^{-1})$

$X_{met, i}$ Concentration of ith intracellular metabolite

Y_{ij} Yield coefficient $(mmoles \, j \, (mmole \, i)^{-1})$

Y_{ij}^{true} The true yield coefficient $(mmoles \, j \, (mmole \, i)^{-1})$

Y_{xATP} ATP requirement for cell growth $(mmoles \, ATP \, (g \, DW)^{-1})$

$Y_{xATP, growth}$ ATP requirement for cell synthesis $(mmoles \, ATP \, (g \, DW)^{-1})$

$Y_{xATP, lysis}$ ATP requirement for cell growth dissipated due to cell lysis $(mmoles \, ATP \, (g \, DW)^{-1})$

$Y_{xATP, leak}$ ATP requirement for leaks and futile cycles $(mmoles \, ATP \, (g \, DW)^{-1})$

Z The phenomenological stoichiometry given by eq. (14.48)

GREEK LETTERS

α_{ji} Stoichiometric coefficient for the ith substrate in the jth reaction

$\mathbf{A}$ Matrix containing the stoichiometric coefficients for the substrates

β_{ji} Stoichiometric coefficient for the ith metabolic product in the jth reaction

$\mathbf{B}$ Matrix containing the stoichiometric coefficients for the metabolic products

χ Force ratio given by eq. (14.49)

χ_i Parameters in eq. (11.78)

δ Vector of measurement errors

ε Vector of residuals given by eq. (4.20)

$\varepsilon_{X_j}^i$ Elasticity coefficient given by eq. (11.11)

ϕ_i^j Metabolite amplification factor given by eq. (11.103)

Φ_i Dissipation function (kJ mole^{-1})

γ_{ji} Stoichiometric coefficient for the ith macromolecular pool in the jth reaction

Γ Matrix containing the stoichiometric coefficients for the macromolecular pools

η_{th} Thermodynamic efficiency

κ Generalized degree of reduction

μ Specific growth rate (h^{-1})

μ_i Chemical potential of the ith compound (kJ mole^{-1})

μ_i^0 Chemical potential of the ith compound at the reference state (kJ mole^{-1})

$\pi_{p_j}^i$ Parameter elasticity coefficient given by eq. (11.19)

τ Characteristic time (h)

CHAPTER 1

The Essence of Metabolic Engineering

The concept of metabolic pathway manipulation for the purpose of endowing microorganisms with desirable properties is a very old one indeed. We have many outstanding examples of this strategy in the areas of amino acids, antibiotics, solvents, and vitamin production. These methods rely heavily on the use of chemical mutagens and creative selection techniques to identify superior strains for achieving a certain objective. Despite widespread acceptance and impressive successes, the genetic and metabolic profiles of mutant strains were poorly characterized and mutagenesis remained a random process where science was complemented with elements of art.

The development of molecular biological techniques for deoxyribonucleic acid (DNA) recombination introduced a new dimension to pathway manipulation. Genetic engineering allowed the precise modification of specific enzymatic reaction(s) in metabolic pathways and, hence, the construction of well-defined genetic backgrounds. Shortly after the feasibility of recombinant DNA technology was established, various terms were coined to represent the potential applications of this technology to *directed* pathway modification. Some of the terms suggested were molecular breeding (Kellogg *et al.*, 1981), *in vitro* evolution (Timmis *et al.*, 1988), (microbial or metabolic) pathway engineering (MacQuitty, 1988; Tong *et al.*, 1991), cellular engineering (Nerem, 1991), and metabolic engineering (Stephanopoulos and Vallino, 1991; Bailey, 1991). Although the exact definition varies from author to author, all convey similar meanings with respect to the general goals and means of metabolic engineering. Here we define metabolic engineering as *the directed improvement of product formation or cellular properties through the modification of specific biochemical reaction(s) or the introduction of new one(s) with the use of recombinant DNA technology.* An essential characteristic of the preceding definition is the *specificity* of the particular biochemical reaction(s) targeted for modification or to be newly introduced. Once such reaction targets are identified, established molecular biological techniques are applied in order to amplify, inhibit or delete, transfer, or deregulate the corresponding genes or enzymes. DNA recombination in a broader sense is routinely employed at various steps toward these ends.

Although a certain sense of direction is inherent in all strain improvement programs, the directionality of effort is a strong focal point of metabolic engineering compared to random mutagenesis, as it plays a dominant role in enzymatic target selection, experimental design, and data analysis. On the other hand, *direction* in cell improvement should not be interpreted as *rational* pathway design and modification, in the sense that it is totally decoupled from random mutagenesis. In fact, strains that are obtained by random mutation and exhibit superior properties can be the source of critical information about pathway configuration and control, extracted via *reverse metabolic engineering.*

As with all traditional fields of engineering, metabolic engineering too encompasses the two defining steps of *analysis* and *synthesis.* Because metabolic engineering emerged with DNA recombination as the enabling technology, attention initially was focused, almost exclusively, on the synthetic side of this field: expression of new genes in various host cells, amplification of endogenous enzymes, deletion of genes or modulation of enzymatic activity, transcriptional or enzymatic deregulation, etc. As such, metabolic engineering was, to a significant extent, the technological manifestation of applied molecular biology with very little engineering content. Bioprocess considerations do not qualify as metabolic engineering. A more significant engineering component can be found in the analytical side of metabolic engineering: How does one identify the important parameters that define the physiological state? How does one utilize this information to elucidate the control architecture of a metabolic network and then propose rational targets for modification to achieve a certain objective? How does one further assess the true biochemical impact of such genetic and enzymatic modifications in order to design the next round of pathway modifications and so on until the goal is attained? Instead of the mostly *ad hoc* target selection process, can one prescribe a rational process to identify the most promising targets for metabolic manipulation? These are some of the questions that the analytical side of metabolic engineering would address.

On the synthetic side, another novel aspect of metabolic engineering is the focus on *integrated* metabolic pathways instead of individual reactions. As such, it examines complete biochemical reaction networks, concerning itself with issues of pathway synthesis and thermodynamic feasibility, as well as pathway flux and its control. We thus are witnessing a paradigm shift away from individual enzymatic reactions and toward systems of interacting biochemical reactions. In this regard, the notion of the *metabolic network* is of central importance in the sense that an enhanced perspective of metabolism and cellular function can be obtained by considering a system of reactions in its entirety rather than reactions in isolation from one another. Through metabolic engineering, attention is shifted to the whole system instead of its constituent parts. In this regard, metabolic engineering seeks to synthesize and design using techniques and information developed from extensive reductionist research. In turn, observations about the behavior of the overall system are the best guide for further rational decomposition and analysis.

Although metabolism and cell physiology provide the main context for analyzing reaction pathways, it should be pointed out that results of flux determination and control have broader applicability. Thus, besides the analysis of material and energy fluxes through metabolic pathways, the concepts of metabolic engineering are equally applicable to the analysis of *information fluxes* as those encountered in signal transduction pathways. Because the latter have not yet been well-defined, the main focus of this book is on applications to metabolic pathways. However, once the concepts of information pathways have crystallized, we expect that many of the ideas and tools presented herein will find good use in the study of the interactions of signal transduction pathways and the elucidation of the complex mechanisms by which external stimuli control gene expression.

Perhaps the most significant contribution of metabolic engineering is the emphasis it places on metabolic fluxes and their control under *in vivo* conditions. To be sure, the concept of metabolic flux *per se* is not new. Metabolic flux and its control have occupied the attention of a small but forward thinking group of researchers in biochemistry for approximately 30 years. As a result of their work, ideas on metabolic control matured and were rigorously defined, although they were not always broadly embraced by traditional biochemists. Metabolic engineering, initially conceived as the *ad hoc* pathway manipulation, quickly became the natural outlet for the analytical skills of engineers who saw the opportunity for introducing rigor in this process by utilizing the available platform of metabolic control analysis. *The combination of analytical methods to quantify fluxes and their control with molecular biological techniques to implement suggested genetic modifications is the essence of metabolic engineering.* When practiced in an iterative manner, it provides a powerful method for the systematic improvement of cellular properties over a broad range of contexts and applications.

The *flux* is a fundamental determinant of cell physiology and the most critical parameter of a metabolic pathway. For the linear pathway of Fig. 1.1a, its flux, J, is equal to the rates of the individual reactions at *steady state*. Obviously, a steady state will be reached when the intermediate metabolites adjust to concentrations that make all reaction rates equal $(v_1 = v_2 = \cdots v_i \cdots = v_L)$. During a transient, the individual reaction rates are not equal and pathway flux is variable and ill-defined (usually by the time varying rate of substrate uptake or product formation). For the branched

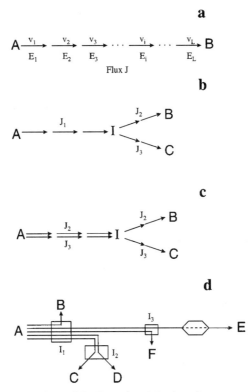

FIGURE 1.1   Examples of simple pathways.

pathway of Fig. 1.1b splitting at intermediate I, we have two additional fluxes for each of the branching pathways, related by $J_1 = J_2 + J_3$ at steady state. The flux of each branch is equal to the rates of the individual reactions at the corresponding branch. It is often convenient to think of flux J_1 as the superposition of the linear pathway fluxes J_2 and J_3, as shown in Fig. 1.1c. In this way, a complex network like the one depicted schematically in Fig. 1.1d can be decomposed into a number of linear pathways, each with its own flux as shown. It should be noted that, for all pathways of Fig. 1.1, a necessary condition to be able to reach steady state is that the rates of the initial and final reactions (or, equivalently, the concentrations of the initial and final metabolites, A and B, C, etc., respectively) must be constant. This is usually accomplished by constant extracellular metabolite concentrations in a continuous bioreactor, often referred to as a chemostat.

As metabolic pathways and their fluxes are at the core of metabolic engineering, it is important to elaborate a little more on their definition and meaning. We define *a metabolic pathway to be any sequence of feasible and observable biochemical reaction steps connecting a specified set of input and output metabolites.* The pathway flux is then defined as the rate at which input metabolites are processed to form output metabolites. The importance of *feasibility* and *observability* should be noted. First, it would be of little value to create nonsense reaction sequences comprising enzymes that are not present in a cell. Similarly, no more valuable is the enumeration of feasible reaction sequences between substrates and products that, however, cannot be observed experimentally. This is a very important point in light of the diversity and complexity of the metabolic maps that have been constructed as result of pioneering research in biochemistry during the previous 50 years. Although there is often more than one bioreaction sequence between speci-fied input and output metabolites, if the fluxes of these sequences cannot be determined independently, their inclusion provides no additional informa-tion. In many ways it is better if these reaction sequences are lumped together in fewer pathways whose fluxes can be observed. In the example of Fig. 1.1d, if the flux of each branch leading to the formation of metabolite E cannot be measured experimentally or otherwise determined, the two branches must be lumped into a single pathway shown by the dashed line. Clearly, the use of more informative measurements, such as those that are able to differentiate between the preceding two branches, should be encour-aged as they enhance the resolution of biochemical pathways that can be observed through them. The determination of metabolic fluxes *in vivo* has been termed metabolic flux analysis (MFA) and is of central importance to metabolic engineering.

In this framework of metabolic pathways and fluxes, a fundamental objective of metabolic engineering is to elucidate the factors and mechanisms responsible for the *control of metabolic flux.* A better understanding of the control of flux provides the basis for rational modification of metabolic pathways. There are three steps in the process for the systematic investigation of metabolic fluxes and their control. The first is to develop the means to observe as many pathways as possible and to measure their fluxes. To this end, one starts with simple material balances based on the measurements of concentrations of extracellular metabolites. The measurement of metabolites A-F of the network of Fig. 1.1d allows one to determine the five indicated fluxes but not the pathway split before F. If, upon administration of a labeled precursor A, metabolite F is labeled differently when it is formed by a particular branch, then this method could provide information about the split flux ratio at the branch point before F. This is one of several techniques that can be applied to provide additional information about branching pathways, and they are discussed at considerable length in this book. It is essential to emphasize that the flux of a metabolic pathway is not the same as the enzymatic activity of one or more of the enzymes in the pathway. In fact, enzymatic assays provide no information about the actual flux of the pathway other than that the corresponding enzyme is present and active under the *in vitro* assay conditions. Their inclusion in metabolic studies has often been misinterpreted to imply a metabolic flux of similar magnitude, which is certainly incorrect to generally conclude.

The second step is to introduce well-defined perturbations to the bioreaction network and to determine the pathway fluxes after the system relaxes to its new steady state. Because all flux control investigations quickly focus on a particular metabolic branch point, it is convenient to think of flux control in terms of the schematic branched pathway of Fig. 1.2. Three perturbations are required, in general, to study this branch point, each one originating in each of the corresponding branches. The ideal perturbation would involve a chemostat, where the activity of an enzyme is suddenly perturbed (through, for example, the use of an inducible promoter) after the system has reached a steady state. This arrangement is most applicable to microbial and cell culture systems. Different experimental configurations will be needed for

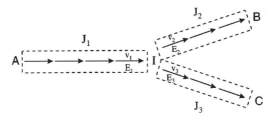

FIGURE 1.2   Branched pathway where the individual reactions are grouped.

other types of systems such as plants and organ function *in vivo*. Other types of perturbations that are easier to implement, such as the addition of a pulse of substrate or switching to a different carbon source, can be also informative. Perturbations should be targeted toward enzymes close to the branch point, although any other perturbation with an appreciable effect on the corresponding branch flux should be acceptable. Finally, one should note that one perturbation can provide information about more than one branch point, and this is useful for minimizing the number of experiments required to elucidate the control structure of a realistic metabolic network.

The third and final step in flux control determination is analysis of the flux perturbation results. Clearly, perturbation of the flux of each of the three branches of Fig. 1.2 allows one to probe the flexibility of the particular branch point. For example, if a large perturbation of flux J_1 has no appreciable effect on the magnitude or distribution of the other two fluxes (flux split ratio), then obviously one is dealing with a branch point that is *rigid* with respect to upstream perturbations. In such a case it will be futile to attempt to alter the downstream fluxes by changing the activity of the upstream enzymes. The other two branches can be analyzed similarly. It is important to understand how a flux perturbation, initiated through either a change in the rate of a reaction (induction), the introduction of a larger flux at the entire pathway (substrate pulse), or other methods, will propagate through the branch point and, in turn, through the network. Perturbations initiated at one of the branches propagate through the network metabolites. For example, the flux increase of J_1 may or may not cause a significant change in the concentration of the branch point metabolite I, reflecting, in turn, the degree of control exercised by the flux on that metabolite's level. On the other hand, even if I changes, there may be little effect on the fluxes J_2 and J_3 if the latter are tightly regulated or weakly affected by the concentration of metabolite I at the particular steady state. This is one of many scenarios that may unfold in the analysis of flux control of branch points and metabolic networks. The importance of elucidating flux control stems from the fact that each flux control architecture will dictate a different strategy for flux amplification or change of the flux split ratio.

The understanding of metabolic flux control is a key objective of metabolic engineering. An important aid in this endeavor is the framework of metabolic control analysis (MCA) (Kacser and Burns, 1973; Heinrich and Rapoport, 1974) developed in the 1970s for the quantitative representation of the degree of flux control exercised by the pathway enzymatic activities, metabolites, effectors, and other parameters. We will provide a comprehensive coverage of MCA in this book. Furthermore, we will present extensions of the basic theory to complex metabolic networks mainly through the top-down

MCA and the grouping of reactions. As illustrated in Fig. 1.2, the main tenet of this approach is to *group* individual reactions around a branch point metabolite (such as I) and then describe the degree of control exercized by the groups of reactions through the introduction of *group control coefficients*. Critical to the success of this approach is to eliminate cross-interactions among reaction groups by restricting all communication through the common branch point metabolite I. One therefore needs to be careful about the way in which the various reaction groups are defined, and guidelines to facilitate the top-down analysis are provided in this book. It should be noted that the successful application of reaction grouping and top-down MCA can be instrumental in locating the principal determinants of metabolic flux control within a well-defined reaction group comprising a much smaller number of enzymes than that which one is typically required to contend. Furthermore, this focused investigation arises as the result of a rational analysis and not an *ad hoc* approach that is largely influenced by single-reaction considerations and ignores the *systemic* characteristics of the metabolic network.

After the key parameters of flux control have been determined, one needs to implement those changes that are likely to be most effective in achieving a certain objective. Various tools are available for this purpose. Although one may think primarily of genetic modifications, the *bioreactor control equivalents* of genetic alterations should not be overlooked. If, for example, metabolite B of Fig. 1.2 is the product of interest and branch point I is *flexible* with respect to flux changes of the branch leading to metabolite C, then a better producer of B is obtained by eliminating the first enzyme (E_3) in the C branch. This genetic change will yield a C auxotroph, and metabolite C will have to be fed under control to optimally balance the rates of cell growth and production. An alternative is to retain a small activity of enzyme E_3 so that metabolite C is synthesized endogenously and the need for control is eliminated. Regardless of the route that one takes in the implementation of a flux control strategy, a combination of genetic and environmental modifications usually is required for optimal results.

It is clear that metabolic engineering is a highly multidisciplinary field (Cameron and Tong, 1993). Biochemistry has provided the basic metabolic maps and a wealth of information about the mechanisms of biochemical reactions, as well as their stoichiometry, kinetics, and regulation. Additionally, the whole area of metabolic control analysis started and took root in biochemistry before gaining acceptance in engineering circles. Genetics and molecular biology provide the tools applied in the construction of well-characterized genetic backgrounds, an important step in studies of flux control. Furthermore, these two disciplines are instrumental in executing the genetic changes necessary for the construction of superior production strains. One should recall the importance of recombinant DNA technology in the emergence of metabolic engineering. In fact, as an enabling technology it provided the main impetus for the definition and advancement of this field. Cell physiology has provided a more integrated view of cellular metabolic function, thus defining the platform for the study of metabolic rates and representation of physiological states. Finally, chemical engineering is the most suitable conduit for the application of the engineering approach to the study of biological systems. *In a general sense, this approach infuses the concepts of integration, quantitation, and relevance in the study of biological systems.* More specifically, it provides the tools and experience for the analysis of systems where rate processes are limiting, a field in which chemical engineering has strongly contributed and frequently excelled.

The multidisciplinary nature of metabolic engineering is certainly an advantage in meeting its goals. At the same time, it makes it imperative to identify those elements that distinguish metabolic engineering from related fields. In our view, such unique characteristics are the concept of a metabolic pathway, its flux, and the factors that control metabolic fluxes in both simple pathways and more realistic metabolic networks. In addition, one must also include experimental and theoretical methods that can be employed for the determination of metabolic fluxes *in vivo*, including material balances, use of isotopic labels, and application of spectroscopic methods (like nuclear magnetic resonance) and their variations (like gas chromatography-mass spectrometry) for the measurement of isotopic enrichment and/or molecular weight distributions of key metabolites. As these measurements are dependent upon the magnitude of the pathway fluxes contributing to their formation, one could use this information to infer metabolic fluxes. The concept of pathway flux, methods for the determination of fluxes *in vivo*, and conclusions that can be reached from the systematic study of metabolic fluxes have been collectively termed metabolic flux analysis. They occupy a considerable place in the overall study of flux control and are presented in detail in this book.

1.1. IMPORTANCE OF METABOLIC ENGINEERING

Metabolic engineering is a field of broad fundamental and practical importance. Basic contributions of metabolic engineering are in the measurement and understanding of the control of flux *in vivo*. As mentioned previously, metabolic maps, despite their elaborate form, convey very little information about the actual fluxes of carbon, nitrogen, and energy through various pathways. The result is that, although these maps provide a picture of possible routes for the conversion of nutrients into products, energy, and reducing equivalents, they reveal little about the *actual* metabolic pathways that are active under a particular set of conditions. *Metabolic flux analysis reveals the degree of pathway engagement in the overall metabolic process.* Furthermore, elucidation of the control of flux provides a mechanistic basis for rationalizing observed fluxes and flux distribution at key metabolic branch points. As these fluxes are determined under *in vivo* conditions, metabolic flux analysis also allows valid comparisons to be made between *in vivo* and *in vitro* enzymatic behavior. Finally, metabolic fluxes provide a generic basis of comparison of strain variants. Even if the fermentation characteristics of such strains differ, these differences may be relatively unimportant if the flux distributions around key branch points have not been altered.

Metabolic engineering offers one of the best ways for meaningfully engaging chemical engineers in biological research, for it allows the direct application of the core subjects of kinetics, transport, and thermodynamics to the analysis of the reactions of metabolic networks. In this context, the latter can be viewed as chemical plants whose units are the individual enzymes, with similar issues of design, control, and optimization. Thus, in attempting to scale up the flux of carbon processed by a microorganism through selected enzyme amplification, one can benefit from accepted scale up principles of chemical plants through the coordinated enlargement of a few key processing units. Similarly, yield optimization depends on byproduct minimization achieved by optimal flux distribution. Of course, flux optimization is possible only after the factors that control flux have first been identified and are well-understood.

Another contribution of metabolic engineering derives from the particular concern of this field about integration and quantitation. An important objective of metabolic engineering is to understand the function of metabolic pathways in their entirety, preferably through the integration of their building blocks, namely, the constituent biochemical reactions. Systemic behavior, however, is not simply the sum of its parts, and one has to deal with issues of complexity as one attempts to reconstruct cellular behavior from genetic and enzymatic information about single genes and reactions. We foresee increasing attention to the need to provide comprehensive cellular descriptions by integrating the plethora of individual pieces of information that have resulted from many years of reductionist research. This need will become more pronounced with the anticipated explosion of information from genomics research. Metabolic engineering provides a valuable forum in this context for upgrading the quality of biological information and synthesizing it for the purpose of developing useful products and processes. In the same vein, other issues of fundamental biochemical and metabolic importance that will benefit from further inquiry into this field are questions of general biological control architecture, hierarchical regulatory structures, and enzymatic reaction channeling effects.

The intellectual framework provided by metabolic engineering for studying cellular metabolism needs to be complemented with the appropriate measurements to achieve maximum results. As such, metabolic engineering plays a strategic role in defining measurements of critical importance for deciphering metabolic or other reaction networks. In the context of metabolic networks this has already contributed to increased attention to problems related to flux measurement *in vivo*. For information networks, it can similarly define analytical needs for elucidating and quantifying the fluxes of information processed through protein-mediated reaction cascades. As the benefits of such measurements become clear in the framework provided, they may drive the development of such instrumentation.

As mentioned in the previous section, elucidation of the flux control structure of metabolic pathways offers tremendous opportunities for the rational design of the optimal enzymatic profile of a cellular catalyst. This activity should be viewed as complementary to molecular biological toolboxes for implementing gene transfers and other similar modifications. In fact, the latter has advanced very rapidly in recent years to the point that rational analysis of metabolic pathways for the identification of target genes and enzymes *is the limiting component* in the directed optimization of cellular function. Evidence for this assertion is the observation that presently, almost 20 years after the pioneering developments in genetic engineering, we have very few significant applications of modern biotechnology in the areas of fuels, chemicals, or materials production (with the possible exception of industrial enzymes). This is so despite the fact that two of the first four biotechnology companies initially focused on these areas as their main business target. The subsequent shift of interest to medical applications was caused by the identification of technically easier and economically more attractive opportunities in health care. It can be argued that the landscape of biotechnological applications is changing rapidly and that future opportunities will include many applications in manufacturing. There are three driving forces supporting this assertion: two are economical and one is technical.

The first driving force behind manufacturing applications is the continuing increase in the production volume of carbohydrate raw materials worldwide. A most natural use of this resource is for the production of derivative products by biotechnology. Some of these products have developed markets, but many others, particularly in the field of materials, will be entirely new applications presenting exciting business opportunities. The second driving force is the continuing decline in the manufacturing cost of biotechnologically produced products compared to an increasing trend in the cost of products manufactured by chemical processes. The reasons for these trends are not entirely clear; however, the realization of economies of scale in large-scale fermentation manufacturing and the increasing burden of environmental compliance for chemical processes have contributed to these trends. We note that not all biotechnological processes are yet fully competitive with chemical ones; however, there is a widespread sense that the costs of the two are changing with different slopes and a crossover will occur soon. According to industrial scientists, the single most important factor accelerating this trend is product selectivity improvements brought about in biotechnological bioprocesses by metabolic engineering.

The last manufacturing driving force is the power of the technologies developed by modern molecular biology. These technologies have not yet been matched by a commensurate capability in locating critical enzymes that need to be manipulated. Progress in recent years, however, allows one to be optimistic regarding the development of methodologies that will permit the identification of the critical enzymes in metabolic networks, as well as the type of modifications needed to bring about significant shifts in the yields of desired products.

Specific areas of industrial production where metabolic engineering can make significant contributions are the production of presently petroleum-derived thermoplastics [poly(hydroxyalkanoates) biosynthesis] by fermentation as well as by expression in whole plants, the production of new materials, and the production of new biologically active agents such as polyketides. The production of gums, solvents, polysaccharides, proteins, diverse antibiotics, foods, biogas, oligopeptides, alcohols, organic acids, vitamins and amino acids, bacterial cellulases, glutathione derivatives, lipids, oils, and pigments are a partial list of product classes that have been produced biologically and presently are the target of metabolic manipulations mainly in microorganisms. Breakthroughs in the principles of metabolic engineering will have a direct impact on the efficiency and economics of these processes.

It should be emphasized that, in an industrial context, the ultimate practical goal of metabolic engineering is the design and creation of *optimal biocatalysts*, optimal in terms of maximizing the yield and productivity of desired products. In this sense, metabolic engineering is equivalent to catalysis in the chemical processing industry, and, although it may be difficult to predict accurately the near-term directions of the field, the preceding analogy makes it easier to envision the long-term impact of metabolic engineering. This impact will be derived primarily from the development of a whole new industry around the fundamental core of, and enabling technologies derived from, applied molecular biology. Just as chemical engineering emerged at the turn of the century as the field implementing industrial applications centered around chemistry, one can envision a new field of biochemical (or biological-metabolic) engineering evolving for the purpose of developing the industrial applications of molecular biology. The central paradigm of metabolic engineering is modeled after that of chemical engineering. In this regard, metabolic engineering, aiming at the development of biocatalysts for process optimization, will play the same role in biological processes that catalysis has for many years in chemical processes. Just as many chemical processes became a reality only after suitable catalysts were developed, the enormous potential of biotechnology will be realized when process biocatalysts become available, to a significant extent through metabolic engineering. The current research activity on sequencing the genomes of many different microbial and other species brings these possibilities much closer to becoming reality. It is on this basis that the long-term potential of metabolic engineering for industrial applications should be assessed.

In addition to the manufacturing applications mentioned previously, metabolic engineering will have a significant impact on the medical field. The main focus here is on the design of new therapies by identifying specific targets for drug development and by contributing to the design of gene therapies. Such approaches presently target a specific single enzymatic step implicated in a particular disease. There is no assurance, however, that the manipulation of a single reaction will translate to systemic responses in the human body. In this regard, medical applications are no different than the ones mentioned earlier in an industrial context, and as such they will benefit from developments in metabolic engineering through a better analysis of experimental results and applications to the rational selection of targets for medical treatment.

Recently, a class of new medicines are increasingly produced containing several chiral centers in their chemical formula. As such molecules are exceedingly difficult to synthesize by organic chemical synthesis, enzymes have been used to carry out one or more difficult steps in an overall chemical synthesis process. It would be desirable to integrate all such steps in a single microorganism by transferring genes expressing the corresponding enzymes needed for overall product synthesis. The identification and transfer of such genes from different organisms, as well as their expression in a host organism, is a theme attracting increasing attention in pharmaceutical manufacturing today. This application is another manifestation of the principles of metabolic engineering in the synthesis of reaction pathways to carry out a particular task instead of employing individual enzymes separated in space. The development of such technologies through the input of metabolic engineering will facilitate the manufacture of drugs that may be needed in significant amounts.

Finally, another application of the concepts and tools of metabolic engineering in the medical field is in the analysis of the function and general metabolism of tissues and whole organs *in vivo*. Perhaps these possibilities can best be described via the analysis of liver function. Although the systemic functions of liver have been studied extensively and are, in general, well-understood, the contributions of the various intrahepatic metabolic pathways to the organ's functions have not been elucidated. For example, it is known that liver metabolism plays an important role in gluconeogenesis and urea production as a mechanism of ammonia removal and, in addition, that there is a complex profile of amino acid uptake and production in the course of normal liver activity. Furthermore, these functions change drastically as a result of burn injuries or in response to administration of cytokines and classic stress hormones such as glucagon, hydrocortisone, and epinephrine. The liver is not a passive participant in the overall metabolic process. On the contrary, it plays an active role in regulating nitrogen disposal from the body, generating glucose, and maintaining a physiological redox and energetic state. The overall biochemical reaction network by which these functions are carried out has been known in sufficient detail; however, the participation of specific pathways in processing carbon skeletons and nitrogen is not understood. Specifically, the various factors controlling the distribution of carbon and nitrogen fluxes through the main metabolic pathways are not known. Because such factors play an important role in maintaining liver function under normal conditions, as well as in response to various perturbations such as injury, it is important that they be identified and sufficiently characterized. Research in this area would, therefore, try to characterize the fluxes through the major hepatic pathways contributing to the processes of gluconeogenesis, urea production, and amino acid processing under normal conditions. This can be done by emulating a continuous system in a perfused liver arrangement under controlled environmental conditions and subject to feeding of physiological media. The fluxes of this network and, especially, variations caused by burn injuries and the introduction of mediators such as cytokines and hormones must be explained through enzyme kinetics and metabolite modulation. Finally, key metabolic pathways in hepatocyte cell culture systems should be identified under normal conditions, as well as conditions of perturbations stimulated by the use of cytokines and combinations of stress hormones. Differences between the two cases would provide invaluable information about *in vivo* enzymatic kinetics and metabolite modulation of the function of the entire organ. It is through this type of research that the actual integral function of organs like the liver can be better understood, both in the context of a broad system and as the overall reconstruction of individual biochemical reactions.

1.2. GENERAL OVERVIEW OF THE BOOK

The approach toward elucidating metabolism and metabolic engineering presented in this book is of general applicability. However, some systems are, by their nature, more amenable to the type of analysis suggested here than others. Also, we note a divergence between the direction suggested here and that followed in other disciplines toward similar objectives, such as the determination of metabolic fluxes or flux control coefficients. In other fields, efforts to obtain estimates of metabolic fluxes, control coefficients, enzymatic elasticities, and other parameters make use of genetically simplified biological systems. Such mutants are deficient in competing pathways, which allows fluxes to be measured directly from the time rate of change of a few extracellular metabolites. Although this approach has yielded some interesting results, it is, in general, rather limited for a number of reasons. The simplified mutants are not always easy to construct, they often exhibit different behavior, and, last, but not least, it is by no means clear that the pathways observed in these altered systems are in any way related to those of the original system. The approach followed here is to study the biological system unaltered, but complemented by as many measurements as current instrumentation allows. In this way, metabolic fluxes and their derivative quantities are obtained from the reconstruction of the metabolic network so as to best describe the rate and label metabolite measurements. By building a significant degree of redundancy into the flux determination process, the level of confidence that they are, indeed, representative of the actual fluxes *in vivo* is significantly enhanced.

In some ways, one can draw an analogy between flux determination and material structure characterization. In material science, there is no single method that will provide the full structure of the material in question. Instead, a number of different techniques are applied, and the structure of the material is determined as the one that gives the best agreement between experimental measurements and their reconstruction from assumed material structures. Similarly, one begins with an assumed biochemical network and determines fluxes so that a large number of diverse, independent, and multidimensional measurements are in good agreement with predictions from the assumed biochemical structure.

It is important to describe at the outset the experimental system where the preceding approach is most applicable. Such a system consists of a continuous bioreactor in which the microbial or cell culture of interest grows and reaches a steady state. Measurement of metabolites in the feed and at the exit of the reactor produces accurate estimates of the metabolic rates of production or consumption of major metabolites. Furthermore, one can introduce directly into the bioreactor well-defined pulses of labeled substrates that are taken up by the cells and reappear, in various forms, in the secreted products. The analysis of the degree of enrichment in the final products, as well as the fine structure of the metabolite peaks in nuclear magnetic resonance (NMR) spectra, provides a powerful combination for the further elucidation of metabolic fluxes. A very important aspect of this experimental system is the concept of the steady state, which is eventually attained after a number of residence times in a flow through reactor. Under such conditions, metabolic fluxes reach a steady state where all of the important response quantities prescribed by metabolic control analysis can be determined. In order to shorten the duration of these experiments, batch or fed-batch cultures may occasionally replace a continuous flow system. However, one should keep in mind that, in this case, only a fraction of the total number of data points is useful in the context of metabolic flux analysis, namely, data collected during the period when the environment remained relatively unchanged. Furthermore, there is limited flexibility in systematically altering the experimental conditions. Clearly, a flow reactor system is suitable primarily for the study of microorganisms and cell cultures. Creative alternatives should be sought to facilitate the investigation of other systems such as plants and organs. The perfusion of whole organs suggested earlier in the context of a liver function analysis is one such possibility that can use the framework of a steady state chemostat to yield valuable information about the integrated function of the liver.

There are two parts in this book. Because there is significant diversity in the backgrounds of readers, Part I provides a general overview of metabolism along with a framework for its comprehensive quantitative description. We do not intend here to replace biochemistry fundamentals that can be found in other excellent texts. The goal rather is to integrate such knowledge in the context of overall metabolism and to provide a systematic quantitative representation that makes use of concepts from chemical reaction engineering. In Chapter 2, cellular reactions are reviewed beginning with transport processes and proceeding to the basic biochemical pathways involved in catabolism and anabolism. An important component is the energy gain obtained from catabolic reactions and energetic cost associated with anabolic processes because they yield valuable information about the true energetic yields of biomass formation and product production. Furthermore, these balances allow one to integrate many different parts of the overall metabolism through the contribution they make to the production or consumption of currency metabolites.

Chapter 3 presents a comprehensive framework for the modeling of cellular reactions. This includes stoichiometric considerations and dynamic material balances used in the determination of reaction rates. The introduction of additional assumptions regarding the use of energy for cellular functions leads to the derivation of useful coefficients and rate equations that have also been observed empirically. Their derivation from first principles and reasonable assumptions provides a rational basis for the explanation of experimental data. Rate equations are, in general, the source of the greatest uncertainty in the analysis of cellular models. It is, therefore, useful to investigate the amount of information that can be extracted from elemental and energetic balances alone without making use of any assumptions regarding biochemical reaction kinetics. This is the subject of Chapter 4, which leads to the derivation of some constraints that rate measurements must satisfy. These constraints are derived from fundamental elemental balances and can be used to test the consistency of the measurements and the assumed biochemistry. In the event that these balances are not satisfied, they can also be used in order to identify the probable sources of error in measurements or in the assumed biochemistry. Part I ends with an overview of the regulation of metabolic pathways. Both transcriptional and enzymatic regulation are discussed in a hierarchical presentation of metabolic controls from the single-enzyme or gene level to the level of the operon and the whole cell. Models are also introduced to describe quantitatively the effect of effectors and inhibitors in regulation at the enzymatic level.

Part II begins with a broad overview of applications of pathway manipulation. This is the subject of Chapter 6, which provides several examples in which metabolic engineering has been profitably applied. In presenting these examples, we follow in part the classification of metabolic engineering applications suggested in the work of Cameron and Tong (1993). By using function as the main criterion, examples are classified into those that lead to the improved production of chemicals already produced by the host organism, those where the range of substrate for growth and product formation was extended, those where new catabolic activities for degradation of toxic chemicals were added, those where chemicals new to the host organism were produced, and situations that contributed to a drastic modification of the overall cellular properties. In most cases, detailed pathways are provided along with a summary of the main problem and the approach that was followed in resolving these problems by metabolic engineering. The list of examples is quite long, with the intention being to provide a broad spectrum that can serve as a guide in future applications with other similar systems.

An important goal of metabolic engineering is to suggest alternative pathways for the biosynthesis of specific products. This can be accomplished by the complete enumeration of all possible pathways that connect a set of specified products with a set of specified reactants within a particular enzymatic database. The problem can be very complex and can lead to the generation of a large number of pathways. An algorithm that can generate all such pathways, which is complete and, at the same time, feasible to implement within a reasonable period of computer time, is presented in Chapter 7.

Metabolic flux analysis is described in detail in Chapter 8. The theoretical background is presented first along with a discussion of the amount and type of information needed in order to solve systems of increasing resolution and complexity. Depending on the number of pathways included and the number of measurements, one may have an exactly-, over-, or under-determined system. A different type of analysis follows each case. Extracellular measurements are the main input in the intracellular flux determination methods presented in Chapter 8. Obviously the resolution of fluxes that can be determined by extracellular methods alone is limited. Additional fluxes and/or flux split ratios can be observed by introducing more measurements, particularly those obtained by making use of isotopic labels. Several examples that illustrate how material balances, radioactive labels, spectroscopic methods, and measurements from gas chromatography-mass spectrometry (GC-MS) can be applied for the experimental determination of fluxes are discussed in Chapter 9. The subject of metabolic flux analysis is completed with two detailed case studies, discussed in Chapter 10. These case studies come from the extended experience of the authors with the systems of amino acid production by glutamic acid bacteria and mammalian cell cultures.

Flux control can best be analyzed within a framework of sensitivity analysis that quantitatively describes the degree of control exercised on each pathway flux. Metabolic control analysis, (MCA), provides the means of describing the extent of (enzymatic) local control and (systemic) global control exercised by a single enzyme or factors affecting enzymatic activity. Within the framework of MCA, it is possible to relate *local* enzymatic kinetics with *global* pathway flux control and thus reconstruct the systemic function of a metabolic network from the kinetic and regulatory properties of the constituent individual reactions. A review of the basic concepts of MCA, along with a comprehensive description of MCA results for linear and branched pathways, is presented in Chapter 11. Much of the discussion revolves around the use of control coefficients as measures of control of flux or metabolites. Depending on the amount of available information, one can opt for the rigorous and quantitative MCA approach or the more qualitative assessment of metabolite and nodal rigidity presented in Chapter 5. Chapter 11 concludes with a presentation of the theory of large perturbations. This is an important extension of MCA that facilitates the determination of flux control coefficients from realistic large perturbation experiments, as opposed to the initial attempts to calculate these coefficients from infinitesimal perturbation experiments suggested by their mathematical definition.

An important limitation in the application of MCA to complex metabolic networks is the large number of reactions involved in such systems. Reaction grouping, as part of a top-down approach, is presented in Chapter 12, where grouping rules are developed for the estimation of reliable *group* control coefficients. The latter are defined in this chapter and, along with group elasticities, provide the metrics for the systematic dissection and analysis of complex reaction networks. These ideas are developed and illustrated with the aid of simulated experiments using a surrogate model of a metabolic network. The use of surrogate cell models is a very useful method as it allows one to bypass time-consuming and expensive experiments in the derivation of flux optimization strategies. The chapter illustrates the systematic application of reaction grouping rules to identify the critical branch points in a metabolic network. The subsequent grouping of reactions around these branch points and determination of the corresponding group control coefficients provide measures of the degree of control exercised by each reaction group on a flux or metabolite of interest. This approach allows the localization of flux control in complex networks. Directed flux amplification strategies using these results are presented in Chapter 13. The goal here is to attempt network flux enhancement through the coordinated amplification of selected reactions, as opposed to the popular single-bullet strategy. It is noted that the approach presented in Chapters 12 and 13 is based entirely on group flux measurements and the Theory of Large Perturbations, which involve several assumptions. It is therefore important to ensure that these assumptions are satisfied, and internal tests are developed for this purpose and presented in Chapter 13.

We conclude this work with Chapter 14, in which several important thermodynamic concepts of cellular pathways are discussed. First, relevant thermodynamic principles of biochemical reactions are reviewed. Then the concept of thermodynamic feasibility based on the magnitude and sign of the standard Gibbs free energy change for individual reactions is extended to reaction pathways. It is shown that there is a limit to the extent that positive values of standard Gibbs free energy changes can be overcome by metabolite concentration differentials. As the number of reactions with positive ΔG° increases, the overall concentration differential necessary to overcome the large positive ΔG°'s increases as well. If certain limits are imposed on the minimum concentrations allowed within a metabolic pathway, then reaction steps or a series of steps can be identified as thermodynamically infeasible, thus creating either a localized or a more distributed thermodynamic bottleneck in metabolic pathways. Although thermodynamic bottlenecks are not necessarily related to kinetic limitations, it is possible to obtain some information about kinetic bottlenecks as well from a thermodynamic analysis. This is done through the use of concepts from thermokinetics and irreversible thermodynamics reviewed in this chapter.

REFERENCES

Bailey, J. E. (1991). Towards a science of metabolic engineering. *Science* **252**, 1668-1674.

Cameron, D. C. & Tong, I.-T. (1993). Cellular and metabolic engineering. *Applied Biochemistry Biotechnology* **38**, 105-140.

Heinrich, R. & Rapoport, T. A. (1974). A linear steady-state treatment of enzymatic chains. *European Journal Biochemistry*. **42**, 89-95.

Kacser, H. & Burns, J. A. (1973). The control of flux. *Symposium Society of Experimental Biology* **27**, 65-104.

Kellogg, S. T., Chatterjee, D. K. & Charkrabarty, A. M. (1981). Plasmid-assisted molecular breeding: new technique for enhanced biodegradation of persistent toxic chemicals. *Science* **214**, 1133-1135.

MacQuitty, J. J. (1988). Impact of biotechnology on the chemical industry. *ACS Symposium Series* **362**, 11-29.

Nerem, R. M. (1991). Cellular engineering. *Annals of Biomedical Engineering* **19**, 529-545.

Stephanopoulos, G. & Vallino, J. J. (1991). Network rigidity and metabolic engineering in metabolite overproduction. *Science* **252**, 1675-1681.

Timmis, K. N., Rojo, F. & Ramos, J. L. (1988). In Environmental Biotechnology, pp. 61-79. Edited by G. S. Omenn. New York, NY: Plenum Press.

Tong, I.-T., Liao, H. H. & Cameron, D. C. (1991). 1,3-Propanediol production by *Escherichia coli* expression genes from the *Klebsiella pneumoniae dha* regulon. *Applied and Environmental Microbiology*. **57**, 3541-3546.

CHAPTER 2

Review of Cellular Metabolism

Formulation of the stoichiometry of metabolic pathways (Chapter 3) is the basis for the quantitative treatment of cellular metabolism. This requires an appreciation of some basic biochemical processes along with an overview of the different pathways normally present in living cells. In this chapter, we review the basic metabolic functions of living cells. We focus primarily on the metabolism of bacteria and fungi, but aspects of the biochemistry of higher eukaryotes are also included. For a more comprehensive discussion of general biochemical concepts and metabolic processes, the reader is referred to a few of many excellent biochemistry textbooks [see, for example, Zubay (1988), Stryer (1995), or Voet and Voet (1995)]. The objective of this chapter is not to substitute but rather to complement a formal course on biochemistry by synthesizing biochemical concepts in the overall framework of cellular metabolism. In this regard, the chapter, although self-contained, is rather dense and best appreciated by readers with some minimal biochemical background, such as that obtained from a first college course in biochemistry.

2.1. AN OVERVIEW OF
CELLULAR METABOLISM

A living cell comprises a large number of different compounds and metabolites. Of these, water is the most abundant component, accounting for approximately 70% of the cellular material. The rest of the cellular mass, usually referred to as the *dry cell weight biomass*, is distributed mainly among the macromolecules DNA, ribonucleic acid (RNA), proteins, lipids, and carbohydrates (Table 2.1). Synthesis and organization of these macromolecules into a functioning cell occur by several independent reactions. The precursors for the synthesis of these macromolecules are small, rapidly used pools of low-molecular-weight compounds that are constantly replenished by biochemical synthesis from metabolites ultimately derived from glucose or other carbon sources (Fig. 2.1). On the basis of their primary function in the overall cell synthesis process, these different reactions can be classified as follows (Neidhardt *et al.*, 1990):

- **Assembly reactions** carry out chemical modifications of macromolecules, their transport to prespecified locations in the cell, and, finally, their association to form cellular structures such as cell wall,

TABLE 2.1   Overall Macromolecular Composition of an Average Cell of
Escherichia coli[a]

Macromolecule	Percentage of total dry weight	Different kinds of molecules
Protein	55.0	1050
RNA	20.5	
rRNA	16.7	3
tRNA	3.0	60
mRNA	0.8	400
DNA	3.1	1
Lipid	9.1	4
Lipopolysaccharide	3.4	1
Peptidoglycan	2.5	1
Glycogen	2.5	1
Soluble pool	3.9	

[a] The data are taken from Ingraham *et al.* (1983).

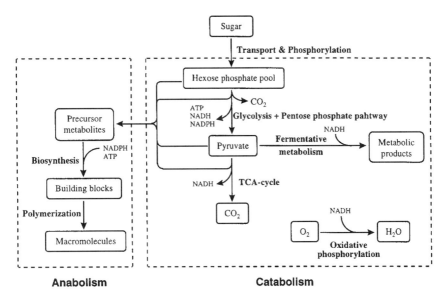

Anabolism   **Catabolism**

FIGURE 2.1   Overall structure of cell synthesis from sugars. The sugar is transported into the cell where it is first phosphorylated and then enters the hexose monophosphate pool. Phosphorylation may occur independently or in conjunction with the transport process. The hexose monophosphates undergo glycolytic reactions, whereby they are converted to pyruvate, or they are used in the synthesis of carbohydrates. Pyruvate, in turn, may be oxidized to carbon dioxide in the respiratory cycle or converted to metabolic products via fermentative pathways. For aerobes, the reducing equivalents in the form of NADH generated in glycolysis and the TCA cycle may be oxidized to NAD^+ in oxidative phosphorylation, whereas for anaerobes, regeneration of NAD^+ occurs in the fermentative pathways. Some of the intermediates in glycolysis and the TCA cycle serve as precursor metabolites for the biosynthesis of building blocks. These building blocks are polymerized into macromolecules, which are finally assembled into the different cellular structures.

membranes, nucleus, etc. These reactions will not be treated further in the present text.

- **Polymerization reactions** represent directed, sequential linkage of activated molecules into long (branched or unbranched) polymeric chains. These reactions form macromolecules from a moderately large number of *building blocks*.

- **Biosynthetic reactions** produce the building blocks used in the polymerization reactions. They also produce coenzymes and related metabolic factors, including signal molecules. There are several biosynthetic reactions, which occur in functional units called biosynthetic pathways, each consisting of one to a dozen sequential reactions leading to the synthesis of one or more building blocks. Pathways are easily recognized and are often controlled *en bloc*. In some cases their reactions are catalyzed by enzymes made from a single piece of mRNA transcribed from a set of contiguous genes forming an operon (see Chapter 5). All biosynthetic pathways begin with one of *12 precursor metabolites*. Some pathways begin directly with such a precursor metabolite, others indirectly by branching from an intermediate or an end product of a related pathway.

- **Fueling reactions** produce the 12 precursor metabolites needed for biosynthesis. Additionally, they generate Gibbs free energy in the form of ATP, which is used for biosynthesis, polymerization, and assembling reactions. Finally, the fueling reactions produce reducing power needed for biosynthesis. The fueling reactions include all biochemical pathways referred to as *catabolic pathways* (degrading and oxidizing substrates).

The different biochemical pathways are connected by virtue of metabolites that participate in more than one pathway by branching, thereby connecting one reaction sequence with another. Another level of pathway integration is mediated by cofactor molecules, such as the global currency metabolites ATP, NADH, and NADPH, which are indispensable for many reactions. Because of their central role in fueling biosynthetic reactions, the continuous formation and utilization of these cofactors connects individual reactions both within the same pathway, and between different pathways as depicted in Fig. 2.1.

Biochemical pathways are organized chemically through the sequential conversion of metabolites in a long chain of reactions and also physically as they may operate in different parts and structures of the cell. This physical organization is most apparent in eukaryotes, where membrane-bound structures provide visible evidence for the localization of sequences or individual biochemical reactions. Thus, DNA and RNA synthesis occur in the nucleus, and many fueling and biosynthetic reactions occur in the mitochondria. This structural organization may have a significant effect on the overall metabolism, but due to the complexity and lack of detailed information about local concentration variations, we will not consider further the structural aspects of metabolic pathway organization.

Besides the chemical and physical (structural) organization of metabolic reactions and pathways, another very useful basis of classification is their dynamics as expressed by their characteristic times. The many different reactions involved in cell metabolism contribute to the dynamics of the various pathways and the overall growth process. As various reactions operate on different time scales, when considering a reaction pathway, only reactions with comparable time scales need be included. Faster reactions can be assumed to be at equilibrium upstream of a slow reaction or give rise to steady states for the metabolites downstream of a slow reaction step. Meanwhile, much slower reactions can be ignored as they operate on a completely different time scale and their impact is minimal within the time frame of interest. Reaction relevance within a given time frame is usually assessed by comparing the *relaxation times* of the various reactions, which are defined as the characteristic times of the reactions approximated as a first-order process (Box 2.1). Figure 2.2 provides a schematic comparison of the relaxation times of different processes operating in a living cell. Processes with much larger relaxation times than that of the system of interest can generally be considered as frozen, *e.g.*, mutations are normally slow compared to cell growth and therefor they can be neglected in studies of cellular growth. On the contrary, processes that have relaxation times much smaller than that of the system can normally be considered as being in pseudo-equilibrium. Thus, enzyme-catalyzed reactions have a relaxation time of milliseconds, which is very fast compared with the relaxation time of cellular growth (normally on the order of hours). Enzymatic reactions therefore generally respond rapidly to new environmental conditions, and a pseudo-steady state in their reaction rates and related metabolites is obtained. A process can normally be considered frozen if its relaxation time is 10 times the relaxation time of the system, and if the former is one-third of the relaxation time of the system, the process can be considered to be at pseudo-steady state. This leads to the following very important conclusion: *Metabolic processes can be simplified significantly by ignoring reactions and pathways operating on time scales outside the time range of interest.*

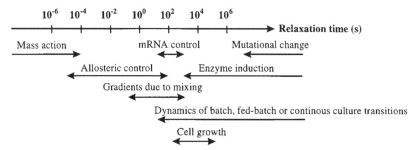

FIGURE 2.2 Relaxation times of different cellular processes in comparison with the relaxation time of bioreactor operation.

BOX 2.1

Relaxation Times

For a chemical reaction with a first-order reaction rate in the reactant,

$$r = kc \tag{1}$$

the relaxation time is given as the reciprocal of the reaction rate constant

$$\tau = 1/k \tag{2}$$

The relaxation time is equal to the characteristic time of the process. If a system is perturbed from its steady state, the relaxation time is the time needed for the system to reach $(1 - 1/e) = 0.63$ of the distance between the old and the new steady state values following the perturbation. For processes that are not first-order, the relaxation time is, by extension, defined as

$$\tau = c/r(c) \tag{3}$$

and represents the characteristic time the process would have if it was a first-order process.

For processes that are not first-order, the relaxation time is obviously a function of the reactant concentration (and, more generally, a function of all state variables). Consequently, the relaxation time is not constant. However, in reality the variation of metabolite concentrations is limited within certain intervals, which confines the actual relaxation times within a reasonable time range.

For a monomolecular reaction

$$A \underset{k_{-1}}{\overset{k_1}{\leftrightarrow}} B \tag{4}$$

the dynamics are described by the material balance:

$$\frac{dc_A}{dt} = -v_1(c_A) + v_{-1}(c_B); \qquad c_A + c_B = \text{constant} \tag{5}$$

By linearizing about the system equilibrium one can write for the deviation of the concentration of A from its equilibrium concentration $c_{A,0}$, i.e., $\sigma_A = c_A - c_{A,0}$:

$$\frac{d\sigma_A}{dt} = -\left(\frac{dv_1}{dc_A} + \frac{dv_{-1}}{dc_B}\right)\sigma_A \Rightarrow \sigma_A(t) = \sigma_A(t_0)\exp\left(-\frac{(t - t_0)}{\tau}\right) \quad (6)$$

with the characteristic time τ given by

$$\tau = \left(\frac{dv_1}{dc_A} + \frac{dv_{-1}}{dc_B}\right)^{-1} \quad (7)$$

or with first-order kinetics ($v_1 = k_1 c_A$ and $v_2 = k_{-1} c_B$):

$$\tau = (k_1 + k_{-1})^{-1} \quad (8)$$

In a similar manner, for a bimolecular reaction

$$A + B \overset{k_1}{\underset{k_{-1}}{\leftrightarrow}} C + D \quad (9)$$

the characteristic time is

$$\tau = \left(k_1(c_{A,0} + c_{B,0}) + k_{-1}(c_{C,0} + c_{D,0})\right)^{-1} \quad (10)$$

where 0 refers to equilibrium concentration.

2.2. TRANSPORT PROCESSES

Species may be transported across the plasma membrane by three different mechanisms: (1) free diffusion, (2) facilitated diffusion, and (3) active transport. The first two mechanisms are passive processes which, in principle, require no supply of Gibbs free energy to occur. In free and facilitated diffusion, species are transported down a concentration gradient. The difference between the two mechanisms is that facilitated diffusion is carrier-mediated, i.e., a specific carrier or transmembrane protein is involved in the transport, whereas free diffusion is molecular transport, driven by a differential of chemical potentials. The third mechanism allows species to be transported against a concentration gradient by an active process requiring the input of Gibbs free energy. Active transport resembles facilitated diffusion in that specific membrane-localized proteins-or permeases-mediate the process.

2.2.1. PASSIVE TRANSPORT

Transport of a compound by free diffusion across a lipid membrane involves three steps: (1) transfer of the compound from the extracellular medium to the membrane phase; (2) diffusion of the compound through the lipid bilayer; and (3) transfer of the compound from the membrane phase to the cytosol. Normally, the cytosol and the extracellular medium have similar physicochemical properties and steps 1 and 3 are, therefore, quite similar. The transfer of a compound between the two phases (*i.e.*, from the extracellular medium to the membrane or from the membrane to the cytosol) will generally be very fast compared with the diffusion process, and it can be assumed to be at equilibrium. The concentration of the transported compound at the interphase of the lipid layer is therefore equal to the product of the concentration in the water phase and the partition coefficient K_{par} (defined as the ratio of the solubility of the compound in the lipid bilayer and the solubility of the compound in water). The mass flux by molecular diffusion follows Fick's first law, and the resulting diffusive flux (mol m^{-2}) is described by

$$r_{tran} = \frac{D_{mem} K_{par}}{d_{mem}} (c_a - c_c) = P(c_a - c_c) \qquad (2.1)$$

where D_{mem} is the diffusion coefficient of the compound in the lipid bilayer (m^2 s^{-1}), d_{mem} is the thickness of the membrane (m), and c_a and c_c are the concentrations of the compound in the abiotic phase and in the cytosol, respectively. The three parameters in eq. (2.1) D_{mem}, d_{mem}, and K_{par}, are normally collected into one parameter $[(D_{mem} K_{par}) / d_{mem}]$ called the permeability coefficient P (Stein, 1990). In eq. (2.1), the mass flux is defined based on the cell surface area, but it can be converted to mass flux per gram dry weight by multiplying with the specific surface area of the cell a_{cell} [m^2 (g DW)$^{-1}$].

Among the compounds transported by passive diffusion are carbon dioxide, oxygen, NH_3, fatty acids, and some alcohols. In their dissociated form, organic acids are practically insoluble in the lipid membrane, whereas the undissociated form of many organic acids is quite soluble. Thus, the permeability coefficients for undissociated lactic acid and acetic acid are 5.0×10^{-7} and 6.9×10^{-5} m s^{-1}, respectively, which means that these compounds diffuse quite rapidly across the plasma membrane. Because it is mainly the undissociated form that diffuses freely, the total transport of compounds like lactic acid and acetic acid is very sensitive to the degree of dissociation and, hence, the pH of both the cytosol and extracellular medium. For many organisms there is a pH gradient across the cytoplasmic membrane (with a higher pH intracellularly), and this may result in a net influx of protons into the cells. In order to maintain the lower intracellular proton concentration, these protons have to be pumped out via a plasma membrane ATPase at the expense of ATP, and the presence of organic acids may therefore result in net ATP consumption (illustrated in Example 2.1).

EXAMPLE 2.1

Decoupling of Energy Generation and Consumption by Organic Acids

The effect of organic acids on the drain of ATP in living cells has been illustrated in a study of Verduyn *et al.* (1992), who analyzed the influence of benzoic acid on the respiration of *Saccharomyces cerevisiae*. They found that the biomass yield on glucose decreased with increasing concentrations of the acid. At the same time the specific uptake rates of glucose and oxygen increased. Thus, there is a less efficient utilization of glucose for biomass synthesis, and this is explained by the drain of ATP due to the proton decoupling effect of benzoic acid. In another study, Schulze (1995) analyzed the influence of benzoic acid on the ATP costs for biomass synthesis in anaerobic cultures of *S. cerevisiae*. He found that the ATP costs for biomass synthesis increased linearly with the benzoic acid concentration, also a consequence of the proton decoupling effect of this acid. Henriksen *et al.* (1998) derived a set of equations that allows quantification of the ATP costs resulting from uncoupling of the proton gradient by organic acids. The aim of the study was to quantify the uncoupling effect of phenoxyacetic acid, a precursor for penicillin V production, on the proton gradient in *Penicillium chrysogenum*. Both forms of this acid may diffuse passively across the plasma membrane, but the undissociated acid has a much larger solubility, *i.e.*, a larger partition coefficient, and is therefore transported much faster. To describe the mass flux of the two forms across the plasma membrane, Henriksen *et al.* (1998) applied eq. (2.1) in the form of

$$r_{tran,i} = P_i a_{cell}(c_{a,i} - c_{c,i}) \tag{1}$$

where the index i refers to either the undissociated or the dissociated form of the acid. The specific cell area is about 2.5 m^2 (g DW)$^{-1}$ for *P. chryso-genum*, and the permeability coefficients for the undissociated and dissociated forms of phenoxyacetic acid have been estimated to be 3.2 x 10^{-6} and 2.6 x 10^{-10} m s^{-1}, respectively (Nielsen, 1997).

Because the undissociated and dissociated forms of the acid are in equilibrium on each side of the cytoplasmic membrane (HA $\leftrightarrow$ H$^+$ + A$^-$) with equilibrium constant K_a, the correlation between the two forms and the total acid concentration are given by

$$c_{undiss} = c_{diss} 10^{pK_a - pH} = \frac{c_{total}}{1 + 10^{pH - pK_a}} \tag{2}$$

where the pK_a for phenoxyacetic acid is 3.1. At pseudo-steady state conditions, the net influx of undissociated acid will equal the net outflux of the dissociated form of the acid:

$$r_{undiss, in} = r_{diss, out}, \text{ or } P_{undiss} a_{cell} (c_{undiss, a} - c_{undiss, c})$$

$$= P_{diss} a_{cell} (c_{diss, c} - c_{diss, a}) \tag{3}$$

where subscript a and c indicate the abiotic and cytosolic side of the membrane, respectively. By substituting for the undissociated and dissociated acid concentrations on the abiotic and cytosolic sides of the membrane in terms of the total concentrations on the abiotic and cytosolic sides from eq. (2) and rearranging, we obtain the following equation for the ratio of the total concentrations on the two sides of the membrane:

$$\frac{c_{c, tot}}{c_{a, tot}} = \frac{P_{undiss} \dfrac{1 + 10^{pH_c - pK_a}}{1 + 10^{pH_a - pK_a}} + P_{diss} \dfrac{1 + 10^{pH_c - pK_a}}{1 + 10^{pK_a - pH_a}}}{P_{diss} 10^{pH_c - pK_a} + P_{undiss}} \tag{4}$$

Because the permeability coefficient for the undissociated form of the acid is orders of magnitude greater than that of the dissociated form, this equation can be reduced to

$$\frac{c_{c, tot}}{c_{a, tot}} = \frac{1 + 10^{pH_c - pK_a}}{1 + 10^{pH_a - pK_a}} \tag{5}$$

Now, the intracellular pH is usually greater than the typical pH of the medium in penicillin cultivations. Equation (5) then indicates that there is a higher total concentration of the acid inside the cells than in the extracellular medium. Using this equation Henriksen et al. (1998) calculated the concentration ratio at different extracellular pH values and an intracellular pH of 7.2. For an extracellular pH of 6.5 the accumulation is low (about 2.3-fold), whereas at an extracellular pH of 5.0 the accumulation is high (about 100-fold).

For a given total extracellular acid concentration, the concentrations of both forms of the acid on each side of the cytoplasmic membrane can be calculated using eq. (2), from which the mass flux of acid across the membrane can be calculated using eq. (1). Because the net outflux of dissociated acid equals the net influx of undissociated acid, the result of acid transport is a net influx of protons, which have to be re-exported by the plasma membrane ATPase in order to maintain a constant intracellular pH. If it is assumed that the export of each proton requires the expenditure of one ATP by the ATPase reaction, Henriksen *et al.* (1997) calculated that the ATP consumption resulting from this futile cycle amounts to 0.15 mmol of ATP $(g \text{ DW})^{-1} h^{-1}$ at an extracellular pH of 6.5 and an intracellular pH of 7.2. This is a low value compared with other non-growth-associated processes that also consume ATP (see Section 2.6). However, at an extracellular pH of 5.0 this value becomes about 7 mmol of ATP $(g \text{ DW})^{-1}$ (again with an intracellular pH of 7.2), which is a significant ATP drain (see Section 2.6). It is thus seen how the maintenance of acid concentration gradients across the plasma membrane contributes to the decoupling of ATP generation and consumption strictly for biosynthetic demands.

2.2.2. FACILITATED DIFFUSION

Many compounds are transported at extremely slow rates by free diffusion because of their very low solubility in the plasma membrane. The transport of such compounds can be significantly enhanced by the mediation of carrier molecules embedded in the cytoplasmic membrane. This transport occurs passively, *i.e.*, requires no energy expenditure, and is referred to as *facilitated diffusion*. It is used for the transport of many compounds in eukaryotes, whereas glycerol transport in *Escherichia coli* and transport of glucose into *Zymomonas mobilis* and *Streptococcus bovis* appear to be the only examples of facilitated diffusion in prokaryotes (Moat and Foster, 1995). Facilitated diffusion resembles free diffusion as transport occurs only in the downhill direction of a concentration gradient. However, the compound can enter the membrane phase only if there is a free carrier molecule available, and the rate of the transport follows typical saturation-type kinetics like Michaelis-Menten kinetics for enzyme-catalyzed reactions (see Example 2.2). Thus, at low concentrations of the substrate the rate is first-order with respect to the substrate concentration, whereas it is zero-order at high substrate concentrations.

In fungi, many sugars are transported by facilitated diffusion. In *S. cerevisiae*, glucose is transported by facilitated diffusion, and the presence of both a glucose-repressed high-affinity system with a K_m of 1 mM and a constitutive low-affinity system with K_m of 20 mM have been reported

(Bisson and Fraenkel, 1983; Bisson *et al.*, 1993). Recently, the complexity of glucose transport in *S. cerevisiae* has been further demonstrated by the cloning of a whole family of genes coding for hexose transporters [see Kruckeberg (1996) for a review]. Also, for filamentous fungi, different facilitated sugar transporters have been reported with K_m values in the range of 8-25 mM (Nielsen, 1997).

EXAMPLE 2.2

Saturation Kinetics of Facilitated Transport

In many biological systems, it has been observed that the facilitated transport rate of compounds follows Michaelis-Menten saturation-type kinetics. Such a rate expression can be derived from the following mechanism. First, a substrate molecule S binds with a molecule of the carrier C to form the complex (CS), which then diffuses freely through the membrane. Steady state conditions require that the number of carrier molecules used to shuttle substrate molecules into the cell rediffuse freely from the cytosolic to the abiotic side of the membrane. The concentration profiles of the substrate-carrier complex and free carrier for this type of mechanism, along with the assumed equilibria and transport processes, are depicted in the following schematic:

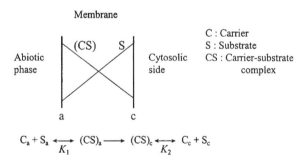

The total transport rate can be written in terms of the complex or free carrier membrane concentrations as follows:

$$r_{transport} = \frac{D_{complex}}{d_{mem}}(c_{cs,a} - c_{cs,c}) = \frac{D_{carrier}}{d_{mem}}(c_{c,c} - c_{c,a}) \qquad (1)$$

where the subscribed parameter D is the diffusion coefficient of the sub-strate-carrier complex and free carrier in the membrane, d_{mem} is the membrane thickness, and c_c and c_{cs} are the carrier and complex membrane concentrations, respectively. By assuming equilibrium for the complex formation and dissociation reactions, as indicated in the preceding figure, the complex concentrations can be eliminated to yield:

$$D_{complex}\left(K_1 c_{c,a} c_{s,a} - K_2^{-1} c_{c,c} c_{s,a}\right) = D_{carrier}\left(c_{c,c} - c_{c,a}\right) \qquad (2)$$

This equation, combined with the conservation of total carrier equation:

$$c_{c,a} + c_{c,c} + K_1 c_{c,a} c_{s,a} + K_2^{-1} c_{c,c} c_{s,c} = c_{total} \qquad (3)$$

can be solved for the free carrier concentrations $c_{c,a}$ and $c_{c,c}$, which, upon substitution into eq. (1), yield an expression for the transport rate. An interesting limiting case arises when the intracellular substrate concentration is zero, $c_{s,c} = 0$, due to rapid catabolic reactions converting the substrate once it has been transported into the cell. Under these conditions, the carrier concentration at the abiotic side of the membrane is found to be

$$c_{c,a} = c_{total}\left(2 + \left(\frac{D_{complex}}{D_{carrier}} + 1\right) K_1 c_{s,a}\right)^{-1} \qquad (4)$$

If we further assume that the diffusivities of the free carrier and carrier-substrate complex are approximately equal, substitution of eq. (4) into eq. (1) yields the transport rate:

$$r_{transport} = \frac{D_{complex}}{d_{mem}} \frac{C_{total}}{2} \frac{K_1 S_a}{1 + K_1 S_a} \qquad (5)$$

The preceding equation exhibits the observed saturation behavior with respect to the substrate concentration in the abiotic phase. Furthermore, it is interesting to note that the preceding transport mechanism is equivalent to the usual Michaelis-Menten (M-M) enzyme-catalyzed reaction mechanism, where the transport step has replaced product formation as the slow step in the overall process. As a consequence, the reaction rate constant of the M-M mechanism has been replaced by the diffusivity multiplied by one-half of the total amount of carrier, which is the effective amount of carrier available for transport at any point in time (the other half being complexed with the substrate).

Several qualitative mechanisms of facilitated diffusion have been described in the literature. Some of them involve actual transport of carrier molecules, such as the one described in this example. Others describe situations where the carrier is a protein spanning the entire membrane and, as such, is relatively stationary. In the latter case, transport of small molecules occurs through intramembrane channels formed by the carrier protein. The transport mechanism still involves some form of substrate-carrier protein binding, as previously described. However, unless a slow step is involved, saturation-type kinetics cannot be obtained. The slow step can be diffusive or reactive in nature and is a critical component in the overall transport process. Equilibrium processes alone, whereby molecules of the carrier protein bind the substrate and turn it over immediately to the inside of the cell under another equilibrium reaction, are insufficient to describe the saturation-type transport rates observed experimentally.

2.2.3. ACTIVE TRANSPORT

Active transport resembles facilitated diffusion because specific membrane-localized proteins-called *permeases*-mediate the transport process. In contrast to facilitated diffusion, active transport can occur against a concentration gradient and therefore is a free energy-consuming process. The free energy required for the transport process may be provided by high-energy phosphate bonds in, for example, ATP (*primary* active transport). Alternatively, the transport process may be coupled to another transport process with a downhill concentration gradient (*secondary* active transport). The net free energy generation in the latter case counterbalances the free energy consumption to drive the active transport process. Finally, in a special type of active transport processes called *group translocation*, the transported compound is converted to a (phosphorylated) derivative that cannot be transported through the membrane in the reverse direction.

An important group of primary active transport systems is the ATPases, which are involved in the excretion of protons at the expense of ATP. Some of these enzymes may work in both directions such that ATP may be generated upon influx of protons, and in prokaryotes this is an important element in oxidative phosphorylation (see Section 2.3.3). Other primary active transport systems exist and involve specific binding proteins that bind the compound to be transported and transfer it to a compatible membrane-bound complex (Moat and Foster, 1995). This transfer triggers ATP hydrolysis, which, in turn, leads to the opening of a pore that allows unidirectional diffusion of the substrate into the cytoplasm. Among the compounds transported by these so-called *traffic ATPases* are histidine, maltose, arabinose, and galactose in *E. coli* (Moat and Foster, 1995).

In secondary active transport, the transport of the compound is coupled with the transport of another compound along a favorable concentration gradient. If the compounds are transported in the same physical direction, the transport is called *symport*, *e.g.*, proton symport, which is one of the most important mechanisms for secondary active transport, and if the compounds are transported in opposite physical directions, the transport is called *antiport*. With proton symport there is an influx of protons that must be re-exported (*e.g.*, by means of an ATPase) in order to maintain the intracellular pH (Fig. 2.3). Even though proton symport is the most common secondary transport mechanism, symport and antiport may also occur with Na^+, K^+, and Mg^{2+}.

In *group translocation*, the transport process is coupled with a subsequent conversion of the transported compound. The best established example of the group translocation systems is the *phosphotransferase system* (PTS), by which certain sugars are transported in bacteria. The system is rather complex, involving the participation of at least four different proteins that function within the cell as phosphocarriers of the high-energy phosphate group from phosphoenolpyruvate (PEP) to the incoming sugar (see Fig. 2.4).

EXTRACELLULAR MEDIUM

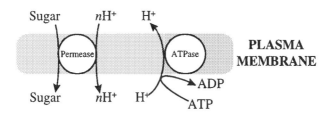

CYTOPLASM

FIGURE 2.3   Transport of sugars by proton symport. The specific membrane-localized permeases transport the sugars into the cell, the whole process being driven by the simultaneous import of n protons. For the lactose permease in *E. coli*, $n = 1$ (Stein, 1990), and this probably also holds for most other sugar permeases. In order to maintain the intracellular pH at a constant level, protons must be pumped out of the cells. In anaerobic bacteria this may occur in the electron transport chain, whereas in fungi it is done under expenditure of ATP by the action of the plasma membrane ATPase. The H^+/ATP stoichiometry of the plasma membrane ATPase is 1 for several eukaryotes, including *N. crassa* (Perlin *et al.*, 1986) and *S. cerevisiae* (Malpartida and Serrano, 1981). The intracellular pH is normally fairly constant over a range of extracellular pH values and is generally higher than the extracellular pH (Cartwright *et al.*, 1989).

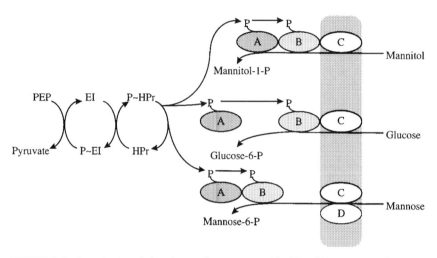

FIGURE 2.4   Organization of phosphotransferase systems. The EI and HPr are general proteins for all PTS, whereas there are many specific EII proteins. The EII proteins consist of three domains (A, B, and C), which may be combined in a single membrane-bound protein (as illustrated for the mannitol PTS) or split into two or more proteins. In the glucose PTS, the B and C domains are combined into one protein called IIBC, which is membrane-bound, whereas the A domain is soluble. In the mannose PTS the A and B domains are combined into one soluble protein called IIAB, whereas the C domain is membrane-bound.

Two of the proteins are soluble, cytoplasmic proteins; these are enzyme I and the histidine protein (HPr) coded by *ptsI* and *ptsH* in *E. coli*, respectively. These two proteins are common to all PTS carbohydrates and therefore are referred to as *general PTS proteins*. In contrast, enzymes II are carbohydrate specific. They consist of three domains (A, B, and C) that may be combined into a single membrane-bound protein or split into two or more proteins called EIIA, EIIB, and EIIC. In the PTS the phosphate group of PEP is transferred to the incoming sugar via phosphorylated intermediates of EI, HPr, EIIA, and EIIB. The C domain of EII (the EIIC protein) forms the translocation channel and at least part of the specific carbohydrate binding site. On the basis of sequence homologies, the EII proteins are grouped into four classes: mannitol, glucose, mannose, and lactose (Moat and Foster, 1995).

The ATP consumption for active transport is a significant fraction of the overall ATP requirements for cell synthesis (see Section 2.6). For transport by proton symport, the ATP consumption depends on the stoichiometry of the transport process and the stoichiometry of proton re-export. Among the best studied systems is the lactose permease in *E. coli*, where one proton is transported together with one lactose (see Fig. 2.3). A similar stoichiometric ratio holds for the transport of many other compounds, *i.e.*, transport of many amino acids in both prokaryotes and eukaryotes and transport of sugars like glucose, fructose, and galactose by high-affinity systems in filamentous fungi (see Table 2.2). For re-export of protons there are some differences between prokaryotes and eukaryotes. In aerobic bacteria, the electron transport chain is located in the plasma membrane and protons may be re-exported from the cytosol by the electron transport chain (see Section 2.3.3). Protons may, however, also be re-exported by the F_0F_1-*ATPase*, which in prokaryotes is located in the plasma membrane. This ATPase is mainly involved in the synthesis of ATP in oxidative phosphorylation, but it is reversible and may pump protons out of the cells at the expense of ATP. The stoichiometric ratio for the F_0F_1-ATPase is not exactly known, but for *E. coli* a stoichiometric ratio of $2H^+/ATP$ is often used. In eukaryotes the F_0F_1-ATPase is located in the mitochondria, but they also possess a *plasma membrane ATPase* belonging to another class of ATPases. This ATPase is probably working only in the direction of ATP hydrolysis, and it normally has a stoichiometric ratio of $1\ H^+/ATP$. On the basis of the preceding stoichiometries it is possible to calculate the ATP costs for transport of different compounds by active transport, and Table 2.2 summarizes the results.

TABLE 2.2   Overview of Active Transport of Some Compounds in Bacteria and Fungi

Species	Compound	Mechanism	ATP cost
Bacteria	Amino acids	H^+ symport	0.5
	Gluconate	H^+ symport	0.5
	Organic acids	H^+ symport	0.5
	Phosphate	H^+ symport	0.5
	Sugar	H^+ symport	0.5
		PTS	0^a
Fungi	Amino acids	H^+ symport	1
	NH_4^+	Chemiosmosis	1
	Phosphate	H^+ symport and Na^+ antiport	2
	Sugars	H^+ symport	1
		Facilitated diffusion	0
	Sulfate	Ca^{2+} and H^+ symport	1

^{a} By the PTS systems the sugar is phosphorylated in conjunction with transport. Even though a high-energy phosphate group present in phosphoenolpyruvate is used, the high-energy bond is conserved in the sugar phosphate and the ATP expenditure is therefore 0.

2.3. FUELING REACTIONS

The fueling reactions serve three purposes: (1) generation of Gibbs free energy, mainly in the form of ATP, which is used to fuel other cellular reactions; (2) production of reducing power (or reducing equivalents), mainly in the form of the cofactor NADPH, required in biosynthetic reactions; and (3) formation of precursor metabolites required in the biosynthesis of building blocks. The substrate that supplies the carbon skeletons for the biosynthesis of building blocks is usually called the *carbon source*, whereas the substrate that supplies the Gibbs free energy and reducing power needed for biosynthesis is called the *energy source*. Many substrates may serve as both *carbon and energy source*, as is the case with some of the most frequently used substrates in industrial processes, such as the sugars glucose, fructose, galactose, lactose, sucrose, or maltose. The catabolism of sugars starts with glycolysis, the end product of which is pyruvate. Pyruvate is subsequently processed further, either in fermentative pathways, anaplerotic pathways, transamination pathways for amino acid formation, the tricarboxylic acid cycle (TCA cycle), or others. All of the precursor metabolites needed for the biosynthesis of building blocks are generated in glycolysis and the TCA cycle, but to replenish the drain of precursor metabolites for biosynthesis, a set of anaplerotic pathways is required. In industrial media (*e.g.*, molasses, cornsteep liquor) there are often additional carbon and energy sources, *e.g.*, amino acids, organic acids, or fats, and many organisms have the ability to catabolize these compounds either in the presence of a sugar or as the sole carbon and energy source.

Due to the importance of the fueling pathways in central carbon metabolism and biosynthesis, in this section we review the basics of the glycolytic, fermentative, and anaplerotic pathways, as well as the TCA cycle and oxidative phosphorylation and the catabolism of fats, oils, and organic acids.

2.3.1. GLYCOLYSIS

Glycolysis is the sum total of all biochemical reactions by which glucose is converted into pyruvate. This can be accomplished by more than one pathway (or metabolic routes). The most frequently encountered pathways are (1) the *Embden–Meyerhof–Parnas* pathway (EMP), (2) the *pentose phosphate* pathway (PP), and (3) the *Entner–Doudoroff* pathway (ED). The common entrance of sugars to the glycolytic pathways is through the three hexose monophosphates, glucose 1-phosphate (G1P), glucose 6-phosphate

(G6P), and fructose 6-phosphate (F6P), which are readily interconvertible by the action of the enzymes phosphoglucomutase (for G1P and G6P) and phosphohexoisomerase (G6P and F6P). These enzymes normally are present in excess, and, under these conditions, the three compounds are therefore at equilibrium constituting a single metabolite pool. The pool can be replenished by the generation of any of its three components (Fig. 2.5), and at equilibrium the pool consists of approximately 3% glucose-1-phosphate, 65% glucose-6-phosphate, and 32% fructose-6-phosphate (Zubay, 1988).

Intracellular glucose and fructose enter directly into the pool of hexose monophosphates by phosphorylation at the C-6 position. In bacteria, where sugars are mainly taken up by PTSs, the phosphorylation occurs in conjunction with transport, whereas in eukaryotes phosphorylation is catalyzed by hexokinases. In *Saccharomyces cerevisiae*, there are three different enzymes involved in the phosphorylation of glucose and fructose: hexokinase A, hexokinase B, and a glucokinase (Gancedo and Serrano, 1989). The two hexokinases phosphorylate both glucose and fructose, whereas the glucokinase is specific toward glucose. Galactose, arising from extracellular galactose or from hydrolysis of intracellular lactose, enters the pool of hexose monophosphates in a more complicated way. If it is transported by a proton-driven permease, phosphorylation occurs via the *Leloir pathway*.

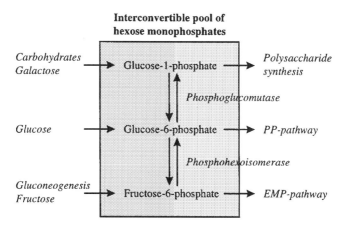

FIGURE 2.5 The pool of hexose monophosphates and their roles in different pathways. Glucose-1-phosphate is formed from phosphorolysis of storage carbohydrates or as an end result of galactose assimilation via the Leloir pathway, and it is used for polysaccharide synthesis. Glucose-6-phosphate is formed from phosphorylation of glucose by ATP, and it is used in the PP pathway (or the ED pathway). Fructose-6-phosphate is formed by gluconeogenesis or by phosphorylation of fructose by ATP, and it is used in the EMP pathway. Thus, each hexose monophosphate in the pool serves both as an entry point and as an exit.

Here galactose is phosphorylated in the C-1 position and then reacts with UDP-glucose, resulting in the formation of glucose-1-phosphate and UDP-galactose. Other reactions provide for the regeneration of UDP-glucose from UDP-galactose. If galactose is transported by a PTS, it is phosphorylated at the C-6 position and metabolism occurs via the *tagatose pathway*. Here galactose-6-phosphate is converted to tagatose-6-phosphate, which is phosphorylated to tagatose-1,6-bisphosphate with consumption of one ATP. Finally, tagatose-1,6-bisphosphate is cleaved to dihydroxyacetonephosphate and glyceraldehyde 3-phosphate.

In the *EMP pathway*, 1 mol of fructose-6-phosphate is converted into 2mol of pyruvate (Fig. 2.6). One of the reactions, the conversion of fructose-6-phosphate to fructose-1,6-bisphosphate, requires free energy in the form of ATP, but there is an overall gain of 3 mol of ATP per mole of glucose-6-phosphate catabolized in the EMP pathway. In reaction (6) of Fig. 2.6 (the oxidation of glyceraldehyde-3-phosphate to 1,3-diphosphoglycerate), NADH is also produced. Thus, the overall stoichiometry for the conversion of glucose to pyruvate in the EMP pathway is

$$2\text{pyruvate} + 2\text{ATP} + 2\text{NADH} + 2H_2O + 2H^+ -$$

$$\text{glucose} - 2\text{ADP} - 2 \sim P - 2\text{NAD}^+ = 0 \qquad (2.2)$$

In the PP pathway, glucose-6-phosphate is oxidized to 6-phosphogluconate, which is further converted to ribulose-5-phosphate and carbon dioxide (Fig. 2.6). In each of these two reactions, 1 mol of NADPH is formed per mole of glucose-6-phosphate entering the pathway. In the subsequent steps, ribulose-5-phosphate is converted into ribose-5-phosphate or erythrose-4-phosphate, which are both precursors for the biosynthesis of building blocks, *i.e.*, aromatic amino acids and nucleotides. In a different sequence of reactions, ribulose-5-phosphate may also be converted back to fructose-6-phosphate and glyceraldehyde-3-phosphate, thus re-entering the EMP pathway (Fig. 2.6).

The individual reactions of the *PP pathway* are listed here:

$$6\text{-P-gluconate} + \text{NADPH} + H^+ - \text{glucose 6-P} - \text{NADP}^+ - H_2O = 0 \quad (2.3a)$$

$$CO_2 + \text{ribulose 5-P} + \text{NADPH}$$

$$+ H^+ - 6\text{-P-gluconate} - \text{NADP}^+ - H_2O = 0 \qquad (2.3b)$$

$$\text{ribose 5-P} - \text{ribulose 5-P} = 0 \qquad (2.3c)$$

$$\text{xylulose 5-P} - \text{ribulose 5-P} = 0 \qquad (2.3d)$$

$$\text{glyceraldehyde 3-P} + \text{sedoheptulose 7-P}$$

$$- \text{xylulose 5-P} - \text{ribose 5-P} = 0 \qquad (2.3e)$$

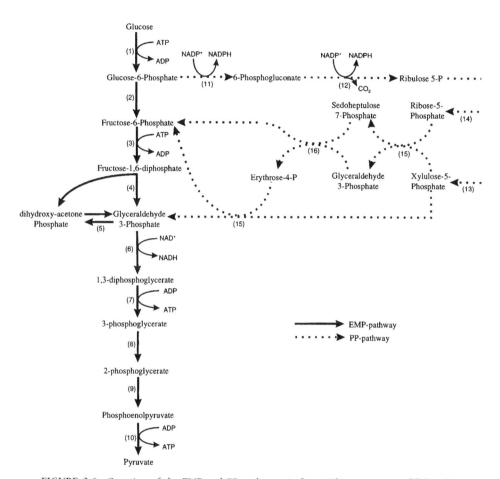

FIGURE 2.6   Overview of the EMP and PP pathways in fungi. The enzymes are (1) hexokinase; (2) phosphohexoisomerase; (3) phosphofructokinase; (4) aldolase; (5) triosephosphate isomerase; (6) 3-phosphoglyceraldehyde dehydrogenase; (7) 3-phosphoglycerate kinase; (8) phosphoglycerate mutase; (9) enolase; (10) pyruvate kinase; (11) glucose-6-phosphate dehydrogenase; (12) 6-phosphogluconate dehydrogenase; (13) ribulosephosphate-3-epimerase; (14) ribosephosphate isomerase; (15) transketolase; (16) and transaldolase.

fructose 6-P + erythrose 4-P - sedoheptulose 7-P - glyceraldehyde 3-P = 0

$$(2.3f)$$

fructose 6-P + glyceraldehyde 3-P - xylulose 5-P - erythrose 4-P = 0   (2.3g)

The overall stoichiometry of the PP pathway depends on the extent to which carbon entering the PP pathway is recycled back into the EMP pathway oxidized into carbon dioxide (with simultaneous production of reducing power in the form of NADPH), or consumed for the formation of biosynthetic precursors (such as five-carbon sugars for ribonucleotide synthesis). For this reason, the PP pathway has been recognized to serve an oxidative as well as an anaplerotic function, each described by the following overall stoichiometries:

anaplerotic PP function:

6ribose 5-P + 5ADP + 4H$_2$O + 4 ~ P - 5glucose 6-P - 5ATP = 0   (2.4)

oxidative PP function:

12NADPH + 12H$^+$ + 6CO$_2$ + ~ P - glucose 6-P - 12NADP$^+$ - 7H$_2$O = 0

$$(2.5)$$

In the *ED pathway*, 6-phosphogluconate is converted to 2-keto-3-deoxy-6-phosphogluconate (KDPG) by the enzyme 6-phosphogluconate dehydratase, and KDPG is subsequently cleaved to form glyceraldehyde-3-phosphate and pyruvate by the enzyme 2-keto-3-deoxy-6-phosphogluconate aldolase (Conway, 1992). Thus, the overall stoichiometry for the conversion of glucose to pyruvate in the ED pathway is

2pyruvate + ATP + NADPH + NADH + 2H$_2$O + 2H$^+$

- glucose - ADP - 1 ~ P - NAD$^+$ - NADP$^+$ = 0   (2.6)

Note that only 1 mol each of ATP and NADPH are formed, in contrast to twice that (NADH instead of NADPH) formed in the glycolytic pathway. The ED pathway also has a nonphosphorylative pathway equivalent, where gluconate is converted to pyruvate and glyceraldehyde via 2-keto-3-deoxy-gluconate. This pathway is, however, only active when the cells metabolize gluconate (Conway, 1992).

Three intermediates of the EMP pathway (glyceraldehyde-3-phosphate, 3-phosphoglycerate, and phosphoenolpyruvate) and two intermediates of the PP pathway (ribose-5-phosphate and erythrose-4-phosphate) serve as precursor metabolites for the biosynthesis of amino acids and nucleic acids. The relative flux through the two glycolytic pathways depends on the requirements of Gibbs free energy, reducing power in the form of NADH and NADPH, and the preceding precursor metabolites. The flux distribution between the two glycolytic pathways has been determined experimentally for a number of species either by the so-called respirometric method where the formation of $^{14}CO_2$ ^{14}C-labeled glucose is quantified (Blumenthal, 1965) or by measuring the fractional enrichment of ^{13}C in intracellular metabolites by NMR (see Chapter 9). Table 2.3 lists the distribution between the EMP and PP pathways for some species. Most of the findings indicate that the EMP pathway is the major pathway. However, for species overproducing metabolites requiring NADPH for product synthesis, e.g., lysine production by *Corynebacterium glutamicum*, the flux through the PP pathway may be larger than the flux through the EMP pathway (see also Chapter 10). The relative flux through the PP pathway in general depends on the specific growth rate and the medium composition, and in *Aspergillus nidulans* the relative PP pathway flux increases with the dilution rate in a steady state chemostat (Fig. 2.7a). This is consistent with an observed decrease (as dilution rate increases) in the *relative* activity of an EMP enzyme (aldolase) and an increase in the *relative* activity of a PP enzyme (glucose-6-phosphate dehydrogenase) (Fig. 2.7a). (Note that the relative activities of the preceding enzymes have been determined with respect to the activity of the hexokinase enzyme supplying the total carbon to both EMP and PP.) The increased activity of the PP pathway at a high specific growth rate is explained by an increased demand at a high specific growth rate for NADPH and for precursor metabolites generated in the PP pathway, especially ribose-5-phosphate for RNA biosynthesis (Fig. 2.7b). It is, however, noted that such a correspondence between fluxes and enzyme activity measurements is not *generally found* because the latter represents v_{max} measurements determined *in vitro*, which may be completely different from the *in vivo* activity due to enzyme regulation or a variable degree of engagement of a particular pathway.

TABLE 2.3 Relative Flux Distribution (in Percentage) between the EMP and PP Pathways in Different Species

Species	EMP	PP	Conditions	Reference
A. niger	78		Batch culture	Shu *et al.* (1954)
C. glutamicum	32	66	Chemostat	Marx *et al.* (1996)
P. chrysogenum	77	23	Batch culture	Wang *et al.* (1958)
	56–70		Batch culture	Lewis *et al.* (1954)
P. digitatum	83	17	Batch culture	Wang *et al.* (1958)
	77	23	Batch culture	Reed and Wang (1959)

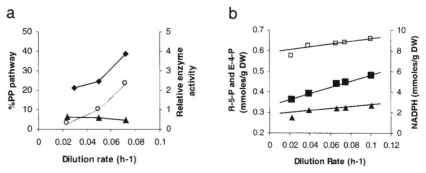

FIGURE 2.7 The relative flux through the PP pathway as function of the dilution rate in a glucose-limited continuous culture. (a) Experimentally determined relative flux through the PP pathway (♦) in *A. nidulans* determined by the respirometric method. Also shown are the relative activities of the EMPenzyme aldolase (▲) and the PP enzyme glucose-6-phosphate dehydrogenase (○) with respect to the activity of hexokinase. The hexokinase activity increased linearly with the specific growth rate. The data are taken from Carter and Bull (1969). (b) Calculated requirements for NADPH (□), ribose-5-phosphate (■), and erythrose-4-phosphate (▲) in *P. chrysogenun* as a function of the specific growth rate (all in millimoles per gram dry weight). The data are taken from Nielsen (1997).

The major control points of the EMP and PP pathways are at the entrances to the pathways, *i.e.*, at the phosphofructokinase and at the glucose-6-phosphate dehydrogenase reactions (Zubay, 1988). Glucose-6-phosphate dehydrogenase is regulated by the $NADPH/NADP^+$ ratio, whereas phosphofructokinase, which is a complex allosteric enzyme, has several effectors. In *S. cerevisiae* it is activated by AMP, ammonia, phosphate, and fructose-2,6-bisphosphate (a regulatory metabolite that is formed by phosphorylation of fructose-6-phosphate by ATP), and it is inhibited by ATP (Gancedo and Serrano, 1989). Binding of ATP leads to a decrease in the affinity for fructose-6-phosphate, whereas binding of the modifier fructose-2,6-bisphosphate results in a large increase in the affinity of the enzyme for fructose-6-phosphate (Zubay, 1988). Thus, fructose-2,6-bisphosphate has a stimulatory effect on the flux through the EMP pathway and ATP has an inhibitory effect. This regulatory profile allows phosphofructokinase to increase its activity to accommodate increasing concentrations of fructose-6-phosphate (which cause the level of fructose-2,6-bisphosphate to also increase); however, at high levels of the energy charge, further production of ATP is interrupted by inhibiting the activity of phosphofructokinase. This key role of phosphofructokinase becomes a central point in the analysis of complex metabolic pathways presented in Chapter 12.

The cofactors NADH and NADPH serve two different purposes in cellular metabolism. In aerobes, NADH is mainly involved in the generation of Gibbs free energy through the oxidative phosphorylation reaction (Section 2.3.3), whereas NADPH is mainly used in the biosynthesis of building blocks (Sections 2.4). Thus, NAD^+ serves as a substrate in the fueling reactions and NADPH serves as a substrate in the biosynthetic reactions, and the ratios $NADH/NAD^+$ and $NADPH/NADP^+$ are therefore regulated at different levels. In bacteria the ratio NADH/NAD is 0.03-0.08, whereas the ratio $NADPH/NADP^+$ is 0.7-1.0 (Ingraham et al., 1983), and in yeasts the two ratios are 0.25-0.30 and 0.58-0.75, respectively. The two coenzymes are, however, interconvertible by the action of the enzyme nicotinamide nucleotide transhydrogenase, which catalyzes the transfer of hydrogen between NAD^+ and $NADP^+$ according to the reaction:

$$NADH + NADP^+ - NAD^+ - NADPH = 0 \qquad (2.7)$$

This enzyme is present in bacteria and in mammalian cells (Hoek and Rydström, 1988), whereas it has not been identified in yeasts (Lagunas and Gancedo, 1973; Bruinenberg et al., 1985) and in filamentous fungi (Eagon, 1963). In mammalian cells the enzyme is located at the inner mitochondrial membrane, where it provides a mechanism for linking the transhydrogenase reaction to other energy-related processes in the mitochondrial membrane (Hoek and Rydström, 1988). The physiological role of this enzyme is not known, but it could provide a protective buffer against the dissipation of either the cellular redox power or the mitochondrial energy supply. In general, if it is present, it is not likely to play an important role in the overall cellular metabolism under normal growth conditions.

2.3.2. FERMENTATIVE PATHWAYS

Pyruvate, which is the last metabolite in glycolysis, may be further converted by several routes depending on the redox and energetic state of the cells. In aerobes, most of the pyruvate enters the TCA cycle (via acetyl-CoA) where it is oxidized completely to carbon dioxide and water (see Section 2.3.3). However, under oxygen-limited conditions or in anaerobic organisms, pyruvate may be converted into metabolic products like lactic acid, acetic acid, and ethanol via fermentative pathways.

The simplest fermentative pathway is the conversion of pyruvate to lactic acid by lactate dehydrogenase. The stoichiometry for this reaction is

$$\text{lactic acid} + NAD^+ - \text{pyruvate} - NADH - H^+ = 0 \qquad (2.8)$$

In this scheme, the NADH that is formed in the EMP pathway from the oxidation of glyceraldehyde-3-phosphate is consumed for the reduction of pyruvate to lactate, so that the overall conversion of glucose to lactic acid involves no net production of NADH. This pathway is active in higher eukaryotes, *e.g.*, muscle cells, under reduced oxygen (hypoxic) conditions. It is also active in many bacteria. In lactic acid bacteria it is the major, and in some cases the only, active fermentative pathway.[1]

Additional fermentative pathways can be operative in lactic acid bacteria that may lead to the formation of several different metabolic products, such as acetic acid, ethanol, formic acid, and carbon dioxide, besides lactic acid (see Example 8.3, where the mixed acid fermentation of lactic acid bacteria is analyzed). In *E. coli*, reaction (2.8) is important under anaerobic conditions, but it always operates together with other fermentative pathways in a mixed acid fermentation to yield several different metabolic products (see Example 3.1, where the mixed acid fermentation of *E. coli* is analyzed). Species of the Gram-positive bacteria *Clostridia*, of which *Clostridium acetobutylicum* is the most important industrially, have a complex fermentative metabolism (Fig. 2.8), leading to many different metabolic products. In this case, besides the conversion of acetyl-CoA to ethanol and acetic acid, two acetyl-CoA molecules may react to form acetoacetyl-CoA. This compound may be converted to butyryl-CoA in a sequence of three reactions (Fig. 2.8). In analogy with the conversion of acetyl-CoA to acetate and ethanol, butyryl-CoA may be converted to butyrate and butanol via butyryl-P and butyraldehyde, respectively. Alternatively, in another branch of the pathway, acetoacetate may be decarboxylated to acetone, which may be further reduced to isopropanol.

The fermentative metabolism of yeasts deviates somewhat from that of bacteria. The major end product is ethanol, and for this reason the process is known as *alcohol fermentation*. However, some acetate and minor amounts of succinate are also formed. The fermentative pathway leading to ethanol and acetate starts with decarboxylation of pyruvate to acetaldehyde, which is further reduced to ethanol by alcohol dehydrogenase or oxidized to acetate (Fig. 2.9). A major difference is that this fermentative metabolism does not proceed via acetyl-CoA, as in bacteria. Four isoenzymes of the alcohol dehydrogenase enzyme (ADHI, ADHII, ADHIII, and ADHIV) have been identified. The cytosolic ADHI is constitutively expressed during anaerobic growth on glucose and is responsible for the formation of ethanol. ADHII is cytosolic and glucose-repressed and is mainly associated with aerobic growth on ethanol. The function of the mitochondrial ADHIII, which is also glucose-repressed, is not known, but it may work as a shuttle of redox equivalents between the cytosol and the mitochondria (Nissen *et al.*, 1997).

---

[1] If the lactate dehydrogenase-catalyzed reaction is the only active pathway, the metabolism is often referred to as homofermentative.

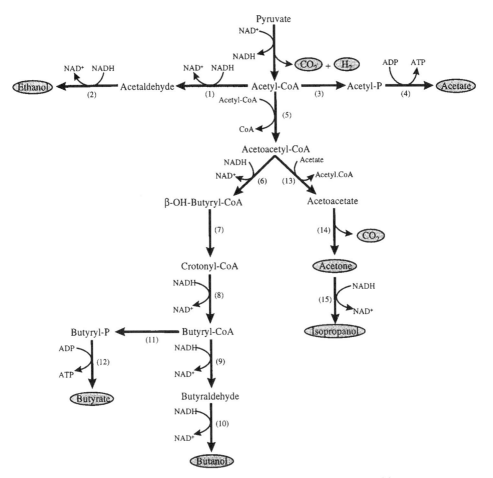

FIGURE 2.8 Mixed fermentation in *C. acetobutylicum*. The enzymes are (1) acetaldehyde
dehydrogenase; (2) ethanol dehydrogenase; (3) phosphotransacetylase; (4) acetate kinase; (5)
acetyl-CoA-acetyl transferase; (6) L(+)-β-hydroxybutyryl-CoA dehydrogenase; (7) 1,3-hydroxy-
acyl-CoA hydrolase; (8) butyryl-CoA dehydrogenase; (9) butyraldehyde dehydrogenase; (10)
butanol dehydrogenase; (11) phosphotransbutyrylase; (12) butyrate kinase; (13) CoA transferase;
(14) acetoacetate decarboxylase; (15) isopropanol dehydrogenase.

Two isoenzymes of aldehyde dehydrogenase have been identified. One may
use both NAD^+ and $NADP^+$ as cofactor, whereas the other is specific for
$NADP^+$. Thus, formation of acetate may lead to the formation of NADPH,
and this may be a key source of NADPH supply in yeast.

  In the overall conversion of glucose to ethanol by *S. cerevisiae*, there is no
net generation of NADH. However, because the synthesis of precursor

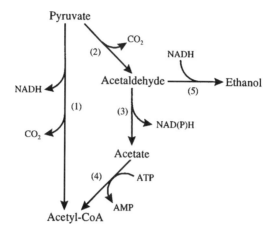

FIGURE 2.9 Fermentative metabolism of yeast. The enzymes are (1) pyruvate dehydrogenase; (2) pyruvate decarboxylase; (3) aldehyde dehydrogenase; (4) acetyl-CoA synthetase; (5) alcohol dehydrogenase. Reaction (1) runs in the mitochondria, whereas all of the other reactions run in the cytosol.

metabolites and building blocks used in biosynthesis causes a net formation of NADH (for example, formation of pyruvate from glucose leads to the generation of one NADH for each pyruvate formed), the cell needs a metabolic route for dissipating the excess NADH generated. One such possibility is glycerol formation from dihydroxyacetonephosphate (DAP). The overall stoichiometry for this pathway is

$$\text{glycerol} + \text{NAD}^+ + \sim P - \text{DAP} - \text{NADH} - H + = 0 \qquad (2.9)$$

The overall conversion of glucose to glycerol therefore yields a net consumption of NADH, and the preceding pathway may compensate for NADH generated in the synthesis of precursor metabolites and building blocks and not utilized in biosynthesis or other NADPH-requiring reactions.

2.3.3. TCA Cycle and Oxidative Phosphorylation

The first step in the complete oxidation of pyruvate is an oxidative decarboxylation leading to formation of acetyl-CoA (Fig. 2.10). This reaction is the result of a complex sequence of events catalyzed by a cluster of three enzymes, collectively called pyruvate dehydrogenase complex (PDC). In eukaryotic cells this enzyme complex is located in the mitochondria, as are

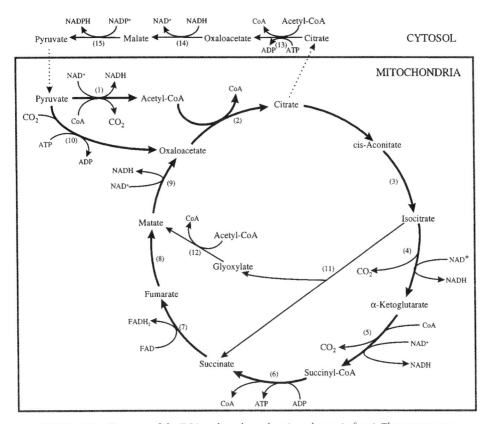

FIGURE 2.10   Overview of the TCA cycle and anaplerotic pathways in fungi. The enzymes are (1) pyruvate dehydrogenase complex; (2) citrate synthase; (3) aconitase; (4) isocitrate dehydrogenase; (5) α-ketoglutarate dehydrogenase; (6) succinate thiokinase; (7) succinate dehydrogenase; (8) fumarase; (9) malate dehydrogenase; (10) pyruvate carboxylase; (11) isocitrate lyase; (12) malate synthase; (13) ATP-citrate lyase; (14) malate dehydrogenase; (15) malic enzyme. In eukaryotic cells, the pyruvate dehydrogenase complex is membrane-bound, such that the conversion of pyruvate to acetyl-CoA is associated with the transport of pyruvate into the mitochondria (there is no carrier for pyruvate in the mitochondrial membrane). Furthermore, the *glyoxylate cycle*, i.e., reactions catalyzed by enzymes (2), (3), (11), (12), and (9), functions not in the mitochondria (as might be suggested from the figure) but in microbodies called *glyoxysomes*, where there is a net synthesis of succinate from acetyl-CoA. Succinate produced in the glyoxylate cycle is then transported to the mitochondria where it enters the TCA cycle. In prokaryotes there is no physical organization of the TCA cycle reactions all occurring in the cytosol, and reactions (13), (14), and (15) consequently are not needed.

all other enzymes relevant to reactions of aerobic catabolism beyond pyruvate. In the oxidative decarboxylation of pyruvate, the coenzyme NAD^+ serves as electron acceptor.

Carbon entry into the TCA cycle occurs by acetyl-CoA condensing with oxaloacetate to form citrate in a reaction catalyzed by citrate synthase (Fig. 2.10). Citrate is subsequently converted to its isomer isocitrate by aconitase. This reaction occurs via cis-aconitate, which has no other function in cellular metabolism, and it is, for this reason, disregarded in the consideration of the TCA cycle. The overall equilibrium constant for the conversion of citrate to isocitrate is close to 1, meaning equivalent amounts of citrate and isocitrate at equilibrium. These two metabolites are often lumped into a single metabolite pool. The next two steps of the TCA cycle are oxidative decarboxylations. First, isocitrate is converted to α-ketoglutarate by isocitrate dehydrogenase. Next, α-ketoglutarate is converted to succinyl-CoA by α-ketoglutarate dehydrogenase in a reaction sequence analogous to that catalyzed by the pyruvate dehydrogenase complex. In fact, the two complexes use the same dehydrogenase subunit. In the next reaction, succinyl-CoA is hydrolyzed to succinate with simultaneous liberation of Gibbs free energy from the hydrolysis of the CoA ester bond, which is recovered through the phosphorylation of GDP to GTP. In the next step succinate is dehydrogenated to fumarate. This reaction requires a strong oxidizing agent, and FAD is reduced to $FADH_2$. FAD is integrated in the succinate dehydrogenase, a flavoprotein enzyme, which catalyzes the reaction. Fumarate is converted to L-malate by fumarase, and, finally, the cycle is completed with the formation of oxaloacetate from L-malate by malate dehydrogenase.

The major regulatory sites of the TCA cycle are at citrate synthase, isocitrate de-hydrogenase, and α-ketoglutarate dehydrogenase. The activity of all three enzymes is favored by a low level of the $NADH/NAD^+$ ratio, while the isocitrate dehydrogenase is strongly regulated by this ratio. Furthermore, in yeast, isocitrate dehydrogenase is activated by AMP and inhibited by ATP (Zubay, 1988).

The overall stoichiometry for the complete oxidation of pyruvate in the TCA cycle is

$$3CO_2 + GTP + 4NADH + FADH_2 + 4H^+$$

$$- \text{pyruvate} - 3H_2O - GDP - 2 \sim P - 4NAD^+ - FAD = 0 \quad (2.10)$$

Thus, 4 mol of NADH and 1 mol of $FADH_2$ are formed for each mole of pyruvate oxidized. Obviously, the TCA cycle can continue to work only if these two cofactors are reoxidized to NAD^+ and FAD. Under aerobic conditions, electrons are transferred from these coenzymes to free oxygen via a chain of electron acceptors. Most of the electron acceptors are organized in large complexes that are embedded in the plasma membrane of prokaryotes and the inner mitochondrial membrane of eukaryotes (Fig. 2.11). The movement of electrons from NADH to oxygen through the electron transport chain is driven by a large, negative Gibbs free energy, *i.e.*, $\Delta G° = -220$ kJ (mol NADH)$^{-1}$, and some of this free energy is captured in the form of ATP. How these two completely different processes-electron transport and ATP synthesis-are coupled is explained by the chemiosmotic theory, which was proposed in 1961 by Peter Mitchell (Mitchell, 1961; Mitchell and Moyle, 1965; Senior, 1988). This theory proposes that when electrons are transported through the chain, protons are pumped out of the mitochondria (or out of the cell in prokaryotes), and thereby both a pH gradient (about 0.05 pH unit, greater inside the cell or the mitochondrion) and an electrical potential gradient (about -0.15 V) are created across the inner mitochondrial membrane (or plasma membrane in prokaryotes) (Fig. 2.11). When protons re-enter the mitochondrial matrix (or the cytosol) via the F_1F_0-ATPase complex (or the ATP-synthase), ATP is produced. For eukaryotes, the theoretical stoichiometry of oxidative phosphorylation (the so-called P/O ratio) is 3 mol of ATP synthesized for each mole of $NADH_2$ oxidized or 2 mol of ATP synthesized for each mole of succinate (or $FADH_2$) oxidized (see Fig. 2.11). For prokaryotes, protons are transported at only two locations (two protons at each location), and because the stoichiometry of the prokaryotic F_1F_0-ATPase is $1ATP/2H^+$, the theoretical stoichiometry of the oxidative phosphorylation yields 2 mol of ATP synthesized for each mole of $NADH_2$ oxidized. Due to incomplete coupling of the oxidation and phosphorylation processes, the so-called operational P/O ratio is much below the theoretical values (see Example 13.6). This is the result of extraneous processes driven by the proton gradient across the membrane. Thus, many compounds are driven by proton symport, as discussed in Section 2.2.3, and this affects the proton gradient across the plasma membrane in prokaryotes. Similarly for eukaryotes, the inner mitochondrial membrane has specific carriers for phosphate, ATP, ADP, and the metabolites pyruvate, citrate, isocitrate, succinate, and malate, and the transport of these compounds is also driven by the proton gradient (LaNoue and Schoolwerth, 1979; Zubay, 1988).

The inner mitochondrial membrane is impermeable to NADH, and, therefore, oxidation of cytosolic NADH in eukaryotes the requires transport of electrons to the electron transport chain by means other than the NADH dehydrogenase, which is specific toward mitochondrial NADH. For this purpose, fungal mitochondria are equipped with a NADH dehydrogenase, which faces the outer surface of the inner mitochondrial membrane and accepts electrons from cytosolic NADH (von Jagow and Klingenberg, 1970;

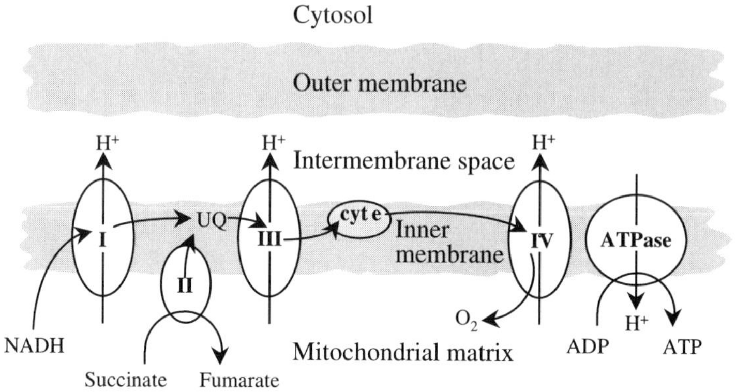

FIGURE 2.11 The electron transport chain and oxidative phosphorylation in eukaryotes. Electrons are transported from NADH or succinate, through the electron transport chain, to oxygen. The elements of the electron transport chain are organized in large complexes located in the inner mitochondrial membrane. Electrons from NADH are first donated to complex I, the NADH dehydrogenase, which is a flavoprotein containing flavin mononucleotide. Electrons from succinate are first donated to FAD, which is integrated in complex II, the succinate dehydrogenase (one of the TCA cycle enzymes). Electrons from complexes I and II (or other flavoproteins located in the inner mitochondrial membrane) are transferred to ubiquinone (UQ), which diffuses freely in the lipid membrane. From UQ, electrons are passed on to the cytochrome system. First, the electrons are transferred to complex III, which consists of two b-type cytochromes ($b566$ and $b562$) and cytochrome $c1$. Electrons are then transferred to complex IV via cytochrome c, which is only loosely attached to the outside face of the membrane. Complex IV (or cytochrome oxidase), which contains cytochromes a and $a3$, finally delivers the electrons to oxygen. Complexes I, III, and IV span the inner mitochondrial membrane, and when two electrons are transported through these complexes, protons (four at each complex) are released into the intermembrane space. These electrons may be transported back into the mitochondrial matrix by a proton-conducting ATP-synthase (or the F_1F_0-ATPase complex). In this complex, one ATP is generated when three protons pass through the ATPase. One additional proton is transported into the mitochondrial matrix in connection with the uptake of ADP and $\sim$ P and the export of ATP, so that a total of four protons are required per ATP generated by this mechanism. Thus, for each electron pair transferred from complex I all the way to complex IV, 12 protons (four protons for each of the three complexes) are pumped from the mitochondrial matrix into the intermembrane space. Upon re-entry of the protons into the mitochondrial matrix through the ATP synthase, 3 mol of ATP are generated (12 protons/4 protons per ATP). The theoretical stoichiometry of the oxidative phosphorylation is, therefore, 3 mol of ATP synthesized per mole NADH oxidized and 2 mol of ATP synthesized per mole of succinate oxidized.

Watson, 1976). This dehydrogenase probably delivers its electrons to the common acceptor UQ, because only two phosphorylation sites are involved in the oxidation of cytoplasmic NADH (Watson, 1976). Another mechanism by which electrons from cytosolic NADH may be processed in the electron transport chain is by means of so-called shuttle systems, the simplest of which involves the reduction of dihydroxyacetonephosphate to glycerol-3-phosphate (DHAP + NADH → G3P + NAD$^+$) in the cytosol, followed by the reoxidation of glycerol-3-phosphate to dihydroxyacetonephosphate by a mitochondrial glycerol-3-phosphate dehydrogenase. The catalytic site of the mitochondrial glycerol-3-phosphate dehydrogenase is on the outside surface of the inner mitochondrial membrane, so that glycerol-3-phosphate does not have to pass through the inner mitochondrial membrane in order to be reoxidized. The mitochondrial glycerol-3-phosphate dehydrogenase is a flavoprotein containing FAD, which feeds electrons to the electron transport chain at the site of UQ. When electrons are transported via this shuttle system, the P/O ratio for oxidation of NADH therefore will be similar to that for oxidation of succinate.

2.3.4. ANAPLEROTIC PATHWAYS

Two of the TCA cycle intermediates, α-ketoglutarate and oxaloacetate serve as precursor metabolites for the biosyntheses of amino acids and nucleotides.[2] There is no net synthesis of these two organic acids in the TCA cycle, and their removal for other cellular functions must be compensated for by other means. Reaction sequences that fulfill this role are summarily referred to as *anaplerotic pathways*. The anaplerotic pathways include (see Fig. 2.10) the following: (1) carboxylation of pyruvate by pyruvate carboxylase; (2) carboxylation of phosphoenolpyruvate by PEP carboxylase; (3) oxidation of malate to pyruvate by the malic enzyme; and (4) the glyoxylate cycle.

The most important anaplerotic pathways are carbon dioxide fixation by either pyruvate carboxylase or PEP carboxylase, leading to the formation of oxaloacetate (Fig. 2.10). Pyruvate carboxylase is activated at a high ATP/ADP ratio and by acetyl-CoA and is inhibited by L-aspartate. Thus, the regulation of this enzyme is almost completely the reverse of that of the pyruvate dehydrogenase complex, which is inhibited by high ATP/ADP or NADH/NAD$^+$ ratios and high acetyl-CoA concentrations (Zubay, 1988). In

---

[2] Succinyl-CoA serves as precursor metabolite in the synthesis of heme and hemelike compounds, but these will not be treated here. Succinyl-CoA also serves as a cofactor in the synthesis of some building blocks. Here succinate is later released and the net expenditure is therefore ATP for regeneration of succinyl-CoA from succinate.

S. cerevisiae, pyruvate carboxylase activity is found in both the cytosol and the mitochondria (Haarasilta and Taskinen, 1977). PEP carboxylase is quite active in many prokaryotes, whereas it has not been identified in fungi. It is regulated similar to pyruvate carboxylase, *i.e.*, inhibited by L-aspartate and activated by acetyl-CoA (Jetten *et al.*, 1994).

Acetyl-CoA cannot be transported through the inner mitochondrial membrane into the cytosol where it is needed as a key precursor for amino acid and lipid biosynthesis. In eukaryotes, if there is an imbalance between the cellular needs for energy and carbon skeletons for biosynthesis, acetyl-CoA synthesis in the cytosol must take place. This may occur by two different pathways: (1) The first utilizes the free transport of citrate through the mitochondrial membrane and involves the enzyme citrate lyase (reaction (13) in Fig. 2.10), which cleaves cytoplasmic citrate to oxaloacetate and acetyl-CoA with concurrent hydrolysis of ATP to ADP. Normally, the requirement for acetyl-CoA exceeds that for oxaloacetate, and the excess oxaloacetate generated by this reaction is converted to L-malate by a cytoplasmic malate dehydrogenase. L-Malate may then either re-enter the mitochondria or be oxidatively decarboxylated to pyruvate by the malic enzyme (reaction (15) in Fig. 2.10). In the oxidation of L-malate by the malic enzyme, NADPH is formed, and this reaction may be a major source for NADPH synthesis in the cytosol of eukaryotes. (2) The second pathway for generation of cytoplasmic acetyl-CoA is via acetate. First, pyruvate is oxidatively decarboxylated to yield acetate

$$\text{acetate} + CO_2 + NADH + H + - \text{pyruvate} - NAD^+ = 0 \quad (2.11)$$

and, thereafter, acetate is converted to acetyl-CoA by acetyl-CoA synthase:

$$\text{acetyl-CoA} + H_2O + AMP + PP_i - \text{acetate-CoA} - ATP = 0 \quad (2.12)$$

This pathway is used by *S. cerevisiae* (Frenkel and Kitchens, 1977), but in *A. nidulans* acetyl-CoA synthase is induced by acetate and repressed by glucose (Kelly and Hynes, 1982). Reaction (2.12) therefore is mainly active during growth on acetate.

In the glyoxylate cycle (also referred to as *glyoxylate shunt*), isocitrate is cleaved by isocitrate lyase [reaction 11 in Fig. (2.10)] to form succinate and glyoxylate, which may react with acetyl-CoA by the action of malate synthase [reaction (12)] to form L-malate. L-Malate may then be converted to isocitrate in a sequence of reactions identical to those of the TCA cycle [reactions (9) and (2) in Fig. 2.10]. The net result of the glyoxylate cycle is, therefore, synthesis of the four-carbon succinate from 2 molecules of acetyl-CoA, and this pathway is important during the metabolism of acetate and fatty acids where acetyl-CoA is a common intermediate.

In eukaryotes, the supply of α-ketoglutarate to the cytosol for biosynthesis is important. This precursor metabolite may be transported across the inner mitochondrial membrane by a specific permease, but it may also be synthesized in the cytosol by a cytosolic isocitrate dehydrogenase. Two different kinds of isocitrate dehydrogenases have been identified in fungi: one NAD^+ dependent and one linked to $NADP^+$. The NAD^+-isocitrate dehydrogenase is always associated with the mitochondria, and in A. nidulans the $NADP^+$ dependent enzyme is repressed by glucose and is mainly active during growth on acetate (Kelly and Hynes, 1982). This may offer an explanation for the existence of two isocitrate dehydrogenases: They allow for the generation of NADPH during growth on acetate, where it is energetically expensive to form NADPH via the PP pathway. For other organisms, activity of the $NADP^+$-linked enzyme is also found during growth on glucose, and here it may play an important role in the supply of NADPH.

2.3.5. CATABOLISM OF FATS, ORGANIC ACIDS, AND AMINO ACIDS

The catabolism of larger fatty acids starts with the activation of the fatty acid by reaction with CoA and ATP to yield fatty acyl-CoA:

$$RCO\text{-}CoA + AMP + PP_i - RCOOH - ATP - CoA = 0 \qquad (2.13)$$

This reaction is catalyzed by acyl-CoA ligase (or thiokinase), and in mammalian cells it takes place in the cytosol. The fatty acyl-CoA is oxidized at β-carbon and cleaved to yield acetyl-CoA and a fatty acyl-CoA with two fewer carbons in a sequence of four reactions. This proceeds until the fatty acyl-CoA has been completely degraded to acetyl-CoA. The overall stoichiometry for the degradation is

$$(n + 1)\text{acetyl-CoA} + n\text{NADH} + n\text{FADH}_2 + n\text{H}^+$$

$$- CH_3(CH_2)_{2n}CO\text{-}CoA - n\text{NAD}^+ - n\text{FAD} - n\text{CoA} = 0 \quad (2.14)$$

In mammalian cells these reactions take place in the mitochondria, and the fatty acyl-CoA is transported there as an acylcarnitine derivative. The electrons captured in $FADH_2$ are transferred to UQ in the electron transport chain. In yeasts, where the oxidation of fatty acids occurs in microbodies, the electrons are transferred from $FADH_2$ directly to free oxygen under the evolution of hydrogen peroxide, which is degraded thereafter by catalase (Tanaka and Fukui, 1989).

Acetyl-CoA is a common intermediate in the metabolism of acetic acid and fatty acids. If these compounds are the *only* carbon source, special pathways are required for the synthesis of some of the precursor metabolites normally synthesized in glycolysis, *e.g.*, pyruvate, glyceraldehyde-3-phosphate, and the hexose-6-phosphates. This occurs by the glyoxylate shunt, whereby succinate is formed from acetyl-CoA, *and* by *gluconeogenesis*, which can be thought of as the reverse of the EMP pathway. During growth on acetate and fatty acids, gluconeogenesis is initiated with oxaloacetate (formed from succinate), which is decarboxylated and phosphorylated by GTP to phosphoenolpyruvate by phosphoenolpyruvate carboxykinase. Phosphoenolpyruvate is then converted to fructose-1,6-bisphosphate by the EMP pathway enzymes. Finally, fructose-1,6-bisphosphate is hydrolyzed to fructose-6-phosphate, a constituent of the hexose monophosphate pool, by fructose bisphosphate phosphatase. During growth on lactate,[3] after lactate uptake and conversion to pyruvate, the first step in gluconeogenesis is the formation of oxaloacetate from pyruvate by pyruvate carboxylase (Fig. 2.10). Oxaloacetate is then converted to phosphoenolpyruvate and the other glycolytic intermediates by the same gluconeogenic reactions as the preceding ones. Of course, the gluconeogenesis pathway can proceed only when the two key reactions of glycolysis, namely, phosphofructokinase and pyruvate kinase, are not active in the forward (*i.e.*, glycolytic) direction. This is accomplished by controlling the *activity* of these enzymes (which are always present) at low levels through the absence of fructose-2,6-bisphosphate (activator of phosphofructokinase) and the low level of fructose-1,6-bisphosphate (activator of pyruvate kinase) (Gancedo and Serrano, 1989). During growth on glucose, the two specific enzymes of gluconeogenesis are repressed by glucose.

Many organisms synthesize and secrete extracellular *proteases*, which hydrolyze proteins to low-molecular-weight peptides and/or amino acids. Peptides do not have to be hydrolyzed completely to amino acids before uptake as many organisms may transport small oligopeptides, *i.e.*, up to pentapeptides. Once inside the cells, these oligopeptides are hydrolyzed by intracellular *proteases* or *peptidases*. Both intracellular and extracellular proteases are normally repressed by ammonia, and synthesis of proteases is often repressed by excess of carbon, sulfur, and phosphate. After protein hydrolysis, catabolism of most resulting amino acids begins with transamina-

---

[3] Lactate metabolism in fungi starts with the conversion to pyruvate by lacate dehydrogenase. The fungal lactate dehydrogenase is a flavoprotein which cannot catalyze the reverse reaction and lactate is therefore not produced by fungi. In yeast the lactate dehydrogenase is located at the intermembrane space of the mitochondria and the enzyme transfers the electrons directly to cytochrome *c* and it contains FAD as coenzyme. The enzyme is repressed by glucose and induced by lactose.

tion where the α-amino nitrogen is transferred to α-ketoglutarate by gluta-mate transaminase:

$$\text{glutamate} + \alpha\text{-keto acid} - \alpha\text{-ketoglutarate} - \text{L-amino acid} = 0 \quad (2.15)$$

Glutamate is subsequently oxidatively deaminated by a NAD-linked gluta-mate dehydrogenase, and hereby α-ketoglutarate is regenerated.

$$\alpha\text{-ketoglutarate} + NH_3 + NADH + H + - \text{glutamate} - NAD^+ - H_2O = 0$$
$$(2.16)$$

The NAD-linked glutamate dehydrogenase mainly serves a catabolic function and its activity is generally low during growth on a minimal medium not containing amino acids. After deamination, the resulting carbon skeleton is broken down to pyruvate, acetyl-CoA or an intermediate of the TCA cycle (Table 2.4). For some amino acids there are many steps in the breakdown process, *e.g.* the breakdown of tryptophan to acetyl-CoA involves 12 steps, whereas others are converted directly to their final metabolite upon deamina-tion.

2.4. BIOSYNTHETIC REACTIONS

The number of building blocks, coenzymes, and prosthetic groups needed for cellular synthesis is about 75-100, and these are all synthesized from the 12 precursor metabolites formed in the biosynthetic pathways (Ingraham *et al.*, 1983). In the overview of biosynthetic reactions provided in this section, the focus is on the role of these biosynthetic pathways in *cell growth* and on the resources required for their operation. In this regard, no detailed description

TABLE 2.4   Overview of the Breakdown of Amino Acids[a]

Breakdown product	Amino acids
Pyruvate	Alanine (1), serine (1), cysteine (3), and glycine (2)
Acetyl-CoA	Threonine (1), lysine (10), leucine (8), tyrosine (7), phenylalanine (8), and tryptophan (12)
α-Ketoglutarate	Glutamate (1), glutamine (2), proline (3), arginine (4) and histidine (5)
Succinyl-CoA	Methionine (9), isoleucine (9), and valine (8)
Oxaloacetate	Aspartate (1) and asparagine (2)

[a] The figure in the parenthesis is the number of steps in the breakdown pathway in the higher eukaryotes. For most of the amino acids where this number is one the breakdown product is directly formed by deamination according to eq. (2.16).

of the individual reaction steps in any of the pathways is given, and only the major biosynthetic pathways are considered, *i.e.*, biosynthesis of amino acids, nucleotides, sugars, amino sugars, and lipids. More details can be found in standard textbooks on biochemistry and in many extensive reviews on these subjects [see, for example, Umbarger (1978) for amino acid biosynthesis in fungi, Jones and Fink (1982) for biosynthesis of amino acids and nucleotides in fungi, and Neidhardt *et al.* (1987) for an extensive treatment of the biosynthesis of building blocks in bacteria].

2.4.1. BIOSYNTHESIS OF AMINO ACIDS

Amino acids are best known as the building blocks of proteins, and this is, indeed, the main function of the 20 L-amino acids most commonly found in cells. However, amino acids also serve as precursors for the biosynthesis of other building blocks and important secondary metabolites, *e.g.*, penicillin. The first step in amino acid biosynthesis is nitrogen assimilation, whereby nitrogen, in the form of ammonia, is fixed and incorporated into organic molecules. This primarily occurs through the biosynthesis of the amino acid L-glutamate from α-ketoglutarate:

$$\text{L-glutamate} + \text{NADP}^+ + \text{H}_2\text{O} - \alpha\text{-ketoglutarate}$$

$$- \text{NH}_3 - \text{NADPH} - \text{H}^+ = 0 \tag{2.17}$$

This reaction is catalyzed by a NADP-linked glutamate dehydrogenase (GDH), which is a key enzyme in cellular metabolism. This enzyme is different from the NAD-GDH, which catalyzes the reverse reaction (2.16), and the two enzymes are under different types of regulation. The NADP-GDH is re-pressed by L-glutamate and is present at high activity during growth on glucose, whereas the NAD-GDH is repressed by glucose.

Another route for L-glutamate biosynthesis is via the so-called GS-GOGAT pathway, which consists of two steps. In the first step, L-glutamine is used as an amino donor to aketoglutarate and the result is the formation of two L-glutamates:

$$2\text{L-glutamate} + \text{NADP}^+ - \alpha\text{-ketoglutarate} - \text{L-glutamine}$$

$$- \text{NADPH} - \text{H}^+ = 0 \tag{2.18}$$

This reaction is catalyzed by glutamate synthase (abbreviated GOGAT from its previous trivial name: glutamine amide-2-oxoglutarate aminotransferase).

The second step is the regeneration of L-glutamine by the reaction:

$$\text{L-glutamine} + \text{ADP} + \ \sim \text{P} - \text{L-glutamate} - \text{NH}_3 - \text{ATP} = 0 \quad (2.19)$$

which is catalyzed by glutamine synthase (GS). The sum of the two reactions (2.18) and (2.19) is the net synthesis of L-glutamate from α-ketoglutarate, similar to reaction (2.17), but an important difference is that now energy is required, *i.e.*, one ATP is hydrolyzed for each L-glutamate formed. The GS-GOGAT pathway is a high-affinity system for ammonia assimilation, and it is mainly active at low ammonia concentrations because glutamate synthase is repressed by ammonia. L-Glutamine serves as an ammonia (nitrogen) donor in the biosynthesis of several nitrogen-containing compounds, and it represents an important branch point in the overall cellular metabolism. The glutamine synthase is tightly regulated: It is repressed by L-glutamine, and it is inhibited by many metabolic pathway end products whose origin can be traced to L-glutamine (adenosine monophosphate, guanosine triphosphate, L-glycine, and L-histidine). Many organisms may also utilize nitrate or nitrite as the sole nitrogen source. Before assimilation occurs, however, these compounds are converted into ammonia, which is thus a central compound in the overall nitrogen metabolism. Reduction of nitrate to ammonia proceeds over nitrite, hyponitrate (N_2O_2), nitrous oxide (N_2O), and hydroxylamine (NH_2OH). The hydrogen donor in these reductive reactions is NADPH. The uptake of nitrate and nitrite is probably coupled to the reduction of these compounds in the cytosol.

The biosynthetic pathways for all 20 common amino acids have been elucidated in many eukaryotes and prokaryotes, and an overview of the pathways is given in Fig. 2.12. There are only a few deviations in the biosynthetic routes, of which lysine biosynthesis is the most important. In bacteria and higher plants, lysine is synthesized from pyruvate and aspartic-β-semialdehyde via diaminopimelic acid (an important building block for bacteria cell wall synthesis), whereas in fungi lysine is synthesized from α-ketoglutarate via α-aminoadipic acid.

Table 2.5 summarizes the "metabolic costs" for the biosynthesis of amino acids in bacteria and fungi. The biosynthesis of L-methionine and L-histidine requires transfer of a one-carbon group, which is donated by N^5-methyltetrahydrofolate and 10-formyltetrahydrofolate, respectively (which are both converted to tetrahydrofolate). These different forms of tetrahydrofolate are interconvertible, and to analyze the metabolic costs it is convenient to use a common basis. In Table 2.5, the conversion of 5,10-methylenetetrahydrofolate (5,10-MTHF) to tetrahydrofolate (THF) is used as the common basis, because this conversion is associated with the synthesis of L-glycine from L-serine. The amount of 5,10-MTHF generated during the biosynthesis of

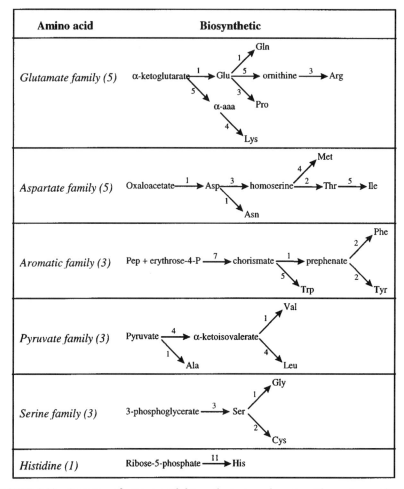

FIGURE 2.12 Overview of amino acid biosynthesis in eukaryotes. The amino acids are classified into five families according to the specific precursor metabolite or amino acid that serves as the starting point for their synthesis. L-Histidine, which has a complex biosynthetic pathway, does not group with any of the other amino acids. The numbers indicate the reaction steps in the pathway. Except for L-lysine, these numbers are the same for bacteria. In bacteria L-lysine is synthesized from aspartate via diaminopimelic acid (an important building block for bacteria cell wall) in a sequence of nine reactions.

TABLE 2.5 Metabolic Costs for Biosynthesis of the Twenty Amino Acids in Bacteria and Fungi

Amino acid	Precursor metabolites[a]	ATP[b]	NADH	NADPH	1-C[c]	NH_3	S[d]
L-Alanine	1 pyr	0	0	-1	0	-1	0
L-Arginine	1 α kg	-7	1	-4	0	-4	0
L-Asparagine	1 oaa	-3	0	-1	0	-2	0
L-Aspartate	1 oaa	0	0	-1	0	-1	0
L-Cysteine[e]	1 pga	-4	1	-5	0	-1	-1
L-Glutamate	1 α kg	0	0	-1	0	-1	0
L-Glutamine	1 α kg	-1	0	-1	0	-2	0
L-Glycine	1 pga	0	1	-1	1	-1	0
L-Histidine	1 penP	-6	3	-1	-1	-3	0
L-Isoleucine	1 oaa, 1 pyr	-2	0	-5	0	-1	0
L-Leucine	2pyr, 1 acCoA	0	1	-2	0	-1	0
L-Lysine (fungi)	1 akg, 1 acCoA	-2	2	-4	0	-2	0
L-Lysine	1 pyr, 1 oaa	-3	0	-4	0	-2	0
L-Methionine	1 oaa	-7	0	-8	-1	-1	-1
L-Phenylalanine	2 pep, 1 eryP	-1	0	-2	0	-1	0
L-Proline	1 α kg	-1	0	-3	0	-1	0
L-Serine	1 pga	0	1	-1	0	-1	0
L-Threonine	1 oaa	-2	0	-3	0	-1	0
L-Tryptophan	1 pep, 1 eryP, 1 penP	-5	2	-3	0	-2	0
L-Tyrosine	2 pep, 1 eryP	-1	1	-2	0	-1	0
L-Valine	2 pyr	0	0	-2	0	-1	0

[a] acCoA, acetyl-CoA; eryP, erythrose-4-phosphate; fruP, fructose-6-phosphate; gluP, glucose-6-phosphate; α kg, α-ketoglutarate; glyP, glyceraldehyde-3-phosphate; oaa, oxaloacetate; penP, ribose-5-phosphate; pep, phosphoenolpyruvate; pga, 3-phosphoglycerate; pyr, pyruvate.
[b] For those reactions where ATP is hydrolyzed to AMP, it is assumed that two ATPs are used.
[c] 5,10-Methylene tetrahydrofolate is used as one-carbon donor which is converted to tetrahydrofolate. Other forms of tetrahydrofolate used in the biosynthesis of 1-methionine and 1-histidine are converted to this basis.
[d] Sulfate is used as S-source, which is reduced to H_2S before assimilation.
[e] Direct sulfhydrylation of L-serine is assumed.

L-glycine is, however, generally not sufficient to meet the requirements for one-carbon transfer groups. The most likely source of the additionally required one-carbon transfer group is the α-carbon of glycine, which is degraded according to reaction (2.20) by glycine oxidase.

$$CO_2 + NH3 + 5,10\text{-MTHF} + NADH + H^+ - \text{L-glycine}$$

$$- THF - NAD^+ = 0 \qquad (2.20)$$

2.4.2. BIOSYNTHESIS OF NUCLEIC AIDS, FATTY ACIDS, AND OTHER BUILDING BLOCKS

Nucleotides in the form of ribonucleotides and deoxyribonucleotides are the building blocks of RNA and DNA, respectively. They also serve, however, many other functions in the cell. They are major constituents of a number of cofactors, *e. g.*, NADH, NADPH, FAD, and CoA, and some nucleotides, *e. g.*, ATP, serve specific purposes in the overall cellular metabolism. Nucleotides are composed of three parts: (1) a heterocyclic nitrogenous base, termed a purine or a pyrimidine; (2) a sugar (ribose in RNA and 2-deoxyribose in DNA); and (3) a phosphoryl group. The bases of DNA are the purines, adenine (A) and guanine (G), and the pyrimidines, thymine (T) and cytosine (C). Three of these bases are also found in RNA, but the fourth, thymine, is replaced by uracil (U).

The deoxyribonucleotides (dAMP, dGMP, dUMP, and dCMP) are formed from the corresponding ribonucleotides (AMP, GMP, UMP, and CMP) by replacement of the 2'-OH group with hydrogen (donated by NADPH), and dTMP is formed by methylation of dUMP. The nucleotides are synthesized from ribose-5-phosphate and 3-phosphoglycerate (for purine biosynthesis) or oxaloacetate (for pyrimidine biosynthesis). Table 2.6 lists the "metabolic costs" for the biosynthesis of nucleotides.

Lipids are a heterologous group of compounds including (1) acylglycerols (2) phospholipids, and (3) sterols. The major building blocks for these lipid components are fatty acids, which are present in acylglycerols, phospholipids, and sterol esters (the predominant pool of sterols). The major fatty

TABLE 2.6 Metabolic Costs for Biosynthesis of Nucleotides

Nucleotide	Precursor metabolite[a]	ATP	NADH	NADPH	1-C	NH$_3$
AMP	1 pga, 1 penP	-9	3	-1	-1	-5
GMP	1 pga, 1 penP	-11	3	0	-1	-5
UMP	1 oaa, 1 penP	-5	0	-1	0	-2
CMP	1 oaa, 1 penP	-7	0	-1	0	-3
dAMP	1 pga, 1 penP	-9	3	-2	-1	-5
dGMP	1 pga, 1 penP	-11	3	-1	-1	-5
dTMP[b]	1 oaa, 1 penP	-5	0	-3	-1	-2
dCMP	1 oaa, 1 penP	-7	0	-2	0	-3

[a] See Table 2.4 for nomenclature.
[b] The costs for dTMP biosynthesis are the same as for synthesis from dUMP. dTMP may also be synthesized from dCMP with higher costs, *i.e.*, nine ATPs. For *E. coli*, Ingraham *et al.* (1983) state that 75% of the dTMP is synthesized from dCMP and 25% from dUMP.

acids are palmitic acid (C16:0), palmitoleic acid (C16:1), stearic acid (C18:0), oleic acid (C18:1), linoleic acid (C18:2), and linolenic acid (C18:3). In fungi, palmitic, oleic, and linoleic acids account for more than 75% (Ratledge and Evans, 1989), whereas the most frequent fatty acids in bacteria are palmitic, palmitoleic, and oleic acids (Ingraham *et al.*, 1983). Biosynthesis of saturated fatty acids occurs by successive additions of two carbon units to an activated form of acetyl-CoA. The carbon units are donated by malonyl-CoA, which is formed by carboxylation of acetyl-CoA (Walker and Woodbine, 1976). In yeasts (and probably in other fungi), the end product is a fatty acyl-CoA ester rather than the fatty acid (Ratledge and Evans, 1989). The overall stoichiometry for the biosynthesis of an n-chain fatty acyl-CoA is

$$CH_3(CH_2)_{n-2}CO\text{-CoA} + \frac{n-2}{2}\, CoA + \frac{n}{2}\, H_2O + \frac{n-2}{2}\, ADP + \frac{n-2}{2}$$

$$\sim P + \frac{n-2}{2}\, NADP^+ - \frac{n}{2}$$

$$acetyl\text{-CoA} - \frac{n-2}{2}\, ATP - (n-2)\, NADPH - (n-2)\, H^+ = 0 \quad (2.21)$$

In bacteria, synthesis of monounsaturated fatty acids occurs by the anaerobic pathway. As the name implies, this pathway is functional in the absence of oxygen. The pathway operates as mentioned earlier until four malonyl-CoAs have been added to the growing chain. The compound formed, β-hydroxydecanoyl-ACP, forms a branch point between the synthesis of saturated and monounsaturated fatty acids. By the action of the enzyme β-hydroxydecanoyl thioester dehydrase, a β,γ-cis-double bond is inserted, and the β,γ-unsaturated acyl is subsequently elongated to yield palmitoleic acid. In eukaryotes, the double bond at the ninth carbon atom is introduced after the C_{16} or C_{18} saturated fatty acyl-CoA has been synthesized (Walker and Woodbine, 1976; Ratledge and Evans, 1989). This conversion is performed by a specific enzyme system associated with the endoplasmic reticulum, and the reaction requires NADH and molecular oxygen according to the following stoichiometry:

$$oleoyl\text{-CoA} + 2H_2O + NAD^+ - stearoyl\text{-CoA} - O_2 - NADH - H^+ = 0 \quad (2.22)$$

Bacteria do not possess polyunsaturated fatty acids, whereas eukaryotes produce a large variety of these. Synthesis of polyunsaturated fatty acids may occur by a reaction similar to eq. (2.22). Alternatively, the monounsaturated fatty acid may be further desaturated after incorporation into a phospholipid (Walker and Woodbine, 1976; Ratledge and Evans, 1989).

Other important building blocks of lipid components are (1) glycerol-3-phosphate, which is the backbone of phospholipids and triacylglycerol; (2) the alcohol moiety of phospholipids; and (3) sterols. Glycerol-3-phosphate is derived directly from dihydroxy acetonephosphate of the EMP pathway. The most common alcohols of phospholipids in fungi are choline, ethanolamine, and inositol (Rose, 1976). Thus, in S. cerevisiae, the phospholipids phosphatidylcholine (PC), phosphatidylethanolamine (PE), and phosphatidylinositol (PI) account for more than 90% of the total phospholipids (Ratledge and Evans, 1989). Ethanolamine and choline cannot be synthesized in free form, and the corresponding phospholipids are therefore formed from other phospholipids. Thus, PE is formed by direct decarboxylation of phosphatidylserine (a phospholipid present in low amounts and that contains L-serine as the alcohol), and PC is formed by successive methylation of PE. PI is formed by incorporation of free inositol, which comes from glucose-6-phosphate via a two-step pathway (Umezawa and Kishi, 1989). First, inositol-phosphate synthase converts glucose-6-phosphate to inositol-1-phosphate using NAD^+ as electron acceptor, and then inositol-1-phosphate is converted to inositol by inositol-1-phosphatase. Of the sterols, ergosterol accounts for more than 90% in S. cerevisiae. The pathway to ergosterol proceeds via squalene, which is a common intermediate in sterol biosynthesis. The steps leading to squalene have been completely elucidated in S. cerevisiae, whereas the details of the further conversion to ergosterol (and other sterols) are not fully understood. The most important classes of phospholipids in bacteria are phosphatidylethanolamine (75-85%), phosphatidylglycerol (10-20%), and cardiolipin (5-15%). All three phospholipids share the same biosynthetic route up to the formation of CDP-diacylglycerol, after which the pathway branches. Phosphatidylethanolamine is synthesized by decarboxylation of phosphatidylserine, which is formed by the replacement of CDP with L-serine. Phosphatidylglycerol is formed from CDP-diacylglycerol by the replacement of CDP with glycerol-3-phosphate and subsequent cleavage of the phosphate from the glycerol moiety. Finally, cardiolipin is formed by condensation of two phosphatidylglycerols with a release of glycerol.

The major building block for the synthesis of *storage carbohydrates* is UDP-glucose, which is also a building block in the synthesis of the lipopolysaccharide layer of *E. coli* and other Gram-negative bacteria and the cell wall of fungi. UDP-glucose is synthesized from glucose-1-phosphate by pyrophosphorylase. When the hexose is incorporated into a polymer UDP is released, and the overall cost of adding a hexose monophosphate to an extending carbohydrate chain is therefore one UTP (equivalent to one ATP). Biosynthesis of the peptidoglycan, which forms the cell wall of bacteria, requires five monomers: UDP-N-acetylglucosamine (UDP-NAG), UDP-Nacetylmuramic acid (UDP-NAM), alanine (both L- and D-forms), di-aminopimelate, and glutamate. UDP-N-acetylglucosamine, which also serves as a building block for chitin synthesis in fungi, is synthesized from fructose-6-phosphate and acetyl-CoA with L-glutamine as the amino donor. The overall stoichiometry is

$$UDP\text{-}NAG + L\text{-}glutamate + CoA + \sim PP \text{ - fructose-6-phosphate}$$

$$\text{- acetyl-CoA - L-glutamine - UTP} = 0 \qquad (2.23)$$

When UDP-Glc-NAc is incorporated into peptidoglycan or chitin, UDP is released and the overall energetic cost for synthesis of a chitin monomer is therefore UTP. The requirements for the biosynthesis of carbohydrate building blocks and lipids are shown in Table 2.7.

TABLE 2.7 Metabolic Costs for Biosynthesis of Lipids and Carbohydrate Building Blocks

Building block	Precursor metabolite[a]	ATP	NADH	NADPH	1-C	NH_3
Glycerol-3-phosphate	1 glyP	0	-1	0	0	0
Palmitoyl-CoA	8 acCoA	-7	0	-14	0	0
Palmitoleoyl-CoA[b]	8 acCoA	-7	0	-14	0	0
Stearoyl-CoA	9 acCoA	-8	0	-16	0	0
Oleoyl-CoA	9 acCoA	-8	1	-16	0	0
Linoleoyl-CoA	9 acCoA	-8	2	-16	0	0
Linolenoyl-CoA	9 acCoA	-8	3	-16	0	0
Ethanolamine[c]	1 pga	0	1	-1	0	-1
Choline	1 pga	0	1	-1	-3	-1
Inositol	1 gluP	0	1	0	0	0
Ergosterol	18 acCoA	-18	0	-13	0	0
UDP-Glucose	1 gluP	-1	0	0	0	0
UDP-Galactose	1 gluP	-1	0	0	0	0
UDP-NAG	1 fruP, 1 acCoA	-2	0	0	0	-1
UDP-NAM	1 fruP, 1 pep, 1 acCoA	-2	0	-1	0	-1
Diaminopimelate	1 oaa, 1 pyr	-2	0	-3	0	-2

[a] See Table 2.4 for nomenclature.
[b] The costs are for the biosynthesis of palmitoleoyl-CoA via the anaerobic pathway.
[c] Because the pathway from squalene to ergosterol is not known in detail, the metabolic cost for the biosynthesis of ergosterol is taken to be the same as that for squalene.

2.5. POLYMERIZATION

The macromolecules of cellular biomass can be grouped into (1) RNA; (2) DNA; (3) proteins; (4) carbohydrates; (5) aminocarbohydrates; and (6) lipids. The macromolecular composition varies from species to species, and within a given species the composition changes with the specific growth rate and environmental conditions. Figure 2.13 shows the content of the most significant macromolecular pools as a function of the specific growth rate for a bacterium and a fungus. In bacteria, protein and RNA account for more than 80% of the biomass and the content of stable RNA increases with the specific growth rate, whereas the protein content decreases with the specific growth

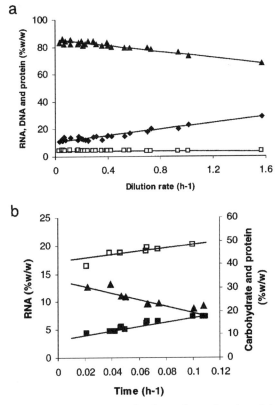

FIGURE 2.13 The most important macromolecular pools as a function of the specific growth rate in a bacterium and a fungus. (a) The pools of protein (▲), RNA (■), and DNA (□) in *E. coli*. The data are taken from Ingraham *et al.* (1983). (b) The pools of protein (□), RNA (■), and carbohydrates (▲) in *P. chrysogenum*. The data are taken from Nielsen (1997).

rate. The content of the other pools, *i.e.*, DNA, lipids, lipopolysaccharide, aminocarbohydrates (peptidoglycan), and carbohydrates (mainly glycogen), is approximately constant. In fungi there are some major differences. The total carbohydrate pool (sum of carbohydrates and aminocarbohydrates) represents a significant fraction of the biomass, *i.e.*, between 20 and 30%, and the fraction decreases with the specific growth rate. The carbohydrate pool is found mainly in the cell wall, which accounts for a decreasing fraction of the biomass with increasing specific growth rate. As for bacteria, the RNA content increases with the specific growth rate, but, in contrast to bacteria, the protein content also increases. This linear increase in the content of stable RNA in both bacteria and fungi is found for many different species, and a slow increase in the protein content with the specific growth rate is also reported for the yeast *S. cerevisiae.*

The total pool of cellular RNA consists of messenger RNA (mRNA), ribosomal RNA (rRNA), and transfer RNA (tRNA). In *E. coli* the composition is 5% mRNA, 18% tRNA, and 77% rRNA at a specific growth rate of 1.0 h^{-1} (Ingraham *et al.*, 1983). At lower specific growth rates, the relative tRNA content increases at the expense of the rRNA content. A similar observation has been made in *Neurospora crassa*, where the number of tRNA molecules per ribosome decreases with increasing specific growth rate (Alberghina *et al.*, 1979). However, even at low specific growth rates, rRNA accounts for more than 75% of the stable RNA, *i.e.*, the sum of rRNA and tRNA. The pool of the stable RNA therefore is a good measure of the ribosome level in the cell. In *A. niger* the ribosomes consist of 53% rRNA and 47% protein (Berry and Berry, 1976), and in *E. coli* the ribosomes consist of about 60% rRNA and 40% protein. Ribosomes play an important role in protein-synthesis, and together with a few enzymes they make up the so-called protein synthesizing system (PSS) (Ingraham *et al.*, 1983).

Because protein synthesis is energetically expensive for the cell (see Table 2.8), there is tight control of the PSS in cells. Thus, the ribosomal level is adjusted to the requirements, and this explains how the pool of stable RNA and the specific growth rate are positively correlated, as seen in Fig. 2.13. A similar correlation between the content of stable RNA and the specific growth rate is observed during dynamic changes in the environmental conditions and shift down experiments in the specific growth rate: The synthesis of rRNA is observed to stop until the level of the ribosomes has adjusted to the new growth conditions (Sturani *et al.*, 1973). At low specific growth rates, the ribosome level of *N. crassa* approaches a constant level, but the ribosome efficiency decreases with the specific growth rate (Alberghina *et al.*, 1979). Consequently, the net efficiency of the PSS continues to decrease as the specific growth rate tends to zero. A similar observation is made for *A. nidulans* (Bushell and Bull, 1976) and for *E. coli* (Ingraham

TABLE 2.8   Amino Acid Composition (in Mole %) of Cellular Protein in
Different Species

Amino acid	E. coli[a]	P. chrysogenum[b]	S. cerevisiae[c]
L-Alanine	9.6%	10%	4.7%
L-Arginine	5.5%	4.8%	5.6%
L-Asparagine/L-Aspartate	9.0%	9.6%	12.7%
L-Cysteine	1.7%	1.4%	0.9%
L-Glutamate/L-Glutamine	9.8%	14.9%	13.4%
L-Glycine	11.5%	9.2%	5.6%
L-Histidine	1.8%	2.4%	9.1%
L-Isoleucine	5.4%	4.3%	5.2%
L-Leucine	8.4%	7.5%	7.9%
L-Lysine	6.4%	5.6%	3.1%
L-Methionine	2.9%	1.7%	4.7%
L-Phenylalanine	3.5%	3.4%	4.0%
L-Proline	4.1%	4.7%	4.3%
L-Serine	4.0%	6.1%	5.2%
L-Threonine	4.7%	5.3%	1.4%
L-Tryptophan	1.1%		4.0%
L-Tyrosine	2.6%	2.6%	6.5%
L-Valine	7.9%	6.4%	1.7%

[a] The data for E. coli are taken from Ingraham et al. (1983).
[b] The data for P. chrysogenum are taken from Henriksen et al. (1996).
[c] The data for S. cerevisiae are taken from Cook (1958).

et al., 1983). Thus, at low specific growth rates, the cells possess some
unused or inefficiently used metabolic machinery that may be activated rapidly
when the environmental conditions change.

Translation of mRNA into proteins takes place mainly on cytoplasmic 80S
ribosomes, which are composed of two subunits of 60S and 40S. In eukaryotes, some proteins, e.g., the F_1F_0-ATPase, cytochrome b, and some of the
TCA cycle enzymes, are synthesized in the mitochondria, which contains
both DNA and ribosomes. The translation process consists of three discrete
phases: initiation, elongation, and termination. The initiation is catalyzed by
a relatively large number of specific protein factors (at least 10), which
results in the formation of an 80S initiation complex. After formation of the
initiation complex, elongation of the peptide chain occurs by the successive
addition of amino acids to the extending peptide chain. The elongation
process consists of four steps: (1) linking of each amino acid to its cognate
tRNA; (2) binding of the aminoacyl-tRNA to the ribosome; (3) formation of
the peptide bond; and (4) translocation, resulting in the movement of the
peptidyl-tRNA in the ribosome and the concomitant movement of the mRNA
by one codon. Activation and formation of aminoacyl-tRNA require the
hydrolysis of ATP to AMP, binding of the aminoacyl-tRNA to the ribosome

requires the hydrolysis of GTP to GDP, and the translocation of the peptidyl-tRNA in the ribosome requires the hydrolysis of GTP to GDP. Thus, a total of four ATP equivalents are required for the addition of an amino acid to an extending peptide chain. In addition, some free energy is required for mRNA synthesis, proofreading, etc., and this amounts to about 0.3 ATP per amino acid incorporated (Ingraham *et al.*, 1983). With the amino acid composition of microbial cells (see Table 2.8), the molecular mass is about 110 g/mol protein-bound amino acid, and the overall stoichiometry for polymerization of 1 g of protein is (with all the stoichiometric coefficients in millimoles)

$$\text{"1 g of protein"} + 39.1\ ADP + 39.1 \sim P - 9.1\ \text{amino acid} - 39.1\ ATP = 0$$

$$(2.24)$$

Transcription of nuclear DNA into RNA is performed by three RNA polymerases: RNA polymerase A, which synthesizes rRNA; RNA polymerase B, which synthesizes mRNA; and RNA polymerase C, which carries out synthesis of tRNA. Before they are incorporated into RNA, the monophosphate nucleotides have to be converted into triphosphate nucleotides, and this requires two ATPs for each nucleotide. Furthermore, in the synthesis of RNA, a segment of the primary transcript is removed and hydrolyzed to monophosphate nucleotides after transcription, and reactivation of these nucleotides also requires two ATPs. Ingraham *et al.* (1983) suggests that approximately 20% of the primary transcript is discarded, and the energetic cost for the incorporation of one ribonucleotide (as monophosphate) is therefore 2.4 ATPs. The nucleotide composition of total RNA on a molar basis is 25.6% AMP, 26.2% UMP, 28.6% GMP, and 19.6% CMP in *S. cerevisiae* (Mounolou, 1975), and the molecular mass for RNA is therefore 323 g/mol RNA-bound nucleotide. The overall stoichiometry for the synthesis of 1 g of RNA is consequently (with all the stoichiometric coefficients in millimoles)

$$\text{"1 g of RNA"} + 7.44\ ADP + 7.44 \sim P$$

$$- 0.79\ AMP - 0.81\ UMP - 0.89\ GMP - 0.61\ CMP - 7.44\ ATP = 0 \quad (2.25)$$

Replication of DNA is a complex process. Before the DNA polymerase can replicate the two strands of DNA, the double-helix structure has to be unwound. This is performed by a special group of enzymes, and the process requires the input of Gibbs free energy to break the hydrogen bonds between the complementary bases. Separation of strands wound in a helix generates loops, termed supercoiled twists, in the single strands. To prevent the generation of supercoiled twists in the unwound DNA strands, the enzyme DNA gyrase periodically breaks a phosphodiester bond in one of the strands and thereby allows free rotation of the opposite strand. After rotation, the same enzyme reforms the bond. The total energetic cost of unwinding the double helix is two ATPs per base pair, *i.e.*, one ATP per nucleotide incorporated into DNA (Ingraham *et al.*, 1983). Once the strands are separated, the DNA polymerases start replication, *i.e.*, the assembly of nucleotides (activated as triphosphates). The replication of DNA is remarkably precise, *i.e.*, approximately only one mistake is made per 10^{10} nucleotide copies. This precision is accomplished partly by built-in exonuclease activity of one of the DNA polymerases. This DNA polymerase removes its own mismatches by moving backward, and it does not catalyze replication in the forward direction unless there is a properly matched nucleotide pair behind it. The energetic cost for proofreading is estimated to be 0.4 ATP per nucleotide (Ingraham *et al.*, 1983). Thus, the overall energetic cost for the incorporation of a deoxyribonucleotide (as monophosphate) is 3.4 ATPs. The composition of cellular DNA varies only slightly between different species, and it is approximately 24.5% dAMP, 24.5% TMP, 25.5% GMP, and 25.5% CMP. This gives a molecular mass for DNA of 310 g/mol DNA-bound nucleotide, and the overall stoichiometry for synthesis of 1 g of DNA bound nucleotides is therefore (with all the stoichiometric coefficients in millmoles)

$$\text{"1 g of DNA"} + 11.0 \text{ ADP} + 11.0 \sim P - 0.79 \text{ dAMP} - 0.79 \text{ dTMP}$$

$$- 0.82 \text{ dGMP} - 0.82 \text{ dCMP} - 11.0 \text{ ATP} = 0 \tag{2.26}$$

The biosynthesis of phospholipids starts with the binding of fatty acyl-CoA to the 1- and 2-carbons of glycerol-3-phosphate to form phosphatidic acid, which is subsequently activated by CTP to CDP-diacylglycerol. Finally, the alcohol is bound to the phosphate group and CMP is released. Thus, the energetic cost for phospholipid synthesis from its building blocks is the conversion of CTP to CMP, corresponding to two ATPs. To calculate the overall metabolic costs for phospholipid biosynthesis, the relative contents of the different classes and the relative contents of the fatty acids need to be known. The phospholipid composition in *S. cerevisiae* is approximately 50% PC, 20% PI, and 30% PE, and in *E. coli* nearly 95% of the phospholipids are PE and phosphatidylglycerol, which are present in approximately equal amounts. The total fatty acid composition varies between species, and it also depends on the environmental conditions [see Ratledge and Evans (1989) for a review of the fatty acid composition of yeasts]. In *S. cerevisiae* a typical composition is 15.6% C16:0, 31.4% C16:1, 5.1% C18:0, 32.0% C18:1, and 13.4% C18:2. In *P. chrysogenum* the total fatty acid composition is 8% C16:0, 7% C18:0, 24% C18:1, 59% C18:2, and 2% C18:3 (Meisgeier *et al.*, 1990), whereas in *E. coli* the composition is about 43% C16:0, 33% C16:1, and 24% C18:1 (Ingraham *et al.*, 1983).

Sterol esters are formed by direct addition of the sterol (typically ergosterol) to the fatty acyl-CoA, and triacylglycerols are formed by replacement of the phosphate group of phosphatidic acid with a fatty acyl-CoA. Phosphatidic acid is formed by the subsequent addition of fatty acyl-CoA to the nonphosphorylated carbons of glycerol-3-phosphate.

2.6. GROWTH ENERGETICS

Growth energetics describe the relationship between generation and consumption of Gibbs free energy (generally in the form of ATP), and it, therefore, links transport, fueling reactions, biosynthetic reactions, and macromolecular synthesis reactions. It originates from the classical yield studies, where a mass yield coefficient, Y_{sx} (see Section 3.4), was determined from simultaneous measurement of substrate consumption and biomass formation (Monod, 1942). Because the role of the energy source is to supply Gibbs free energy in the form of ATP to drive the biosynthetic and polymerization reactions, the ATP yield coefficient Y_{xATP}, i.e., the amount of ATP consumed for the formation of biomass [mmol ATP (g DW)$^{-1}$], is a more fundamental property of the cell than the mass yield coefficient Y_{sx}. For more than a decade following the initial work of Bauchop and Elsden (1960), who introduced the ATP yield coefficient Y_{xATP}, the latter was regarded as a universal constant with the value 95 mmol of ATP (g DW)$^{-1}$ (Forrest and Walker, 1971). In 1973 Stouthamer and Bettenhaussen, however, proposed the following linear relationship between ATP production and consumption:

$$r_{ATP} = Y_{xATP} \mu + m_{ATP} \qquad (2.27)$$

Today this equation is the authoritative frame for interpreting growth energetics. It is based on the assumption of a pseudo-steady state for the ATP level. This assumption is quite reasonable in light of the strong regulation of the energy level of the cell and the higher turnover rate of the ATP pool, i.e., relaxation times on the order of seconds compared to relaxation times on the order of hours for cellular growth. Thus, even for sudden environmental perturbations, e.g, addition of a glucose pulse to a steady state continuous culture of S. cerevisiae, a new steady ATP level is reached within a few minutes after the perturbation (Theobald et al., 1993).

Equation (2.27) states that the rate of ATP generation, r_{ATP} [mmol of ATP (g DW h)$^{-1}$], balances the rate of ATP consumption, the latter expressed as the sum of two terms: one to account for the ATP expenditures for growth and another for non-growth-associated processes that consume ATP, such as maintenance, given by m_{ATP}. The second term is, in fact, more general and comprises all nongrowth ATP consumption, including futile cycles and other unaccountable ATP uses (see Box 2.2). The growth-associated ATP consump-

BOX 2.2
_____

ATP Consumption for Maintenance

Many cellular reactions require the consumption of ATP without contributing to a net synthesis of biomass, and these reactions are usually referred to as maintenance reactions (or processes). Some of these are associated with growth, *e. g.*, to maintain the electrochemical gradients across the plasma membrane, whereas others are independent of the specific growth rate of the cells. We therefore distinguish between growth- and non-growth-associated maintenance. The ATP costs for growth-associated maintenance are included in the two terms $Y_{xATP, lysis}$ and $Y_{xATP, leak}$. It is difficult to group different processes as growth- and nongrowth-associated maintenance, but some of the most important maintenance processes can be mentioned:

- **Maintenance of concentration gradients and electrical potential gradients.** In order to ensure proper function, cells need to maintain various concentration gradients and electrochemical potential gradients across the plasma membrane and, in eukaryotes, also across the mitochondrial membrane. These processes require Gibbs free energy, but they do not lead to the synthesis of new biomass and they are, therefore, typical examples of maintenance processes. Part of these processes are growth-associated, *i.e.*, when the cell expands the gradients need to be maintained over an increasing area/volume, but there is also a non-growth-associated component, *i.e.*, the gradients also need to be maintained when the cells do not grow. Expenditure of ATP for maintenance of gradients has been estimated to account for up to 50% of the total ATP produced (Stouthamer, 1979).
- **Futile cycles.** Inside the cells there are sequences of reactions whose net result is hydrolysis of ATP. Such an example is the conversion of fructose-6-phosphate to fructose-1,6-bisphosphate by phosphofructokinase and the subsequent degradation of fructose-1,6-bisphosphate to fructose-6-phosphate by fructosebisphosphate phosphatase. These two reactions result in a net consumption of ATP. This type of cycling was originally regarded as an imperfection in metabolic control (hence the name futile cycle), but it is now considered an important control mechanism of metabolism as the presence of both enzymes allows rapid adjustments to new environmental conditions.

> • **Turnover of macromolecules**. To maintain the ability to control the metabolic function of the cells, it is necessary that many molecules are degraded and synthesized continuously. Thus, mRNA has a typical half-life of a few minutes. This continued degradation and repolymerization of macromolecules results in net ATP consumption without the generation of new biomass, and therefore it can also be considered a maintenance process.

tion can be further subdivided into three terms (Benthin *et al.*, 1994):

$$Y_{xATP} = Y_{xATP,\,growth} + Y_{xATP,\,lysis} + Y_{xATP,\,leak} \tag{2.28}$$

where $Y_{xATP,\,growth}$ is the ATP consumption for transport, biosynthesis, and polymerization, $Y_{xATP,\,lysis}$ is the ATP consumption for repolymerization of degraded macromolecules, etc., and $Y_{xATP,\,leak}$ embodies all other ATP consumption, *i.e.*, the ATP consumption due to leaks, futile cycles, and growth-associated maintenance.

On the basis of estimates of the metabolic costs for transport, biosynthesis, and polymerization, it is possible to calculate the theoretical value of $Y_{xATP,\,growth}$, but it is not possible to theoretically evaluate the two other contributions to Y_{xATP} and no theoretical estimate of the overall ATP requirement for growth therefore can be given. However, through a combination of stoichiometry and precise measurements of the substrate uptake rates and metabolic product formation, one may calculate the ATP production. When this is done as a function of the specific growth rate, experimental values of Y_{xATP} and m_{ATP} may be determined (see Example 3.3). Under aerobic growth, this requires knowledge of the operational P/O ratio, whereas under anaerobic growth it is normally possible to estimate the ATP production quite precisely from measurements of the metabolic product formation. Table 2.9 lists some experimentally determined values of Y_{xATP} and m_{ATP}.

With the estimates of metabolic costs for transport, biosynthesis, and polymerization provided in the previous sections, it is possible to calculate the total ATP and NADPH requirements for cell synthesis. The basis for this calculation is the carbon source (typically glucose), whereas the basis in the previous sections was the 12 precursor metabolites. These are formed from the carbon source in the fueling reactions, as discussed in Section 2.3, and

TABLE 2.9 Experimentally Determined Y_{xATP} and m_{ATP}

Species	Y_{xATP}[a]	m_{ATP}[b]	Reference
Aerobacter aerogenes	71	6.8	Stouthamer and Bettenhaussen (1976)
	57	2.3	Stouthamer and Bettenhaussen (1976)
Escherichia coli	97	18.9	Hempfling and Mainzer (1975)
Lactobacillus casei	41	1.5	de Vries *et al.* (1970)
Lactobacillus delbruckii	72	0	Major and Bull (1985)
Lactococcus cremoris	73	1.4	Otto *et al.* (1980)
	53	—	Brown and Collins (1977)
	15-50	7-18	Benthin *et al.* (1993)[c]
Lactococcus diacetilactis	47	—	Brown and Collins (1977)
Saccharomyces cerevisiae	71-91	< 1	Verduyn *et al.* (1990)

[a] The unit is mmol ATP $(g\ DW)^{-1}$.
[b] The unit is mmol ATP $(g\ DW\ h)^{-1}$.
[c] In their analysis, Benthin *et al.* (1993) found a large variation in the energetic parameters depending on the medium composition.

TABLE 2.10 Stoichiometry for the Formation of the 12 Precursor Metabolites from Glucose

Precursor metabolite	Glucose[a]	ATP	NADH	NADPH	CO_2
Glucose-6-phosphate	-1	-1	0	0	0
Fructose-6-phosphate	-1	-1	0	0	0
Ribose-5-phosphate	-1	-1	0	2	1
Erythrose-4-phosphate	-1	-1	0	4	2
Glyceraldehyde-3-phosphate	-0.5	-1	0	0	0
3-Phosphoglycerate	-0.5	0	1	0	0
Phosphoenolpyruvate	-0.5	0	1	0	0
Pyruvate	-0.5	1	1	0	0
Acetyl-CoA[b]	-0.5	1	2	0	1
Acetyl-CoA + oxaloacetate[c]	-1	0	3	0	0
Acetyl-CoA[d]	-0.5	-1	1	1	1
Acetyl-CoA[e]	-0.5	-1	2	0	1
α-Ketoglutarate[f]	-1	1	4	0	1
Succinyl-CoA	-1	1	5	0	2
Oxaloacetate[g]	-0.5	0	1	0	-1

[a] All of the stoichiometric coefficients are per mole of glucose.
[b] Formation by the pyruvate dehydrogenase complex (active in prokaryotes).
[c] Formation by citrate lyase.
[d] Formation by citrate lyase and recycling of excess oxaloacetate.
[e] Formation from pyruvate via acetate.
[f] Formation by an NADH-linked isocitrate dehydrogenase.
[g] Formation by pyruvate carboxylase.

Table 2.10 lists the stoichiometric coefficients for their synthesis from 1 mol of glucose.

When the metabolic costs for the formation of precursor metabolites, building blocks, and macromolecules are known, the ATP requirements for synthesis of a whole cell can be calculated, and Table 2.10 shows the results for a typical bacterial cell and a typical fungal cell. It is observed that the ATP costs are almost the same for the two cells. The ATP cost for the synthesis of protein, RNA, DNA, and carbohydrates is almost the same in the bacterium and the fungus, whereas the ATP cost for lipid biosynthesis is significantly higher in the eukaryote. This is due to the more complex biosynthesis of special phospholipids in these cells. It is also noted that the costs for transport are lower in the bacterium (see also Section 2.2.3), and the net result is a lower ATP requirement for the synthesis of a bacterial cell than

TABLE 2.11   Typical Composition [g (g DW)$^{-1}$] and ATP Requirements [mmol of ATP (g DW)$^{-1}$] for Synthesis of a Bacterial Cell and a Fungal Cell in a Minimal Medium[a]

Macromolecule	E. coli		P. chrysogenum	
	Content	ATP	Content	ATP
Protein	0.52	21.88	0.45	19.92
RNA	0.16	4.37	0.08	3.32
DNA	0.03	1.05	0.01	0.39
Lipid	0.09	0.14	0.05	
Phospholipid			0.035	1.65
Sterol esters			0.010	0.81
Triacylglycerols			0.005	0.30
Carbohydrates	0.17	2.06	0.25	
Chitin			0.22	2.90
Glycogen			0.03	0.37
Soluble pool[b]			0.08	
Amino acids			0.04	
Nucleotides			0.02	
Metabolites, etc.			0.02	
Ash	0.03		0.08	
Transport				
Ammonia		4.24		7.10
Sulfate				0.14
Phosphate		0.77		2.12
Total	1.00	34.71	1.00	39.02

[a] The data for E. coli are from Stouthamer (1979), and the data for P. chrysogenum are from Nielsen (1997). Both calculations are based on growth on a minimal medium, i.e., glucose as the carbon and energy source and inorganic salts as the source of other elements.

[b] The ATP costs for synthesis of the soluble pool of metabolites are included in the biosynthesis of the macromolecules, e.g., the ATP costs for biosynthesis of free amino acids are included in the ATP cost for protein synthesis.

for the synthesis of a fungal cell. The total ATP costs given in Table 2.11 are valid for the macromolecular composition specified, and the ATP costs change with variations in the macromolecular composition (see Fig. 2.13). Thus, in *P. chrysogenum*, the ATP cost changes from about 35 mol of ATP (g DW)$^{-1}$ at low specific growth rates to about 40 mmol of ATP (g DW)$^{-1}$ (given in Table 2.11) at high specific growth rates (Nielsen, 1997).

REFERENCES

Alberghina, L., Sturani, E., Costantini, M. G., Martegani, E. & Zippel, R. (1979). Regulation of macromolecular composition during growth of *Neurospora crassa* In *Fungal Walls and Hyphal Growth*, pp. 295-318. Edited by J. H. Bunnett & A. P. J. Trinci. Cambridge, UK: Cambridge Univ.

Bauchop, T. & Elsden, S. R. (1960). The growth of microorganisms in relation to their energy supply. *Journal of General Microbiology* 23, 35-43.

Benthin, S., Schulze, U., Nielsen, J. & Villadsen, J. (1994). Growth energetics of *Lactococcus cremoris* FD1 during energy-, carbon- and nitrogen-limitation in steady state and transient cultures. *Chemical Engineering Science* 49, 589-609.

Berry, D. R. & Berry, E. A. (1976). Nucleic acid and protein synthesis in filamentous fungi. In *The Filamentous Fungi*, Vol. II, pp. 238-291. Edited by J. E. Smith and D. R. Berry. London: Edward Arnold.

Bisson, L. F. & Fraenkel, D. G. (1983). Involvement of kinases in glucose and fructose uptake by *Saccharomyces cerevisiae*. *Proceedings of the National Academy of Science USA* 80, 1730-1734.

Bisson, L. F., Coons, D. M., Kruckeberg, A. L. & Lewis, D. A. (1993). Yeast sugar transporters. *Critical Reviews in Biochemistry and Molecular Biology*. 28, 259-308.

Blumenthal, H. J. (1965). Carbohydrate metabolism. 1. Glycolysis. In *The Fungi*, Vol. II, pp. 229-268. Edited by G. C. Ainsworth & A. S. Sussman London. Edward Arnold.

Brown, W. V. & Collins, E. B. (1977). End product and fermentation balances for lactis Streptococci grown aerobically on low concentrations of glucose. *Applied Environmental Microbiology* 59, 3206-3211.

Bruinenberg, P. M., Jonker, R., van Dijken, J. P. & Scheffers, W. A. (1985). Utilization of formate as an additional energy source by glucose limited chemostat cultures of *Candida utilis* CBS621 and *Saccharomyces cerevisiae* CBS8066. Evidence for the absence of transhydrogenase activity in yeasts. *Archives of Mikrobiology*. 142, 302-306.

Bushell, M. E. & Bull, A. T. (1976). Growth rate dependent ribosomal efficiency of protein synthesis in the fungus *Aspergillus nidulans*. *Journal of Applied. Chemistry and Biotechnology* 26, 339-340.

Carter, B. L. A. & Bull, A. T. (1969). Studies of fungal growth and intermediary carbon metabolism under steady state and non-steady state conditions. *Biotechnology and Bioengineering* 11, 785-804.

Cartwright, C. P., Rose, A. H., Calderbank, J. & Keenan, M. H. J. (1989). Solute transport. In *The Yeasts*, Vol. 3, pp. 5-56. Edited by A. H. Rose & J. S. Harrison. London: Academic Press.

Conway, T. (1992). The Entner-Doudoroff pathway. History, physiology and molecular biology. *FEMS Microbioligal Reviews* 103, 1-28.

Cook, A. H. (1958). The Chemistry and Biology of Yeasts. New York: Academic Press.

de Vries, W., Kapteijn, W. M. C., van der Beek, E. G. & Stouthamer, A. H. (1970). Molar growth yields and fermentation balances of *Lactobacillus casei* L3 in batch cultures and in continuous cultures. *Journal of General Microbiology* **63**, 333-345.

Eagon, R. G. (1963). Rate limiting effects of pyridine nucleotides on carbohydrate catabolic pathways in microorganisms. *Biochemistry and Biophysics Research Communications* **12**, 274-279.

Forrest, W. W. & Walker, D. J. (1971). The generation and utilization of energy during growth. *Advances in Microbial Physiology* **5**, 213-274.

Frenkel, E. P. & Kitchens, R. L. (1977). Purification and properties of acetyl coenzyme A synthetase from bakers yeast. *Journal of Biological Chemistry* **30**, 760-761.

Gancedo, C. & Serrano, R. (1989). Energy-yielding metabolism. In *The Yeasts*, Vol. 3, pp. 205-259. Edited by A. H. Rose and J. S. Harrison. London: Academic Press, London, UK.

Haarasilta, S. & Taskinen, L. (1977). Location of three key enzymes of glyconeogenesis in baker's yeast. *Archives of Mikrobiology* **113**, 159-161.

Hempfling, W. P. & Mainzer, S. E. (1975). Effects of varying the carbon source limiting growth on yield and maintenance characteristics of *Escherichia coli* in continuous culture. *Journal of Bacteriology* **123**, 1076-1087.

Henriksen, C. M., Christensen, L. H., Nielsen, J. & Villadsen, J. (1996). Growth energetics and metabolic fluxes in continuous cultures of *Penicillium chrysogenum*. *Journal of Biotechnology*. **45**, 149-164.

Henriksen, C. M., Nielsen, J. & Villadsen, J. (1998). Modelling the protonophoric uncoupling by phenoxyacetic acid of the plasma membrane potential of *Penicillium chrysogenum*. *submitted*.

Hoek, J. B., Rydström, J. (1988). Physiological roles of nicotinamide nucleotide transhydrogenase. *Biochemical Journal* **254**, 1-10.

Ingraham, J. L., Maaløe, O. & Neidhardt, F. C. (1983). Growth of the bacterial cell. Sunderland: Sinnauer Associated.

Jetten, M. S. M., Pitoc, G. A., Follenttie, M. T. & Sinskey, A. J. (1994). Regulation of phospho(enol)-pyruvate- and oxaloacetate-converting enzymes in *Corynebacterium glutamicum*. *Applied Microbiology and Biotechnology* **41**, 47-52.

Jones, E. W. & Fink, G. R. (1982). Regulation of amino acid and nucleotide biosynthesis in yeast. In *The Molecular Biology of the yeast* Saccharomyces. *Metabolism and Gene Expression*. pp. 181-299. Edited by J. N. Starhern, E. W. Jones & J. R. Broach. Cold Spring Harbor, NY: Cold Spring Haibor Laboratory Press.

Kelly, J. M. & Hynes, M. J. (1982). The regulation of NADP-linked isocitrate dehydrogenase in *Aspergillus nidulans*. *Journal of General Microbiology* **128**, 23-28.

Kruckeberg, A. L. (1996). The hexose transporter family of *Saccharomyces cerevisiae*. *Archives of Microbiology* **166**, 283-292.

Lagunas, R. & Gancedo, J. M. (1973). Reduced pyridine-nucleotides balance in glucose-growing *Saccharomyces cerevisiae*. *European Journal of Biochemistry*. **37**, 90-94.

LaNoue, K. F., Schoolwerth, A. C. (1979). Metabolite transport in mitochondria. *Annual Reviews in Biochemistry and Biophysics* **52**, 93-109.

Lewis, K. F., Blumenthal, H. J., Wenner, C. E. & Weinhouse, S. (1954). Estimation of glucose catabolism pathways. *Federation Proceedings* **13**, 252.

Major, N. C. & Bull, A. T. (1985). Lactic acid productivity of a continuous culture of *Lactobacillus delbrueckii*. *Biotechnology Letters*. **7**, 401-405.

Malpartida, F. & Serrano, R. (1981). Proton translocation catalyzed by the purified yeast plasma membrane ATPase reconstituted in liposomes. *FEBS Letters* **131**, 351-354.

Marx, A., de Graaf, A. A., Wiechert, W., Eggeling, L. & Sahm, H. (1996). Determination of the fluxes in the central metabolism of *Corynebacterium glutamicum* by nuclear magnetic resonance spectroscopy combined with metabolite balancing. *Biotechnology Bioengineering* 49, 111-129.

Meisgeier, G., Müller, H., Ruhland, G. & Christner, A. (1990). Qualitative und quantitative zusammensetzung der Fettsäurespektren von selektanten des *Penicillium chrysogenum*. *Zentralblatt für Mikrobiologie* 145, 183-186.

Mitchell, P. (1961). Coupling of phosphorylation to electron and hydrogen transfer by a chemisomotic type of mechanism. *Nature* 191, 144-148.

Mitchell, P. & Moyle, J. (1965). Stoichiometry of proton translocation through the respiratory chain and adenosine triphosphatase systems or rat liver mitochondria. *Nature* 208, 147-151.

Moat, A. G. & Foster, J. W. (1995). *Microbial physiology*. New York: Wiley-Liss.

Monod, J. (1942). Recherches sur la Croissance des Cultures Bacteriennes. Paris: Hermann et Cie.

Mounolou, J. C. (1975). The properties and composition of yeast nucleic acids. In *the Yeasts*, Vol. 2, pp. 309-334. Edited by A. H. Rose and J. S. Harrison. London: Academic Press.

Neidhardt, F. C., Ingraham, J. L., Low, K. B., Magasanik, B., Schaechter, M. & Umbarger, H. E. (1987). *Escherichia coli* and *Salmonella typhimurium*. In *Cellular and Molecular Biology*. Washington, DC: ASM.

Neidhardt, F. C., Ingraham, J. L. & Schaechter, M. (1990). Physiology of the bacterial cell. A molecular approach. Sunderland: Sinauer Associates.

Nielsen, J. (1997). Physiological engineering aspects of *Penicillium chrysogenum*. Singapore: World Scientific Publishing Co.

Nielsen, J. & Villadsen, J. (1994). *Bioreaction Engineering Principles*. New York: Plenum Press.

Nissen, T., Schulze, U., Nielsen, J. & Villadsen, J. (1997). Flux distribution on anaerobic, glucose-limited continuous cultures of *Saccharomyces cerevisiae*. *Microbiology* 143, 203-218.

Otto, R., Sonnenberg, A. S. M., Veldkamp, H. & Konings, W. N. (1980). Generation of an electrochemical proton gradient in *Streptococcus cremoris* by lactate efflux. *Proceedings of the National Academy of Science USA* 77, 5502-5506.

Perlin, D. S., San Fransisco, M. J. D., Slayman, C. W. & Rosen, B. P. (1986). H^+/ATP stoichiometry of proton pumps from *Neurospora crassa* and *Escherichia coli*. *Archives of Biochemistry and Biophysics* 248, 53-61.

Ratledge, C. & Evans, C. T. (1989). Lipids and their metabolism. In *The Yeasts*, Vol. 3, pp. 367-455. Edited by A. H. Rose and J. S. Harrison. London: Academic Press.

Reed, D. J. & Wang, C. H. (1959). Glucose metabolism in *Penicillium digitatum*. *Canadian Journal of Microbiology*. 16, 157-167.

Rose, A. H. (1976). Chemical nature of membrane components. In *The filamentous Fungi*, Vol. II, pp. 308-327. Edited by J. E. Smith and D. R. Berry. London: Edward Arnold.

Schulze, U. (1995). Anaerobic physiology of *Saccharomyces cerevisiae*. Ph.D. Thesis, Technical University of Denmark.

Senior, A. E. (1988). ATP synthesis by oxidative phosphorylation. *Physiological Reviews* 68, 177-231.

Shu, P., Funk, A. & Neish, A. C. (1954). Mechanism of citric acid formation from glucose by *Aspergillus niger*. *Canadian Journal of Biochemistry and Physiology* 32, 68-80.

Stein, W. D. (1990). Channels, Carriers and Pumps. An Introduction to Membrane Transport. San Diego: Academic Press.

Stouthamer, A. H. (1979). The search for correlation between theoretical and experimental growth yields. In *International Review of Biochemistry: Microbial Biochemistry*, Vol. 21, pp. 1-47. Edited by J. R. Quayle. Baltimore: University Park Press.

Stouthamer, A. H. & Bettenhaussen, C. (1973). Utilization of energy for growth and maintenance in continuous and batch cultures of microorganisms. *Biochimica Biophysica Acta* **301**,
53-70.

Stryer, L. (1995). *Biochemistry*, 4th ed., San Fransisco: W. H. Freeman and Company.

Sturani, E., Martegani, F. & Alberghina, F. A. M. (1973). Inhibition of ribosomal RNA synthesis
during a shift down transition of growth in *Neurospora crassa*. *Biochimica Biophysica Acta*
319, 153-164.

Tanaka, A. & Fukui, S. (1989). Metabolism of n-alkanes. In *The Yeasts*, Vol. 3, pp. 261-287.
Edited by A. H. Rose and J. S. Harrison. London: Academic Press.

Theobald, U., Mailinger, W. Reuss, M. & Rizzi, M. (1993). *In vivo* analysis of glucose-induced
fast changes on yeast adenine nucleotide pool applying a rapid sampling technique. *Analytical Biochemistry* **214**, 31-37.

Umbarger, H. E. (1978). Amino acid biosynthesis and its regulation. *Annual Reviews in
Biochemistry* **47**, 1127-1162.

Umezawa, C., Kishi, T. (1989). Vitamin metabolism. In *The Yeasts*, Vol. 3, pp. 457-488. Edited
by A. H. Rose and J. S. Harrison. London: Academic Press.

Verduyn, C., Postma, E., Scheffers, W. A. & van Dijken, J. P. (1990). Energetics of Saccharomyces cerevisiae in anaerobic glucose limited chemostat cultures. *Journal of General
Microbiology* **136**, 405-412.

Verduyn, C., Postma, E., Scheffers, W. A. & van Dijken, J. P. (1992). Effect of benzoic acid on
metabolic fluxes in yeast. A continuous culture study on the regulation of respiration and
alcoholic fermentation. *Yeast* **8**, 501-517.

Voet, D., Voet, J. G. (1995). *Biochemistry*, 2nd ed. New York: John Wiley & Sons.

von Jagow, G.; Klingenberg, M. 1970. Pathways of hydrogen in mitochondria of *Saccharomyces
carlsbergensis*. *European Journal of Biochemistry* **12**, 583-592.

Walker, P. & Woodbine, M. (1976). The biosynthesis of fatty acids. In *The Filamentous Fungi*,
Vol. II, pp. 137-158. Edited by J. E. Smith and D. R. Berry. London: Edward Arnold.

Wang, C. H., Stern, I., Gilmour, C. M., Klungsoyr, S., Reed, D. J., Bialy, J. J., Christensen, B. E.
& Cheldelin, V. H. (1958). Comparative study of glucose catabolism by radiorespirometric
method. *Journal of Bacteriology* **76**, 207-216.

Watson, K. (1976). The biochemistry and biogenesis of mitochondria. In *The Filamentous Fungi*,
Vol. II, pp. 92-120. Edited by J. E. Smith and D. R. Berry. London: Edward Arnold.

Zubay, G. (1988). *Biochemistry*, 2nd ed. New York: Macmillan.

CHAPTER 3

Comprehensive Models for Cellular Reactions

In the previous chapter, we reviewed the main metabolic pathways by which carbon-energy sources are catabolized for the production of free energy and carbon skeletons needed for metabolite synthesis and cell growth. To complete the picture of intermediary metabolism, the complex pathways of Chapter 2 need to be complemented with information about the stoichiometry, kinetics, and thermodynamics of the individual reactions. Thermodynamics, a subject dealing primarily with issues of reaction and pathway feasibility, will be discussed at length in Chapter 14. Information on reaction kinetics, especially under *in vivo* conditions, is particularly hard to come by, yet such information is necessary for understanding the mechanisms of metabolic flux control, a central issue of metabolic engineering and, indeed, of this textbook. Stoichiometry is readily available for individual reactions from extensive studies conducted on intermediary metabolism. Extension of these reaction stoichiometries to metabolic pathways provides the basis for setting up energy and material balances, as well as accounting for interactions among coupled reactions and pathways. Pathway stoichiometry imposes constraints that must be satisfied by cell cultivation measurements, raising issues of data consistency discussed in the next chapter.

In this chapter we review methods by which the sum total of cellular reactions are *lumped* together to provide comprehensive models of cellular metabolism. Clearly, only a few of all possible cellular reactions are considered explicitly, as many of them are not even known or their inclusion would lead to an overwhelmingly large system. At the same time it is not clear to what extent the fidelity or usefulness of cellular models might be enhanced by increasing the detail of metabolic representation. As there are many ways in which reaction lumping can be executed, it is important that such lumping be carried out rationally and in a consistent manner. We therefore provide a framework for developing the stoichiometry of cellular models, which is also applicable to the analysis of very large reaction networks.

3.1. STOICHIOMETRY OF CELLULAR REACTIONS

The overall result of the totality of cellular reactions is the conversion of substrates into free energy and metabolic products (*e.g.*, primary metabolites), more complex products (such as secondary metabolites), extracellular proteins, and constituents of biomass, *e.g.*, cellular proteins, RNA, DNA, and lipids. These conversions occur via a large number of metabolites, including precursor metabolites and building blocks in the synthesis of macromolecular pools. In order to analyze this vast set of reactions, we have to specify their stoichiometry, and we will do this within a general formalism. For this purpose it is important to distinguish between substrates, metabolic products, intracellular metabolites, and biomass constituents, defined in this text as follows:

- A *substrate* is a compound present in the sterile medium that can be further metabolized by, or directly incorporated into, the cell. With this broad definition the number of substrates is normally large, ranging from the carbon, nitrogen, and energy sources to various minerals essential for cell function. In most cases only the carbon, nitrogen, and energy sources are considered, and, with glucose usually serving as the carbon and free energy source, the number of substrates considered is generally small (glucose, ammonia, and perhaps oxygen).
- A *metabolic product* is a compound produced by the cells that is excreted to the extracellular medium. Thus, it may be either a com-

pound produced in the primary metabolism, *e.g.*, carbon dioxide, ethanol, acetate, or lactate, or a more complex one, *e.g.*, a secondary metabolite or a heterologous protein secreted to the extracellular medium.

- *Biomass constituents* are pools of macromolecules (or a single macro-molecule) that make up biomass. This group includes true cellular constituents like the macromolecular pools of RNA, DNA, protein, lipids, carbohydrates, etc., as well as macromolecular products accumu-lating inside the cell, *e.g.*, a polysaccharide, a biopolymer, or a nonse-creted heterologous protein.

- Finally, *intracellular metabolites* is a class that includes all other com-pounds within the cell. Thus, this includes both intermediates in the different cellular pathways, *e.g.*, glycolytic intermediates, and building blocks used for macromolecular synthesis, *e.g.*, amino acids.

The preceding distinction between biomass constituents and intracellular metabolites differentiates intracellular compounds according to the time scale of their turnover in cellular reactions. Thus, small metabolites like glycolytic intermediates (but also amino acids) have a very rapid turnover of their pool compared with macromolecules (see Example 8.1), and a pseudo-steady state assumption is generally applicable for them (see Chapter 8). On the other hand, biomass constituents can change slowly with time during transient growth. In some cases it may be difficult to decide whether a compound is to be considered as an intracellular metabolite or as a metabolic product. Pyruvate, for example, may accumulate and be excreted at very high glycolytic fluxes in *Saccharomyces cerevisiae*, but when the glycolytic flux decreases as a result of decreasing glucose concentration, it may be taken up again and further metabolized. Obviously pyruvate is a pathway intermedi-ate, but because it is excreted it can also be considered as a metabolic product. In this text, we will consider such compounds as metabolic prod-ucts. Thus, all compounds that can be measured in the extracellular medium at any time throughout a cultivation experiment should be treated as either a substrate or a metabolic product.

The following two examples illustrate the preceding classification. The first is the uptake of glucose by the phosphotransferase system (PTS) active in many bacteria. The actual PTS transport mechanism involves a number of enzymes (see Section 2.2.3); however, for the purpose of this discussion, the PTS is summarized through the following overall stoichiometry:

$$\text{glucose - PEP + glucose-6-phosphate + pyruvate} = 0 \qquad (3.1)$$

The other example is ammonia assimilation by the NADPH-linked glutamate dehydrogenase (see Section 2.4.1), for which the stoichiometry is

$$\alpha\text{-ketoglutarate} - NH_3 - NADPH - H^+$$

$$+ \text{L-glutamate} + NADP^+ + H_2O = 0 \qquad (3.2)$$

In the preceding reactions, the stoichiometry is written such that a compound used in the forward reaction (as reactant) has a negative stoichiometric coefficient, and a compound formed in the forward reaction has a positive stoichiometric coefficient. Besides the sign, stoichiometric coefficients in these reactions supply us with important information, *e.g.*, that 1 mol of PEP is used in the uptake and phosphorylation of 1 mol of glucose, and at the same time 1 mol of pyruvate is formed.

For generalization, we now term the stoichiometric coefficients for substrates α, the stoichiometric coefficients for metabolic products β, and the stoichiometric coefficients for intracellular metabolites g. Notice that the stoichiometric coefficients for substrates are generally negative and the stoichiometric coefficients for metabolic products positive, whereas the stoichiometric coefficients for the intracellular metabolites may be either positive or negative. Although glutamate and α-ketoglutarate are components of cellular biomass, they are, according to the definition given earlier, considered as intracellular metabolites in reaction (3.2). No biomass compounds are considered in these examples, but in the general formalism their stoichiometric coefficients are designated by γ. An overall stoichiometry for the synthesis of macromolecular constituents of biomass was derived in Section 2.4.2, and it is the stoichiometric coefficients in these reactions that are designated by γ to differentiate them from intracellular metabolites that are participating in other metabolic reactions.

With these general definitions of stoichiometric coefficients for substrates, metabolic products, intracellular metabolites, and biomass constituents, we return to the two reactions (3.1) and (3.2). Obviously glucose and ammonia are substrates, and except for protons and water all the other compounds are intracellular metabolites. Thus, we have that both α_{glc} and g_{PEP} are -1 in reaction (3.1) and that g_{glut} is 1 in reaction (3.2). Protons and water may be considered as metabolic products, although they generally are not included in reaction stoichiometries due to their preponderance among other cellular constituents.

We now proceed to formulate a general stoichiometry for cellular reactions. For this purpose we consider a system where N substrates are converted to M metabolic products and Q biomass constituents. The conversions are carried out in J reactions in which K intracellular metabolites also participate as pathway intermediates. A two-numbered index on the stoichiometric coefficients will indicate the reaction number and the compound, *e.g.*, α_{ji} is the stoichiometric coefficient for the ith substrate in the jth reaction. In the generalized stoichiometry, we introduce stoichiometric coefficients for all substrates, metabolic products, intracellular metabolites, and biomass constituents in each of the reactions. Many of the stoichiometric coefficients therefore become zero, as the corresponding compound does not participate in a reaction, *e.g.*, the stoichiometric coefficient for glucose in reaction (3.2) is zero. The substrates are termed S_i, the metabolic products P_i and the biomass constituents $X_{macro, i}$. Furthermore, the K pathway intermediates are termed $X_{met, i}$. With these definitions, the stoichiometry for the jth cellular reaction can be specified as

$$\sum_{i=1}^{N} \alpha_{ji} S_i + \sum_{i=1}^{M} \beta_{ji} P_i + \sum_{i=1}^{Q} \gamma_{ji} X_{macro, i} + \sum_{i=1}^{K} g_{ji} X_{met, i} = 0 \qquad (3.3)$$

In a metabolic model there will be an equation like (3.3) for each of the J cellular reactions, and it is therefore convenient to write the stoichiometry for all J cellular reactions in a compact form using matrix notation:

$$\mathbf{AS} + \mathbf{BP} + \boldsymbol{\Gamma} \mathbf{X}_{macro} + \mathbf{GX}_{met} = 0 \qquad (3.4)$$

where the matrices $\mathbf{A}$, $\mathbf{B}$, $\boldsymbol{\Gamma}$, and $\mathbf{G}$ are stoichiometric matrices containing stoichiometric coefficients in the J reactions for the substrates, metabolic products, biomass constituents, and pathway intermediates, respectively. In matrices $\mathbf{A}$, $\mathbf{B}$, $\boldsymbol{\Gamma}$, and $\mathbf{G}$, rows represents reactions and columns metabolites. Thus, the element in the jth row and the ith column of $\mathbf{A}$ specifies the stoichiometric coefficient for the ith substrate in the jth reaction. As discussed earlier, the stoichiometric coefficients may be positive, negative, or zero. With a stoichiometric formulation of the general type of eq. (3.4), a large number of the stoichiometric coefficients become zero, and one may find it cumbersome to specify stoichiometric coefficients for all compounds and all reactions considered in the model. However, the advantage is that the general matrix formulation facilitates much of the subsequent analysis because matrix operations can be used (for those who are not familiar with matrix notation and matrix operations, we refer to Box 4.2). Furthermore, when the stoichiometry has been formulated in matrix notation, it is very easy to get an overview of the participation of different compounds in the various reactions as one only has to look at the column for this compound in the appropriate matrix. Stoichiometric matrix columns collect the stoichiometric coefficients for a particular compound in all J reactions. This is illustrated in Example 3.1, where the stoichiometry for the mixed acid fermentation of *Escherichia coli* is developed following the formalism of eqs. (3.3) and (3.4).

In cellular reactions there are a number of cofactor pairs, with ATP/ADP, NAD^+/NADH, and $NADP^+$/NADPH being the most important. For the two compounds in a cofactor pair, the stoichiometric coefficients will normally be the same in magnitude but of opposite sign, *e.g.*, in reaction (3.2) the stoichiometric coefficients for NADPH and $NADP^+$ are -1 and 1, respectively. The intracellular concentration of the different forms of these cofactors is, therefore, closely related, *e.g.*, the sum of concentrations of NADPH and $NADP^+$ is constant in any reaction. If both cofactors are included in the stoichiometry, linearly dependent columns will arise in the stoichiometric matrix G, *i.e.*, one of the columns will be a linear combination of the other columns. This creates problems in later analysis, and, because cofactor pairs are directly related, it is not necessary to consider them both. In all such cases we include only one of the cofactors in the pair. It is, however, important to note that, if only one of the compounds in a cofactor pair is considered, the same basis for their interconversion should be used, *e.g.*, interconversion of ATP and ADP. If in some reactions another compound is formed from one of the cofactors, *e.g.*, formation of AMP from ATP, it is necessary to take this into account, in analogy with the energetic cost calculations of Section 2.6, where the conversion of ATP to ADP is used as basis. In some cases the identification of conserved moieties like cofactor pairs is not straightforward, but they can be identified by checking for linearly dependent rows in the stoichiometric matrix and thereafter eliminated.

EXAMPLE 3.1

Mixed Acid Fermentation by *E. coli*

E. coli is a facultative anaerobe that mediates a relatively complex fermentation normally referred to as mixed acid fermentation. Seven metabolic products are produced, and with the exception of succinate, which is made from phosphoenolpyruvate, all metabolites are formed from pyruvate. Succinate is formed via oxaloacetate, which undergoes transamination with glutamate to yield aspartate (one NADPH and one ammonium are used to regenerate glutamate from α-ketoglutarate, so they appear as reactants in Fig. 3.1). Aspartate is then deaminated to form fumarate, which is finally reduced to succinate by fumarate dehydrogenase (which is different from the succinate dehydrogenase that functions in the opposite direction). Figure 3.1 gives an overview of the key steps in the mixed acid fermentation, and Table

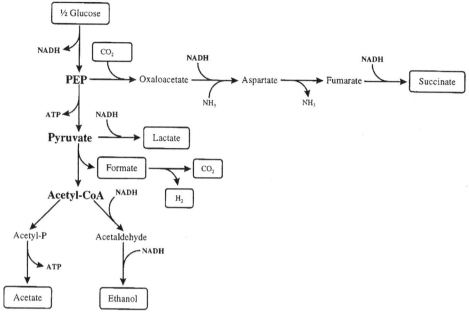

FIGURE 3.1   Mixed acid fermentation by *E. coli*. The substrate (glucose) and the seven metabolic products are circled. Intracellular metabolites and cofactors included in the metabolic model are marked in boldface type. The transamination reaction where aspartate is formed from oxaloacetate is shown as a direct amination, *i.e.*, glutamate and 2-ketoglutarate are not shown in order to reduce the complexity.

3.1 lists some typical yields. Our goal in setting up a stoichiometric model is to account for the net change of metabolites in the medium in the context of catabolic reactions operative in a typical *E. coli* cell.

We now set up a simple metabolic model for this mixed acid fermentation. In order to minimize unnecessary complexity, we do not consider all

TABLE 3.1   Typical Yields of the Mixed Acid *E. coli* Fermentation[a]

Metabolic product	Moles formed per 100 mol of glucose fermented
Formate	2.4
Acetate	36.5
Lactate	79.5
Succinate	10.7
Ethanol	49.8
CO_2	88.0
H_2	75.0

[a] The data are taken from Ingraham *et al.* (1983).

glycolytic intermediates. Consequently, the conversion of glucose to phosphoenolpyruvate (PEP) is lumped into an overall reaction with the following stoichiometry:

$$-\tfrac{1}{2}\text{glucose} + \text{PEP} + \text{NADH} = 0 \tag{1}$$

Intermediates in the pathways leading to succinate, i.e., oxaloacetate, aspartate, and fumarate, are also omitted. Furthermore, NADH and NADPH have been lumped in a single pool of reducing equivalents, implying, in essence, that in the regeneration of glutamate, which is used in the conversion of oxaloacetate to aspartate, we take NADH (instead of NADPH) to be the cofactor. Consequently, the overall stoichiometry for the conversion of PEP to succinate is

$$-\text{PEP} - CO_2 - 2\text{NADH} + \text{succinate} = 0 \tag{2}$$

The stoichiometry for the conversion of PEP to pyruvate and the stoichiometries for further conversion of this compound to lactate, formate, and acetyl-CoA are taken directly from Fig. 3.1:

$$-\text{PEP} + \text{pyruvate} + \text{ATP} = 0 \tag{3}$$

$$-\text{pyruvate} - \text{NADH} + \text{lactate} = 0 \tag{4}$$

$$-\text{pyruvate} + \text{acetyl-CoA} + \text{formate} = 0 \tag{5}$$

Similarly, for the hydrolysis of formate to carbon dioxide and hydrogen:

$$-\text{formate} + CO_2 + H_2 = 0 \tag{6}$$

Finally, we need to describe the stoichiometry of the pathways leading to acetate and ethanol. Here we also leave out the intermediates (acetyl-P and acetaldehyde) whereby the overall stoichiometry becomes:

$$-\text{acetyl-CoA} + \text{acetate} + \text{ATP} = 0 \tag{7}$$

$$-\text{acetyl-CoA} - 2\text{NADH} + \text{ethanol} = 0 \tag{8}$$

In the preceding formulation, we have described the mixed acid fermentation of E. coli with eight overall reactions. It is interesting to note that what we have done is to leave out all intermediates along linear segments of a pathway, so that only metabolites at branch points are considered. As discussed at length in Chapter 8, this is a consequence of pseudo-steady state assumptions for all intracellular metabolites.

The preceding eight equations can be grouped following the matrix notation of eq. (3.4). Glucose is identified as a substrate and succinate, carbon dioxide, lactate, formate, hydrogen, acetate, and ethanol as metabolic products, with phosphenolpyruvate (PEP), pyruvate, acetyl-CoA, ATP, and NADH as intracellular metabolites. No biomass constituents are included in the model, because no anabolic reactions are involved and only fueling reactions are considered. From this and the stoichiometry of reactions (1) - (8), we get

$$
\begin{array}{cc}
\mathbf{A} & \mathbf{B}
\end{array}
$$

$$
\begin{pmatrix} -\frac{1}{2} \\ 0 \\ 0 \\ 0 \\ 0 \\ 0 \\ 0 \\ 0 \end{pmatrix} S_{glc} +
\underbrace{\begin{pmatrix} 0 & 0 & 0 & 0 & 0 & 0 & 0 \\ 1 & -1 & 0 & 0 & 0 & 0 & 0 \\ 0 & 0 & 0 & 0 & 0 & 0 & 0 \\ 0 & 0 & 1 & 0 & 0 & 0 & 0 \\ 0 & 0 & 0 & 1 & 0 & 0 & 0 \\ 0 & 1 & 0 & -1 & 1 & 0 & 0 \\ 0 & 0 & 0 & 0 & 0 & 1 & 0 \\ 0 & 0 & 0 & 0 & 0 & 0 & 1 \end{pmatrix}}_{}
\begin{pmatrix} P_{suc} \\ P_{CO_2} \\ P_{lac} \\ P_{for} \\ P_{H_2} \\ P_{ac} \\ P_{et} \end{pmatrix}
$$

$$
+ \underbrace{\begin{pmatrix} 1 & 0 & 0 & 0 & 1 \\ -1 & 0 & 0 & 0 & -2 \\ -1 & 1 & 0 & 1 & 0 \\ 0 & -1 & 0 & 0 & -1 \\ 0 & -1 & 1 & 0 & 0 \\ 0 & 0 & 0 & 0 & 0 \\ 0 & 0 & -1 & 1 & 0 \\ 0 & 0 & -1 & 0 & -2 \end{pmatrix}}_{\mathbf{G}}
\begin{pmatrix} X_{PEP} \\ X_{Pyr} \\ X_{AcCoA} \\ X_{ATP} \\ X_{NADH} \end{pmatrix} =
\begin{pmatrix} 0 \\ 0 \\ 0 \\ 0 \\ 0 \\ 0 \\ 0 \\ 0 \end{pmatrix}
\qquad (9)
$$

Equation (9) provides a convenient overview of the reactions considered. Thus, by looking at the fourth column of the last stoichiometric matrix (**G**), it can be seen that ATP is produced only in two reactions, namely, the conversion of PEP to pyruvate [reaction (3)] and the conversion of acetyl-CoA to acetate [reaction (7)]. Because the fluxes of these two reactions are measurable (the flux to acetate can be measured directly as the formation rate of acetate, whereas the flux from PEP to pyruvate can be measured from the sum of the rates of formation of all the metabolic products except succinate or from the difference between the glucose uptake rate and the rate of succinate formation), we can obtain information on the total rate of ATP synthesis. As there are no other sources of ATP supply under anaerobic conditions, the latter is also an estimate of the consumption rate of ATP for growth and maintenance. Furthermore, because the redox balance (or, equivalently, the NADH balance) has to close for the conversion of sugar to the different metabolites, we see that the uptake of glucose must be related to the formation of succinate, lactate, and ethanol. Thus, using the data of Table 3.1 and the last column of the stoichiometric matrix for the intracellular metabolites we find that the consumption of NADH per 100 mol of glucose is

$$2 \times 10.7 + 79.5 + 2 \times 49.8 = 200.5$$

which matches the 200 mol of NADH produced per 100 mol of glucose converted to PEP very well.

3.2. REACTION RATES

The stoichiometry specified in section 3.1 defines the relative amounts of the compounds produced or consumed in each of the J intracellular reactions, but does not allow one to calculate the rates or the relative amounts at which metabolic products are secreted in the medium. This can be done by introducing the rates of the individual reactions and further coupling them to determine the overall rates of product secretion. The rate of a chemical reaction is defined as the forward rate (or velocity) v, which specifies that a compound that has a stoichiometric coefficient β is formed at the rate βv. Normally, one of the stoichiometric coefficients in each reaction is arbitrarily set to be 1, whereby the forward reaction rate becomes equal to the consumption or production rate of this compound in the particular reaction, e.g., the rate of reaction (3.1) is equal to the glucose uptake rate by the PTS. For this reason, the forward reaction rate is normally specified with units mol h^{-1}, or if we want to specify that a certain volume element is considered the unit mol $(L\ h)^{-1}$ is used. For cellular reactions we often use the biomass as reference to define the so-called *specific rates*, which have the unit mol $(g\ DW\ h)^{-1}$. We now collect the forward reaction rates of the J reactions considered in section 3.1 in the rate vector **v**. Thus, $\beta_{ji} v_j$ specifies the specific rate of formation of the ith metabolic product in the jth reaction. Because the stoichiometric coefficients for the substrates, i.e., the elements of **A**, are generally negative, the specific conversion rate of the ith substrate in the jth reaction is given by $-\alpha_{ji} v_j$. When we want to calculate the overall production or consumption of a compound, we have to sum the contributions from the different reactions. This can be illustrated by considering the mixed acid fermentation of E. coli in Example 3.1, where carbon dioxide is formed in one reaction (decarboxylation of formate) and used in another reaction (carboxylation of PEP). The total production rate of carbon dioxide is obviously determined by the relative rates of these two reactions. We can therefore write the net specific uptake rate for the ith substrate as the sum of its consumption rate in all J reactions:

$$r_{s,i} = - \sum_{j=1}^{J} \alpha_{ji} v_j \qquad (3.5)$$

and similarly for the net specific rate of formation of the ith metabolic product:

$$r_{p,i} = \sum_{j=1}^{J} \beta_{ji} v_j \qquad (3.6)$$

Equations (3.5) and (3.6) specify very important relationships between what can be directly measured, namely, the specific uptake rates of substrates and the specific formation rate of products on one hand and the rates of the various intracellular reactions on the other. We often use the term *fluxes* for the latter to indicate that they are rates through pathways rather than rates of single reactions. If a compound is formed in only one reaction, a measure of its rate of consumption or formation is an indirect measure of the rate of this reaction. Thus, if we return to the mixed acid fermentation of *E. coli* in Example 3.1, we see that, for example, the rate of acetate formation equals the flux from acetyl-CoA to acetate via acetyl-P. In other words, the rate (or flux) of the seventh reaction of this example can be estimated from measurements of the acetate formation rate.

Similar to eqs. (3.5) and (3.6), for the biomass constituents and the intracellular metabolites we can write

$$r_{macro,i} = \sum_{j=1}^{J} \gamma_{ji} v_j \qquad (3.7)$$

$$r_{met,i} = \sum_{j=1}^{J} g_{ji} v_j \qquad (3.8)$$

These rates are not as easy to determine experimentally as the specific substrate uptake rates and the specific product formation rates. The rates in eqs. (3.7)-(3.8) are net specific formation rates and can be quantified from measurements of intracellular components. Thus, a compound may be consumed in one reaction and produced in another, and the rates specified on the left-hand sides of eqs. (3.7) and (3.8) are the net result of consumption and production of that compound in all J intracellular reactions. If the rate $r_{met,i}$ is positive there is a net formation of the ith intracellular metabolite, and if it is negative there is a net consumption of this metabolite. If the rate is zero, the rates of formation in the J reactions exactly balance the rates of consumption (this balancing is the basis for metabolic flux analysis discussed in detail in Chapters 8-10).

The summation equations (3.5) - (3.8) can be formulated in matrix notation as

$$\mathbf{r}_s = -\mathbf{A}^T \mathbf{v} \tag{3.9}$$

$$\mathbf{r}_p = \mathbf{B}^T \mathbf{v} \tag{3.10}$$

$$\mathbf{r}_{macro} = \mathbf{\Gamma}^T \mathbf{v} \tag{3.11}$$

$$\mathbf{r}_{met} = \mathbf{G}^T \mathbf{v} \tag{3.12}$$

where the specific rate vector $\mathbf{r}_s$ contains the N specific substrate uptake rates, $\mathbf{r}_p$ the M specific product formation rates, etc.

EXAMPLE 3.2

Specific Rates of the Mixed Acid Fermentation of E. coli

We now specify the rates of the eight reactions considered in Example 3.1 as v_1-v_8. This immediately shows us that the specific glucose uptake rate is

$$r_{glc} = \tfrac{1}{2}v_1 \tag{1}$$

which, of course, could also have been found by using eq. (3.9):

$$r_{glc} = -\left(-\tfrac{1}{2} \quad 0 \quad 0 \quad 0 \quad 0 \quad 0 \quad 0 \quad 0\right) \begin{pmatrix} v_1 \\ v_2 \\ v_3 \\ v_4 \\ v_5 \\ v_6 \\ v_7 \\ v_8 \end{pmatrix} = \tfrac{1}{2}v_1 \tag{2}$$

Thus, the flux to PEP, which is given by v_1, can be determined as 2 times the specific glucose uptake rate.

The specific rates of metabolic product formation are found similarly, e.g., for carbon dioxide:

$$r_{CO_2} = v_6 - v_2 \tag{3}$$

Because the specific rate of hydrogen formation must equal v_6 and the specific rate of succinate formation must equal v_2, the rate of carbon dioxide formation should be equal to the difference between these two rates, and this constraint can be used to check the consistency of the data. Note that the data of Table 3.1 do not seem to be consistent as there is more carbon dioxide formed than hydrogen per 100 mol of glucose metabolized. Either there is an experimental error or the model is too simple.

In a simple model like this it is easier to derive correlations between the individual reaction rates and the specific substrate uptake and product formation rates directly (without using the matrix notation). This is, however, not the case for the intracellular metabolites, as these may take part in several different reactions. Thus, for the five intracellular metabolites considered, we find

$$
\begin{pmatrix} r_{PEP} \\ r_{Pyr} \\ r_{AcCoA} \\ r_{ATP} \\ r_{NADH} \end{pmatrix} = \begin{pmatrix} 1 & -1 & -1 & 0 & 0 & 0 & 0 & 0 \\ 0 & 0 & 1 & -1 & -1 & 0 & 0 & 0 \\ 0 & 0 & 0 & 0 & 1 & 0 & -1 & -1 \\ 0 & 0 & 1 & 0 & 0 & 0 & 1 & 0 \\ 1 & -2 & 0 & -1 & 0 & 0 & 0 & -2 \end{pmatrix} \mathbf{v}
$$

$$
= \begin{pmatrix} v_1 - v_2 - v_3 \\ v_3 - v_4 - v_5 \\ v_5 - v_7 - v_8 \\ v_3 + v_7 \\ v_1 - 2v_2 - v_4 - 2v_8 \end{pmatrix} \tag{4}
$$

As discussed in Example 3.1, r_{ATP} is seen to be given as the sum of fluxes from PEP to pyruvate and from pyruvate to acetate.

3.3. DYNAMIC MASS BALANCES

In the previous section, we derived equations that relate the intracellular reaction rates with the rates of substrate uptake and metabolic product formation. The latter may be obtained experimentally from the measurement of the concentrations of substrates and metabolic products. Before we turn to how the specific rates are obtained from such measurements (which is treated at the end of this section), it is necessary to consider the dynamics of the bioreactor in which such measurements are usually made. Figure 3.2 is a

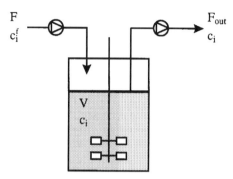

FIGURE 3.2 Bioreactor with the addition of fresh, sterile medium and removal of spent medium. c_i^f is the concentration of the ith compound in the feed and c_i is the concentration of the ith compound in the spent medium. The bioreactor is assumed to be very well mixed (or ideal), so that the concentration of each compound in the spent medium becomes identical to its concentration in the bioreactor.

general representation of a bioreactor. It has volume V (L) and it is fed with a stream of fresh, sterile medium with a flow rate F_{in} (L h^{-1}). Spent medium is removed with a flow rate F_{out} (L h^{-1}). The medium in the bioreactor is assumed to be completely (or ideally) mixed, i.e., there are no spatial variations in the concentration of the different medium compounds. For small volume bioreactors (< 1 L) (including shake flasks), this can generally be achieved through aeration and some agitation, whereas for laboratory, stirred tank bioreactors (1-10 L), special designs may have to be introduced in order to ensure a homogeneous medium (Sonnleitner and Fiechter, 1988; Nielsen and Villadsen, 1993). Not all bioreactors involve the continuous flow of medium, but they may be operated in several different modes, the three most common of which are considered:

- **Batch**, where $F = F_{out} = 0$, i.e., the volume is constant. This is the classical fermentation operation, and it is used by many life scientists, because it requires a relatively simple experimental setup. Batch experiments have the advantage of being easy to perform and can produce large volumes of experimental data in a short period of time. The disadvantage is that the experimental data are difficult to interpret as there are dynamic variations throughout the experiment, i.e., the environmental conditions experienced by the cells vary with time. By using well-instrumented bioreactors at least some variables, e.g., pH and dissolved oxygen tension, may, however, be controlled at a constant level.

- **Continuous**, where $F = F_{out} \neq 0$, i.e., the volume is constant. A typical operation of the continuous bioreactor is the so-called *chemostat*, where the added medium is designed such that there is a single rate-limiting substrate. This allows for controlled variation in the specific growth rate of the biomass. The advantage of the continuous bioreactor is that a steady state can be obtained, which allows for precise experimental determination of specific rates under well-defined environmental conditions. These conditions can be further varied by changing the feed flow rate to the bioreactor. This allows valuable information concerning the influence of the environmental conditions on cellular physiology to be obtained. The disadvantage of the continuous bioreactor is that it is laborious to operate as large amounts of fresh, sterile medium have to be prepared and requires long periods of time for a steady state to be achieved. Despite the advantages of continuous operation, it is rarely used in industrial processes because it is sensitive to contaminations, e.g., via the feed stream, and to genetic instability that may lead to the formation of fast-growing mutants that out-compete the production strain. Other examples of continuous operation are the *pH-stat*, where the feed flow is adjusted to maintain constant pH in the bioreactor, and the *turbidostat*, where the feed flow is adjusted to maintain the biomass concentration at a constant level.
- **Fed-batch** (or semibatch), where $F \neq 0$ and $F_{out} = 0$, i.e., the volume increases. This is probably the most common operation in industrial practice, because it allows for control of the environmental conditions, e.g., maintaining the glucose concentration at a certain level, and it enables formation of much higher titers (up to several hundreds grams per liter of some metabolites), which is of importance in the subsequent downstream processing. At the same time, the fed-batch operation is a convenient experimental system for maintaining steady environmental conditions to facilitate physiological studies, albeit for a limited period of time.

With the bioreactor of Fig. 3.2, the dynamic mass balances for the ith substrate are given by (see Box 3.1 for derivation):

$$\text{Accumulation} = - \text{Rate of substrate} + \text{Rate of substrate} - \text{Rate of substrate}$$
$$\text{of substrate} \quad \text{consumption} \qquad \text{addition} \qquad \text{removal}$$

$$\frac{dc_{s,i}}{dt} = -r_{s,i}x + D\left(c^f_{s,i} - c_{s,i}\right) \tag{3.13}$$

where $c_{s,i}$ is the concentration of the ith substrate in the bioreactor (mol L^{-1}), $c^f_{s,i}$ is the concentration of the ith substrate in the feed (mol L^{-1}), $r_{s,i}$ is the specific substrate consumption rate of the ith substrate, x is the biomass concentration (g DW L^{-1}), and D is the so-called *dilution rate* (h^{-1}), which for a batch reactor is zero and for a chemostat and a fed-batch reactor is given by:

$$D = \frac{F}{V} \tag{3.14}$$

The first term on the right-hand side of eq. (3.13) is the volumetric rate of substrate consumption, which is given as the product of the specific rate of substrate consumption multiplied by the biomass concentration. The second term accounts for the addition and removal of substrate from the bioreactor. The left-hand side of eq. (3.13) is the accumulation term, given by the time rate of change of the substrate, which, in a batch reactor, equals the volumetric rate of substrate consumption. At steady state the accumulation term is equal to zero, so that the volumetric rate of substrate consumption becomes equal to the product of the dilution rate multiplied by the difference in the substrate concentrations between the inlet and outlet of the reactor.

In analogy with the dynamic mass balances for the substrates, we have for the metabolic products:

$$\frac{dc_{p,i}}{dt} = r_{p,i}x + D\left(c^f_{p,i} - c_{p,i}\right) \tag{3.15}$$

where the first term on the right-hand side is the volumetric formation rate of the ith metabolic product. Normally, metabolic products are not present in the sterile feed to the bioreactor, so that $c^f_{p,i}$ is zero. In these cases the volumetric rate of product formation at steady state is equal to the dilution rate multiplied by the concentration of the metabolic product in the bioreactor (which is equal to that at the outlet).

For the biomass constituents we normally use biomass as reference, *i.e.*, concentrations are given with biomass as the basis. In this case the mass balances are given by (see Box 3.1 for derivation)

$$\frac{dX_{macro,i}}{dt} = r_{macro,i} - \mu X_{macro,i} \tag{3.16}$$

where $X_{macro,i}$ is the concentration of the ith biomass constituent and μ is the specific growth rate (h^{-1}). Different units may be applied for the concentrations of the biomass constituents, but they are normally given as g (g DW)$^{-1}$. With these units it, can be shown that the sum of all biomass

BOX 3.1

Derivation of Dynamic Mass Balances

In this note we show the derivation of the mass balances for the substrates and biomass constituents. The mass balances for the metabolic products and intracellular metabolites can be similarly derived.

Substrates.

We consider the i'th substrate which is added to the bioreactor via the feed and is consumed by the cells present in the bioreactor. The mass balance for this compound is:

$$\frac{d(c_{s,i}V)}{dt} = -r_{s,i}xV + Fc_{s,i}^{f} - F_{out}c_{s,i} \qquad (1)$$

where $r_{s,i}$ is the specific consumption rate of compound i (moles (g DW h)$^{-1}$), $c_{s,i}$ is its concentration in the bioreactor (assumed to be equal to the concentration in the outlet, moles L^{-1}), $c_{s,i}^{f}$ is the concentration of the same compound in the feed (moles L^{-1}), and x is the biomass concentration in the bioreactor (g DW L^{-1}). The first term in eq. (1) is the accumulation term, the second term is the consumption (or reaction term), the third term is accounting for the supply, and the last term is accounting for the removal rate of the compound. Rearrangement of eq. (1) gives:

$$V\frac{dc_{s,i}}{dt} + c_{s,i}\frac{dV}{dt} = -r_{s,i}xV + Fc_{s,i}^{f} - F_{out}c_{s,i} \qquad (2)$$

or upon division by V and rearrangement:

$$\frac{dc_{s,i}}{dt} = -r_{s,i}x + \frac{F}{V}c_{s,i}^{f} - \left(\frac{F_{out}}{V} + \frac{1}{V}\frac{dV}{dt}\right)c_{s,i} \qquad (3)$$

Since for a fed-batch reactor:

$$F = \frac{dv}{dt} \qquad (4)$$

(continues)

(*continued*)

and $F_{out} = 0$ the term inside the parenthesis becomes equal to the dilution rate. For a continuous or batch reactor the volume is constant and $F = F_{out}$, so that for these bioreactor modes too the term inside the parenthesis becomes equal to the dilution rate. Eq. (2) therefore reduces to the mass balance (3.13) for any type of operation.

Biomass constituents.

For the biomass constituents the mass balance is given by (sterile feed is assumed):

$$\frac{d(X_{macro,i}\,xV)}{dt} = r_{macro,i}\,xV - F_{out}\,X_{macro,i}\,x \tag{5}$$

where $X_i\,x$ is the concentration of the i'th biomass component in the bioreactor (g L^{-1}) and $r_{macro,i}$ is the specific, net rate of formation of the i'th biomass constituent. Rearrangement of (5) gives:

$$\frac{dX_{macro,i}}{dt} = r_{macro,i} - \left(\frac{F_{out}}{V} + \frac{1}{x}\frac{dx}{dt} + \frac{1}{V}\frac{dV}{dt} \right) X_{macro,i} \tag{6}$$

Again we have that for any mode of bioreactor operation:

$$D = \frac{F_{out}}{V} + \frac{1}{V}\frac{dV}{dt} \tag{7}$$

which together with the mass balance (3.20) for the total biomass concentration gives the mass balance of eq. (3.16).

In order to derive eq. (3.18) we use eq. (3.17) which implies that:

$$\sum_{i=1}^{Q} \frac{dX_{macro,i}}{dt} = \sum_{i=1}^{Q} r_{macro,i} - \mu \sum_{i=1}^{Q} X_{macro,i} = 0 \tag{8}$$

which directly gives eq. (3.18).

constituent concentrations equals unity:

$$\sum_{i=1}^{Q} X_{macro,i} = 1 \qquad (3.17)$$

Furthermore, the preceding unit is consistent with the experimentally determined macromolecular composition of cells where mass fractions are generally used (see Table 2.11). In eq. (3.16) it is observed that the mass balance for the biomass constituents is independent of the mode of bioreactor operation as the dilution rate does not appear explicitly in the mass balance. However, indirectly there is a coupling via the last term, which accounts for dilution of the biomass constituents when the biomass expands due to growth. Thus, if there is no net synthesis of a macromolecular pool ($r_{macro,i} = 0$) but the biomass still grows ($\mu > 0$), the intracellular level decreases. By using eq. (3.17) the specific growth rate can be shown to be given by the sum of the net formation rates of all biomass constituents (see Box 3.1):

$$\mu = \sum_{i=1}^{Q} r_{macro,i} \qquad (3.18)$$

For intracellular metabolites it is not convenient to use the same unit of concentration as for the biomass constituents. These metabolites are dissolved in the cytosol of the cell and it is therefore more appropriate to express their concentration in moles per liquid cell volume. This unit choice also allows for direct comparison of intracellular metabolite concentrations with enzyme affinities, typically quantified by their K_m values, which are normally given in moles L^{-1}. If the concentration is known in one unit it is, however, easily converted to another unit if the density of the biomass (in the range of 1 g cell per mL cell) and the water content (in the range of 0.67 mL water per mL cell) are known. Despite the use of different units for the concentration of intracellular metabolites, biomass is still the basis and the mass balances for the intracellular metabolites take the same form as for the biomass constituents:

$$\frac{dc_{met,i}}{dt} = r_{met,i} - \mu c_{met,i} \qquad (3.19)$$

where $c_{met,i}$ is the concentration of the i'th intracellular metabolite. It is important to distinguish between concentrations of intracellular metabolites given in moles per liquid reactor volume and in moles per liquid cell volume.

If concentrations are given in the first of the above units, the mass balance will be completely different.

Equations (3.13), (3.15), (3.16) and (3.19) specify the mass balances for the four different types of species we consider in cellular stoichiometry. Additionally, the total biomass balance is important, and this is given by:

$$\frac{dx}{dt} = (\mu - D)x \qquad (3.20)$$

From this mass balance it is easily seen that at steady state the specific growth rate equals the dilution rate. Thus, by varying the dilution rate (or, equivalently, the feed flow rate) in a continuous culture, different specific growth rates can be obtained.

The mass balances of eps. (3.13), (3.15), (3.16), and (3.19) form the basis for any type of quantitative treatment of cell cultivation processes, ranging from calculation of specific (or volumetric) rates to design and simulation of bioprocesses. For design and simulation of bioprocesses, it is necessary to specify kinetic expressions for the individual reaction rates in v, *i.e.*, provide functional relationships between the reaction rates and the variables of the system. The combination of stoichiometric equations and kinetic expressions for the reaction rates constitutes a *kinetic model*. Such models allow one to simulate the process, *e.g.*, determine how the process behaves at different operating conditions. Application of kinetic models, both simple, so-called unstructured models and more advanced structured models, to design and simulate cultivation processes is treated extensively in several biochemical engineering textbooks [see, for example, Bailey and Ollis (1986) or Nielsen and Villadsen (1994)], and we will not consider it further here. Instead, we turn to the question of how to use experimental data to determine specific rates for metabolites and biomass constituents.

Table 3.2 gives an overview of the different equations used to calculate the specific rates for substrate uptake, metabolic product formation, and net formation of biomass components (both biomass constituents and intracellular metabolites) in bioreactors under different modes of operation. In all cases we see that the biomass concentration is an important variable. If reliable biomass concentration data are not available, one may decide to use volumetric rates rather than specific rates. The disadvantage of volumetric rates, however, is that they are not directly comparable between experiments. Thus, the growth rate of biomass may be 2.0 g DW L^{-1} h^{-1} in one experiment and 1.0 g DW L^{-1} h^{-1} in another, even though the specific growth rate may be the same, *e.g.*, 0.2 h^{-1} with a biomass concentration of 10 g DW L^{-1} in the first and 5 g DW L^{-1} in the second experiment.

Another observation made from Table 3.2 is that, except for a continuous bioreactor at steady state conditions, the determination of specific rates

TABLE 3.2 Equations Used to Calculate Specific Rates from Experimental Data[a]

Bioreactor	Substrates	Metabolic products	Intracellular species
Batch	$r_s = -\dfrac{1}{x}\dfrac{dc_s}{dt}$	$r_p = \dfrac{1}{x}\dfrac{dc_p}{dt}$	$r_{macro} = \dfrac{dX_{macro}}{dt} + \mu X_{macro}$
Continuous			
Dynamic conditions	$r_s = \dfrac{1}{x}\left(D\left(c_s^f - c_s\right) - \dfrac{dc_s}{dt}\right)$	$r_p = \dfrac{1}{x}\left(Dc_p + \dfrac{dc_p}{dt}\right)$	$r_{macro} = \dfrac{dX_{macro}}{dt} + \mu X_{macro}$
Steady state	$r_s = \dfrac{1}{x}D\left(c_s^f - c_s\right)$	$r_p = \dfrac{1}{x}Dc_p$	$r_{macro} = \mu X_{macro}$
Fed-batch	$r_s = \dfrac{1}{x}\left(D\left(c_s^f - c_s\right) - \dfrac{dc_s}{dt}\right)$	$r_p = \dfrac{1}{x}\left(Dc_p^f + \dfrac{dc_p}{dt}\right)$	$r_{macro} = \dfrac{dX_{macro}}{dt} + \mu X_{macro}$

[a] For intracellular species, only the expressions for biomass constitutents are given. The expressions are similar for intracellular metabolites.

requires calculation of the time derivatives of the concentrations. These time derivatives may be obtained from the slope of the tangent to the curve of concentration measurements. It should be recognized, however, that good estimates of time derivatives are difficult to obtain by this approach because the data are generally sparse and noisy. This clearly underlines the advantage of steady state continuous cultures, whereby accurate estimates of the specific rates can be obtained. Much of the information on cellular physiology is, however, obtained from batch cultures and continuous cultures at dynamic conditions, and it is therefore important to apply robust procedures for extracting information on the specific rates from this type of experiments.

Another approach is to carry out a functional representation of the data, e.g., polynomial splining, and to calculate the derivatives and specific rates from the fitted functions. This approach too can give rise to large fluctuations in the specific rates, because it is difficult to find good functional representations of experimental cultivation data (notice that, in order to get the specific rates, functional relationships have to be derived both for the compound of interest and for the biomass concentration). This is due to the exponential behavior of most cellular processes, which polynomial splining is not well-suited to represent. A better approach, especially for the analysis of batch cultivation data, is to use a simple model, e.g., the Monod model,[1] fit the parameters in the model, and then calculate the specific rates directly

[1] In the Monod model, the specific growth rate is described by $\mu = \mu_{max} c_s / c_s + K_s$, where μ_{max} is the maximum specific growth rate and K_m is the Monod constant. c_s is the concentration of the limiting substrate. In the model it is assumed that the biomass yield on the limiting substrate (see Section 3.4) is constant, and the specific substrate uptake rate therefore is given by $r_s = Y_{xs} \mu$ (see also Section 3.4).

from the model. Because simple models like the Monod model are empirical in nature, this approach is, in principle, the same as searching for a proper functional representation of the data. However, because these models generally are well-suited to fit experimental data from batch cultures, they yield satisfactory results.

The preceding approaches to the functional representation of experimental data (either completely empirical or by using simple kinetic models) are applicable in those cases where sufficient data are available. Intracellular compounds, however, are measured sparsely, which makes it difficult to obtain good estimates of their time derivatives. For intracellular metabolites the assumption of a pseudo-steady state can, in general, be applied (see Section 8.1). Furthermore, during the exponential growth phase of batch cultures, so-called *balanced growth* conditions normally prevail. This means that the intracellular composition of biomass does not change, *i.e.*, the concentrations of the biomass constituents are at steady state, and the elements of r_{macro} therefore can be determined alone from one (or a few) measurement of the macromolecular composition and a good estimate of the specific growth rate.

3.4. YIELD COEFFICIENTS AND LINEAR RATE EQUATIONS

As mentioned in the introductory comments of this chapter, metabolic models are the basis for the quantitative analysis of cell physiology. It is often the case, however, that one is interested not in the full details of intracellular reactions but, rather, in a more macroscopic assessment of the overall distribution of metabolic fluxes, *e.g.*, how much carbon in the glucose substrate is recovered in the metabolite of interest. This overall distribution of fluxes is normally represented by the so-called *yield coefficients*. These are usually defined as overall fluxes with respect to a reference compound, often the carbon source or the biomass. Yield coefficients are, therefore, dimensionless and take the form of unit mass of metabolite per unit mass of the reference, *e.g.*, moles of lysine produced per moles of glucose consumed. However, other references may be of interest in various situations. Thus, the moles of carbon dioxide produced per mole of oxygen consumed, called the respiratory quotient (RQ), frequently is used to characterize aerobic cultivations. Unfortunately, several different nomenclatures have been used in the literature for the yield coefficients. Here we will use the formulation of Nielsen and Villadsen (1994) [which was introduced by Roels (1983)], where the yield coefficients are designated with a double subscript, Y_{ij}, to denote the mass of j formed or consumed per mass of i formed or consumed. Taking the ith substrate as the reference, the yield coefficients are then defined as

$$Y_{s_i s_j} = \frac{r_{s,j}}{r_{s,i}}; \qquad j = 1, \ldots, N \tag{3.21}$$

$$Y_{s_i p_j} = \frac{r_{p,j}}{r_{s,i}}; \qquad j = 1, \ldots, M \tag{3.22}$$

$$Y_{s_i x} = \frac{\mu}{r_{s,i}} \tag{3.23}$$

With this formulation, it is obvious that

$$Y_{ij} = \frac{1}{Y_{ji}} \tag{3.24}$$

The yield coefficients are extremely valuable for evaluating experimental data, and they are often among the first quantitative values to be extracted from raw fermentation data. The fundamental significance of the yield coefficients is that they are the vehicle for introducing some useful relationships among the intrinsic rates defining them. For example, if a yield coefficient is found or assumed to be constant, this immediately implies that a certain relationship exists between the two rates. Such relationships may be entirely empirical or derived from other fundamental biochemical and physiological relationships, as discussed in the following.

In the classical description of cellular growth introduced by Monod (1942), the yield coefficient Y_{sx} was taken to be constant, and all cellular reactions were lumped into a single overall growth reaction whereby substrate is converted to biomass. In 1959, however, Herbert (1959) showed that the yield of biomass on substrate is not constant. In order to describe this, he introduced the concept of *endogenous metabolism* and specified a fraction of the total substrate consumption for this process, in addition to that for biomass synthesis. In the same year, Luedeking and Piret (1959a,b) found that lactic acid bacteria produce lactic acid under non-growth conditions, which was consistent with an endogenous metabolism of the cells. Their results suggested a linear correlation between the specific lactic acid production rate and the specific growth rate:

$$r_p = a\mu + b \tag{3.25}$$

In 1965, Pirt introduced a similar linear correlation between the specific substrate uptake rate and the specific growth rate, and he suggested use of the term *maintenance*, which now is the most commonly used term to describe endogenous metabolism. The linear correlation of Pirt takes the following form:

$$r_s = Y_{xs}^{true}\mu + m_s \tag{3.26}$$

where Y_{xs}^{true} is referred to as the true growth yield coefficient and m_s is the maintenance coefficient. With the introduction of these linear correlations, yield coefficients obviously can no longer be constants. Thus, the yield of biomass on substrate becomes

$$Y_{sx} = \frac{\mu}{Y_{xs}^{true}\mu + m_s} \tag{3.27}$$

which shows that the actual yield, Y_{sx}, can be smaller than the true growth yield when an increasing fraction of the substrate is used to meet the maintenance requirements of the cell, as is the case at low specific growth rates. For large specific growth rates, the yield coefficient approaches the reciprocal of Y_{xs}^{true}, i.e., the actual growth yield coefficient approaches the true growth yield coefficient under fast growth as most of the substrate is consumed for growth under such conditions.

The empirically derived linear correlations are very useful to correlate growth data, especially those obtained from steady state continuous cultures, where linear correlations similar to eq. (3.26) can be derived for most of the important specific rates. This is illustrated in Fig. 3.3, where the specific glucose uptake rate, the specific carbon dioxide production rate, and the specific oxygen uptake rate are plotted as a function of the specific growth rate of the filamentous fungus *Penicillium chrysogenum* grown in steady state continuous cultures. Similar linear correlations are found in the literature. The remarkable robustness and general validity of such linear yield correlations suggest a possible fundamental basis from which they are all derived. One possibility is the balance between a continuous supply and consumption of ATP due to the tight coupling of ATP production and depletion in all cells. Under this hypothesis, the role of the energy-producing substrate is to generate sufficient ATP to drive both the biosynthetic and polymerization reactions of the cell, as well as the different maintenance processes. This coupling is captured by the linear relationship given in eq. (2.27):

$$r_{ATP} = Y_{xATP}\mu + m_{ATP} \tag{3.28}$$

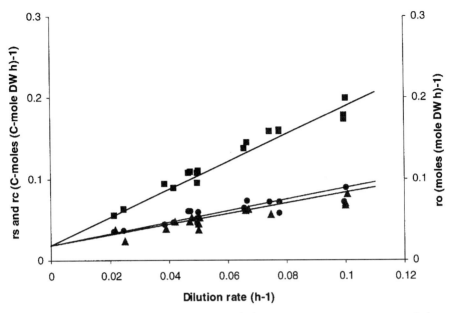

FIGURE 3.3   The specific glucose uptake rate r_s (■), the specific oxygen uptake rate r_o (▲), and the specific carbon dioxide production rate r_c (●) as a function of the specific growth rate (equal to the dilution rate) in glucose-limited continuous cultures of *P. chrysogenum*. The rates are given in C-mol (or mol) per C-mole of biomass per hour. The bioreactor was fed with a defined medium containing glucose, ammonium, inorganic salts, and phenoxyacetate. The data are taken from Nielsen (1997).

which is a formal analogue of the linear correlation of Pirt (1965). As discussed in Section 2.6, eq. (3.28) states that the ATP produced is exactly balanced by that consumed for growth and maintenance. Furthermore, if the ATP yield on the energy-producing substrate is constant, *i.e.*, r_{ATP} is proportional to r_s, it is quite obvious that the linear correlation eq. (3.26) can be derived directly from eq. (3.28). Notice that Y_{xATP} in eq. (3.28) actually is a true ATP yield coefficient, but it is normally specified without the superscript *true*.

The concept of balancing ATP production and consumption can be extended to other cofactors, *e.g.*, NADH and NADPH, whereby it becomes possible to derive linear rate equations for three different cases (Nielsen and Villadsen, 1994): (1) anaerobic growth, where ATP is generated by substrate level phosphorylation; (2) aerobic growth without metabolite formation; and (3) aerobic growth with metabolite formation. The two first cases are each illustrated with the following examples (Examples 3.3 and 3.4).

EXAMPLE 3.3

Metabolic Model of *Penicillium chrysogenum*

To illustrate the derivation of the linear rate equations for an aerobic process without metabolite formation, we consider a simple metabolic model for the filamentous fungus *P. chrysogenum* as presented by Nielsen (1997). The stoichiometric model summarizes the overall cellular metabolism, and by employing pseudo-steady state assumptions for ATP, NADH, and NADPH metabolites, it is possible to derive linear rate equations where the specific uptake rates for glucose and oxygen and the specific carbon dioxide formation rate are expressed in terms of the specific growth rate. By evaluating the parameters in these linear rate expressions, which can be done from a comparison with experimental data, information on key energetic parameters may be extracted. In the analysis, formation of metabolites (both primary metabolites like gluconate and metabolites related to penicillin biosynthesis) was neglected, because the carbon flux to these products was small compared with the flux to biomass and carbon dioxide.

The overall stoichiometry for synthesis of the constituents of a *P. chrysogenum* cell can be summarized as (Nielsen, 1997)

$$\text{biomass} + 0.139CO_2 + 0.458NADH - 1.139CH_2O - 0.20NH_3 -$$

$$0.004H_2SO_4 - 0.010H_3PO_4 - Y_{xATP} \text{ ATP} - 0.243NADPH = 0 \quad (1)$$

This stoichiometry is valid for a cell with composition specified in Table 2.8 when the substrate is glucose and inorganic salts, with ammonia as the N source, sulfate as the S source, and phosphate as the P source. The stoichiometry is given on a C-mole basis, *i.e.*, glucose is specified as CH_2O, and the elemental composition of biomass, as calculated from the macromolecular composition, was found to be $CH_{1.81}O_{0.58}N_{0.20}S_{0.004}P_{0.010}$ (Nielsen, 1997). This agrees quite well with experimentally determined compositions reported in the literature (see Table 4.1). Note that ADP, NAD^+, and $NADP^+$ are not included in the stoichiometry as they are coupled cofactors (as discussed in Section 3.1.) The required ATP and NADPH for biomass synthesis are supplied by the catabolic pathways, and excess NADH formed in the biosynthetic reactions is, together with NADH formed in the catabolic pathways, reoxidized by transfer of electrons to oxygen via the electron transport chain.

Reactions (2)-(4) summarize the overall stoichiometry for the catabolic pathways. Reaction (2) specifies NADPH formation by the pentose phosphate pathway, where glucose is completely oxidized to CO_2; reaction (3) is the overall stoichiometry for the combined EMP pathway and the TCA cycle.

Finally, reaction (4) is the overall stoichiometry for oxidative phosphorylation. Compartmentation is not considered in the stoichiometry *i.e.*, no differentiation is made between cytosolic and mitochondrial NADH and $FADH_2$ formed in the TCA cycle is pooled together with NADH. The P/O ratio in reaction (4) therefore is the overall (or operational) P/O ratio for oxidative phosphorylation.

$$CO_2 + 2NADPH - CH_2O = 0 \tag{2}$$

$$CO_2 + 2NADH + 0.667ATP - CH_2O = 0 \tag{3}$$

$$P/OATP - 0.5O_2 - NADH = 0 \tag{4}$$

Finally, consumption of ATP for maintenance is included simply as a reaction where ATP is used:

$$- ATP = 0 \tag{5}$$

Note that, with the stoichiometry defined on a C-mole basis, the stoichiometric coefficients extracted from the biochemistry, *e.g.*, formation of 2 mol ATP per mole of glucose in the EMP pathway, are divided by 6, because 1 mol of glucose contains 6 C-mol.

The preceding stoichiometry is written in the form of eq. (3.3), but we can easily convert it to the more compact matrix notation of eq. (3.4):

$$
\begin{pmatrix} -1.139 & 0 \\ -1 & 0 \\ -1 & 0 \\ 0 & -0.5 \\ 0 & 0 \end{pmatrix}
\begin{pmatrix} S_{glc} \\ S_{o_2} \end{pmatrix}
+
\begin{pmatrix} 0.139 \\ 1 \\ 1 \\ 0 \\ 0 \end{pmatrix} P_{CO_2}
+
\begin{pmatrix} 1 \\ 0 \\ 0 \\ 0 \\ 0 \end{pmatrix} X
$$

$$
+
\begin{pmatrix} -Y_{xATP} & 0.458 & 0.243 \\ 0 & 0 & 2 \\ 0.667 & 2 & 0 \\ P/O & -1 & 0 \\ -1 & 0 & 0 \end{pmatrix}
\begin{pmatrix} X_{ATP} \\ X_{NADH} \\ X_{NADPH} \end{pmatrix}
=
\begin{pmatrix} 0 \\ 0 \\ 0 \\ 0 \\ 0 \end{pmatrix} \tag{6}
$$

where X represents the biomass. If we introduce the forward reaction rates for reactions (1) - (5) in the rate vector **v**:

$$
\mathbf{v} = \begin{pmatrix} \mu \\ v_{PP} \\ v_{EMP} \\ v_{OP} \\ m_{ATP} \end{pmatrix} \tag{7}
$$

the production and consumption of the three cofactors ATP, NADH, and NADPH can be balanced, in analogy with eq. (3.28), to yield the three equations:

$$-Y_{xATP}\mu + 0.667v_{EMP} + P/Ov_{OP} - m_{ATP} = 0 \tag{8}$$

$$0.458\mu + 2v_{EMP} - v_{OP} = 0 \tag{9}$$

$$-0.243\mu + 2v_{PP} = 0 \tag{10}$$

Notice that these balances correspond to zero net specific formation rates for the three cofactors. Consequently, these balances can also be derived using eq. (3.12):

$$\mathbf{r}_{met} = \begin{pmatrix} r_{ATP} \\ r_{NADH} \\ r_{NADPH} \end{pmatrix} = \mathbf{G}^T\mathbf{v}$$

$$= \begin{pmatrix} -Y_{xATP} & 0 & 0.667 & P/O & -1 \\ 0.458 & 0 & 2 & -1 & 0 \\ -0.243 & 2 & 0 & 0 & 0 \end{pmatrix} \begin{pmatrix} \mu \\ v_{PP} \\ v_{EMP} \\ v_{OP} \\ m_{ATP} \end{pmatrix} = \begin{pmatrix} 0 \\ 0 \\ 0 \end{pmatrix} \tag{11}$$

In addition to the three steady state balances [eps. (8)-(10)], we also have the relationships for the specific glucose and oxygen uptake rates, as well as the specific carbon dioxide production rate, in terms of the five reaction rates [eps. (7)]. These relationships are given by eqs. (3.5) and (3.6) or, by using the matrix notation of eqs. (3.9) and (3.10):

$$\begin{pmatrix} r_{glc} \\ r_{O_2} \end{pmatrix} = -\begin{pmatrix} -1.139 & -1 & -1 & 0 & 0 \\ 0 & 0 & 0 & -0.5 & 0 \end{pmatrix} \begin{pmatrix} \mu \\ v_{PP} \\ v_{EMP} \\ v_{OP} \\ v_{ATP} \end{pmatrix}$$

$$= \begin{pmatrix} 1.139\mu + v_{PP} + v_{EMP} \\ 0.5v_{OP} \end{pmatrix} \tag{12}$$

$$r_{CO_2} = (0.139 \quad 1 \quad 1 \quad 0 \quad 0) \begin{pmatrix} \mu \\ v_{PP} \\ v_{EMP} \\ v_{OP} \\ m_{ATP} \end{pmatrix} = 0.139\mu + v_{PP} + v_{EMP} \tag{13}$$

By eliminating the three reaction rates v_{EMP}, v_{PP}, and v_{OP} among the combined set of eqs. (11)-(13), the linear rate eqs. (14)-(16) are derived.

$$r_{glc} = (a + 1.261)\mu + b = Y_{xs}^{true}\mu + m_s \tag{14}$$

$$r_{CO_2} = (a + 0.261)\mu + b = Y_{xc}^{true}\mu + m_c \tag{15}$$

$$r_{O_2} = (a + 0.229)\mu + b = Y_{xo}^{true}\mu + m_o \tag{16}$$

The two common parameters a and b are obtained as a function of the energetic parameters Y_{xATP} and m_{ATP} and the P/O ratio, according to eqs. (17) and (18):

$$a = \frac{Y_{xATP} - 0.458\text{P}/\text{O}}{0.667 + 2\text{P}/\text{O}} \tag{17}$$

$$b = \frac{m_{ATP}}{0.667 + 2\text{P}/\text{O}} \tag{18}$$

Equation (14) is seen to be the same as the linear rate eq. (3.26) introduced by Pirt, with the difference that the yields of that correlation are now obtained in terms of basic cellular energetic parameters. This is true for all parameters in the preceding linear correlations as they are coupled via the ATP, NADH, and NADPH balances. It is thus seen that the three true yield coefficients cannot take any value but that they are actually coupled through these balances. Furthermore, the maintenance coefficients are the same. This is due to the choice of the unit C-moles per C-mole of biomass per hour for the specific rates. If other units were used for the specific rates, the maintenance coefficients would not take the same values, but they would still be related. This coupling of the parameters shows that there are only two degrees of freedom in the system [equivalent to defining parameters a and b in eqs. (14-16)], and one actually has to determine only one yield coefficient and one maintenance coefficient; the other parameters can be calculated using the three eqs. (14)-(16).

The derived linear rate equations certainly are useful for correlating experimental data, but they also allow the evaluation of the key energetic parameters Y_{xATP} and m_{ATP} and the operational P/O ratio. Thus, if the true yield and maintenance coefficients of eqs. (14)-(16) are estimated (by linearly regressing experimental data for the glucose, oxygen, and carbon dioxide rates against the specific growth rate), the values of a and b can be determined in turn and, through them, the three energetic parameters using eqs. (17) and (18). By using the data of Fig. 3.3, a and b were found to be 0.436 C-mol (C-mol DW)$^{-1}$ and 0.018 C-mol (C-mol h)$^{-1}$, respectively (Nielsen, 1997). The value of a is an average value obtained by using the

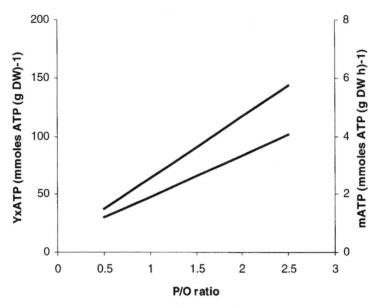

FIGURE 3.4 Energetic parameters Y_{xATP} and m_{ATP} as a function of the operational P/O ratio in *P. chrysogenum*.

three yield coefficients (the relative standard deviation is 2.7%). With only two equations it is not possible to estimate all three energetic parameters Y_{xATP} and m_{ATP} and the operational P/O ratio, but if one of the parameters is known the two others can be calculated. Figure 3.4 depicts Y_{xATP} and m_{ATP} as a function of the operational P/O ratio. The exact operational P/O ratio is not known for *P. chrysogenum*, but it is probably in the range 1.0-2.0 (Nielsen, 1997). For an operational P/O ratio of 1.5, Y_{xATP} is about 2.29 mmol of ATP (C-mol biomass)$^{-1}$ [or 80 mmol (g DW)$^{-1}$ with a molecular mass of 26.33 g (C-mol)$^{-1}$ and an ash content of 8%]. This is in the same range reported for *S. cerevisiae* (see Table 2.6).

EXAMPLE 3.4

Energetics of *S. cerevisiae*

During anaerobic growth, most cells generate ATP by substrate level phosphorylation with the simultaneous conversion of the energy source to one or more metabolites. With growth on a complex medium, there is no (or little) synthesis of NADH and no (or little) consumption of NADPH in the biosynthetic reactions, as these metabolites can be derived from amply supplied medium components. *The redox level of the energy source therefore is conserved in the metabolites formed in the catabolic pathways, i.e.,* all NADH formed in glycolysis is consumed in the conversion of pyruvate to various metabolites. This was illustrated in Example 3.1 for the mixed acid fermentation of *E. coli*, and it is also an important consideration in the homofermentative lactic acid fermentation, where pyruvate is converted to lactate by lactate dehydrogenase. In these processes, it is only ATP that couples the catabolism and anabolism, so that it is possible to derive the linear rate eq. (3.25) of Leudeking and Piret from ATP balance alone [see also Nielsen and Villadsen (1994)].

With growth on a minimal medium there is a substantial NADH production and NADPH consumption associated with the biosynthetic reactions (see Section 2.4). These cofactors have to be regenerated in the fueling reactions, resulting in a tight coupling between catabolism and anabolism, as illustrated in Example 3.3. In anaerobic growth of *S. cerevisiae* this coupling gives rise to the formation of glycerol in addition to the primary metabolite, which in this case is, of course, ethanol. Glycerol is formed by the reduction of dihydroxyacetonephosphate to glycerol-3-phosphate (catalyzed by glycerol-3-P dehydrogenase), which is then dephosphorylated to glycerol by glycerol-1-phosphatase [see reaction (2.6)]. In the first reaction NADH is oxidized to NAD^+, and the overall conversion of glucose to glycerol therefore results in NADH consumption. Because there is no regeneration of ATP in the dephosphorylation of glycerol-3-phosphate, the conversion of glucose to glycerol is an ATP-consuming process overall.

The anaerobic physiology of *S. cerevisiae* has been studied extensively by Schulze (1995), and in this example we will analyze some of his results by using a simple metabolic model. From an analysis of the macromolecular composition of *S. cerevisiae*, he calculated the stoichiometric coefficients for NADH and NADPH connected with biomass synthesis. His figures can be used to set up an overall stoichiometry for cell growth:

$$CH_{1.82}O_{0.58}N_{0.16} + 0.105CO_2 + 0.355NADH - 1.105CH_2O$$

$$- 0.16NH_3 - Y_{xATP}ATP - 0.231NADPH = 0 \qquad (1)$$

The elemental composition is taken from Table 4.1, and the substrate is glucose and ammonia. Sulfate and phosphate are included in the ash content, which is taken to be 8%. As in Example 3.3, the stoichiometry is given on a C-mole basis, *i.e.*, glucose is specified as CH_2O. The required NADPH for biosynthesis is assumed to be formed in the PP pathway only, and as in Example 3.3 we therefore have

$$CO_2 + 2NADPH - CH_2O = 0 \tag{2}$$

ATP necessary for biosynthesis is supplied by conversion of glucose to ethanol, for which the stoichiometry can be specified as

$$0.5CO_2 + CH_3O_{\frac{1}{2}} + 0.5ATP - 1.5CH_2O = 0 \tag{3}$$

The NADH formed in connection with biosynthesis is removed by the conversion of glucose to glycerol, as previously mentioned. Because 2 mol of NADH is oxidized per mole of glucose converted to glycerol and 2 mol of ATP is used to form fructose-1,6-bisphosphate from glucose (with no ATP regeneration in glycerol formation), the overall stoichiometry for glycerol formation is

$$CH_{8/3}O - 0.333NADH - 0.333ATP - CH_2O = 0 \tag{4}$$

Finally, consumption of ATP for maintenance is included as in Example 3.3:

$$-ATP = 0 \tag{5}$$

Following the previous procedure, we collect the forward reaction rates for the preceding five reactions in the rate vector $\mathbf{v}$:

$$\mathbf{v} = \begin{pmatrix} \mu \\ v_{PP} \\ r_{EtOH} \\ r_{gly} \\ m_{ATP} \end{pmatrix} \tag{6}$$

With the stoichiometric coefficients and intracellular reaction rates defined, we can set up balances for the three cofactors. The balance for NADPH gives a direct coupling between v_{PP} and the specific growth rate:

$$-0.231\mu + 2v_{PP} = 0 \tag{7}$$

and, similarly, the NADH balance yields a correlation between the specific growth rate and the specific rate of glycerol formation:

$$0.355\mu - 0.33r_{gly} = 0 \tag{8}$$

Equation (8) corresponds to a yield of glycerol on biomass of 1.066 C-mol of glycerol (C-mol biomass)$^{-1}$. In his analysis, Schulze (1995) found a value of 10.01 mmol of glycerol (g DW)$^{-1}$, which, with the elemental composition of biomass given earlier and an ash content of 8%, corresponds to 0.827 C-mol of glycerol (C-mol biomass)$^{-1}$. This is substantially below the value calculated from the metabolic model. Moreover, secretion of succinate was observed experimentally, which is accompanied by further substantial NADH formation (5 mol of NADH per mole of succinate). The succinate production therefore necessitates the production of even more glycerol to accommodate the increased NADH production. The yield of succinate on biomass was found to be about 0.25 mmol of succinate (g DW)$^{-1}$, and this will correspond to the formation of 1.25 mmol of glycerol (g DW)$^{-1}$. Thus, only 8.76 mmol (g DW)$^{-1}$, or 0.724 C-mol of glycerol (C-mol biomass)$^{-1}$, of the formed glycerol is due to the biosynthesis, which makes the deviation from the metabolic model even larger. Either the NADH production in connection with biosynthesis given in eq. (1) is likely to be too large or there is some transhydrogenase activity that allows conversion of NADH to NADPH.

The ATP balance gives

$$0.5r_{EtOH} - 0.333r_{gly} = Y_{xATP}\mu + m_{ATP} \tag{9}$$

Insertion of eq. (8) for the rate of glycerol yields

$$r_{EtOH} = 2(Y_{xATP} + 0.355)\mu + 2m_{ATP} \tag{10}$$

which is equivalent to the Luedeking-Piret model applied to ethanol formation. In his analysis, Schulze found the true yield coefficient of ethanol on biomass to be 85.05 mmol (g DW)$^{-1}$, which corresponds to 4.69 C-mol of ethanol (C-mol biomass)$^{-1}$. Thus, Y_{xATP} is 1.95 mol of ATP (C-mol biomass)$^{-1}$ or about 71 mmol ATP (g DW)$^{-1}$. If eq. (9) is used directly to calculate the ATP production and Y_{xATP} is determined from a plot of r_{ATP} versus μ, a slightly higher value of 73 mmol of ATP (g DW)$^{-1}$ is obtained (Schulze, 1995). This difference is due to the overestimation of glycerol formation in the metabolic model, as discussed earlier.

For the glucose uptake rate we find

$$r_s = 1.105\mu + 1.5r_{EtOH} + r_{gly} \tag{11}$$

which can be reformulated as eqs. (12) by inserting eq. (8) and (10):

$$r_s = \left(1.105 + 3.0(Y_{xATP} + 0.355) + \frac{0.355}{0.333}\right)\mu + 2m_{ATP} \tag{12}$$

Again we see that the linear rate eq. (3.26) can be derived and that the true yield coefficient Y_{xs}^{true} is determined by key parameters for cell synthesis, *i.e.*, the ATP and NADPH costs and the production of NADH. A linear rate equation similar to eq. (12) can be derived for the production of carbon dioxide.

REFERENCES

Bailey, J. E. & Ollis, D. F. (1986). *Biochemical Engineering Fundamentals*, 2nd ed. New York: MacGraw-Hill.

Herbert, D. (1959). Some principles of continuous culture. *Recent Progress in Microbiology* 7, 381-396.

Ingraham, J. L., Maaløe, O. & Neidhardt, F. C. (1983). *Growth of the Bacterial Cell*. Sunderland: Sinnauer Associated.

Luedeking, R. & Piret, E. L. (1959a). A kinetic study of the lactic acid fermentation. Batch process at controlled pH. *Journal of Biochemical and Microbiological Technology and Engineering* 1, 393-412.

Luedeking, R. & Piret, E. L. (1959b). Transient and steady state in continuous fermentation. Theory and experiment. *Journal of Biochemical and Microbiological Technology and Engineering* 1, 431-459.

Monod, J. (1942). *Recherches sur la Croissance des Cultures Bacteriennes*. Paris: Hermann et Cie.

Nielsen, J. (1997). *Physiological Engineering Aspects of* Penicillium chrysogenum. Singapore: World Scientific Publishing Co.

Nielsen, J. & Villadsen, J. (1993). Bioreactors: Description and modelling. In *Biotechnology*, 2nd ed., vol. 3, Chapter 5 pp. 77-104. Edited by H.-J. Rehm & G. Reed (volume editor G. Stephanopoulos). VCR Verlag.

Nielsen, J. & Villadsen, J. (1994). *Bioreaction Engineering Principles*. New York: Plenum Press.

Pirt, S. J. (1965). The maintenance energy of bacteria in growing cultures. *Proceedings of The Royal Society* London. Series B **163**, 224-231.

Roels, J. A. (1983). *Energetics and Kinetics in Biotechnology*. Amsterdam: Elsevier Biomedical Press.

Schulze, U. (1995). Anaerobic physiology of *Saccharomyces cerevisiae*. Ph.D. Thesis, Technical University of Denmark.

Sonnleitner, B. & Fiechter, A. (1988). High performance bioreactors: A new generation. *Analytica Chimica Acta* **213**, 199-205.

CHAPTER 4

Material Balances and Data Consistency

Quantitative analysis of metabolism requires experimental data for the determination of metabolic fluxes, flux distributions, and measures of flux control (see Chapter 11), among other parameters. As such, these calculations exemplify methods and procedures for *upgrading the information content* of primary fermentation data. Whereas our focus in this book is on metabolism and its control, the basic philosophy of information content upgrade is applicable throughout the life sciences, so long as quantitative measurements are available.

Because these methods of information upgrade are data-driven, it is of the utmost importance to ensure the reliability of the data used. This can be done by applying the usual methods of random error minimization, *i.e.* use of repeat experiments, multiple sensors, careful calibrations, etc. An additional consideration (which is the subject of this chapter) is the introduction of *data redundancy* for the validation of both the actual measurements, and the broader mechanistic framework within which such measurements are analyzed. For example, in the context of metabolic analysis, flux calculations are based on the measurement of the specific rates for substrate uptake and product formation, which represent the fluxes in and out of the cells. Before any such derivative calculations are carried out, it is important that the consistency of the data be confirmed, for example, the closure of the carbon balance.

Data redundancy is introduced when multiple sensors are employed for the measurement of the same variable or when certain constraints must be satisfied by the measurements so obtained, such as closure of material balances. Obviously, the greater the redundancy, the higher the degree of confidence for the data and their derivative parameters. Furthermore, redundancies can be employed for the systematic detection of the source of gross measurement errors or the identification of a particular element of the framework (i.e. model) most likely responsible for any observed inconsistencies. We demonstrate these ideas in this chapter in the context of fluxes, metabolism, and material balances. For this purpose, experimental data that are to be used for quantitative analysis must be

- *Complete.* This does not mean that *all* substrates and metabolic products must be measured, but those present in sufficient amounts should be quantified to allow the validation of the carbon and nitrogen (and, in some cases, also sulfur and phosphorus) balances. This requirement necessitates the use of defined, minimal media and essentially eliminates complex media from systematic metabolic studies.

- As much as possible, *noise free.* As discussed in Section 3.3, specific rates are derived from measurements of the concentration profiles, which make rate calculations difficult if these data are noisy. An important aspect of quantitative analysis of cellular metabolism therefore is the development of reliable and accurate analytical techniques, generally computerized high-performance bioreactors, where the most important culture variables are monitored on-line.

There are two approaches in assessing the consistency of experimental data. The first is based on a very simple metabolic model, the so-called *black box model*, where all cellular reactions are lumped into a single one for the overall cell biomass growth, and the method basically consists of validating elemental balances. It is rather easy to apply as the only information needed is that of the elemental composition of the substrates, metabolic products, and biomass, together with a set of fluxes in and out of the cell. The second approach recognizes far more biochemical detail in the overall conversion of substrates into biomass and metabolic products. As such, it is mathematically more involved, but, of course, it provides a more realistic depiction of the actual degrees of freedom than a black box model. We develop such metabolic models in connection with our discussion of metabolic flux analysis (in Section 8.3). Because our focus in this chapter is mostly on methodology development for consistency analysis, we have eliminated unnecessary complications due to metabolic complexity and instead use a black box model.

4.1. THE BLACK BOX MODEL

In the black box model, cell biomass is the black box exchanging material with the environment, as depicted in Fig. 4.1, and processing it through many cellular reactions lumped into one, that of biomass growth. The fluxes in and out of the black box are given by the specific rates (in grams or moles of the compound per gram or mole of biomass and unit time). These are the specific substrate uptake rates (elements of r_s) and the specific product formation rates (elements of r_p). Additionally, there is accumulation of biomass within the box, which is represented as a flux with the specific rate μ. Because all cellular reactions are lumped into one overall reaction, the stoichiometric coefficients in this overall reaction are given by the yield coefficients introduced in Section 3.4:

$$X + \sum_{i=1}^{M} Y_{xp_i} P_i - \sum_{i=1}^{N} Y_{xs_i} S_i = 0 \qquad (4.1)$$

where the specific rate of biomass formation is used as reference. Because the stoichiometric coefficient for biomass is 1, the forward reaction rate is given by the specific growth rate of the biomass, which, together with the yield coefficients of eq. (4.1), completely specifies the system. In the application of

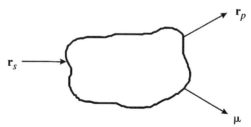

FIGURE 4.1 Representation of the black box model. The cell is considered as a black box, and fluxes in and out of the cell are the only variables measured. The fluxes of substrates into the cell are elements of the vector r_s, and the fluxes of metabolic products out of the cell are elements of the vector r_p. Some of the mass originally present in the substrates accumulates within the black box as formation of new biomass with the specific rate μ.

the black box model for analyzing data consistency, one may use either (1) a set of yield coefficients given in eq. (4.1) together with the specific growth rate; (2) a set of yield coefficients with respect to another reference, *e.g.*, one of the substrates, along with the specific rate of formation/consumption of this reference compound; (3) a set of specific rates for *all* substrates and products, including biomass; or 4) a set of all volumetric rates that are the product of the specific rates by the biomass concentration. All of these variables supply the same information. In the following, we will use either the yield coefficients given in eq. (4.1) or the set of specific rates that we collect in the total rate vector **r**, given by

$$\mathbf{r} = \begin{pmatrix} \mu \\ \mathbf{r}_p \\ -\mathbf{r}_s \end{pmatrix} = \begin{pmatrix} \mu & r_{p,1} & \cdots & -r_{s,1} & \cdots \end{pmatrix}^T \qquad (4.2)$$

EXAMPLE 4.1

A Simple Black Box Model

Consider the aerobic cultivation of the yeast *Saccharomyces cerevisiae* on a defined, minimal medium, *i.e.*, glucose is the carbon and energy source and ammonia is the nitrogen source. During aerobic growth, the yeast oxidizes glucose completely to carbon dioxide. However, at very high glycolytic fluxes, a bottleneck in the oxidation of pyruvate leads to ethanol formation. Thus, at high glycolytic fluxes, both ethanol and carbon dioxide should be considered as metabolic products. Finally, water is formed in the cellular pathways, and this is also included as a product in the overall reaction. Thus, the black box model for this system is

$$X + Y_{xe} \text{ ethanol} + Y_{xc}CO_2 + Y_{xw}H_2O - Y_{xs} \text{ glucose}$$

$$- Y_{xo} O_2 - Y_{xN} NH_3 = 0 \qquad (1)$$

which can be represented with the specific rate vector:

$$\mathbf{r} = \begin{pmatrix} \mu & r_e & r_c & r_w & -r_s & -r_o & -r_N \end{pmatrix}^T \qquad (2)$$

Obviously, the stoichiometric (or yield) coefficients of eq. (1) are not constant, as Y_{xe} is zero at low specific growth rates (corresponding to low glycolytic fluxes) and greater than zero for higher specific growth rates.

4.2. ELEMENTAL BALANCES

In the black box model, we have $M + N + 1$ variables: M yield coefficients for the metabolic products, N yield coefficients for the substrates, and the forward reaction rate μ [see eq. (4.1)] or the $M + N + 1$ specific rates of eq. (4.2). Because mass is conserved in the overall conversion of substrates to metabolic products and biomass, the $(M + N + 1)$ rates of the black box model are not completely independent but must satisfy several constraints. Thus, the elements flowing into the system must balance the elements flowing out of the system, e.g., the carbon entering the system via the substrates has to be recovered in the metabolic products and biomass. Each element considered in the black box obviously yields one constraint. Thus, a carbon balance gives

$$1 + \sum_{i=1}^{M} h_{p,i} Y_{xp_i} - \sum_{i=1}^{N} h_{s,i} Y_{xs_i} = 0 \qquad (4.3)$$

where $h_{s,i}$ and $h_{p,i}$ represent the carbon content (C-moles mole^{-1}) in the ith substrate and the ith metabolic product, respectively. In this equation, the elemental composition of biomass is normalized with respect to carbon, i.e., it is represented by the form $CH_aO_bN_c$. The elemental composition of biomass depends on its macromolecular content and, therefore, on the growth conditions and the specific growth rate [e.g., the nitrogen content is much lower under nitrogen-limited conditions than under carbon-limited conditions (see Table 4.1)]. However, except for extreme situations, it is reasonable to use the general composition formula $CH_{1.8}O_{0.5}N_{0.2}$ whenever the biomass composition is not known exactly.

The carbon balance of eq. (4.3) may also be formulated in terms of the specific rates. Thus, after multiplying eq. (4.3) by μ and applying the definition of the yield coefficients:

$$\mu + \sum_{i=1}^{M} h_{p,i} r_{p,i} - \sum_{i=1}^{N} h_{s,i} r_{s,i} = 0 \qquad (4.4)$$

Often, we normalize the elemental composition of substrates and metabolic products with respect to their carbon content, e.g., glucose is specified as CH_2O. Equation (4.3) is then written on per C-mole basis as

$$1 + \sum_{i=1}^{M} Y_{xp_i} - \sum_{i=1}^{N} Y_{xs_i} = 0 \qquad (4.5)$$

TABLE 4.1 Elemental Composition of Biomass for Several Microorganisms[a]

Microorganism	Elemental composition	Ash content (w/w%)	Conditions
Candida utilis	$CH_{1.83}O_{0.46}N_{0.19}$	7.0	Glucose limited, D = 0.05 h^{-1}
	$CH_{1.87}O_{0.56}N_{0.20}$	7.0	Glucose limited, D = 0.45 h^{-1}
	$CH_{1.83}O_{0.54}N_{0.10}$	7.0	Ammonia limited, D = 0.05 h^{-1}
	$CH_{1.87}O_{0.56}N_{0.20}$	7.0	Ammonia limited, D = 0.45 h^{-1}
Klebsiella aerogenes	$CH_{1.75}O_{0.43}N_{0.22}$	3.6	Glycerol limited, D = 0.10 h^{-1}
	$CH_{1.73}O_{0.43}N_{0.24}$	3.6	Glycerol limited, D = 0.85 h^{-1}
	$CH_{1.75}O_{0.47}N_{0.17}$	3.6	Ammonia limited, D = 0.10 h^{-1}
	$CH_{1.73}O_{0.43}N_{0.24}$	3.6	Ammonia limited, D = 0.08 h^{-1}
Saccharomyces cerevisiae	$CH_{1.82}O_{0.58}N_{0.16}$	7.3	Glucose limited, D = 0.08 h^{-1}
	$CH_{1.78}O_{0.60}N_{0.19}$	9.7	Glucose limited, D = 0.255 h^{-1}
	$CH_{1.94}O_{0.52}N_{0.25}P_{0.025}$	5.5	Unlimited growth
Escherichia coli	$CH_{1.77}O_{0.49}N_{0.24}P_{0.017}$	5.5	Unlimited growth
	$CH_{1.83}O_{0.50}N_{0.22}P_{0.021}$	5.5	Unlimited growth
	$CH_{1.96}O_{0.55}N_{0.25}P_{0.022}$	5.5	Unlimited growth
	$CH_{1.93}O_{0.55}N_{0.25}P_{0.021}$	5.5	Unlimited growth
Pseudomonas fluorescens	$CH_{1.83}O_{0.55}N_{0.26}P_{0.024}$	5.5	Unlimited growth
Aerobacter aerogenes	$CH_{1.64}O_{0.52}N_{0.16}$	7.9	Unlimited growth
Penicillium chrysogenum	$CH_{1.70}O_{0.58}N_{0.15}$		Glucose limited, D = 0.038 h^{-1}
	$CH_{1.68}O_{0.53}N_{0.17}$		Glucose limited, D = 0.098 h^{-1}
Aspergillus niger	$CH_{1.72}O_{0.55}N_{0.17}$	7.5	Unlimited growth
Average	$CH_{1.81}O_{0.52}N_{0.21}$	6.0	

[a] The P content is given only for some microorganisms. The composition for *P. chrysogenum* is taken from Christensen *et al.* (1995), whereas the other data are taken from Roels (1983).

In eq. (4.5), the yield coefficients have the unit C-moles per C-mole biomass. Conversion to this unit from other units is illustrated in Box 4.1. Equation (4.5) (or eq. (4.4)) is very useful for checking the consistency of experimental data. Thus, if the sum of carbon in the biomass and the metabolic products does not equal the sum of carbon in the substrates, there is an inconsistency in the experimental data.

EXAMPLE 4.2

Carbon Balance in a Simple Black Box Model

We now return to the black box model of Example 4.1 and rewrite the conversion eq. (1) using the elemental composition of the substrates and metabolic products specified. For biomass we use the elemental composition of $CH_{1.83}O_{0.56}N_{0.17}$, and therefore have

$$CH_{1.83}O_{0.56}N_{0.17} + Y_{xe}CH_3O_{0.5}$$

$$+ Y_{xc}CO_2 + Y_{xw}H_2O - Y_{xs}CH_2O - Y_{xo}O_2 - Y_{xN}NH_3 = 0 \qquad (1)$$

Some may find it difficult to identify $CH_3O_{0.5}$ as ethanol, but the advantage of using the C-mole basis becomes apparent immediately when we look at the carbon balance:

$$1 + Y_{xe} + Y_{xc} - Y_{xs} = 0 \qquad (2)$$

This simple equation is very useful for checking the consistency of experimental data. Thus, by using the classical data of von Meyenburg (1969), we find $Y_{xe} = 0.713$, $Y_{xc} = 1.313$, and $Y_{xs} = 3.636$ at a dilution rate of $D = 0.3$ h^{-1} in a glucose-limited continuous culture. Obviously the data are not consistent as the carbon balance does not close. A more elaborate data analysis (Nielsen and Villadsen, 1994) suggests that the missing carbon is ethanol, which could have evaporated as a result of ethanol stripping due to intensive aeration of the bioreactor.

By analogy to eq. (4.4), we find that a nitrogen balance gives

$$Y_{xN} = 0.17 \qquad (3)$$

or in terms of specific rates:

$$r_N = 0.17\mu \qquad (4)$$

If the measured rates of ammonia uptake and biomass formation do not conform with eq. (4), an inconsistency is identified in one of these two measurements or the nitrogen content of the biomass is different from that specified.

Similar to eq. (4.3), balances can be written for all other elements participating in the conversion [eq. (4.1)]. These balances can be conveniently written by collecting the elemental compositions of biomass, substrates, and metabolic products in the columns of a matrix E, where the first

---

BOX 4.1
_____

Calculation of Yields with Respect to C-mole Basis

Yield coefficients are typically described as mol (g DW)$^{-1}$ or g (g DW)$^{-1}$. To convert the yield coefficients to a C-mol basis, information on the elemental composition and the ash content of biomass is needed. To illustrate the conversion, we calculate the yield of 0.5 g DW biomass (g glucose)$^{-1}$ on a C-mol basis. First, we convert the grams dry weight biomass to an ash-free basis, i.e., determine the amount of biomass that is made up of carbon, nitrogen, oxygen, and hydrogen (and, in some cases, also phosphorus and sulfur). With an ash content of 8% we have 0.92 g ash-free biomass (g DW biomass)$^{-1}$, which gives a yield of 0.46 g ash-free biomass (g glucose)$^{-1}$. This yield can now be directly converted to a C-mol basis using the molecular masses in g C-mol^{-1} for ash-free biomass and glucose. With the standard elemental composition for biomass of $CH_{1.8}O_{0.5}N_{0.2}$, we have a molecular mass of 24.6 ash-free biomass C-mol^{-1}, and therefore we find a yield of $0.46/24.6 = 0.0187$ C-mol biomass (g glucose)$^{-1}$. Finally, by multiplication with the molecular mass of glucose on a C-mol basis (30 g C-mol^{-1}), a yield of 0.56 C-mol biomass (C-mol glucose)$^{-1}$ is found.

---

column contains the elemental composition of biomass, columns 2 through $M + 1$ contain the elemental compositions of the M metabolic products, and columns $M + 2$ through $M + N + 1$ contain the elemental composition of the N substrates. If we consider I elements (normally four, namely, C, H, O and N), there are I rows in matrix E and the I elemental balances are represented by an equal number of algebraic equations similar to eq. (4.3), which can be summarized as

$$Er = 0 \qquad (4.6)$$

With $N + M + 1$ specific rates (or volumetric rates) and I constraints, the degree of freedom is $F = M + N + 1 - I$. If exactly F rates are measured, it may be possible to calculate the other rates by using the I algebraic equations given by eq. (4.6), but, in this case, there are no redundancies left to check the consistency of the data. For this reason, it is advisable to strive for more measurements than the degrees of freedom of the system.

EXAMPLE 4.3

Elemental Balances in a Simple Black Box Model

We return to the black box model of Examples 4.1 and 4.2. With the elemental composition of biomass given previously, we can write the elemental composition matrix E as

$$
E = \begin{pmatrix} 1 & 1 & 1 & 0 & 1 & 0 & 0 \\ 1.83 & 3 & 0 & 2 & 2 & 0 & 3 \\ 0.56 & 0.5 & 2 & 1 & 1 & 2 & 0 \\ 0.17 & 0 & 0 & 0 & 0 & 0 & 1 \end{pmatrix} \begin{matrix} \leftarrow \text{carbon} \\ \leftarrow \text{hydrogen} \\ \leftarrow \text{oxygen} \\ \leftarrow \text{nitrogen} \end{matrix} \tag{1}
$$

where the rows indicate the content of carbon, hydrogen, oxygen, and nitrogen, respectively, and the columns give the elemental composition of biomass, ethanol, carbon dioxide, water, glucose, oxygen, and ammonia, respectively. By using eq. (4.6), where r is replaced by a vector specifying the yield coefficients, we find

$$
\begin{pmatrix} 1 & 1 & 1 & 0 & 1 & 0 & 0 \\ 1.83 & 3 & 0 & 2 & 2 & 0 & 3 \\ 0.56 & 0.5 & 2 & 1 & 1 & 2 & 0 \\ 0.17 & 0 & 0 & 0 & 0 & 0 & 1 \end{pmatrix} \begin{pmatrix} 1 \\ Y_{xe} \\ Y_{xc} \\ Y_{xw} \\ -Y_{xs} \\ -Y_{xo} \\ -Y_{xN} \end{pmatrix}
$$

$$
= \begin{pmatrix} 1 + Y_{xe} + Y_{xc} - Y_{xs} \\ 1.83 + 3Y_{xe} + 2Y_{xw} - 2Y_{xs} - 3Y_{xN} \\ 0.56 + 0.5Y_{xe} + 2Y_{xc} + Y_{xw} - Y_{xs} - 2Y_{xo} \\ 0.17 - Y_{xN} \end{pmatrix} = \begin{pmatrix} 0 \\ 0 \\ 0 \\ 0 \end{pmatrix} \tag{2}
$$

The first and last rows are identical to the balances derived in Example 4.2 for carbon and nitrogen, respectively. The balances for hydrogen and oxygen introduce two additional constraints. However, because the rate of water formation is impossible to measure, one of these equations must be used to calculate this rate (or yield). This leaves only one additional constraint from these two balances.

Equation (4.6) summarizes the balances for all elements. As discussed in Example 4.3, there is actually one fewer constraint as either the hydrogen or oxygen balance must be used to calculate the (nonmeasurable) water production rate. Obviously, water can be excluded by eliminating the yield coefficient for water between the O and H balances. A more elegant approach is to use the so-called *generalized degree of reduction* balance, which is derived as a linear combination of the elemental balances of eq. (4.6). This balance was introduced by Roels (1983) as a generalization of the earlier work of Erickson *et al.* (1978). It is generated by adding the elemental balances after multiplying them by a certain factor. By choosing appropriate multiplication factors, the yield coefficients (or rates) for *water, carbon dioxide,* and *nitrogen source* are eliminated from the resulting equation. To illustrate the procedure, consider the elemental balances given in Example 4.3, where we multiply the carbon balance by 4, the hydrogen balance by 1, the oxygen balance by -2, and the nitrogen balance by -3 to obtain the following:

$$
\begin{array}{llllll}
4 & +4Y_{xe} & +4Y_{xc} & -4Y_{xs} & & = 0 \\
1.83 & +3Y_{xe} & 2Y_{xw} & -2Y_{xs} & -3Y_{xN} & = 0 \\
(-2)\,0.56 + & (-2)0.5Y_{xe} + & (-2)2Y_{xc} + & (-2)Y_{xw} - (-2)Y_{xs} - (-2)2Y_{xo} & & = 0 \\
(-3)\,0.17 & & & & -(-3)Y_{xN} & = 0 \\
\hline
4.20 & +6Y_{xe} & & -4Y_{xs} & +4Y_{xo} & = 0
\end{array}
$$

The resulting equation is the generalized degree of reduction balance for the system. Of course, this balance is not independent of the other elemental balances. It normally replaces either the oxygen or hydrogen balance, while the other is used to calculate the rate of water formation. Finally, one has the carbon, nitrogen, and degree of reduction balances to use for consistency analysis or for the calculation of unmeasured rates. In the original formulation of Erickson *et al.* (1978), the multiplication factor of each of the C, H, O, and N balances was interpreted as the number of free electrons available in C, H, O, and N, respectively, for transfer to oxygen upon the combustion of each element to *water, carbon dioxide,* and *ammonia* (as the nitrogen source). For the nitrogen balance the multiplier was always taken to be -3, as this is the predominant valency of nitrogen in biomass. In the generalized concept of Roels, the multiplication factors are arbitrary coefficients free to be chosen such that the resulting coefficients for water, carbon dioxide, and nitrogen source vanish. In this way, if another nitrogen source is used, *e.g.,* ammonium nitrate, a different multiplication factor is selected for the nitrogen balance in order to eliminate the yield coefficient for the nitrogen source from the generalized degree of reduction balance.

The coefficient multiplying the yield in the generalized degree of reduction balance is called the *degree of reduction* of the corresponding compound. For the preceding system, the degree of reduction is 4.2 for biomass, 6 for ethanol, 4 for glucose, 0 for water, ammonia, and carbon dioxide, and -4 for oxygen. With the generalization of Roels, the degree of reduction of the nitrogen-containing compounds depends on the nitrogen source used. In most cases ammonium is used as either the sole nitrogen source or in combination with another nitrogen source, yielding the following general expression for the degree of reduction κ of a compound with the elemental composition $CH_aO_bN_c$:

$$\kappa = 4 + a - 2b - 3c \tag{4.7}$$

Table 4.2 of the following section lists the degrees of reduction for compounds typically encountered in fermentation processes. Roels (1983), or the more recent publication of Nielsen and Villadsen (1994), can be consulted for further elaboration on the concept of the degree of reduction. With the introduction of a compound's degree of reduction κ, the generalized degree of reduction balance for any system is given by

$$\kappa_x + \sum_{i=1}^{M} \kappa_{p,i} Y_{xp_i} - \sum_{i=1}^{N} \kappa_{s,i} Y_{xs_i} = 0 \tag{4.8}$$

This balance is very useful as it is simple to set up and, together with the carbon and nitrogen balances, contains all the constraints imposed by the four elemental balances.

EXAMPLE 4.4

Analysis of Data Consistency in Anaerobic Yeast Cultivations

To illustrate the application of the generalized degree of reduction to the analysis of data consistency, we consider data from the anaerobic continuous cultures of *S. cerevisiae* obtained by Schulze (1995). Yield coefficients (all in C-moles or moles per C-mole biomass), obtained under conditions of glucose limitation, are listed for glucose (Y_{xs}), ethanol (Y_{xe}), carbon dioxide (Y_{xc}), and glycerol (Y_{xg}):

Dilution rate h⁻¹	Y_{xs}	Y_{xe}	Y_{xc}	Y_{xg}
0.1	7.81	3.88	2.13	0.67
0.2	8.06	4.00	2.26	0.73

First, the carbon balance

$$1 + Y_{xe} + Y_{xc} + Y_{xg} - Y_{xs} = 0 \qquad (1)$$

is satisfied within 2% at $D = 0.1$ h^{-1} and within 1% at $D = 0.2$ h^{-1}. Such deviations (estimated relative to the carbon supply in the form of glucose) are very satisfactory.

The generalized degree of reduction balance gives

$$\kappa_x + 6Y_{xe} + 4.67Y_{xg} - 4Y_{xs} = 0 \qquad (2)$$

The elemental composition of yeast was determined to be $CH_{1.78}O_{0.60}N_{0.19}$. Thus, the degree of reduction of biomass is 4.01, and, upon substitution of the yield coefficients from the preceding table, the generalized degree of reduction balance is found to close within 3% in both cases (again the deviation is given relative to glucose). It is interesting to note that for $D = 0.1$ h^{-1} the degree of reduction of the "missing carbon" is found (by invoking the carbon balance as well) to be very close to 6, indicating that the ethanol measurement may be underestimated. This is a general problem in yeast and other cultivations where volatile compounds are produced, and a loss of carbon in the neighborhood of less than 2% in these cultivations generally is considered acceptable.

4.3. HEAT BALANCE

In the conversion of substrates to metabolic products and biomass, part of the Gibbs free energy in the substrates is dissipated to the surrounding environment as heat. Especially under aerobic conditions, the energy dissipation may be substantial. Energy dissipation is determined by the difference between the total Gibbs free energy in the substrates and the total Gibbs free energy recovered in the metabolic products and biomass. The energy dissipation normally gives rise to changes in both the enthalpy and entropy of the system, and it is difficult to quantify (see also Section 13.1). Attention is, therefore, generally focused on heat production determined by the change in enthalpy, as this heat production has direct consequences for process cooling requirements for temperature control. With the black box model, the heat production Q_{heat} [kJ (C-mol biomass)$^{-1}$] of the overall process can be calculated from

$$Q_{heat} = -\Delta H_c^0 = \sum_{i=1}^{N} Y_{xs_i} \Delta H_{c,i}^0 - \Delta H_{c,x}^0 - \sum_{i=1}^{M} Y_{xp_i} \Delta H_{c,i}^0 \quad (4.9)$$

where $\Delta H_{c,i}^0$ is the heat of combustion [kJ $(C\text{-mol})^{-1}$] of the ith compound at standard conditions (298 K and 1 atm). The yield coefficients in the preceding equation are given on a C-mole basis. Table 4.2 lists the heats of combustion for some compounds typically found in fermentation media. Notice that Q_{heat} is actually a yield rather than a rate. To determine the rate of heat production, the preceding equation is multiplied by the growth rate. Equation (4.9) is useful for calculating the heat production from the yield coefficients and can be used for designing the cooling capacity of a bioreactor, as illustrated in Example 4.5.

EXAMPLE 4.5

Heat Generation at Anaerobic versus Aerobic Growth

We consider the growth of *S. cerevisiae* under anaerobic and aerobic conditions. The black box models for these two growth conditions can be taken to be

$$CH_{1.62}O_{0.53}N_{0.15} + 4.78CH_3O_{0.5} + 2.42CO_2$$
$$+ 0.41H_2O - 8.20CH_2O - 0.15NH_3 = 0 \quad (1)$$

for anaerobic growth and

$$CH_{1.62}O_{0.53}N_{0.15} + 0.67CO_2 + 1.08H_2O$$
$$- 1.67CH_2O - 0.15NH_3 - 0.64O_2 = 0 \quad (2)$$

for aerobic growth. We calculate the heat production for the two reactions to be

$$Q_{anarob} = [(8.20)(467) + (0.15)(383) - 560 - (4.78)(683)]$$
$$= 62.11 \text{ kJ } (C\text{-mol biomass})^{-1} \quad (3)$$

$$Q_{aerob} = [(1.67)(467) + (0.15)(383) - 560]$$
$$= 277.3 \text{ kJ } (C\text{-mol biomass})^{-1} \quad (4)$$

TABLE 4.2   Heats of Combustion for Various Compounds at Standard Conditions
(298 K and 1 atm) and pH 7 [a]

Compound	Formula	Degree of reduction	$\Delta H^o_{c,i}$ (kJ C-mol^{-1})
Acetaldehyde	C_2H_4O	5	583
Acetic acid	$C_2H_4O_2$	4	437
Acetone	C_3H_6O	5.33	597
Ammonia	NH_3		383[c]
Biomass	$CH_{1.8}O_{0.5}N_{0.2}$	4.2	560
n-Butanol	$C_4H_{10}O$	6	669
Butyric acid	$C_4H_8O_2$	5	546
Citric acid	$C_6H_8O_7$	3	327
Ethane	C_2H_6	7	780[c]
Ethanol	C_2H_6O	6	683
Formaldehyde	CH_2O	4	571[c]
Formic acid	CH_2O_2	2	255
Fructose	$C_6H_{12}O_6$	4	469
Fumaric acid	$C_4H_4O_4$	3	334
Galactose	$C_6H_{12}O_6$	4	468
Glucose	$C_6H_{12}O_6$	4	467
Glycerol	$C_3H_8O_3$	4.67	554
Isopropanol	C_3H_8O	6	673
Lactic acid	$C_3H_6O_3$	4	456
Lactose	$C_{12}H_{22}O_{11}$	4	471
Malic acid	$C_4H_6O_5$	3	332
Methane	CH_4	8	890[c]
Methanol	CH_4O	6	727
Oxalic acid	$C_2H_2O_4$	1	123
Palmitic acid	$C_{16}H_{32}O_2$	5.75	624[b]
Propane	C_3H_8	6.67	740[c]
Propionic acid	$C_3H_6O_2$	4.67	509
Succinic acid	$C_4H_6O_4$	3.5	373
Sucrose	$C_{12}H_{22}O_{11}$	4	470
Urea	CH_4ON_2		632
Valeric acid	$C_5H_{10}O_2$	5.2	568

[a] The heat of combustion is given with the reference being CO_2, H_2O and N_2.
[b] Solid form.
[c] Gaseous form.

We see that much more heat is generated in the aerobic process [corresponding to 165 kJ (C-mol glucose metabolized)$^{-1}$] than in the anaerobic process [corresponding to about 8 kJ (C-mol glucose)$^{-1}$]. Thus, in the aerobic process, a large fraction of the free energy originally present in glucose dissipates as heat, whereas in the anaerobic process it is retrieved in ethanol. To illustrate the cooling requirements of a large-scale bioreactor, we calculate the total heat production for a typical industrial baker's yeast fermentation. We use a bioreactor volume of 100 m^3 and a biomass concentration of 50 g L^{-1} (corresponding to about 1.96 C-mol L^{-1}). For the batch phase, of such a process the specific growth rate is approximately equal to 0.25 h^{-1}. Using these data we first find the specific rate of heat production:

$$r_q = Q_{aerob}\, \mu = \left[277.3 \text{ kJ (C-mol biomass)}^{-1}\right](0.25 \text{ h}^{-1})$$

$$= 69 \text{ kJ (C-mol biomass)}^{-1} \text{ h}^{-1} \tag{5}$$

and from here we find the total heat production to be

$$\left(69 \text{ kJ (C-mol biomass)}^{-1} \text{ h}^{-1}\right)\left(1.96 \text{ C-mol L}^{-1}\right)\left(100.000 \text{ L}\right) = 3.8 \text{ MW} \tag{6}$$

This large heat production clearly illustrates the requirement for large amounts of cooling water to maintain a constant temperature in the bioreactor.

If the heat production rate can be measured accurately, such as by using a calorimeter [as illustrated in several publications; see, for example Larsson et al. (1991) and von Stockar and Birou (1989)] or measuring the temperature change in the bioreactor, the heat balance [eq. (4.9)] may be used to supply an additional redundancy along with the elemental balances of Section 4.2. If, however, the heat production rate cannot be measured (as would be the case of an anaerobic process where the heat production is very small), the introduction of an additional equation does not change the degrees of freedom due to the additional unknown variable (Q_{heat}). For aerobic processes, it is generally found that the rate of heat production is proportional to the oxygen uptake rate:

$$Q_{heat} = aY_{xo} \tag{4.10}$$

Equation (4.10) is empirically found to be valid for microbial growth on different substrates with a proportionality constant approximately equal to 460 kJ per mol O$_2$ [see Table 4.3 and Example 4.5, where it was found to be

TABLE 4.3   Comparison of Y_{xo} and Q_{heat} for Bacteria Grown on Different Carbon Sources[a]

Substrate	Y_{xo} [mmol O_2 (g DW)$^{-1}$]	Q_{heat} [kJ (g DW)$^{-1}$]	Q/Y_{xo} [kJ (mol O_2)$^{-1}$]
Malate	30.6	14.0	458
Acetate	44.6	19.9	446
Glucose	21.3	10.0	469
Methanol	71.0	34.9	492
Ethanol	51.2	23.2	453
Isopropanol	135.8	56.5	416
n-Paraffins	62.5	26.2	419
Methane	156.3	68.6	439

[a] The data are taken from Abbott and Clamen (1973).

equal to 433 kJ (mol O_2)$^{-1}$]. Equation (4.10) may also be derived from a generalized degree of reduction balance, where the reference for nitrogen-containing compounds is taken to be N_2 (Roels, 1983; Nielsen and Villadsen, 1994). A consequence of eq. (4.10) is that the measurement of the rate of heat production is well-suited for checking the measurements of the oxygen uptake rate or as an alternative to this measurement.

4.4. ANALYSIS OF OVERDETERMINED SYSTEMS—IDENTIFICATION OF GROSS MEASUREMENT ERRORS

If there are more measurements available than the degrees of freedom F, the system is generally called *overdetermined*. In this case the redundancy of the measurements can be used to (a) calculate the rates of nonmeasured metabolites; (b) increase the accuracy of the available measurements through the application of essentially a least squares calculation; and (c) identify the most likely source of gross measurement errors or even the source of inconsistencies in the formulation of the black box framework. This can be carried out in a straightforward manner. For example, if only one rate is not measured, we can use the carbon balance to calculate that rate and the remaining (nitrogen and degree of reduction) balances to check the overall consistency of the data. A more effective analysis is based on the *simultaneous* use of all balances, elemental and otherwise, for the calculation of the nonmeasured rates as well as for data consistency analysis. This is best carried out through the use of matrix manipulations. We follow this procedure here; however, we have reduced the use of matrix operations to a minimum in order to facilitate the review of this material by those with limited exposure to this subject. Furthermore, we have provided a rudimentary review of matrix operations in Box 4.2. Finally, we provide several examples to illustrate these operations when applied to our system of aerobic yeast cultivation without ethanol formation.

We begin our analysis with the elemental balances of eq. (4.6), which we rewrite in the following form by partitioning the rate vector $\mathbf{r}$ into two vectors: One, $\mathbf{r}_m$, collects all measured rates, and another, $\mathbf{r}_c$, collects the remaining rates (that need to be calculated, hence the subscript c):

$$\mathbf{Er} = \mathbf{E}_c \mathbf{r}_c + \mathbf{E}_m \mathbf{r}_m = 0 \qquad (4.11)$$

Similarly, the elemental matrix $\mathbf{E}$ is partitioned by separating the columns with the elemental composition of the compounds that have been measured into one matrix $\mathbf{E}_m$ and the columns of the nonmeasured compounds (that must be calculated from the balances) into matrix $\mathbf{E}_c$. Of course, if exactly F variables are measured, there are just enough equations to determine the nonmeasured rates. In this case $\mathbf{E}_c$ is a square matrix with dimensions (I x I) equal to the number of constraints (or balances; I). If it has full rank, i.e., rank($\mathbf{E}_c$) = I (see Box 4.2), the nonmeasured specific rates of $\mathbf{r}_c$ can be calculated by solving eq. (4.11):

$$\mathbf{r}_c = -\mathbf{E}_c^{-1} \mathbf{E}_m \mathbf{r}_m \qquad (4.12)$$

If $\mathbf{E}_c$ is square and has full rank, the system is called *observable*, as there are exactly enough measurements to determine the unknown rates, i.e., the system is overdetermined. If more rates are measured than the degrees of freedom F, there are more equations available than the minimum needed for the determination of the (now fewer) unknown rates. In this case, a least squares approach is usually employed, whereby the unknown rates are calculated from a combination of the available balances in order to increase the accuracy of the so-obtained estimates. The matrix equivalent of this situation is that the elemental submatrix $\mathbf{E}_c$ now is not square, and its inverse therefore cannot be determined. However, multiplication of eq. (4.11) by the transpose matrix $\mathbf{E}_c^T$ (see Box 4.2) of $\mathbf{E}_c$ yields

$$\mathbf{E}_c^T (\mathbf{E}_c \mathbf{r}_c + \mathbf{E}_m \mathbf{r}_m) = (\mathbf{E}_c^T \mathbf{E}_c) \mathbf{r}_c + \mathbf{E}_c^T \mathbf{E}_m \mathbf{r}_m = 0 \qquad (4.13)$$

$\mathbf{E}_c^T \mathbf{E}_c$ is certainly square (see Box 4.2), and if it has full rank, it can be inverted to give the solution for $\mathbf{r}_c$:

$$\mathbf{r}_c = -\mathbf{E}_c^{\#} \mathbf{E}_m \mathbf{r}_m \qquad (4.14)$$

BOX 4.2

Matrix Operations

A matrix is simply a set of numbers arranged in some array. The arrays that you will use in this text will usually be in the form of either a vector with multiple components or a square/rectangular matrix. In this box we give an introduction to the most simple matrix operations that are used throughout this text [for more details, see Strang (1988)].

Consider the generalized matrix **A** shown with two rows and two columns:

$$\mathbf{A}\begin{pmatrix} A_{1,1} & A_{1,2} \\ A_{2,1} & A_{2,2} \end{pmatrix}$$

Note that in matrix notation $A_{i,j}$ refers to the element of the ith row and the jth column, where $i = 1..n$, and $j = 1..m$. The dimension of a matrix is specified as $n \times m$, and in this case **A** is a 2 x 2 matrix.

Basic Matrix Operations

Consider matrix **A** already shown and another 2 x 2 matrix **B**:

$$\mathbf{B}\begin{pmatrix} B_{1,1} & B_{1,2} \\ B_{2,1} & B_{2,2} \end{pmatrix}$$

The sum and difference of matrices **A** and **B** give 2 x 2 matrixes **C** and **D**, respectively:

$$\mathbf{C} = \mathbf{A} + \mathbf{B} = \begin{pmatrix} A_{1,1} + B_{1,1} & A_{1,2} + B_{1,2} \\ A_{2,1} + B_{2,1} & A_{2,2} + B_{2,2} \end{pmatrix}$$

$$\mathbf{D} = \mathbf{A} - \mathbf{B} = \begin{pmatrix} A_{1,1} - B_{1,1} & A_{1,2} - B_{1,2} \\ A_{2,1} - B_{2,1} & A_{2,2} - B_{2,2} \end{pmatrix}$$

Matrices are multiplied by numerical constants one component at a time:

$$\mathbf{E} = 2\mathbf{A} = 2\begin{pmatrix} A_{1,1} & A_{1,2} \\ A_{2,1} & A_{2,2} \end{pmatrix} = \begin{pmatrix} 2A_{1,1} & 2A_{1,2} \\ 2A_{2,1} & 2A_{2,2} \end{pmatrix}$$

The multiplication operation, of a matrix by a vector or a matrix by a matrix, is less apparent and is illustrated next. Consider a vector $\mathbf{v}$, where $\mathbf{v}$ has the same number of elements as the number of columns of $\mathbf{A}$:

$$\mathbf{v} = \begin{pmatrix} v_1 \\ v_2 \end{pmatrix}$$

The product of $\mathbf{A}$ and $\mathbf{v}$ is then specified as

$$\mathbf{F} = \mathbf{Av} = \begin{pmatrix} A_{1,1} & A_{1,2} \\ A_{2,1} & A_{2,2} \end{pmatrix} \begin{pmatrix} v_1 \\ v_2 \end{pmatrix} = \begin{pmatrix} A_{1,1}v_1 + A_{1,2}v_2 \\ A_{2,1}v_1 + A_{2,2}v_2 \end{pmatrix}$$

Note that in essence each element of $\mathbf{F}$ represents the sum of the products of the corresponding rows of $\mathbf{A}$ and $\mathbf{v}$. For the general case of a matrix $\mathbf{A}$ having dimensions m x n, and a vector $\mathbf{v}$ with dimensions r x 1, it is obviously necessary that r must be equal to n; ($i.e.$, the number of columns of $\mathbf{A}$), and the product matrix will have dimensions m x n.

In a similar fashion, one can multiply two matrices of compatible dimensions as shown next, where in this case each row of $\mathbf{A}$ is multiplied by the corresponding column of $\mathbf{B}$ to give the 2 x 2 matrix $\mathbf{G}$:

$$\mathbf{G} = \mathbf{AB} = \begin{pmatrix} A_{1,1} & A_{1,2} \\ A_{2,1} & A_{2,2} \end{pmatrix} \begin{pmatrix} B_{1,1} & B_{1,2} \\ B_{2,1} & B_{2,2} \end{pmatrix}$$

$$= \begin{pmatrix} A_{1,1}B_{1,1} + A_{1,2}B_{2,1} & A_{1,1}B_{1,2} + A_{1,2}B_{2,2} \\ A_{2,1}B_{1,1} + A_{2,2}B_{2,1} & A_{2,1}B_{1,2} + A_{2,2}B_{2,2} \end{pmatrix}$$

Note that matrix multiplication is associative, i.e., $(\mathbf{AB})\mathbf{C} = \mathbf{A}(\mathbf{BC})$, and distributive, $i.e.$, $\mathbf{A}(\mathbf{B} + \mathbf{C}) = \mathbf{AB} + \mathbf{BC}$, but not commutative, $i.e.$, $\mathbf{AB} \neq \mathbf{BA}$.

(continues)

(continued)

Example 1

Consider matrices **A** and **B** and vector **v** with the following numerical values:

$$\mathbf{A} = \begin{pmatrix} 0 & 1 \\ 2 & 3 \end{pmatrix}, \quad \mathbf{B} = \begin{pmatrix} 4 & 3 \\ 2 & 2 \end{pmatrix}, \quad \mathbf{v} = \begin{pmatrix} 3 \\ 5 \end{pmatrix}$$

$$\mathbf{A} + \mathbf{B} = \begin{pmatrix} 4 & 4 \\ 4 & 5 \end{pmatrix} \quad \mathbf{A} - \mathbf{B} = \begin{pmatrix} -4 & -2 \\ 0 & 1 \end{pmatrix}$$

$$2\mathbf{A} = \begin{pmatrix} 0 & 2 \\ 4 & 6 \end{pmatrix} \quad \mathbf{Av} = \begin{pmatrix} 5 \\ 21 \end{pmatrix} \quad \mathbf{AB} = \begin{pmatrix} 2 & 2 \\ 14 & 12 \end{pmatrix}$$

The reader is strongly encouraged to reproduce these results as an exercise.

Matrix Transpose

The transpose of matrix **A**, denoted as $\mathbf{A}^T$, is a matrix whose columns are taken directly from the rows of **A**, *i.e.*, row i of **A** becomes column i of $\mathbf{A}^T$. Thus, for the general case:

$$\mathbf{A}^T = \begin{pmatrix} A_{1,1} & A_{1,2} \\ A_{2,1} & A_{2,2} \end{pmatrix}^T = \begin{pmatrix} A_{1,1} & A_{2,1} \\ A_{1,2} & A_{2,2} \end{pmatrix}$$

Note: The transpose of **AB** is $(\mathbf{AB})^T = \mathbf{A}^T \mathbf{B}^T$.

Matrix Inverse

The inverse of an $n \times n$ matrix **A**, denoted as $\mathbf{A}^{-1}$, is another $n \times n$ matrix **B**, so that $\mathbf{AB} = \mathbf{BA} = \mathbf{I}$, where **I** is the so-called identity matrix that contains 1's on its diagonal and 0's everywhere else. For example when $n = 2$, **I** is

$$\mathbf{I} = \begin{pmatrix} 1 & 0 \\ 0 & 1 \end{pmatrix}$$

The inverse of the general matrix **A** is calculated as follows:

$$\mathbf{A}^{-1} = \begin{pmatrix} A_{1,1} & A_{1,2} \\ A_{2,1} & A_{2,2} \end{pmatrix} = \frac{1}{\det(\mathbf{A})} \begin{pmatrix} A_{2,2} & -A_{1,2} \\ -A_{2,1} & A_{1,1} \end{pmatrix}$$

Where, det(A) is what is known as the *determinant* of A and is defined as

$$\det(A) = \det\begin{pmatrix} A_{1,1} & A_{1,2} \\ A_{2,1} & A_{2,2} \end{pmatrix} = \begin{vmatrix} A_{1,1} & A_{2,1} \\ A_{1,2} & A_{2,2} \end{vmatrix}$$

$$= A_{1,1}A_{2,2} - A_{1,2}A_{2,1}$$

To find the determinant of a matrix with dimensions larger than 2 x 2, we refer to textbooks on linear algebra [see, for example Strang (1988)]. It is important to note that the inverse of A does not exist when its determinant equals zero. Such matrices that cannot be inverted are commonly referred to as *singular matrices*. Note also that the transpose of A^{-1}, *i.e.*, $(A^{-1})^T$, is equal to $(A^T)^{-1}$.

Another important matrix property is the *rank* of a matrix (r), which corresponds to the *number of genuinely independent rows* in a matrix. For an n x n square matrix A, when $r = n$, it can be proven that (1) A has an inverse and (2) this inverse is unique.

Example 2

For the matrices given in Example 1, the determinants are calculated to be $\det(A) = -2$ and $\det(B) = 2$, which indicates that these are nonsingular matrices that should have an inverse. These are calculated as follows:

$$A^{-1} = \begin{pmatrix} A_{1,1} & A_{1,2} \\ A_{2,1} & A_{2,2} \end{pmatrix}^{-1} = \begin{pmatrix} -1.5 & 0.5 \\ 1 & 0 \end{pmatrix} \text{ and}$$

$$B^{-1} = \begin{pmatrix} B_{1,1} & B_{1,2} \\ B_{2,1} & B_{2,2} \end{pmatrix}^{-1} = \begin{pmatrix} 1 & -1.5 \\ -1 & 2 \end{pmatrix}$$

Such 2 x 2 systems are rather easy to handle on paper; however, for matrices with larger dimensions, software packages such as MATLAB, MATHCAD, or MATHEMATICA should be employed to facilitate these operations.

where $E_c^{\#}$ is the so-called pseudo-inverse (or the Moore-Penrose inverse) of the matrix given by

$$E_c^{\#} = \left(E_c^T E_c\right)^{-1} E_c^T \tag{4.15}$$

Equation (4.14) is essentially the least squares estimate of the nonmeasured rates contained in the vector r_c, where all balances have been employed for their determination. It can be shown that if E_c has full rank (*i.e.*, there are at least as many linearly independent balances as the number of nonmeasured rates), then $E_c^T E_c$ also has full rank and the pseudo-inverse therefore can be found.

In the case of an overdetermined system, after the nonmeasured rates (r_c) have been determined by eq. (4.14), one may still be left with unused balances that can be employed to check the overall consistency of measured and calculated rates. To accomplish this, eq. (4.14) is inserted into eq. (4.11) to yield

$$R r_m = 0 \tag{4.16}$$

with R being the so-called *redundancy matrix* (van der Heijden *et al.*, 1994a,b) given by

$$R = E_m - E_c\left(E_c^T E_c\right)^{-1} E_c^T E_m \tag{4.17}$$

The rank of the redundancy matrix specifies the number of independent equations that must be satisfied by the measured and calculated [per eq. (4.14)] rates, and therefore it contains I - rank(R) *dependent* rows. If the dependent rows are removed, we obtain rank(R) independent equations relating the measured variables, *i.e.*

$$R_r r_m = 0 \tag{4.18}$$

where R_r is the reduced redundancy matrix containing only the independent rows of R. Equation (4.16) is the basis for our further analysis of gross error identification, but before we proceed, we first illustrate the preceding concepts and the method of determining R_r.

EXAMPLE 4.6

Analysis of Aerobic Yeast Cultivation without Ethanol Formation

We return to the case of aerobic yeast cultivation, which was also discussed in Examples 4.1-4.3, but now we consider the situation where there is no ethanol formation. In this case Y_{xe} is zero, and therefore we will not include ethanol in the black box model. Thus, we have μ, r_c, r_w, r_s, r_o, and r_N as the rates in the black box model. With measurements of the specific glucose uptake rate, specific oxygen uptake rate, specific carbon dioxide production rate, and specific growth rate (equal to the dilution rate in a steady state chemostat), the elemental matrix is partitioned as follows:

$$\text{Glc } O_2 \text{ } CO_2 \text{ biomass} \qquad NH_3 \text{ } H_2O$$

$$
\mathbf{E}_m = \begin{pmatrix} 1 & 0 & 1 & 1 \\ 2 & 0 & 0 & 1.83 \\ 1 & 2 & 2 & 0.56 \\ 0 & 0 & 0 & 0.17 \end{pmatrix}; \mathbf{E}_c = \begin{pmatrix} 0 & 0 \\ 3 & 2 \\ 0 & 1 \\ 1 & 0 \end{pmatrix} \tag{1}
$$

Note that the four columns of matrix $\mathbf{E}_m$ correspond to the four rates of glucose, oxygen, carbon dioxide, and biomass, respectively, whereas the two columns of the $\mathbf{E}_c$ matrix correspond to ammonia and water rates, respectively. The rows of these above matrices represent, of course, the four elemental balances. With a total of six compounds and four elemental balances there are $F = 2$ degrees of freedom. Because four rates are measured, the system is overdetermined. By using eq. (4.17) the redundancy matrix is found to be

$$
\mathbf{R} = \begin{pmatrix} 1 & 0 & 1 & 1 \\ 0 & -0.286 & -0.286 & 0.014 \\ 0 & 0.572 & 0.572 & -0.028 \\ 0 & 0.858 & 0.858 & -0.042 \end{pmatrix} \tag{2}
$$

with rank($\mathbf{R}$) = 2. It is easily seen that the last two rows of $\mathbf{R}$ are proportional to the second row (the third row is equal to the second row, multiplied by -2, and the fourth row is equal to the second row multiplied by -3). We therefore delete these two rows and thereby obtain the reduced redundancy matrix:

$$
\mathbf{R}_r = \begin{pmatrix} 1 & 0 & 1 & 1 \\ 0 & -0.286 & -0.286 & 0.014 \end{pmatrix} \tag{3}
$$

Equation 3, along with the four measured specific rates, yields the following redundant equations according to eq. (4.18):

$$
\mathbf{R}_r \mathbf{r}_m = \begin{pmatrix} -r_s + r_c + \mu \\ 0.286 r_o - 0.286 r_c + 0.014\mu \end{pmatrix} = \begin{pmatrix} 0 \\ 0 \end{pmatrix} \tag{4}
$$

Obviously the first row is recognized as a carbon balance, but the second row is not that easily identified even though it contains all the information from the constraints of the three other elemental balances.

Normally, experimental data are overlaid with noise, and in some cases there may even be systematic errors. As a consequence of such errors, eq. (4.16) is not, in general, exact. There will be some residuals different from zero when the measured rates (or yields) are multiplied into the reduced redundancy matrix. This is better expressed by recognizing that the *measured* rate vector $\bar{\mathbf{r}}_m$ equals the sum of the *actual rate* vector $\mathbf{r}_m$ and its corrupting general measurement error δ:

$$\bar{\mathbf{r}}_m = \mathbf{r}_m + \delta \tag{4.19}$$

Combination eq. (4.19) into eq. (4.18) yields the following equation for the vector of the residual ε:

$$\varepsilon = \mathbf{R}_r \bar{\mathbf{r}}_m = \mathbf{R}_r (\mathbf{r}_m + \delta) = \mathbf{R}_r \delta \tag{4.20}$$

If the model is correct and if there are no systematic or random errors, *i.e.*, $\delta = 0$, all equations [eq. (4.18)] are satisfied exactly and yield zero values for the residuals. In all data sets, however, there is some noise present in the measurements that makes the residuals vector different from zero. The best rate estimates are those that minimize the magnitude of the residual, and they are determined as follows.

By assuming that the error vector is distributed normally with a mean value of zero and a variance-covariance matrix $\mathbf{F}$,

$$E(\delta) = 0 \tag{4.21}$$

$$\mathbf{F} \equiv E\left[(\bar{\mathbf{r}}_m - \mathbf{r}_m)(\bar{\mathbf{r}}_m - \mathbf{r}_m)^T\right] = E(\delta\delta^T) \tag{4.22}$$

where E is the expected value operator, it can be shown that the residuals also will be distributed normally with a mean of zero

$$E(\varepsilon) = \mathbf{R}_r E(\delta) = 0 \tag{4.23}$$

and a variance-covariance matrix given by

$$\mathbf{P} = E(\varepsilon\varepsilon^T) = \mathbf{R}_r E(\delta\delta^T)\mathbf{R}_r^T = \mathbf{R}_r \mathbf{F} \mathbf{R}_r^T \tag{4.24}$$

The minimum variance estimate of the error vector δ is obtained by minimizing the sum of squared errors scaled according to their variance:

$$\min_{\delta} (\delta^T F^{-1} \delta) \qquad (4.25)$$

The solution is given by

$$\hat{\delta} = FR_r^T P^{-1} \varepsilon = FR_r^T P^{-1} R_r r_m \qquad (4.26)$$

where the circumflex specifies that the value of δ is an *estimate*. Because δ is distributed normally, the function to be minimized in eq. (4.25) is the same for the least squares minimization problem and for the maximum likelihood minimization problem. If the error vector is not distributed normally, the estimate in eq. (4.26) remains valid for the least squares minimization problem but it no longer is the maximum likelihood estimate (Wang and Stephanopoulos, 1983). By using eq. (4.26), the best estimates for the measured rates are obtained as

$$\hat{r}_m = \overline{r_m} - \hat{\delta} = (I - FR_r^T P^{-1} R_r)\overline{r_m} \qquad (4.27)$$

where I is a identity matrix. It can be shown that the estimates of the measured rates given by eq. (4.27) have a smaller standard deviation than the raw measurements (Wang and Stephanopoulos, 1983), and the estimate therefore is likely to be more reliable than the measured data. By using the best estimates for the measured rates, the nonmeasured rates of the black box model can be calculated using eq. (4.14).

EXAMPLE 4.7

Analysis of Aerobic Yeast Cultivation without Ethanol Formation (Continued)

We now continue our analysis of aerobic yeast cultivation that was initiated in Example 4.6, where we derived the reduced redundancy matrix. At a dilution rate of $0.15 \ h^{-1}$, the measured specific rates for glucose, oxygen, carbon dioxide, and biomass are given by

$$\bar{r}_m = \begin{pmatrix} -r_s \\ -r_o \\ \mu \\ r_c \end{pmatrix} = \begin{pmatrix} -0.250 \\ -0.113 \\ 0.113 \\ 0.141 \end{pmatrix} \qquad (1)$$

with all rates in C-moles (C-mole biomass hour)$^{-1}$. We now want to calculate better estimates for the measured rates when it is assumed that there is a 5% error in the biomass and glucose measurements and a 10% error in the gas measurements, $i.e.$, the oxygen and carbon dioxide measurements. With these errors the variance-covariance matrix is given by

$$F = 10^{-3} \begin{pmatrix} 0.1563 & 0 & 0 & 0 \\ 0 & 0.1277 & 0 & 0 \\ 0 & 0 & 0.0319 & 0 \\ 0 & 0 & 0 & 0.1988 \end{pmatrix} \qquad (2)$$

and by using eq. (4.24) we find

$$P = 10^{-3} \begin{pmatrix} 0.3870 & -0.0563 \\ -0.0563 & 0.0267 \end{pmatrix} \qquad (3)$$

(the reduced redundancy matrix is taken from Example 4.6). The error vector for the measured fluxes is then found from eq. (4.26):

$$\hat{\delta} = \begin{pmatrix} -0.0055 \\ 0.0115 \\ -0.0013 \\ 0.0108 \end{pmatrix} \qquad (4)$$

and this leads to better estimates for the measured fluxes:

$$\hat{r}_m = \begin{pmatrix} -0.2445 \\ -0.1245 \\ 0.1143 \\ 0.1302 \end{pmatrix} \qquad (5)$$

Thus, there are only small corrections to the measurements, and the original measurements therefore seem to be good. However, the corrected measurements conform better with the elemental balances, and therefore they are better estimates than the raw measurements.

Normally, the variance-covariance matrix is assumed to be diagonal, meaning that the measurements are uncorrelated. However, specific rates, yield coefficients, and even volumetric rates are seldom measured directly but instead are derived from measurements of the so-called prime variables, which may influence more than one measured rate. An example is the measurement of the oxygen uptake rate and the carbon dioxide production rate, which are both based on the measurement of the gas flow rate through the bioreactor, in conjunction with measurements of the partial pressure of the two gasses in the head space. If there is an error in the measurement of the gas flow rate, this influences both rates, and, therefore, errors in the measured rates are indirectly correlated. The same objection holds for the measurement of other rates as well that are normally obtained by combination of concentration and flow rate measurements. In all such cases with indirect error correlations, it is difficult to specify the true variance-covariance matrix F. Madron *et al.* (1977) describe a simple algorithm by which the true variance-covariance matrix can be found when the properties of the noise of the prime variables are known (see Box 4.3). In many cases, however, we have inadequate information about the noise of even the prime variables, and the true variance-covariance matrix cannot be derived. In these cases, one may decide to neglect covariances and use a diagonal variance-covariance matrix, where reasonable values for the errors are used. Alternatively, one may use the least squares estimate given by

$$\hat{\mathbf{r}}_m = \left(\mathbf{I} - \mathbf{R}_r^T\left(\mathbf{R}_r\mathbf{R}_r^T\right)^{-1}\mathbf{R}_r\right)\bar{\mathbf{r}}_m \qquad (4.28)$$

which is based on the assumption of the same absolute error in all the measured rates. Because absolute values for the errors are used, it is only reasonable to apply eq. (4.28) when the variables are of the same magnitude.

If any constraint residuals are *significantly different from zero*, either a systematic error is present in at least one of the measurements or the model employed is incorrect. To quantify the statement "significantly different from zero," we introduce the test function h given by the sum of *weighted* squares of the residuals:

$$h = \varepsilon^T \mathbf{P}^{-1}\varepsilon \qquad (4.29)$$

When the raw measurements are uncorrelated, the test function h is χ^2 distributed (Wang and Stephanopoulos, 1983), and this was shown to be the case for correlated raw data as well (van der Heijden *et al.*, 1994b). The degrees of freedom of the χ^2 distribution are equal to the rank($\mathbf{P}$) = rank($\mathbf{R}$), i.e., the number of independent constraints. By comparing the calculated value of the test function h with the values of the χ^2 distribution at the degrees of freedom [rank($\mathbf{R}$)], it is possible to detect the presence of a systematic error in the data at a certain confidence level. Thus, if at a high enough confidence level one obtains a test function value that is greater than the value of the χ^2 distribution, then there is something wrong with the data

BOX 4.3

Calculation of the Variance-Covariance Matrix from Errors in Prime Variables

Normally the measured rates are determined from measurements of the so-called prime variables, $e.g.$, the volumetric glucose uptake rate in a steady state chemostat is determined as the difference between the glucose concentration in the feed and that in the bioreactor multiplied by the dilution rate. Specification of the variance-covariance matrix therefore is not straightforward. Madron $et\ al.$ (1977) describe a simple approach to find **F** from the measurement noise of the prime variables. First, the measured rates are specified as functions of the prime variables. When the latter are collected in the vector y, we have for the ith rate:

$$r_{m,i} = f_i(y) \tag{1}$$

Generally the functions f_i are nonlinear, but in order to obtain an approximate estimate of the variances and covariances, these functions are linearized. The error of the measured i'th rate, δ_j, is expressed as a linear combination of the errors δj^* of the prime variables:

$$\delta_i = \sum_{j=1}^{k} \left(\frac{\partial f_i}{\partial y_j} \right) \delta_i^* = \sum_{j=1}^{k} g_{ij} \delta_j^* \tag{2}$$

where K is the number of prime variables and g_{ij} are the sensitivities. If the sensitivities are collected in the matrix **G**, the variance-covariance matrix **F** can be calculated from

$$\mathbf{F} = \mathbf{G}\mathbf{F}^*\mathbf{G}^T \tag{3}$$

where $\mathbf{F}^*$ is a diagonal matrix with the variances of the prime variables. The preceding method is very simple to compute the covariances, but obviously the calculated values are limited by the accuracy of the linear approximation in eq. (2).

TABLE 4.4   Values of the χ^2 Distribution

Degrees of freedom	Confidence level					
	0.500	0.750	0.900	0.950	0.975	0.990
1	0.46	1.32	2.71	3.84	5.02	6.63
2	1.39	2.77	4.61	5.99	7.38	9.21
3	2.37	4.11	6.25	7.81	9.35	11.30
4	3.36	5.39	7.78	9.49	11.10	13.30
5	4.35	6.63	9.24	11.10	12.80	15.10

or the model. Table 4.4 gives values of the χ^2 distribution at different confidence levels and different degrees of freedom.

EXAMPLE 4.8

Analysis of Aerobic Yeast Cultivation without Ethanol Formation (Continued)

We now continue our analysis of aerobic yeast cultivation that was initiated in Example 4.6 and analyzed further in Example 4.7. From the matrices derived in Example 4.7, we calculate the residuals using eq. (4.20):

$$\varepsilon = \begin{pmatrix} 0.0040 \\ -0.0064 \end{pmatrix} \tag{1}$$

and then the test function can be calculated using eq. (4.29):

$$h = \varepsilon^T P^{-1} \varepsilon = 1.87 \tag{2}$$

Because there are two independent rows in the reduced redundancy matrix, its rank is 2, i.e., the test function has to be compared with the χ^2 distribution with two degrees of freedom. From Table 4.4 it is seen that the test function is lower than the χ^2 distribution even at a confidence level of 0.75. Thus, it is only at a very low confidence level that it can be concluded that the data contain gross errors; hence, the data quality is satisfactory.

The finding of a large value for the test function, $h > \chi^2$, at a given confidence level does not allow one to conclude whether the unsatisfactorily large errors are due to systematic errors in the data or due to large random errors. One approach that can be applied to this end is to eliminate one measurement at a time from the given set of data and use one of the constraints to calculate such a measurement. The remaining constraints are then used for consistency analysis to recalculate the test function h and compare it again with the χ^2 statistic for one fewer degree of freedom. If a significantly lower value is obtained for the test function upon elimination of a certain measurement, this is strong evidence for the presence of gross (systematic) errors in the measurement that was eliminated. The same can be applied to constraints other than the elemental balances, such as those arising from application of the steady state hypothesis to intracellular metabolites. This approach of error diagnosis requires that the system be overdetermined by at least two measurements, i.e., rank(R) ≥ 2. This overdetermination allows for one constraint to be used for the calculation of the eliminated measurement while the other is used for the recalculation of the test function. The procedure of measurement elimination is very simple, as illustrated in Example 4.9, and it allows for the rapid determination of the probable source of a systematic error.

EXAMPLE 4.9

Error Diagnosis in Yeast Cultivation Measurements

For aerobic growth of S. cerevisiae with glucose as the carbon source and ammonia as the nitrogen source, the specific rates of glucose uptake, oxygen uptake, biomass growth, and carbon dioxide formation are measured to be

$$\mathbf{r} = \begin{pmatrix} -r_s \\ -r_o \\ \mu \\ r_c \end{pmatrix} = 0.008 \begin{pmatrix} -2.1 \\ -3.8 \\ 1 \\ 1.4 \end{pmatrix} \tag{1}$$

at a specific growth rate of 0.008 h^{-1} [all rates in C-moles (C-mole biomass hour)$^{-1}$]. The elemental composition of the biomass is assumed to be the same as in Examples 4.6-4.8. The measurement errors are 6%, 11.7%, 5%, and 11.1% for glucose, oxygen, biomass, and carbon dioxide, respectively. There are no covariances. We now want to examine whether there are any experimental errors. Because the stoichiometry is the same and the same rates are measured as in Examples 4.6-4.8, we can use the reduced redundancy matrix derived in Example 4.6. Furthermore, with the given errors the variance-covariance matrix is found to be

$$F = 10^{-4} \begin{pmatrix} 0.0102 & 0 & 0 & 0 \\ 0 & 0.1265 & 0 & 0 \\ 0 & 0 & 0.0016 & 0 \\ 0 & 0 & 0 & 0.0155 \end{pmatrix} \qquad (2)$$

When the test function is calculated as illustrated in Example 4.8, it is found to be

$$h = 35.06 \qquad (3)$$

which shows that even at a confidence level of 0.99 there is a measurement error. From inspection of the measured rates it seems likely that this error is in either the oxygen or the carbon dioxide measurement, as the respiratory quotient ($RQ = r_c/r_o$) is less than the 1, the normal value for an aerobic culture of *S. cerevisiae* growing at low specific growth rates. To identify the measurement error, we do, however, eliminate each of the four reactions - one at a time - and then calculate the test function. The result of this gives

Compound eliminated	h
Glucose	27.06
Oxygen	2.12
Biomass	26.43
Carbon dioxide	34.96

Clearly, if any of the three measurements glucose, biomass, or carbon dioxide is eliminated, there is still a measurement error. Only when the oxygen measurement is eliminated does the value of the test function drop to a low value, and by comparing it with the χ^2 distribution with one degree of freedom it is seen that at a confidence greater than 90% it cannot be concluded that there are gross measurement errors. Thus, it is very likely that the oxygen measurement is erroneous.

If the oxygen measurement is left out, it is possible to calculate both better estimates for the three measured rates and best estimates for the three nonmeasured rates (including oxygen). First, by using eq. (4.27), we find for the measured rates

$$\hat{r}_m = \begin{pmatrix} -r_s \\ \mu \\ r_c \end{pmatrix} = 0.008 \begin{pmatrix} -2.21 \\ 0.98 \\ 1.23 \end{pmatrix} \qquad (4)$$

and thereafter we find the nonmeasured rates (ammonia uptake, water formation, and oxygen uptake) using eq. (4.14) [with eq. (4) inserted for the measured rates]:

$$\hat{r}_c = \begin{pmatrix} -r_N \\ r_w \\ -r_o \end{pmatrix} = 0.008 \begin{pmatrix} -0.17 \\ 1.56 \\ -1.18 \end{pmatrix} \tag{5}$$

Thus, we see that the oxygen uptake rate is corrected drastically. Furthermore, with the estimated rates the RQ is found to be 1.04, which is a much more realistic value.

REFERENCES

Abbott, B. J. & Clamen, A. (1973). A. The relationship of substrate, growth rate, and maintenance coefficients to single cell protein production. *Biotechnology and Bioengineering* **15**, 117-127.

Christensen, L. H. & Henriksen, C. M.; Nielsen, J. & Villadsen, J. (1995). Continuous cultivation of *Penicillium chrysogenum*. Growth on glucose and penicillin production. *Journal of Biotechnology* **42**, 95-107.

Erickson, L. E., Minkevich, I. G. & Eroshin, V. K. (1978). Application of mass and energy balance regularities in fermentation. *Biotechnology and Bioengineering* 20:1595-1621

Larsson, C., Blomberg, A. & Gustafsson, L. (1991). Use of microcalorimetric monitoring in establishing continuous energy balances and in continuous determinations of substrate and product concentrations of batch grown *Saccharomyces cerevisiae*. *Biotechnology and Bioengineering*. **38**, 447-458.

Madron, F., Veverka, V. & Vanecek, V. (1977). Statistical analysis of material balance of a chemical reactor. *AIChE Journal* **23**, 482-486.

Nielsen, J. & Villadsen, J. (1994). *Bioreaction engineering principles*. New York: Plenum Press.

Roels, J. A. (1983). *Energetics and Kinetics in Biotechnology*. Amsterdam: Elsevier Biomedical Press.

Schulze, U. (1995). Anaerobic physiology of *Saccharomyces cerevisiae*. Ph.D. Thesis, Technical University of Denmark.

van der Heijden, R. T. J. M., Heijnen, J. J., Hellinga, C., Romein, B. & Luyben, K. Ch. A. M. (1994a). Linear constraint relations in biochemical reaction systems: I. Classification of the calculability and the balanceability of conversion rates. *Biotechnology and Bioengineering* **43**, 3-10.

van der Heijden, R. T. J. M., Heijnen, J. J., Hellinga, C., Romein, B. & Luyben, K. Ch. A. M. (1994b). Linear constraint relations in biochemical reaction systems: II. Diagnosis and estimation of gross measurement errors. *Biotechnology and Bioengineering*. **43**, 11-20.

von Meyenburg, K. (1969). Katabolit-Repression und der Sprossungszyklus von *Saccharomyces cerevisiae*. Ph.D. Thesis, ETH Zürich.

von Stockar, U. & Birou, B. (1989). The heat generated by yeast cultures with a mixed metabolism in the transition between respiration and fermentation. *Biotechnology and Bioengineering*. **34**, 86-101.

Wang, N. S. & Stephanopoulos, G. (1983). Application of macroscopic balances to the identification of gross measurement errors. *Biotechnology and Bioengineering*. **25**, 2177-2208.

CHAPTER 5

Regulation of Metabolic Pathways

In living systems, control of biological function occurs at the cellular and molecular levels. These controls are implemented by the regulation of concentrations of species taking part in biochemical reactions. Such regulatory variables include concentrations of enzymes (E), substrates (S), products (P), and regulatory molecules (R), so that the rate of an enzymatic reaction can be generally expressed as

$$v = v\left(c_e, c_s, c_p, c_r\right)$$

The preceding molecular species are present in vast numbers in a typical cell and participate in specific interactions with one another. Despite their immense complexity, biochemical systems are characterized by their ability to reach stable steady states, and this is believed to be the outcome of special interrelationships among this molecular cast of characters, often described as "controls." Investigation of metabolic regulation attempts to elucidate the specific mechanisms by which the activity of individual enzymes is modulated or even controlled at the molecular level (Atkinson, 1970; Khoshland, 1970; Khoshland and Neet, 1968; Van Dam *et al.*, 1993). Once the regulatory characteristics of enzymes are understood at the local level, global control of pathway flux can be deduced by making use of the results of metabolic control analysis (Chapter 11).

For certain pathways, individual reactions are organized such that the metabolic intermediates constituting the pathway are always bound to the enzyme surface and the products are handed on as substrates for the subsequent reaction (Hofmeyr, 1991). One such example is the β-oxidation pathway for the conversion of fatty acyl-CoA derivatives to acetyl-CoA. In such cases, intermediates typically participate in one specific reaction and are not detected in solution. On the other hand, metabolic pathways such as the EMP pathway, gluconeogenesis, the TCA cycle, and the pentose phosphate (PP) pathway are not organized in the same way. The intermediates of these pathways, although present in low concentrations in the cell, are indeed detectable. The interrelated nature of metabolism requires that a number of compounds function as intermediates and/or precursors for more than one pathway (*e.g.*, G6P, PEP, PYR, OAA). Such metabolites cannot be enzyme-bound because they always have to be available for more than one pathway, and, more importantly, the rates of these pathways are likely to vary under different conditions. As the nature of metabolism begins to unfold, it is becoming apparent that the cell utilizes some of these metabolites for the purpose of metabolic regulation, so that certain compounds function both as metabolic intermediates and as metabolic *regulators* of key enzymes.

The biosynthetic reaction sequences leading to the formation of cellular building blocks, such as amino acids, purine and pyrimidine nucleosides, and steroids, originate from simpler cell materials that arise from the metabolism of the carbohydrates or fatty acids (Chapter 2). Each metabolic pathway is associated with a *committed step*, *i.e.*, the reaction that produces the first metabolic intermediate destined for the formation of the end product of the particular sequence of reactions. In most cases, the committed step is exergonic (*i.e.*, $\Delta G^{o\prime} < 0$, $K' > 1$) so that the reaction is almost irreversible. It is widely accepted that the metabolic control intended to regulate the formation of the end product functions most satisfactorily at such metabolic nodes, while control of intermediate steps is exercised less frequently.

When posed with an involved metabolic network, an important question is identification of the relative control exerted by the various intermediary reactions. Despite their high degree of structural and functional complexity, metabolic systems portray rather simple kinetics and unique stable steady states. This has been attributed to the hierarchical structure of enzymes, especially with respect to their characteristic time scales that span more than

15 orders of magnitude (de Koning and van Dam, 1992; Heinrich and Sonntag, 1982; Hiromi, 1979). At the fast end are the kinetics of metabolic reactions (10^{-2}-10^{4}), and at the slowest end an genetic regulation and evolution (see Section 2.1 and Box 2.1).

Table 5.1 summarizes the relaxation times for glycolytic reactions in erythrocytes. Based on these values, one can differentiate two groups of enzyme-catalyzed reactions. One group includes HK, PFK, DPGM, DPGP, PK, ATPase, and which catalyze slow reactions, whereas the rest of the reactions have relaxation times 2-6 orders of magnitude faster. For these fast reactions the concentrations of substrates and products are essentially at equilibrium, and their role is relatively unimportant for the dynamic response of the overall metabolic system. In other words, the dynamic response of the system can be represented essentially by the slow reactions, which in turn considerably reduces the number of variables that need to be considered and the complexity of the overall system.

The probability of instability rises with the number of variables (see Chapter 1). Thus, in essence, time hierarchy may be an important factor in stabilizing metabolic systems by reducing the number of dynamically important variables. Therefore, despite the immense complexity of biological systems at a microscopic level, metabolic systems can be successfully lumped and modeled by relatively simple expressions. Otherwise, the theoretical

TABLE 5.1 Relaxation Times of Glycolytic Reactions from Erythrocytes[a]

Enzymatic reaction	Relaxation time (s)
Hexokinase (HK)	> 1100
Phosphofructokinase (PFK)	> 75
2,3-Bisphosphoglycerate mutase (DPGM)	4
2,3-Bisphosphoglycerate phosphatase (DPGP)	34,000
Pyravate kinase (PK)	28
ATPase	1800
Phosphoglucoisomerase (PGI)	$\sim 10^{-2}$
Aldolase (Ald)	$\sim 10^{-2}$
Triose phosphatisomerase	$\sim 10^{-2}$
Glyceraldehydephosphate dehydrogenase (GAPD)	$\sim 10^{-2}$
Phosphoglycerate kinase (PGK)	$\sim 10^{-2}$
Phosphoglycerate mutase (PGM)	$\sim 10^{-2}$
Enolase	$\sim 10^{-2}$
Lactate dehydrogenase (LDH)	$\sim 10^{-2}$
Adenylate kinase (AD)	$\sim 10^{-2}$

[a] From Rappoport et al., 1974; Schuster et al., 1989.

analysis of biological systems would be unattainable for all practical purposes.

A fundamental distinction between the numerous mechanisms of metabolic regulation is whether they directly modulate the activity of the enzyme present in the cell (see Box 5.1) or whether they entail secondary responses that alter the net rate of synthesis of the particular protein. Living cells employ a plethora of control mechanisms that regulate enzyme activity. At the extremes of the spectrum one finds complete deactivation/activation by covalent modifications. Phosphorylation is a common mechanism for this, but addition or removal of adenylyl, acetyl, methyl, and other moieties has also been observed. What is more common though is the gradual variation of enzymatic activity by reversible association with another molecule, such as a reaction product. Control of enzyme levels can also take several forms, which include regulation of enzyme synthesis and/or its degradation. In general, differentiated cells of higher plants and animals display fewer and smaller enzymatic level modulations due to their specialized nature. On the other hand, unicellular organisms, such as bacteria, heavily rely on enzymatic level modifications to coordinate different reactions, in addition to enzymatic activity adjustments.

Mechanisms of enzymatic regulation are reviewed in the following sections. Section 5.1 describes the various modes by which cells regulate the activity of enzymes present in the cell, while Section 5.2 deals with the regulation of the actual enzyme concentration, which takes place at the genetic level. Sections 5.3 and 5.4 address regulation issues at the cellular level: For instance, how the cell regulates sets of related enzymatic reactions and, ultimately, how systemic or global regulation takes place.

5.1. REGULATION OF ENZYMATIC ACTIVITY

By far, the most common form of metabolic regulation takes place by regulating the level of enzymatic activity. Feedback inhibition and activation are extremely rapid responses for regulating enzyme activity in all types of cells. Repression and induction (or derepression) of enzyme synthesis represent slower, long-term regulatory mechanisms whereby the amount of a particular enzyme in a cell is optimized. For instance, 2-D protein gels of cell extracts prepared from *Escherichia coli* grown at 20 and 37°C are indistinguishable, despite the enormous adjustment made by the cell to accommodate the physiological differences at the two growth temperatures (Ingraham, 1987). Such adjustments must be achieved mainly by altering the enzymatic activities without significant changes in enzyme concentrations. Adjustments

BOX 5.1
_____

Modes of Feedback Inhibition / Activation (Neidhardt *et al.*, 1990)

In sequential feedback inhibition, regulation of the various branches is decoupled and the branch point metabolite acts as a regulator for the first enzyme of the pathway. More complex pathways involving multiple common metabolites entail more complicated control patterns that can combine both activation and inhibition in order to coordinate the metabolic fluxes. Examples of these patterns of regulation are found throughout the bacterial world.

The presence of isozymes, for example, or the production of multiple end products by a particular pathway (either sequentially or divergently) requires a combination of the regulatory mechanisms discussed in this section. Isofunctional enzymes that catalyze the first reaction of a common step are usually sensitive to one of the end products of the pathway. This mechanism (cooperative or synergistic inhibition) ensures that end product inhibition does not starve the cells of metabolites derived from the same metabolic pathway. Cumulative (partial) feedback inhibition is a more sophisticated form of the previous mechanism, whereupon a single enzyme contains multiple allosteric sites responding to each of the various end products of the pathway (see Fig. 5.1).

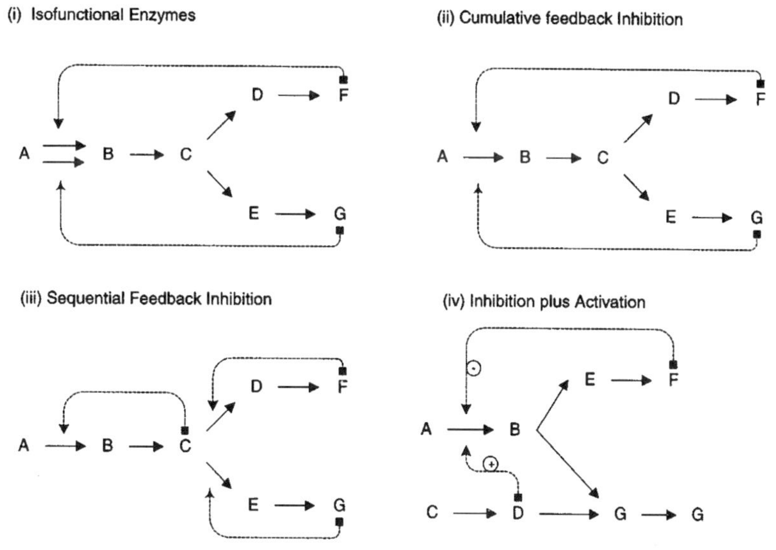

FIGURE 5.1   Regulatory networks involving feedback inhibition and activation.

are mediated by metabolites that may be a substrate or product of a reaction sequence or key global metabolites such ATP or NAD(P)H. Regulatory metabolites can act as either activators or inhibitors, and they are usually termed as effectors or ligands, respectively.

5.1.1. OVERVIEW OF ENZYME KINETICS

The earliest attempt to analyze the kinetics of enzyme-catalyzed reactions was made by Michaelis and Menten, whose work led to the introduction of two fundamental kinetic parameters of enzymatic reaction rates (Dixon and Webb, 1979; Laidler and Bunting, 1973; Schulz, 1994; Segel, 1993). The first of these parameters, called v_{max}, is defined as the maximum velocity (rate) that the reaction can attain in the presence of a given amount of enzyme, i.e., the reaction rate when the enzyme is saturated with substrate. This is a measure of the enzyme efficiency and it is directly related to its turnover number, defined as the number of substrate molecules converted to product per second. Turnover numbers range from several hundred thousands per second (e.g. for carbonic anhydrase the turnover number is 600,000 s^{-1}) down to less than one molecule per second (e.g., 0.5 s^{-1} for lysozyme).

The other important kinetic parameter is the Michaelis constant or K_m, defined as the substrate concentration at which the reaction velocity is $v_{max}/2$. An important distinction between the two parameters is the fact that the value of K_m is independent of enzyme concentration. The numerical value of K_m is important for a number of reasons: (1) It provides a useful indicator of the relative affinity of an enzyme for its substrate (lower values imply higher affinity). (2) It establishes an approximate value for the intracellular level of substrate - no physiological sense for $c_s \gg K_m$ because v would be close to v_{max} and v would be insensitive to changes in c_s, also, at $c_s \ll K_m$ it implies that $v < v_{max}$ and, thus, most of the catalytic activity of the enzyme would be wasted. (3) By investigating the effects of various compounds on K_m it is possible to identify potential allosteric effectors for the particular enzyme. (4) Knowledge of the K_m value enables the selection of appropriate substrate concentrations in enzymatic assay. (5) K_m indicates the relative "suitability" of alternate substrates of a particular enzyme (substrates with highest v_{max}/K_m are the most favorable).

The relationship between reaction rates and substrate concentration in terms of the two parameters v_{max} and K_m is depicted in Fig. 5.2. Mathematically it can expressed as follows:

$$v = \frac{v_{max} c_s}{K_m + c_s} \qquad (5.1)$$

It should be noted that, although the Michaelis-Menten equation was proposed initially as an empirical expression describing experimental measurements, it was later justified on the basis of specific molecular mechanisms. One such mechanism assumes the formation of an enzyme-substrate complex (ES), which decomposes to yield the enzyme and product (P):

$$E + S \overset{k_1, k_{-1}}{\leftrightarrow} (ES) \overset{k_2}{\rightarrow} E + P \tag{5.2}$$

The assumption of equilibrium for the enzyme-substrate complex formation step allows one to derive the rate of eq. (5.1) from the reaction mechanism of eq. (5.2). It was later determined that the equilibrium assumption was incorrect and that the enzyme-substrate complex is, instead, at steady state. However, a pseudo-steady state assumption for the complex (ES) yields the same rate eq. (5.1), although the rate constant K_m is now a different function of the rate constants k_1, k_{-1}, and k_2 of reaction (5.2) (for derivations see Box 5.2).

Lineweaver and Burk converted the preceding equation into a linear form that enables a more accurate determination of the K_m and v_{max} values:

$$\frac{1}{v} = \frac{1}{v_{max}} + \left(\frac{K_m}{v_{max}} \right) \frac{1}{c_s} \tag{5.3}$$

Thus, a plot of $1/v$ versus $1/c_s$ should give a straight line with a slope K_m/v_{max}, a y-intercept (i.e., at $1/c_s = 0$) $1/v_{max}$, and an x-intercept (i.e., at $1/v = 0$) $-1/K_m$ (Fig. 5.2b).

5.1.2. SIMPLE REVERSIBLE INHIBITION SYSTEMS

Any molecule that reduces the velocity of an enzymatic reaction is considered to be an inhibitor. Many naturally occurring and synthetic molecules are known to interfere with enzymatic activity, and they can do so either reversibly or irreversibly (Webb, 1963). This section summarizes the four basic types of reversible inhibition, namely, substrate, competitive, noncompetitive, and uncompetitive inhibition for enzymes that have a single substrate and a single inhibitor.

Substrate Inhibition

In many instances where a large amount of substrate is present, the enzyme-catalyzed reaction is negatively affected as a result of excess substrate. As indicated in Fig. 5.3, the reaction rate goes through a maximum as

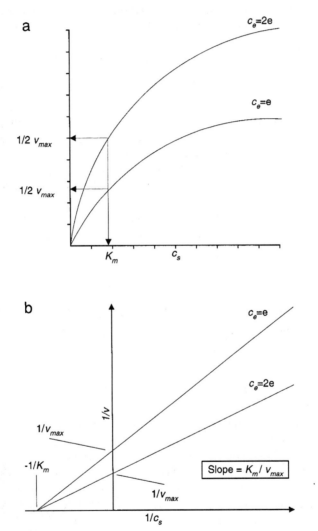

FIGURE 5.2   (a) Velocity versus substrate concentration at two enzyme concentrations $c_e = e$ and $c_e = 2e$. (b) Double-reciprocal ($1/v$ versus $1/c_s$) Lineweaver-Burk plots are generally utilized for determining v_{max} and K_m values. Note that v_{max} is a macroscopic property that varies with c_e, whereas K_m is an intrinsic property that does not depend on c_e.

BOX 5.2

Derivation of Equation (5.1)

The simplest enzyme-catalyzed reaction involves a single substrate (S) converted to a single product (P) via a *central complex* (ES). This is usually referred to as a Uni-Uni reaction system and can be represented as follows:

$$E + S \overset{k_1, k_{-1}}{\leftrightarrow} (ES) \overset{k_p}{\to} E + P$$

The velocity eq. (5.1) can be derived in either of two ways: (1) the rapid equilibrium approach or (2) the steady-state (or Briggs and Haldane) approach.

1. Rapid Equilibrium

This is a simple approach that assumes, as the name implies, conditions of rapid equilibrium, *i.e.*, E, S, and ES equilibrate very rapidly compared with the rate at which ES is converted to E + P. The instantaneous velocity will, therefore, depend only on the concentration of ES:

$$v = k_p c_{es} \tag{1}$$

where k_p is called the catalytic rate constant. The total enzyme is distributed between E and ES:

$$c_{et} = c_e + c_{es} \tag{2}$$

If eq. (1) is divided by eq. (2):

$$\frac{v}{c_{et}} = \frac{k_p c_{es}}{c_e + c_{es}} \tag{3}$$

On the basis of the equilibrium assumption, c_{es} can be expressed in terms of c_s, c_e, and K_s, where K_s is the dissociation constant of the ES complex:

$$K_s = \frac{c_e c_s}{c_{es}} = \frac{k_{-1}}{k_1} \qquad \text{Therefore,} \qquad c_{es} = \frac{c_s}{K_s} c_e \tag{4}$$

(*continues*)

(continued)

Substitute the expression for c_{es} from eq. (4) into eq. (3):

$$\frac{v}{c_{et}} = \frac{k_p \dfrac{c_s c_e}{K_s}}{c_e + \dfrac{c_s c_e}{K_s}}$$ Cross multiply k_p and eliminate c_e:

$$\frac{v}{k_p c_{et}} = \frac{\dfrac{c_s}{K_s}}{1 + \dfrac{c_s}{K_s}}$$ Let $k_p c_{et} = v_{max}, \Rightarrow$ $\dfrac{v}{v_{max}} = \dfrac{\dfrac{c_s}{K_s}}{1 + \dfrac{c_s}{K_s}}$ (5)

Equation (5) upon rearrangement yields the more familiar Henri-Michaelis-Menten equation:

$$\frac{v}{v_{max}} = \frac{c_s}{K_s + c_s} \qquad (6)$$

Thus, the preceding equation provides the instantaneous or initial velocity relative to v_{max} at any substrate concentration. It is important to bear in mind that this expression is valid only if v is measured over a short enough time so that c_s remains essentially constant or, in other words, no more than approximately 5% of the substrate will be utilized over the assay period.

2. Steady State Approach (Briggs and Haldane) (Haldane, 1965; Walter, 1965)

If we assume that the rate at which ES is converted to product is rapid compared with the rate at which ES dissociates back to E + S, then shortly after E and S are mixed, a *steady state* will be established in which the concentration of ES remains invariant with time. By following a procedure similar to that in the case of rapid equilibrium (part 1), a velocity equation can be derived based on a steady state assumption. Similar to the previous derivation:

$$E + S \overset{k_1, k_{-1}}{\leftrightarrow} (ES) \overset{k_p}{\rightarrow} E + P$$

$$v = k_p c_{es} \qquad (7)$$

$$\frac{v}{c_{et}} = \frac{k_p c_{es}}{c_e + c_{es}} \qquad (8)$$

Because c_{es} is assumed to be time invariant, the rate of its formation is then equal to the rate of its decomposition:

$$k_1 c_e c_s = k_{-1} c_{es} + k_p c_{es} = \left(k_{-1} + k_p \right) c_{es} \qquad \Rightarrow c_{es} = \frac{k_1 c_{es}}{\left(k_{-1} + k_p \right)} \tag{9}$$

We define

$$K_m = \frac{\left(k_{-1} + k_p \right)}{k_1} \qquad \Rightarrow c_{es} = \frac{c_s c_{et}}{K_m + c_s} \tag{10}$$

and then substitute eq. (10) into eq. (7) and rearrange::

$$\Rightarrow \frac{v}{v_{max}} = \frac{c_s}{Km + c_s} \tag{11}$$

Thus, the *form* of the velocity equation is the same in both cases, but the *rate constants* are different [K_s in eq. (6) and K_m in eq. (11)]. Note that in the cases where $k_p \ll k_{-1}$, K_m reduces to K_s. The physical significance of K_m (or K_s) is that it corresponds to the substrate concentration that yields half-maximal velocity, *i.e.*, when $c_s = K_m$:

$$v = \left(\frac{K_m v_{max}}{K_m + K_m} \right) = \frac{1}{2} v_{max} \tag{12}$$

substrate concentration is increased. Beyond this maximum, an increase in substrate concentration causes a decrease in the enzymatic velocity.

A model can be derived according to the rapid equilibrium approach (Box 5.2), assuming that when S binds the ES complex an unreactive intermediate results. At equilibrium:

$$E + S \overset{k_1, \, k_{-1}}{\leftrightarrow} ES \qquad \text{(Dissociation constant } K_1 = k_{-1}/k_1)$$

$$ES + S \leftrightarrow ES_2 \qquad \text{(Dissociation constant } K_2)$$

$$ES \overset{k_p}{\to} E + P \qquad \text{(Slow Step)}$$

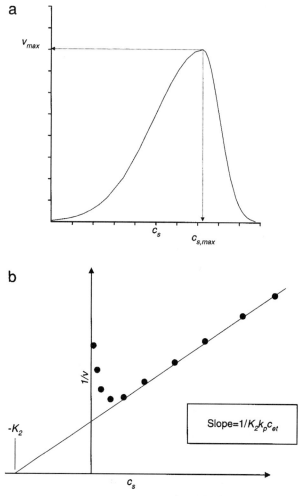

FIGURE 5.3 Substrate inhibition. (a) At substrate levels greater than $c_{s,\,max}$, an increasing substrate concentration results in lower reaction velocities. (b) Plot of $1/v$ versus c_s enables the determination of K_2 (see text).

On the basis of the analysis presented in Box 5.2, the velocity equation is

$$v = \frac{k_p c_{et}}{1 + K_1/c_s + c_s/K_2}$$

Experimentally, K_2 can be determined as indicated in Fig. 5.3 from the x-intercept of the linearized plot of $1/v$ versus c_s. Parameter K_1 can then be evaluated from the expression shown:

$$c_{S,\max} = \sqrt{K_1 K_2}$$

Note: $dv/dt = 0$ at $c_s = c_{s,\max}$.

Competitive Inhibition

A competitive inhibitor is a molecule that shares common characteristics with one of the substrates, thus allowing it to compete with that substrate for available enzymatic active sites. Reaction products, non metabolizable substrate analogues, derivatives of the substrate, or alternate substrates can act as competitive inhibitors. Malonic acid is a well-known competitive inhibitor of succinic dehydrogenase, which catalyzes the oxidation of succinate to fumarate (succinate + FAD $\Leftrightarrow$ fumarate + FADH$_2$). As shown, in Fig. 5.4, competitive inhibition of succinate dehydrogenase by malonate, glutarate, and oxalate is primarily due to structural similarity with the primary substrate succinate. The inhibition of the hexokinase-catalyzed reaction between glucose and ATP by fructose or mannose is another example of competitive inhibition by alternate substrates.

Because the binding of the inhibitor to the enzyme is reversible, an equilibrium is established between free and enzyme-bound inhibitor molecules according to the reaction scheme shown in Fig. 5.5. The rate

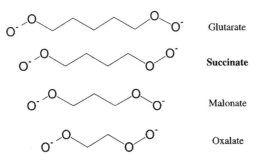

FIGURE 5.4   Structural similarities between substrate (succinate) and competitive inhibitors of succinate dehydrogenase.

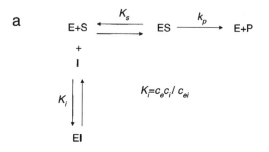

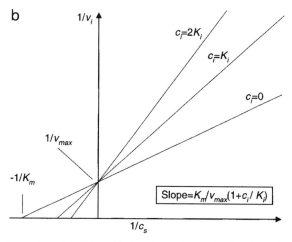

FIGURE 5.5 Competitive inhibition. (a) Reaction scheme. In the classical model S and I compete for the same binding site, and I must resemble S structurally. $K_i = c_e c_i / c_{ei}$, $K_s = c_e c_s / c_{es}$, and k_p is the rate constant for the breakdown of ES to E + P. (b) Reciprocal plots in the presence of different fixed concentrations $(0, K_i, 2K_i)$ of a competitive inhibitor.

equation for competitive inhibition can be derived from this reaction mechanism and is given by

$$\frac{1}{v} = \frac{K_m}{v_{max}}\left(1 + \frac{c_i}{K_i}\right)\frac{1}{c_s} + \frac{1}{v_{max}} \qquad (5.4)$$

A competitive inhibitor acts only to increase the apparent K_m for the substrate in a linear fashion ($K_{m,app} = K_m(1 + c_i/K_i)$), whereas the v_{max} remains unchanged. In this scheme the magnitude of the inhibition parameter K_i equals the inhibitor concentration that doubles the slope of the reciprocal plot and is *not* equivalent to the c_i that yields 50% inhibition. Because at low substrate concentrations relative to c_i the active sites are more likely to be occupied by the inhibitor and vice versa, reaction rates can be enhanced by raising c_s; the inhibition can be overcome with very high substrate concentrations as the probability of the enzyme encountering an inhibitor molecule becomes small under these conditions.

The principles of competitive inhibition have been exploited both in the medical field and in biochemical processes. For instance, drugs such as sulfanilamide are designed on the basis of their structural resemblance to key bacterial substrates. Sulfanilamide closely resembles the structure of *p*-aminobenzoic acid, a substance that bacteria utilize as a precursor for folic acid synthesis. Thus, patients with bacterial infections can be treated effectively by this compound that blocks folic acid synthesis by competitive inhibition of the corresponding bacterial enzyme, without any adverse effects on humans (who do not have a pathway for folic acid synthesis).

Noncompetitive Inhibition

This class of inhibitors also binds noncovalently and reversibly to enzymes. These inhibitors bind on cognate regulatory sites on the enzyme molecule (away from the active site) and trigger conformational changes that interfere with the enzyme's catalytic efficiency. Noncompetitive inhibitors have no effect on substrate binding, and, in contrast with their competitive counterpart, S and I bind randomly and independently at different sites of the enzyme. Such inhibition is common in steady state multireactant systems. A mechanism of noncompetitive inhibition is depicted in Fig. 5.6, which shows the reaction mechanism along with the corresponding reciprocal plots. As shown, I binds to E and to ES; S binds to E and to EI. The binding of one ligand has no effect on the dissociation constant of another, but the resulting ESI complex is inactive.

An expression relating v, v_{max}, c_s, K_m, c_i, and K_i in the presence of a noncompetitive inhibitor can be derived from rapid equilibrium assumptions:

$$\frac{1}{v} = \frac{K_m}{v_{max}}\left(1 + \frac{c_i}{K_i}\right)\frac{1}{c_s} + \frac{1}{v_{max}}\left(1 + \frac{c_i}{K_i}\right) \tag{5.5}$$

The preceding equation indicates that both the slope and the *y*-intercept of the usual $1/v$ versus $1/c_s$ plot are functions of c_i. Mechanistically, this is due to the dual effect of a noncompetitive inhibitor on both the substrate binding as well as the catalytic efficiency. Because both the *y*-intercept and slope increase by the same factor $(1 + c_i/K_i)$, the *x*-intercept remains unchanged and equals to $-1/K_m$ (Fig. 5.6). As predicted, the only effect of a

a

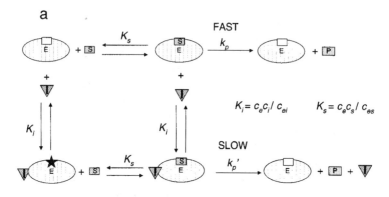

SLOW

$$K_i = c_e c_i / c_{ei} \qquad K_s = c_e c_s / c_{es}$$

★ Enzymatic Site Inactivated by Inhibitor

b

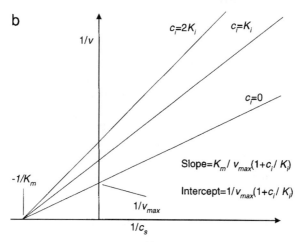

$$\text{Slope} = K_m / v_{max}(1 + c_i / K_i)$$

$$\text{Intercept} = 1/v_{max}(1 + c_i / K_i)$$

FIGURE 5.6 Noncompetitive inhibition. (a) The inhibitor and the substrate bind reversibly, randomly, and independently at different sites. (b) The $1/v$ versus $1/c_s$ plot in the presence of different fixed concentrations $(0, K_i, 2K_i)$ of a noncompetitive inhibitor. A classic noncompetitive inhibitor decreases v_{max} but has no effect on the K_m value.

noncompetitive inhibitor is to decrease v_{max} while K_m remains unchanged. It is important to note that k_p is also unchanged, and it is the steady state concentration of ES that is actually decreased as part of ES is converted to the unreactive ESI form.

Because noncompetitive inhibitors typically do not resemble any particular substrate, a single type of inhibitor can affect a wide spectrum of enzymes. For example, chelating agents, such as EDTA and cyanide that bind magnesium and iron, respectively, belong to this class of inhibitors. Because iron-containing proteins are important in redox reactions and magnesium ions are required for all reactions involving ATP, such inhibitors can have detrimental physiological effects.

Uncompetitive Inhibition

Uncompetitive inhibitors are molecules that reversibly bind to an enzyme-substrate complex, yielding an inactive ESI complex. This type of inhibition, common in multireactant systems, is depicted in Fig. 5.7, which yields, under the usual equilibrium assumptions, the following velocity equation:

$$\frac{1}{v} = \frac{K_m}{v_{max}} \frac{1}{c_s} + \frac{1}{v_{max}} \left(1 + \frac{c_i}{K_i} \right) \tag{5.6}$$

Uncompetitive inhibitors will lower both v_{max} and the apparent K_m by the same factor [$v_{max,i} = v_{max}/(1 + c_i/K_i)$, and $K_{m,app} = K_m/(1 + c_i/K_i)$]. The decrease in K_m results from the reaction $ES + I \Leftrightarrow ESI$, which reduces the amount of ES and causes the $E + S \Leftrightarrow ES$ equilibrium to shift to the right. The reaction equilibria and reciprocal plots characteristic of this type of inhibition are shown in Fig. 5.7. As c_i increases, the y-axis intercept increases, yielding a series of parallel curves.

5.1.3. IRREVERSIBLE INHIBITION

Irreversible inhibitors are molecules that bind covalently to enzymes and permanently disrupt their catalytic activity. A classic irreversible inhibitor is diisopropyl phosphofluoridate (DIPF), which inactivates enzymes by attaching to the hydroxyl group of serine moieties. In other cases, irreversible inhibitors are very selective, as in the case of the antibiotic penicillin, which selectively inhibits enzymes that participate in bacterial cell wall synthesis. A substance that irreversibly inactivates an enzyme can be misinterpreted as a noncompetitive inhibitor because v_{max} is reduced. Irreversible inhibition and reversible noncompetitive inhibition may be distinguished by plotting v_{max} versus c_{et} (total units of enzyme activity added to the assay). As indicated in Fig. 5.8 for a reversible noncompetitive inhibitor, the "plus inhibitor" curve will have a smaller slope than the control curve and both will pass through the origin. For an irreversible inhibitor, on the other hand, the "plus inhibitor" curve will have the same slope as the control curve, but will

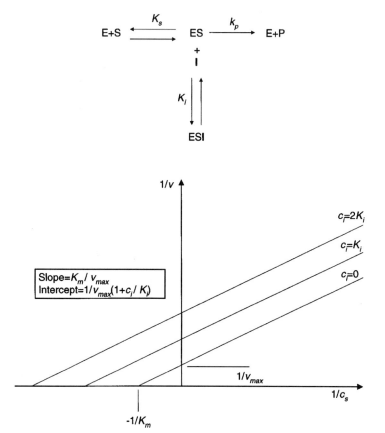

FIGURE 5.7 Uncompetitive inhibition. (a) Reaction scheme: The classical uncompetitive inhibitor reversibly binds to the enzyme-substrate complex, yielding an inactive ESI complex. (b) The reciprocal plot indicates that the slope is still K_m/v_{max} but the $1/v$-axis intercept is increased by the factor $(1 + c_i/K_i)$. Thus, increasing the inhibitor concentration yields a series of parallel curves, with the y-axis intercept increasing with increasing inhibitor concentration.

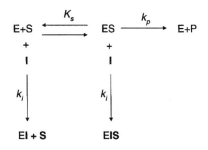

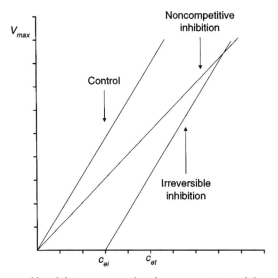

FIGURE 5.8   Irreversible inhibition compared with noncompetitive inhibition. c_{et} represents the amount of enzyme titrated by the irreversible inhibitor.

intersect the x-axis at a position equivalent to the amount of the enzyme that is irreversibly inactivated.

EXAMPLE 5.1

Biosynthetic Network Control by Enzyme Activity Regulation; Feedback Control Architecture in the Aspartate Amino Acid Pathways (Umbarger, 1978)

Members of the aspartate family of amino acids (lysine, methionine, threonine, and isoleucine) are produced by the branched pathway depicted in

Fig. 5.9. The regulation of this type of branched pathways is usually complex, because it requires an architecture where excess of one product does not accidentally shut off the entire pathway. The regulatory mechanisms for this particular pathway in E. coli, for example, are very complicated indeed as these also involve the participation of isozymes. As indicated here, each product inhibits and/or represses the first enzyme of the common pathway, aspartate kinase. However, so that one product does not shut down all of the kinase activity, three different isozymes (different enzymes that catalyze the same chemical reaction) of aspartate kinase are present: one sensitive to lysine inhibition, one sensitive to threonine inhibition, and one repressed mainly by excess methionine. In addition, E. coli is able to adjust the presence of an unbalanced mixture of pathway products. For example, excess methionine but low levels of the other products (lysine, isoleucine,

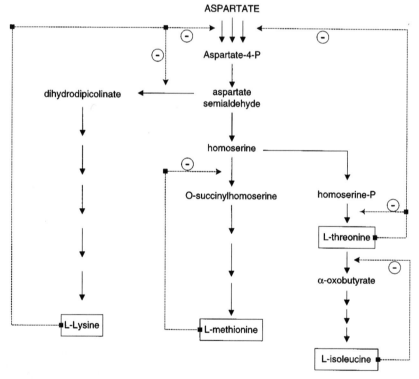

FIGURE 5.9 Regulation of the aspartate family of amino acids in enteric bacteria. L-Lysine inhibits aspartate kinase III and dihydropicolinate synthase; L-methionine inhibits O-suc-cinylhomoserine synthetase; L-threonine inhibits aspartate kinase I and homoserine dehydrogenase; and L-isoleucine inhibits threonine deaminase.

and threonine) cannot be corrected by regulating the aspartate kinase alone as it will not lead to the proper ratio of methionine to the other three amino acids. Thus, each amino acid usually controls the first enzyme of its own particular branch. Methionine and isoleucine regulate the pathway through both feedback inhibition and repression mechanisms, whereas lysine and threonine participate in more complex regulatory patterns.

Luckily for the purposes of metabolic engineering, numerous microorganisms have much simpler regulatory mechanisms, presumably because these organisms do not often encounter mixtures of amino acids of unbalanced composition in their environment. If an organism usually lives in an environment that is poor in amino acids, the major function of the regulation of amino acid biosynthesis is to adjust its rate in response to the growth rate of the microorganism. The major regulation of the branched pathway can then be achieved by a system in which only one or a few of the products inhibit the first common enzyme. This type of simple regulatory scheme indeed is found in the soil bacterium *Corynebacterium glutamicum* and other members of the "coryneform cluster" of amino acid-producing organisms. The simplicity of the regulatory schemes in the *Brevibacterium-Corynebacterium* group makes it possible to utilize these organisms for the overproduction of lysine by simply isolating either *auxotrophic or regulatory mutants*.

In such mutants, the activity of an enzyme-catalyzing a step in the biosynthesis of a metabolite has been disrupted. Auxotrophic mutants require supplementation of said metabolite in the medium for growth. For example, a homoserine dehydrogenase (HDH, the enzyme converting aspartate semialdehyde to homoserine) deficient mutant is homoserine or the threonine and methionine auxotroph. Such *C. glutamicum* auxotrophs overproduce lysine and none of the other amino acids for two reasons: (a) all aspartokinase flux is diverted to lysine formation because of the HDH disruption, and (b) the inhibition of aspartate kinase by a mixture of lysine and threonine at around lmM is close to 94%, whereas lysine alone only marginally inhibits this enzyme (12-20%) at around 1 mM. Because such mutants require precise supplementation of the growth medium with methionine and threonine, regulatory mutants are more useful for commercial production of lysine.

Mutant strains of *Brevibacterium*, altered in the regulation of lysine biosynthesis, typically have been obtained by using the lysine analogue S-aminoethylcystine (AEC). Like lysine, this analogue inhibits the activity of aspartate kinase and, hence, inhibits the growth of the wild-type bacteria. On the other hand, AEC-resistant mutants are likely to have an alteration in their aspartate kinase such that the altered allosteric regulatory site of the enzyme has a lower affinity for AEC and also for lysine. These mutant enzymes are likely to bind lysine with lower affinity and to be defective in the feedback regulation by lysine. Through a series of selection of regulatory mutants, commercial strains are now available that can overproduce lysine in excess of 60 g/l.

5.1.4. ALLOSTERIC ENZYMES: COOPERATIVITY

The previous sections described enzymes that possess multiple, but independent, substrate binding sites, i.e., the binding of one molecule has no effect on the intrinsic associations/dissociations of the vacant sites. A second fundamentally different class of regulatory enzymes includes those enzymes where binding of one molecule induces structural or electronic changes that result in altered affinities for the remaining vacant sites of the enzyme. Such interactions can be either homotropic, i.e., involving only the substrate (simple cooperativity), or heterotropic, where the interactions involve substrate and inhibitor or substrate and activator molecules. This group of enzymes is more commonly known as *allosteric* enzymes. Allosteric enzymes typically are composed of *multiple subunits* and are also characterized by *sigmoidal* velocity curves (Harford and Weitzman, 1975; Kurganov, 1982; Monod *et al.*, 1963; Sanwal, 1969, 1970; Stadtman, 1966).

There is a significant advantage to such kinetics compared to the hyperbolic response of enzymes that obey Michaelis-Menten (M-M) kinetics in controlling pathway flux. The sigmoidal shape of kinetic curves obtained for these enzymes indicates that, at low c_s, small changes in c_s cause the enzyme to become more efficient. Positive cooperativity causes an enzyme's activity to increase faster than normal with increasing c_s, whereas negative cooperativity causes activity to increase slower than normal with increasing c_s. Cooperativity is physiologically important, because it provides enzymes with varying degrees of sensitivity to fluctuations of the levels of either substrates or allosteric regulators. Thus, the sigmoidal response acts in essence as an "off-on switch" at low or high concentrations, respectively, whereas at intermediate concentrations the sigmoidal response provides a much more sensitive control of the reaction rate. Another interesting property of such regulation is the existence of a concentration threshold below which there is no catalytic activity. Such a threshold phenomenon is extremely important in cases of hormonal responses, neural signals, or cell differentiation.

The term "allosteric" was originally used by Monod, Changeux, and Jacob to classify enzymes that display altered kinetic characteristics (typically in K_m) in the presence of molecules that have *no structural resemblance to the substrate* (Monod *et al.*, 1963). It was later extended to also include enzymes that can exist in different conformations as a result of substrate or effector molecule binding.

Such a type of response is of vital importance in metabolic control. For example, in a cascade of enzymatic steps, the cell can regulate the flux through the pathway by on-off feedback inhibition of the first step by the end product. This ensures an economical way of avoiding the overproduction and accumulation of cellular building blocks. General important properties of allosteric enzymes are summarized:

- They are composed of polypeptide subunits assembled into multimeric complexes. Typical allosteric enzymes consist of 2, 4, 6, or more subunits. The subunits can differ, with one type having catalytic and the other regulatory function, or be identical, with both functions on the same subunit.

- Effectors and substrates are usually of very different chemical structure and bind the enzyme at distinct sites (regulatory versus catalytic sites).

- Allosteric effectors are thought to work by modifying the conformational state of the enzyme in such a way that its activity may increase or decrease. This implies that the enzyme can exist in different conformations, hence the name allosteric. The binding of the ligand induces the change, either by effecting a change from one conformational state of the enzyme to the other or by altering the equilibrium between the two conformations. In either case, the enzyme changes from a less active to a more active state, and thus the effector can modulate its activity.

- A clear characteristic of allosteric enzymes is the deviation of their kinetic properties from classic Michaelis-Menten kinetics: a plot of v versus c_s results in a sigmoidal curve instead of the M-M hyperbola. Due to the fact that most allosteric enzymes display sigmoidal kinetics, allosterism has become synonymous with sigmoidal responses. Strictly speaking, however, this is not true because not all sigmoidal curves result from allosteric interactions. For instance a random bireactant reaction system (two substrates or substrate plus effector) where there is a preferred but kinetic pathway to a ternary complex would yield a similar response.

Noncooperative Sites

As an introduction to allosterism, let us consider the case of a dimeric enzyme in which both sites are identical and independent as represented in

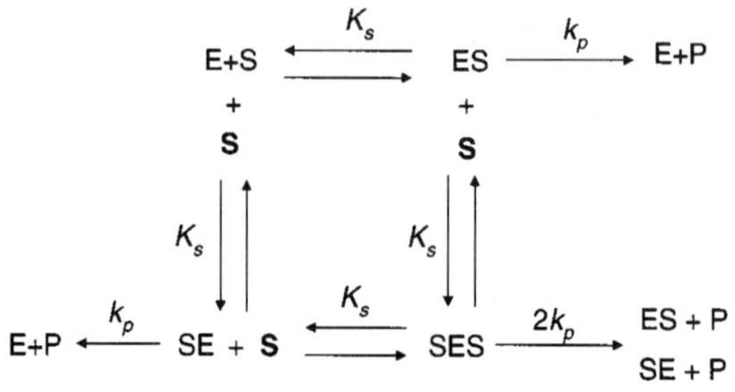

FIGURE 5.10 Substrate binding sequence for a noncooperative dimeric enzyme. Dimer (two-site) model in which both sites are identical and independent. K_s is called an *intrinsic* dissociation constant.

Fig. 5.10. For such a mechanism the rate equation is given by

$$\frac{v}{v_{max}} = \frac{\dfrac{c_s}{K_s} + \dfrac{c_s^2}{K_s^2}}{1 + \dfrac{2c_s}{K_s} + \dfrac{c_s^2}{K_s^2}} = \frac{\dfrac{c_s}{K_s}\left(1 + \dfrac{c_s}{K_s}\right)}{\left(1 + \dfrac{c_s}{K_s}\right)} \tag{5.7}$$

where: $K_s = \dfrac{c_s c_e}{c_{es}}$, $v_{max} = 2k_p c_{et}$.

In general, the velocity equation for an enzyme with n identical and independent sites is given by

$$\frac{v}{v_{max}} = \frac{\dfrac{c_s}{K_s}\left(1 + \dfrac{c_s}{k_s}\right)^{n-1}}{\left(1 + \dfrac{c_s}{K_s}\right)^{n}} \quad \text{or,} \quad \frac{v}{v_{max}} = \frac{c_s}{K_s + c_s} \tag{5.8}$$

It thus can be seen that, irrespective of the value n, the v versus c_s plots are always hyperbolic, and consequently it is impossible to determine n from kinetic data. In other words, it is not possible to distinguish between 1 mol of enzyme with n identical sites versus n mol of an enzyme with one site.

Cooperative Binding

The essence of allosteric interaction, on the other hand, is that the binding of one substrate molecule induces structural and/or electronic changes that affect the binding characteristics at vacant sites. In this type of enzyme, the binding of the first molecule facilitates the binding of additional molecules by increasing the affinities of vacant binding sites for substrate molecules. This phenomenon is also known as *positive cooperativity* or *positive homotropic response*, meaning that it involves one type of binding molecule, *i.e.*, substrate. Interactions involving pairs of distinct ligands (*e.g.*, substrate and activator, substrate and inhibitor, inhibitor and activator) are termed heterotropic responses and they can be either negative or positive.

Let us consider again an enzyme with two binding sites, but in this case the binding of the first molecule will influence the binding of the second. This is represented in Fig. 5.11, where the dissociation constant of the second vacant site changes to αK_s (where $\alpha < 1$ for positive cooperativity and $\alpha > 1$ for negative cooperativity).

This mechanism yields the following kinetic expression for the reaction rate. Unlike the noncooperative case, eq. (5.9) does not reduce to the simple M-M equation and exhibits the characteristic sigmoidal shape with respect to c_s.

$$\frac{v}{v_{max}} = \frac{\dfrac{c_s}{K_s} + \dfrac{c_s^2}{\acute{\alpha}K_s^2}}{1 + \dfrac{2c_s}{K_s} + \dfrac{c_s^2}{\acute{\alpha}K_s^2}} \tag{5.9}$$

where: $K_s = \dfrac{c_s c_e}{c_{es}}$, $v_{max} = 2k_p c_{et}$

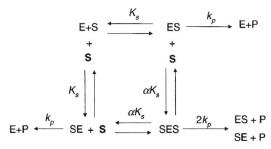

FIGURE 5.11 Substrate binding sequence for allosteric enzyme with dual binding sites. Binding at one substrate site (leading to ES) changes the intrinsic dissociation constant K_s by a factor α, where $\alpha < 1$.

For the more elaborate case of a tetramer, the rate equation becomes

$$\frac{v}{v_{max}} = \frac{\dfrac{c_s}{K_s} + \dfrac{3c_s^2}{\alpha K_s^2} + \dfrac{3c_s^3}{\alpha^2\beta K_s^3} + \dfrac{c_s^4}{\alpha^3\beta^2\gamma K_s^4}}{1 + \dfrac{4c_s}{K_s} + \dfrac{6c_s^2}{\alpha K_s^2} + \dfrac{4c_s^3}{\alpha^2\beta K_s^e} + \dfrac{c_s^4}{\alpha^3\beta^2\gamma K_s^4}} \qquad (5.10)$$

Note that in this model the effects are progressive and cumulative, i.e., the binding of the second molecule changes the dissociation constant of the remaining two sites by a factor β to $\alpha\beta K_s$, and the binding of the third molecule changes the dissociation of the last site by a factor γ to $\alpha\beta\gamma K_s$.

The Hill Equation: Simplified Velocity for Allosteric Enzymes

For the case of an enzyme exhibiting significant cooperativity the factors α, β, γ, etc. are very small, and consequently the velocity equation will be dominated by the $(c_s)^n$ term (i.e., the concentrations of all enzyme-substrate complexes containing less than n molecules of substrate will be negligible at any c_s that is higher than K_s). For example, the equation for the four-site enzyme in this case reduces to

$$\frac{v}{v_{max}} = \frac{\dfrac{c_s^4}{\alpha^3\beta^2\gamma K_s^4}}{1 + \dfrac{c_s^4}{\alpha^3\beta^2\gamma K_s^4}} = \frac{\dfrac{c_s^4}{K'}}{1 + \dfrac{c_s^4}{K'}} = \frac{c_s^4}{c_s^4 + K'} \qquad (5.11)$$

For an enzyme with n equivalent substrate binding sites, the following kinetic expression (Hill equation) has been proposed:

$$\frac{v}{v_{max}} = \frac{c_s^n}{K' + c_s^n} \qquad (5.12)$$

where: $K' = \alpha^{n-1}\beta^{n-2}\gamma^{n-3} \ldots K_s^n$. The constant K' in eq. (5.12) equals the substrate concentration raised to the power n (c_s^n), where $v = 0.5v_{max}$. It comprises the intrinsic dissociation constant K_S, as well as the interaction factors (< 1 indicates positive cooperativity, and > 1 indicates negative cooperativity).

The Hill equation [eq. (5.12)] can be rewritten in a more useful form:

$$\log\left(\frac{v}{v_{max} - v}\right) = n_{app}\log c_s - \log K' \qquad (5.13)$$

Thus, a plot of the left-hand side versus $\log(c_s)$ should produce a straight line with slope equal to n_{app}. If cooperativity is not high, then the calculate values of n will be less than the actual number of sites, with the next higher integer above the n_{app} value representing the minimum number of actual sites.

5.2. REGULATION OF ENZYME CONCENTRATION

In the previous section we discussed the various modes of modulation of enzymatic activity. These mechanisms are responsible for modifying the rate of enzyme molecules already present in a cell. They involve reversible or irreversible reactions that in essence affect the number of active enzyme molecules of the total available. As a result, the action of these mechanisms yields rapid responses with very small characteristic times. Of course, an alternate method by which the rate of an enzymatic reaction can be modified is by altering the *total* amount of enzyme. As enzymes are the products of a gene expression process their amount essentially can be controlled at two levels: DNA transcription and RNA translation. Most prokaryotic genes that are under regulation are controlled at the transcriptional level, which prevents the wasteful production of unnecessary mRNA.

5.2.1. CONTROL OF TRANSCRIPTION INITIATION

Transcription initiation takes place when RNA polymerase binds the promoter region upstream of the encoding gene (see Box 5.3). The most obvious place for regulating transcription would be at or around the promoter region of a gene. By controlling the rate of RNA polymerase binding to the promoter, the cell can regulate the amount of mRNA produced that ultimately determines the level of enzyme synthesized. The sequences adjacent to the actual coding region (structural gene) that are involved in this control are termed regulatory regions. These include, in addition to the promoter, an operator region where a regulatory molecule can bind. Regulatory proteins can either prevent transcription (negative control) or increase transcription (positive control) (Hoopes and McClure, 1987).

An operon is defined as several distinct genes arranged in tandem that are controlled by a common regulatory region. The mRNA produced from an operon contains information for *all* of the structural genes comprising the operon (polycistronic). This allows for coordinated regulation of all gene products that typically participate in a common biochemical pathway.

BOX 5.3

Bacterial Promoters

Promoters are segments of DNA, typically found upstream of certain genes, where RNA polymerase binds to initiate mRNA synthesis. The level of activity of bacterial promoters can be modulated in various ways. Positive regulation is one of the best studied mechanisms, and it involves the binding of an activator near the -35 region of the promoter. A well-studied activator is the cAMP binding protein (CAP), an allosteric protein that changes conformation upon binding cAMP. The activity of some protein activators can also be modulated by covalent modifications, as in the case of the NR_1 protein, which, upon phosphorylation, activates transcription from the Ntr (nitrogen-regulated) promoters. A second form of positive regulation involves the association of various sigma factors with the RNA polymerase holoenzyme. Promoters that utilize this type of regulation have characteristic sequences (usually in the -10 region), and modulation of transcription is thought to occur by changes in the cellular levels of the relevant sigma factors.

Mechanisms of down-regulating transcription initiation are also commonly found. A basic mechanism involves the binding of repressors to operators to decrease promoter activity. Binding regions for repressor molecules vary in position from upstream of the -35 region to downstream of the -10 region of the promoter. Some E. coli operons (gal, deo, and ara) have two, well-separated, repressor binding sites. Like activators, repressors are allosteric proteins that change conformation upon binding of their cognate ligand, whereas covalent modifications are less frequent. Other less well-understood types of negative regulation involve methylation of DNA in the promoter region and supercoiling. Autogenous regulation is another interesting mechanism whereby gene regulation takes place by the gene's own product. This pattern of control, following the theme of enzymatic feedback inhibition, provides a feedback mechanism by which repressor/activator can prevent its own overproduction. Representative examples include the genes crp (global regulation of carbon metabolism), araC (regulator of operons of L-arabinose metabolism), trpR (repressor of Trp biosynthetic operon), and glnG (regulator of ammonia assimilation). Figure 5.13 shows portions of selected promoters from E. coli and its phages, outlining common characteristics. Examination of more than 100 E. coli promoters has revealed the following frequency of occurrence for the Pribnow box (5-10 bases to the left of mRNA start):

$T_{80}A_{95}T_{45}A_{60}A_{50}T_{96}$. This region is thought to orient RNA polymerase, such that synthesis proceeds from left to right, and it is also the place where the double helix opens.

Further examination of the promoter region indicates that a significant number contain a second important section, to the left of the Pribnow box, that contains certain preserved features. This six-base sequence, which is called the -35 sequence and has the consensus TTGACA, is believed to be the initial site of binding of RNA polymerase. The difference between weak and strong promoters lies in the sequences of the -35 and -10 regions.

In eukaryotic cells, promoter sequences vary with the type of RNA polymerase involved. Promoters for protein-coding genes, which are transcribed by RNA polymerase II, usually contain the consensus sequence TATAAA located about 30 bases upstream from the transcription start site. This sequence, called the TATA box, is present in about 80% of the genes transcribed by RNA polymerase II. This sequence, which resembles the -10 prokaryotic region, plays a crucial role in positioning RNA polymerase II, and its removal abolishes transcription initiation. In addition to (or instead of) the TATA box, 10-15% of the genes transcribed by RNA polymerase II contain two other types of control sequences, typically located 60-120 bases upstream from the transcription start site: one has the consensus sequence CCAAT (CCAAT box) and the other GGGCG (GC box).

Individual Operons-The *lac* operon

One of the earliest and most fully characterized operons is the *lac* operon (Müller-Hill, 1996). The complete regulation of this operon is quite complex; however, in this section we provide a rather fundamental description of its features. This operon, carrying the structural genes for the utilization of lactose as a carbon source, is depicted in Fig. 5.12. Lactose, a disaccharide of glucose and galactose, is a central regulatory molecule of this operon.[1] The promoter and operator for the *lac* operon are *lacP* and *lacO*, respectively. The *lacI* gene codes for the repressor protein, which in the absence of lactose will bind the operator region, thus physically hindering the binding of RNA

---

[1] More precisely, the regulatory molecule is a lactose isomer, allolactose, that acts as an inducer. This species is formed *in situ* by β-galactosidase, which transforms lactose to allolactose. Thus, β-galactosidase, which is actually under the control of the *lac* operon, also has to be expressed constitutively at a basal level as a primer of the induction sequence.

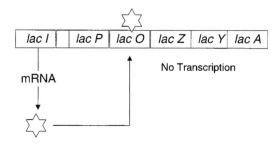

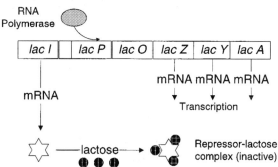

FIGURE 5.12 Transcriptional control of the *lac* operon. In the absence of lactose (top), the *lac* repressor binds to the *lac* operator DNA, blocking the transcription of the *lac* genes by RNA polymerase. When lactose is present (bottom), it binds to the *lac* repressor, forming an inactive complex that cannot bind to the *lac* operator. This allows RNA polymerase to bind the *lac* promoter region and transcribe mRNAs for the three enzymes of the *lac* operon: β-galactosidase (*lacZ*), lactose permease (*lacY*), and galactoside transacetylase (*lacA*).

polymerase at the promoter region. Induction of the *lac* operon occurs in the presence of lactose. Binding of lactose (inducer) at specific sites of the repressor molecule allosterically diminishes its DNA binding affinity with the *lac* operator. Therefore, RNA polymerase can carry out its function of transcribing the polycistronic mRNA that codes for β-galactosidase (*lacZ*), lactose permease (*lacY*), and galactoside transacetylase (*lacA*). As a typical operon, these gene products participate in related functions. The permease ensures the entrance of lactose into the cell, whereas β-galactosidase cleaves the β-1,4-linkage of lactose, releasing the free monosaccharides. Thus, the *lac* operon is a negatively controlled inducible system.

CGTATAA**T**GTGTGG<u>A</u>
GGTACGA**T**GTACCAC<u>A</u>
AGTAAGA**T**ACAAATC<u>G</u>
GTG ATAA**T**GGTTGC<u>A</u>
CTTATAA**T**GGTTAC<u>A</u>
CGTATGT**T**GTGTGG<u>A</u>

FIGURE 5.13   Segments of promoters found in *E. coli* and its phages. Common sequences are in bold letters (Pribnow box). The "conserve T" is underlined, and the start point for mRNA synthesis is indicated by double underlining.

The *lac* operon also has an additional, positive regulatory control system. The purpose of this is to prevent wasteful energy generation in the presence of excess glucose. Because the enzymes for glucose utilization in *E. coli* are constitutive, it would be pointless for the cell to produce enzymes for lactose metabolism when glucose and lactose are both present. When *E. coli*, for example, is grown on glucose, cAMP levels are low, but when grown on an alternate carbon source, cAMP levels tend to be high. It is believed that cAMP provides the positive regulating signal for the *lac* operon. At high levels (that indicate absence of glucose) cAMP binds a special protein (CAP, catabolite activator protein), and the CAP-cAMP complex in turn binds the CAP binding site of the *lac* promoter. This was found to promote helix destabilization farther downstream, which in turn facilitates the binding of RNA polymerase, thus increasing the efficiency of the promoter.

Once initiated, transcription should proceed until polymerase encounters a DNA stop sequence (GC-rich followed by series of U residues in RNA) or the paused polymerase is exposed to a transcription terminator molecule, such as the Rho factor. Although such events typically take place at the end of an operon, in certain instances they can occur prematurely, thus serving a regulatory function. The formation of a hairpin loop, for example, in the nascent mRNA causes the pausing of the RNA polymerase, which 80-90% of the time results in the ejection of the enzyme from the transcribed bubble, thereby terminating elongation. This mechanism of regulated transcription termination is called *attenuation*, and it operates by a special coupling between translation and transcription. A classic example for this type of regulation is the tryptophan operon of *E. coli*. When the supply of tryptophan (more precisely tryptophanyl-tRNA) is low, an early termination loop near the beginning of the *trp* transcript cannot form and *trp* mRNA is completely transcribed. Conversely, high levels of tryptophan trigger an attenuation mechanism that halts the synthesis of tryptophan biosynthetic genes.

5.2.2. CONTROL OF TRANSLATION

Genetic control of certain enzymes can also take place at the translational level in what is referred to as post-transcriptional regulation. This regulatory mechanism controls the number of times a completed mRNA molecule will be translated. The first system elucidated was the production of an enzyme in *E. coli* infected with phage R17. This phage encodes only three gene products-two structural proteins and the enzyme replicase. Because far more coat protein is needed, the mRNA molecule has a binding site for the coat protein between the termination codon of the coat protein gene and the AUG codon of the replicase gene. As the coat protein is synthesized, this binding site gradually is filled with protein molecules, blocking the ribosome from translating the replicase region. When the necessary coat particles for packaging a virus particle have been synthesized, they are released to form a new virus and replicase translation ensues. In general, translational regulation is achieved by regulating one of the following: (1) mRNA stability; (2) probability of translation initiation; or (3) overall rate of protein synthesis (Iserentant and Fiers, 1980; Stormo *et al.*, 1982).

Translational regulation is common in several of the operons that encode ribosomal proteins. The growth rates of all bacterial species vary with the composition of the growth medium. In minimal medium with an efficiently utilized carbon source such as glucose, *E. coli* cells divide roughly every 45 min at 37°C, compared with about every 500 min for a poorer carbon source such as proline. In rich medium containing glucose, amino acids, purines, pyrimidines, vitamins, and fatty acids, a cell does not have to synthesize these building blocks and hence it will grow very rapidly, with a generation time of less than 30 min. Because ribosomes have a limited capacity for protein synthesis (15 amino acids/s at 37°C), at different specific growth rates (thus, different rates of protein synthesis) the number of ribosomes per cell varies accordingly. For instance, *E. coli* growing at a doubling rate of 25 min contains ca. 70,000 ribosomes per cell versus 2000 ribosomes at a doubling time of 300 min. Furthermore "downshift" experiments (*i.e.*, transferring bacterial cells from rich to minimal medium) indicate that the ribosome content of each cell decreases accordingly by pausing the synthesis of rRNA.

A constant ratio of rRNA to ribosomes and r-proteins to ribosomes, at different growth rates, is regulated by two feedback mechanisms. In the first, ribosomes are made in slight excess and free nontranslating ribosomes inhibit the synthesis of rRNA (ribosome feedback regulation). In the second, certain r-proteins inhibit translation of mRNA that encodes the one or more r-proteins (translational repression). Studies indicate that translational repression is a result of binding of one particular r-protein to a base sequence near the ribosomal binding site. On the other hand, when translational repression is not occurring, each mRNA is completely translated, yielding equal numbers of each r-protein encoded in a particular mRNA species. This mechanism ensures that r-proteins are synthesized at the same rate and coupled to the synthesis of ribosomes.

Another important regulating scheme for rRNA is the so-called stringent response. If a growing culture is depleted for an amino acid, this results in the formation of uncharged tRNA, which in turn triggers the induction of the relA gene. The product of the relA gene is a protein known as the stringent factor (localized exclusively in the 50S ribosome), which is responsible for the production of ppGpp. Production of ppGpp signals the cessation of protein synthesis (due to amino acid deprivation), which in turn triggers the termination of rRNA synthesis.

EXAMPLE 5.2

Genetic Regulatory Network: Cholesterol Synthesis and Elimination (Hardie, 1992; Ku, 1996)

A classic example of a condition that involves altered gene expression is hypercholesterolemia, which is manifested as high cholesterol levels. Chronic diseases such as coronary vascular disease, obesity, and diabetes have metabolic components that affect the expression of specific genes in this pathway. Cholesterol enters the bloodstream principally as a very low density lipoprotein (VLDL) that is formed in the liver and small intestines. Although cholesterol synthesis requires several enzymes, 3-hydroxy-3-methylglutaryl-CoA (HMG-CoA) reductase is believed to be the controlling enzyme under physiological conditions. Among the drugs that lower cholesterol levels is a class of inhibitors that blocks HMG-CoA reductase. The pathway for cholesterol biosynthesis with its principal feedback mechanisms is depicted in Fig. 5.14.

Mevalonic acid exerts negative control over HMG-CoA reductase at two distinct levels. Mevalonate or a related metabolite decreases the rate of translation of mRNA, which encodes HMG-CoA reductase by up to 5-fold, and in the meantime it also increases the degradation rate of HMG-CoA reductase. Both effects result in a reduction of HMG-CoA reductase levels, thus protecting the cell against the accumulation of toxic levels of intermediate metabolites of this pathway. Elevated cholesterol levels were found to inhibit the transcription of the gene by up to 8-fold. These negative feedback

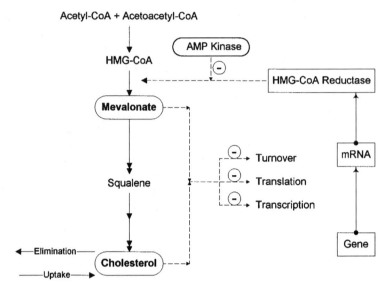

FIGURE 5.14   Pathway for cholesterol biosynthesis and its regulation.

loops produce a dynamic range of 200-fold for the concentration of HMG-CoA reductase. Further negative control on this pathway is exerted by inhibition due to the phosphorylation of HMG-CoA reductase by an AMP-dependent protein kinase.

5.3. GLOBAL CONTROL: REGULATION AT THE WHOLE CELL LEVEL

The elucidation of the *lac* operon as a detailed example of gene regulation some 30 years ago sparked intense interest in investigating additional operons. These efforts led to the discovery of many new operons, whose systematic study and analysis yielded the general realization that operons do not function in isolation, but rather they are members of higher level regulatory networks (Neidhardt *et al.*, 1990). This type of global regulation is an integral part of the capacity of living organisms to maintain the ability to grow in different and changing environments, such as shifts from rich to minimal medium, changes of carbon source, amino acid limitations, shifts between aerobic and anaerobic environments (facultative only), heat shock, and numerous starvations such as of phosphate, nitrogen, and carbon source.

Global control refers to the inherent ability of cells to utilize regulatory signals in order to control multiple aspects of cell physiology resulting in a coordinated response to any of the shifts mentioned in the preceding paragraph. Such regulatory networks include sets of operons and regulons with apparently unrelated functions and physical locations that are discontinuous on the chromosomal map. In a global regulatory network, sets of operons, scattered physically throughout the bacterial genome and sometimes representing disparate functions, are coordinately controlled. A global *regulon* is defined as a network of operons under the control of a common regulatory protein in a global regulatory network. However, most commonly, even a simple environmental change can induce several regulons. Therefore, a *stimulon* is used to refer to the entire sets of regulons responding to a single environmental stimulus. In order to emphasize the intricacy and overlapping nature of global regulatory networks, the term *modulon* was introduced and defined as a group of operons and/or regulons under different specific controls by sharing a *pleiotropic* (producing more than one effect) regulatory protein. It has become obvious in recent years that global regulation is a common theme in the prokaryotic world.

There are at least two reasons for the existence of such global regulatory networks (Fig. 5.15). First, certain cellular processes involve more genes than can be accommodated in a single operon. Take, for example, the bacterial translational machinery, a group of at least 150 gene products that includes rRNA, ribosomal proteins, initiation, elongation, and termination factors, aminoacyl-tRNA synthetases, and tRNAs. Coordinated regulation of these numerous interrelated components is important to the overall efficiency of cell growth, yet it would be almost impossible to incorporate all these genes into a single operon. The second reason is that, although some genes need to be regulated independently, it is also crucial to have an overriding coordinating control system. Examples of such cases are commonly found in genes encoding catabolic enzymes involved in the utilization of carbon and energy sources. Glucose is the premium substrate for most bacteria, and when present in the growth medium it overrides the expression of enzymes required for the metabolism of secondary or redundant substrates (catabolic repression). However, in the absence of the premium substrate, each operon must have the ability of independent induction in the presence of its cognate substrate. These examples illustrate the need for a regulatory organization that supersedes that of the single operon.

It is estimated that bacterial cells have evolved several hundreds of multigene systems, few of which have been studied in depth to date. Some examples from various bacteria are shown in Table 5.2, with the entries arranged in three broad categories for cellular responses to nutrient limitations, redox variations, or unfavorable environmental conditions. It is evia

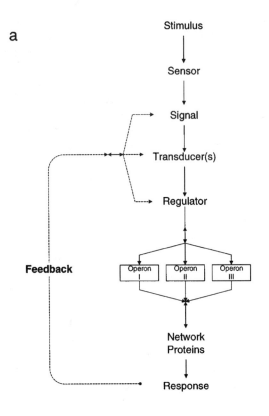

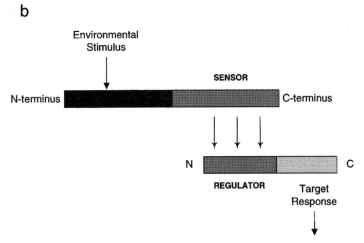

dent that bacteria have evolved diverse mechanisms of global regulation that are just beginning to be elucidated. It is not surprising that components of such networks are borrowed from operon regulation, such as protein repressors, activators, or sigma factors. Heat shock and sporulation systems utilize sigma factors to reprogram RNA polymerase so that it recognizes promoters of the member operons. SOS, oxidation damage, and anaerobic electron transport systems utilize protein repressors/activators that recognize a particular sequence common to controlling regions of constituent operons. On the other hand, one of the most pervasive and baffling operons, the stringent control system, has no apparent protein modulator, but instead uses the nucleotide guanosine tetraphosphate (ppGpp) as a global regulatory molecule.

A critical aspect of multigene systems is the ability to process information, a function typically carried out through protein-protein interactions. Genetic and biochemical analyses of the response of bacteria to environmental stimuli have revealed that signal transduction in response to such stimuli typically takes place through the communication between pairs of two classes of proteins. Each pair is composed of a sensor protein that can detect an environmental change and subsequently transmit a signal to its partner protein, the regulator(y) protein. In turn, the sensitized regulator will affect the transcription initiation of certain sets of operons, either positively or negatively. Significant homologies have been observed between members of each class of signal transducing proteins. In most cases, the conserved domain of the sensor protein extends to over 250 amino acids at the C-terminus, and for the regulator protein to about 120 amino acids at the N-terminus (Fig. 5.15b).

Signal Transduction by Phosphorylation

By the mid-1980s, it became clear that many multigene systems contained a pair of proteins, one from each of two families (Wanner, 1992). In a typical sensor-regulator two-component regulatory system, a transmembrane domain separates the sensor into the periplasmic domain that senses a specific

---

FIGURE 5.15 Global controls. (a) Signals generated in response to environmental conditions (e.g., temperature, pH, nutrient status) directly or indirectly (through one or more transducers) control the rate of synthesis of a regulatory molecule (regulator). In turn, the regulator controls the rate of protein synthesis of member operons whose function would promote growth or enable cellular survival. A feedback control mechanism allows the system to return to a new equilibrium state. (b) Signal transduction between sensor and regulator. Conserved domains are indicated by similar shading (cross).

TABLE 5.2 Bacterial Multigene Regulatory Systems[a]

Stimulus	System	Organism	Regulatory genes	Regulated genes
Nutrient limitations				
Carbon	Catabolic repression	Enteric bacteria	*crp*: encodes activator *cya*: adenylate cyclase	catabolic genes, including *lac*, *mal*, *gal*, *ara*, *tna*, *dsd*, *hut*, and many others
Amino acid or ATP	Stringent response	Enteric and other bacteria	*RelA* and *spoT*: encode enzymes of (p)ppGpp metabolism	Genes (> 200) for ribosomes, other translational proteins, and biosynthetic enzymes
Ammonia	Ntr system	Enteric bacteria	*GlnB*: protein P_{II} *glnD*: UR/UTase *glnG*: protein NR_I *glnL*: protein NR_{II}	*glnA*, *hut*, others
Ammonia	Nif system	*Klebsiella aerogens*, many other	Multiple genes including those controlling ammonia assimilation	Multiple genes encoding nitrogenase
Phosphate	Pho system	Enteric bacteria	*PhoB*: modulator of PhoB activity *phoU*: sensor of P transport	*phoA*, and others
Starvation	Sporulation	*B. subtilis*	*SpoOA*: activator *spoOF*: modulator	Many (> 100) genes for spore formation
Starvation or inhibition	Stationary phase	All bacteria	Unknown	Hundreds of genes
Energy metabolism				
Oxygen supply	Arc system (aerobic)	*E. coli*	*ArcA*: repressor *arcB*: modulator of ArcA activity	Many genes of aerobic metabolism
Electron acceptors other than O_2	Anaerobic respiration	*E. coli*	*fnr*: encodes activator	Genes for nitrate reductase and other enzymes of anaerobic respiration
Absence of electron acceptors	Fermentative metabolism	*E. coli* and other facultatives	Unknown	Many (> 20) genes for fermentative enzymes
Stress Response				
UV and other DNA damagers	SOS response	*E. coli* and others	*LexA*: repressor *recA*: modulator of LexA activity	About 20 genes for repair of UV-damaged DNA

(*continues*)

TABLE 5.2 (*continued*)

Stimulus	System	Organism	Regulatory genes	Regulated genes
Stress Response				
DNA alkylation	Ada system	*E. coli* and others	*Ada*: activator	Four genes for removal of alkylated bases from DNA
H_2O_2 or similar oxidants	Oxidation response	Enteric bacteria	*OxyR*: repressor	About 12 genes for oxidant protection
Shift to high T	Heat shock	*E. coli* and other bacteria	*HtpR* σ^{32}	About 20 genes involved in macromolecular synthesis, processing, and degradation
Shift to low T	Cold shock	*E. coli*	Unknown	Several genes for macromolecular synthesis
High osmolarity	Porin response	*E. coli* and others	*EnvZ*: sensor *ompR*: DNA binding protein	Genes for porins

[a] From Neidhardt, 1987; Neidhardt *et al.*, 1990.

environmental stimulus and subsequently induces autophosphorylation of its C-terminus (compare to Fig. 5.15b). In a second step, the phosphate group is then transferred to the N-terminus of the corresponding regulator. Members of the first family are protein kinases (PKs), which can become phosphorylated by transfer of phosphoryl groups from ATP to histidine residues in the kinase. In the second step, these phosphoryl groups are transferred to aspartate residues in the members of the second group, known as the phosphorylated response regulators (PRRs). In a final step, the phosphoryl group can be removed from the PRR-phosphoaspartate residues by hydrolysis. The phosphorylation process thus can be summarized as

$$ATP + PK\text{-}His \rightarrow ADP + PK\text{-}His\text{-}P$$

$$PK\text{-}His\text{-}P + PRR\text{-}Asp \rightarrow PK\text{-}His + PRR\text{-}Asp\text{-}P$$

$$PRR\text{-}Asp\text{-}P \rightarrow PRR\text{-}Asp + P_i$$

It is estimated that there may be as many as 50 of these transduction systems in a typical microbe. Signaling by phosphorylation has since been found to be widespread in the bacterial world, controlling diverse functions such as regulation of gene expression, chemotaxis, or developmental pathways (Table 5.3). Members of the histidine protein kinase family have conserved carboxy termini, whereas response regulators have a conserved domain of about 100 amino acids near the amino terminus. The significance of this conservation lies in the fact that in several cases it allows interaction (or cross-talk) between different systems. For example, in *phoR* mutants (*i.e.*, cells with a disrupted *phoR* gene and, hence, deficient in phoR activity) the product of *phoM* gene (apparently another PK) can partially substitute for the defective PhoR in modifying PhoB (RRP). As a result of this cross-talk, activation of one adaptive response may allow the cell to partially activate a homologous system by cross-phosphorylation.

Phosphorylation is the preferred mechanism for gene regulation in eukaryotes. Protein phosphatases present in the cytoplasm of mammalian cells have been classified into two types. Type 1 phosphatases are inhibited by nanomolar concentrations of two thermostable proteins, termed inhibitors 1 and 2, and dephosphorylate the β-subunit of phosphorylase kinase preferentially. Type 2 protein phosphatases, on the other hand, are unaffected by inhibitors 1 and 2 and preferentially dephosphorylate the α-subunit of phosphorylase kinase. Various protein kinases have been identified that phosphorylate enzymes that regulate glucogenolysis, glycogen synthesis, glycolysis, gluconeogenesis, aromatic amino acid breakdown, fatty acid synthesis, cholesterol synthesis, and protein synthesis in mammalian skeletal muscle and liver.

Although protein-protein interactions are very important in regulating networks, the ways in which groups of different operons are coordinated are quite diverse. For example, the heat shock response in E. coli is controlled by the cellular level of an RNA polymerase sigma factor, σ^{32}, whose elevation induces a set of about 20 genes. The SOS system that induces about 20 DNA repair proteins is induced by proteolytic cleavage of the repressor of the member operons. In a third mechanism, a group of 12 or so genes is induced by oxidative damage. Induction is mediated by activation of the positive regulator OxyR.

TABLE 5.3   Examples of Homologous Phosphotransferases[a]

System	Protein kinase	Response regulator
Nitrogen regulator (Ntr)	NR_{II}	NR_1
Phosphate regulator (Pho)	PhoR	PhoB
Aerobic respiration (Arc)	ArcB	ArcA
Porin regulator (Omp)	EnvZ	OmpR
Sporulation(Spo)	SpoIIJ	SpoOA/SpoOF

[a] From Neidhardt *et al.*, 1990.

EXAMPLE 5.3

Cross-Regulation: The Pho Regulon

Cross-regulation may be especially important in the control of pathways of central carbon metabolism. The best known example of cross-regulation is the phosphate (Pho) regulon of *E. coli* (Lee *et al.*, 1990; Shin and Seo, 1990; Wanner and Wilmes-Riesenberg, 1992). This regulon contains the gene *phoA* (bacterial alkaline phosphatase, BAP) and several other genes encoding enzymes for the degradation and uptake of phosphorus sources. Extracellular P_i is the preferred P source for growth and is taken up via the PstSCAB system. Intracellular P_i is then incorporated into ATP via one of several distinct routes in central pathways (Fig. 5.16).

Three of the controls that act on the Pho regulon are shown in Fig. 5.17, two of which involve cross-regulation. One control makes use of the sensor

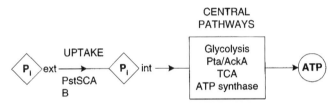

FIGURE 5.16 P_i uptake and assimilation into ATP. P_i is incorporated into ATP via glyceralde-hyde-3-P dehydrogenase and phosphoglycerate kinase in glycolysis via the phosphotransacety-lase and acetate kinase (Pta-AckA), via succinyl-CoA synthetase in the TCA cycle during aerobic growth, or via the F_1F_0-ATP synthase.

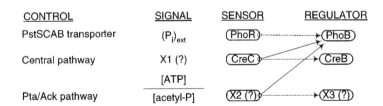

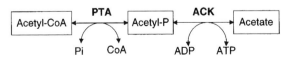

FIGURE 5.17 Controls on the Pho regulon. Dashed arrows indicate interactions between partner proteins, whereas solid arrows show cross-regulation among nonpartner proteins.

CreC, which is induced by glucose, and the second is induced by growth on pyruvate. The latter control probably detects acetyl-P and activates PhoB through an unknown sensor either directly by acetyl-P or through a phosphokinase. Thus, the Pho regulon control is associated with the Pta-AckA pathway via acetyl-P. Acetyl-P is made by Pta and further utilized by AckA during growth on glucose, with the converse being true for growth on acetate or pyrovate. It has been demonstrated that, in the presence of pyruvate, the precursor, or acetyl-CoA, the Pho regulon is induced. Furthermore, control via acetyl-P is abolished in *pta ack* deletion mutants.

5.4. REGULATION OF METABOLIC NETWORKS

It is the essence of regulation that reaction rates, as the controlled variable, respond to signals (typically, the concentrations of metabolites, effectors, or other regulatory molecules). An essential property of such control systems is that their actual output is determined solely by the signal and is, therefore, independent of the properties of individual components of the control system. In analogy to a valve controlling liquid flow, the capacity of a feedback-regulated enzymatic reaction is larger than the maximum possible flux, and flux itself is determined only by the signal. While the total amount of enzyme and signal determine the maximum and actual flux magnitudes, respectively, the dynamic properties of the control are a function of the particular regulatory mechanisms, as discussed in the following.

In a metabolic pathway comprising more than one enzyme, the pathway flux is determined by the kinetics and regulation of the individual enzymes in the pathway. For pathways in the central carbon metabolism (such as glycolysis and TCA cycle), although the enzymes are not physically organized, they are highly organized in a chemical framework. In certain cases, the pathway flux is controlled by the total amount of an enzyme present that is low enough to qualify as a rate-controlling step (see Chapter 11). In other cases pathway flux control is distributed among several enzymes, whereas other reactions in the pathway occur very rapidly and become limited by the substrate or cofactor concentrations. In such a metabolic framework, catalytic activity is not the only function of enzymes, as they also play a significant role in the regulation of the pathway flux as well. Although not necessarily in physical contact, each enzyme is informed of the state of the pathway by specific chemical signals (such as concentrations of substrate, product, or specific regulators, *e.g.*, ATP, NADH) that control the enzyme rate and, by extension, the pathway flux. In a functional cell, reaction rates must be modulated very precisely. It is thus easy to visualize allosteric enzymes as dynamic entities in which the structure of the enzyme is responsive to relatively small changes in the level of both substrate and allosteric activators or inhibitors. These structural changes affect the enzyme activity by altering the affinity of the enzyme for its substrate; this differs markedly from the inactivation of enzyme by covalent binding of inhibitors.

Identification of enzymes that catalyze nonequilibrium reactions may provide clues as to the regulatory control of the network. One of the difficulties in understanding metabolic control lies in the definition of a *regulatory* enzyme, as to some extent the activities of *all* the enzymes in a pathway need to be coordinated as the flux through the pathway changes. The crucial question then becomes whether an enzyme exists that first responds to the original metabolic signal (master enzyme) and thereby initiates secondary responses in the activities of the remaining enzymes (slave enzymes). A series of criteria developed to allow the identification of such regulatory enzymes is outlined:

- The enzyme should catalyze a nonequilibrium ("irreversible") reaction. A "nonequilibrium" enzyme possesses low catalytic activity, which thus can limit the flux through the pathway. On the other hand, an "equilibrium enzyme" is considered to be in excess, and fluxes should not respond significantly to its perturbation.
- It should possess regulatory properties, such as allosteric sites.

It is also possible that no single regulatory enzyme exists, but rather the activities of several enzymes in a pathway change in tandem in response to changing conditions and the ensuing metabolic signals.

It is important to establish the relationship between a pathway flux J and the amount (or activity) of an enzyme in a pathway, c_e. Although the exact relationship depends on many parameters, including the intrinsic kinetics of the pathway enzymes and the influence of products and allosteric modulators, a general hyperbolic relationship of the following type has been observed in the majority of cases:

$$J = \frac{J_{\max} c_e}{(c_e + K)} \tag{5.14}$$

where J is the observed flux, J_{max} is the flux obtained when infinite enzyme is present, c_e is assayable maximum *in vitro* activity for the specific enzyme, and K is a constant defined as the value of c_e when $J = J_{max}/2$ (similar to K_m). Thus, at small activities the pathway flux increases linearly with the

BOX 5.4

Regulation in Eukaryotes versus Prokaryotes

The regulatory systems of prokaryotes and eukaryotes are quite distinct, a reflection of their distinct "lifestyles." Although many functions carried out by prokaryotic and eukaryotic cells are quite similar, they differ with respect to several structural and genetic features. Prokaryotes are generally free-living unicellular organisms that grow and divide indefinitely under appropriate environmental and nutritional conditions. Thus, regulation of prokaryotic systems is geared toward maximizing growth while utilizing nutrients as efficiently as possible. Due to the lack of a nucleus, prokaryotic DNA is continuously exposed to regulatory signals from the cytoplasm; hence, the on-off control of protein synthesis is often implemented at the transcriptional level. Eukaryotes, on the other hand, are typically multicellular (with the exception of yeast, algae, and protozoa), larger, and structurally more complex (Towle, 1995). Cellular differentiation in particular requires a special type of regulation, as cells of different tissues have different requirements (Hofestadt *et al.*, 1996; Krauss and Quant, 1996). For instance, in an embryo a cell not only must produce progeny cells but must also undergo considerable morphological and biochemical changes that need to be maintained indefinitely. This type of permanent switching requires other regulatory strategies to be used in such cell types, such as gene loss, gene inactivation, gene amplification, and gene rearrangement.

The first clear evidence that mammalian metabolism could be regulated at the genetic level by nutritional substrates and hormones was provided by the discovery that tryptophan and adrenal glucocorticoid hormones regulated the activity of tryptophan dioxygenase in liver. This enzyme initiates the utilization of tryptophan when the amino acid is present in excess over protein synthesis requirements. This pathway is essential for tryptophan utilization in gluconeogenesis and for endogenous synthesis of nicotinamide-containing cofactors. In contrast to the rapid increase in lactose utilization by bacteria, studies with tryptophan dioxygenase demonstrated that steroids such as cortesol take 8-12 h to attain steady state enzyme and the corresponding mRNA levels. This is a general phenomenon: enzyme induction in bacteria reaches maximal levels within 30-40 min, whereas the process in mammals requires several hours and in some cases may require several days to complete.

This sharp distinction between unicellular and multicellular organisms, which represents an important aspect in the understanding of gene regulation, is further elaborated in Table 5.4. On the basis of the size of the two genes, the *lac* operon can be transcribed in under 2 min, whereas the tryptophan dioxygenase gene requires about 6 min for a round of transcription. This time difference is still small compared to the significant difference in the rate of transcription initiation, which thus becomes the rate-controlling step. Additionally, the translational efficiency of prokaryotes is about 3-fold greater than that of eukaryotes. These effects, in conjunction with the fact that the typical cell volume of a growing *E. coli* cell is about 1 μm^3 compared with 11,000 μm^3 for a typical liver cell, allows prokaryotes to achieve a high concentration of catalysts very rapidly. In contrast to the relatively long half-lives of mammalian mRNAs, which range from 20 min to 100 h, half-lives of bacterial mRNAs are typically 1-3 min. Most proteins are relatively stable in bacteria, but they become diluted within each cell as a result of cell division. Thus, the effective protein half-life in bacteria corresponds to the generation time, which lies in the range of 20-60 min. Overall, genetic control mechanisms are intrinsically more suitable for prokaryotes, allowing them to promptly adjust enzyme concentration to meet changing cellular demands.

As mentioned earlier, some fundamental differences exist between eukaryotic and prokaryotic systems. Some of the differences pertaining to genetics and regulation between the two systems are summarized here:

- Eukaryotic cells translate mRNA only to a single polypeptide chain, thus excluding the presence of operons common in prokaryotes.
- Unlike prokaryotic DNA, only a small fraction of eukaryotic DNA is bare: it is mostly bound to histones to form chromatin.
- A large fraction of DNA sequences of higher eukaryotes consists of untranslated regions (introns).
- Eukaryotes possess mechanisms for rearranging DNA segments in a controlled fashion, thus allowing them to increase the number of genes when required.

(continues)

(continued)

- Prokaryotic transcriptional regulatory sites are small, near to, and usually upstream from the promoter, and binding of proteins to such sites directly stimulates binding of RNA polymerase. In contrast, corresponding eukaryotic sites are much larger and may be hundreds of bases away from promoters. Although they bind proteins, due to special limitations they are unable to interact with the promoter simultaneously.
- In eukaryotes RNA is synthesized in the nucleus and must be transported through the nuclear membrane to the cytoplasm, where it is translated.

Overall, mRNA production in eukaryotes is much more complicated than in prokaryotes, and hence many points of control can be envisioned. For example, promoter availability could be determined by the state of the chromatin, the activity or concentration of transcription factors, or enhancer availability. Production of mRNA from primary transcripts could be regulated by control of splicing and polyadenylation. Regulation of mRNA lifetime is rarely encountered in prokaryotes because of the frequent need to switch the synthesis of regulated gene products on and off fairly quickly. On the other hand, eukaryotes that normally have more time to respond employ the strategy of extending the lifetime of mRNA when a large amount of a particular protein is needed.

availability of the enzyme, but eventually the pathway flux will be limited by other factors, such as substrate availability, the activity of another enzyme in the pathway, or the accumulation of an inhibiting product.

For controlling enzymes that are present in "basal" levels, small increases in enzyme concentrations are, thus, likely to increase linearly the flux through the pathway. However, an increase in enzyme concentration beyond a certain level is ineffective, and other factors need to be considered in order to optimize and coordinate fluxes. As a rule of thumb, it is not effective in most cases to increase transcriptional levels of a particular enzyme more than 10-fold. This is also important from a cellular economy perspective, as cellular resources are limited and there are competing demands among numerous pathways for the same pools of precursors. For example, if the requirement for a building block declines suddenly, it is economical to reduce both the enzyme level that produces that particular monomer as well

TABLE 5.4   Comparison of Transcriptional and Translational
Time Scales between *E. coli* and Mammalian Cells[a]

	E. coli	Mammals
Initiation	1 s^{-1}	10 min^{-1}
Transcription	2500 nt min^{-1}	3000 nt min^{-1}
RNA processing		ca. 10 min
Nuclear RNA $t_{1/2}$		ca. 10 min
Nucleocytoplasmic transport		ca. 10 min
mRNA $t_{1/2}$	1-3 min	1–20 h
mRNA translation	2700 nt min^{-1}	720 nt min^{-1}
Protein $t_{1/2}$	20-60 min	2–100 h

[a] nt = nucleotides.

as its activity. For cells grown in minimal medium, it can be shown using radiolabeled substrates that within seconds of the addition of an amino acid all carbon flow toward its biosynthetic pathway ceases. These issues of pathway flux control will be revisited in Chapter 11 in the context of metabolic control analysis.

5.4.1. BRANCH POINT CLASSIFICATION

Often, in biological processes, poor product yields are attributed to limitations caused by enzymatic levels or activities. Although overall activity of the enzymes in the product branch determines *productivities*, product *yields* are, however, functions of the flux split ratios at intermediate branch points. The latter, also referred to as nodes, are points in a metabolic network where a reaction sequence bifurcates between two or more different pathways. For example, the yield of product P_1 in Fig. 5.18 depends on the flux split ratio at the I node and not on the activity of the I to P_1 branch *per se*. Because the pathways that synthesize primary metabolites typically can support substantial fluxes, metabolic engineering efforts should be focused on altering flux partitioning at selected metabolic branch points to enhance product yields.

Although a network may consist of a large number of nodes, it is generally believed that the split ratios at relatively few nodes actually affect product yield. These nodes are referred to as *principal nodes*. Although the metabolic control structure used by different organisms to regulate flux varies markedly,

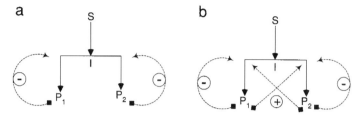

FIGURE 5.18   Metabolic nodes: (a) flexible or weakly rigid node; (b) strongly rigid node. Dashed lines indicate (negative or positive) feedback control mechanisms.

it is nevertheless possible to classify them into three general categories based on their branch point rigidity (Vallino, 1991).

- *Flexible nodes* (markedly subsensitive to regulation): Nodes at which flux partitioning will change readily in response to cellular metabolic requirements. Such nodes are characterized by competing enzymes that have similar substrate affinities and similar reaction velocities. By definition a flexible node will not limit product yield, as any branch can attain a split ratio of 100% whenever required.
- *Weakly rigid nodes* (moderately sensitive to regulation): Flux partitioning at these nodes is dominated by one branch. The dominating enzyme(s) is characterized by high specific activity and/or substrate affinity, together with a lack of feedback inhibition.
- *Strongly rigid nodes* (highly sensitive to regulation): Flux partitioning at these nodes is tightly controlled at one or more of its branches by a combination of feedback control and enzyme transactivation by metabolites from a competing branch.

In a flexible node, the flux split ratio will solely depend on the cellular demands for the two products P_1 and P_2 (Fig. 5.18a). A weakly rigid node is similar to a flexible node with the difference that one branch may exhibit significant dominance over the other(s) due to greater affinity of the corresponding enzyme for the common metabolite and/or greater overall activity over that of the subordinate branch. In this case, complete deregulation of the subordinate branch will not have a notable effect on the flux. On the other hand, if the activity of the dominant branch were to be attenuated, this would significantly increase the flux through the subordinate branch. A strongly rigid node is highly regulated by end product activation or inhibition, as shown in Fig. 5.18b. In this example, each product metabolite acts as an activator of the competing branch, as well as an inhibitor of its own formation. A rigid node can have a controlling effect on pathway product yield, and its deregulation can be more complex than enzymatic attenuation.

Let us assume that the desired goal is to increase the P_1 branch flux split ratio, and, furthermore, that the steady state concentration of P_1 is high enough to inhibit its own synthesis in the absence of the activator, P_2. In such a case, increasing the split ratio of the P_1 branch would require that the flux to P_2 be attenuated. However, in a strongly rigid node such an approach would result in lower levels of P_2, and consequently the P_1 branch will lose its activation by P_2. In short, attenuation of one branch results in the attenuation of the competing branch (overall nodal collapse), leaving the split ratios for each branch relatively unchanged.

Metabolic networks may not always exhibit the binary response to flux partitioning implied by the previous classification, as a result of the complex and diverse control architectures and enzyme kinetics (Srere, 1994). Thus, attenuation of one branch of a strongly rigid node may result only in partial attenuation of the competing branch, so that some improvement in product yield may be attained in a dependent network that harbors a rigid node. It is also possible that attenuating unwanted pathways may cause the excretion of intermediate metabolites instead of flux reduction or flux redirection.

Although other possible control architectures could be postulated that would produce a rigid node, the key aspect to any such control is the presence of a feedback mechanism between competing branches (cross-talk) that will have the tendency to maintain a constant flux split ratio. Not only can a rigid node limit product yield, but its effects are not easily mitigated via simple enzyme activity attenuation.

The glyoxylate shunt shown in Fig. 5.19 is a thoroughly investigated metabolic branch point. Isocitrate lyase (IL) is induced when cells are grown on acetate, and it serves an important role in supplying anaplerotic carbon

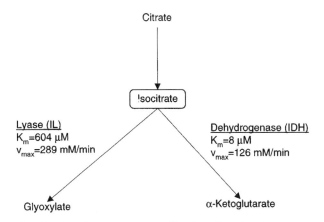

FIGURE 5.19   Kinetic parameters of isocitrate branch point enzymes.

(see Chapter 2) that would otherwise be lost as CO_2 in the TCA cycle. Under growth on acetate, isocitrate dehydrogenase (IDH) is primarily (ca. 80%) in a phosphorylated state that renders it inactive. When cells are presented with a more advantageous carbon source, such as glucose, there is rapid dephosphorylation of IDH, leading to its activation.

The kinetic properties of the two competing enzymes are summarized Fig. 5.19 for growth on acetate (glucose increases IDH v_{max} 5-fold). An important feature of this branch point is the large discrepancy in the respective K_m values: IL has a 75-fold larger K_m. At physiological concentrations of isocitrate (about 160 μM), IL is first-order with respect to isocitrate while IDH is saturated by the common substrate. As a result of the mutual disposition of the kinetics of these two enzymes, the flux through the shunt is ultrasensitive to changes in the level of isocitrate concentration. Introduction of glucose into an acetategrown culture induces both an increase in the v_{max} of IDH as well as a corresponding decrease in the carbon flow through citrate synthase, with an overall effect of lowering the intracellular isocitrate level (about 170-fold). Because the IL has a significantly higher K_m for isocitrate than IDH, the isocitrate concentration drop decreases the flux though the shunt to a great extent (ca. 99%). Thus, although there is no allosteric modulator known to act directly on IL, the flux through it can shift from 30% of the total down to virtually zero. This control has the interesting feature that the enzyme most affected, IL, is not subject to direct control, and this has been referred to as the "branch point effect."

5.4.2. COUPLED REACTIONS AND THE ROLE OF GLOBAL CURRENCY METABOLITES

Biosynthetic reactions in general rely on the supply of ATP, and in cases where this is low biosynthesis suffers. ATP and its related molecules, ADP and AMP, play a much more important role in the regulation of the fueling reactions of central carbon metabolism than just being mere participants in metabolic reactions. These central catabolic processes (glycolysis and TCA cycle) supply the starting materials for biosynthesis, as illustrated in Fig. 5.20 (see also Section 2.1). These pathways also serve an essential anabolic function by providing not only the precursor metabolites but also free energy and cofactors. Therefore, such pathways are termed *amphibolic* to indicate their dual roles in catabolism and anabolism (Sanwal, 1969, 1970). Regulation of these pathways must reflect this dual function, *i.e.*, to ensure correct and coordinated flow of carbon skeletons into biosynthetic pathways and into the energy-generating pathways. Strictly biosynthetic or catabolic path-

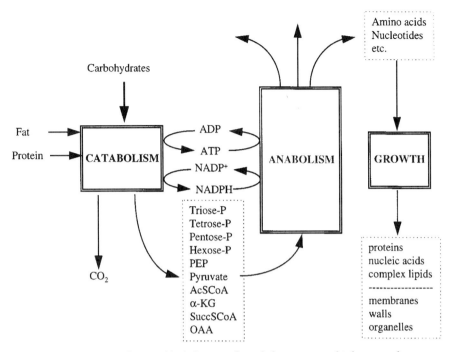

FIGURE 5.20   Schematic block diagram of metabolism in an aerobic heterotroph.

ways differ mainly in the nature of the *regulatory signal* utilized for flux control (Vaulont and Kahn, 1994).

In both cases feedback inhibition is exerted on the enzyme(s) catalyzing the first step(s) of the pathway. If glycolysis, or the TCA cycle, were regulated by the energy charge alone, the flux through these pathways would be severely decreased under conditions of plentiful energy supply. Along with a decline in ATP regeneration, however, such a feedback mechanism would also limit the availability of growth intermediates, thus impairing growth. Such a response would be suboptimal as growth would be limited when the supply of resources is plentiful.

In order to satisfactorily regulate a process that simultaneously serves two functions, a sophisticated mechanism has evolved that requires two types of inputs: one that is a measure of the level of biosynthetic intermediates and another indicative of the energy level (Anthony, 1988; Atkinson, 1970, 1977; Dietzler *et al.*, 1979; Nicholls, 1992). *Signal 1:* The first signal is a specific pathway end product and is typically found in strictly anabolic processes. Numerous examples are now available, showing that the regulation of most biosynthetic pathways involves a simple feedback mechanism in which the ultimate end product inhibits the first step involved in its synthesis, *e.g.*, threonine inhibits homoserine dehydrogenase, isoleucine inhibits threonine deaminase, histidine inhibits phosphoribosyl-ATP pyrophosphorylase, tryptophan inhibits anthranilase synthetase, etc. (see Example 5.1) *Signal 2:* The second signal is characteristic of strictly catabolic pathways and is the ultimate/intermediate product of energy metabolism (*i.e.*, P_i, AMP, ADP, ATP, NADH, NADPH), *e.g.*, biodegradative threonine deaminase of *E. aerogens* activated by AMP or biodegradative histidase of *P. aeruginosa* inhibited by P_i and inhibition relieved by AMP and GDP.

Amphibolic pathways use both types of flux control (*i.e.*, inhibition by end product and control of the activity by metabolites that are indicators of the energetic state of the cell). When either signal is low, glycolysis should proceed; when both are within a normal range, the functions of glycolysis are adequately served and the flux through this pathway is reduced. Of course glycolysis is never totally turned off in a cell that is metabolizing glucose, because ATP is required continuously. Balancing of the formation rates of the 12 precursor metabolites with respect to one another and also to the biosynthetic needs involves a highly sophisticated regulatory mechanism. The need for coordination goes beyond carbon-containing metabolites; fueling reactions must supply energy and reducing power (both NADH and NADPH), and it is not surprising that these compounds are themselves allosteric effectors at numerous points of the network.

$$\text{energy charge (EC)} = \frac{c_{ATP} + \dfrac{C_{ADP}}{2}}{c_{ATP} + c_{ADP} + c_{AMP}} \qquad (5.15)$$

The coordinated control of flux by the energy charge given by eq. (5.15) and the availability of biosynthetic precursors is depicted schematically in Fig. 5.21.

The energy charge that is determined by experimental measurements of the adenylate pool can vary between 0 (all AMP) and 1 (all ATP). In practice this value ranges from 0.87 to 0.95 for bacteria and does not change significantly with the specific growth rate. In general, high levels of energy charge typically inhibit energy-replenishing enzymes while they activate ATP-utilizing enzymes (such as biosynthetic enzymes) and vice versa.

An analogous situation exists for oxidation-reduction reactions that are regulated similarly by a "reduction charge." Pyridine nucleotides serve as carriers of H ions or reducing power, and whereas NAD functions primarily in redox reactions of fueling pathways, NADP is commonly used in biosynthetic reactions. In fueling reactions, *oxidized* NAD is the reactant whereas

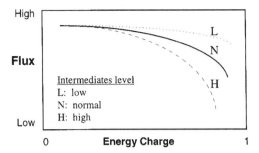

FIGURE 5.21 Interaction between energy charge and the concentration of metabolic intermediates in regulating the rates of amphibolic pathways.

biosynthetic reactions use the *reduced* form of NADP. It thus is not surprising that under physiological conditions most NAD is present intracellularly in the oxidized form and NADP is in the reduced form. Useful indexes of the oxidoreductive state are the catabolic reduction charge (ca. 0.03-0.07) and the anabolic reduction charge (ca. 10-fold higher):

$$\text{catabolic reduction charge (CRC)} = \frac{c_{NADH}}{c_{NADH} + c_{NAD}} \qquad (5.16)$$

$$\text{anabolic reduction charge (ARC)} = \frac{c_{NADPH}}{c_{NADPH} + c_{NADP}} \qquad (5.17)$$

Both ARC and CRC play an important role in coordinating fueling and biosynthetic reactions, although it is less clear than the role of the energy charge. In most cells four fueling reactions generate NADPH production (two in PPP, and one each in isocitrate dehydrogenase and malic enzyme) that can be used in biosynthesis. Adjustments of the CRC and ARC are believed to take place by the transhydrogenase enzyme that catalyzes the reversible reaction:

$$NADP + NADH_2 \Leftrightarrow NADPH_2 + NAD$$

This is a membrane-bound enzyme that is driven by protonmotive force; thus, it can only proceed at an energized membrane, and the equilibrium of this reaction is believed to be shifted by the protonmotive force (see Chapter 2).

REFERENCES

Anthony, C. (1988). *Bacterial Energy Transduction.* pp. 517. San Diego, CA: Academic Press.

Atkinson, D. E. (1970). Enzymes as control elements in metabolic regulation. In *The Enzymes*, pp. 461. Edited by P. Boyer. San Diego, CA: Academic Press.

Atkinson, D. E. (1977). *Cellular Energy Metabolism and Its Regulation.* New York: Academic Press, Inc.

de Koning, W. & van Dam, K. (1992). A method for the determination of changes of glycolytic metabolites in yeast on a subsecond time scale using extraction at neutral pH. *Analytical Biochemistry* **204**, 118-123.

Dietzler, D. N., Leckie, M. P., Lewis, J. W., Porter, S. E., Taxman, T. L. & Lais, C. J. (1979). Evidence for new factors in the coordinate regulation of energy metabolism in *Escherichia coli. Journal of Biological Chemistry* **254**, 8295-8307.

Dixon, M. & Webb, E. C. (1979). *Enzymes*, 3rd ed. New York: Academic Press.

Haldane, J. B. S. (1965). *Enzymes.* Cambridge, MA: MIT Press.

Hardie, D. G. (1992). Regulation of fatty acid and cholesterol metabolism by the AMP-activated protein kinase. *Biochimica Biophysica Acta* **1123**, 231-238.

Harford, S. & Weitzman, P. D. J. (1975). Evidence for isosteric and allosteric nucleotide inhibition of citrate synthase from multiple-inhibition studies. *Biochemical Journal* **151**, 455-458.

Heinrich, R. & Sonntag, I. (1982). Dynamics of non-linear biochemical systems and the evolutionary significance of time hierarchy. *Biosystems* **15**, 301-316.

Hiromi, K. (1979). *Kinetics of Fast Enzyme Reactions: Theory and Practice.* New York: John Wiley & Sons.

Hofestadt, R. M., Collado-Vides, J. & Loffler, M. (1996). Modeling and simulation of metabolic pathways, gene regulation and cell differentiation. *BioEssays* **18**, 33.

Hofmeyr, J.-H. S. (1991). Metabolite channelling and metabolic regulation. *Journal of Theoretical Biology.* **152**, 101.

Hoopes, B. C. & McClure, W. R. (1987). Strategies in regulation of transcription initiation. In *Escherichia coli and Salmonella typhimurium*, pp. 1231-1240. Edited by F. C. Neidhardt. Washington, DC: American Society for Microbiology.

Ingraham, J. (1987). Effect of temperature, pH, water activity, and pressure on growth. In *Escherichia coli and Salmonella typhimurium*, pp. 1543-1554. Edited by F. C. Neidhardt. Washington, DC: American Society for Microbiology.

Iserentant, D. & Fiers, W. (1980). Secondary structure of mRNA and efficiency of translation initiation. *Gene* **9**, 1-12.

Khoshland, D. E. J. (1970). The molecular basis for enzyme regulation. In *The Enzymes*, pp. 461. Edited by P. Boyer. San Diego, CA: Academic Press.

Khoshland, D. E. J. & Neet, K. E. (1968). The catalytic and regulatory properties of enzymes. *Annual Reviews of Biochemistry* **37**, 359.

Krauss, S. & Quant, P. A. (1996). Regulation and control in complex, dynamic metabolic systems: Experimental application of the top-down approaches of metabolic control analysis to fatty acid oxidation and ketogenesis. *Journal of Theoretical Biology* **182**, 381.

Ku, E. C. (1996). Regulation of fatty acid biosynthesis by intermediates of the cholesterol biosynthetic pathway. *Biochemical and Biophysical Research Communications* **225**, 173.

Kurganov, B. I. (1982). *Allosteric Enzymes: Kinetic Behaviour.* New York: John Wiley & Sons.

Laidler, K. J. & Bunting, P. S. (1973). *The Chemical Kinetics of Enzyme Action.* London: Oxford University Press.

Lee, T.-Y., Makino, K., Shinagawa, H. & Nakata, A. (1990). Overproduction of acetate kinase activates the phosphate regulon in the absence of the *pho*R and *pho*M function in *Escherichia coli*. *Journal of Bacteriology* **172**, 2245-2249.

Monod, J., Changeux, J.-P. & Jacob, F. (1963). Allosteric proteins and cellular control systems. *Journal of Molecular Biology* **6**, 306-329.

Müller-Hill, B. (1996). *The lac Operon: A Short History of a Genetic Paradigm*. Berlin: Walter de Gruyter & Co.

Neidhardt, F. C. (1987). Multigene systems and regulons. In *Escherichia coli and Salmonella typhimurium*, pp. 1313-1317. Edited by F. C. Neidhardt. Washington, DC: American Society for Microbiology.

Neidhardt, F. C., Ingraham, J. L. & Schaechter, M. (1990). *Physiology of the Bacterial Cell: A Molecular Approach*. Sunderland, MA: Sinauer Associates, Inc.

Nicholls, D. G. (1992). *Bioenergetics 2*, 2nd ed. San Diego: Academic Press Limited.

Rapoport, T. A., Heinrich, R., Jacobasch, G. & Rapoport, S. (1974). A linear steady-state treatment of enzymatic chains: A mathematical model of glycolysis of human erythrocytes. *European Journal of Biochemistry* **42**, 107-120.

Sanwal, B. D. (1969). Regulatory mechanisms involving nicotinamide adenine nucleotides as allosteric effectors. I. Control characteristics of malate dehydrogenase. *Journal of Biological Chemistry* **244**, 1831-1837.

Sanwal, B. D. (1970). Allosteric controls of amphibolic pathways in bacteria. *Bacteriological Reviews* **34**, 20-39.

Schulz, A. R. (1994). *Enzyme Kinetics: From Diastase to Multi-Enzyme Systems*. Cambridge: Cambridge University Press.

Schuster, R., Jacobasch, G. & Holshütter, H.-G. (1989). Mathematical modeling of metabolic pathways affected by an enzyme deficiency. Energy and redox metabolism of glucose-6-phosphate-dehydrogenase-deficient erythrocytes. *European Journal of Biochemistry* **182**, 605-612.

Segel, I. H. (1993). *Enzyme Kinetics. Behavior and Analysis of Rapid Equilibrium and Steady-State Enzyme Systems*. New York: John Wiley & Sons, Inc.

Shin, P. K. & Seo, J.-H. (1990). Analysis of *Escherichia coli* phoA-lacZ fusion gene expression inserted into a multicopy plasmid and host cell's chromosomes. *Biotechnology and Bioengineering* **36**, 1097-1104.

Srere, P. (1994). Complexities of metabolic regulation. *Trends in Biochemical Sciences*. **19**, 519.

Stadtman, E. R. (1966). Allosteric regulation of enzyme activity. In *Advances in Enzymology and Related Subjects of Biochemistry*, pp. 41-154. Edited by F. F. Nord. New York: John Wiley & Sons.

Stormo, G. D., Schneider, T. D. & Gold, L. M. (1982). Characterization of translational initiation sites in *E. coli*. *Nucleic Acids Research* **10**, 2971-2996.

Towle, H. C. (1995). Metabolic regulation of gene transcription in mammals. *Journal of Biological Chemistry* **270**, 23235.

Umbarger, H. E. (1978). Amino acid biosynthesis and its regulation. *Annual Reviews of Biochemistry* **47**, 533-606.

Vallino, J. J. (1991). Identification of branch-point restrictions in microbial metabolism through metabolic flux analysis and local network perturbations. In *Chemical Engineering*, pp. 394. Cambridge, MA: Massachusetts Institute of Technology.

Van Dam, K., Jansen, N., Postma, P., Richard, P., Ruijter, G., Rutgers, M., Smits, H. P., Teusink, B., van der Vlag, J. & Walsh, M. (1993). Control and regulation of metabolic fluxes in microbes by substrates and enzymes. *Antonie van Leeuwenhoek* **63**, 315-322.

Vaulont, S. & Kahn, A. (1994). Transcriptional control of metabolic regulation genes by carbohydrates. *FASEB Journal* **8**, 28.

Walter, C. (1965). *Steady State Applications in Enzyme Kinetics*. The Ronald Press.

Wanner, B. L. (1992). Is cross regulation by phosphorylation of two-component response regulator proteins important in bacteria? *Journal of Bacteriology* 174, 2053-2058.

Wanner, B. L. & Wilmes-Riesenberg, M. R. (1992). Involvement of phosphotransacetylase, acetate kinase, and acetyl phosphate synthesis in control of the phosphate regulon in *Escherichia coli*. *Journal of Bacteriology* 174, 2124-2130.

Webb, J. L. (1963). *Enzyme and Metabolic Inhibitors*. San Diego, CA: Academic Press.

CHAPTER 6

Examples of Pathway Manipulations: Metabolic Engineering in Practice

Nature has provided a remarkable array of metabolic pathways as witnessed by the diversity of extant microorganisms. In certain cases, the assembly and kinetic coordination of such pathways in a particular organism are suitable for a useful commercial application. In most cases, however, genetic improvements are required for the optimization of the conversion reactions and kinetic properties of a cell to render it suitable for practical use. These improvements are guided by the current understanding of microbial metabolism and molecular genetics and implemented by molecular biological techniques and recombinant DNA technology. The rational transfer of conversion pathways has produced new and desirable functionalities in cells, thus benefiting the pharmaceutical, agricultural, food, chemical, and environmental sectors.

In this chapter we review applications of metabolic pathway manipulation. We follow the classification of Cameron and Tong (1993), slightly modified, in organizing the large number of examples in basically five groups, *i.e.*, applications aiming at (a) yield and productivity improvement of products made by microorganisms, (b) expansion of the range of substances that can be metabolized by an organism, (c) formation of new and novel products, (d) general improvement of cellular properties, and (e) xenobiotic degradation. We have three goals in reviewing these applications in the context of this book. First, we provide a sample of the truly enormous range of possibilities for biocatalyst improvement afforded by pathway manipulation and metabolic engineering. We note that this review is focused, almost exclusively, on industrial applications. Except for the introduction (Chapter 1) and a few passing references, very little is included on the broad applications of metabolic engineering in the medical field. The reason is, simply, that most of the work in this field is current and not sufficiently crystallized for our review purposes. Reported progress, however, leaves little doubt about the impact of the tools and methodologies of metabolic engineering in analyzing tissue and organs *in vivo* and *in vitro*, as well as providing fundamentals for the rational analysis of the organization and function of signal transduction pathways.

The second goal of this review is to provide the reader with a sense of the complexity of metabolic pathways, along with their regulation and coordination with the overall metabolism. An important corollary of the admittedly complex structure of these pathways is that the methods for their systematic analysis may not always be as simple as one might desire. This point will become clearer as we delve into the methods of metabolic flux determination (Chapters 8 and 9) and issues of control and flux amplification in complex metabolic networks (Chapters 11 and 12). In the same vein, representative assays of the state of cell metabolism, metabolic production pathways, or signal transduction pathways may require a *multidimensional array of measurements*, in contrast to current practice. The recognition of truly distinguishing patterns in large volumes of measurements will be a challenge, and methods of flux analysis could provide a useful framework to this end.

The third goal of this review is to underscore the methods used for effecting desired changes in cellular systems for industrial use or medical reasons. It will become apparent that most successful applications require coordinated modification of more than one enzymatic step in a metabolic network. This is almost necessary for those applications that extend beyond a simple product-forming pathway and involve the complex structures of central carbon metabolism. We believe that this will become a generic requirement as research focus gradually shifts from the simplicity of highly reduced model systems to the realm of realistic industrial or medical situations. As with complex measurement interpretation mentioned earlier, metabolic engineering can have an impact in the rational design of metabolic pathway modifications.

6.1. ENHANCEMENT OF PRODUCT YIELD AND PRODUCTIVITY

A large number of mainly industrial applications can be classified in this group. We note that, although often interchangeable in a loose sense, yield and productivity represent different figures of merit that also require different strategies for their enhancement. Yield impacts primarily the cost of raw materials and is affected by *redirection of metabolic fluxes* toward the formation of the desired product. Productivity, on the other hand, is the key determinant of the capital cost of bioprocessing equipment and can be improved by *amplification of metabolic fluxes*. Admittedly, overall process optimization must include both yield and productivity concerns, although in certain cases decoupling of the two may be possible. Productivity depends, first and foremost, on the specific rate of substrate uptake, which for most industrial organisms ranges between 0.2 and 0.5 of substrate per gram of biomass per hour. If such uptake rates are realized under process conditions, then productivity can be economically acceptable provided that byproduct formation is minimized. Under these conditions, yield and productivity optimization methods may indeed converge. If, on the other hand, uptake rates are too low, then productivity optimization should begin with the amplification of the substrate transport system, followed by flux redirection as dictated by strategies of yield optimization.

Yield and productivities obviously are more important in large volume, low cost industrial operations. We next review efforts aimed at the improvement of yield and productivity of ethanol, amino acids, and solvents by metabolic engineering.

6.1.1. ETHANOL

Ethanol is an important industrial chemical with emerging potential as a biofuel to replace vanishing fossil fuels. Additionally, it may have significant environmental impact as ethanol combustion is less polluting, and it may serve as feedstock for the production of oxygenated fuels. Also, because its production is mainly based on agricultural products, it will enhance the "carbon cycle" for atmospheric CO_2 removal. According to current estimates, the United States will be importing more than 50% of its crude oil and refined products to meet energy needs in the year 2010. From an economic point of view, ethanol (and other biofuels), by providing domestic resources to meet part of this demand, can play a major role in stabilizing energy prices, improving national energy security, and ensuring rural and regional economic development.

Ethanol can be made from a number of renewable feedstocks, including sugar crops such as sugar-cane, starch-containing grains such as corn, or lignocellulosic materials including agricultural residues, herbaceous crops, and wood. The economics of the ethanol process is determined by the cost of sugar. Almost all of the U.S. fuel ethanol production of 2.3 billion liters was made from corn, and it is estimated that an additional 20 billion liters of ethanol per year could be made with surplus corn. Over the past decade, the cost of ethanol has dropped from more than $1.0 L^{-1} to approximately $0.3-0.5 L^{-1}, with a projected cost of less than $0.25 L^{-1} in the near future.

Lignocellulosic materials are such an abundant and inexpensive resource that existing supplies could support the sustainable production of liquid transportation fuels on the same scale as the total US consumption. The National Renewable Energy Laboratory (NREL) has estimated the current cost of ethanol production from lignocellulose to be about $0.32 L^{-1}, assuming a feedstock cost of $42 per dry ton (National Renewable Energy Laboratory, 1996). This average biomass cost amounts to approximately $0.06 kg^{-1} of sugar or a contribution to the feedstock costs for ethanol production of as low as $0.10 L^{-1}. Lignocellulosic crops considered to be suitable raw materials for fuel ethanol production are fast-growing wood, agricultural and forestry residues and various kinds of wastes, e.g., pulping waste, newsprint, and municipal solid waste. Efficient utilization of the hemicellulose component of lignocellulosic feedstocks (25% dry weight of hardwood and predominantly D-xylose) offers an opportunity to reduce the cost of producing fuel ethanol by 25% (Bull, 1990). Whereas lignocellulose is inexpensive because it cannot be digested and therefore does not compete as a food, its inability to be digested also makes it difficult to convert to fermentable sugars. Furthermore, lignocellulose is a complex structure with three major components (cellulose, hemicellulose, and lignin), each of which must be processed separately to make the best use of the high efficiencies inherent in biological processes. A general process schematic for the conversion of lignocellulose to ethanol is shown in Fig. 6.1. The hydrolysate, resulting after prehydrolysis and hydrolysis, contains varying amounts of monosaccharides, in the form of both pentoses (D-xylose and L-arabinose) and hexoses (Table 6.1) and a broad range of substances either derived from the raw material or formed as reaction byproducts from the pretreatment stage of the process (sugar and lignin degradation products). Xylose is the most abundant sugar in the hemicellulose of hardwoods and crop residues, whereas mannose is more abundant in the hemicellulose of softwoods. Furthermore, xylose is second only to glucose in natural abundance. Microbial conversion of the sugar residues present in wastepaper and yard trash from U.S. landfills alone could

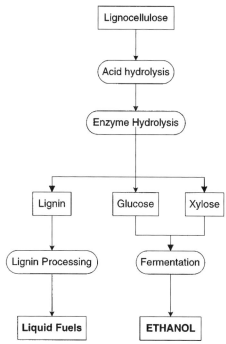

FIGURE 6.1   Conversion of lignocellulose to ethanol. Crystalline cellulose, the largest (50%) and most difficult fraction, is hydrolyzed by a combination of acid and enzymatic processes. During these steps 95-98% of the xylose and glucose is recovered. These monosaccharides subsequently are converted to ethanol by appropriate microorganisms.

TABLE 6.1   Carbohydrate Structural Polymers in Lignocellulose for a Typical Softwood, Such as Common Beech

Polymer	Monomer(s)	Typical % Total
Cellulose	Glucose	40
Hemicellulose	Xylose	30
	Arabinose	
	Mannose	
	Glucose	
	Galactose	
Lignin	Phenylpropane	25
Pectin	Uronic acids	5

provide more than 400 billion liters of ethanol (Lynd *et al.*, 1991), 10 times the corn-derived ethanol burned annually as a 10% blend with gasoline (Keim and Venkatasubramanian, 1989).

The fermentation organism must be able to ferment all monosaccharides present and, in addition, withstand potential inhibitors in the hydrolysate. The most commonly used ethanol producer, *Saccharomyces cerevisiae*, cannot *ferment pentoses*, which may constitute 8-28% of the raw material (Ladisch *et al.*, 1983). Yeasts produce ethanol efficiently from hexoses by the pyruvate decarboxylase-alcohol dehydrogenase (PDC-ADH) system. However, during xylose fermentation the byproduct xylitol accumulates, thereby reducing the yield of ethanol. Furthermore, yeasts are reported to ferment L-arabinose only very weakly. The efficient fermentation of xylose and other hemicellulose constituents may prove essential for the development of an economically viable process to produce ethanol from biomass.

Pentose-fermenting microorganisms are found among bacteria, yeasts, and fungi, with the yeasts *Pichia stipitis*, *Candida shehatae*, and *Pachysolen tannophilus* being the most promising naturally occurring microorganisms. Only a handful of bacterial species are known that do possess the important PDC-ADH pathway to ethanol. Among these, *Zymomonas mobilis* has the most active PDC-ADH system; however, it is incapable of dissimilating pentose sugars. During recent years, the application of metabolic engineering resulted in recombinant bacteria (Alterhum and Ingram, 1989; Feldmann *et al.*, 1989; Ingram *et al.*, 1987; Ohta *et al.*, 1991a,b; Tolan and Finn, 1987a,b) and yeasts (Hallborn *et al.*, 1991; Kötter *et al.*, 1990; Tantirungkij *et al.*, 1993) as competent ethanol producers. A recent study that compared the performance of various ethanol producers, natural and recombinant, in pentose-rich corn cob hydrolysate concluded that the recombinant ethanologenic *Escherichia coli* KO11 (*E. coli* carrying *Z. mobilis pdc* and *adhB* integrated on the chromosome) is currently the best fermentation organism (Hahn-Hägerdal *et al.*, 1993).

Initial studies were only partially successful in redirecting fermentative metabolism in *Erwinia chrysanthemi* (Tolan and Finn, 1987a,b), *Klebsiella planticola* (Tolan and Finn, 1987a,b) and *E. coli* (Brau and Sahm, 1986). The first generation of recombinant organisms amplified the PDC activity only and depended on endogenous levels of ADH activity to couple the further reduction of acetaldehyde to the oxidation of NADH (see Fig. 6.2). Because ethanol is just one of a number of fermentation products normally produced by these enteric bacteria, a deficiency in ADH activity together with NADH accumulation contributed to the formation of various unwanted byproducts. This problem was solved by amplifying the ADH activity through overexpression of the *Z. mobilis adhB* gene yielding recombinants of *E. coli* (Ingram *et al.*, 1987) and *K. oxytoca* (Ohta *et al.*, 1991a,b; Wood and Ingram, 1992)

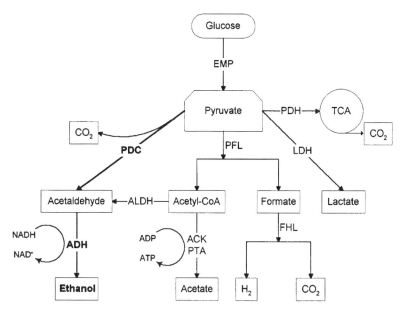

FIGURE 6.2 Competing pathways at the pyruvate branch point. Abbreviations: EMP, Emb-den-Meyerhof-Parnas enzymes and intermediates; PDC, pyruvate decarboxylase; ADH, alcohol dehydrogenase; PFL, pyruvate formate-lyase; ACK/PTA, phosphotransacetylase and acetate kinase; ALDH, acetaldehyde dehydrogenase; FHL, formate hydrogen lyase; LDH, lactate dehydrogenase; PDH, pyruvate dehydrogenase.

that efficiently ferment a variety of sugars to ethanol. This was accomplished by assembling both *Z. mobilis* genes (*i.e.*, *pdc* and *adhB*) into an artificial operon to produce a portable genetic element for ethanol production (PET operon). *E. coli* is an advantageous host organism, especially because it can grow efficiently on a wide range of carbon substrates that *includes five-carbon sugars.*

Pyruvate decarboxylase catalyzes the nonoxidative decarboxylation of pyruvate to produce acetaldehyde and carbon dioxide (Fig. 6.2). Two alcohol dehydrogenase isozymes are present in *E. coli* that catalyze the reduction of acetaldehyde to ethanol during fermentation accompanied by the oxidation of NADH to NAD^+. In the recombinant *E. coli*, both enzymes [pyruvate decarboxylase, (PDC), and alcohol dehydrogenase (ADH)], required to divert pyruvate metabolism to ethanol are present at high levels. The combined effect of high PDC levels and low apparent K_m (Table 6.2) of this enzyme for pyruvate effectively is to divert carbon flow to ethanol even in the presence of native fermentation enzymes like lactate dehydrogenase.

TABLE 6.2   Comparison of Apparent K_m Values for Pyruvate for Selected
E. coli and Z. mobilis Pyruvate-Acting Enzymes[a]

Organism	Enzyme	K_m Pyruvate	K_m NADH
E. coli	PDH	0.4 mM	0.18 mM
	LDH	7.2 mM	> 0.5 mM
	PFL	2.0 mM	
	ALDH		50 μM
	NADH-OX		50 μM
Z. mobilis	PDC	0.4 mM	
	ADH II		12 μM

[a] Abbreviations: PDH, pyruvate dehydrogenase; LDH, lactate dehydrogenase; PFL, pyruvate formate lyase; ALDH, aldehyde dehydrogenase; NADH-OX, NADH oxidase; PDC, pyruvate decarboxylase; ADH II, alcohol dehydrogenase II.

Significant amounts of ethanol were produced in recombinant E. coli containing the pet operon under both aerobic and anaerobic conditions (Table 6.3). Under aerobic conditions, wild-type E. coli metabolizes pyruvate through PDH and PFL (K_m 0.4 and 2.0 mM, respectively, Table 6.3), with main products CO_2 and acetate (formed by the conversion of excess acetyl-CoA). The apparent K_m for the Z. mobilis PDC is similar to that of PDH and lower than those of PFL and LDH, thereby facilitating acetaldehyde production. NAD^+ regeneration under aerobic conditions primarily results from biosynthesis and from the NADH oxidase (coupled to the electron transport system). Again, because the apparent K_m for Z. mobilis ADH II is over 4-fold lower than that for E. coli NADH oxidase, the heterologous ADH II effectively competes for endogenous pools of NADH, allowing the reduction of

TABLE 6.3   Comparison of Fermentation Products during Aerobic and Anaerobic
Growth of Wild-Type and Recombinant E. coli[a]

Growth	Plasmid	Fermentation Product (mM) Ethanol	Lactate	Acetate	Succinate
Aerobic					
	None	0	0.6	55	0.2
	PLO1308-10 (PET)	337	1.1	17	4.9
Anaerobic					
	None	0.4	22	7	0.9
	PLO1308-10 (PET)	482	10	1.2	5.0

[a] From Ingram and Conway, 1988.

acetaldehyde to ethanol. Under anaerobic conditions, wild-type *E. coli* metabolizes pyruvate primarily via LDH and PFL. As indicated again in Table 6.2, the apparent K_m values for these two enzymes are 18-fold and 5-fold higher, respectively, than that for *Z. mobilis* PDC. Furthermore, the apparent K_m values for primary native enzymes involved in NAD^+ regeneration are also considerably higher in *E. coli* than those of *Z. mobilis* ADH. Overall, overexpressed ethanologenic *Z. mobilis* enzymes in *E. coli* are quite competitive with respect to the native enzymes in channeling carbon (pyruvate) and reducing power (NADH) into ethanol.

In March 1991, the University of Florida was awarded U.S. Patent No. 5,000,000 for the ingenious microbe created at its Institute of Food and Agricultural Sciences. The fermentation characteristics of the recombinant *E. coli* strain have been reported in numerous studies. Typical final ethanol concentrations are in excess of 50 g L^{-1} [*e.g.*, 54.4. and 41.6 g L^{-1} were obtained from 10% glucose and 8% xylose, respectively (Ohta *et al.*, 1991a,b)] at nearmaximum theoretical yields of 0.5 g of ethanol/g of sugar (sugar → 2ethanol + $2CO_2$). Published volumetric and specific ethanol productivities with xylose in simple batch fermentations are 0.6 g of ethanol (L h)$^{-1}$ and 1.3 g of ethanol (g DW h)$^{-1}$, respectively (Alterhum and Ingram, 1989). Further improvements have resulted in volumetric productivities of as high as 1.8 g of ethanol (L h)$^{-1}$ (Ohta *et al.*, 1991). The production cost of ethanol from pentoses (*e.g.*, willow or pine) using *E. coli* KO11 is estimated at around $0.13 L^{-1} (von Sivers and Zacchi, 1995; von Sivers *et al.*, 1994), which can easily bring the final cost of ethanol well below the $0.18 L^{-1} target for the year 2000. In addition, this ethanologenic *E. coli* also has the ability to ferment - besides xylose - all other sugar constituents of lignocellulosic material: glucose, mannose, arabinose, and galactose. When the recombinant strain was grown on mixtures of sugars typically present in hemicellulose hydrolysates, sequential utilization was observed with glucose consumed first, followed by arabinose and xylose, to produce near-maximum theoretical yields of ethanol (Takahashi *et al.*, 1994).

Recently, Ohta *et al.* investigated the expression of the *pdc* and *adh* genes of *Z. mobilis* in a related enteric bacterium, *Klebsiella oxytoca* (Ohta *et al.*, 1991a,b). In *Klebsiella* strains, two additional fermentation pathways are present compared with *E. coli* (Fig. 6.2), converting pyruvate to succinate and butanediol. As in the case of *E. coli*, it was possible to divert more than 90% of the carbon flow from sugar catabolism away from the native fermentative pathways and toward ethanol. Overexpression of recombinant PDC alone produced only about twice the ethanol level of the parental strain. However, when both PDC and ADH were elevated in *K. oxytoca* M5A1, ethanol production was both very rapid and efficient: volumetric productivities > 2.0 g $(L\ h)^{-1}$, yields 0.5 g of ethanol/g of sugar, and final ethanol of 45 g L^{-1} for both glucose and xylose carbon sources were obtained.

6.1.2. AMINO ACIDS

Amino acids have a wide spectrum of commercial use as food additives, feed supplements, infusion compounds, therapeutic agents, and precursors for the synthesis of peptides or agrochemicals. Most microorganisms have the metabolic machinery to synthesize all essential amino acids from carbon and nitrogen sources (Fig. 6.3). It is also possible that certain microorganisms can overproduce one or a group of amino acids. In the mid-1950s, for example, Japanese scientists isolated a novel bacterium that excreted large quantities of L-glutamate, giving rise to a new era of amino acid production by fermentation (Kinoshita et al., 1957). Before that the main sources of amino acids had been natural proteins and, to a lesser extent, chemical synthesis. This bacterium, later known as Corynebacterium glutamicum, is a Gram-positive, short aerobic rod capable of excreting very large amounts of glutamate into the medium, close to 100 g L^{-1} under certain conditions.

The success in the industrial production of glutamic acid stimulated further interest in isolating producing strains for other amino acids. Wild-type strains of glutamic acid bacteria are capable of producing only a few amino acids extracellularly, such as glutamic acid, valine, proline, glutamine, and alanine. For the extracellular accumulation of a desired amino acid, changes in the cellular metabolism and/or regulatory controls are required. For many years following the discovery of these bacteria, attempts were made to induce auxotrophic and regulatory mutants (see Example 5.1). The rationale of utilizing auxotrophic mutants is to bypass feedback control (see Chapter 5) by minimizing the intracellular accumulation of feedback inhibitors or repressors or by modifying the inhibitor binding site, thus rendering the enzyme insensitive to the presence of the inhibitor. For example, an ornithine producer was isolated by using an arginine auxotrophic mutant followed, a year later, by homoserine auxotrophic mutants for successful lysine fermentation (Kinoshita et al., 1957). Most amino acids are produced today by use of strains that contain combinations of auxotrophic and regulatory mutations. More than 500,000 tons of L-glutamate are produced annually with C. glutamicum, whereas while its auxotrophic mutant is responsible for about 400,000 tons of L-lysine per year. The demand for amino acids is constantly increasing.

In light of the commercial importance of the aspartate family of amino acids, particularly lysine, intense strain improvement programs were carried out to isolate strains with superior properties. These programs initially

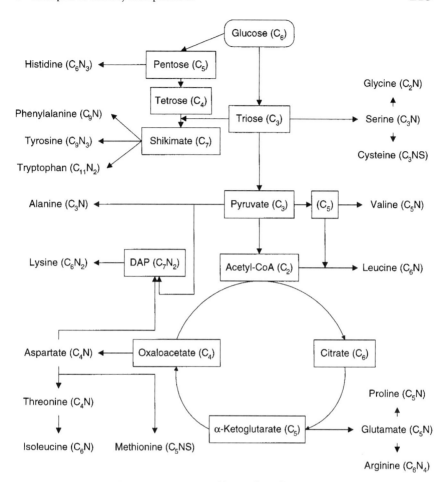

FIGURE 6.3   Amino acid biosynthesis from glucose.

employed random mutation and auxotrophic selection procedures utilizing the wellunderstood metabolic pathways of amino acid biosynthesis and regulation. Wild-type *Corynebacteria* do not accumulate lysine due to concerted feedback inhibition of aspartokinase by threonine and lysine (see also Section 10.1.1 and Fig. 10.1). Thus, the first lysine-producing mutant was a homoserine dehydrogenase (HDH) deficient and, hence, homoserine-auxotrophic strain; incapable of homoserine synthesis and, therefore, requiring homoserine or threonine and methionine supplementation in the medium. The latter is provided at a controlled rate so as to satisfy protein synthesis needs and allow lysine to accumulate to high concentrations at the same time. Further improvement involved strains resistant to S-(2-aminoethyl)-L-cysteine (AEC), a lysine analogue. Because AEC resembles lysine, it elicits similar inhibitory effects, such as inhibition of aspartokinase and arrest of lysine synthesis. AEC-resistant strains apparently involve deregulated aspartokinases that are not inhibited by lysine and, as such, can accumulate large amounts of lysine in the medium. Subsequent efforts focused on central carbon metabolism in an attempt to divert increased amounts of carbon away from the respiratory and into the anaplerotic pathways. For this purpose, mutants with citrate-synthase attenuated activity were isolated and found to offer further improvement in lysine yield, especially in combination with the previous two phenotypes of HDH deficiency and AEC resistance. This theme has since been applied with many variations, leading to fluoropyruvate-sensitive (i.e., pyruvate dehydrogenase-attenuated) mutants, alanine auxotrophs, and many others. The exact nature of the mechanism(s) responsible for any claimed improvements in these strains is not known due to the poor characterization of random mutations and also the incomplete understanding of the function and regulation of the anaplerotic pathways in C. glutamicum.

In recent years, even more potent producing strains have been obtained by further pathway manipulation, e.g. by eliminating the ability of the production strain to degrade the product or by improving cell permeability to favor excretion of the final product. Several research groups have independently initiated research programs focusing on the development of metabolic engineering tools for Corynebacterium species. Essential prerequisites are the availability of vectors derived from endogeneous plasmids and efficient DNA transfer systems. Small, cryptic plasmids were isolated in various Corynebacterium strains, and new and efficient transformation techniques have been developed in the past few years. This facilitated the isolation of amino acid biosynthetic genes from Corynebacteria, which currently number around 50 or so [see the review by Jetten and Sinskey (1995)]. These genes, either individually or in combination, can be utilized to improve production strains by raising the activities of enzymes or by removing the feedback regulation of critical enzymes. Use of these genes also allows for specific probes to be utilized in order to elucidate the biochemistry of amino acid synthesis and central carbon metabolism (CCM) in general.

Tryptophan

Tryptophan synthesis in E. coli is highly regulated by a complex set of feedback mechanisms. By transducing each mutation one at a time, researchers combined a long list of alterations to these mechanisms within a single strain, thus creating a tryptophan overproducer (Aiba et al., 1980; Shio, 1986). The first step of the aromatic pathway, the conversion of erythrose-4-P and PEP to 3-deoxy-D-arabino-heptulosonate-7-P (DAHP), is catalyzed by three isofunctional enzymes (AroF, AroG, and AroH) regulated by tyrosine, tryptophan, and phenylalanine, respectively. One of the initial approaches was to simplify the regulation of the system by deleting *aroG* and *aroH*. Furthermore, the tyrosine-regulated enzyme (AroF) was rendered insensitive to feedback inhibition by mutation (*aroF394*), and the repression of this gene was removed by inactivating the repressor gene (*tyrR*). Other modifications included the removal of branches leading to tyrosine and phenylalanine (*tyrA* and *pheA*), inactivation of the gene for tryptophanase (*tna*) to prevent the possible degradation of the synthesized tryptophan, alleviation of the feedback inhibition of the tryptophan branch by making anthranilate synthetase insensitive to tryptophan (*trpE382*), inactivation of the tryptophan repressor (*trpR*); and destruction of the cell's attenuation control by mutating the gene for tryptophanyl-tRNA synthetase (*trpS*). The industrial *E. coli* strain (NST100) produces about 6.2 g L^{-1} tryptophan when cultured in a medium containing 5% glucose for 24 h. Higher tryptophan yields are possible with the addition of anthranilate to the cultivation medium.

Recently, a *C. glutamicum* strain able to produce 18 g L^{-1} tryptophan has been altered to produce large amounts of tyrosine (26 g L^{-1}) by overexpressing deregulated 3-deoxy-D-arabino-heptulosonate-7-phosphate (DAHP) synthase and chorismate mutase (Ikeda and Katsumata, 1992). Overexpression of an additional gene in the previous construct, prephenate dehydratase, led to the predominant production of phenylalanine (28 g L^{-1}).

Alanine

Uhlenbusch *et al.* were able to construct a *Z. mobilis* alanine overproducer by introducing the gene for alanine dehydrogenase (*alaD*) from *Bacillus sphaericus* (Uhlenbusch *et al.*, 1991). Alanine yield reached 10 mmol per 280 mmol of glucose, which was later increased to 41 mmol by the addition of 85 mM ammonia that was apparently limiting before. At this production rate growth ceased, presumably due to the strong competition for pyruvate between pyruvate decarboxylase (PDC) and alanine dehydrogenase. Starvation for the PDC cofactor thiamine-PP resulted in further growth inhibition and higher alanine yields (84 mmol in 25 h).

Threonine

Whereas lysine and methionine can be manufactured economically for use as feed additives, the demand for threonine cannot yet be filled due to the low yields of existing processes. Recently, however, significant progress was

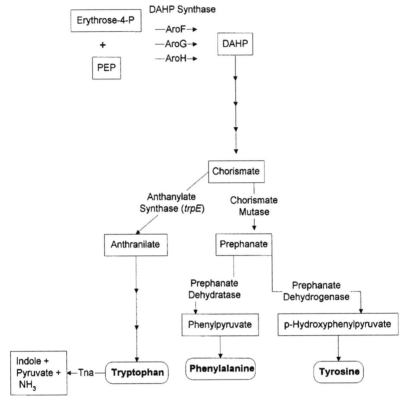

FIGURE 6.4 Aromatic acid biosynthesis. In *E. coli*, the structural genes form an operon (*trpEDCBA*) under a common operator. The regulator gene, which is situated away from this operon, allows for feedback inhibition of enzyme formation (repression) by the end product tryptophan.

made in the efforts for the construction of efficient threonine-producing strains by metabolic engineering. The *C. glutamicum* genes encoding the first two enzymes in the threonine pathway, homoserine dehydrogenase (HD) and homoserine kinase (HK), were isolated by complementation of the *E. coli thrB* mutant (Follettie *et al.*, 1988). These two genes form an operon that is expressed from a single promoter upstream of the *hom* gene (Peoples *et al.*, 1988), which is regulated at the transcriptional level by methionine via a unique attenuation system (Jetten *et al.*, 1993). The final step in the threonine pathway involves the conversion of homoserine phosphate to threonine by the constitutive enzyme threonine synthase (TS). The gene encoding TS (*thrC*) was obtained recently by complementation of a *C. glutamicum* auxotroph (Han *et al.*, 1990).

Threonine production is determined by the distribution of metabolic fluxes at the common substrate aspartate-β-semialdehyde (ASA) to the divergent threonine and lysine biosynthetic pathways (Fig. 6.5). This flux distribution is controlled by the relative affinities of the competing enzymes, homoserine dehydrogenase and dihydropicolinate synthase, for the common

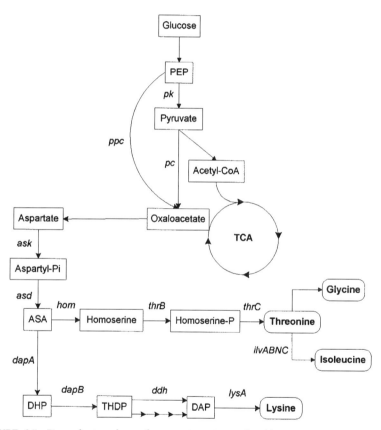

FIGURE 6.5   Biosynthetic pathways for aspartate amino acids. Abbreviations: *ppc*, PEP carboxylase; *pc*, pyruvate carboxylase; *pk*, pyruvate kinase; *ask*, aspartokinase; *asd*, ASA dehydrogenase; *dapA*, DHP synthase; *dapB*, DHP reductase; *ddh*, DAP dehydrogenase; *lysA*, DAP decarboxylase; *hom*, homoserine dehydrogenase; *thrB*, homoserine kinase; *thrC*, threonine synthase; *ilvA*, threonine dehydratase (deaminase); *ilvBN*, acetohydroxy acid synthase, acetohydroxy acid isomeroreductase, dihydroxy acid dehydratase; *ilvC*, transaminase C.

substrate ASA. The activity of homoserine dehydrogenase is highly sensitive to allosteric inhibition by L-threonine (K_i = 0.16 mM), and, therefore, under nominal growth conditions, ASA predominantly enters the lysine pathway. Critical to the understanding of the molecular basis of threonine inhibition of homoserine dehydrogenase, as well as to the construction of threonine overproducers, has been the isolation of feedback-resistant HD (HD^{dr}). Two groups have independently isolated and characterized genes encoding deregulated, feedback-resistant homoserine dehydrogenase (Archer et al., 1991; Reinscheid et al., 1991). Archer et al. (1991) determined that the hom^{dr} mutation is due to a single base deletion that radically alters the structure of the carboxy terminus, leading to 10 amino acid changes and deletion of the last 7 residues relative to the wild type. These residues apparently are part of the threonine binding site, and their removal from the HD^{dr} mutant renders the HD activity insensitive to feedback inhibition by threonine.

In order to study the regulation of carbon fluxes around the ASA node, well-defined recombinant strains were constructed (Colón et al., 1993; Eikmanns et al., 1991; Reinscheid et al., 1994). Amplification of the threonine genes in wild-type C. glutamicum 13032 did not yield any threonine secretion, presumably due to the feedback inhibition of aspartokinase and homoserine dehydrogenase by threonine (Eikmanns et al., 1991). Amplification of the deregulated hom^{dr} alone (plasmid pJD4, Table 6.4) yielded an approximately equal flux split between the lysine and threonine pathways, along with intracellular accumulation of threonine (100 mM) and homoserine (74 mM), which led to the conclusion that threonine production is probably limited by either its efflux and/or a possible lack of balance between the activities of HD and HK. In order to prevent an intracellular buildup of homoserine, Colón et al. fused the thrB gene to the tac promoter (plasmid pGC42, Table 6.4) and regulated hom^{dr}/Ptac-thrB expression by

TABLE 6.4   Threonine Production by C. glutamicum Recombinant Strains[a]

C. glutamicum strain	Excreted amino acids (g L^{-1})				
	Lysine	Threonine	Homoserine	Glycine	Isoleucine
ATCC 21799(pM2)[b]	22.0 ± 1.0	< 0.1	< 0.1	< 0.1	< 0.1
ATCC 21799(pJD4)	4.5 ± 0.2	5.4 ± 0.2	2.0 ± 0.1	2.0 ± 0.1	1.3 ± 0.1
ATCC 21799(pGC42)[c]					
No induction	0.9 ± 0.1	5.6 ± 0.3	6.7 ± 0.3	1.3 ± 0.1	1.0 ± 0.1
1.5 mmol of IPTG	0.9 ± 0.1	11.8 ± 0.6	< 0.1	4.6 ± 0.2	1.9 ± 0.2

[a] From Colón et al., 1995a,b.
[b] E. coli-C. glutamicum shuttle vector; pJD4, Km^R hom^{dr}-thrB operon; pGC42, Km^R Ap^R laqI^q hom^{dr} tac-thrB.
[c] ATCC 21799(pGC42) induced by given amount of IPTG.

the addition of IPTG (Colón et al., 1993; Colón et al., 1995a,b). By increasing the activity of homoserine kinase relative to that of homoserine dehydrogenase, homoserine secretion essentially was eliminated and the final threonine titer was increased by about 120% (Table 6.4).

As indicated in Table 6.4, a significant fraction of threonine is either converted to isoleucine or further degraded to glycine. The conversion of threonine to isoleucine was prevented by the construction of defined ilvA mutants via marker exchange mutagenesis (Colón et al., 1997). At this point emphasis is placed on blocking the degradation of threonine to glycine. This is a more challenging problem that involves more than one pathway, the genes of which have not yet been characterized in Corynebacteria.

Isoleucine

The biosynthesis of isoleucine in C. glutamicum starts with the conversion of L-threonine to α-ketobutyrate by L-threonine deaminase (LTD, ilvA), followed by the condensation of this molecule with α-acetolactate catalyzed by acetohydroxy acid synthase (AHAS, ilvB-ilvN). This pathway also provides the precursors for the synthesis of the other two branched-chain amino acids (BCAA), namely, valine and leucine. Even though up to five different AHAS isozymes have been reported in enterobacteria, only one enzyme is known in C. glutamicum. Both LTD and AHAS are inhibited by isoleucine, whereas AHAS is also inhibited by leucine and valine. Furthermore, all three BCAA repress the expression of AHAS. The identification and characterization of genes involved in BCAA biosynthesis in Corynebacteria have been the subject of intensive investigation in the last few years (Colón et al., 1995; Cordes et al., 1992; Keilhauer et al., 1993).

Overproduction of isoleucine was achieved through the amplification of the ilvA gene (plasmid pGC77, Table 6.5), in combination with the hom^{dr} and thrB genes of plasmid pGC42 (see the threonine case). The Corynebacterium ilvA gene encoding threonine dehydratase was isolated from a pM2-

TABLE 6.5   Isoleucine Production by C. glutamicum Recombinant Strains[a]

C. glutamicum strain	Excreted amino acids (g L^{-1})				
	Lysine	Threonine	Homoserine	Glycine	Isoleucine
ATCC 21799(pM2)	22.0 ± 1.0	< 0.1	< 0.1	< 0.1	< 0.1
ATCC 21799(pGC42)	0.9 ± 0.1	11.8 ± 0.6	< 0.1	4.6 ± 0.2	1.9 ± 0.2
ATCC 21799(pGC77)[b]	0.4 ± 0.1	< 0.1	< 0.1	0.5 ± 0.1	15.1 ± 0.2

[a] From Colón et al., 1995.
[b] Derived from pGC42 (Table 6.4) Km^R Ap^R laqI^q hom^{dr} ilvA tac-thrB.

based genomic *C. glutamicum* library by heterologous complementation of an *E. coli ilvA* mutant. The resulting plasmid (pGC77), when inserted in the lysine strain (ATCC 21799), resulted in about 15g L^{-1} isoleucine, along with small amounts of lysine and glycine. A carbon balance indicates that the majority of carbon previously converted to threonine, lysine, glycine, and isoleucine (21799/pGC42) was incorporated into isoleucine by the new strain (21799/pGC77).

6.1.3. SOLVENTS

The history of acetone, butanol, and ethanol (ABE) industrial fermentation processes dates back to the beginning of the century. Due to the shortage of natural rubber, the English firm Strange and Graham investigated the possibility of manufacturing synthetic rubber. It was then determined that the synthetic rubber precursors butadiene and isoprene could be best produced from butanol or isoamyl alcohol. It was in this situation that Professor Perkins and his assistant Chaim Weizmann (later to become the first president of Israel) were recruited to study the chemical production of rubber precursors. Despite his chemistry education, Weizmann soon concluded that the key to the success of such a process was through fermentation; thus, he retrained himself as a microbiologist. Between 1912 and 1914 he screened several bacterial strains and succeeded in isolating one, initially termed BY, that was later termed *Clostridium acetobutylicum*, which gave the highest yields of acetone and butanol from starch.

The subsequent development of ABE fermentation processes was accelerated rapidly by the outbreak of World War I, due to the demand for acetone as the colloidal solvent for nitrocellulose. World War II resulted in a further demand for acetone and the substrate was changed from maize to molasses, which was relatively inexpensive and abundant in the 1930s. After World War II, ABE fermentation declined and virtually ceased in the United States and United Kingdom with the advent of solvent production from petroleum and an escalation in molasses prices.

The demise of ABE fermentation was due to a number of intrinsic system limitations, such as low final concentrations, yields, and productivities, undesirable solvent ratios, and relatively high substrate costs. Genetic engineering of microbial solvent-producing strains can potentially revive ABE fermentation processes by addressing the following challenges:

- increase product yields and introduce alternative substrates derived from lignocellulose- or waste-based feedstocks
- develop of a strain that exhibits high productivities in continuous and immobilized cell systems

- develop of a strain that gives higher final product concentrations and exhibits enhanced endproduct tolerance
- develop a strain that will easily allow the manipulation of solvent ratios

The induction of several solventogenic enzymes at the onset of solvent formation suggests that the genetic control of solvent formation is important (Bennett and Rudolph, 1995; Sauer and Duerre, 1995). However, despite the cloning and sequencing of several acid- and solvent-associated genes, the understanding of metabolic flux regulation still remains elusive. Genetic tools, such as plasmid vectors, have been developed for a number of clostridial strains. These include (1) cloning vectors that utilize the broad-host-range conjugal pAMβ1 replicon of *Enterococcus faecalis* or the pIM13 replicon of *Bacillus subtilis*; (2) suitable selection markers are erythromycin and clarinthromycin, which is stable at the low prevailing pH; (3) the highly expressed *Clostridium* ferredoxin promoter has been exploited in the construction of an expression vector; (4) the conjugative transposons Tn916, Tn925, and Tn1545 function in clostridial strains, so that transposon mutagenesis may be possible. Numerous clostridial genes have been cloned and studied in *E. coli* (Bennett and Rudolph, 1995; Durre *et al.*, 1995; Papoutsakis and Bennett, 1991), waiting to be used in gene-inactivation physiological studies in *Clostridium* strains.

The first successful cloning of acid and solvent formation genes in *C. acetobutylicum* was reported in 1992 for heterologous overexpression of acetoacetate decarboxylase (*adc*) and phosphotransbutyrylase (*ptb*) in strain ATTC 824 (Mermelstein *et al.*, 1992) (see Fig. 2.8). This was possible with the development of a *B. subtilis/C. acetobutylicum* shuttle vector (pFNK1), in conjunction with an improved electrotransformation protocol. Acetoacetate decarboxylase (AADC) is the terminal enzyme in the pathway for acetone production, converting acetoacetate to acetone and CO_2. Phosphotransbutyrylase (PTB) is the branch point enzyme for butyrate production, converting butyryl-CoA and inorganic phosphate into butyryl phosphate (subsequently converted to butyrate by butyrate kinase) and reduced CoA. Alternatively, butyryl-CoA is converted into butanol in two enzymatic steps. AADC activity in the recombinant strain increased by over 9-fold in the exponential phase and over 33-fold in the stationary phase, whereas PTB activities of the engineered strain increased by over 20-fold in the exponential phase and 40-fold in the stationary phase. The transformed strain showed an increase of 95, 37, and 90% in the levels of acetone, butanol, and ethanol, respectively. Furthermore, acid concentrations at the end of the fermentation were considerably lower in the engineered strain (22-fold) than in the control, and the solvent yield from glucose increased by about 50% in the redesigned strain.

In a different study, the feasibility of genetic manipulation in *Clostridia* was demonstrated by altering the substrate range of *C. acetobutylicum* NCIMB 8052: An artificial operon containing the *celC* and *celA* genes from *C. thermocellum* was transferred to the NCIMB 8052 strain. The resulting transformant was able to grow on lichenan (a β-glycan) as the sole carbon source.

1,3-Propanediol

1,3-Propanediol (1,3-PD) is an intermediate in chemical and polymer synthesis, *e.g.*, in the synthesis of polyurethanes and polyesters. It is currently derived from petroleum and is very expensive to produce relative to similar diols. Tong *et al.* (1991) recently constructed an *E. coli* propanediol-producing strain carrying genes from the *Klebsiella pneumoniae dha* regulon. The *dha* regulon in *Klebsiella pneumoniae* enables the organism to grow anaerobically on glycerol and produce 1,3-PD. *Escherichia coli*, which does not have a *dha* system, is unable to grow anaerobically on glycerol without an exogeneous electron acceptor and does not produce 1,3-PD. In the first step (see Fig. 6.6), glycerol is converted into 3-hydroxypropionaldehyde by a

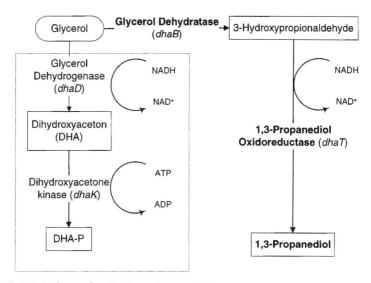

FIGURE 6.6 Pathways for the dissimilation of glycerol in *K. pneumoniae*. Cloning of *K. pneumoniae dhaB* and *dhaT* genes in *E. coli* yielded a recombinant strain that converts glycerol into the industrially useful product 1,3-propanediol.

coenzyme B_{12} dependent dehydratase, which is then reduced to 1,3-propanediol by an NAD dependent oxidoreductase.

A genomic library of *K. pneumoniae* ATCC 25955 constructed in *E. coli* AG1 was enriched for the ability to grow anaerobically on glycerol and dihydroxyacetone and was screened for the production of 1,3-PD. The cosmid pTC1 was isolated from a 1,3-PD-producing strain of *E. coli* and found to possess enzymatic activities associated with four genes of the *dha* regulon: glycerol dehydratase (*dhaB*), 1,3-PD oxidoreductase (*dhaT*), glycerol dehydrogenase (*dhaD*), and dihydroxyacetone kinase (*dhaK*) (see Fig. 6.6). All four activities were inducible by the presence of glycerol. When *E. coli* AG1/pTC1 was grown on complex medium plus glycerol, the yield of 1,3-PD from glycerol was 0.46 mol mol^{-1}. The major fermentation byproducts were formate, acetate, and D-lactate. The 1,3-PD fermentation provides a useful model system for studying the interaction of a biochemical pathway in a foreign host and for developing strategies for metabolic pathway engineering.

Further progress in this area is needed in order to minimize byproduct formation, eliminate the need for glycerol supplementation, and also extend the substrate range of the pathway to more abundant renewable compounds. An analysis of the maximum theoretical yield of 0.875 mol 1,3-PD per mol from glycerol indicates that the yield can be improved. No microorganism is available that can convert glycerol entirely to 1,3-PD and CO_2 due to the need for the regeneration of reducing power. Cells regenerate reducing power (NADH) by forming a mixture of byproducts, such as acetate and formate, that in essence reduces the maximal theoretical yield to 0.667 mol mol^{-1}. In principle, it is possible to provide an alternative source of reducing power by supplementing glycerol fermentations with pentoses or hexoses. Theoretical yields of such processes are summarized (moles of 1,3-PD per mole of glycerol) (Tong and Cameron, 1992):

$$- 2glycerol - glucose + 2\ 1,3\text{-PD} + 2acetate$$
$$+ 2formate = 0\ \left(\text{theoretical yield} = 1.0\ \text{mol mol}^{-1}\right)$$

$$- 5glycerol - 3xylose + 5\ 1,3\text{-PD} + 5acetate$$
$$+ 5formate = 0\ \left(\text{theoretical yield} = 1.0\ \text{mol mol}^{-1}\right)$$

Pilot runs with the *E. coli* strain carrying the *K. pneumoniae dha* regulon indeed resulted in enhanced yields of 1,3-PD from glycerol with cosubstrate feed: The yield was improved from 0.46 mol mol^{-1} with glycerol alone to 0.63 mol mol^{-1} with glycerol + glucose and 0.55 mol mol^{-1} with glycerol + xylose. Such improvements are important economically as the prices of glucose and xylose are significantly lower than that of glycerol.

6.2. EXTENSION OF SUBSTRATE RANGE

Most of the work in this area focused on engineering organisms to use xylose, the primary five-carbon sugar in hemicellulosic biomass, and lactose, a major byproduct of the dairy industry. Other efforts have examined the utilization of other plentiful carbon sources, such as whey, starch, and cellulose. In general, expansion of the ability of microbial strains to utilize a spectrum of carbonenergy sources provides increased flexibility in the design and improves the economic feasibility of fermentation processes. This is particularly true for large-volume commodity operations in which the cost of the substrate may contribute a very large fraction of the total production cost (60-65% for ethanol, 40-45% for lysine, and 25-35% for antibiotics and industrial enzymes). As most microorganisms share a large number of common metabolic pathways, extension of the substrate range usually involves the addition of only a few enzymatic steps. Occasionally, however, such steps need to be coordinated with downstream reactions, and it is in these cases where the tools of metabolic engineering are very useful indeed.

6.2.1. METABOLIC ENGINEERING OF PENTOSE METABOLISM FOR ETHANOL PRODUCTION

Along with the introduction of ethanol genes in enteric bacteria, parallel efforts were also undertaken to incorporate pentose-metabolizing pathways in natural ethanol producers such as *S. cerevisiae* and *Z. mobilis*. Microorganisms, in general, metabolize xylose to xylulose through two separate routes (Fig. 6.7). The one-step pathway catalyzed by xylose isomerase is typical in bacteria, whereas the two-step reaction involving xylose reductase and xylitol dehydrogenase is usually found in yeast. Xylulose is subsequently phosphorylated by ATP and catabolized via the pentose phosphate pathway and the EMP pathway (or the ED pathway in organisms such as *Z. mobilis*). During the last few years, the genes encoding the enzymes of xylose utilization have been cloned and characterized in *E. coli* and some other bacteria (Lawlis et al., 1984; Rygus et al., 1991). Efforts to isolate natural ethanologenic microbes that can utilize xylose have not been successful. These provided the impetus for introducing xylose utilization genes into organisms, especially those used for ethanol production, and, as such, had the advantage of high ethanol tolerance.

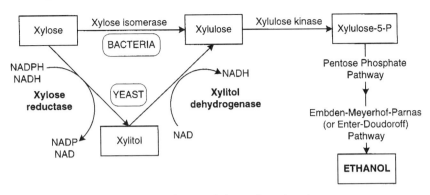

FIGURE 6.7   Xylose metabolism in bacteria and yeast.

Yeast

Even though certain types of yeast such as *Pachysolen tannophilus*, *Pichia stipitis*, or *Candida shehatae* are xylose-fermenting, they have poor ethanol yields and low ethanol tolerance compared with the common glucose-fermenting yeasts, such as *S. cerevisiae*. Early attempts to introduce the one-step pathway by cloning the xylose isomerase gene from either *E. coli* (Sarthy *et al.*, 1987) or *B. subtilis* (Hollenberg and Sahm, 1988) in *S. cerevisiae* were unsuccessful due to the inactivity of the heterologous protein in the recombinant host cell.

In most yeasts and fungi, xylose reductase and xylitol dehydrogenase are dependent on NADPH and NAD, respectively (Fig. 6.8). However, examples of yeast xylose reductases exist that have dual coenzyme specificity (*i.e.*, NADPH and NADH), such as those from *P. stipitis* and *C. shehatae*. Such a type of enzyme has the advantage of preventing imbalances of the NAD/NADH redox system, especially under oxygen-limiting conditions. Recently, the *P. stipitis* genes for xylose reductase and xylitol dehydrogenase were introduced in *S. cerevisiae* (strain H) (Kötter *et al.*, 1990; Tantirungkij *et al.*, 1993). Whereas *P. stipitis* converts xylose primarily to ethanol under anaerobic conditions, ethanol production in the recombinant *S. cerevisiae* strain (strain H) was marginal (2.7 g L^{-1}), accompanied by the accumulation of considerable amounts of xylitol (35 g L^{-1}). The observation that ethanol yield and productivity were higher in aerobic conditions was explained on the basis of improved NAD regeneration from NADH, which in turn stimulates xylitol dehydrogenase. Additional limitations of xylose utilization in *S. cerevisiae* were also attributed to the inefficient capacity of the nonoxidative PPP, as indicated by the accumulation of sedoheptulose-7-P.

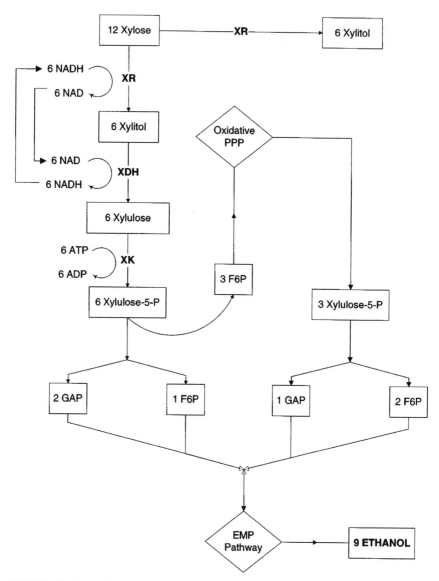

FIGURE 6.8   Anaerobic xylose utilization and cofactor regeneration in recombinant *S. cerevisiae*. Abbreviations: EMP, Embden-Meyerhof-Parnas; PPP, pentose phosphate pathway; XR, xylose reductase; XDH, xylitol dehydrogenase; XK, xylulose kinase.

Further improvement of strain H was attempted via random mutagenesis and selection for cells that grow rapidly on xylose (Tantirungkij *et al.*, 1994). Interestingly, strain IM2, which grew 3 times faster in xylose medium than strain H, showed lower specific activities of both xylose reductase and xylitol dehydrogenase, but 1.5 times higher specific activity of xylulose kinase. Despite the higher growth rate, however, ethanol production by strain IM2 was improved marginally to about 4.2 g L^{-1} at a yield of 0.08 g g^{-1}.

Zymomonas mobilis

Xylose also could be a useful carbon source for the ethanol producer *Z. mobilis*. This is a bacterium that has been used as a natural fermentative agent in alcoholic beverage production and has been shown to have ethanol productivity superior to that of yeast. Overall, it demonstrates many of the desirable traits sought in an ideal biocatalyst for ethanol, such as high ethanol yield, selectivity and specific productivity, as well as low pH and high ethanol tolerance. In glucose medium, *Z. mobilis* can achieve ethanol levels of at least 12% (w/v) at yields of up to 97% of the theoretical value. When compared to yeast, *Z. mobilis* exhibits 5-10% higher yields and up to 5-fold greater volumetric productivities. The notably high yield of this microbe is attributed to reduced biomass formation during fermentation, apparently limited by ATP availability. Note that this organism produces only 1 mol of ATP per mole of glucose through the ED pathway [see reaction (2.6) and Box 6.1] compared with 2 mol for yeast (EMP pathway). As a matter of fact, *Zymomonas* is the only genus identified to date that exclusively utilizes the Entner-Doudoroff pathway anaerobically. Furthermore, glucose can readily cross the cell membrane of this organism by facilitated diffusion, efficiently be converted to ethanol by an overactive pyruvate decarboxylase/alcohol dehydrogenase system, and is generally recognized as a safe (GRAS) organism for use as an animal feed. As discussed earlier, the main drawback of this microorganism is that it can only utilize glucose, fructose, and sucrose and thus is unable to ferment the widely available pentose sugars.

This led Zhang *et al.* at the National Renewable Energy Laboratory (Golden, CO) to attempt to introduce a pathway for pentose metabolism in *Z. mobilis* (Zhang *et al.*, 1995). Early attempts by other groups using the xylose isomerase (*xylA*) and xylulokinase (*xylB*) genes (Fig. 6.7) from either *Klebsiella* or *Xanthomonas* were met with limited success, despite the functional expression of these genes in *Z. mobilis*. It soon became evident that such failures were due to the absence of detectable transketolase and transaldolase activities in *Z. mobilis*, which are necessary to complete a

---

BOX 6.1
_____

Theoretical Ethanol Yield on Xylose by Recombinant *Zymomonas* Strain

    The stoichiometry of ethanol production in this recombinant organism (Fig. 6.8) can be summarized as follows (neglecting the NAD(P)H balances):

$$3\text{xylose} + 3\text{ADP} + 3\text{P}_i \rightarrow 5\text{ethanol} + 5\text{CO}_2 + 3\text{ATP} + 3\text{H}_2\text{O}$$

Thus, the theoretical yield on ethanol is 0.51 g of ethanol/g of xylose (1.67 mol mol^{-1}). It is important to note that the metabolically engineered pathway yields only 1 mol of ATP from 1 mol of xylose, compared with 5/3 mol typically produced through a combination of the pentose phosphate and EMP pathways. The energy limitation results in less biomass formation and thus a more efficient conversion of substrate to product.

---

functional pentose metabolic pathway (Fig. 6.9). After the transketolase *E. coli* gene was cloned and introduced in *Z. mobilis*, a small conversion of xylose to CO_2 and ethanol occurred (Feldmann *et al.*, 1992). The next step was to introduce the transaldolase reaction, as this strain accumulated significant amounts of sedoheptulose-7-P intracellularly. Sophisticated cloning techniques therefore were applied for the construction of a chimeric shuttle vector (pZB5) that carries two independent operons: the first encoding the *E. coli xylA* and *xylB* genes and the second expressing transketolase (*tktA*) and transaldolase (*tal*) again from *E. coli*. The two operons comprising the four xylose assimilation and nonoxidative pentose phosphate pathway genes were expressed successfully in *Z. mobilis* CP4. The recombinant strain was capable of fast growth on xylose as the sole carbon source, and moreover it efficiently converted glucose and xylose to ethanol with 86 and 94% of the theoretical yield from xylose and glucose respectively. This represents a complementary approach to the previously discussed expression of the *Z. mobilis* PET operon in *E. coli* for ethanol production.

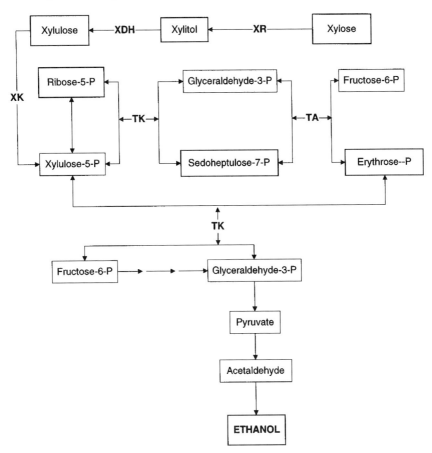

FIGURE 6.9   Ethanol production from pentose sugars in metabolically engineered *Z. mobilis*. Appreviations: XR, xylose reductase; XDH, xylulose dehydrogenase; TK, transketolase; TA, transaldolase.

6.2.2. CELLULOSE - HEMICELLULOSE DEPOLYMERIZATION

It would be desirable if ethanol-producing microbes from lignocellulose also had means to depolymerize cellulose, hemicellulose, and associated carbohydrates. Many plant pathogenic bacteria (soft-rot bacteria), such as *Erwinia carotovora* and *Erwinia chrysanthemi*, have evolved sophisticated systems of hydrolases and lyases that aid the solubilization of lignocellulose and allow them to macerate and penetrate plant tissue (Kado, 1992). Genetic engineering of these bacteria for ethanol production represents an attractive alternative to the solubilization of lignocellulosic biomass by chemical or enzymatic means. *E. carotovora* SR38 and *E. chrysanthemi* EC16 were genetically engineered with the PET operon and shown to produce ethanol and CO_2 efficiently as primary fermentation products from cellobiose, glucose, and xylose (Beall and Ingram, 1993). Both ethanologenic *Erwinia* strains produced about 50 g L^{-1} ethanol from 100 g L^{-1} cellobiose in less than 48 hour with a maximum volumetric productivity of 1.5 g of ethanol L^{-1} hour. This rate is over twice that reported for the cellobiose-utilizing yeast, *Brettanomyces custersii*, in batch culture (Spindler *et al.*, 1992).

Along similar lines, the incorporation of saccharifying traits to ethanol-producing microorganism was also attempted. The gene encoding for the xylanase enzyme (*xynZ*) from the thermophilic bacterium *C. thermocellum* was expressed at high cytoplasmic levels in ethanologenic strains of *E. coli* KO11 and *K. oxytoca* M5A1(pLOI555) (Burchhardt and Ingram, 1992). This is a *temperature stable* enzyme that depolymerizes xylan to its primary monomer (99%) xylose. In order to increase the amount of xylanase in the medium and facilitate xylan hydrolysis, a two-stage, cyclical process was employed for the fermentation of polymeric feedstocks to ethanol by a single, genetically engineered microorganism. Cells containing xylanase were harvested and added to a xylan solution at 60°C, thereby lysing and releasing xylanase for saccharification. After cooling to 30°C, the hydrolysate was fermented to ethanol, in the meantime replenishing the supply of xylanase for the subsequent saccharification. *K. oxytoca* was found to be a superior strain for such an application, because, in addition to xylose (metabolizable by *E. coli*), it can also consume xylobiose and xylotriose. Even though the maximum theoretical yield of M5A1(pLOI555) is in excess of 48 g L^{-1} ethanol from 100 g L^{-1} xylose, about one-third of that was achieved in this process because xylotetrose and longer oligomers remained unmetabolized by this strain. The yield appeared to be limited by the digestibility of commercial xylan rather than by the lack of sufficient xylanase activity or by ethanol toxicity.

6.2.3. LACTOSE AND WHEY UTILIZATION

Whey is a nutrient-rich byproduct of the dairy industry that can provide an inexpensive carbon and nitrogen source in biotechnological processes. It reaches an annual production of 10^{11} kg, with high lactose (75% of dry matter) and protein contents (12-14%), as well as small amounts of organic acids, minerals, and vitamins. Whereas its protein content is separated and

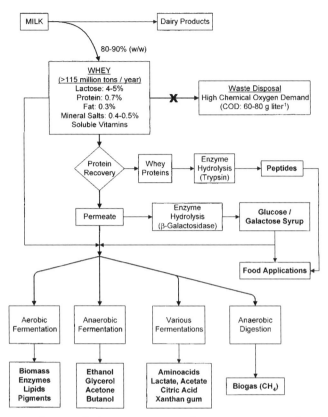

FIGURE 6.10   Utilization of whey components in fermentation processes (von Stockar and Marison, 1990).

concentrated for food purposes (see Fig. 6.10), the lactose and salts in the permeate have lower values and are typically discarded. In addition to the loss of valuable nutrients, disposal also requires expensive sewage treatment. Efforts therefore are intensifying to find useful applications for whey in general and for the permeate in particular. Some examples of fermentation processes that can utilize whey byproducts as feedstocks are summarized in Fig. 6.10.

Although a variety of microbes can utilize whey, some of the industrially most prominent organisms such as *S. cerevisiae*, *Z. mobilis*, and *Alcaligenes eutrophus* are unable to do so. Utilization of lactose requires the presence of the catabolic enzyme β-galactosidase (coded by the *lacZ* gene) for the hydrolysis of the lactose disaccharide into its constituent sugars, glucose and galactose. Additionally, an efficient lactose transport system, along with glucose- and galactose- catabolizing pathways, is needed. These requirements were evident in the early work of introducing the lactose transposon, Tn 951 from *Yersinia enterocolitica* (harbors *lacI, lacZ* and *lacY* genes: see Chapter 5 section on the *lac* operon), in *Z mobilis* (Carey *et al.*, 1983; Goodman *et al.*, 1984). Although the *E. coli* β-galactosidase was successfully expressed in this strain, ethanol yields were much lower than theoretical values for at least two reasons: lactose is cleaved by β-galactosidase to glucose and galactose; however, only glucose can be fermented to ethanol by the recombinant *Z. mobilis* strains, with galactose accumulating to inhibitory concentrations (Yanase *et al.*, 1988). This indicates that the galactose operon is needed in addition to the lactose operon. Secondly, the poor ethanol productivity was attributed to the slow uptake of lactose. Later studies utilized the Tn 951 to express β-galactosidase in *Pseudomonas saccharophila* and *Alcaligenes eutrophus* (Pries *et al.*, 1990). This allowed transconjugants of *P. saccharophila* to grow slowly on lactose mineral medium whereas the parent strain did not grow at all. Plasmid pPL76, harboring an *A. eutrophus* promoter-*lacZ* fusion, enabled *A. eutrophus* not only to express β-galactosidase but also to grow slowly on lactose. Subsequently, the *E. coli gal* operon was also transferred to these strains to allow galactose utilization.

The *E. coli lacZY* operon (coding for β-galactosidase and lactose permease) was also integrated into the *Pseudomonas aeruginosa* chromosome, an important producer of rhamnolipid biosurfactants (Koch *et al.*, 1988). The transconjugants grew well in lactose-based media (minimal medium and whey) and produced rhamnolipid during the stationary phase. The *E. coli lacZY* gene, under the control of phage φLO promoter, was also inserted in *Xanthomonas campestris*, a bacterium that causes tremendous agricultural losses worldwide but also used in the production of xanthan gum (Fu and Tseng, 1990). For the production of xanthan, glucose, sucrose, or starch media are normally used. The recombinant strain, however, expressed high levels of β-galactosidase and grew well in a medium containing lactose as the sole carbon source. Production of xanthan gum in lactose or diluted whey by the engineered strain was evaluated, and it was found to produce as much xanthan gum using these substrates as did cells in a glucose medium. These examples illustrate attractive processes for treating industrial waste materials while producing useful compounds at the same time.

An alternative strategy is to construct a strain that secretes β-galactosidase into the medium or into the periplasm wherein lactose is freely diffusible. This approach was applied in yeast, where a lactose utilizing *S. cerevisiae* was constructed by expressing the gene for a *secreted*, thermostable β-galactosidase (*lacA*) from *Aspergilus niger* (Kumar *et al.*, 1992). This study demonstrated that 40% of the total recombinant protein was secreted into the medium, allowing *S. cerevisiae* to grow on whey permeate (4% w/v lactose) with a doubling time of 1.6 h. This approach offers significant advantages over the earlier processes for the fermentation of whey by *S. cerevisiae*, which used either β-galactosidase prehydrolyzed whey or yeast co-immobilized with β-galactosidase.

The entire *E. coli* lactose operon was also inserted into the amino acid producer *C. glutamicum* R163 (Brabetz *et al.*, 1991). Recombinant *C. glutamicum* strains carrying their *lac* genes under the control of a strong promoter grew rapidly in defined media with lactose as the sole carbon source (3% w/v^{-1}). The growth characteristics, which were indistinguishable from those in glucose media, depended on the presence of the *lacY* gene (lactose permease) in addition to *lacZ*. Furthermore, enzymatic assays indicated that all β-galactosidase activity was intracellular. Again, the main drawback of this system is the inability of the cells to utilize the second monosaccharide of lactose, namely, galactose.

6.2.4. SUCROSE UTILIZATION

Sucrose (a disaccharide of glucose and fructose) is another abundant and inexpensive carbon source found in cane molasses, for example. Even though certain *E. coli* strains can utilize sucrose, *E. coli* K-12, a potentially useful industrial organism for amino acid production, is unable to grow on sucrose. Various researchers have attempted to express the sucrose utilization system (Scr$^+$) from other bacterial species, but are unable to stably maintain the Scr$^+$ phenotype in *E. coli*. Recently, a successful attempt came about from the cloning of the *scrA* gene coding for sucrase from *E. coli* B-62 onto a plasmid and then transferring the cloned DNA fragment onto the chromosome of *E. coli* K-12. Tryptophan producer derivatives of *E. coli* K-12 expressing the *scrA* gene grew well in sucrose medium and excreted amounts of tryptophan (5.7 g L^{-1}) comparable to these from similar strains grown on glucose (Tsunekawa *et al.*, 1992).

6.2.5. STARCH - DEGRADING MICROORGANISMS

Starch, derived from renewable resources such as corn and cereals, is a very important carbon and energy source in biotechnological processes. Substitution of glucose with starch not only can reduce fermentation feedstock costs but also can minimize or eliminate negative physiological effects associated with glucose, such as catabolic repression or acidogenesis.

Starch is a mixture of linear and branched homopolymers of D-glucose that are connected by $\alpha(1 \rightarrow 4)$ linkages and at branch points by $\alpha(1 \rightarrow 6)$ linkages. It is formed as a carbohydrate reserve in plants and is present in significant amounts in potato tubers and in the seeds of wheat, corn, barley, and sorghum. The linear component, *amylose*, consists of chains of α-1,4-D-glucopyranose ranging in degree of polymerization from about 10^2 to 4×10^5. In the branched component, *amylopectin*, shorter chains (17-23 units long) of α-1,4-D-glucopyranose are linked together by α-1,6 bonds to form a branched structure with a degree of polymerization ranging from 10^4 to 4×10^7. Four types of starch-decomposing enzymes are of importance: α-amylase, β-amylase, pullulanase or isoamylase (debranching enzymes), and glucoamylase. α-Amylase is an endoglucanase that randomly cleaves $\alpha(1 \rightarrow 4)$ linkages, converting starch to dextrins, maltose, and glucose. It is produced by bacteria and fungi, notably by *Bacillus* species, *Pseudomonas*, and *Lactobacilli* and by *Aspergillus* species. β-Amylase is an exoglucanase typically found in plants that successively removes maltose units from the nonreducing end of starch. Pullulanase and isoamylase belong to the debranching group of enzymes that hydrolyze $\alpha(1 \rightarrow 6)$ linkages. Glucoamylase is a fungal enzyme that removes glucose residues from the nonreducing end of starch.

Because most microorganisms are unable to degrade this glucose biopolymer, work has focused on cloning genes for enzymatic starch hydrolysis into various organisms (Kennedy *et al.*, 1988). This approach offers an attractive alternative to current processes that first convert starch enzymatically into glucose and some oligosaccharides and then use them as carbon sources in a separate fermentation step. Along these lines, a *S. cerevisiae* strain was constructed that contained a glucoamylase gene from *Aspergillus* sp. (Innis *et al.*, 1985). The recombinant strain was able to grow on amylodextrins, albeit at a lower rate than the case where glucoamylases are added added to the fermentation medium.

Useful applications of such a recombinant strain include brewing and baking. In the case of brewing, the malting process, the partial hydrolysis of barley starch, results in a considerable amount of dextrins that cannot be fermented by the yeast *S. cerevisiae*. These dextrins are of high caloric content and have to be removed for the production of light beer, presently achieved by the external addition of glucoamylase. Therefore, an engineered strain with amylolytic properties would offer a suitable alternative for brewing, especially for the production of a low-calorie product. Also, such a strain would eliminate the need for α-amylase-enriched flour in certain types of bread manufacturing.

Strains with the proceding desirable characteristics recently were constructed by expressing the yeast *Schwanniomyces occidentalis* α-amylase (AMY1) and glucoamylase (GAM1) genes in *S. cerevisiae* (Hollenberg and Strasser, 1990). Comparative enzymatic studies illustrated that the engineered amylolytic system is as effective as the original *S. occidentalis* strain. During fermentation of ground, liquefied wheat, this recombinant strain showed the same ethanol production rate as a conventional distillery yeast with saccharifying enzymes added prior to fermentation.

6.3. EXTENSION OF PRODUCT SPECTRUM AND NOVEL PRODUCTS

This is an area with immense potential for metabolic engineering. Rational expression of heterologous genes can extend existing pathways of the host organism for the overproduction of both known and novel compounds with attractive chemical and/or physical properties.

6.3.1. ANTIBIOTICS

Antibiotic production by microorganisms is one of their more interesting features, particularly from a medical and commercial point of view. More than 10,000 antibiotics and similar bioactive metabolites have been isolated from microbes, with approximately 500 new classes of low-molecular-weight compounds published every year. In monetary terms, antibiotics currently are the most important group of microbial biotechnological products, with an estimated world sales in excess of $15 billion. The primary classes are cephalosporins, penicillins, and tetracyclines, and the majority of these agents are produced by *Streptomyces* (and other actinomycetes) and various *Bacillus* species. Their primary use is in the treatment of human infectious diseases, although a significant number have agricultural and veterinary applications.

Antibiotics are made by secondary metabolic pathways that use common metabolites in less specific and, sometimes, more intricate ways than primary metabolism. The polyketide antibiotics, for example, are made from simple fatty acids by a pathway that superficially resembles the one used to make long-chain fatty acids, but the resulting compounds exhibit a range of structural complexity far surpassing the simple hydrocarbon framework of the essential fatty acids. Recently, it has become apparent that yields of secondary metabolites, including antibiotics, can also be improved by overcoming rate-controlling biosynthetic steps through genetic techniques. In addition, metabolic engineering techniques are applied in order to modify known antibiotics to improve their properties and also to synthesize new forms of antibiotics (Summers *et al.*, 1992). For years antibiotic production in filamentous fungi and *Streptomyces* was improved by random mutation/screening and, to a lesser extent, by selecting mutants that over-produced primary metabolic precursors of antibiotics. Work spanning four decades developed production strains less amenable to improvement by these traditional techniques.

The application of recombinant DNA technology was based on the development of genetic transformation systems for β-lactam-producing organisms and cloning of biosynthetic genes. The ability to transform industrially important organisms (such as *P. chrysogenum* and *C. acremonium*) provided a powerful tool for the precise manipulation of biosynthetic pathways and an avenue for practical applications, such as gene dosage studies for possible limiting steps or gene disruptions to alter final products. The discovery that many antibiotics genes are clustered, and also that certain genes of related pathways exhibit cross-hybridization, has opened new avenues in this area (Charter, 1990). Gene clustering facilitates cloning, and the fact that these genes are often positively regulated increases the possibility of improving production through overexpression of the genetic regulatory molecule. Overexpression of such regulatory genes caused, for instance, the overproduction of streptomycin, undecylprodigiosin, and actinorhodin in wild-type strains (Charter, 1990).

Streptomyces rank near the top among microorganisms of industrial importance, especially as antibiotic producers. Actinorhodin biosynthesis genes were transferred from *Streptomyces coelicitor*, the only species with well-established genetics, to *Streptomyces lividans*, enabling the latter strain to produce actinorhodin. Later on, clustered erythromycin genes from *Streptomyces erythreus* were transferred to *S. lividans*, allowing the recombinant strain to produce erythromycin A. Transformation of the fungi *Neurospora crassa* and *Aspergillus niger* with a cosmid containing P*enicillium chrysogenum* penicillin biosynthetic genes resulted in the production of penicillin V by these strains.

Yield improvements through metabolic engineering have been demonstrated for a number of systems. For example, the production of cephalosporin C by *Cephalosporium acremonium* was increased by 15% by overexpressing the *cefEF* gene (Skatrud *et al.*, 1989). This gene codes for a bifunctional protein with two sequentially acting activities: deacetoxycephalosporin C synthetase and deacetylcephalosporin C synthetase (DACS). The recombinant strain with a 2-fold increase in DACS activity, was able to convert penicillin N, a precursor normally excreted in large quantities, into the final product cephalosporin C. This work also identified DAC acetyltransferase, the final enzyme in the cephalosporin C pathway, as a potentially controlling step, as a substantial amount of deacetylcephalosporin (DAC) was observed in the medium.

Recombinant DNA techniques can also be utilized to engineer hybrid or even novel antibiotics. The main inherent obstacle in such applications is, of course, the fact that the producing organism must be resistant to the hybrid antibiotic in order to achieve high yields. Genes for biosynthetic steps in different organisms can be combined in the same organism, thus extending the diversity of natural antibiotics. In an early attempt, part of the cloned pathway for actinorhodin from *Streptomyces coelicitor* was transformed into a *Streptomyces* strain that produces the compound medermycin (Hopwood *et al.*, 1985). The recombinant strain produced a hybrid antibiotic identified as *mederrhodin*. Conversion of the native medermycin to mederrhodin involves a β-hydroxylation step postulated to be catalyzed by heterologous β-hydroxylation activity of an enzyme with a broad substrate specificity. McAlpine *et al.* have used a similar strategy to transform a mutant of *Saccharopolyspora erythraea*, which is blocked in an early step of erythromycin biosynthesis, with a DNA library from the oleandromycin producer *Streptomyces antibioticus* (McAlpine *et al.*, 1987). One of the recombinant strains produced an antibiotic with a novel structure, called 2-norerythromycin. A greater challenge in generating novel antibiotics goes beyond single-group substitutions and involves the alteration of their backbone structure. *Streptomyces galilaeus* normally produces aclacinomycin A and B. Following its transformation with polyketide synthase genes, clones were obtained that produced anthraquinone (Bartel *et al.*, 1990). These exciting results provide the foundation for ongoing efforts to rationally design and synthesize novel antibiotics.

6.3.2. POLYKETIDES

Polyketides are found in most organisms and are especially abundant in a class of filamentous bacteria, the actinomycetes. The polyketide family is a rich source of bioactive molecules with antibiotic (such as tetracycline and erythromycin) and pharmacological (*e.g.*, cancer agents and immunosuppressants) properties. Synthesis of these molecules involves giant modular enzymes known as polyketide synthases (PKS). Polyketides are made from simple fatty acids by a pathway that superficially resembles the one used to make long-chain fatty acids, but the resulting compounds exhibit a range of structural complexity far surpassing the simple hydrocarbon framework of biological fatty acids. A major distinction of fatty acid synthesis is the fact that the initial condensations (β-keto acid reduction, dehydration, hydrogenation) do not occur in a regular fashion, but rather depend on the modular structure of the *given* polyketide synthase. In fatty acid synthesis, an acetyl group is added at each round of synthesis to produce a long unbranched chain, and the carbonyl group introduced at each round is reduced to CH_2. In the biosynthesis of polyketides, the unit added is often larger than an acetyl (*e.g.*, malonyl-CoA), yet each condensation step adds two carbons to the elongating chain in a way that the remaining part of the unit extends from the main chain as a branch. Some of the carbonyl groups are not reduced at all, and others are reduced only to the level of CHOH.

Reasons that make polyketides an attractive study model for metabolic engineering include the following: (1) their complex structure results from simple units combined in diverse ways; (2) the modular construction of the enzymatic catalyst (PKS) allows control of enzyme structure and, hence, polyketide type at the genetic level. Recent progress in this area has established the groundwork to generate novel polyketide structures through genetic engineering of polyketide synthases and, at the same time to derive knowledge that elucidates the structure-function relationship in polyketide synthases (Kao, 1997; McDaniel *et al.*, 1993). Moreover, this system provides an opportunity to bridge the fields of genetics and chemistry and, above all, promises to enable scientists to rationally design novel molecules at the level of DNA.

Erythromycin Production by *Saccharopolyspora erythrea*

The production of this polyketide in *S. erythrea* involves only three giant genes, each of which codes for a protein of more than 300 kDa (Cane *et al.*, 1983, 1987). Each protein is in turn made up of two inexact repeats that can be divided into six modules, as shown in Fig. 6.11. Each module contains *combinations* of at least six monofunctional polypeptides, each responsible for one *single reaction step*: acyl transferase (AT), acyl-carrier proteins (ACP), β-ketoacyl-ACP-synthase (KS), β-ketoacyl-ACP-reductase (KR), dehydrase (DH), and enoyl reductase (ER) (Fig. 6.11). An interesting and perhaps anticipated observation is the fact that some of these genes are highly homologous with genes of the fatty acid biosynthesis pathway. What is fascinating about this type of organization is the fact that novel molecules can be generated by using different combinations and permutations of these basic modules, as well as by introducing point mutations within functional domains. Nature has already produced a vast diversity of polyketides by this same technique. A major contribution of metabolic engineering in this area is to design chemical structures of potentially useful molecules, currently known or not, by "genetic design."

In the last few years, Khosla and colleagues have focused their attention on deciphering the rules of polyketide synthetases by developing a *Streptomyces* host-vector system for the expression of recombinant polyketide synthases (PKS) (Kao, 1997; McDaniel *et al.*, 1993). This work has led to the concept of a *minimal polyketide forming-system* containing the condensing enzyme, the acyl carrier protein, and the malonyl-CoA transferase (McDaniel

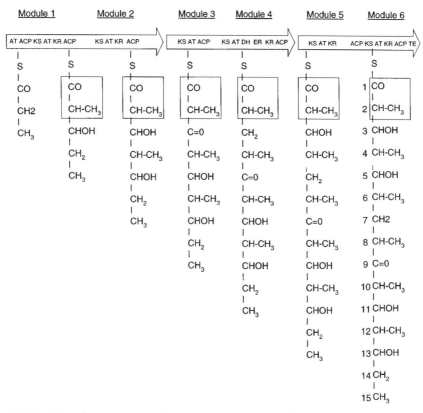

FIGURE 6.11 The organization of genes for erythromycin A biosynthesis in *Saccharospora erythrea*. The DNA region is divided into three open reading frames (ORFs), each coding for a large, complex enzyme molecule. In turn, each enzyme can be subdivided into two modules, with each successive module adding a new propionic acid unit (box) to the growing chain. Subunits: acyl transferase (AT), acyl-carrier proteins (ACP), β-ketoacyl-ACP-synthase (KS), β-ketoacyl-ACP-reductase (KR), dehydrase (DH), and enoyl reductase (ER).

et al., 1994). Additional proteins may then function either as chain length factors, which determine the extent of elongation, or as cyclases, which direct the mode of cyclization (Hutchinson and Fujii, 1995). A number of polyketide molecules were produced recently by *Streptomyces* strains transformed with various combinations of PKS genes comprising minimal systems, thus paving the way for combinatorial biosynthetic approaches (Shen and Hutchinson, 1996; Tsoi and Khosla, 1995). Characterization of these metabolites has provided new insights into the programming aspects of PKS genes (Box 6.2).

BOX 6.2
─────────

Examples of Minimal Polyketide-Forming Systems and Programming Rules of PKS

S. coelicolor A3(2) is an actinomycete with well-developed genetics that produces the blue-pigmented polyketide actinorhodin. The *act* PKS gene cluster has been cloned and completely sequenced, and, in addition, a *S. coelicolor* strain (CH999) was constructed by deleting the entire *act* cluster through homologous recombination. This mutant strain was transformed by plasmid carrying combinatorial "minimal" gene clusters of various PKS genes in order to elucidate the mechanisms by which PKSs achieve their high degree of specificity.

For example, the recombinant strain CH999/pRM37 expresses a "minimal" *act* PKS gene cluster together with a minimal gene set for tetracenomycin (*tcm*) PKS (see figure; AT, acyl transferase; ACP, acyl-carrier proteins; KS, β-ketoacyl-ACP-synthase; KR, β-ketoacyl-ACP-reductase; CYC, cyclase; OMT, *o*-methyltransferase). The actinorhodin (*act*) PKS catalyzes chain termination after nine condensation cycles, whereas *tcm* (PKS) does so after nine cycles. This particular strain was found to produce two novel aromatic polyketides, whose structures were determined by ^{1}H and ^{13}C NMR. Similar experiments have been repeated using various minimal gene clusters followed by the identification and structural analysis of the resulting polyketide(s).

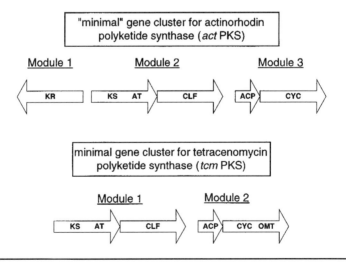

Some of the primary conclusions from such studies are summarized (McDaniel *et al.*, 1993):

- The chain length is, at least in part, dictated by a specific protein, which has been given the name "chain length determining factor" (CLF)
- Some heterologous ketosynthase/putative acyl transferase (KS/AT) CLF pairs give rise to functional PKSs, but other pairs are monofunctional.
- Acyl-carrier proteins (ACP) could be interchanged among different synthases without affecting product structure.
- A specific ketoreductase (KR) reduced polyketide chains of different lengths, probably after the complete polyketide chain had been synthesized.
- Regardless of chain length, this ketoreductase reduces the C-9 carbonyl from the carboxyl end.
- The regiospecificity of the first cyclization is controlled by the KS/AT and/or the CLF.
- A specific cyclase (CYC), responsible for catalyzing the second cyclization reaction, evidently can discriminate between intermediates of different chain lengths and degrees of reduction.

Such findings form the basis for the systematic exploration of structure-function relationships in these complex systems and the rational design of novel polyketides that are based on minimal polyketide-forming systems.

With the further elucidation of PKS strategies, it is envisioned that combinatorially generated PKS systems will allow the synthesis of polyketide libraries that contain thousands of new molecules. These libraries could then be screened for molecules with any type of property, ranging from pharmacological to materials. Clearly, the upcoming years of modular PKS research promise to be very exciting ones, especially when one considers the richness and engineering potential of these fascinating enzyme systems.

6.3.3. VITAMINS

Vitamin C

Established commercial production of the vitamin C (ascorbic acid) precursor 2-keto-L-gluconic acid (2-KLG) involves a two-stage fermentation

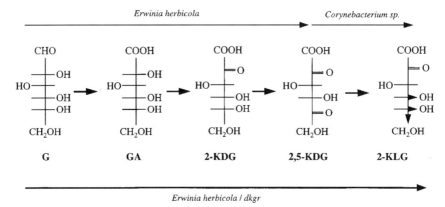

FIGURE 6.12 Biological conversion of glucose (G) to 2-keto-L-gluconic acid (2-KLG). Comparison of a two-stage process involving *Erwinia herbicola* and *Corynebacterium* sp. with a single step process based on *E. herbicola* expressing heterologous diketo-D-gluconate reductase (DKGR). Intermediates: GA, gluconic acid; 2-KDG, 2-keto-D-gluconic acid; 2,5-DKG, 2,5-diketo-D-gluconic acid.

process. The first utilizes *Erwinia herbicola* to convert glucose to 2,5-diketo-D-gluconate, which is subsequently converted to 2-KLG in a fermentation step by a *Corynebacterium* sp. In an effort to change this into a one-stage process, the *E. herbicola* was genetically transformed with the *Corynebacterium* gene that encodes 2,5-DKG reductase (DKGR), which catalyes the 2,5-DKG to 2-KLG conversion (Anderson *et al.*, 1985; Grindley *et al.*, 1988) (Fig. 6.12). After optimizing the culture conditions, these recombinant strains of *Erwinia* produced about 120 g L^{-1} 2-KLG within 120 with a molar yield from glucose of about 60%. Followup studies illustrated the potential economic advantages of the metabolically engineered strain for vitamin C production and have led to a number of U.S. patents (Anderson *et al.*, 1991; Hardy *et al.*, 1990).

Biotin

Biotin is an essential nutrient for many microorganisms and animals. It acts as cofactor for enzymes involved in fatty acid and carbohydrate metabolism and is used in animal feed and as an additive in industrial fermentation processes. Currently, biotin is produced by a complicated and expensive chemical synthesis method. Even though current economics favor chemical synthesis, further improvements in microbial biotin production processes could make bioconversions competitive with existing technologies. The metabolic pathway for biotin synthesis from pimelic-CoA was first described in *E. coli* (Barker and Campbell, 1980), and all enzymes involved in biotin synthesis from pimelic acid in *B. sphaericus* were identified (Izumi *et al.*, 1981). The finding that *B. sphaericus* secreted significant quantities of biotin pathway intermediates led to the isolation of the *bio* genes from this organism. The genes involved in biotin synthesis, organized in two clusters *bioXWF* and *bioDAYB*, have recently been cloned on *E. coli* vectors (Gloecker *et al.*, 1990). *E. coli* transformed with these genes produced up to 457 mg L^{-1} of biotin and 350 mg L^{-1} biotin intermediates (Sabatié *et al.*, 1991).

Vitamin A

Another example of the application of metabolic engineering to convert native metabolic intermediates to desirable endproducts is the production of β-carotene precursor for vitamin A. In the past, many species of algae and fungi (*e.g.*, *Neurospora crassa*, *Penicillium sclerotiorum*, *Phycomyces blakesleeanus*) and also yeasts (*Rhodotorula*) were considered for use in β-carotene production, but were found to be unsuitable (Ninet and Renaut, 1979). Because the precursor for carotenoid biosynthesis, geranylgeranyl pyrophosphate, exists in many organisms for the synthesis of sterols, hopanoids, and terpenes, it can be utilized with appropriate genetic engineering to produce β-carotene. Recently, the *Erwinia uredovora* genes for the biosynthesis of cyclic carotenoids including β-carotene have been cloned and analyzed (Misawa *et al.*, 1990). Following the genetic transformation of *Z. mobilis* and *Agrobacterium tumefaciens* with four of the β-carotene genes, yellow colonies were obtained on agar plates (Misawa *et al.*, 1991). Even though neither strain is a native producer of β-carotene, the transconjugants produced 220-350 mg DW of the vitamin A precursor in liquid culture. It is also suggested that β-carotene-producing *Z. mobilis* strains, which is used on a large scale for ethanol production, can be subsequently used as an animal feedstock due to its enhanced nutritional value.

In a related effort, again involving heterologous expression of six of the *Erwinia* carotenoid genes, an array of geranylgeranyl pyrophosphate byproducts was obtained in *S. cerevisiae*. One or more of the following products was detected, depending on the number of genes in the linear pathway that were actually expressed: phytoene, lycopene, β-carotene, zeaxanthin, and zeaxanthin diglucoside (Ausich *et al.*, 1991).

6.3.4. BIOPOLYMERS

Improvement of polymer production by organisms (*e.g.*, xanthan gum and bacterial cellulose), as well as the production of new biological polymers, is yet another major application of metabolic engineering (Peoples and Sinskey, 1990). Approximately 93% of fossil resources consumed in the world is for energy production, while only 7% is used by industries for the production of a variety of organic chemicals, including solvents and plastics (Eggersdorfer et al., 1992). Replacement of a fraction of synthetic plastics with biodegradable polymers produced from renewable resources is, thus, likely to have only a marginal impact on the overall consumption of fossil fuels. Greater use of biodegradable plastics could, however, significantly contribute to solving problems associated with environmental pollution and waste management. The same intrinsic qualities of durability and resistance to degradation that have made plastics ideal industrial and consumer materials are now regarded as a source of environmental and waste management problems. In contrast, biodegradable polymers are either partly or fully composed of material that can be degraded either by nonenzymatic hydrolysis or by the action of enzymes secreted by microorganisms. Although some polymers, such as blends of starch and polyethylene, are only partly biodegradable, polymers such as poly(3-hydroxybutyric acid) [P(3HB)] are 100% biodegradable as they can be converted into carbon dioxide and energy by microorganisms, such as bacteria, fungi, and algae. More than a dozen biodegradable plastics are now on the market, representing a range of properties suitable for various consumer products, with estimates of the current global market for these biodegradable plastics of up to 1.3 billion kg per year (Lindsay, 1992).

Poly(hydroxyalkanoate)s

Among the various biodegradable plastics available, poly(hydroxyalkanoate)s (PHAs) are attracting growing interest. PHAs are a class of intracellular carbon and energy storage materials accumulated by numerous bacteria in response to environmental limitations (e.g., oxygen or nitrogen deprivation and sulfate or magnesium limitation). Changes in environmental conditions often cause dramatic shifts in intermediary metabolism. Many of these shifts are controlled by global regulatory networks capable of coordinated induction or repression of enzyme repertoires (Chapter 5). These polymers have recently attracted considerable attention because of their potential use as biodegradable thermoplastics. By changing the carbon source and/or the bacterial strain used in the fermentation process, it is possible to produce biomaterials having properties ranging from stiff and brittle plastics to rubbery polymers.

Poly(hydroxybutyrate) was first discovered in 1926 as a constituent of the bacterium Bacillus megaterium. Since then, PHB and related PHAs have been shown to occur in over 90 genera of bacteria. The majority of PHAs are composed of R-(-)-3-hydroxyalkanoic acid monomers ranging from 3 to 14

$$\begin{array}{cc} R & O \\ | & || \\ \end{array}$$
$$[\text{-O-CH-CH}_2\text{-C-}]_x$$

R=hydrogen	3-hydroxypropionate	(3HP)
R=methyl	3-hydroxybutyrate	(3HB)
R=ethyl	3-hydroxyvalerate	(3HV)
R=propyl	3-hydroxycaproate	(3HC)
R=butyl	3-hydroxyheptanoate	(3HH)
R=pentyl	3-hydroxyoctanoate	(3HO)
R=hexyl	3-hydroxynonanoate	(3HN)
R=heptyl	3-hydroxydecanoate	(3HD)
R=octyl	3-hydroxyundecanoate	(3HUD)
R=nonyl	3-hydroxydodecanoate	(3HDD)

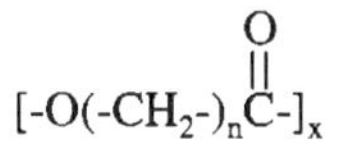

$$\begin{array}{c} O \\ || \\ \end{array}$$
$$[\text{-O(-CH}_2\text{-)}_n\text{C-}]_x$$

n=3	4-hydroxybutyrate	(4HB)
n=4	5-hydroxyvalerate	(5HV)

FIGURE 6.13   Structures of major biological poly(hydroxyalkanoate)s.

carbons in length (Fig. 6.13). PHAs synthesized by bacteria can be broadly subdivided in two groups: short-chain PHAs with C3-C5 monomers (*e.g.*, *Alcaligenes eutrophus*) and medium-chain PHAs with C6-C14 monomers (*e.g.*, *Pseudomonas oleovorans*). Over 40 different PHAs have been characterized, with some polymers containing unsaturated bonds or other functional groups (Steinbüchel, 1991).

PHB is the most widespread and thoroughly characterized PHA. Most knowledge of PHB biosynthesis has been obtained from the bacterium *Alcaligenes eutrophus*, which derives PHB from acetyl-CoA by the sequential action of three enzymes (Fig. 6.14). The first enzyme of the pathway, 3-ketothiolase (or β-ketothiolase), catalyzes the reversible condensation of two acetyl-CoA molecules to form acetoacetyl-CoA. Acetoacetyl-CoA reductase then reduces acetoacetyl-CoA to R-(-)-3-hydroxybutyryl-CoA, which is then polymerized by the action of PHA synthase to form PHB. Molecular studies have revealed that the genes for these three enzymes are organized in a single operon. PHA typically is produced as a polymer of 10^3-10^4 monomers, which accumulates up intracellularly as inclusion of 0.2-0.5 μm in diameter. In *A. eutrophus*, PHB inclusions can typically accumulate to 80% of the dry weight when bacteria are grown in media containing excess

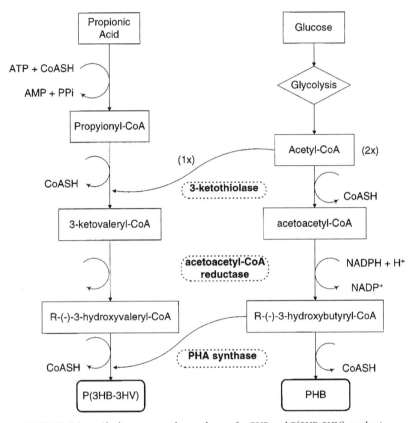

FIGURE 6.14   *Alcaligenes eutrophus* pathways for PHB and P(3HB-3HV) synthesis.

carbon, such as glucose, but limited in one essential nutrient, such as nitrogen or phosphate. Under these conditions, PHB synthesis acts as a carbon reserve and an electron sink. When growth conditions are restored by the addition of phosphate or nitrogen, PHB is catabolized to acetyl-CoA and PHB returns to preinduction levels.

Induction studies on 3-ketothiolase and acetoacetyl-CoA reductase revealed that both enzymatic activities increase markedly in response to PHB-stimulating limitations. These experiments indicate that the PHB pathway may exhibit a mode of transcriptional control that resembles these of other metabolic pathways that are induced by environmental stress. Examples of such global regulatory networks include the heat shock regulon, the *pho* regulon, and the carbon starvation regulon (see Section 5.4). Recently, the *A. eutrophus* PHB biosynthetic genes (*phaA, phaB, phaC*) were cloned and expressed in *E. coli* (Peoples and Sinskey, 1989; Schubert *et al.*, 1988;

Slater *et al.*, 1988) and various species of *Pseudomonas* (Timm and Steinbüchel, 1990). The *A. eutrophus* PHB pathway was found to be functional in all recombinant strains, with PHB accumulation representing a significant portion of the cellular dry matter when growth took place in excess carbon source under nitrogen limitation. Interestingly, *E. coli* clones produced PHB to approximately 50% of the level achieved in *A. eutrophus* H16, while expressing reductase levels that were less than 2% of reductase levels in *A. eutrophus* H16. Further subcloning identified two distinct forms of *A. eutrophus* 3-ketothiolase, one believed to serve a biosynthetic role and the other a catabolic role. The high levels of PHB achieved in certain recombinant *E. coli* strains (up to 90% of cell dry weight) are indicative of either a high degree of transcriptional versatility or a high degree of transcriptional homology between the various strains. Another interesting result was the fact that recombinant *P. aeruginosa* strains possess three different pathways for the synthesis of poly(hydroxyalkanoate)s. When these cells are grown on glucose, they accumulate a polymer consisting β-hydroxybutyrate, β-hydroxydecanoate, and β-hydroxydodecanoate as the main constituents and of β-hydroxyoctanoate and β-hydroxyhexanoate as minor constituents (Timm and Steinbüchel, 1990).

Copolymers

At present, the PHA copolymer of greatest industrial interest is poly(3-hydroxybutyric-*co*-3-hydroxyvaleric acid) [P(3HB-*co*-3HV)] due to its enhanced flexibility over the homopolymer [P(3HB)]. Addition of propionic acid or valeric acid to the growth medium containing glucose leads to the production of a random copolymer composed of 3-hydroxybutyrate and 3-hydroxyvalerate (Fig. 6.14). This random copolymer is currently produced commercially under the brand name Biopol by Monsanto by fermentation of the bacterium *A. eutrophus* on glucose and propionic acid. Incorporation of various C3-C5 units in PHA is possible because of the broad specificity of the bacterial enzymes involved in PHA synthesis. For instance, two 3-ketothiolases have been detected in *A. eutrophus*, which together accept from C4 to at least C10 3-ketoacyl-CoAs, and the acetoacetyl-CoA reductase has been shown to be active with C4-C6 3-ketoacyl-CoAs. By altering the intermediate metabolism, Slater *et al.* have constructed an *E. coli* strain that produces this copolymer at high titers (Slater *et al.*, 1992). The strategy was based on genetic elimination of the transcriptional regulation of *E. coli* genes of the propionate pathway (constitutive expression), thus resulting in a strain that can efficiently take up propionate and incorporate it into the copolymer [P(3HB-*co*-3HV)]. Furthermore, this strategy introduced the ability to control the ratios of the two polymers in P(3HB-*co*-3HV) by manipulating propionate and/or glucose concentrations in the growth medium.

Substrates used to produce the biodegradable polymer poly(hydroxy-butyrate) (PHB) in *A. eutrophus* include fructose, glucose, acetic or butyric acid, and a mixture of H_2 and CO_2. *A. eutrophus* does not normally utilize ethanol as a carbon source, but an ethanol-utilizing strain was engineered by expressing the gene for ethanol dehydrogenase that converts ethanol to acetaldehyde, which then enters the acetyl-CoA pool, a precursor of PHB. More importantly, expression of the same gene allowed for the utilization of propanol, which leads to the formation of the copolymer poly(hydroxy-butyrate-valerate) (PHBV), a compound with a reduced melting point and an improved polymer processibility compared with PHB (Alderete *et al.*, 1993). Up to 74% PHB by weight was obtained with 63 g L^{-1} dry cell mass. The copolymer content increased with a higher fraction of propanol in the feed and reached a maximum of 35.2 mol % from pure propanol.

A central issue of metabolic engineering in biopolymer production is the maximization of polymer formation through the coordinated amplification of the thiolase, reductase, and polymerase enzymes. Simple maximization of the activity of all three is not the solution to this problem. PHB synthesis depends on acetyl-CoA supply and NADPH availability. The first is maximized by increasing the rate of glycolysis; however, an increase in glycolytic flux will reduce the flux into the pentose phosphate pathway and, hence, NADPH generation. Therefore, maximization of PHA production is more an issue of optimally balancing the flux distribution at the G6P branch point rather than straightforward amplification of the three enzymes. This balance, incidentally, may depend on the growth conditions, as the need for reduction power and carbon may shift as cells pass from the growth to the production phase.

Another important point to be made is that the relative activities of the preceding three enzymes (thiolase, reductase, and polymerase) have a profound impact on the quality of the product: Increasing the polymerase activity while keeping the other two enzyme activities constant yielded PHB with lower molecular weight, a counterintuitive result that can be explained when one considers the actual mechanism of polymerization. Finally, the production of the copolymer can be accomplished by feeding propionic acid in the fermentation or by engineering the organism to provide its own supply of propionic acid, such as through the threonine degradation pathway.

Plant Poly(hydroxyalkanoate)s

Recently the poly(hydroxyalkanoate) pathway has also been expressed in crop plants, a very attractive system for such a purpose, with the potential for producing large amounts of several chemicals at low cost. Synthesis of

PHB in plants initially was explored with the expression of PHB biosynthetic genes of the bacterium *A. eutrophus* in the plant *Arabidopsis thaliana* (Poirier *et al.*, 1992). Although of no agricultural importance, *A. thaliana* was chosen because of its extensive use as a model system for genetic and molecular studies in plants, and because it is closely related to the oil-producing crop rapeseed, a target crop for PHB production on an agricultural scale. Of the three enzymes required for PHB synthesis, only 3-ketothiolase is present in plants. In order to complete the pathway, *A. eutrophus* genes encoding the acetyl-CoA reductase and PHA synthase were expressed in transgenic *A. thaliana*, with activity found to be localized in the cytoplasm. This initial attempt resulted in only 0.14% dry weight yield (approximately 2 orders of magnitude below commercially attractive levels) and had an adverse effect on cell growth.

A second generation of transgenic plants producing PHB proved to be more successful. In plants, biosynthesis of fatty acids from acetyl-CoA occurs in the plastid. The plastid is therefore a site of high carbon flux through acetyl-CoA. This flux is particularly enhanced in the seeds of oil-accumulating plants, such as *Arabidopsis*, where up to 40% of the seed dry weight is triglycerides. Furthermore, the plastid is also the site of starch accumulation, and as such it can accommodate large amounts of inclusions without disruption of organelle function. Expression of the PHB pathway in the plastid recently has been demonstrated in transgenic *A. thaliana* (Nawrath *et al.*, 1994). Plastid expression was achieved by fusing the transit peptide of the small subunit of ribulose bisphosphatase carboxylase to the N-terminus of 3-ketothiolase and acetoacetyl-CoA reductase. The PHB content gradually increased over the life span of the plant, reaching a maximum of 10 mg/g fresh weight, representing approximately 14% dry weight. Thus, redirection of the PHB from the cytoplasm to the plastid resulted in a 100-fold increase in PHB production.

Fructan

Fructan is a poly(fructose) molecule naturally produced as a storage compound in a limited number of plants and characterized by a low degree of polymerization (5–60 units). Such polymers can be hydrolyzed enzymatically or chemically to yield fructose, which is becoming an increasingly popular sweetener in many food products. Because oligofructose molecules are sweet, fructans themselves can be utilized directly as natural sweeteners. Also, the human digestive system has no enzymes that can degrade the $\beta(2 \rightarrow 1)$ or $\beta(2 \rightarrow 6)$ glycosidic linkages found in fructan, making this sugar attractive as a low-calorie food ingredient. Besides plants, microorganisms are capable of producing fructans of very high molecular weight ($> 100,000$ units). For example, in *Bacilli*, *Pseudomonas*, and *Streptococci*, extracellular fructosyltransferase converts sucrose to bacterial fructan, often called levan. The main reaction for fructan biosynthesis is nGF (sucrose) $\rightarrow$ G-Fn (fructan) $+ n - 1G$. For this purpose the *SacB* gene of *B. subtilis*, which encodes the fructosyltransferase enzyme commonly known as levansucrase, was modified and introduced into tobacco plants, resulting in transgenic plants that can accumulate fructans. Production levels achieved a range from 3 to 8% dry weight, and the size and properties of this fructan were found to be similar to those of *B. subtilis*. An important feature of the recombinant fructans is their stability in plants, which makes this work attractive for applications in food and nonfood products.

Xanthan Gum

Xanthan is an extracellular polysaccharide produced by the Gram-negative bacterium *Xanthomonas campestris*. Its unique rheological properties, such as high viscosity and pseudoplasticity, account for its extensive use in a variety of food and industrial applications. The chemical structure consists of a cellulosic $\beta(1 \rightarrow 4)$-glucose backbone with trisaccharide side chains composed of two mannose residues and one glucuronic acid residue attached to alternate glucose molecules on the backbone. Typically, the mannose sugars are acetylated and pyruvylated at specific sites, but to various degrees. Many of the genes involved in exopolysaccharide synthesis are often clustered. A cluster of genes essential for xanthan synthesis has also been isolated in *X. campestris* (Barrere *et al.*, 1986). Recent studies have illustrated the potential of recombinant DNA technology for altering the structure and properties of xanthan gum. A plasmid containing several xanthan biosynthetic genes increased the production of xanthan by 10%. Furthermore, by cloning and overexpressing the gene for the enzyme ketal pyruvate transferase, the extent of pyruvylation of the xanthan side chains was increased by up to 45% (Harding *et al.*, 1987). Conversely, using transposon mutagenesis, a strain could be constructed that formed xanthan with a severely reduced pyruvate content (Marzocca *et al.*, 1991). Such studies are illustrative of the promise of metabolic engineering in both the manipulation of product structure and the elucidation of xanthan synthesis.

6.3.5. BIOLOGICAL PIGMENTS

Indigo

One of the classic examples of metabolic engineering is the production of indigo in genetically engineered *E. coli* carrying the naphthalene dioxygenase gene from *Pseudomonas putida*, which catalyzes the final step in indigo biosynthesis (Ensley, 1985) (see Fig. 6.15). Indigo, or indigotin, occurs as a glucoside in many plants and has been used throughout history as a blue dye. For the past century, indigo manufacturing has been carried out by chemical synthesis leading to indoxyl, which is finally oxidized to indigo. By using selective cultivation techniques, a soil organism (*Pseudomonas indoloxidans*) was isolated in 1927 that could also decompose indole (Fig. 6.4) with the formation of blue crystals later identified as indigotin. Even though several other microbes were isolated that were also able to produce indigo from indole, none was actually put to use in the large-scale microbial synthesis of indigo due to (i) low availability of the precursor indole and (ii) low activity of naphthalene dioxygenase (NDO), the final indigo biosynthetic enzyme. Indigo production in these early strains required the cofeeding of tryptophan or free indole, whose costs limited their use in commercial processes.

Early attempts focused on enhancing the conversion of indole by overexpressing NDO from *Pseudomonas putida*. Superior enzyme activity was obtained by combining the first four genes comprising the naphthalene dioxygenase enzyme system into a multicopy plasmid under the control of the strong λp_L promoter. Further genetic manipulations were necessary to improve the stability of heterologous NDO in *E. coli*. This work resulted in high-level expression of stable NDO that can be utilized for indigo biosynthesis.

In the meantime, parallel efforts investigated synthesis of the indigo precursor, indole, directly from glucose. Normally, indole is present in the cell either in the form of indole 3-glycerol phosphate (IGP) or as a tryptophan moiety. In recombinant *E. coli*, tryptophan biosynthesis is carried out by the five gene products of the tryptophan operon (Figs. 6.16 and 6.4). For both bacteria and fungi, chorismate is the major branch point compound for aromatic amino acid biosynthesis that includes phenylalanine, tyrosine, and tryptophan. Also present in the cell is the enzyme tryptophanase, which degrades tryptophan and releases indole. Early attempts, however, to stimulate indole production by overexpression of tryptophanase did not prove to be very successful due to apparent limitation by the total tryptophan synthesis flux. To correct for this, enzymes in the tryptophan synthesis pathway were amplified. In particular, the *trpB* moiety was modified by site specific mutagenesis which increased indole biosynthesis by more than an order of magnitude (Murdock *et al.*, 1993). An *E. coli* production strain was finally developed that combined both enhanced indole production (mutated *trpB*) and enhanced indole conversion to indigo (NDO), which can be used for *de novo* indigo biosynthesis from glucose.

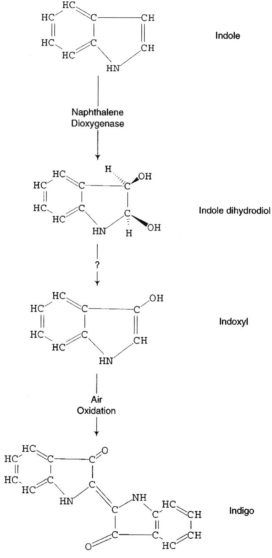

FIGURE 6.15 Indigo biosynthesis from indole by naphthalene dioxygenase (NDO). Indole (see Fig. 6.4) is oxidized by NDO to form either an unstable dihydrodiol or indoxyl, which is further condensed to form indigo by air oxidation.

6.3.6. HYDROGEN

Hydrogen could be the ultimate substitute for polluting and irreplaceable fossil fuels for transportation, direct heating, or electricity generation. Because hydrogen combustion emits only water vapor and small quantities of nitrogen oxides, hydrogen-fueled cars and other devices do not impact global warming (carbon dioxide) and contribute only insignificantly to air pollution. Prototype vehicles produced by manufacturers such as Mercedes Benz show that hydrogen-powered cars can be practical in terms of performance, comfort, and safety. Hydrogen is no more hazardous than methane or gasoline, and it is, in fact, used routinely in industrial processes. Also, the fact that the U.S. space program has relied on hydrogen as the fuel of choice for nearly 40 years provides a knowledge base for extended uses of this fuel.

Conventional electrolysis techniques for extracting hydrogen from water, though inexhaustible, still require slightly more energy than hydrogen would yield upon combustion. Other means of hydrogen production have been proposed, such as those utilizing wind or solar power (photovoltaic technology), gasification, and pyrolysis.

Fermentation or enzymatic techniques are also currently being investigated as potential biological routes to hydrogen production (Kitani and Hall, 1989; Taguchi *et al.*, 1995) (Fig. 6.17). It is well-known that selected microorganisms can efficiently produce hydrogen as an end product of metabolism. Metabolic engineering techniques will prove very useful in redirecting cellular metabolism toward hydrogen production above physiological levels. Several genes involved in hydrogen synthesis have been isolated thus far, such as the hydrogenase gene from *Citrobacter freudii* cloned in *E. coli* (Kanamayia *et al.*, 1988).

Recently, the possibility of *in vitro* hydrogen production has also been investigated (Woodward *et al.*, 1996). This system consists of two enzymes, namely, glucose dehydrogenase (GDH) isolated from *Thermoplasma acidophilum* and hydrogenase from *Pyrococcus furiosus*. GDH catalyzes the oxidation of glucose to glucono-δ-lactone, which further hydrolyzes to gluconic acid using NADH or NADPH as a cofactor. Even though bacterial hydrogenases rarely interact with NADPH because of its insufficiently low potential, hydrogenases from *P. furiosus* and *A. eutrophus* have been shown to use NADPH as an electron donor. Woodward *et al.* (1996) demonstrated that the combination of GDH and hydrogenase was capable of hydrogen production from glucose *in vitro* (Fig. 6.17). Stoichiometric yields of hydrogen were produced from glucose with continuous recycling of the cofactor

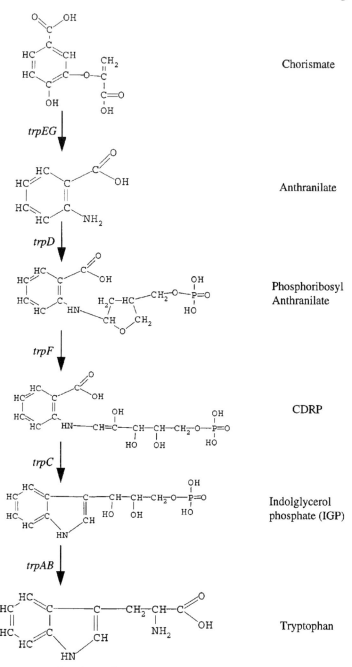

FIGURE 16 Tryptophan biosynthesis. Structural gene designations: *trpAB*, tryptophan synthase; *trpC*, indole glycerol phosphate synthase; *trpD*, anthranilate phosphoribosyl transferase; *trpEG*, anthranilate synthase.

(at least 20 times). This newly discovered pathway seems to be an attractive alternative for hydrogen production from renewable sources without the immediate formation of waste gases such as CO_2 and CO. One limitation is the need to identify further uses for the enormous amounts of gluconic acid that would be produced as a byproduct of even a small-scale hydrogen production plant. Future generations of such a process could involve immobilized forms of these enzymes for continuous hydrogen synthesis.

Green algae, such as *Chlamydomonas reinhardtii*, could provide an appropriate mode of capturing light energy in fuel hydrogen (Cinco *et al.*, 1993; Lee *et al.*, 1995). In a recent study (Greenbaum *et al.*, 1995), it was shown that mutation of the membrane-bound photosystem I reaction center does not disable the complete photosynthesis. This original finding contradicts the premise that both photosystems I and II are required for the conversion of light energy to chemical energy, and it also doubles the maximum theoretical conversion rate of light to chemical energy from 10 to 20%.

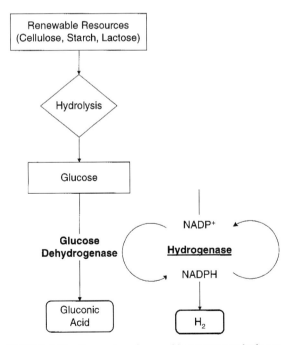

FIGURE 6.17   Conversion of renewable resources to hydrogen.

6.3.7. PENTOSES: XYLITOL

Xylitol, a sugar alcohol, is a good anticariogenic sweetener that does not require insulin for its digestion by diabetics (Emodi, 1978). In nature xylitol is found in certain fruits and vegetables in small amounts, making its quantitative extraction difficult and uneconomical. Currently xylitol is manufactured chemically in alkaline conditions by catalytic reduction of xylose derived from hemicellulose hydrolysates. Lately, much attention has focused on the microbial production of xylitol from D-xylose. Xylitol has been reported to be produced by yeasts, especially species of genus *Candida*, such as *C. pelliculosa* (Nishio *et al.*, 1989), *C. boidinii* (Vongsuvanlert and Tani, 1989), *C. guillliermondi* (Meyrial *et al.*, 1991), and *C. tropicalis* (Gong *et al.*, 1981), *Petromyces albertensis* (Dahiya, 1991), by bacteria such as *Enterobacter liquefaciens* (Yoshitake *et al.*, 1973), *Corynebacterium* sp. (Yoshitake *et al.*, 1971), and *Mycobacterium smegmatis* (Izumori and Tuzaki, 1988). Yeasts generally possess the first two enzymes needed for the metabolism of xylose: xylose reductase (XR) and xylitol dehydrogenase (XDH) (Fig. 6.7). Efforts to increase xylitol productivities and yields through culture optimization have yielded moderate results (0.32-2.67 g L^{-1} h^{-1}) (Horitsu *et al.*, 1992). Therefore, emphasis has been placed on genetic modifications that would enhance both the yield and productivity of xylose conversion to xylitol.

In an initial study, the *P. stipitis* XR was chosen for the construction of xylitol-producing recombinant yeasts (Hallborn *et al.*, 1991). The choice was based on the high specific activity of XR and the fact that this particular XR uses both NADH and NADPH as cofactors (Verduyn *et al.*, 1985). Due to the lack of XDH (needed for NADH regeneration), the recombinant strain was studied on medium containing both glucose and xylose. Xylose utilization commenced after glucose exhaustion and proceeded to give about 97% theoretical yield conversion of xylose to xylitol at a specific productivity of 0.08 g (gcells h)$^{-1}$. A later study, aimed at expanding the substrate utilization range of ethanologenic *S. cerevisiae* by overexpressing the *P. stipitis* genes for XR and XDH, fortuitously resulted in the production of 35 g L^{-1} xylitol (Tantirungkij *et al.*, 1993).

6.4. IMPROVEMENT OF CELLULAR PROPERTIES

This type of metabolic engineering is aimed at the organism as a whole; thus, it is also referred to as cellular engineering. There are already a number of successful applications of cellular engineering that involve a wide range of organisms, from bacteria to animal cells. Such applications have been targeted at improving specific growth rates and growth yields, providing resistance to toxic compounds, improving the secretion of a specific product, enhancing drought and salt tolerance of plant cells, and altering glycosylation sequences of recombinant polypeptides. It is an area of great challenges and also vast opportunities.

6.4.1. ALTERATION OF NITROGEN METABOLISM

An early success of metabolic engineering was the alteration of the nitrogen assimilation pathway of the methylotrophic bacterium *Methylophilus methylotrophus* to enhance the yield of single-cell protein (SCP). *M. methylotrophus* was the industrial choice for the production of SCP from methanol due to its high carbon conversion efficiency, methanol tolerance, and nutritional profile. However, the ammonia assimilation pathway of this organism has a major drawback: it utilizes the glutamine synthase (GS) and glutamate synthase system (GOGAT) that requires mol of ATP for every mole of ammonia transported into the cell (see Section 2.4.1). In contrast, the corresponding *E. coli* nitrogen assimilation pathway uses glutamate dehydrogenase (GDH), which requires no ATP consumption (Fig. 6.18). *M. methylotrophus* probably uses the energetically suboptimal pathway for ammonia assimilation because it evolved in an environment of low ammonia concentrations, as GS has a much higher affinity for ammonia than GDH. The glutamate dehydrogenase gene (*gdh*) of *E. coli* cloned on a shuttle vector was shown to complement *gs* mutants of *M. methylotrophus* (Windass *et al.*, 1980). As a result, the engineered organism exhibited higher methanol conversion into cellular carbon, presumably because ammonia utilization is more energy-efficient via GDH than via the coupled GS/GOGAT pathway. The efficiency of carbon conversion was increased by 4-7%.

This work, which was one of the first industrial applications of metabolic engineering, illustrated that the properties of organisms that evolved to maximize the chances for survival in their natural habitat are not necessarily optimal in the artificial environment of a large-scale bioreactor.

6.4.2. ENHANCED OXYGEN UTILIZATION

A common engineering challenge in large-scale aerobic fermentations is ensuring an adequate level of dissolved oxygen to achieve the desired cell growth and productivity. Under microaerobic conditions that can arise, for

A. GS/GOGAT Pathway

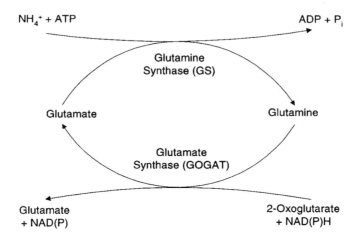

B. GDH Pathway

$$2\text{-Oxoglutarate} + NH_4^+ + NAD(P)H \quad \text{-----GDH}\longrightarrow \quad \text{Glutamate} + NAD(P)$$

FIGURE 6.18   Pathways of bacterial ammonia assimilation.

example, from nonideal mixing, elements of both respiration and fermentative metabolism are active and compete for accomplishing energy synthesis and redox balance. Such oxygen fluctuations inevitably lead to undesirable physiological events, such as the elicitation of global and specific oxygen regulation responses that activate or repress key enzymes of central carbon metabolism. Such survival responses usually are manifested as alterations in growth rate and product formation (Onken and Leifke, 1989).

A recent finding that may in part alleviate the undesirable consequences of oxygen fluctuation and hypoxic environments is the cloning of the hemoglobin gene (*vgb*) from the bacterium *Vitreoscilla* (VHb) (Khosla and Bailey, 1988a,b). Although its precise *in vivo* function has not yet been established, VHb is thought to serve as an oxygen-binding protein enabling *Vitreoscilla* to survive in microaerobic conditions characteristic of its natural habitat. Recent studies indicate that VHb increases the number of protons extruded per reduced oxygen atom across the cytoplasmic membrane and enhances the ATPase-catalyzed ATP synthesis rate in microaerobic *E. coli* (Chen and Bailey, 1994; Kalio *et al.*, 1994) (see Section 2.3.3). Flux distribution analysis (Chapter 8) also revealed that VHb^+ cells have a smaller ATP synthesis rate from substrate-level phosphorylation, but a larger overall ATP production rate under microaerobic conditions (Tsai *et al.*, 1995a-c). Furthermore, results from *cyo/cyd* (aerobic/microaerobic terminal oxidases) mutants suggest that the expression of VHb in *E. coli* increases the level and activity of terminal oxidases and thereby improves the efficiency of microaerobic respiration and growth (Tsai *et al.*, 1995). On-line culture NAD(P)H fluorescence and redox potential measurements (CRP) suggested that Vhb buffers intracellular redox potential perturbations caused by intracellular DO fluctuations (Tsai *et al.*, 1995).

Motivated by the hypothesis that this enzyme could possibly be beneficial for growth in oxygen-limiting environments, the *vgb* gene was transferred in various industrially important microbes. Heterologous expression of this protein in a wide range of hosts has demonstrated that it elicits *in vivo* effects of reduced oxygen starvation, improved cell growth, and enhanced product formation (DeModena *et al.*, 1993; Khosla and Bailey, 1988a,b; Khosravi *et al.*, 1990; Magnolo *et al.*, 1991). For example, *E. coli* that carried a single copy of this gene integrated on its chromosome synthesized total cell protein more rapidly than an isogenic wild-type strain under oxygen-limiting conditions. Furthermore, coexpression of VHb increases the expression of cloned β-galactosidase, chloramphenicol acetyltransferase (CAT), and α-amylase by 1.5- to 3.3-fold relative to controls in oxygen-limiting *E. coli* cultures. In other cases, expression of this protein achieved a 13-fold increase in the production of an antibiotic in *Streptomyces* and a 1.2-fold increase in the production of amino acids in *Coryneform* bacteria. Exogene, the company founded to explore this technology, has also successfully expressed Vhb in *Penicillium* and mammalian cells. Furthermore, in the field of bioremediation, transformation of *Xanthomonas maltophilia* with *vgb* resulted in an enhanced efficiency of benzoic acid conversion to biomass (Liu *et al.*, 1996).

6.4.3. PREVENTION OF OVERFLOW METABOLISM

Currently, one of the current major technical challenges in recombinant protein production processes employing *E. coli* is to maintain high intracellular product levels at high cell concentrations. This dual goal is difficult to achieve due to the accumulation of inhibitory culture byproducts. During both aerobic and anaerobic growth, carbon and reductant fluxes are balanced by the excretion of acidic byproducts, the most abundant of which is acetate. This weak acid is a well-known growth inhibitor. Most importantly, it reduces the cellular efficiency for the expression of recombinant products and affects the quality of intracellular proteins, apparently by interfering with disulfide bond formation.

Organic acids have been shown to influence cell growth at concentrations that are low in comparison to inhibitory levels of mineral acids. The undissociated forms of short-chain fatty acids produced intracellularly, such as acetic acid, can freely permeate the cell membrane and accumulate in the medium. Subsequently, a fraction of the undissociated acid that is present extracellularly re-enters the cell, where it dissociates given the relatively higher intracellular pH. This means that, in effect, weak acids act as proton conductors (see Section 2.2.1 and Example 2.1). If this process continues undiminished, the intracellular pH will approach the external pH and hence the ΔpH component of the protonmotive force (see Section 2.33 and Example 13.6) will collapse (Diaz-Ricci et al., 1990; Slonczewski et al., 1981). In addition, a low external pH (< 5) can cause almost complete growth stasis (without cell lysis) presumably due to the irreversible denaturation of DNA and protein (Cherrington et al., 1991).

In addition to the preceding effects on cellular energetics, there are many other factors contributing to the inhibitory nature of weak organic acids that make minimization of acetate excretion a prerequisite for optimizing the production yields of recombinant processes (Yee and Blanch, 1992; Zabriskie and Arcuri, 1986). The chemostat data of Jensen and Carlsen (1990), who studied the effects of acetate on the production of human growth hormone in E. coli, clearly illustrate the significance of acetate in this recombinant system. By varying the amount of acetate in the feed, it was determined that acetate levels of 40 mM reduce recombinant protein yields by approximately 35% without having any effect on the biomass yield. This result agrees with the general observation that the acetate threshold that influences recombinant protein yields is usually lower than that that causes notable growth inhibition. In the same study, increasing the acetate level to 100 mM caused a reduction in biomass yield by more than 70%, whereas recombinant product yield declined by a factor of 2. Several other investigators have implicated acetate as an important factor in the deterioration of recombinant process productivities (Brandes et al., 1993; Brown et al., 1985; Curless et al., 1988; Luli and Strohl, 1990; Starrenburg and Hugenholtz, 1991).

It is widely accepted that acetate excretion results from an imbalance between the glycolytic flux and the cell's actual requirements for metabolic precursors and energy. Pyruvate, which is the end product of glycolysis as well as the precursor of acetate, provides a suitable junction for effecting acetate accumulation. The strategy involves the introduction of a heterologous enzyme to catalyze the redirection of surplus carbon flux to a less harmful byproduct than acetate. The *B. subtilis* acetolactate synthase (ALS) enzyme was selected for this purpose on the basis of the fermentation characteristics of this group of microorganisms (Johansen *et al.*, 1975). Table 6.6 compares the amounts of fermentation byproducts excreted by mixed acid producers, such as *E. coli*, with those of members of the butanediol family, namely, *Bacillus subtilis* and *Aerobacillus polymyxa*. Evidently, members of the second group form very small amounts of acids compared with *E. coli* while they convert glucose primarily to the neutral compound 2,3-butanediol.

As shown in Fig. 6.19, butanediol producers normally have two distinct enzymes that convert pyruvate to acetolactate: the "pH 6" acetolactate synthase (ALS) and the acetohydroxy acid synthetase (AHAS). AHAS, an anabolic enzyme found in many microorganisms, also catalyzes the initial steps from pyruvate in the formation of the branched-chain amino acids valine, leucine, and isoleucine. AHAS is also a flavoprotein that is regulated by end product feedback inhibition, such as by valine. On the other hand, ALS does not require FAD for activity, nor is it inhibited by the presence of branched amino acids (Störmer, 1968).

The *alsS* gene from *B. subtilis* encoding the acetolactate synthase enzyme was successfully expressed in *E. coli* (Aristidou *et al.*, 1994a,b). This enzyme

TABLE 6.6 Comparison of Mixed Acid and Butanediol Fermentations[a]

Products[b]	Mixed Acid		Butanediol	
	E. coli	*P. formicans*	*B. subtilis*	*A. polymyxa*
2,3-Butanediol	0.26	-	54.60	65.1
Acetoin	0.19	-	1.56	2.8
Glycerol	0.32	-	56.80	-
Ethanol	50.5	64.0	7.65	66.2
Formate	86.0	105.0	1.32	-
Acetate	38.7	62.0	0.16	2.9
Lactate	70.0	43.0	17.61	-
Succinate	14.8	22.0	1.08	-
CO_2	1.75	-	117.8	199.6
H_2	0.26	-	0.16	70.9
C recovery (%)	94.7	96.0	98.0	101.6
O/R balance	0.91	1.02	0.99	0.99

[a] From Wood, 1961.
[b] mmoles/100 mmoles of fermented glucose.

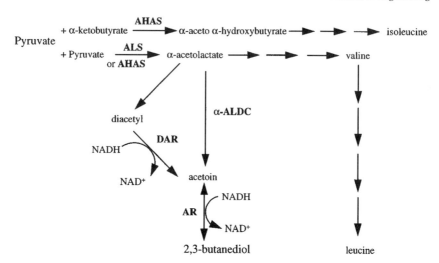

FIGURE 6.19 Comparison of ALS and AHAS enzymes and their roles in branched-chain amino acid synthesis and butanediol formation. Abbreviations: ALS, acetolactate synthase; AHAS, acetohydroxy acid synthetase; α-ALDC, α-acetolactate decarboxylase; AR, acetoin reductase (or 2,3-butanediol dehydrogenase); DAR, diacetyl reductase.

acts at the pyruvate branch point, redirecting excess carbon flux away from acetate and toward the noninhibitory byproduct α-acetolactate. Characterization of the resulting strain indicated that acetate excretion can be maintained below 20mM even in dense cultures employing rich glucose medium. Moreover, the engineered strain is a more efficient host for the production of recombinant proteins (Aristidou et al., 1994): the volumetric expression of recombinant β-galactosidase was found to increase by about 50% in batch cultivations and by about 220% in high cell density fed-batch cultivations. These results demonstrate the successful application of metabolic engineering for the improvement of cellular characteristics.

6.4.4. ALTERATION OF SUBSTRATE UPTAKE

Nutrient uptake by transport through the cell membrane is an important task of all living organisms. In addition to free diffusion, transport mechanisms fall into three general categories, namely, (i) facilitated diffusion, (ii) group translocation, and (iii) active transport (see Section 2.2). Hexoses, like glucose, mannitol, and fructose are primarily transported into the cell by the phosphoenolpyruvatedependent carbohydrate:phosphotransferase system (PTS), which is a group translocation process (Dills et al., 1980; Postma *et al.*, 1993; Saier *et al.*, 1988). The overall process catalyzed by PTS can be summarized as

$$- (PEP)_{IN} - (carbohydrate)_{OUT} + (pyruvate)_{IN} + (carbohydrate\text{-}P_i)_{IN} = 0$$

regardless of the carbohydrate or microorganism. The PTS accomplishes both the translocation and phosphorylation of the carbohydrate in a series of steps that involves a number of cytoplasmic as well as membrane-bound proteins (Section 2.2.3).

In addition to serving an important role in sugar uptake, PEP is also an essential precursor for several specialty chemicals, including aromatic amino acids, indigo, enterobactin, and melanin. Thus, by providing a non-PTS sugar uptake alternative, one should, in principle, save 1 mol of PEP for every glucose consumed. Some of the approaches utilized to improve the availability of PEP for biosynthetic purposes include the use of non-PTS carbon sources, pyruvate recycling to PEP by PEP synthase overproduction (Patnaik and Liao, 1994; Patnaik *et al.*, 1995), and inactivation of pyruvate kinase (Mori *et al.*, 1987). A major breakthrough in this area resulted from deletion of the glucose PTS genes *ptsH*, *ptsI*, and *crr* of an *E. coli* strain so that glucose could no longer be transported into the cell by the PEP-consuming system (Flores *et al.*, 1996). "Revertant" strains subsequently were selected in chemostat cultures, which were later characterized to possess a galactosepermease gene that is able to transport glucose efficiently. The stable, rapidly growing engineered strain NF9 subsequently was used in a genetic background of elevated DAHP synthase (the enzyme that condenses PEP and erythrose-4-P into DAHP), where it was illustrated that PEP saved during glucose transport was redirected into the aromatic pathway.

6.4.5. Maintenance of Genetic Stability

During recent years, the use of plasmid cloning vectors as carriers of genes whose products are of scientific or commercial interest has become common. Some of the desirable characteristics of a plasmid vector include small size, high or controllable copy number depending on the application, strong and controllable promoter for high-level gene expression, and, most importantly, structural and genetic stability. Genetic instability is a primary impediment to industrial utilization of recombinant microorganisms. There can be various reasons leading to such instability, and in general expression vectors that are most effective in directing protein production are usually more unstable. This is perhaps a consequence of the elevated metabolic burden imposed on the cell and can lead to a plasmid-free population within a few generations.

In addition to segregational instability, structural instability through homologous recombination events can lead to plasmid derivatives that no longer produce the desired product.

Prevention of segregational and structural instability has been the subject of intense research in the last 15-20 years. Structural instability generally is minimized by deletion of the *recA* gene from the host cell. This deletion should severely limit homologous recombination between extrachromosomal and chromosomal DNA. In their early work, Csonka and Clark (1979) showed that the Δ(*srl-recA*)306 mutation reduced the rate of recombination by a factor of 36,000. Additionally, Laban and Cohen (1981) have shown that a *recA* point mutation lowers the frequency of recombination events within a plasmid by 100-fold. Segregational instability, on the other hand, is a more challenging issue that requires more elaborate solutions.

Several strategies to obtain genetically stable plasmids have been devised. The most primitive solution seems to be the addition of antibiotics to the growth medium thus selecting for cells harboring the plasmid vector with antibiotic resistance. This approach has a number of obvious drawbacks, *e.g.* the use of costly and contaminating antibiotics. A more pragmatic approach was suggested by Skogman and Diderichsen (Diderichsen, 1986; Skogman *et al.*, 1983). These authors used a special host-vector system where a specific chromosomal mutation giving rise to an auxotrophic host is complemented by a corresponding functional gene on the plasmid vector of interest. Although a powerful approach, it has applicability limitations due to the requirement of special host strains carrying specific mutations.

A technique of wider applicability involves the *hok/sok* locus (formerly *parB*) of plasmid R1. This system was discovered through its ability to mediate efficient stabilization of a variety of plasmids in Gram-negative bacteria (Gerdes, 1988; Gerdes *et al.*, 1986). Later it was illustrated that the increased plasmid maintenance was a consequence of the selective killing of cells that at the point of division lose the *hok/sok*-bearing plasmid. The *hok/sok* locus codes for two RNAs, Hok (host killing) mRNA and Sok (suppression of killing) RNA. The *hok* product is a potent cell-killing agent that destroys bacterial cells from within by damaging the cell membrane. The *sok* product is a *trans*-acting *antisense* RNA (Sok-RNA, see Box 6.3) that represses *hok* gene expression at a post-transcriptional level. The rapid and selective killing of plasmid-free segregates can be explained on the basis that hok mRNA is extremely stable ($t_{1/2} \sim 20$ min), whereas sok mRNA decays very rapidly. Hence, in a plasmid-carrying cell, the Sok RNA prevents synthesis of the Hok protein and the cell remains viable. On the other hand, when a plasmid-free segregate appears, the unstable Sok RNA molecules decay, thereby rendering the stable Hok RNA accessible to translation.

BOX 6.3

Antisense RNA

Natural antisense RNAs are small (15-50 nucleotides), untranslated, regulatory RNA molecules that inhibit the function of their target RNAs, to which they are complementary. Antisense RNAs regulate such diverse functions as plasmid replication, conjugation, and maintenance, transposition, and lysis-lysogeny decisions in bacteriophages. Antisense RNAs recognize their target RNA via an initial reversible contact between a single-stranded loop in the antisense RNA and a complementary loop in the target RNA. The initial contact usually is followed by formation of a thermodynamically very stable duplex between the two RNA molecules.

Antisense RNAs may inhibit the function of their target RNAs by several different mechanisms. For example, the transposable gene of Tn10 is regulated by RNA-OUT, an antisense complementary to the translational initiation region (TRI) of the transposable mRNA. In this case, it has been shown that hybridization of the antisense RNA to the mRNA physically blocks entry of the ribosome at the transposable gene and, thus, is an example of direct target RNA inhibition mediated by an antisense RNA.

Indirect target RNA inhibition by antisense RNAs has been described in several cases. The classic example is perhaps the replication control circuit of plasmid ColE1. The antisense RNA (RNA I) interacts with RNA II, the primer of replication initiation, and thereby induces a secondary structure change several hundred nucleotides downstream from the hybridization site. The change in secondary structure prevents the ribonuclease (RNase) H-dependent cleavage of the primer RNA that is a prerequisite for replication initiation.

Consequently, the Hok protein is synthesized, thus leading to the death of the plasmid-free segregates.

6.5. XENOBIOTIC DEGRADATION

Natural processes, both biological and geochemical, produce enormous amounts of diverse organic compounds, and, over the eons of evolution, different microbes have developed the ability to degrade nearly all such natural compounds as a source of carbon and/or energy. However, many of the tens of thousands of artificial organic compounds produced by humans for industrial or agricultural purposes have no apparent counterparts in the microbial world. Such synthetic novel compounds are called *xenobiotics*, from the Greek word *xenos*, which means "foreign." Xenobiotics (such as the PCBs discussed later) are stable compounds that are also fat-soluble; thus, they become increasingly concentrated as they travel up the food chain.

Xenobiotic degradation is a rapidly developing area that holds great opportunities for metabolic engineering. It involves primarily the utilization of genetically engineered microbes for the degradation of pollutants and is commonly known as *bioremediation*. Bioremediation currently is being used to decrease the organic chemical waste content of soils, ground water, and effluent from chemical plants and food processing and oil sludge from petroleum refineries.

Early attempts for the isolation of xenobiotic-degrading organisms focused on chemostat selection methods. This technique is based on inoculating chemostats with microbial samples from various toxic waste dump sites and selecting for desirable strains. By using this approach, it was possible to isolate a *Pseudomonas* strain capable of degrading halogenated compounds, such as the herbicide 2,4,5-trichlorophenoxyacetic acid (Kilbane *et al.*, 1983). The isolated strain was successful in removing 98% of the pollutant from soil within a week, after which it died rapidly, as it was unable to compete with endogenous strains.

In addition to chemostat selection, rational approaches for designing useful catabolic pathways have also been reported in the last decade, some of which are discussed next.

6.5.1. POLYCHLORINATED BIPHENYLS (PCBS)

PCBs are a class of 209 distinct man-made compounds carrying 1-10 chlorines attached to a biphenyl (Fig. 6.20). From 1929 to 1978, approximately 700,000 tons of PCBs were produced, with several hundred thousand

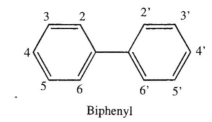

Biphenyl

FIGURE 6.20   General polychlorobiphenyl (PCB) structure. Numbers indicate possible chlorinated sites. Empirical formula: $C_{12}H_{12-n}Cl_n$ where n = 1-10.

tons of that released into the environment. Although there are naturally occurring microbes that can oxidize the mono-, di-, and trichlorinated PCBs, this is not the case for the higher chlorinated PCBs. Cometabolism is probably the most common method of PCB degradation in nature, but this is a slow process as the initial transformations do not provide carbon or energy for growth.

The genes for PCB-degrading enzymes (*bph*A, -B, -C, and -D) have been isolated from the *Pseudomonas* strain LB400 (Fig. 6.21) and subsequently overexpressed in an *E. coli* strain. The PCB-degrading ability of the recombinant strain was comparable to that of LB400 in terms of both specificity and the extent of degradation.

By introducing specific genes into *Pseudomonas* strains it has been possible to construct strains that are able to degrade mixtures of aromatic compounds, such as 3-chloro- or 4-methylbenzoate, which are frequently present in industrial wastes (Rojo *et al.*, 1987). Another interesting enzyme system for the degradation of various aromatic compounds is the ligninase of the white rot fungus *Phanerochaete chrysosporium*, which can decompose a broad range of xenobiotics, even including such diverse structures as benzopyrene and triphenylmethane dyes (Bumpus and Brock, 1988).

6.5.2. BENZENE, TOLUENE, AND *p*-XYLENE MIXTURES (BTX)

This group of petroleum-derived contaminants are water-soluble and, thus, can pollute water resources, posing serious health threats to humans. A number of studies have focused on the development of genetically engineered strains that would convert such pollutants, either individually or

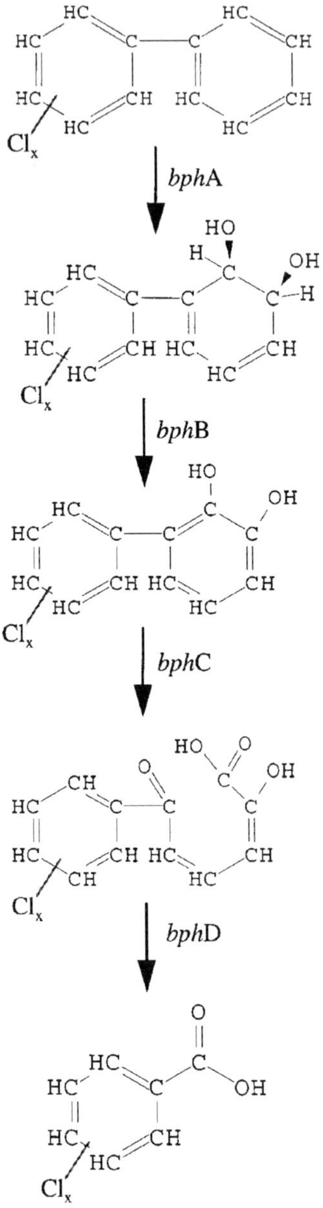

FIGURE 6.21 Degradation of chlorobiphenyl by enzymes of the 2,3-dioxygenase pathway of *Pseudomonas* strain LB400. Gene designations: *bph*A, biphenyl 2,3-dioxygenase; *bph*B, dihydro-diol dehydrogenase; *bph*C, 2,3-dihydroxybiphenyl dioxygenase; *bph*D, 2-hydroxy-6-oxo-6-phen-ylhexa-2,4-dienoic acid hydrase.

collectively, to less harmful compounds, such as pyruvate. Two such examples are discussed next.

TOL (pWW0) Catabolic Plasmid

Catabolic plasmids typically contain a complete set of genes required for a certain degradative pathway and are particularly ubiquitous in *Pseudomonas*. They are self-transmissible, and many have a broad host range making them particularly useful in nature, where such catabolic pathways can be made available to different species. Many catabolic plasmids isolated thus far carry pathways for the degradation of aromatic compounds, presumably because the benzene ring is second in natural abundance to glucose as a building block.

The TOL (pWW0) catabolic plasmid from *Pseudomonas putida* has been shown to confer on its host the capacity to degrade toluene, as well as *m*- and *p*-xylene and other benzene derivatives. The genes are organized in two operons (see Table 6.7 and Fig. 6.22): (1) *xylCAB*, which encodes the degradation of toluene and xylenes to benzoate and toluates, respectively (Upper pathway), and (2) *xylXYZLEGFJKIH*, which encodes the degradation of benzoate and toluates to acetaldehyde and pyruvate (Lower pathway). The branching at 2-hydroxymuconic semialdehyde broadens the substrate range

TABLE 6.7 Genes and Corresponding Products of the TOL Catabolic Plasmid pWW0

Gene	Enzyme or function
Upper pathway operon	
xylA	Xylene oxygenase
xylB	Benzyl alcohol dehydrogenase
xylC	Benzaldehyde dehydrogenase
Lower pathway operon	
xylX,Y,Z	Toluene dioxygenase
xylE	Catechol 2,3-dioxygenase
xylF	2-Hydroxymuconic semialdehyde hydrolase
xylG	2-Hydroxymuconic semialdehyde dehydrogenase
xylH	4-Oxalocrotonate tautomerase
xylI	4-Oxalocrotonate decarboxylase
xylJ	2-Oxopent-4-enoate hydratase
xylK	2-Oxo-4-hydroxypentenoate
xylL	Dihydroxycyclohexadiene carboxylase dehydrogenase
Regulation	
xylR	Transcription regulation
xylS	Transcription regulation

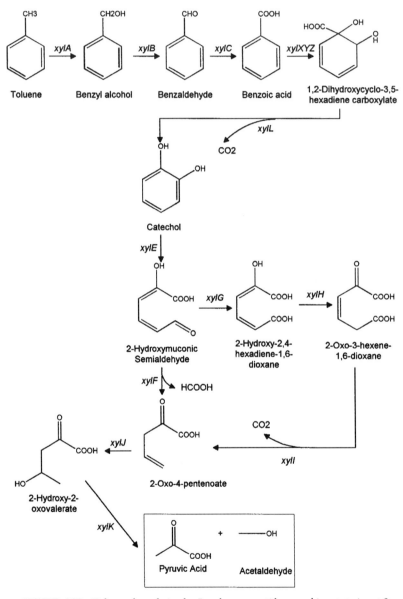

FIGURE 6.22 Toluene degradation by *Pseudomonas putida* recombinant strain mt-2.

that can be degraded by such pathway. For example, toluate is degraded by the *xylF* branch, whereas benzoate and *p*-toluate are degraded by the *xylGHI* branch.

Genetic engineering techniques have been applied in order to extend the range of substrates that an organism can utilize. For example, *P. putida* carrying the TOL plasmid is able to grow on a variety of alkylbenzoates, such as benzoate, 3- and 4-methylbenzoate (3MB and 4MB), 3,4-dimethylbenzoate (34DMB), and 3-ethylbenzoate (3EB), but not on the closely related compound 4-ethylbenzoate (4EB). A series of experiments to explain this concluded that (1) 4EB did not activate a key regulatory protein (*xylS*), and as a result, genes of the benzoate catabolic pathway (*xylABC*, Fig. 6.22) remained uninduced, and (2) the native enzyme catechol 2,3-dioxygenase (*xylE*, Fig. 6.22) was unable to use 4-ethylcatechol as substrate (but could use 4-methyl-catechol, for instance) (Ramos and Timmis, 1987). Thus, 4EB degradation by this pathway would require (1) a *xylS* regulator with a broader effector specificity and (2) generation of catechol 2,3-dioxygenase mutant enzymes capable of degrading 4-ethylcatechol. Both strategies have been pursued, leading to the isolation of mutated *xylS* and *xylE* genes, whose products behave in accordance with the preceding requirements (*xylS'* and *xylE'*). *P. putida* cells containing a TOL plasmid with both the *xylS'* and *xylE'* genes were indeed able to grow on all the usual substrates mentioned previously as well as on 4EB.

Simultaneous Biodegradation of BTX Mixtures

Because BTX compounds typically are discharged into the environment as a mixture of the three aromatics, effort was placed on developing a strain that could simultaneously degrade all three pollutants (Lee *et al.*, 1994). To this end, a *P. putida* strain was developed that contains two catabolic pathways: the TOL pathway discussed previously and the TOD pathway. Enzymes of the TOD pathway act on all three aromatics; however, the end products of this pathway (benzene-, toluene-, and *p*-xylene-*cis*-glycol) cannot be metabolized further. On the other hand, enzymes of the TOL pathway can act only on toluene and xylene; however, its end products can be utilized for energy and carbon intermediates. The general form of the combined catabolic pathways is summarized in Fig. 6.23.

The original strain was constructed by including all enzymes of TOL and TOD. Characterization studies indicated, however, that this strain accumulated significant amounts of TOD end products, *i.e.*, benzene-, toluene-, and *p*-xylene-*cis*-glycols. In order to alleviate this, the final step of the TOD pathway, toluene-*cis*-glycol dehydrogenase, was blocked as indicated in Fig. 6.23. The final strain was able to degrade all three aromatics at

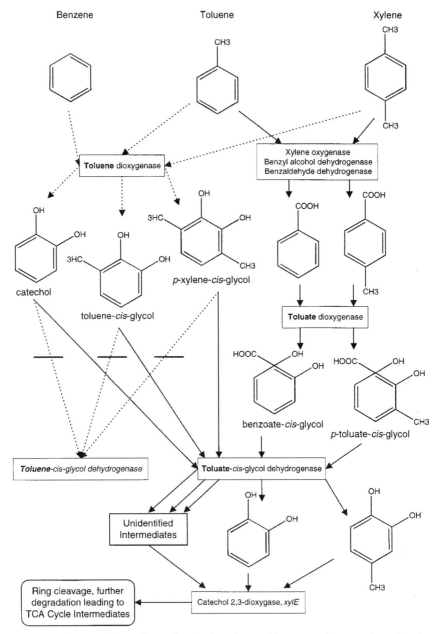

FIGURE 6.23 Metabolic pathways for the degradation of benzene, toluene, and *p*-xylene in *Pseudomonas putida*. Dashed lines represent the TOD pathway and solid lines the TOL pathway (see also Table 6.7 and Fig. 6.22). (−) represents genetic blockage in the TOD pathway (toluene-*cis*-glycol dehydrogenase).

significant rates: 0.27, 0.86, and 2.89 mg (mg biomass h)$^{-1}$ for benzene, toluene, and *p*-xylene, respectively.

This is an interesting illustration of how existing catabolic plasmids (that constitute complete pathways) can be combined judiciously to generate new organisms with novel properties, such as wider substrate utilization range or better product formation.

REFERENCES

Aiba, S., Imanaka, T., & Tsunekawa, H. (1980). Enhancement of tryptophan production in *Escherichia coli* as an application of genetic engineering. *Biotechnology Letters* **2**, 525.

Alderete, J. E., Karl, D. W., & Park, C.-H. (1993). Production of poly(hydroxybutyrate) homopolymer and copolymer from ethanol and propanol in a fed-batch culture. *Biotechnology Progress* **9**, 520-525.

Alterhum, F., & Ingram, L. O. (1989). Efficient ethanol production from glucose, lactose and xylose by recombinant *Escherichia coli*. *Applied and Environmental Microbiology* **55**, 1943-1948.

Anderson, S., Marks, C. B., Lazarus, R., Miller, J., Stafford, K., Seymour, J., Light, D., Rastetter, W. & Estell, D. (1985). Production of 2-keto-L-gluconate, an intermediate in L-ascorbate synthesis by a genetically modified *Erwinia herbicola*. *Science* **230**, 144-149.

Anderson, S., Lazarus, R. & Miller, J. (1991). Metabolic pathway engineering to increase production of ascorbic acid intermediates. US Patent 5,032,514.

Archer, J. A. C., Solowcordero, D. E. & Sinskey, A. J. (1991). A C-terminal deletion in *Corynebacterium glutamicum* homoserine dehydrogenase abolishes allosteric inhibition by L-threonine. *Gene* **107**, 53.

Aristidou, A. A., Bennett, G. N. & San, K.-Y. (1998). Improvement of biomass yield and recombinant gene expression in *Escherichia coli* by using fructose as the primary carbon source. *Journal of Biotechnology,* in press

Aristidou, A. A., Bennett, G. N. & San, K.-Y. (1994b). Modification of the central metabolic pathways of *Escherichia coli* to reduce acetate accumulation by heterologous expression of the *Bacillus subtilis* acetolactate synthase gene. *Biotechnology and Bioengineering* **44**, 944-951.

Ausich, R. L., Proffitt, J. H. & Yarger, J. G. (1991). Biosynthesis of carotenoids in genetically engineered hosts. *Chemical Abstracts* **116**, 35644d.

Barker, D. F. & Campbell, A. M. (1980). Use of the *bio lac* fusion strains to study regulation of biotin synthesis in *Escherichia coli*. *Journal of Bacteriology* **143**, 789-800.

Barrere, G. C., Barber, C. E. & Daniels, M. J. (1986). Molecular cloning of genes involved in the production of the extracellular polysaccharide xanthan by *Xanthomonas campestris* pv. *campestris*. *International Journal of Biological Macromolecules* **8**, 372-374.

Bartel, P. L., Zhu, C.-B., Lampel, J. S., Dosch, D. C., Connors, N., Strohl, W. R., Beale, J. M. & Floss, H. G. (1990). Biosynthesis of anthraquinones by interspecies cloning of actinorhodin biosynthesis genes in *Streptomyces*: Clarification of actinorhodin gene functions. *Journal of Bacteriology* **172**, 4816-4826.

Beall, D. S. & Ingram, L. O. (1993). Genetic engineering of soft-rot bacteria for ethanol production from lignocellulose. *Journal of Industrial Microbiology* **11**, 151-155.

Bennett, G. N. & Rudolph, F. B. (1995). The central metabolic pathway from acetyl-CoA to butyryl-CoA in *Clostridium acetobutylicum*. *FEMS Microbiology Reviews* **17**, 241.

Brabetz, W., Liebl, W. & Schleifer, K.-H. (1991). Studies on the utilization of lactose by *Corynebacterium glutamicum*, bearing the lactose operon of *Escherichia coli*. *Archives of Microbiology* **155**, 607-6012.

Brandes, L., Wu, X., Bode, J., Rhee, J. & Schügerl, K. (1993). Fed-batch cultivation of recombinant *Escherichia coli* JM103 and production of the fusion protein SPA::EcoRI in a 60-L working volume airlift tower loop reactor. *Biotechnology and Bioengineering* **42**, 205-214.

Brau, B. & Sahm, H. (1986). Cloning and expression of the structural gene for pyruvate decarboxylase of *Zymomonas mobilis* in *Escherichia coli*. *Archives in Microbiology* **144**, 296-301.

Brown, S. W., Meyer, H.-P. & Fiechter, A. (1985). Continuous production of human leukocyte interferon with *Escherichia coli* and continuous cell lysis in a two stage chemostat. *Applied Microbiology and Biotechnology* **23**, 5-9.

Bull, S. R. (1990). *Energy from Biomass & Wastes XIV*. Chicago: Institute of Gas Technology.

Bumpus, J. A. & Brock, B. J. (1988). Biodegradation of crystal violet by the white rot fungus *Phanerochaete chrysosporium*. *Applied and Environmental Microbiology* **54**, 1143-1150.

Burchhardt, G. & Ingram, L. O. (1992). Conversion of xylan to ethanol by ethanologenic strains of *Escherichia coli* and *Klebsiella oxytoca*. *Applied and Environmental Microbiology* **58**, 1128-1133.

Cameron, D. C. & Tong, I. T. (1993). Cellular and metabolic engineering: An overview. *Applied Biochemistry and Biotechnology* **38**, 105-140.

Cane, D. E., Hasler, H., Taylor, P. B. & Liang, T.-C. (1983). Macrolide biosynthesis - II: Origin of the carbon skeleton and oxygen atoms of the erythromycins. *Tetrahedron* **39**, 3449-3455.

Cane, D. E., Hasler, H., Taylor, P. B. & Liang, T.-C. (1987). Macrolide biosynthesis - IV. Intact incorporation of a chain-elongation intermediate into erythromycin. *Journal of the American Chemical Society* **109**, 1255-1257.

Carey, V. C., Walia, S. K. & Ingram, L. O. (1983). Expression of a lactose transposon (Tn951) in *Zymomonas mobilis*. *Applied and Environmental Microbiology* **465**, 1163-1168.

Charter, K. F. (1990). The improving prospects for yield increase by genetic engineering in antibiotic-producing *Streptomyces*. *Bio/Technology* **8**, 115-121.

Chen, R. & Bailey, J. E. (1994). Energetic effect of *Vitreoscilla* hemoglobin expression in *Escherichia coli*: An on-line ^{31}P NMR saturation transfer study. *Biotechnology Progress* **10**, 360-364.

Cherrington, C. A., Hinton, M., Pearson, G. R. & Chopra, I. (1991). Short-chain organic acids at pH 5.0 kill *Escherichia coli* and *Salmonella spp.* without causing membrane perturbation. *Journal of Applied Bacteriology* **70**, 161-165.

Cinco, R. M., MacInnis, J. M. & Greenbaum, E. (1993). The role of carbon dioxide in light-activated hydrogen production by *Chlamydomonas reinhardtii*. *Photosynthesis Research* **38**, 27.

Colón, G. E., Aristidou, A. A., Jetten, M. S. M., Yeni-Komshian, H., Sinskey, A. J. & Stephanopoulos, G. (1997). Disruption of *ilvA* in a *Corynebacterium lactofermentum* threonine producer results in increased threonine production. *Biotechnology Progress,* in press.

Colón, G. E., Follettie, M. T., Jetten, M. S. M., Stephanopoulos, G. & Sinskey, A. J. (1993). Redirections of carbon flux at a *Corynebacterium glutamicum* threonine metabolic branch point by controlled enzyme overexpression. In *Annual ASM Meeting*, pp. 320.

Colón, G. E., Jetten, M. S. M., Nguyen, T. T., Gubler, M. E., Sinskey, A. J. & Stephanopoulos, G. (1995). Effect of inducible *thrB* expression on amino acid production in *Corynebacterium lactofermentum* 21799. *Applied and Environmental Microbiology* **61**, 74-78.

Colón, G. E., Nguyen, T. T., Jetten, M. S. M., Sinskey, A. J. & Stephanopoulos, G. (1995b). Production of isoleucine by overexpression of *ilvA* in a *Corynebacterium lactofermentum* threonine producer. *Applied Microbiology and Biotechnology* **43**, 482.

Cordes, C., Möckel, B., Eggeling, L. & Sahm, H. (1992). Cloning, organization and functional analysis of *ilvA*, *ilvB* and *ilvC* genes from *Corynebacterium glutamicum*. *Gene* **112**, 113.

Csonka, L. N. & Clark, A. J. (1979). Deletions generated by the transposon *Tn10* in the *srl recA* region of the *Escherichia coli* K-12 chromosome. *Genetics* 93, 321-343.

Curless, C. E., Forrer, P. D., Mann, M. B., Fenton, D. M. & Tsai, L. B. (1988). Chemostat study of kinetics of human lymphokine synthesis in recombinant *Escherichia coli*. *Biotechnology and Bioengineering* 34, 415-421.

Dahiya, J. S. (1991). Xylitol production by *Petromyces albertensis* grown on medium containing D-xylose. *Canadian Journal of Microbiology* 37, 14-18.

DeModena, J. A., Gutierrez, S., Velasco, J., Fernandez, F. J., Fachini, R. A. & Galazzo, J. L. (1993). The production of cephalosporin C by *Acremonium chrysogenum* is improved by the intracellular expression of a bacterial hemoglobin. *Bio/Technology* 11, 926-929.

Diaz-Ricci, J. C., Hiltzmann, B., Rinas, U. & Bailey, J. E. (1990). Comparative studies of glucose catabolism by *Escherichia coli* grown in complex medium under aerobic and anaerobic conditions. *Biotechnology Progress* 6, 326-332.

Diderichsen, B. (1986). A genetic system for stabilization of cloned genes in *Bacillus subtilis*. In *Bacillus Molecular Genetics and Biotechnology Applications*, pp. 35-46. Edited by A. T. Ganesan & J. A. Hoch. Orlando, FL: Academic Press.

Dills, S. S., Apperson, A., Schmidt, M. R. & Saier, M. H., Jr. (1980). Carbohydrate transport in bacteria. *Microbiological Reviews* 44, 385-418.

Durre, P., Fischer, R.-J., Kuhn, A., Lorenz, K., Schreiber, W., Sturzenhofecker, B., Ullmann, S., Winzer, K. & Sauer, U. (1995). Solventogenic enzymes of *Clostridium acetobutylicum*: Catalytic properties, genetic organization, and transcriptional regulation. *FEMS Microbiology Reviews* 17, 251.

Eggersdorfer, M., Meyer, J. & Eckes, P. (1992). Use of Renewable Resources for Non-Food Materials. *FEMS Microbiology Reviews* 103, 355-364.

Eikmanns, B., Metz, M., Reinscheid, D., Kricher, M. & Sahm, H. (1991). Amplification of three threonine biosynthetic genes in *Corynebacterium glutamicum* and its influence on carbon flux in different strains. *Applied Microbiology and Biotechnology* 34, 617.

Emodi, A. (1978). Xylitol: Its properties and food applications. *Food Technology* 32, 20-32.

Ensley, B. D. (1985). Microbial production of indigo. US Patent 4,520,103.

Feldmann, S., Sprenger, G. A. & Sahm, H. (1989). Ethanol production from xylose with a pyruvate-formate-lyase mutant of *Klebsiella planticola* carrying a pyruvate-decarboxylase gene from *Zymomonas mobilis*. *Applied Microbiology and Biotechnology* 31, 152-157.

Feldmann, S., Sahm, H. & Sprenger, G. A. (1992). Pentose metabolism in *Zymomonas mobilis* wild type and recombinant strains. *Applied Microbiology and Biotechnology* 38, 354.

Flores, N., Xiao, J., Berry, A., Bolivar, F. & Valle, F. (1996). Pathway engineering for the production of aromatic compounds in *Escherichia coli*. *Nature Biotechnology* 14, 620-623.

Follettie, M. T., Shin, H. K. & Sinskey, A. J. (1988). Organization and regulation of the *Corynebacterium glutamicum hom-thrB* and *thrB* loci. *Molecular Microbiology* 2, 53.

Fu, J.-F. & Tseng, Y.-H. (1990). Construction of lactose-utilizing *Xanthomonas campestris* and production of xanthan gum from whey. *Applied and Environmental Microbiology* 56, 919-923.

Gerdes, K. (1988). The *par*B (*hok/sok*) locus of plasmid R1: A general purpose plasmid stabilization system. *Bio/Technology* 6, 1402-1405.

Gerdes, K., Rasmussen, P. B. & Molin, S. (1986). Unique type of the plasmid maintenance function: Postsegregational killing of plasmid-free cells. *Proceedings of the National Academy of Science* 83, 3116-3120.

Gloecker, R., Ohsawa, I., Speck, D., Ledoux, C., Bernand, S., Zinsius, M., Villeval, D., Kisou, T., Kamogawa, K. & Lemoine, Y. (1990). Cloning and characterization of the *Bacillus sphaericus* genes controlling the bioconversion of pimelate into desthiobiotin. *Gene* 87, 63-70.

Gong, C. H., Chen, L. F. & Tsao, G. T. (1981). Quantitative production of xylitol from D-xylose by a high-xylitol producing yeast mutant *Candida tropicalis* HXP2. *Biotechnology Letters* 3, 130-135.

Goodman, A. E., Strzelecki, A. T. & Rogers, P. L. (1984). Formation of ethanol from lactose by *Zymomonas mobilis*. *Journal of Biotechnology* 1, 219-228.

Greenbaum, E., Lee, J. W., Tevault, C. V., Blankinship, S. L. & Mets, L. J. (1995). CO_2 fixation and photoevolution of H_2 and O_2 in a mutant of *Chlamydomonas* lacking photosystem I. *Nature* 376, 438.

Grindley, J. F., Payton, M. A., van de Pol, H. & Hardy, K. G. (1988). Conversion of glucose to 2-keto-L-gluconate, an intermediate in L-ascorbate synthesis, by a recombinant strain of *Erwinia citreus*. *Applied and Environmental Microbiology* 54, 1770-1775.

Hahn-Hägerdal, B., Jeppsson, H., Olsson, L. & Mohagheghi, A. (1993). An Interlaboratory Comparison of the Performance of Ethanol-Producing Microorganisms in a Xylose-Rich Acid Hydrolysate. *Applied Microbiology and Biotechnology* 41, 62-72.

Hallborn, J., Walfridsson, M., Airaksinen, U., Ojamo, H., Hahn-Hägerdal, B., Penttilä, M. & Keränen, S. (1991). Xylitol Production by Recombinant *Saccharomyces cerevisiae*. *Bio/Technology* 9, 1090-1095.

Han, K.-S., Archer, J. A. C. & Sinskey, A. J. (1990). The molecular structure of the *Corynebacterium glutamicum* threonine synthase gene. *Molecular Microbiology* 4, 1693.

Harding, N. E., Cleary, J. M., Cabana, D. K., Rosen, I. G. & Kang, K. S. (1987). Genetic and physical analyses of a cluster of genes essential for xanthan gum biosynthesis in *Xanthomonas campestris*. *Journal of Applied Bacteriology* 169, 2854-2861.

Hardy, K. G., van de Pol, H., Grindley, J. F. & Payton, M. A. (1990). Production of vitamin C precursor using genetically modified organisms. U.S. Patent 4,945,052.

Hollenberg, C. P. & Sahm, H. (1988). Biosensors and environmental biotechnology. In *Biotech*, pp. 150. New York: Gustav Fischer.

Hollenberg, C. P. & Strasser, A. W. M. (1990). Improvement of baker's and brewer's yeast by gene technology. *Food Biotechnology* 4, 527-534.

Hopwood, D. A., Malpartida, F., Kieser, H. M., Ikeda, H., Duncan, J., Fujii, I., Rudd, B. A. M., Floss, H. G. & Omura, S. (1985). Production of hybrid antibiotics by genetic engineering. *Nature* 314, 642-646.

Horitsu, H., Yahashi, Y., Takamizawa, K., Kawai, K., Suzuki, T. & Watanabe, N. (1992). Production of xylitol from D-xylose by *Candida tropicalis*: Optimization of production rates. *Biotechnology and Bioengineering* 40, 1085-1091.

Hutchinson, C. R. & Fujii, I. (1995). Polyketide synthase gene manipulation: A structure-function approach in engineering novel polyketides. *Annual Reviews in Microbiology* 49, 201-238.

Ikeda, M. & Katsumata, R. (1992). Metabolic engineering to produce tyrosine or phenylalanine in a tryptophan-producing *Corynebacterium glutamicum* strain. *Applied and Environmental Microbiology* 58, 781-785.

Ingram, L. O. & Conway, T. (1988). Expression of different levels of ethanolgenic enzymes from *Zymomonas mobilis* in recombinant strains of *Escherichia coli*. *Applied and Environmental Microbiology* 54, 397-404.

Ingram, L. O., Conway, T., Clark, D. P., Sewell, G. W. & Preston, J. F. (1987). Genetic engineering of ethanol production in *Escherichia coli*. *Applied and Environmental Microbiology* 53, 2420-2425.

Innis, M. A., Holland, M. J., McCabe, P. C., Cole, G. E., Wittman, V. P., Tal, R., Watt, K. W. K., Gelfand, D. H., Holland, J. P. & Meade, J. H. (1985). Expression, glycosylation and secretion of an *Aspergillus* Glucoamylase by *Saccharomyces cerevisiae*. *Science* 228, 21-26.

Izumi, Y., Kano, Y., Inagaki, K., Tani, Y. & Yamada, H. (1981). Characterization of biotin biosynthetic enzymes of *Bacillus sphaericus*: A dethiobiotin producing bacterium. *Agricultural and Biological Chemistry* 45, 1983-1989.

Izumori, K. & Tuzaki, K. (1988). Production of xylitol from D-xylose by *Mycobacterium smegmatis*. *Journal of Fermentation Technology* **66**, 33-36.

Jensen, E. B. & Carlsen, S. (1990). Production of recombinant human growth hormone in *Escherichia coli*: Expression of different precursors and physiological effects of glucose, acetate, and salts. *Biotechnology and Bioengineering* **36**, 1-11.

Jetten, M. S. M. & Sinskey, A. J. (1995). Recent advances in the physiology and genetics of amino acid-producing bacteria. *Critical Reviews in Biotechnology* **15**, 73.

Jetten, M. S. M., Gubler, M. E., McCormick, M. M., Colón, G. E., Follettie, M. T. & Sinskey, A. J. (1993). Molecular organization and regulation of the biosynthetic pathway for aspartate derived amino acids in *Corynebacterium glutamicum*. In *Industrial Microorganisms: Basic and Applied Molecular Genetics*, pp. 97. Edited by R. H. Batltz, G. Hegeman & P. L. Skatrud. Washington, D.C.: ASM.

Johansen, L., Bryn, K. & Störmer, F. C. (1975). Physiological and biochemical role of the butanediol pathway in *Aerobacter (Enterobacter) aerogens*. *Journal of Bacteriology* **123**, 1124-1130.

Kado, C. I. (1992). Plant pathogenic bacteria. In *The Prokaryotes*, pp. 659-674. Edited by A. Balows, H. G. Truper, M. Dworkin, W. Harder & K.-H. Scheifer. New York: Springer-Verlag.

Kalio, P. T., Kim, D. J., Tsai, P. S. & Bailey, J. E. (1994). Intracellular expression of *Vitreoscilla* hemoglobin alters *Escherichia coli* energy metabolism under oxygen-limited conditions. *European Journal of Biochemistry* **219**, 201-208.

Kanamayia, H., Sode, K., Yakamoto, T. & Kurube, I. (1988). *Biotechnology and Genetic Engineering Reviews* **6**, 379-401.

Kao, C. M.-F. (1997). Structure, function and engineering of modular polyketide synthases. Ph.D. Thesis, Department of Chemical Engineering, Stanford University, Palo Alto, CA.

Keilhauer, C., Eggeling, L. & Sahm, H. (1993). Isoleucine synthesis in *Corynebacterium glutamicum*: Molecular analysis of the *ilvB-ilvN-ilvC* operon. *Journal of Bacteriology* **175**, 5595.

Keim, C. R. & Venkatasubramanian, K. (1989). Economics of current biotechnological methods of producing ethanol. *Trends in Biotechnology* **7**, 22-29.

Kennedy, J. F., Cabalda, V. M. & White, C. A. (1988). Enzymatic starch utilization and genetic engineering. *Trends in Biotechnology* **6**, 184-189.

Khosla, C. & Bailey, J. E. (1988). Heterologous expression of a bacterial hemoglobin improves the growth properties of recombinant *Escherichia coli*. *Nature* **331**, 633-635.

Khosla, C. & Bailey, J. E. (1988b). The *Vitreoscilla* hemoglobin gene: Molecular cloning, nucleotide sequence, and genetic expression in *Escherichia coli*. *Molecular and General Genetics* **214**, 158-161.

Khosravi, M., Webster, D. A. & Stark, B. C. (1990). Presence of bacterial hemoglobin gene improves α-amylase production of a recombinant *Escherichia coli* strain. *Plasmid* **24**, 190-194.

Kilbane, J. J., Chatterjee, D. K. & Chakrabarty, A. M. (1983). Detoxification of 2,4,5-trichloro-phenoxyacetic acid from contaminated soil by *Pseudomonas cepacia*. *Applied and Environmental Microbiology* **45**, 1697-1700.

Kinoshita, S., Nakayama, K. & Udaka, S. (1957). Studies on the amino acid fermentations. Part I. Production of L-glutamic acid by various microorganisms. *Journal of General and Applied Microbiology* **3**, 193-205.

Kitani, O. & Hall, C. W. (1989). *Biomass Handbook*. New York: Gordon and Breach Science Publishers.

Koch, A. K., Reiser, J., Käppeli, O. & Firchter, A. (1988). Genetic construction of lactose-utilizing strains of *Pseudomonas aeruginosa* and their application in biosurfactant production. *Bio/Technology* **6**, 1335-1339.

Kötter, P., Amore, R., Hollenberg, C. P. & Ciriacy, M. (1990). Isolation and characterization of the *Pichia stipitis* xylitol dehydrogenase gene, XYL2, and construction of a xylose-utilizing *Saccharomyces cerevisiae* transformant. *Current Genetics* 18, 493-500.

Kumar, V., Ramakrishnan, S., Teeri, T. T., Knowles, J. K. C. & Hartey, B. S. (1992). *Saccharomyces cerevisiae* cells secreting an *Aspergilus niger* β-galactosidase grown on whey permeate. *Bio/Technology* 10, 82-85.

Laban, A. & Cohen, A. (1981). Interplasmidic and intraplasmidic recombination in *Escherichia coli* K-12. *Molecular and General Genetics* 184, 200-207.

Laboratory, National Renewable Energy (1996). Ethanol production from lignocellulose. In *The Eighteenth Annual Meeting on Fuels & Chemical From Biomass*. Edited by B. Davison & C. Wayman. Gatlinburg, Tennessee.

Ladisch, M. R., Lin, K. W., Voloch, M. & Tsao, G. T. (1983). Process considerations in the enzymatic hydrolysis of biomass. *Enzyme and Microbial Technology* 5, 82-102.

Lawlis, V. B., Dennis, M. S., Chen, E. Y., Smith, D. H. & Henner, D. J. (1984). Cloning and sequencing of the xylose isomerase and the xylulose kinase genes of *Escherichia coli*. *Applied and Environmental Microbiology* 47, 15-21.

Lee, J. W., Tevault, C. V., Blankinship, S. L., Collins, R. T. & Greenbaum, E. (1994a). Photosynthetic water splitting: In situ photoprecipitation of metallocatalysts for photoevolution of hydrogen and oxygen. *Energy & fuels: An American Chemical Society Journal* 8, 770-773.

Lee, J.-Y., Roh, J.-R. & Kim, H.-S. (1994b). Metabolic engineering of *Pseudomonas putida* for the simultaneous biodegradation of benzene, toluene, and p-xylene mixture. *Biotechnology and Bioengineering* 43, 1146-1152.

Lee, J. W., Blankinship, S. L. & Greenbaum, E. (1995). Temperature effect on production of hydrogen and oxygen by *Chlamydomonas* cold strain CCMP1619 and wild type 137c. *Applied Biochemistry and Biotechnology* 51-52, 379.

Lindsay, K. F. (1992). "Truly degradable" resins are now truly commercial. *Modern Plastics* 2, 62-64.

Liu, S.-C., Webster, D. A., Wei, M.-L. & Stark, B. C. (1996). Genetic engineering to contain the *Vitreoscilla* hemoglobin gene enhances degradation of benzoic acid by *Xanthomonas maltophilia*. *Biotechnology and Bioengineering* 49, 101-105.

Luli, G. W. & Strohl, W. R. (1990). Comparison of growth, acetate production and acetate inhibition of *Escherichia coli* strains in batch and fed-batch fermentations. *Applied and Environmental Microbiology* 56, 1004-1011.

Lynd, L. R., Cushman, J. H., Nichols, R. J. & Wyman, C. E. (1991). Fuel Ethanol from cellulosic biomass. *Science* 251, 1318-1323.

Magnolo, S. K., Leenutaphong, D. L., DeModena, J. A., Curtis, J. E., Bailey, J. E. & Galazzo, J. L. (1991). Actinorhodin production by *Streptomyces coelicitor* and growth of *Streptomyces lividans* are improved by expression of a bacterial hemoglobin. *Bio/Technology* 9, 473-476.

Marzocca, M. P., Harding, N. E., Petroni, E. A., Cleary, J. M. & Ielpi, L. (1991). Location and cloning of the ketal pyruvate transferase gene of *Xanthomonas campestris*. *Journal of Bacteriology* 173, 7519-7524.

McAlpine, J. B., Tuan, J. S., Brown, D. P., Grebner, K. D., Whitern, D. N., Bako, A. & Katz, L. (1987). New antibiotics from genetically engineered *Actinomycetes*. I. 2-Nonerythromycins, isolation and structural determination. *Journal of Antibiotics* 40, 1115-1122.

McDaniel, R., Ebert-Khosla, S., Hopwood, D. A. & Khosla, C. (1993). Engineering biosynthesis of novel polyketides. *Science* 262, 1546-50.

McDaniel, R., Ebert-Khosla, S., Hopwood, D. A. & Khosla, C. (1994). Engineering biosynthesis of novel polyketides: Influence of a downstream enzyme on the catalytic specificity of a minimal aromatic polyketide synthase. *Proceedings of the National Academy of Science USA* **91**, 11542-11346.

Mermelstein, L. D., Welker, N. E., Bennett, G. N. & Papoutsakis, E. T. (1992). Expression of cloned homologous fermentative genes in *Clostridium acetobutylicum* ATCC 824 for. *Bio/Technology* **10**, 190-195.

Meyrial, V., Delgenes, J. P., Moletta, R. & Navarro, J. M. (1991). Xylitol production from D-xylose by *Candida guillermondii*: Fermentation behaviour. *Biotechnology Letters* **13**, 281-286.

Misawa, N., Nakagawa, M., Kobayashi, K., Yamano, S., Izawa, Y., Nakamura, K. & Harashima, K. (1990). Elucidation of the *Erwinia uredovora* carotenoid biosynthesis pathway by functional analysis of gene products expressed in *Escherichia coli*. *Journal of Bacteriology* **172**, 6704-6712.

Misawa, N., Yamano, S. & Ikenaga, H. (1991). Production of β-carotene in *Zymomonas mobilis* and *Agrobacterium tumefaciens* by introduction of the biosynthetic genes from *Erwinia uredovora*. *Applied and Environmental Microbiology* **57**, 1847-1849.

Mori, M., Yokota, A., Sugitomo, S. & Kawamura, K. (1987). Process for the isolation of a strain with reduced pyruvate kinase activity or completely lacking it. Japanese Patent JP 62,205,782.

Murdock, D., Ensley, B. D., Serdar, C. & Thalen, M. (1993). Construction of metabolic operons catalyzing the *de novo* biosynthesis if indigo in *Escherichia coli*. *Bio/Technology* **11**, 381-386.

Nawrath, C., Poirier, Y. & Somerville, C. (1994). Targeting of the polyhydroxybutyrate biosynthetic pathway to the plastids of *Arabidopsis thaliana* results in high levels of polymer accumulation. *Proceedings of the National Academy of Sciences of the USA* **91**, 12760.

Ninet, L. & Renaut, J. (1979). Carotenoids. In *Microbial Processes*, pp. 529-544. Edited by H. J. Peppler & D. Perlman. New York: Academic Press, Inc.

Nishio, N., Sugawa, K., Hayase, N. & Nagai, S. (1989). Conversion of D-xylose into xylitol by immobilized cells of *Candida pelliculosa* and *Methanobacterium* sp. HU. *Journal of Fermentation and Bioengineering* **67**, 356-360.

Ohta, K., Beall, D. S., Mejia, J. P., Shanmugam, K. T. & Ingram, L. O. (1991a). Genetic improvement of *Escherichia coli* for ethanol production: Chromosomal integration of *Zymomonas mobilis* genes encoding pyruvate decarboxylase and alcohol dehydrogenase II. *Applied and Environmental Microbiology* **57**, 893-900.

Ohta, K., Beall, D. S., Mejia, J. P., Shanmugam, K. T. & Ingram, L. O. (1991b). Metabolic engineering of *Klebsiella oxytoca* M5A1 for ethanol production from xylose and glucose. *Applied and Environmental Microbiology* **57**, 2810-2815.

Onken, U. & Leifke, E. (1989). Effect of total and partial pressure (oxygen and carbon dioxide) on aerobic microbial processes. *Advances in Biochemical Engineering / Biotechnology* **40**, 137-169.

Papoutsakis, E. T. & Bennett, G. N. (1991). Cloning, structure and expression of acid and solvent pathway genes of *Clostridium acetobutylicum*. In *The Clostridia and Biotechnology*. Edited by D. R. Wood. Oxford, UK: Butterworths.

Patnaik, R. & Liao, J. C. (1994). Engineering of *Escherichia coli* central metabolism for aromatic metabolite production with near theoretical yield. *Applied and Environmental Microbiology* **60**, 3903-3908.

Patnaik, R., Spitzer, R. G. & Liao, J. C. (1995). Pathway engineering for production of aromatics in *Escherichia coli*: Confirmation of stoichiometric analysis by independent modulation of AroG, TktA and Pps activities. *Biotechnology and Bioengineering* **46**, 361-370.

Peoples, O. P. & Sinskey, A. J. (1989). Poly-β-hydroxybutyrate biosynthesis in *Alcaligenes eutrophus* H16. Identification and characterization of the PHB polymerase gene (*phbC*). *Journal of Biological Chemistry* **264**, 15298-15303.

Peoples, O. P. & Sinskey, A. J. (1990). Novel biodegradable microbial polymers. In Proceedings of the NATO Advanced Research Workshop on New Biosynthetic, Biodegradable Polymers of Industrial Interest from Microorganisms. Edited by E. A. Dawes. Dordrecht: Kluwer Academic Publishers.

Peoples, O. P., Liebl, W., Bodis, M., Maeng, P. J., Follettie, M. T., Archer, J. A. C. & Sinskey, A. J. (1988). Nucleotide sequence and fine structural analysis of the *Corynebacterium glutamicum hom-thrB* operon. *Molecular Microbiology* **2**, 63.

Poirier, Y., Dennis, D. E., Klomparens, K. & Somerville, C. (1992). Polyhydroxybutyrate, a biodegradable thermoplastic, produced in transgenic plants. *Science* **256**, 250-253.

Postma, P. W., Lengeler, J. W. & Jacobson, G. R. (1993). Phosphoenolpyruvate:carbohydrate phosphotransferase systems of bacteria. *Microbiological Reviews* **57**, 543-594.

Pries, A., Steinbüchel, A. & Schlegel, H. G. (1990). Lactose- and galactose-utilizing strains of poly(hydroxyalkanoic acid)-accumulating *Alcaligenes eutrophus* and *Pseudomonas saccharophila* obtained by recombinant DNA technology. *Applied Microbiology and Biotechnology* **33**, 410-417.

Ramos, J. L. & Timmis, K. N. (1987). Experimental evolution of catabolic pathways in bacteria. *Microbiological Science* **4**, 228-237.

Reinscheid, D. J., Eikmanns, B. J. & Sahm, H. (1991). Analysis of *Corynebacterium glutamicum hom* gene coding for a feedback resistant homoserine dehydrogenase. *Journal of Bacteriology* **173**, 3228.

Reinscheid, D. J., Eikmanns, B. J. & Sahm, H. (1994). Stable expression of *hom1-thrB* in *Corynebacterium glutamicum* and its effect on the carbon flux to threonine and related amino acids. *Applied and Environmental Microbiology* **60**, 126-132.

Rojo, F., Pieper, D. H., Engesser, K. H., Knackmuss, H. M. & Timmis, K. N. (1987). Assemblage of *ortho*-clevage route for simultaneous degradation of chloro- and methylaromatics. *Science* **235**, 1395-1398.

Rygus, T., Scheler, A., Allmansberger, R. & Hillen, W. (1991). Molecular cloning, structure, promoters and regulatory elements for transcription of the *Bacillus megaterium* encoded regulon for xylose utilization. *Archives in Microbiology* **155**, 535-542.

Sabatié, J., Speck, D., Reymund, J., Hebert, C., Caussin, L., Weltin, D., Gloeckler, R., O'Regan, M., Bernard, S., Ledoux, C., Ohsawa, I., Kamogawa, K., Lemoine, Y. & Brown, S. W. (1991). Biotin formation by recombinant strains of *Escherichia coli*: Influence of the host physiology. *Journal of Biotechnology* **20**, 29-50.

Saier, M. H, Jr., M. H., Yamada, M., Erni, B., Suda, K., Lengeler, J., Ebner, R., Argos, P., Rak, B., Schnetz, K., Lee, C. A., Stewart, G. C., Breidt, F., Waygood, E. B., Peri, K. G. & Doolittle, R. F. (1988). Sugar permeases of the bacterial phosphoenolpyruvate-dependent phosphotransferase system: Sequence comparisons. *FASEB Journal* **2**, 199-208.

Sarthy, A. V., McConaughy, B. L., Lobo, Z., Sundstrom, J. A., Furlong, C. E. & Hall, B. D. (1987). Expression of the *Escherichia coli* xylose isomerase gene in *Saccharomyces cerevisiae*. *Applied and Environmental Microbiology* **53**, 1996-2000.

Sauer, U. & Duerre, P. (1995). Differential induction of genes related to solvent formation during the shift from acidogenesis to solventogenesis in continuous culture of *Clostridium acetobutylicum*. *FEMS Microbiology Letters* **125**, 115.

Schubert, P., Steinbüchel, A. & Schlegel, H. G. (1988). Cloning of the *Alcaligenes eutrophus* genes for synthesis of poly-β-hydroxybutyrate (PHB) and synthesis of PHB in *Escherichia coli*. *Journal of Bacteriology* **170**, 5837-5847.

Shen, B. & Hutchinson, C. R. (1996). Deciphering the mechanism for the assembly of aromatic polyketides by a bacterial polyketide synthase. *Proceedings of the National Academy of Science USA* **93**, 6600-6604.

Shio, I. (1986). Production of individual amino acids: Tryprophan, phenylalanine and tyrosine. In *Biotechnology of Amino Acid Production*, pp. 188-206. Edited by K. Aida, I. Chibata, K. Nakayama, K. Takinami & H. Yamada. Tokyo: Elsevier.

Skatrud, P. L., Tietz, A. J., Ingolia, T. D., Cantwell, C. A., Fisher, D. L., Chapman, J. L. & Queener, S. W. (1989). Use of recombinant DNA to improve production of cephalosporin C by *Cephalosporium acremonium*. *Bio/Technology* **7**, 477-485.

Skogman, G., Nilsson, J. & Gustafsson, P. (1983). The use of a partitioning locus to increase stability of tryptophan-operon bearing plasmids in *E. coli*. *Gene* **23**, 105-115.

Slater, S. C., Voige, W. H. & Dennis, D. E. (1988). Cloning and expression in *Escherichia coli* of the *Alcaligenes eutrophus* H16 poly-β-hydroxybutyrate biosynthetic pathway. *Journal of Bacteriology* **170**, 4431-4436.

Slater, S. C., Gallaher, T. & Dennis, D. E. (1992). Production of poly(3-hydroxybutyrate-co-3-hydroxyvalerate) in a recombinant *Escherichia coli*. *Applied and Environmental Microbiology* **58**, 1089-1094.

Slonczewski, J. L., Rosen, B. P., Alger, J. R. & Macnab, R. M. (1981). pH homeostasis in *Escherichia coli*: Measurement by ^{31}P nuclear magnetic resonance of methylphosphonate and phosphate. *Proceedings of the National Academy of Science USA* **78**, 6271-6275.

Spindler, D. D., Wyman, C. E., Grohmann, K. & Philippidis, G. P. (1992). Evaluation of the cellobiose-fermenting yeast *Brettanomyces custersii* in the simultaneous saccharification and fermentation of cellulose. *Biotechnology Letters* **14**, 403-407.

Starrenburg, M. J. C. & Hugenholtz, J. (1991). Citrate fermentation by *Lactococcus* and *Leuconostoc* spp. *Applied and Environmental Microbiology* **57**, 3535-3540.

Steinbüchel, A. (1991). Polyhydroxyalkanoic acids. In *Novel Biomaterials from Biological Sources*. Edited by D. Byrom. New York: MacMillan.

Störmer, F. C. (1968). The pH 6 acetolactate-forming enzyme from *Aerobacter aerogenes*. II. Evidence that it is not a flavoprotein. *Journal of Biological Chemistry* **243**, 3740-3741.

Summers, R. G., Wendt-Pienkowski, E., Motamedi, H. & Hutchinson, C. R. (1992). Nucleotide sequence of the *tcmII-tcmIV* region of the tetracenomycin C biosynthetic gene cluster of *Streptomyces glaucescens* and evidence that the *tcmN* gene encodes a multifunctional cyclase--dehydratase--O-methyl transferase. *Journal of Bacteriology* **174**, 1810-1820.

Taguchi, F., Mizukami, N., Yamada, K., Hasegawa, K. & Saito-Taki, T. (1995). Direct conversion of cellulosic materials to hydrogen by *Clostridium* sp. strain no. 2. *Enzyme and Microbial Technology* **17**, 147-150.

Takahashi, D. F., Carvalhal, M. L. & Alterhum, F. (1994). Ethanol production from pentoses and hexoses by recombinant *Escherichia coli*. *Biotechnology Letters* **16**, 747-750.

Tantirungkij, M., Nakashima, N., Seki, T. & Yoshida, T. (1993). Construction of xylose-assimilating *Saccharomyces cerevisae*. *Journal of Fermentation and Bioengineering* **75**, 83-88.

Tantirungkij, M., Seki, T. & Yoshida, T. (1994). Genetic improvement of *Saccharomyces cerevisiae* for ethanol production from xylose. *Annals of the New York Academy of Science* **721**, 138-147.

Timm, A. & Steinbüchel, A. (1990). Formation of polyesters consisting of medium-chain-length 3-hydroxyalkanoic acids from gluconate by *Pseudomonas aeruginosa* and other fluorescent pseudomonads. *Applied and Environmental Microbiology* **56**, 3360-3367.

Tolan, J. S. & Finn, R. K. (1987). Fermentation of D-xylose and L-arabinose to ethanol by *Erwinia chrysanthemi*. *Applied and Environmental Microbiology* **53**, 2033-2038.

Tolan, J. S. & Finn, R. K. (1987). Fermentation of D-xylose to ethanol by *Genetically Modified Klebsiella planticola*. *Applied and Environmental Microbiology* **53**, 2039-2044.

Tong, I.-T. & Cameron, D. C. (1992). Enhancement of 1,3-propanediol production by cofermentation in *Escherichia coli* expressing *Klebsiella pneumoniae dha* regulon genes. *Applied Biochemistry and Biotechnology* 34 / 35, 149-158.

Tong, I., Liao, H. H. & Cameron, D. C. (1991). Propanediol production by *Escherichia coli* expressing genes from the *Klebsiella pneumoniae dha* regulon. *Applied and Environmental Microbiology* 57, 3541-3546.

Tsai, P. S., Hatzimanikatis, V. & Bailey, J. E. (1995a). Effect of *Vitreoscilla* hemoglobin dosage on microaerobic *Escherichia coli* carbon and energy metabolism. *Biotechnology and Bioengineering* 49, 139-150.

Tsai, P. S., Nägeli, M. & Bailey, J. E. (1995b). Intracellular expression of *Vitreoscilla* hemoglobin modifies microaerobic *Escherichia coli* metabolism through elevated concentration and specific activity of cytochrome o. *Biotechnology and Bioengineering* 49, 151-160.

Tsai, P. S., Rao, G. & Bailey, J. E. (1995c). Improvement of *Escherichia coli* microaerobic metabolism by vitreoscilla hemoglobin: New insights from NAD(P)H fluorescence and culture redox potential. *Biotechnology and Bioengineering* 47, 347-354.

Tsoi, C. & Khosla, C. (1995). Combinatorial biosynthesis of "unnatural" products: The polyketide example. *Chemical Biology* 2.

Tsunekawa, H., Azuma, S., Okabe, M., Okamoto, R. & Aiba, S. (1992). Acquisition of sucrose utilization system in *Escherichia coli* K-12 derivatives and its application to industry. *Applied and Environmental Microbiology* 56, 2081-2088.

Uhlenbusch, I., Sahm, H. & Sprenger, G. A. (1991). Expression of an L-alanine dehydrogenase gene in *Zymomonas mobilis* and excretion of L-alanine. *Applied and Environmental Microbiology* 57, 1360-1366.

Verduyn, C., van Kleff, R., Schreuder, H., van Dijken, J. P. & Scheffers, W. A. (1985). Properties of the NAD(P)H-dependent xylose reductase from the xylose fermenting yeast *Pichia stipitis*. *Biochemical Journal* 226, 669-677.

Vongsuvanlert, V. & Tani, Y. (1989). Xylitol production by a methanol yeast, *Candida boidinii*. *Journal of Fermentation and Bioengineering* 37, 35-39.

von Sivers, M. & Zacchi, G. (1995). A techno-economical comparison of three processes for the production of ethanol from pine. *Bioresource Technology* 51, 43.

von Sivers, M., Zacchi, G., Olsson, L. & Hahn-Hägerdal, B. (1994). Cost analysis of ethanol production from willow using recombinant *Escherichia coli*. *Biotechnology Progress* 10, 555-560.

von Stockar, U. & Marison, I. M. (1990). Unconventional utilization of whey in Switzerland. In *Bioprocess Engineering*, pp. 343-365. Edited by K. Tarun & E. Ghose.

Windass, J. D., Worsey, M. J., Pioli, E. M., Pioli, D., Barth, P. T., Atherton, K. T. & Dart, E. C. (1980). Improved conversion of methanol to single-cell protein by *Methyllophilus methylotrophus*. *Nature* 287, 396-401.

Wood, W. A. (1961). Fermentation of carbohydrates and related compounds. In *The Bacteria, Vol II: Metabolism*, pp. 59-149. Edited by I. C. Gunsalus & R. Y. Stanier. London: Academic Press.

Wood, B. E. & Ingram, L. O. (1992). Ethanol production from cellobiose, amorphous cellulose, and crystalline cellulose by recombinant *Klebsiella oxytoca* containing chromosomally integrated *Zymomonas mobilis* genes for ethanol production and plasmids expressing thermostable cellulase genes from *Clostridium thermocellum*. *Applied and Environmental Microbiology* 58, 2103-2110.

Woodward, J., Mattingly, S. M., Danson, M., Hough, D., Ward, N. & Adams, M. (1996). *In vitro* hydrogen production by glucose dehydrogenase and hydrogenase. *Nature Biotechnology* 14, 872-874.

Yanase, H., Kurii, J. & Tonomura, K. (1988). Fermentation of lactose by *Zymomonas mobilis* carrying a Lac + recombinant plasmid. *Journal of Fermentation Technology* **66**, 409-415.

Yee, L. & Blanch, H. W. (1992). Recombinant protein expression in high cell density fed-batch cultures of *Escherichia coli*. *Bio/Technology* **10**, 1550-1556.

Yoshitake, J., Ohiwa, H. & Shimamura, M. (1971). Production of polyalcohols by a *Corynebacterium* sp. Part I. Production of pentitol from aldopentose. *Agricultural and Biological Chemistry* **35**, 905-911.

Yoshitake, J., Ishizaki, H., Shimamura, M. & Imai, T. (1973). Xylitol production by an *Enterobacter* species. *Agricultural and Biological Chemistry* **37**, 2261-2267.

Zabriskie, D. W. & Arcuri, E. J. (1986). Factors influencing productivity of fermentations employing recombinant microorganisms. *Enzyme and Microbial Technology* **16**, 933-941.

Zhang, M., Eddy, C., Deandra, K., Finkelstein, M. & Picataggio, S. (1995). Metabolic engineering of a pentose metabolism pathway in ethanologenic *Zymomonas mobilis*. *Science* **267**, 240-243.

CHAPTER 7

Metabolic Pathway Synthesis

Metabolic pathway synthesis deals with the construction of stoichiometrically consistent routes of biochemical (enzyme-catalyzed) reactions that meet certain specifications. There are various types of specifications that can be formulated according to the role of metabolites and bioreactions in metabolic pathways. A common specification is the synthesis of a metabolite product from a designated set of substrates. We focus, in particular, on metabolites and bioreactions as they constitute the fundamental building blocks of metabolic pathways. As such, some metabolites are designated as *required* substrates or *required* final products, whereas most other metabolites are designated as *allowed* reactants or byproducts. Similarly, many bioreactions are allowed to participate in the construction of a pathway, whereas others may be prohibited from participating in the metabolic network.

Systematic enumeration of *all* possible routes leading to a specified product is relevant not only in the context of metabolic pathway analysis but also in the early steps of bioprocess design. First, in the case of bioprocess design, if all possible alternative pathways leading to a desired product are identified, common features shared by different pathways emerge that may be helpful in evaluating the various biosynthetic options in a meaningful way. Such common characteristics are an intermediate metabolite shared by all different pathways, or one or more enzymes that are absolutely necessary for product synthesis or critical for the yield of the entire process. Second, yield calculations for the various alternatives can be carried out to identify the pathway with the most favorable yield among all possible alternatives. Third, knowledge of all possible pathways is important in defining more accurately the phenotype of mutant strains lacking a particular enzyme. One needs to identify those sets of substrates on which the mutant microbe should be able to grow and those sets of substrates on which the microbe should not grow. As the ability to grow depends primarily on the function of suitable pathways that catabolize the allowed carbon and nitrogen sources, pathway enumeration can direct the elimination of specific enzymes to block growth. Finally, knowledge of the various possible ways by which growth-limiting substrates can be catabolized or products can be formed by microorganisms can improve the way we interpret experimental observations in fermentation experiments.

In the context of metabolic pathway analysis and flux determination, pathway enumeration can be of critical importance in reconciling the large volume of metabolite rate and isotopic label data that are obtained in the course of metabolic flux analysis. As mentioned in Chapter 9, metabolic flux estimates must be confirmed by redundant measurements in a trial and error process that may also involve pathway modification. In such a process, it is obviously important to know how to generate all possible pathways that are relevant to the distribution of isotope label and select the one that yields the best agreement between predicted and measured rates and label distributions. It is noted that, in addition to validating flux estimates, pathway enumeration in combination with labeling reconciliation can be a powerful tool for new pathway discovery.

A pathway is not a mere collection of bioreactions, as many distinct pathways can include the same bioreactions but achieve different transformations. For example, the reactions $A \rightarrow B + C$ and $2B + C \rightarrow D$ can form the pathways $2A \rightarrow \rightarrow D + C$ as well as $A + B \rightarrow \rightarrow D$, depending on whether the reactions are combined in 2:1 or 1:1 proportions. A fully specified pathway must include a set of coefficients, one for each bioreaction included in the pathway, indicating the proportions at which the reaction stoichiometries are combined. The set of these coefficients is the *pathway stoichiometry*, whereas the overall transformation achieved by the pathway in terms of net reactants and products is the *molecular stoichiometry* of the pathway. In the next section we define special notation to reflect these differences and, in particular, the pathway stoichiometry.

An intuitive method for the systematic synthesis of metabolic pathways is depicted in Fig. 7.1. It reflects an approach that starts from a given substrate and lists all possible reactions that can deplete this designated substrate. Each of these reactions produces products (B_1, B_2, B_3), which, in turn, become available for use by other reactions. Pathways thus are constructed recursively proceeding from the substrate to the product. For any pathway being expanded, the available substrates for the next round of expansion would include all products that the pathway has produced up to that point. As this approach rather quickly leads to combinatorial explosion, means for the elimination of redundant pathways must be introduced to bring about a partial reduction in the high computational complexity that is created in this manner. Despite various means of bounding the number of possible combinations, this method still has a number of fundamental shortcomings: (a) it is incomplete in the sense that pathways do exist that this method cannot synthesize; (b) the problem is formulated in a rather restrictive manner because it involves only required substrates and products, but makes no provision for allowed reactants and byproducts; (c) the application of the method to large bioreaction networks demands exponentially increasing computer time; and (d) this method is not really an algorithm, but rather a computer program.

In this chapter, we present an approach for the synthesis of biochemical pathways that overcomes the previous problems through a radically different formalism. The application of the algorithm is demonstrated first via a detailed step by step example, followed by a section that summarizes the generic characteristics of the metabolic pathway synthesis algorithm. Finally, a case study is presented that summarizes the intrinsic characteristics of the methodology, along with a sample of the type of results that can be expected from similar exercises on metabolic pathway synthesis.

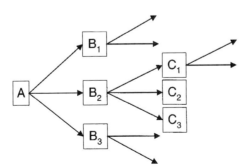

FIGURE 7.1 Schematic diagram of the expansion of three pathways from a designated initial reactant, A. Enumeration of pathways by this method may lead to combinatorial explosion.

7.1. METABOLIC PATHWAY SYNTHESIS ALGORITHM

We defer most of the discussion on the algorithm fundamentals and proofs of accuracy and completeness to the original references of Mavrovouniotis *et al.* (1992a, b). Here we illustrate the procedure by considering the problem of generating all possible biosynthetic routes leading from metabolite A to metabolite L in an enzymatic database comprising the following reactions:

$$
\begin{array}{ll}
\text{(a)} & A \rightarrow B \\
\text{(b)} & B \leftrightarrow C \\
\text{(c)} & C \leftrightarrow D \\
\text{(d)} & C + D \leftrightarrow F + K \\
\text{(e)} & F + K \leftrightarrow H + E \\
\text{(f)} & H + D \leftrightarrow E + F \\
\text{(g)} & A \leftrightarrow E \\
\text{(h)} & E \rightarrow F + G \\
\text{(k)} & F \leftrightarrow G \\
\text{(m)} & G \rightarrow L
\end{array}
$$

All metabolites are designated as excluded reactants and excluded products, with the exception of A, which is a *required reactant*, and L, which is a *required product*. An *excluded reactant or product* cannot participate in a pathway as this would imply that such a reactant must be supplied externally, or the product can accompany the formation of the target metabolite (see Examples 7.1 and 7.2).

We first construct the reverse reactions for reactions (b), (c), (d), (e), (f), (g), and (k). We designate these reverse reactions as (-b), (-c), (-d), (-e), (-f), (-g), and (-k), respectively. Each (forward or reverse) reaction is now considered a one-step partial pathway. To complete the description of the initial state of the problem, we also list separately each metabolite and the pathways in which it participates. Two systems of notation are used here for pathways. To represent only the reactions from which a pathway is constructed, an expression like [2a, 2(-g), b] denotes a pathway that is constructed as a linear combination of the reactions (a), (-g), and (b), with coefficients 2, 2, and 1, respectively. To represent, instead, the overall transformation accomplished by this pathway, the expression $2E \rightarrow \rightarrow B + C$ is used. The two alternative expressions can be combined into $2E \rightarrow [2a, 2(-g), b] \rightarrow B + C$. By using this notation, the initial state of the problem entails the following pathways:

$$A \rightarrow [a] \rightarrow B$$
$$B \rightarrow [b] \rightarrow C$$
$$C \rightarrow [-b] \rightarrow B$$
$$C \rightarrow [c] \rightarrow D$$
$$D \rightarrow [-c] \rightarrow C$$
$$C + D \rightarrow [d] \rightarrow F + K$$
$$F + K \rightarrow [-d] \rightarrow C + D$$
$$F + K \rightarrow [e] \rightarrow H + E$$
$$H + E \rightarrow [-e] \rightarrow F + K$$
$$H + D \rightarrow [f] \rightarrow E + F$$
$$E + F \rightarrow [-f] \rightarrow H + D$$
$$A \rightarrow [g] \rightarrow E$$
$$E \rightarrow [-g] \rightarrow A$$
$$E \rightarrow [h] \rightarrow F + G$$
$$F \rightarrow [k] \rightarrow G$$
$$G \rightarrow [-k] \rightarrow F$$
$$G \rightarrow [m] \rightarrow L$$

The set of metabolites with the pathways in which they participate is as follows:

(A) $[a], [g], [-g]$

(B) $[b], [-b], [a]$

(C) $[b], [-b], [c], [-c], [d], [-d]$

(D) $[c], [-c], [d], [-d], [f], [-f]$

(E) $[e], [-e], [f], [-f], [g], [-g], [h]$

(F) $[d], [-d], [e], [-e], [f], [-f], [h], [k], [-k]$

(G) $[h], [k], [-k], [m]$

(H) $[e], [-e], [f], [-f]$

(K) $[d], [-d], [e], [-e]$

(L) $[m]$

Following the algorithm, the metabolites that participate in fewer pathways must be processed first. Here, L, which participates in only one pathway, is processed first. L is a required product (and an excluded reactant). Because L is produced by one partial pathway and is not consumed by any partial pathway, processing the constraints on this metabolite does not change any pathway. The metabolite that is processed next must be either A or B; the order in which these two metabolites are processed does not affect the results, and we arbitrarily choose B. One new pathway is constructed: [a, b] as a combination of [a] and [b]; this operation is denoted as [a] + [b] = [a, b]. Note that it is not permissible to construct the pathway [b] + [-b], because it would involve the same reaction in both the forward and reverse directions; such a pathway would be a trivial loop, not accomplishing any net transformation. After the combination [a, b] is constructed, the pathways [a], [b], and [-b] are deleted from the database of allowed reactions. This reduces the set of active pathways to the following:

$$A \rightarrow [a, b] \rightarrow C$$
$$C \rightarrow [c] \rightarrow D$$
$$D \rightarrow [-c] \rightarrow C$$
$$C + D \rightarrow [d] \rightarrow F + K$$
$$F + K \rightarrow [-d] \rightarrow C + D$$
$$F + K \rightarrow [e] \rightarrow H + E$$
$$H + E \rightarrow [-e] \rightarrow F + K$$
$$H + D \rightarrow [f] \rightarrow E + F$$
$$E + F \rightarrow [-f] \rightarrow H + D$$
$$A \rightarrow [g] \rightarrow E$$
$$E \rightarrow [-g] \rightarrow A$$
$$E \rightarrow [h] \rightarrow F + G$$
$$F \rightarrow [k] \rightarrow G$$
$$G \rightarrow [-k] \rightarrow F$$
$$G \rightarrow [m] \rightarrow L$$

The updated set of metabolites that still need to be processed becomes

(A) [a, b], [g], [-g]
(C) [a, b], [c], [-c], [d], [-d]
(D) [c], [-c], [d], [-d], [f], [-f]
(E) [e], [-e], [f], [-f], [g], [-g], [h]
(F) [d], [-d], [e], [-e], [f], [-f], [h], [k], [-k]
(G) [h], [k], [-k], [m]
(H) [e], [-e], [f], [-f]
(K) [d], [-d], [e], [-e]

Metabolite A is processed next. Because A is a required reactant and excluded product, a new combination pathway is constructed as [-g] + [a, b] = [-g, a, b], and only pathway [-g] is deleted. For the next step, G is selected arbitrarily from metabolites G, H, and K, which participate in the same number of reactions. In processing G, there are two pathways consuming it, [-k] and [m], and two pathways producing it, [h] and [k]. Hence, four combinations would be constructed, except that [k] cannot be combined with [-k]. Three legitimate combinations remain, namely, [h] + [-k] = [h, -k]; [h] + [m] = [h, m], and [k] + [m] = [k, m]. The original four pathways in which G participated are deleted. After processing of A and G, the active pathways are as follows:

$$A \rightarrow [a, b] \rightarrow C$$
$$C \rightarrow [c] \rightarrow D$$
$$D \rightarrow [-c] \rightarrow C$$
$$C + D \rightarrow [d] \rightarrow F + K$$
$$F + K \rightarrow [-d] \rightarrow C + D$$
$$F + K \rightarrow [e] \rightarrow H + E$$
$$H + E \rightarrow [-e] \rightarrow F + K$$
$$H + D \rightarrow [f] \rightarrow E + F$$
$$E + F \rightarrow [-f] \rightarrow H + D$$
$$A \rightarrow [g] \rightarrow E$$
$$E \rightarrow [-g, a, b] \rightarrow C$$
$$E \rightarrow [h, -k] \rightarrow 2F$$
$$E \rightarrow [h, m] \rightarrow F + L$$
$$F \rightarrow [k, m] \rightarrow L$$

The updated set of metabolites that still need to be processed becomes

(C) [a, b], [c], [-c], [d], [-d], [-g, a, b]
(D) [c], [-c], [d], [-d], [f], [-f]
(E) [e], [-e], [f], [-f], [g], [-g, a, b], [h, m], [h, -k]
(F) [d], [-d], [e], [-e], [f], [-f], [h, m], [k, m], [h, -k]
(H) [e], [-e], [f], [-f]
(K) [d], [-d], [e], [-e]

Metabolite K, participating in four pathways, is processed next. The combinations [d] + [e] = [d, e] and [-e] + [-d] = [-e, -d] are created, and the pathways [d], [-d], [e], and [-e] are deleted. The set of active pathways becomes

$$A \rightarrow [a, b] \rightarrow C$$
$$C \rightarrow [c] \rightarrow D$$
$$D \rightarrow [-c] \rightarrow C$$
$$C + D \rightarrow [d, e] \rightarrow H + E$$
$$H + E \rightarrow [-e, -d] \rightarrow C + D$$
$$H + D \rightarrow [f] \rightarrow E + F$$
$$E + F \rightarrow [-f] \rightarrow H + D$$
$$A \rightarrow [g] \rightarrow E$$
$$E \rightarrow [-g, a, b] \rightarrow C$$
$$E \rightarrow [h, -k] \rightarrow 2F$$
$$E \rightarrow [h, m] \rightarrow F + L$$
$$F \rightarrow [k, m] \rightarrow L$$

The updated set of metabolites that still need to be processed becomes

(C) $[a, b], [c], [-c][d, e], [-e, -d], [-g, a, b]$
(D) $[c], [-c], [d, e], [-e, -d], [f], [-f]$
(E) $[d, e], [-e, -d], [f], [-f], [g], [-g, a, b], [h, m], [h, -k],$
(F) $[f], [-f], [h, m], [k, m], [h, -k]$
(H) $[d, e], [-e, -d], [f], [-f]$

By processing H in a similar fashion, two combination pathways are constructed, namely, $[-f] + [-e, -d] = [-f, -e, -d]$ and $[d, e] + [f] = [d, e, f]$; it should be mentioned again that pathways that involve the same reaction in both the forward and reverse directions are never constructed. The pathways now become

$$A \rightarrow [a, b] \rightarrow C$$
$$C \rightarrow [c] \rightarrow D$$
$$D \rightarrow [-c] \rightarrow C$$
$$C + 2D \rightarrow [d, e, f] \rightarrow 2E + F$$
$$2E + F \rightarrow [-f, -e, -d] \rightarrow C + 2D$$
$$A \rightarrow [g] \rightarrow E$$
$$E \rightarrow [-g, a, b] \rightarrow C$$
$$E \rightarrow [h, -k] \rightarrow 2F$$
$$E \rightarrow [h, m] \rightarrow F + L$$
$$F \rightarrow [k, m] \rightarrow L$$

The updated set of metabolites that still need to be processed becomes

(C) $[a,b],[c],[-c],[d,e,f],[-f,-e,-d],[-g,a,b]$
(D) $[c],[-c],[d,e,f],[-f,-e,-d]$
(E) $[d,e,f],[-f,-e,-d],[g],[-g,a,b],[h,m],[h,-k]$
(F) $[d,e,f],[-f,-e,-d],[h,m],[k,m],[h,-k]$

Because D participates in only four pathways, it is processed next. The fact that the coefficient of D in reactions $[d,e,f]$ and $[-f,-e,-d]$ is 2 must be reflected in the construction of the combinations. The new pathways are constructed as $2[c]+[d,e,f]=[2c,d,e,f]$ and $[-f,-e,-d]+2[-c]=[-f,-e,-d,2(-c)]$, and all four pathways that involve D are deleted. The set of active pathways has now become significantly smaller:

$$A \rightarrow [a,b] \rightarrow C$$
$$3C \rightarrow [2c,d,e,f] \rightarrow 2E + F$$
$$2E + F \rightarrow [-f,-e,-d,2(-c)] \rightarrow 3C$$
$$A \rightarrow [g] \rightarrow E$$
$$E \rightarrow [-g,a,b] \rightarrow C$$
$$E \rightarrow [h,-k] \rightarrow 2F$$
$$E \rightarrow [h,m] \rightarrow F + L$$
$$F \rightarrow [k,m] \rightarrow L$$

Only three metabolites still need to be processed:

(C) $[a,b],[2c,d,e,f],[-f,-e,-d,2(-c)],[-g,a,b]$
(E) $[2c,d,e,f],[-f,-e,-d,2(-c)],[g],[-g,a,b],[h,m],[h,-k]$
(F) $[2c,d,e,f],[-f,-e,-d,2(-c)],[h,m],[k,m],[h,-k]$

As it participates in only four pathways, C is processed next and leads to two combinations: $3[a,b]+[2c,d,e,f]=[3a,3b,2c,d,e,f]$ and $3[-g,a,b]+[2c,d,e,f]=[3(-g),3a,3b,2c,d,e,f]$. After deletion of the original four pathways, the active pathways are as follows:

$$3A \rightarrow [3a,3b,2c,d,e,f] \rightarrow 2E + F$$
$$A \rightarrow [g] \rightarrow E$$
$$E \rightarrow [3(-g),3a,3b,2c,d,e,f] \rightarrow F$$
$$E \rightarrow [h,-k] \rightarrow 2F$$
$$E \rightarrow [h,m] \rightarrow F + L$$
$$F \rightarrow [k,m] \rightarrow L$$

The two metabolites yet to be processed are

(E) $[3a, 3b, 2c, d, e, f], [g], [3(-g), 3a, 3b, 2c, d, e, f], [h, m], [h, -k]$

(F) $[3a, 3b, 2c, d, e, f], [3(-g), 3a, 3b, 2c, d, e, f], [h, m], [k, m], [h, -k]$

The preceding two metabolites participate in the same number of pathways and can be processed in either order to yield the final results. Processing of F first leads to three new combinations of pathways: $[3a, 3b, 2c, d, e, f] + [k, m] = [3a, 3b, 2c, d, e, f, k, m]$; $[h, m] + [k, m] = [h, k, 2m]$; and $[3(-g), 3a, 3b, 2c, d, e, f] + [k, m] = [3(-g), 3a, 3b, 2c, d, e, f, k, m]$. After the original five pathways in which F participated are deleted, the remaining active pathways are

$$3A \rightarrow [3a, 3b, 2c, d, e, f, k, m] \rightarrow 2E + L$$
$$A \rightarrow [g] \rightarrow E$$
$$E \rightarrow [3(-g), 3a, 3b, 2c, d, e, f, k, m] \rightarrow L$$
$$E \rightarrow [h, k, 2m] \rightarrow 2L$$

The remaining unprocessed metabolite, E, participates in all four pathways. Processing of E (and omitting pathways that include the same reactions in opposite directions) leads to the following combinations: $(1/3)[3a, 3b, 2c, d, e, f, k, m] + (2/3)[3(-g), 3a, 3b, 2c, d, e, f, k, m] = [2(-g), 3a, 3b, 2c, d, e, f, k, m]$; $[3a, 3b, 2c, d, e, f, k, m] + 2[h, k, 2m] = [3a, 3b, 2c, d, e, f, 3k, 5m, 2h]$; and the much simpler $[g] + [h, k, 2m] = [g, h, k, 2m]$. It should be noted that in order to obtain smaller integer coefficients for the combination pathway, the fractions $1/3$ and $2/3$ were used instead of 1 and 2 in the construction of the combination. This has the same effect as dividing the resulting pathway by 3.

Clearly, the essence of the transformation and the overall significance of the pathway are not affected by multiplicative constants. Only the molar *proportions* of metabolites and reactions matter. Thus, the final pathways are

$$A \rightarrow [2(-g), 3a, 3b, 2c, d, e, f, k, m] \rightarrow L \qquad [P1]$$
$$3A \rightarrow [3a, 3b, 2c, d, e, f, 3k, 5m, 2h] \rightarrow 5L \qquad [P2]$$
$$A \rightarrow [g, h, k, 2m] \rightarrow 2L \qquad [P3]$$

These three pathways are all feasible solutions to the original synthesis problem. All other pathways are linear combinations of pathways from this set, with positive coefficients. Other methods for pathway synthesis utilize special rules to produce only the basic pathways; for example, they apply a rule stating that, "If an enzyme requires a substrate already used by the pathway to that enzyme, then this substance must be produced by the same set of enzymes used in the pathway." This example shows that the method presented here achieves the construction of basic pathways (and avoids the construction of redundant ones) without additional rules of this type.

One may observe that of the three pathways constructed, only two are linearly independent: pathway [P2] can be obtained as [P1] + 2[P3]. All three pathways are useful, however, because they are *genotypically* independent, *i.e.*, they involve independent sets of enzymes. Although [P2] is the sum of [P1] and 2[P3], the set of enzymes of [P2] is not the union of the respective sets for [P1] and [P3]. Specifically, the enzyme g exists in both [P1] and [P3], but not in [P2].

If a high yield of the product L on the substrate A is desired, it is observed that the pathway [P3] is the best of the three pathways. Thus, in addition to investigating different routes between the substrate(s) and the metabolic product(s), the pathway synthesis algorithm is useful for the evaluation of overall yields when the carbon is processed via different metabolic routes. In many cases the pathways generated by the algorithm will, however, depend on the physiological state of the cell, *e.g.*, the drain of certain metabolites for biosynthesis depends on the growth conditions. For calculation of overall yields, metabolic flux analysis, as described in the following chapters, therefore is often more suitable, as here it is possible to specify constraints for certain pathway fluxes when the overall yield is to be calculated.

7.2. OVERVIEW OF THE ALGORITHM

Metabolites can participate in biochemical pathways in one of the following three capacities: (1) as *net reactants* of the pathway, where there is net depletion of the metabolite; (2) as *net products* of the pathway, where there is net production of the metabolite; and (3) as *intermediates*, in other words, participating without net consumption or production.

Additionally, constraints can be imposed on the various metabolites considered in the construction of metabolic pathways. They can be constrained by the fact that they are designated as follows: (i) *required* reactants, whereby they must be consumed by the pathway and their stoichiometric coefficient is strictly less than zero; (ii) as *allowed* reactants, where they may or may not be consumed by the pathway and their stoichiometric coefficient in the pathway is less than or equal to zero; and (iii) *excluded* reactants (or prohibited reactants), which must not be consumed by the pathway. In most cases, only a few reactants are classified as required, whereas most metabolites assume the default characterization of excluded reactants. In this capacity, they can be neither produced nor consumed by the pathway, and reactions must be provided so that their net rate of change is equal to zero.

In addition to constraints on metabolites, users can also impose constraints on bio-reactions. Thus, one needs to specify which bioreactions are *required*, which are *allowed*, and which are *prohibited* to participate in the pathways. Some of these constraints are easy to satisfy. For example, constraints dictating that certain reactions are excluded can be satisfied right from the start by simply eliminating them from the active database. One should also note that reactions may proceed in the forward or reverse direction. Constraints designating that some bioreactions are excluded in one of the two directions can also be present. Such constraints are derived from thermodynamic or mechanistic arguments on the irreversibility of bioreactions. Bioreaction constraints are not independent. For example, a bioreaction required in the forward direction must be excluded in the backward direction.

All previous constraints are, in essence, stoichiometric constraints. The algorithm illustrated in the previous section basically attempts to satisfy constraints imposed on the *participation* of metabolites in biochemical pathways. The first step in the implementation of the algorithm is to assemble a database of allowed biochemical reactions and their stoichiometry. In this form, the initial set of bioreactions does not, in general, satisfy the constraints, as they require the net consumption of excluded reactants that must not be consumed by the pathway. The algorithm attempts the iterative satisfaction of the various constraints, whereby the initial set of bioreactions is transformed in a stepwise fashion into the final set of pathways that satisfy all imposed constraints.

The main body of the algorithm tackles one constraint at a time. At each iteration stage, the state of the pathway synthesis algorithm is characterized by a set of partial pathways satisfying the constraints that have already been processed and a set of stoichiometric constraints that have yet to be satisfied. At the next pathway expansion step, one more constraint will be satisfied by selecting one of the remaining metabolites as the goal. The most suitable metabolite is chosen to be the one participating in the smallest possible number of active pathways. Metabolites can participate as reactants or products in such pathways. Upon the selection of said suitable metabolite, the set of active pathways is then modified to satisfy the constraint. If, for example, the constraint designates a metabolite as an excluded reactant or excluded product, all possible combination pathways must be constructed by combining one pathway consuming the metabolite and one pathway producing it, such that the metabolite is eliminated from the overall net stoichiometry. Once the combinations are constructed, all pathways consuming or producing the metabolite are deleted because they violate the constraint requiring that there be no net production or consumption of said metabolite.

If the metabolite is an excluded reactant but a required product, the same combination pathways are constructed; however, only the pathways consuming the metabolite are deleted in this case.

The linear nature of the constraints has an important consequence. Once a constraint is satisfied by all surviving pathways, further linear combinations of such pathways will never violate the constraint. Thus, after the processing of each constraint, the new active pathways satisfy all previously processed constraints.

A requirement for application of the pathway synthesis algorithm is a database of possible enzymatic reactions. Such a database with 250 enzymatic reactions and 400 metabolites has been generated by Mavrovouniotis (1989). We illustrate below application of the algorithm with some examples (Examples 7.1, 7.2) produced using this database [see also Mavrovouniotis et al., 1992a, b)].

EXAMPLE 7.1

Synthesis of Serine

The pathway synthesis problem is defined using the following specifications: Enumerate *all* possible pathways within the database of single enzymatic reactions that can lead to the synthesis of serine (required product), with glucose as required reactant and NH_3, and CO_2 as allowed reactant and product, respectively. In addition, oxidation-reduction currency metabolites (NAD, NADH, NADP, NADPH, FAD, and $FADH_2$), direct energy currency metabolites (ATP, ADP, AMP), indirect energy currency metabolites (GTP, GDP, GMP, TTP, TDP, TMP, UTP, UDP, UMP, CTP, CDP, and CMP), and some others [coenzyme A, phosphate (P_i), and pyrophosphate (PP_i)] are introduced as *allowed* metabolites.

With this formulation, the algorithm stalls in the steps processing the constraints on the acetyl-CoA and malate metabolites, which is evidence that these metabolites are quite central in the serine synthesis pathways. For this reason they are specified explicitly as *allowed* metabolites, meaning that, if they appear in the overall serine biosynthetic pathway, additional pathways for their generation will also have to be included. With this addition the algorithm ran to completion. It produced 1526 pathways, the longest of which comprised 26 reactions. Some of these pathways are discussed next.

Figure 7.2 shows the normal pathway for serine biosynthesis. Glucose is first catabolized through glycolysis to 3-phosphoglycerate (3-PG), which is converted to serine through 3-phosphohydroxypyruvate and 3-phosphos-

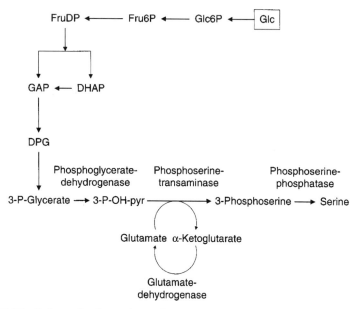

FIGURE 7.2   Pathways for the synthesis of serine from glucose with recovery of glutamate by glutamate dehydrogenase.

erine. The glutamate required for the nitrogen of serine is supplied by glutamate dehydrogenase.

Figure 7.3 depicts an alternative pathway for serine synthesis similar to those of Fig. 7.2, except that the recovery of glutamate is done not through glutamate dehydrogenase but through a set of reactions involving the intermediates fumarate, aspartate, oxaloacetate, and malate. This particular option for converting α-ketoglutarate and ammonia into glutamate is merely one of a number of possibilities. Another analogous possibility is the well-known route through glutamine, using the reactions catalyzed by glutamine synthetase and glutamine synthase.

Thus, the algorithm specifies different routes for glutamate recovery, and this clearly illustrates the complexity of cellular metabolism. In biochemical textbooks one may form the impression that the pathways listed are those running in the cell. However, when it comes to key compounds, like glutamate, there may be several different routes for its recovery, and several of these may act in parallel. This is illustrated by the different pathways generated by the algorithm. In the following chapters on metabolic flux analysis, we will investigate how the relative contribution of these different routes can be quantified.

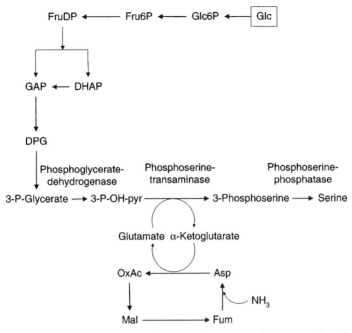

FIGURE 7.3   Synthesis of serine from glucose with recovery of glutamate through a loop involving oxaloacetate, aspartate, fumarate, and malate.

EXAMPLE 7.2

Alanine Synthesis

In the case of alanine synthesis, the initial formulation is to specify glucose as the required reactant and alanine as the required product, with NH_3 and CO_2 as allowed reactant and product, respectively. As with serine, malate and acetyl-CoA are designated as allowed reactants and products, respectively. Figure 7.4 shows what would be considered the usual pathway for alanine synthesis with glucose as the main reactant. As before, glucose is catabolized by glycolysis to pyruvate, which is then converted to alanine by alanine aminotransferase. The glutamate required by this reaction is recovered from α-ketoglutarate by glutamate dehydrogenase.

Alternative pathways can be generated similarly by incorporating different means for glutamate recovery such as that involving the fumarate, aspartate,

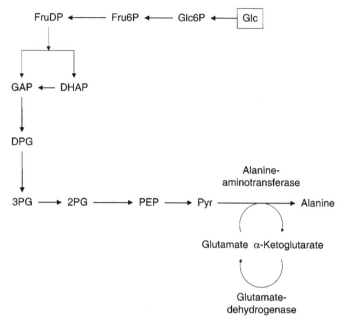

FIGURE 7.4 Pathway for alanine synthesis from glucose with glutamate recovery via glutamate dehydrogenase. Glutamate can be recovered alternatively by the loop of Fig. 7.3. With
alanine dehydrogenase, ammonia is incorporated directly into alanine and all glutamate recovery
reactions are bypassed.

oxaloacetate, and malate loop depicted in Fig. 7.3 for serine. Yet another
pathway can convert pyruvate to alanine not by alanine aminotransferase,
but rather by alanine dehydrogenase. Because the latter reaction uses ammonia directly, there is no need for glutamate dehydrogenase or any other
scheme for the conversion of α-ketoglutarate to glutamate in this case.

7.3. A CASE STUDY: LYSINE BIOSYNTHESIS

A case study of lysine synthesis from glucose and ammonia is presented in
this section in order to demonstrate the different types of results that can be
obtained from the application of the pathway synthesis algorithm. It should
be noted that the aim here is not to provide a complete enumeration of all
possible pathways, as they can easily number several hundred. The objective
is rather to illustrate the construction of generally accepted pathways and

explore alternatives with regard to the participation of key enzymes and metabolites. This process can lead to some fundamental constraints on the structure and yield of lysine-producing pathways.

It is noted that the set of pathways constructed is complete only with respect to the bioreaction database employed. When additional bioreactions are introduced in the database, the results of the search become incomplete; however, the pathways previously constructed still remain valid. By the same token, if the number of pathways constructed is overwhelmingly large, one may consider the judicious reduction of the enzymatic reaction database in order to reduce the number of pathways generated.

By using the same database as for serine and alanine synthesis, the algorithm generated approximately 500 different pathways. In order to simplify the case study, the enzyme α-ketoglutarate dehydrogenase was assumed to be nonfunctional and was excluded from the database. In addition to simplifying the results, this case also illustrates what would be needed for lysine production in a mutant deficient in the preceding enzyme. The resulting pathway is shown in Fig. 7.5, where it can be seen that the glyoxylate shunt has been invoked in order to complement the TCA cycle and make up for the absence of α-ketoglutarate dehydrogenase activity. The overall pathway includes the basic units of glycolysis, TCA cycle, the pathway from oxaloacetate to aspartate, and the sequence of reactions between aspartate and lysine.

One of the applications of the pathway synthesis algorithm is to explore possibilities to bypass a single reaction if it turned out that such a reaction constituted a bottleneck in the overall pathway. For example, if we assumed that malate dehydrogenase was a key limiting enzyme in the pathway, it would be desirable to generate alternative pathways bypassing this enzyme. Figure 7.6a shows one such pathway that excludes malate dehydrogenase. This pathway, in fact, bypasses the whole TCA cycle through the direct carboxylation of pyruvate to oxaloacetate, which can be achieved by either pyruvate carboxylase or oxaloacetate decarboxylase. This pathway also yields a more attractive maximum molar yield. If we neglect constraints from redox and ATP balances, the maximum yield of the pathway of Fig. 7.6a is 100% on a molar basis, compared to a molar yield of only 67% for the pathway of Fig. 7.5.

Smaller perturbations may also be explored if one prefers to bypass only the immediate vicinity of a limiting reaction and retain most of the structure of the original pathway, including the TCA cycle. Such an alternative bypasses malate dehydrogenase with a set of just two reactions converting malate to oxaloacetate:

$$malate + pyruvate \rightarrow oxaloacetate + lactate$$

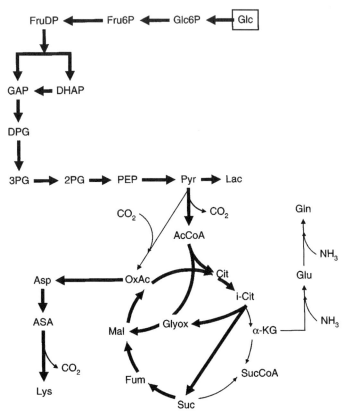

FIGURE 7.5 One possible bioreaction network for lysine synthesis from glucose. Bold arrows indicate the path through the glyoxylate shunt in the absence of α-ketoglutarate dehydrogenase activity. Other possibilities include PEP and pyruvate carboxylation for the anaplerotic reaction. The pathway is for the formation of carbon skeletons only, without energy and redox metabolite regeneration.

by lactate-malate transhydrogenase and

$$\text{lactate} \rightarrow \text{pyruvate}$$

by lactate dehydrogenase in the reverse direction.

Another alternative, shown in Fig. 7.6b, involves the conversion of malate to fumarate by fumarase in the direction opposite that of Fig. 7.6a and conversion of succinate to fumarate by succinate dehydrogenase as in the original pathway. Additionally, fumarate is converted into aspartate through aspartate amino lyase. Because oxaloacetate is used in order to form citrate, half of the aspartate must be recycled into oxaloacetate to close the TCA cycle. In the pathway of Fig. 7.6b, the reaction of aspartate glutamate transaminase converts aspartate to oxaloacetate by operating in the reverse direction of that of the original bioreaction network (Fig. 7.5). A small variation in the previous pathway of Fig. 7.6b is created if a set of two reactions is used for the conversion of aspartate to oxaloacetate (Fig. 7.7). First, glycine oxaloacetate amino transferase converts glyoxylate and aspartate into glycine and oxaloacetate. Second, glycine dehydrogenase recycles glycine into glyoxylate.

7.3.1. THE ROLE OF OXALOACETATE

It can be observed in the previous pathways that oxaloacetate is a central metabolite in all of them. Although oxaloacetate was partly bypassed in the pathway of Fig. 7.6b, a key question is whether this metabolite can be bypassed altogether and whether pathways can be constructed for the production of lysine from pyruvate or glucose without the involvement of oxaloacetate at any point. Upon examination of all the pathways generated, it turned out that, within the enzymatic database used, such a situation is impossible. Also, no single reaction surrounding oxaloacetate is fixed in the sense that it is present in all pathways. Particular reactions consuming and producing oxaloacetate may vary; however, the intermediate itself is always present. Oxaloacetate thus is a key node in the production of lysine.

In the pathway of Fig. 7.6b, aspartate and lysine are not derived directly from oxaloacetate, because fumarate is converted to aspartate by a single enzyme. In fact, aspartate is converted into oxaloacetate, rather than the reverse. Thus, in this case, the metabolism in the neighborhood of aspartate, fumarate, malate, and oxaloacetate is quite different from what one would typically find. This portion of the metabolism may suggest that it is possible to derive aspartate without the intervention of oxaloacetate. It turns out, however, that the necessary TCA intermediates (malate or succinate) cannot be produced from glucose without the intervention of oxaloacetate. This constraint necessitates the presence of oxaloacetate in any pathway leading from glucose to lysine.

To further illustrate this point, assume that in addition to glucose we could use succinate as an allowed reactant. A priori biosynthetic classifications would still entail oxaloacetate as a required intermediate. Inspection of the pathway of Fig. 7.6b, however, indicates that succinate can be converted to fumarate and from then on produce aspartate by aspartate amino lyase without the intervention of malate or oxaloacetate. Thus, with succinate as an additional allowed substrate, it is entirely possible to synthesize lysine with a pathway that does not entail oxaloacetate.

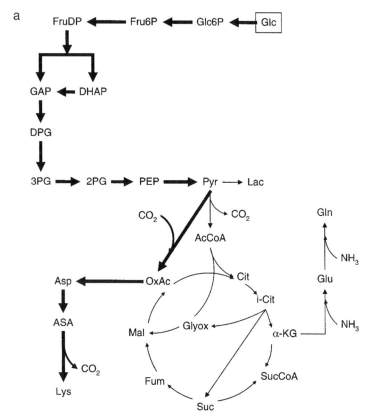

FIGURE 7.6   Alternative pathways for lysine synthesis bypassing malate dehydrogenase, which catalyzes the conversion of malate to oxaloacetate.

7.3.2. OTHER ALTERNATIVES

The lysine pathways examined so far involve either pyruvate dehydrogenase or pyruvate carboxylase as key reactions for the formation of aspartate. There are, however, pathways that bypass pyruvate carboxylase and pyruvate dehydrogenase, pointing to other means by which pyruvate can enter the citric acid cycle. Another pathway that can achieve the carboxylation of pyruvate through methylmalonyl-CoA carboxytransferase and propionyl-CoA carboxylase is shown in Fig. 7.8. Of course, yet another possibility is the direct carboxylation of phosphoenol pyruvate, allowing glycolysis to be connected to the TCA cycle through a route that bypasses pyruvate altogether.

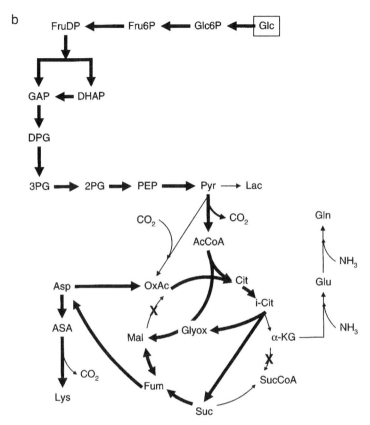

FIGURE 7.6 Continued.

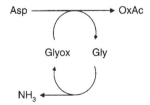

FIGURE 7.7 Alternative pathway for the conversion of aspartate to oxaloacetate.

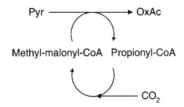

FIGURE 7.8 Alternative pathway for the carboxylation of pyruvate.

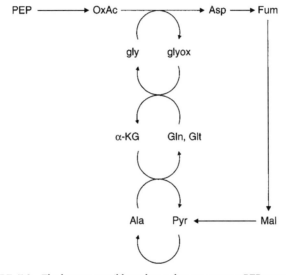

FIGURE 7.9 The longest possible pathway that can convert PEP to pyruvate.

Of all possible alternative pathways produced by the synthesis algorithm, only some of the simplest ones have been presented. It should be remembered, however, that the algorithm can find all pathways and that some among them are very complex. As an example, the simple task of the conversion of PEP to pyruvate (which can be achieved in a single step by pyruvate kinase) was considered. The algorithm produced several pathways for this conversion, the longest of which is depicted in Fig. 7.9.

7.3.3. Restrictions on the Maximum Yield

One of the most intriguing applications of the pathway synthesis algorithm is to hypothesis investigation. For example, it was mentioned earlier that no pathway was found that can produce lysine from glucose without involving oxaloacetate as intermediate. The algorithm also revealed that the maximum yield of the pathway can exceed 67% only through carbon dioxide fixation by a carboxylation reaction. Indeed, if carboxylation reactions are eliminated, the yield is restricted to 67% or less. Without a systematic enumeration of all possibilities, one cannot be certain that other pathways do not exist that can produce the same product with a higher yield. For example, if a pathway were devised to convert 2 mol of pyruvate to 3 mol of acetyl-CoA (without production or consumption of CO_2),

$$2CH_3COCOO^- + 3HS\text{-}CoA + 4H^+ + 2NADH$$
$$\rightarrow 3CH_3CO\text{-}S\text{-}CoA + 3H_2O + 2NAD^+$$

a yield of lysine over glucose greater than 67% would be possible. Considering that pyruvate and oxaloacetate participate in a large number of enzymatic reactions, it is not obvious *a priori* that the existence of such a pathway can be ruled out. The synthesis algorithm shows that within a reasonably complete database, no such pathway exists. It is noted, again, that maximum yields are obtained with respect to carbon consumption only. Such yields are likely to be further reduced when energy and reducing equivalent balances are included in the calculations, as illustrated in Chapters 9-11.

7.4. DISCUSSION OF THE ALGORTIHM

The pathway synthesis algorithm presented in this chapter is very efficient and can process large numbers of stoichiometric constraints in minimal time. The complexity is, in the worst case, exponential in the number of reactions. This, however, occurs only in very specialized cases, such as the one depicted in Fig. 7.10. For each diamond numbered D1, D2, etc., a pathway can follow either the upper or the lower branch. If there are N diamonds, there will be $N - 1$ junctions where these choices occur. Thus, there are $2N - 1$ distinct pathways that are all genotypically independent.

In such cases, because the algorithm will construct all genotypically independent pathways, it would require time and storage that are exponential in the number of reactions. This is, however, the worst case scenario. In practice, metabolism contains long sequences of reactions rather than parallel

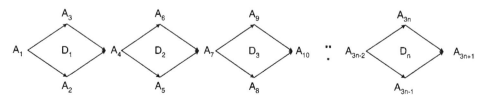

FIGURE 7.10   A reaction set giving rise to an exponential number of pathways.

branches of the type in Fig. 7.10. Such long reaction sequences are processed very efficiently by the method presented and lead to the quick generation of all alternative pathways. One characteristic of the algorithm that contributes to its efficiency is the fact that it does not necessarily start from metabolites that are required reactants (as would be the case with the scheme of Fig. 7.1), but may start from those that participate in constraints that are easiest to satisfy, i.e., those metabolites participating in the smallest number of reactions. This feature, reminiscent of greedy algorithms, greatly facilitates the search through the reaction space for the generation of all possible solutions.

Another advantage of the methodology presented is that in the metabolite-processing phases, metabolites are selected on the basis of the number of reactions in which they participate. In this way, they are eliminated from the constraints and the size of the problem is continuously reduced. If such metabolites participate in long reaction sequences, such sequences are considered only once and, therefore, do not represent a particular burden for the storage and computational resources.

A final point concerns the treatment of common currency metabolites. Such metabolites must be designated as allowed metabolites, otherwise the algorithm may run for a very long time in order to satisfy all stoichiometric constraints. Consider, for example, the currency metabolites involved in the management of the Gibbs energy (P_i, ATP, and ADP) and two classes of subpathways: those acting as futile cycles converting ATP to ADP and P_i and those that achieve the transformation specified by the particular synthesis problem, except that these pathways also produce ATP and consume ADP and P_i. If ATP and ADP are allowed reactants and allowed products, pathways from the second class are acceptable solutions and they need not be expanded further. If, however, ATP and ADP are excluded from appearing in the net stoichiometry, pathways must be combined from the two classes to form acceptable solutions. In other words, a pathway from the second class must be combined with one from the first in order to eliminate ATP and ADP from the stoichiometry. The number of combinations that are created are too numerous, and this leads to serious implementation problems. It is thus recommended that such common currency metabolites be included in the list of allowed reactants and products.

REFERENCES

Mavrovouniotis, M. L. (1989). Computer-aided design of biochemical pathways. Ph.D. Thesis, MIT, Cambridge, MA.

Mavrovouniotis, M. L., Stephanopoulos, G. & Stephanopoulos, G. (1992a). Computer-aided synthesis of biochemical pathways. Biotechnology and Bioengineering 36, 1119-1132.

Mavrovouniotis, M. L., Stephanopoulos, G. & Stephanopoulos, G. (1992b). Synthesis of biochemical production routes. Computers and Chemical Engineering 16, 605-619.

CHAPTER 8

Metabolic Flux Analysis

It was argued in Chapter 1 that metabolic fluxes constitute a fundamental determinant of cell physiology, primarily because they provide a measure of the degree of engagement of various pathways in overall cellular functions and metabolic processes. Accurate quantification of the magnitude of pathway fluxes *in vivo* is, therefore, an important goal of cell physiology and metabolic engineering, especially in the context of metabolite production, where the aim is to convert as much substrate as possible to useful products. A powerful methodology for the determination of metabolic pathway fluxes is metabolic flux analysis (MFA), whereby intracellular fluxes are calculated by using a stoichiometric model for the major intracellular reactions and applying mass balances around intracellular metabolites. A set of *measured* extracellular fluxes is used as input to the calculations, typically uptake rates of substrates and secretion rates of metabolites. The final outcome of flux calculation is a metabolic flux map showing a diagram of the biochemical reactions included in the calculations along with an estimate of the steady state rate (*i.e.*, the flux) at which each reaction in the diagram occurs. This is illustrated in Fig. 8.1, depicting the distribution of fluxes through the catabolic pathways in *Saccharomyces cerevisiae* during anaerobic growth. All fluxes were calculated from material balances around the intracellular metabolites and measurements of the fluxes of compounds exchanged be-

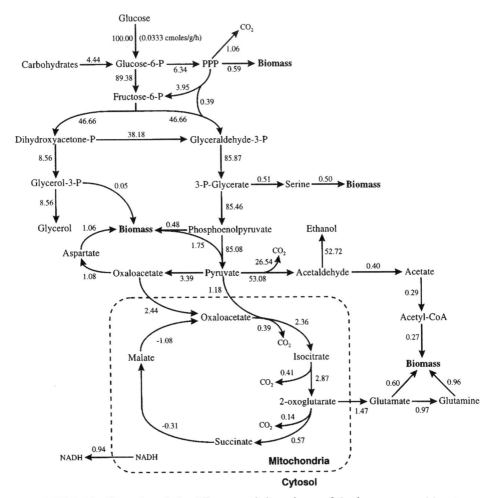

FIGURE 8.1   Fluxes through the different metabolic pathways of *Saccharomyces cerevisiae* at anaerobic growth. The fluxes were calculated using the stoichiometric model of Nissen *et al.* (1997) and measurements of the uptake rates of glucose and ammonia, the production rates of carbon dioxide, acetate, ethanol, glycerol, pyruvate, and succinate, and the rate of synthesis of the key macromolecular pools, *i.e.*, DNA, RNA, protein, lipid, and carbohydrates.

tween the cell and its environment. Metabolic flux maps of the type shown in Fig. 8.1 contain useful information about the *contribution* of various pathways to the overall metabolic processes of substrate utilization and product formation. However, the real value of such metabolic flux maps lies in the *flux differences* that are observed when flux maps obtained with different strains or under different conditions are compared with one another. It is through such comparisons that the impact of genetic and environmental perturbations can be fully assessed and the importance of specific pathways, and reactions within such pathways, accurately described.

In addition to quantification of pathway fluxes, metabolic flux analysis can provide additional insights about other important cell physiological characteristics. We provide here a summary of the type of additional information that may be obtained from MFA along with references to sections in this book where such applications are elaborated upon in more detail.

(1) *Identification of branch point control (nodal rigidity) in cellular pathways.* By comparing changes in branch point flux split ratios resulting from changes in operating conditions and with different mutants, the flexibility or rigidity of branch points can be assessed (Stephanopoulos and Vallino, 1991). In general, rigid branch points resist changes in flux split ratios whereas flexible ones tend to be more accommodating. Knowledge of such characteristics of metabolic networks is important in rationalizing the type of changes that will be most effective for altering product yields. The concept of nodal rigidity was defined in Section 5.4, and in Section 10.1.3 we demonstrate how it can be applied to the study of amino acid biosynthesis.

(2) *Identification of alternative pathways.* Formulation of reaction stoichiometries, which is the basis of MFA, requires detailed knowledge of the *actual biochemical route* by which substrates are converted into products. This, however, may not be clear for many microorganisms, as several alternative pathways have been identified in different organisms and are known to operate under different conditions. It is important to repeat the pathway definition here (see Chapter 1), as the sequence of *feasible and observable* reactions connecting a set of input with a set of output metabolites. MFA is important in identifying pathways that can reproduce the macroscopic flux measurements of extracellular metabolites equally well or, for that matter, can eliminate alternative pathways that are not possible by virtue of their inability to satisfy material balances. Aiba and Matsuoka (1979) demonstrated the first application of this type in a citric acid fermentation (see Example 8.3), and in Section 10.1 we provide another

example of MFA application to the identification of transhydrogenase activity in *C. glutamicum.*

(3) Calculation of nonmeasured extracellular fluxes. Sometimes, the number of extracellular fluxes that can be measured is smaller than what is needed to calculate the unknown intracellular fluxes. In this case, by using flux split ratios determined from previous experiments, it is possible to calculate the nonmeasured extracellular fluxes, *e.g.*, the rates of production of various byproducts, by use of the stoichiometric model and the measured fluxes. If measurements of these fluxes also were to become available, the extent of their agreement with model predictions could be used for model validation or revision, as the case might be (see the discussion on overdetermined systems in Section 8.1.2).

(4) *Calculation of maximum theoretical yields.* The calculation of theoretical yield is based on the stoichiometry of a metabolic network *configured such as to yield maximum product* from a designated substrate. Flux split ratios must be fixed such as to maximize the amount of product formed, and, additionally, constraints on metabolic intermediates and currency metabolites must be observed. *Ad hoc* theoretical yield determination certainly is possible; however, in complex metabolic networks a more formalized approach based on MFA is preferable. Theoretical yield calculations provide a benchmark for real processes and can also identify alternative pathways with attractive features for a given application. Additionally, they provide useful markers (*i.e.*, respiratory quotient) of metabolic activity under conditions of maximum yield that can be pursued in the quest for optimal control strategies. Examples of all methods are provided in Section 10.1.

In this chapter, we present the theory of metabolic flux analysis and illustrate the basis of calculations similar to those leading to the results of Fig. 8.1. One particular aspect of MFA is that the stoichiometric determination of intracellular metabolic fluxes may be based either on an equal number of metabolite balances or on a greater or smaller number of such balances, depending on the number of available measurements. In the first case there is no redundancy, whereas with more measurements the validity of the process can be rigorously assessed. If fewer measurements are available, optimization constraints may be introduced to provide for the closure of equations. These aspects are discussed in Sections 8.2-8.4. In the following chapter, we discuss methods for the experimental determination of metabolic fluxes using isotope-labeled substrates, and in Chapter 10 we demonstrate the application of MFA with specific examples from microbial and cell culture case studies.

8.1. THEORY

The starting point of MFA is the reaction network stoichiometries describing how substrates are converted to metabolic products and biomass constituents (or macromolecular pools). Using the framework of Chapter 3, we consider J intracellular reactions that proceed via K pathway metabolites, for which the mass balances are given by eq. (3.19), or in vector notation:

$$\frac{dX_{met}}{dt} = r_{met} - \mu X_{met} \qquad (8.1)$$

In eq. (8.1), X_{met} is the concentration vector for the intracellular metabolites (or pathway intermediates), and r_{met} is a vector containing the *net rates of formation* of the intracellular metabolites in the J reactions.

It is generally accepted that there is very high turnover of the pools of most metabolites. As a result, the concentrations of the different metabolite pools rapidly adjust to new levels, even after large perturbations in the environment experienced by the cells. It is, therefore, reasonable to assume that the pathway metabolites are at a pseudo-steady state. This implies that there is no metabolite accumulation, or

$$0 = r_{met} - \mu X_{met} \qquad (8.2)$$

The first term of the right-hand side of eq. (8.2) expresses the net synthesis rate of pathway intermediates in the J reactions, i.e., the sum of fluxes leading into and out of a pathway metabolite. The second term describes the dilution of the metabolite pool due to biomass growth. If the dilution effect is considered as an artificial consumption reaction, eq. (8.2) states that the fluxes leading into and out of a pathway metabolite pool match each other, i.e., the fluxes around a given pathway metabolite pool are conserved. Because the intracellular level of most pathway metabolites is very low, the dilution effect is generally small, especially when compared with the other fluxes affecting the same metabolite. Therefore, the last term generally be can be neglected (see Example 8.1). Thus, we arrive at the simpler balance:

$$0 = r_{met} = G^T v \qquad (8.3)$$

where eq. (3.12) has been inserted for the net rate of metabolite synthesis.

EXAMPLE 8.1

Dilution Effect of Pathway Metabolites

To illustrate that the dilution term of eq. (8.2) is negligible, we consider three different types of intracellular compounds: (1) intermediates in the glycolytic pathway; (2) amino acids; and (3) ATP. Table 8.1 collects the relevant data for the three groups of compounds.

In aerobically grown *S. cerevisiae*, the intracellular level of metabolites in the glycolytic pathway has been measured in continuous cultures (dilution rate = 0.1 h^{-1}) to be in the range of 0.05-1.0 μmol (g DW)$^{-1}$ (Theobald *et al.*, 1997). At these growth conditions, the flux through the EMP pathway is about 1.1 mmol (g DW)$^{-1}$ h^{-1}, and it is therefore seen that the flux through a metabolite pool is much higher than the dilution term of eq. (8.2), which is about 0.005-0.1 μmol (g DW)$^{-1}$ h^{-1}. Thus, even for much lower glycolytic fluxes the dilution term is negligible. For mammalian cells, where the intracellular pool levels are about the same but the fluxes are smaller, the difference between turnover rate and dilution rate is smaller, but, in general, a similar conclusion can be drawn.

TABLE 8.1 Typical Intracellular Levels of Different Metabolites

Compound	Concentration (μmol/g DW)	Metabolic route	Microorganism	Reference
Glucose-6-P	0.90	Catabolism (EMP)	*S. cerevisiae*	Theobald *et al.* (1997)[a]
Fructose-6-P	0.20			
Fructose-1,6-dP	0.10			
Glyceraldehyde-3-P	0.065			
Phosphoenolpyruvate	0.052			
Alanine	22.8	Anabolism	*P. chrysogenum*	Nielsen (1997)[b]
Glutamate	44.5			
Lysine	3.9			
Phenylalanine	0.9			
ATP	8.0	Energy metabolism	*S. cerevisiae*	Theobald *et al.* (1997)[a]

[a] The data are from aerobically grown *S. cerevisiae* in a steady state continuous culture ($D = 0.1$ h^{-1}).

[b] The data are from the initial phase of fed-batch cultivations where the specific growth rate was high, i.e., about 0.1 h^{-1}. The pool of free amino acid concentration was not found to change much with the environmental conditions.

In *P. chrysogenum*, the intracellular concentrations of amino acids are in the range of 1-45 μmol (g DW)$^{-1}$. The flux through the pool of amino acids depends on the rate of protein synthesis, which at steady state can be determined from the protein content of the cells and the specific growth rate. With a typical protein content of 45% (w/w) and a specific growth rate of 0.1 h^{-1}, the rate of protein synthesis is 0.045 g of protein (g DW)$^{-1}$ h^{-1}. For the amino acid composition of the protein pool in *P. chrysogenum*, the average molecular mass is 112 g/mol protein-bound amino acid, and the rate of amino acid consumption for protein synthesis is therefore 0.4 mmol (0.045/0.112) of protein bound-amino acids (g DW)$^{-1}$ h^{-1} (Nielsen, 1997). With a content (on a molar basis) of 6.6% alanine, 12.8% glutamate, 5.3% lysine, and 3.7% phenylalanine, the fluxes through these four amino acid pools are 26.4, 51.2, 21.2, and 14.8 μmol (g DW)$^{-1}$ h^{-1} for alanine, glutamate, lysine, and phenylalanine, respectively. Thus, for those few amino acids where the free pool is large, *e.g.*, alanine and glutamate, the dilution term is on the order of 10 times smaller than the fluxes through the pools of free amino acids. Neglect of the dilution term for these amino acids may cause a small error in the calculations, but, if data are available for the size of the pool, it is possible to account for the dilution term by including this as an artificial consumption reaction [see, for example, Reardon *et al.* (1987)]. For the amino acids where the free pool is small (which is the case for most amino acids), the dilution term is negligible as it is about 50 times smaller than the flux through the pool.

The ATP pool in *S. cerevisiae* is about 8.0 μmol (g DW)$^{-1}$. Because ATP is involved in many reactions, it is difficult to assess the flux through the pool. However, from the ATP yield, Y_{xATP}, and the specific growth rate, one can obtain an estimate of the total ATP requirements for cell growth. Using a Y_{xATP} of 70 mmol of ATP/ (g DW)$^{-1}$ (see Example 3.4), we find that the ATP requirement is 7.0 mmol (g DW)$^{-1}$ h^{-1} at a specific growth rate of 0.1 h^{-1}. Thus, the flux through the pool is a factor of 10,000 higher than the dilution term, and this term obviously can be neglected in a balance for ATP. Even for cells that have a higher pool of ATP, *e.g.*, lactic acid bacteria, this conclusion still holds. A similar conclusion can be drawn for other cofactors like NADH and NADPH.

From the preceding analysis, we see that for metabolites and cofactors, for which there is a very high turnover of their intracellular pool, the dilution term can be neglected. This can be expressed in terms of characteristic times (see Section 2.1). Thus, if the characteristic time for the turnover of a pool, which is equal to the ratio of the pool size to the flux through the pool, is much smaller than the characteristic time for growth, which is the reciprocal of the specific growth rate, then the dilution term can be neglected.

A consequence of the pseudo-steady state assumption is that only metabolites positioned at branch points of the metabolic reaction network need to be considered. All pathway intermediates in linear reaction sequences can be eliminated. This is illustrated by the conversion of fructose-6-phosphate to dihydroxyacetonephosphate and glyceraldehyde-3-phosphate in the EMP pathway. This conversion occurs in two reaction steps, for which the stoichiometry is

$$- \text{fructose-6-P} - \text{ATP} + \text{fructose-1,6-bisP} = 0 \qquad (8.4)$$

$$- \text{fructose-1,6-bisP} + \text{dihydroxyacetone-P} + \text{glyceraldehyde-3-P} = 0 \quad (8.5)$$

As discussed in Section 3.1, we have left out ADP formed in the first reaction step. The preceding two reactions are the only cellular reactions where fructose-1,6-bisphosphate is involved, and a steady state mass balance around this metabolite gives

$$v_1 - v_2 = 0 \qquad (8.6)$$

where v_1 and v_2 are the net forward reaction rates[1] for the two reactions, respectively. The rate of the first reaction is, thus, equal to the rate of the second reaction, and we can lump the two reactions into one overall reaction converting fructose-6-phosphate into dihydroxyacetonephosphate and glyceraldehyde-3-phosphate with the forward reaction rate v_1. Notice that this lumping does not change the total degrees of freedom because the removal of one reaction rate (i.e., one unknown) is also accompanied by the elimination of one mass balance (that for fructose-1,6-bisphosphate). A similar reaction lumping can be performed whenever there is a linear sequence of reactions (another example is the conversion of 3-phosphoglycerate to phosphoenolpyruvate in the EMP pathway). Thus, in setting up the stoichiometry of metabolic pathways, only metabolites at network branch points need be considered, and this leads to a significant reduction in the complexity of the stoichiometric model generally employed in metabolic flux analysis. This is demonstrated in Example 8.2.

---

[1] Intracellular reaction rates v in this chapter refer to net forward reaction rates. Such rates equal the forward reaction rate for irreversible reactions and the difference between the forward and reverse reaction rates for reversible reactions. Because macroscopic balances considered in this chapter cannot differentiate between the rates of forward and reverse reactions, v describes the difference between the two or the net rate.

EXAMPLE 8.2

Determination of Metabolic Fluxes in Lysine Biosynthesis

The metabolic network of lysine biosynthesis is described in more detail in Table 10.1 which can be consulted for more information about the specific reactions considered here and their stoichiometries. Figure 8.2 illustrates the rates at which various intracellular reactions proceed, starting with 1 mol of glucose consumed per gram dry weight and unit time. The following overall stoichiometry has been used for the pentose phosphate pathway and TCA cycle, respectively:

$$3Glc6P + 3H_2O + 6NADP \rightarrow 3CO_2 + 6NADPH + 2Fru6P + GAP \quad (1)$$

$$Pyr + 3NAD + NADP + FAD + ADP$$

$$\rightarrow 3CO_2 + NADPH + 3NADH + FADH + ATP \quad (2)$$

The net glycolytic flux is positive in the direction from Glc6P to Fru6P. Products are trehalose (at rate r_T), glycolytic products (i.e., lactate, alanine, and valine, lumped together under the rate r_G), CO_2 (at rate r_{CO_2}), and lysine (at rate r_L), while the biosynthesis uses glucose (at a rate of 1 mol per unit time), ammonia and oxygen. Glucose transport is through the PTS system:

$$Glc + PEP \rightarrow Glc6P + Pyr \quad (3)$$

Separate reactions have been included for the carboxylation of pyruvate (flux v_6), and PEP (flux v_3), and the decarboxylation of OAA (flux v_4). Because these three reactions cannot be observed independently from the preceding metabolite measurements, fluxes v_3, v_4, and v_6 are lumped together as $v_8 = v_3 - v_4 + v_6$. By ignoring the amounts of metabolites consumed for biosynthesis, one then can write the following balances to determine the two unknown flux split ratios v_1 and v_2:

Trehalose	$r_T = 1 - v_1 - v_2$	(4)
Lysine	$r_L = v_8 = v_3 - v_4 + x_6$	(5)
Glycolytic products	$r_G = v_5$	(6)
NADPH	$0 = 2v_1 - 4v_8 + z$	
	(z is the TCA cycle flux from Fig. 8.2)	(7)

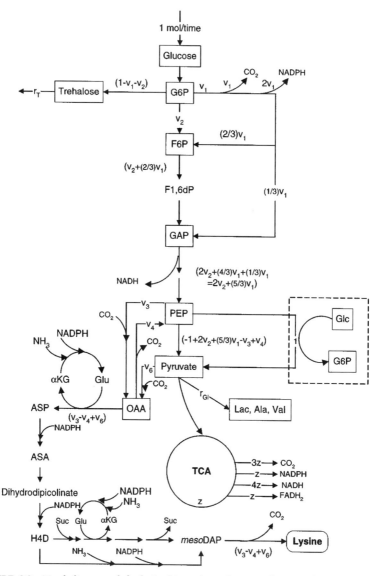

FIGURE 8.2 Metabolic network for lysine biosynthesis depicting the rates of consumption and production of intermediate and currency metabolites. Table 10.1 can be consulted for reaction stoichiometry. Positive flux values indicate that net flux is in the forward direction. Balances are summarized in eqs. (4)-(10) of Example 8.2.

Ammonia $r_{amm} = 2v_8$ (8)

Evolution of CO_2 $r_{CO2} = v_1 + 3z$ (9)

Uptake of O_2 $r_{O2} = (1/2)r_{NADH} = (1/2)(5\,v_1/3 + 2v_2 + 4z)$ (10)

The preceding equations can be solved for the unknowns to give

$$v_1 = (3/5)\{r_G - 2 + 6r_L + 2r_T\} \tag{11}$$

$$v_2 = 1 - r_T - (3/5)\{r_G - 2 + 6r_L + 2\,r_T\} \tag{12}$$

$$v_8 = r_L \tag{13}$$

$$v_5 = r_G \tag{14}$$

whereas the additional measurements, although unable to delineate the carboxylation fluxes, they provide the following constraints on the experimental data:

$$r_{CO2} + 6r_L + 3r_G + 6r_T = 6 \tag{15}$$

$$2r_{O2} + 14r_L + 5r_G + 12r_T = 12 \tag{16}$$

$$r_{amm} = 2r_L \tag{17}$$

Equation (8.3) forms the basis for metabolic flux analysis, i.e., the determination of the unknown pathway fluxes in the intracellular rate vector v. This vector equation represents K linear algebraic balances for the K metabolites with J unknowns (the pathway fluxes). Because the number of reactions (J) is always greater than the number of pathway metabolites (K), there is a certain degree of freedom in the set of algebraic equations given by $F = J - K$. Some of the elements in v therefore have to be measured in order to allow the determination of the remaining. A typical example of a reaction rate that can be measured is the conversion of glucose to glucose-6-phosphate, which may be taken to be equal to the glucose uptake rate. If exactly F fluxes (or reaction rates) in v are measured, the system becomes *determined* and the solution is unique and rather simple to obtain (see the following). In the event that more than F fluxes are measured, the system is *overdetermined*, meaning that extra equations exist that can be used for testing the consistency of the overall balances, the accuracy of the flux measurements, the validity of the pseudo-steady state assumption, and, ultimately, the calculation of more accurate values for the unknown intracellular fluxes. If fewer than F fluxes are measured, the system is *under-determined* and the unknown fluxes can be determined only if additional constraints are introduced or an overall optimization criterion is imposed on the metabolic balances (see Section 8.3).

In a determined system, that is one for which exactly F fluxes (or reaction rates in v) are measured, the remaining fluxes can be calculated by solving the linear system of eq. (8.3). It is very convenient to introduce matrix algebra to describe the various steps. To this end, the solution to eq. (8.3) is found by collecting the measured rates in a new vector, v_m, and the remaining elements of vector v (which are the rates to be calculated) in another vector, v_c. Similarly, the stoichiometric coefficients in matrix G are partitioned by collecting those of the measured reactions in G_m and the remaining in matrix G_c. Equation (8.3) may then be rewritten as

$$0 = G^T v = G_m^T v_m + G_c^T v_c \qquad (8.7)$$

Because exactly $F = J - K$ fluxes are measured, G_c is a square matrix (dimensions $K \times K$) and, if this matrix can be inverted (see the following) the elements of v_c can be found from

$$v_c = -\left(G_c^T\right)^{-1} G_m^T v_m \qquad (8.8)$$

We demonstrate the application of matrix algebra to flux determination in Example 8.3, and we have gone to the length of showing every matrix in the preceding formalism to provide a template for the application of this approach to other systems.

EXAMPLE 8.3

Metabolic Flux Analysis of Citric Acid Fermentation by *Candida lipolytica*

To further illustrate the concept of metabolic flux analysis, we consider a study of *Candida lipolytica* producing citric acid carried out by Aiba and Matsuoka (1979), which is probably the first application of metabolite balancing to fermentation data. In their analysis, Aiba and Matsuoka employed the simplified metabolic network shown in Fig. 8.3, which includes the EMP pathway, the TCA cycle, the glyoxylate shunt, pyruvate carboxylation, and formation of the major macromolecular pools, *i.e.*, proteins, carbohydrates, and lipids. The two anaplerotic routes obviously are necessary to replenish TCA cycle intermediates when citrate and isocitrate are secreted to the extracellular medium.

Notice the structure of the model, in particular some metabolic products such as citrate and isocitrate depicted as pathway intermediates. The intro-

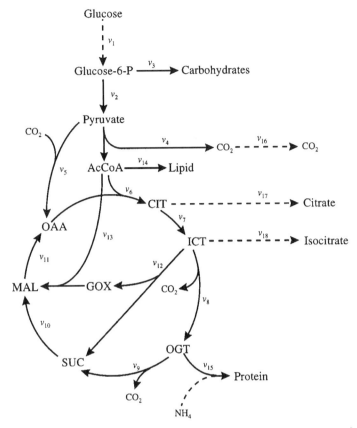

FIGURE 8.3 Simplified metabolic network for *C. lipolytica*. Abbreviations: G6P, glucose-6-phosphate; Pyr, pyruvate; AcCoA, acetyl-CoA; OAA, oxaloacetate; CIT, citrate; ICT, isocitrate; OGT, α-ketoglutarate; SUC, succinate; MAL, malate; GOX, glyoxylate. Broken lines represent transport reactions, whereas solid lines represent intracellular reactions.

duction of secretion reactions (indicated by the dotted lines in Fig. 8.3) allows the application of the pseudo-steady state assumption for all pathway intermediates. Carbon dioxide participates in several reactions, and it is therefore also included as an intracellular compound that is secreted with a rate v_{16}. A balance therefore may also be introduced around this compound. With this structure of the model, some of the fluxes can be measured directly.

The stoichiometry of the individual reactions is quite simple because, except for the conversion of glucose to pyruvate, all stoichiometric coefficients are 1 or -1. By applying the pseudo- steady state assumption to all branch points in the network (circled in Fig. 8.3), we can easily derive eqs. (1)-(11) for the individual reaction rates. All rates are given in moles per gram dry weight per hour of the compound formed in the reaction.

$$\text{G6P: } v_1 - v_2/2 - v_3 = 0 \tag{1}$$

$$\text{Pyr: } v_2 - v_4 - v_5 = 0 \tag{2}$$

$$\text{AcCoA: } v_4 - v_6 - v_{13} - v_{14} = 0 \tag{3}$$

$$\text{CIT: } v_6 - v_7 - v_{17} = 0 \tag{4}$$

$$\text{ICT: } v_7 - v_8 - v_{12} - v_{18} = 0 \tag{5}$$

$$\text{OGT: } v_8 - v_9 - v_{15} = 0 \tag{6}$$

$$\text{SUC: } v_9 - v_{10} + v_{12} = 0 \tag{7}$$

$$\text{MAL: } v_{10} - v_{11} + v_{13} = 0 \tag{8}$$

$$\text{GOX: } v_{12} - v_{13} = 0 \tag{9}$$

$$\text{OOA: } v_5 + v_{11} - v_6 = 0 \tag{10}$$

$$CO_2: v_4 + v_8 + v_9 - v_{16} = 0 \tag{11}$$

Notice that reaction 13 could have been eliminated as the pseudo-steady state assumption for glyoxylate implies that the rate of this reaction is equal to v_{12}.

With 18 reaction rates and 11 balance equations, there are 7 degrees of freedom. Thus, if 7 reaction rates are specified, the other rates can be calculated. In their analysis Aiba and Matsuoka measured six reaction rates in the network: the glucose uptake rate ($r_{glc} = v_1$); the carbon dioxide production rate ($r_c = v_{16}$); the citric acid production rate ($r_{cit} = v_{17}$); the isocitrate production rate ($r_{ict} = v_{18}$); the protein synthesis rate ($r_{prot} = v_{15}$); and the carbohydrate synthesis rate ($r_{car} = v_3$). The rates r_{prot} and r_{car} were found from measurements of the protein and carbohydrate contents, respectively, of the biomass (in a steady state chemostat). In addition to the six measurements, Aiba and Matsuoka imposed a constraint by setting one of the rates in the network equal to zero. Three different cases were examined, reflecting three different models of citric acid biochemistry:

- **Model 1**: The glyoxylate shunt is inactive, i.e., $v_{12} = 0$.
- **Model 2**: Pyruvate carboxylase is inactive, i.e., $v_5 = 0$.
- **Model 3**: The TCA cycle is incomplete, i.e., $v_9 = 0$.

With the preceding six measured rates (r), the system of equations is determined exactly for each of these models and can be solved to determine the unknown reaction rates. This can be done by sequential elimination of the unknown rates, e.g., we rapidly find that

$$v_2 = 2(r_{glc} - r_{car}) \tag{12}$$

and similar expressions can be derived for the other rates as well. A more general approach is to use matrix algebra calculations. The following material illustrates how the unknown fluxes are calculated in the case of model 1. For this purpose, we first rewrite eqs. (1)-(11) in matrix notation using the stoichiometric matrix **G**:

$$
\begin{pmatrix}
1 & -0.5 & -1 & 0 & 0 & 0 & 0 & 0 & 0 & 0 & 0 & 0 & 0 & 0 & 0 & 0 & 0 & 0 \\
0 & 1 & 0 & -1 & -1 & 0 & 0 & 0 & 0 & 0 & 0 & 0 & 0 & 0 & 0 & 0 & 0 & 0 \\
0 & 0 & 0 & 1 & 0 & -1 & 0 & 0 & 0 & 0 & 0 & 0 & -1 & -1 & 0 & 0 & 0 & 0 \\
0 & 0 & 0 & 0 & 0 & 1 & -1 & 0 & 0 & 0 & 0 & 0 & 0 & 0 & 0 & -1 & 0 \\
0 & 0 & 0 & 0 & 0 & 0 & 1 & -1 & 0 & C & 0 & -1 & 0 & 0 & 0 & 0 & -1 \\
0 & 0 & 0 & 0 & 0 & 0 & 0 & 1 & -1 & 0 & 0 & 0 & 0 & -1 & 0 & 0 & 0 \\
0 & 0 & 0 & 0 & 0 & 0 & 0 & 0 & 1 & -1 & 0 & 1 & 0 & 0 & 0 & 0 & 0 \\
0 & 0 & 0 & 0 & 0 & 0 & 0 & 0 & 0 & 1 & -1 & 0 & 1 & 0 & 0 & 0 & 0 \\
0 & 0 & 0 & 0 & 0 & 0 & 0 & 0 & 0 & 0 & 0 & 1 & -1 & 0 & 0 & 0 & 0 \\
0 & 0 & 0 & 0 & 1 & -1 & 0 & 0 & 0 & 0 & 1 & 0 & 0 & 0 & 0 & 0 & 0 \\
0 & 0 & 0 & 1 & -1 & 0 & 0 & 1 & 1 & 0 & 0 & 0 & 0 & 0 & -1 & 0 & 0
\end{pmatrix} \mathbf{v}
$$

$$
= \begin{pmatrix}
0 \\ 0 \\ 0 \\ 0 \\ 0 \\ 0 \\ 0 \\ 0 \\ 0 \\ 0 \\ 0 \\ 0 \\ 0
\end{pmatrix} \tag{13}
$$

When rates v_1, v_3, v_{15}, v_{16}, and v_{17} are measured (indicated by the corresponding subscribed r variables) and there is no isocitrate secretion ($v_{18} = 0$), for the case of model 1 v_{12} is set to zero and columns 1, 3, 12, 15, 16, 17, and 18 are collected in the matrix $\mathbf{G_m}$, whereas the remaining columns are collected in matrix G_c. Equation (8.8) then yields

$$
\begin{pmatrix} v_2 \\ v_4 \\ v_5 \\ v_6 \\ v_7 \\ v_8 \\ v_9 \\ v_{10} \\ v_{11} \\ v_{13} \\ v_{14} \end{pmatrix} = -
\begin{pmatrix}
-0.5 & 0 & 0 & 0 & 0 & 0 & 0 & 0 & 0 & 0 & 0 \\
1 & -1 & -1 & 0 & 0 & 0 & 0 & 0 & 0 & 0 & 0 \\
0 & 1 & 0 & -1 & 0 & 0 & 0 & 0 & 0 & -1 & -1 \\
0 & 0 & 0 & 1 & -1 & 0 & 0 & 0 & 0 & 0 & 0 \\
0 & 0 & 0 & 0 & 1 & -1 & 0 & 0 & 0 & 0 & 0 \\
0 & 0 & 0 & 0 & 0 & 1 & -1 & 0 & 0 & 0 & 0 \\
0 & 0 & 0 & 0 & 0 & 0 & 1 & -1 & 0 & 0 & 0 \\
0 & 0 & 0 & 0 & 0 & 0 & 0 & 1 & -1 & 1 & 0 \\
0 & 0 & 0 & 0 & 0 & 0 & 0 & 0 & 0 & -1 & 0 \\
0 & 0 & 1 & -1 & 0 & 0 & 0 & 0 & 1 & 0 & 0 \\
0 & 1 & -1 & 0 & 0 & 1 & 1 & 0 & 0 & 0 & 0
\end{pmatrix}^{-1}
$$

$$
\times
\begin{pmatrix}
1 & -1 & 0 & 0 & 0 & 0 & 0 \\
0 & 0 & 0 & 0 & 0 & 0 & 0 \\
0 & 0 & 0 & 0 & 0 & 0 & 0 \\
0 & 0 & 0 & 0 & 0 & -1 & 0 \\
0 & 0 & -1 & 0 & 0 & 0 & -1 \\
0 & 0 & 0 & -1 & 0 & 0 & 0 \\
0 & 0 & 1 & 0 & 0 & 0 & 0 \\
0 & 0 & 0 & 0 & 0 & 0 & 0 \\
0 & 0 & 1 & 0 & 0 & 0 & 0 \\
0 & 0 & 0 & 0 & 0 & 0 & 0 \\
0 & 0 & 0 & 0 & -1 & 0 & 0
\end{pmatrix}
\begin{pmatrix} r_{glc} \\ r_{car} \\ 0 \\ r_{prot} \\ r_c \\ r_{cit} \\ r_{ict} \end{pmatrix}
\tag{14}
$$

or

$$
\begin{pmatrix} v_2 \\ v_4 \\ v_5 \\ v_6 \\ v_7 \\ v_8 \\ v_9 \\ v_{10} \\ v_{11} \\ v_{13} \\ v_{14} \end{pmatrix} =
\begin{pmatrix}
2 & -2 & 0 & 0 & 0 & 0 & 0 \\
2 & -2 & 1 & -1 & 0 & -1 & -1 \\
0 & 0 & -1 & 1 & 0 & 1 & 1 \\
-1 & 1 & 0 & 1.5 & 0.5 & 2 & 2 \\
-1 & 1 & 0 & 1.5 & 0.5 & 1 & 2 \\
-1 & 1 & -1 & 1.5 & 0.5 & 1 & 1 \\
-1 & 1 & -1 & 0.5 & 0.5 & 1 & 1 \\
-1 & 1 & 0 & 0.5 & 0.5 & 1 & 1 \\
-1 & 1 & 1 & 0.5 & 0.5 & 1 & 1 \\
0 & 0 & 1 & 0 & 0 & 0 & 0 \\
3 & -3 & 0 & -2.5 & -0.5 & -3 & -3
\end{pmatrix}
\begin{pmatrix} r_{glc} \\ r_{car} \\ 0 \\ r_{prot} \\ r_c \\ r_{cit} \\ r_{ict} \end{pmatrix}
\tag{15}
$$

Notice the complexity of the solution, *i.e.*, the intracellular fluxes are functions of almost all the measured rates. In such cases, solution of the algebraic equations by Gaussian elimination may be quite cumbersome, whereas the matrix equation is easily solved using computer programs like Mathematica, Mable, or Matlab. From the first row of eq. (15) it is seen that the solution for v_2 is the same as that given by eq. (12). Obviously $v_{13} = 0$ and $v_{10} = v_{11}$ when the glyoxylate shunt is inactive. By calculating the unknown rates (or fluxes) at different dilution rates in a steady state chemostat, *i.e.*, for different sets of measured rates, Aiba and Matsuoka (1979) concluded that model 1 describes reasonable values for the fluxes. Furthermore, *in vitro* measurements of the activity of four different enzymes (pyruvate carboxylase, citrate synthase, isocitrate dehydrogenase, and isocitrate lyase) correlated fairly well with the calculated fluxes. When the two other models were tested, it was found that some of the fluxes were negative, *e.g.*, model 2 predicts that α-ketoglutarate is converted to isocitrate. This is not impossible, but most of the reactions are favored thermodymically in the direction specified by the arrows in Fig. 8.3. Thus, the ΔG^0 for the conversion of isocitrate to α-ketoglutarate is -20.9 kJ $(mol)^{-1}$, and a large concentration ratio of α-ketoglutarate to isocitrate would be required to allow this reaction to run in the opposite direction. Furthermore, there is better agreement between the measured enzyme activities and flux predictions by using model 1 than with the two other models. Aiba and Matsuoka (1979) therefore concluded that glyoxylate is inactive or operates at a very low rate in *C. lipolytica* under citric acid production conditions.

Equation (8.8) actually shows the result obtained when K linear algebraic equations are solved using matrix notation (see Box 4.2). A requirement for a unique solution is that the set of algebraic equations be linearly independent, *i.e.*, that none of the algebraic equations is a linear combination of the other equations. This linear independence is necessary for inversion of the G_c matrix, and it is checked most easily by determining the rank of the matrix. Thus, if the matrix has full rank (equal to K), *i.e.*, the $\det(G_c)$ is nonzero, the matrix can be inverted and the nonmeasured fluxes can be calculated. If the rank of G_c is less than K, the matrix is singular meaning that $\det(G_c) = 0$, and a solution cannot be found using eq. (8.8), as we are dealing in essence with an underdetermined system in this case.

The rank of matrix G_c may be smaller than K mainly for two reasons:

1. The set of reaction stoichiometries (*e.g.*, a set of rows) collected in G_c are linearly dependent, *i.e.*, one or more of the rows included in G_c can

be written as a linear combination of the other rows. Whether linearly dependent reaction stoichiometries are present in a reaction network can be checked easily by determining the rank of the total stoichiometric matrix G. If G has full rank, i.e., rank$(G) = K$, at least one square submatrix exists that can be used for the inversion calculations. If rank$(G) < K$, at least one of the rows of G_c is linearly dependent on the other rows. Due to the large number of reactions typically present in metabolic networks, the question of linear dependence of reaction stoichiometries arises frequently (see Example 8.4). In these cases it is necessary to eliminate all linearly dependent reactions and/or include additional information in the solution procedure.

2. Even if G has full rank, it is possible that the set of measurements was chosen in such a way that G_c has a rank that is less than K. In this case some of the measured rates are redundant. They can be used for consistency testing, but only after a different set of measurements is used first to determine the unknown fluxes (Example 8.5).

EXAMPLE 8.4

Linearly Dependent Reaction Stoichiometries

Because most living cells are capable of utilizing a large variety of compounds as carbon, energy and nitrogen sources, many complementary pathways exist that would serve similar functions if they operated at the same time. The inclusion of all such pathways may give rise to observability problems when metabolic flux analysis is carried out, meaning that these pathways normally are not discernible from extracellular measurements alone. This situation usually manifests itself by matrix singularity, whereby the nonobservable pathways appear as linearly dependent reaction stoichiometries. Some of the most often encountered problems are caused by one or more of the following metabolic routes:

- Glyoxylate cycle in prokaryotes
- Nitrogen assimilation via the GS-GOGAT system
- Transhydrogenases
- Isoenzymes

In prokaryotes, the TCA cycle and all anaplerotic reactions, including the glyoxylate cycle, operate in the cytosolic matrix. Often the glyoxylate cycle is considered as a bypass of the TCA cycle because it shares a number of reactions with this cycle (see Fig. 2.10). However, the two pathways serve very different purposes: the TCA cycle has the purpose of oxidizing pyruvate to carbon dioxide, whereas the glyoxylate cycle has the purpose of synthesizing precursor metabolites, *e.g.*, oxaloacetate, from acetyl-CoA. Considered by themselves, the TCA cycle and the glyoxylate cycle are not linearly dependent, but if other anaplerotic pathways, *e.g.*, pyruvate carboxylase, are included, a singularity arises. This may be illustrated by writing lumped reactions for the three pathways (see Fig. 2.10 for an overview of the pathways). In all cases we use pyruvate as the starting point:

$$\text{TCA cycle: - pyruvate} + 3CO_2 + NADH + FADH_2 + GTP = 0 \quad (1)$$

$$\text{Glyoxylate cycle: - 2pyruvate} + 2CO_2 + \text{oxaloacetate}$$
$$+ 4NADH + FADH_2 = 0 \quad (2)$$

$$\text{Pyruvate carboxylase: - pyruvate - ATP - } CO_2 + \text{oxaloacetate} = 0 \quad (3)$$

If ATP and GTP are pooled together (which is often done in the analysis of cellular reactions), it is quite obvious that the glyoxylate cycle is a linear combination of the two other pathways, and all three pathways cannot be determined independently by flux analysis. It may be a difficult task to decide which pathway should be eliminated, and as illustrated in Fig. 8.4, the calculated flux distributions may be radically different when either the TCA cycle or the glyoxylate cycle is included. Fortunately, these pathways rarely operate at the same time as their enzymes are induced differently. Information about induction and regulation of the corresponding enzymatic activities is critical in making an informed decision as to the exact pathway to be considered under a set of environmental conditions. For example, expression of isocitrate lyase (the first enzyme of the glyoxylate cycle) is repressed by glucose in many microorganisms, so that the glyoxylate cycle possibility can be eliminated for growth on glucose. In eukaryotes, presence of the glyoxylate cycle does not give rise to linear dependency due to compartmentation of the different reactions, *i.e.*, the TCA cycle operates in the mitochondria and the glyoxylate cycle either in the cytosol or in microbodies. All three routes were included in the analysis of Aiba and Matsuoka (1979), but in the three models considered, one of the routes was inactive in each model. In our treatment of their model in Example 8.2 we included the stoichiometry for all three pathways in the total stoichiometric matrix, but in deriving the solution, we collected the stoichiometry for one of the pathways (illustrated with the glyoxylate shunt) in the submatrix G_m as this flux was set to zero. Alternatively, we could have eliminated the stoichiometry for this pathway from the network, but this would have resulted in the same solution.

Another example of linear dependent reactions is the two ammonia assimilation routes: the glutamate dehydrogenase reaction and the GS-

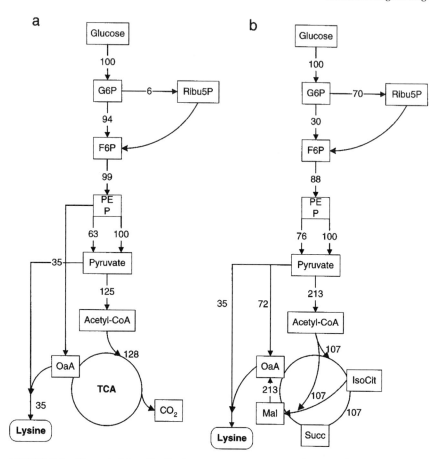

FIGURE 8.4 Theoretical flux distributions necessary to support a lysine molar yield of 0.35 in *Corynebacterium glutamicum*. The fluxes were calculated using the stoichiometric model of Vallino and Stephanopoulos (1993) (see also section 10.1). The two fluxes between PEP and Pyr account for reactions catalyzed by pyruvate kinase (left) and the glucose PTS transport system (right). (a) The flux distribution with active TCA cycle and inactive glyoxylate cycle. (b) The flux distributions with active glyoxylate cycle and inactive TCA cycle. Used with permission from Vallino and Stephanopoulos, 1993. © 1993 John Wiley & Sons, Inc.

GOGAT system (see Section 2.4.1). The stoichiometries of these two routes are

$$\text{GDH: } -\alpha\text{-ketoglutarate} - NH_3 - NADPH + \text{glutamate} = 0 \qquad (4)$$

$$\text{GS-GOGAT: } -\alpha\text{-ketoglutarate} - NH_3 - NADPH - ATP + \text{glutamate} = 0 \qquad (5)$$

Thus, the only difference is that ATP is used in the GS-GOGAT route (which is a high-affinity system) but not in the GDH reaction. The problem here is that an ATP balance is not easy to write due to lack of sufficient information about all ATP-consuming reactions. In the absence of an ATP balance to differentiate between them, the two nitrogen assimilation reactions are linearly dependent and, as such, nonobservable. Because the only difference between the two routes is the consumption of ATP in the GS-GOGAT system, distinction between the two routes may not be important, and they are therefore often lumped into a single reaction in stoichiometric models.

Another reaction that may give rise to a singularity is the transhydrogenase reaction for the interconversion of NADH and NADPH [see eq. (2.4)]. In the presence of this reaction, the balances for the two cofactors are coupled, obviously becoming linearly dependent. Even though transhydrogenases are reported to exist in many organisms, these enzymes normally are not very active under normal growth conditions. They have, however, been assayed in certain cases (see Section 10.1), and they serve a very important role in dissipating the large amounts of NADPH formed under conditions of very high pentose cycle activity. An important point here is that the presence of transhydrogenase activity can be detected by the lack of consistency in the stoichiometric balances, another feature of MFA applied to overdetermined systems.

In many species different isoenzymes exist, catalyzing the same reaction but using different cofactors. A well-known example is the glutamate dehydrogenase reaction, for which many cells have two isoenzymes, one linked to NADH as cofactor (referred to as GDH-NADH) and one linked to NADPH as cofactor (referred to as GDH-NADPH). If both isoenzymes are operational, they will act as a transhydrogenase reaction and thereby give rise to a singularity. However, the GDH-NADH is mainly involved in catabolism of amino acids and therefore is repressed by glucose, whereas the GDH-NADPH is mainly involved in glutamate synthesis and is repressed by glutamate. Depending on the growth conditions, one of the two enzymes therefore can often be left out of the network formulation.

If a singularity arises in the stoichiometric matrix, one has the following two options:

(1) Remove the linearly dependent reaction(s) from the model, invoking information about specific enzyme regulation and induction.
(2) Introduce additional information such as the relative flux of the two pathways. Such information may be derived from measurements of enzyme activities, e.g., the relative activity of key enzymes in the two routes. However, this approach is hampered by the fact that *in vitro* enzyme activity measurements often are found to bear little relationship to actual *in vivo* flux distributions. A more powerful technique is the use of labeled substrates, e.g., ^{13}C-enriched glucose, followed by NMR measurements of the distribution of the ^{13}C label in various

products. Such techniques can supply the missing information about flux distribution through linearly dependent routes (see Chapter 9).

EXAMPLE 8.5

Heterofermentative Metabolism of Lactic Acid Bacteria

Lactic acid bacteria are often grown on complex media that are rich in amino acids (*e.g.*, yeast extract or casein peptone), providing the carbon skeletons for biosynthesis. There is, therefore, no (or very little) net consumption of redox equivalents in the anabolic pathways. This results in a conservation of redox equivalents in the conversion of the energy source (glucose or lactose) to the primary metabolites (lactate, ethanol, acetate, formate, and carbon dioxide). The catabolic pathway utilized by lactic acid bacteria (Fig. 8.5) is, therefore, decoupled from growth and, as such, can be analyzed separately. Because the drain of precursor metabolites for growth is negligible for growth on a complex medium, the EMP pathway can be considered as a linear pathway with no branch points, so that all intermediates between glucose and pyruvate are eliminated as discussed earlier in this chapter. Under conditions of good growth, some species of lactic acid bacteria use the pathway only from glucose to lactate (often called homofer-

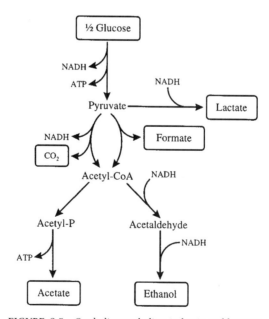

FIGURE 8.5   Catabolic metabolism in lactic acid bacteria.

mentative metabolism as leading to a single product). In this case the redox balance closes exactly, as the NADH formed in the conversion of glucose to pyruvate is regenerated in the conversion of pyruvate to lactate. Under conditions of extreme starvation, however, the cells strive to gain more ATP in the catabolic reactions, and this is done by channeling some pyruvate toward acetate. When this happens, a redox imbalance results, requiring that a fraction of the pyruvate be channeled toward ethanol where NAD^+ is regenerated. Thus, in order to obtain more ATP in the catabolism of glucose, several metabolic products are formed, and the metabolism in this case is called heterofermentative (or mixed acid fermentation).

Pyruvate obviously is a pathway branch point metabolite, and therefore a balance can be set up around this metabolite. In the conversion of pyruvate to metabolic products, three other pathway intermediates are involved: acetyl-CoA, acetyl-P, and acetaldehyde. Of these, only acetyl-CoA is located at a branch point and needs to be considered. Finally, conservation of the redox equivalents in the catabolic pathway yields an additional balance for NADH. One could also set up a balance for NAD^+, but because this compound always is linked to NADH, such a balance would be linearly dependent on the NADH balance and give no additional information. Note that a balance for ATP could be set up, but because only catabolic metabolism is considered, there is no consumption of ATP and this balance therefore would not close. In summary, three pathway metabolites (pyruvate, acetyl-CoA, and NADH), one substrate (glucose), and five metabolic products (lactate, carbon dioxide, formate, acetate, and ethanol) are included in the lactic acid pathway analysis. According to the general eq. (3.4), the stoichiometry of the six reactions considered in the pathway is given by (note that subscribed r variables indicate extracellular rates of accumulation of the corresponding metabolite and that because all fluxes can be measured directly we use the specific rates r_i instead of the intracellular fluxes v_i)

$$
\begin{pmatrix} -\frac{1}{2} \\ 0 \\ 0 \\ 0 \\ 0 \\ 0 \end{pmatrix} S_{glc} +
\begin{pmatrix} 0 & 0 & 0 & 0 & 0 \\ 1 & 0 & 0 & 0 & 0 \\ 0 & 1 & 0 & 0 & 0 \\ 0 & 0 & 1 & 0 & 0 \\ 0 & 0 & 0 & 1 & 0 \\ 0 & 0 & 0 & 0 & 1 \end{pmatrix}
\begin{pmatrix} P_{lac} \\ P_{CO_2} \\ P_{for} \\ P_{ac} \\ P_{et} \end{pmatrix} +
\begin{pmatrix} 1 & 0 & 1 \\ -1 & 0 & -1 \\ -1 & 1 & 1 \\ -1 & 1 & 0 \\ 0 & -1 & 0 \\ 0 & -1 & -2 \end{pmatrix}
\begin{pmatrix} X_{pyr} \\ X_{acCoA} \\ X_{NADH} \end{pmatrix}
$$

$$
= \begin{pmatrix} 0 \\ 0 \\ 0 \\ 0 \\ 0 \\ 0 \end{pmatrix} \tag{1}
$$

By using the stoichiometry of eq. (1) and applying the pseudo-steady state assumption to the three intracellular compounds, pyruvate, AcCoA, and NADH, yields the following linear equation result:

$$
\begin{pmatrix} 0 \\ 0 \\ 0 \end{pmatrix} = \begin{pmatrix} 1 & -1 & -1 & -1 & 0 & 0 \\ 0 & 0 & 1 & 1 & -1 & -1 \\ 1 & -1 & 1 & 0 & 0 & -2 \end{pmatrix} \begin{pmatrix} r_{pyr} \\ r_{lac} \\ r_c \\ r_{for} \\ r_{ac} \\ r_{et} \end{pmatrix}
\tag{2}
$$

The first of the preceding equations (corresponding to the first row in the matrix) is the pyruvate balance, the second is the balance for acetyl-CoA, and the last equation is the balance for NADH. It is noted that, in this example, all fluxes in the pathway can actually be determined directly from the measurement of the substrate uptake and product secretion fluxes. This is, however, rather an exceptional case, as illustrated in Example 8.3 where several of the fluxes could not be directly observed.

There are no linearly dependent reactions in this pathway, and this is confirmed by the rank of matrix **G**, which is found to be equal to 3. Therefore, it is possible to find a set of three measured rates that can be used to solve for the unknown rates. If we choose rate measurements of glucose (from which r_{pyr} is determined as $2r_{glc}$), lactate, and formate, we find by using eq. (8.8)

$$
\begin{pmatrix} 0 \\ 0 \\ 0 \end{pmatrix} = \begin{pmatrix} 1 & -1 & -1 \\ 0 & 0 & 1 \\ 1 & -1 & 0 \end{pmatrix} \begin{pmatrix} r_{pyr} \\ r_{lac} \\ r_{for} \end{pmatrix} + \begin{pmatrix} -1 & 0 & 0 \\ 1 & -1 & -1 \\ 1 & 0 & -2 \end{pmatrix} \begin{pmatrix} r_c \\ r_{ac} \\ r_{et} \end{pmatrix}
\tag{3}
$$

The last matrix $(\mathbf{G}_c^T)$ has a determinant different from zero, and therefore it can be inverted to yield

$$
\begin{pmatrix} r_c \\ r_{ac} \\ r_{et} \end{pmatrix} = - \begin{pmatrix} -1 & 0 & 0 \\ 1 & -1 & -1 \\ 1 & 0 & -2 \end{pmatrix}^{-1} \begin{pmatrix} 1 & -1 & -1 \\ 0 & 0 & 1 \\ 1 & -1 & 0 \end{pmatrix} \begin{pmatrix} r_{pyr} \\ r_{lac} \\ r_{for} \end{pmatrix}
$$

$$
= \begin{pmatrix} 1 & -1 & -1 \\ 0 & 0 & \frac{1}{2} \\ 1 & -1 & -\frac{1}{2} \end{pmatrix} \begin{pmatrix} r_{pyr} \\ r_{lac} \\ r_{for} \end{pmatrix}
\tag{4}
$$

It is observed that the formation of 2 mol of formate is accompanied by the formation of 1 mol of acetate. This fixed ratio between the rates of formation of two extracellular metabolites is explained by the NADH balance: When acetyl-CoA is formed together with formate, it is necessary to regenerate exactly 1 mol NAD^+ per mole of acetyl-CoA formed (namely, the NAD^+ used in the EMP pathway), and the flux from acetyl-CoA is therefore split equally between formation of ethanol (where two NAD^+ molecules are regenerated) and acetate. Similarly, if no formate is produced, the conversion of pyruvate to acetyl-CoA occurs solely via the pyruvate dehydrogenase route, which leads to additional NADH production. In this case, all the acetyl-CoA must be channeled toward ethanol, where the required NAD^+ is regenerated.

With the previous measurements of glucose (equal to one-half the rate of pyruvate), lactate, and formate, there were no difficulties in determining the fluxes of eq. (4). However, if the measurements of glucose, acetate, and formate are selected instead, we have

$$\det\left(G_c^T\right) = \begin{vmatrix} -1 & -1 & 0 \\ 0 & 1 & -1 \\ -1 & 1 & -2 \end{vmatrix} = 0 \tag{5}$$

In other words, the preceding measurements cannot be used to calculate uniquely the remaining three unknown pathway fluxes to acetate, CO_2, and ethanol. The reason, of course, is that with formate and acetate as the only measurements of metabolic products there is no information about the rate of NADH consumption (one of the fluxes r_{lac}, r_c, or r_{et} has to be measured) and the NADH balance is of little use. Therefore, measurement of both acetate and formate fluxes does not provide additional information as one can be determined from the other. Note that it is not necessary to calculate the determinant of the matrix (even though this is easily done using commercial mathematical software packages), as it is observed that the last row can be written as: row3 = row1 + 2row2.

8.2. OVERDETERMINED SYSTEMS

In Section 8.1 we developed the method of flux analysis for exactly determined systems. It is often the case, however, that more flux measurements are available than the degrees of freedom of the system. This, of course, yields an overdetermined system in which, by analogy to our black box model treatment of Section 4.4, redundancies can be exploited to calculate not only better estimates for the nonmeasured fluxes but also better estimates for the measured fluxes. Furthermore, with an overdetermined system it may be possible to check the validity of the steady state assumption for the intracellular metabolites.

In the case of an overdetermined system, the G_c^T matrix of eq. (8.8) is nonsquare and therefore can not be inverted. A simple solution is obtained by use of its Moore-Penrose pseudo-inverse (see also Section 4.4):

$$v_c = -\left(G_c^T\right)^{\#} G_m^T v_m \tag{8.9}$$

where $(G_c^T)^{\#}$ is the pseudo-inverse of G_c^T given by

$$\left(G_c^T\right)^{\#} = \left(G_c G_c^T\right)^{-1} G_c \tag{8.10}$$

Equation (8.9) of course can be used provided that matrix $G_c G_c^T$ is nonsingular, a condition that is satisfied when matrix G_c has full rank. Thus, the requirement for solution of an overdetermined system is the same as that for an exactly determined system. For a quadratic matrix the pseudo-inverse is identical with its inverse, and eq. (8.9) is a general solution to eq. (8.7) when the measurements are greater than or equal to the degrees of freedom.

Equation (8.9) is very useful when there is little noise in the measurements. It *represents the least squares estimate for the nonmeasured fluxes* under the assumption that the constraint residuals arising from measurement noise are uniformly distributed among the system constraints, *i.e.* the pseudo-steady state balances for the intracellular metabolites. However, if a significant amount of noise is present in only some of the measured fluxes, the solution of eq. (8.9) does not satisfy flux conservation around some network nodes, meaning that at some nodes the sum of mass fluxes into the node will be different from the sum of mass fluxes emanating from the node. This is obviously problematic in studies of flux distributions, so that it is desirable to identify gross error-containing flux measurements similar to our treatment of the black box model in Section 4.4. Because eq. (8.7) is analogous to the elemental balances given by eq. (4.10), the analysis of Section 4.4 may be applied here in exactly the same way. The redundancy matrix is calculated first:

$$R = G_m - G_c G_c^{\#} G_m \tag{8.11}$$

and from this the *reduced redundancy matrix* is determined, which is used in calculations similar to those outlined by eqs. (4.15)-(4.17) and illustrated in Example 4.6.

A slightly different procedure [described by Tsai and Lee (1988)], can provide better estimates *for both nonmeasured and measured fluxes*. For this purpose, eq. (8.7) is reformulated to read:

$$\begin{pmatrix} \mathbf{v}_m \\ 0 \end{pmatrix} = \begin{pmatrix} \mathbf{I} & 0 \\ \mathbf{G}^T & \end{pmatrix} \begin{pmatrix} \mathbf{v}_m \\ \mathbf{v}_c \end{pmatrix} = \mathbf{Tv} \qquad (8.12)$$

where $\mathbf{I}$ is the identity matrix of dimension equal to the number of measured fluxes. The first row in eq. (8.12) simply specifies that $\mathbf{v}_m = \mathbf{v}_m$, whereas the second row is identical with eq. (8.7). Most metabolic models in the literature have or easily can be reduced to the form of eq. (8.12). In that form, eq. (8.12) introduces a strict separation between the calculated and measured flux vectors, $\mathbf{v}_c$ and $\mathbf{v}_m$, respectively, and does not allow for any improvement of the measured fluxes through elimination of measurement noise. This is accomplished by a more general partitioning of the flux vector, whereby measured and unmeasured fluxes are not separated:

$$\begin{pmatrix} \mathbf{v}_m \\ 0 \end{pmatrix} = \begin{pmatrix} \mathbf{T}_{11} & \mathbf{T}_{12} \\ \mathbf{T}_{21} & \mathbf{T}_{22} \end{pmatrix} \begin{pmatrix} \mathbf{v}_1 \\ \mathbf{v}_2 \end{pmatrix} \qquad (8.13)$$

The second row of the preceding equation clearly reflects the steady state metabolite balances. Matrices $\mathbf{T}_{11}$ and $\mathbf{T}_{12}$ are defined from the identity matrix and $\mathbf{G}^T$ as illustrated in Example 8.6. Vectors $\mathbf{v}_1$ and $\mathbf{v}_2$ can be selected to be equal to $\mathbf{v}_m$ and $\mathbf{v}_c$, respectively (in which case the method is identical to the previous one), or not so long as matrix $\mathbf{T}_{22}$ can be inverted to yield a solution for vector $\mathbf{v}_2$ in terms of $\mathbf{v}_1$:

$$\mathbf{v}_2 = -\mathbf{T}_{22}^{-1}\mathbf{T}_{21}\mathbf{v}_1 \qquad (8.14)$$

When this equation is inserted in eq. (8.13), the following relationship between the measured rates $\mathbf{v}_m$ and the elements of $\mathbf{v}_1$ is obtained:

$$\mathbf{v}_m = \mathbf{T}_r\mathbf{v}_1 \qquad (8.15)$$

where $\mathbf{T}_r$ is given by

$$\mathbf{T}_r = \mathbf{T}_{11} - \mathbf{T}_{12}\mathbf{T}_{22}^{-1}\mathbf{T}_{21} \qquad (8.16)$$

Equation (8.15) specifies an overdetermined system just as the original balance equations, and if the measurement errors are normally distributed with a mean value of zero and a variance-covariance matrix equal to $\mathbf{F}$, then the maximum likelihood estimate for the elements of $\mathbf{v}_1$ is given by (Madron et al., 1977)

$$\hat{\mathbf{v}}_1 = \left(\mathbf{T}_r^T\mathbf{F}^{-1}\mathbf{T}_r\right)^{-1}\mathbf{T}_r^T\mathbf{F}^{-1}\mathbf{v}_m \qquad (8.17)$$

where the circumflex specifies the best estimate of the corresponding quantity for normally distributed measurement errors. Estimates for the elements of $v_2, \hat{v}_2$, may be found by introducing eq. (8.17) into eq. (8.14):

$$\hat{v}_2 = -T_{22}^{-1}T_{21}\hat{v}_1 \tag{8.18}$$

Equations (8.17) and (8.18) give the best estimates for *all fluxes* in the metabolic model, *i.e.*, both those that are measured and those that are not measured. Often variance and covariance information is not available, but if the errors in the measured fluxes are of the same magnitude and uncorrelated, one may use the least squares estimate for the elements of v_1 given by

$$\hat{v}_1 = \left(T_r^T T_r\right)^{-1} T_r^T v_m \tag{8.19}$$

EXAMPLE 8.6

Flux Determination in an Overdetermined System

We now return to the heterofermentative metabolism of lactic acid bacteria treated in Example 8.5 and consider the situation where four rates are measured, thus yielding an overdetermined system. If measurements of the pyruvate production rate (equal to $2r_{glc}$) and lactic acid (r_{lac}), formate (r_{for}), and ethanol (r_{et}) are available, eq. (8.7) becomes

$$\begin{pmatrix} 0 \\ 0 \\ 0 \end{pmatrix} = \begin{pmatrix} 1 & -1 & -1 & 0 \\ 0 & 0 & 1 & -1 \\ 1 & -1 & 0 & -2 \end{pmatrix} \begin{pmatrix} r_{pyr} \\ r_{lac} \\ r_{for} \\ r_{et} \end{pmatrix} + \begin{pmatrix} -1 & 0 \\ 1 & -1 \\ 1 & 0 \end{pmatrix} \begin{pmatrix} r_c \\ r_{ac} \end{pmatrix} \tag{1}$$

where the first row is the balance for pyruvate, the second row is the balance for acetyl-CoA, and the last row is the balance for NADH. Again, subscribed r variables have been introduced to indicate the extracellular rates of accumulation of the corresponding metabolites. We now calculate the least squares estimates for the nonmeasured rates using both the simple approach of eq. (8.9) and the procedure based on eqs. (8.12) and (8.13).

First we calculate the Moore − Penrose pseudo-inverse of G_c^T from eq. (8.10):

$$(G_c^T)^\# = \left(\begin{pmatrix} -1 & 1 & 1 \\ 0 & -1 & 0 \end{pmatrix} \begin{pmatrix} -1 & 0 \\ 1 & -1 \\ 1 & 0 \end{pmatrix} \right)^{-1} \begin{pmatrix} -1 & 1 & 1 \\ 0 & -1 & 0 \end{pmatrix}$$

$$= \begin{pmatrix} -0.5 & 0 & 0.5 \\ -0.5 & -1 & 0.5 \end{pmatrix} \tag{2}$$

and then the solution for the nonmeasured rates from eq. (8.9):

$$\begin{pmatrix} r_c \\ r_{ac} \end{pmatrix} = -\begin{pmatrix} -0.5 & 0 & 0.5 \\ -0.5 & -1 & 0.5 \end{pmatrix} \begin{pmatrix} 1 & -1 & -1 & 0 \\ 0 & 0 & 1 & -1 \\ 1 & -1 & 0 & -2 \end{pmatrix} \begin{pmatrix} r_{pyr} \\ r_{lac} \\ r_{for} \\ r_{et} \end{pmatrix}$$

$$= \begin{pmatrix} 0 & 0 & -0.5 & 1 \\ 0 & 0 & 0.5 & 0 \end{pmatrix} \begin{pmatrix} r_{pyr} \\ r_{lac} \\ r_{for} \\ r_{et} \end{pmatrix} \tag{3}$$

This solution is consistent with the one found for the determined system in Example 8.3, i.e, if there are no measurement errors the preceding solution is identical with that found in Example 8.3 [try to insert the expression for r_{et} from eq. (4) (Example 8.5) into the preceding equation and compare the resulting expressions for r_c and r_{ac} with those found in Example 8.5]. From eq. (3) it is observed that the estimates for the nonmeasured fluxes are independent of the measurements of the pyruvate flux and the lactate accumulation rates.

In order to update the flux estimates of both nonmeasured and measured metabolites following the method of Tsai and Lee (1988), we rewrite eq. (1) in the form indicated by eq. (8.12):

$$\begin{pmatrix} r_{pyr} \\ r_{lac} \\ r_{for} \\ r_{et} \\ 0 \\ 0 \\ 0 \end{pmatrix} = \begin{pmatrix} 1 & 0 & 0 & 0 & 0 & 0 \\ 0 & 1 & 0 & 0 & 0 & 0 \\ 0 & 0 & 1 & 0 & 0 & 0 \\ 0 & 0 & 0 & 1 & 0 & 0 \\ \hline 1 & -1 & -1 & 0 & -1 & 0 \\ 0 & 0 & 1 & -1 & 1 & -1 \\ 1 & -1 & 0 & -2 & 1 & 0 \end{pmatrix} \begin{pmatrix} r_{pyr} \\ r_{lac} \\ r_{for} \\ r_{et} \\ r_c \\ r_{ac} \end{pmatrix} \tag{4}$$

where the partitioning of the matrix $\mathbf{T}$ is indicated. Notice the appearance of the stoichiometric matrix $\mathbf{G}^T$ (the bottom three rows), the 4 x 4 identity matrix, and the zero matrix. The preceding equation has been written such that $\mathbf{v}_1$ only consists of measured fluxes, but this does not necessarily have to be the case. Nonmeasured rates may well be included in $\mathbf{v}_1$ together with measured rates. With the indicated partitioning, we find

$$
\mathbf{T_r} = \begin{pmatrix} 1 & 0 & 0 \\ 0 & 1 & 0 \\ 0 & 0 & 1 \\ 0 & 0 & 0 \end{pmatrix} - \begin{pmatrix} 0 & 0 & 0 \\ 0 & 0 & 0 \\ 0 & 0 & 0 \\ 1 & 0 & 0 \end{pmatrix} \begin{pmatrix} 0 & -1 & 0 \\ -1 & 1 & -1 \\ -2 & 1 & 0 \end{pmatrix}^{-1} \begin{pmatrix} 1 & -1 & -1 \\ 0 & 0 & 1 \\ 1 & -1 & 0 \end{pmatrix}
$$

$$
= \begin{pmatrix} 1 & 0 & 0 \\ 0 & 1 & 0 \\ 0 & 0 & 1 \\ 1 & -1 & -0.5 \end{pmatrix} \tag{5}
$$

and this gives

$$
\hat{\mathbf{v}}_1 = \begin{pmatrix} \hat{r}_{pyr} \\ \hat{r}_{lac} \\ \hat{r}_{for} \end{pmatrix} = \tfrac{1}{13} \begin{pmatrix} 9 & 4 & 2 & 4 \\ 4 & 9 & -2 & -4 \\ 2 & -2 & 12 & -2 \end{pmatrix} \begin{pmatrix} r_{pyr} \\ r_{lac} \\ r_{for} \\ r_{et} \end{pmatrix} \tag{6}
$$

and

$$
\hat{\mathbf{v}}_2 = \begin{pmatrix} \hat{r}_{et} \\ \hat{r}_{c} \\ \hat{r}_{ac} \end{pmatrix} = \tfrac{1}{13} \begin{pmatrix} 4 & -4 & -2 & 9 \\ 3 & -3 & -8 & 10 \\ 1 & -1 & 6 & -1 \end{pmatrix} \begin{pmatrix} r_{pyr} \\ r_{lac} \\ r_{for} \\ r_{et} \end{pmatrix} \tag{7}
$$

where the rate vector on the right-hand side represents the measured values of the corresponding metabolite fluxes. Notice also that this solution is consistent with the one found from the determined system of Example 8.3 [again try to insert the expression for r_{et} in eq. (4) of example 8.3 and compare the resulting expressions for r_c and r_{ac} with those found from the above equation]. However, here the presence of measurement noise in any of the measured rates affects the best least squares estimates of both nonmeasured and measured rates. Thus, the estimate of the pyruvate flux (r_{pyr}) is a function of all four measured rates.

In the preceding estimation procedure, we indirectly assumed that the steady state assumptions are exactly satisfied and that all noise is distributed only among the measured fluxes. If there is noise also in the pseudo-steady state equations (meaning that small accumulations of the corresponding metabolites are possible), then eq. (8.12) still forms the basis for the estimation of the intracellular fluxes, but the solution is found directly as

$$\hat{v} = \left(T^T F^{-1} T\right)^{-1} T^T F^{-1} \begin{pmatrix} v_m \\ 0 \end{pmatrix} \tag{8.20}$$

where F is the variance-covariance matrix for the residuals of both the measured fluxes and the pseudo-steady state assumptions. Again, if the noise is of the same magnitude, the least square estimate given by eq. (8.21) may be used:

$$\hat{v} = \left(T^T T\right)^{-1} T^T \begin{pmatrix} v_m \\ 0 \end{pmatrix} \tag{8.21}$$

It is generally preferable to use the procedure where noise is assumed to be distributed only among the measured rates, but due to its simple structure eq. (8.21) frequently is used for calculation of metabolic fluxes. One can, however, use eq. (8.20) and specify very low variances for the pseudo-steady state assumptions (Vallino and Stephanopoulos, 1993), as this will give estimates identical to those obtained with the procedure described earlier (see Example 8.7).

EXAMPLE 8.7

Analysis of an Overdetermined System

We now again return to the heterofermentative metabolism of lactic acid bacteria treated in Examples 8.5 and 8.6 and calculate estimates for the intracellular fluxes using eqs. (8.20) and (8.21). First we assume that there are no covariances and that the variances in the measured fluxes are the same (arbitrarily set to 1). The variances of the pseudo-steady state assumptions are given very small values (10^{-9}), effectively corresponding to no noise.

Using eq. (8.20) we then find:

$$
\begin{pmatrix} \hat{r}_{pyr} \\ \hat{r}_{lac} \\ \hat{r}_{for} \\ \hat{r}_{et} \\ \hat{r}_{c} \\ \hat{r}_{at} \end{pmatrix}
= \frac{1}{13}
\left(\begin{array}{cccc|ccc}
9 & 4 & 2 & 4 & 2 & 0 & 2 \\
4 & 9 & -2 & -4 & -2 & 0 & -2 \\
2 & -2 & 12 & -2 & -1 & 0 & -1 \\
4 & -4 & -2 & 9 & -2 & 0 & -2 \\
3 & -3 & -8 & 10 & -8 & 0 & 5 \\
1 & -1 & 6 & -1 & -7 & -13 & 6
\end{array}\right)
\begin{pmatrix} r_{pyr} \\ r_{lac} \\ r_{for} \\ r_{et} \\ 0 \\ 0 \\ 0 \end{pmatrix}
\tag{1}
$$

These estimates are identical to those derived in eqs. (6) and (7) of Example 8.6 (note that only the elements left of the line in the matrix are important due to the multiplication of the elements right of the line with the zero vector). If, on the contrary, we use eq. (8.21) directly, we find the estimates to be given by

$$
\begin{pmatrix} \hat{r}_{pyr} \\ \hat{r}_{lac} \\ \hat{r}_{for} \\ \hat{r}_{et} \\ \hat{r}_{c} \\ \hat{r}_{at} \end{pmatrix}
= \frac{1}{15}
\left(\begin{array}{cccc|ccc}
11 & 4 & 2 & 4 & 2 & 0 & 2 \\
4 & 11 & -2 & -4 & -2 & 0 & -2 \\
2 & -2 & 14 & -2 & -1 & 0 & -1 \\
4 & -4 & -2 & 11 & -2 & 0 & -2 \\
3 & -3 & -9 & 12 & -9 & 0 & 6 \\
1 & -1 & 7 & -1 & -8 & -15 & 7
\end{array}\right)
\begin{pmatrix} r_{pyr} \\ r_{lac} \\ r_{for} \\ r_{et} \\ 0 \\ 0 \\ 0 \end{pmatrix}
\tag{2}
$$

which are quite different from the estimates of eq. (1).

In analogy with our analysis of the overdetermined black box model of Section 4.4, it is possible here too, if the number of measured rates is greater than $F + 1$, to use the measurement redundancy in combination with the metabolic model to perform error diagnosis. If it is assumed that the pseudo-steady state assumptions are correct, this error diagnosis is based on the reduced redundancy matrix ($\mathbf{R}_r$) of eq. (8.11), and, as in Section 4.4, the test function h is calculated from

$$
h = \varepsilon^T \mathbf{P}^{-1} \varepsilon \tag{8.22}
$$

where $\mathbf{P}$ is calculated according to eq. (4.22) and ε is the residual vector given by

$$
\varepsilon = \mathbf{R}_r \mathbf{v}_m \tag{8.23}
$$

Comparison of the test function h with the statistics of the χ^2 distribution described in section 4.4 may identify gross measurement errors at a given confidence level. By eliminating one measurement at a time and subsequently calculating the resulting test function, it may be possible to identify the source of the gross measurement error. If there are indications that one or more of the pseudo-steady state assumptions are violated, error diagnosis may also be carried out to check the consistency of these assumptions. Here the basis for the analysis is eq. (8.12), and, besides flux measurements, the pseudo-steady state assumptions are eliminated one at a time, followed each time by a calculation of the new redundancy matrix, the new test function h [eq. (8.22)], and a similar comparison to the χ^2 distribution statistics.

From the Gibbs rule of stoichiometry, the degrees of freedom in a metabolic model are always less than or equal to the degrees of freedom in a black box model (Nielsen and Villadsen, 1994). Hence, it should be attractive to use a metabolic model for identification of gross measurement errors and error diagnosis as fewer measurements may be needed for the analysis. However, for simple metabolic models the degrees of freedom are generally the same as for the black box model representing the system [in the lactic acid model of Example 8.3 there are three degrees of freedom, and with six metabolites and three elements (C, H, and O) the degrees of freedom for a black box model of this system are also three], whereas for more complex models the analysis may become cumbersome as a result of a large stoichiometric matrix. Error diagnosis therefore is performed preferentially using the black box model, and, after suspect measurements are identified and corrected, the improved set of estimated rates may be used to calculate better estimates for the metabolic fluxes.

8.3. UNDERDETERMINED SYSTEMS— LINEAR PROGRAMMING

If the number of measured fluxes is smaller than the system degrees of freedom, an infinite number of solutions exist for the network fluxes. In this case *linear programming* could be used to determine intracellular flux distributions, provided that a suitable objective function can be specified (such as, for example, maximizing the specific growth rate of the cells). With this approach it is possible to obtain a unique solution for the intracellular fluxes by optimizing the objective function subject to the constraints of the metabolite balances. Before we illustrate the usefulness of this approach in studies of cellular systems, we present a basic description of the method using a very simple example which provide a useful insight into the nature of linear programming problems.

We consider a system of two variables x and y related through the constraint:

$$2x + y = 4 \qquad (8.24)$$

which specifies that the solution lies on the straight line AB (see Fig. 8.6). In addition to eq. (8.24), another constraint common to all linear programming problems is that all variables (in this case x and y) are required to be nonnegative. This violates our definition of metabolic fluxes, which may be either positive or negative. Therefore, application of linear programming for solving metabolic flux balances requires extension of the usual metabolic model to include both forward and reverse fluxes for each of the J reactions in the model, or, at least, for those reactions that may be reversible. With the constraints $x > 0$ and $y > 0$, the solutions of the preceding problem are restricted to those lying in the positive quadrant of (x, y) space. This still represents infinite solutions, which are further reduced to one by applying the requirement that *the solution maximize or minimize a certain cost (or objective) function*. For example, if we want to maximize the objective function $2x + 3y$ in the preceding example, then the problem is to find a solution that lies on the line AB and at the same time maximizes this function. A family of cost lines is shown in Fig. 8.6. The line that yields the largest value for the objective function and at the same time satisfies the

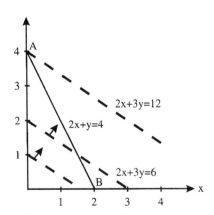

FIGURE 8.6 Illustration of linear programming. The optimal solution is constrained to lie on line AB in the nonnegative quadrant of the solution space. The objective function consists of a family of lines (of which three are shown). The line that maximizes the objective function intercepts the constraint line AB at $(x, y) = (0, 4)$, which represents the solution for this optimization problem.

constraint is given by $2x + 3y = 12$. It intersects with the line AB at $x = 0$ and $y = 4$. A feasible solution is obtained in this example, but one may encounter situations where there is no feasible solution or there are infinite solutions [with the preceding example this will happen if the objective function is the same as or a multiple of eq. (8.24)].

We now return to our metabolic model, where the constraints are given by eq. (8.7). Because all variables (i.e., fluxes) have to be positive in a linear programming formalism, we rewrite the model such that the stoichiometry includes forward and reverse reactions (in those cases involving reversible reactions). Thus, the constraints are summarized by

$$G_{ex}^T v_{ex} = 0 \quad \text{and} \quad v_{ex} \geq 0 \tag{8.25}$$

where G_{ex}^T is the stoichiometric matrix revised to include the stoichiometry of both the forward and reverse reactions and v_{ex} is the corresponding revised flux vector with dimension J_{ex}. In the previous simple example the constraint specified a line in 2-dimensional space. With J_{ex} variables (or fluxes), each constraint specifies a hyperplane in J_{ex}-dimensional space. Because all constraints must be satisfied, eq. (8.25) specifies that a solution has to lie at the intersection of the K hyperplanes in this J_{ex}-dimensional space. If one of the hyperplanes does not intersect with the others, a feasible solution for the fluxes does not exist (it is said in this case that the set of feasible solutions is empty). This situation, however, rarely is encountered in practice. The next step is to specify an objective function that is a linear function of the fluxes, i.e., a function of the elements of vector v_{ex}. The optimal solution is then found by solving the corresponding maximization or minimization problem:

$$\text{Minimize/maximize } a v_{ex}, \text{ subject to constraints in eq. (8.25)} \tag{8.26}$$

where a is a row vector containing weights of the individual variables specifying the influence of the individual fluxes on the objective function.

Finding the optimal solution to the problem of eq. (8.26) may be complicated due to the potentially large number of feasible solutions that satisfy all the constraints. However, as with the previous simple example, the optimal solution is located at the end (or corner) of the set of feasible solutions, (a result of the linearity of both the constraints and the objective functions). Thus, we could (in principle) find the solution by, first, enumerating all solution corners and, second, evaluating the objective function at each point of the space of feasible solutions. This approach is practically infeasible due to the very large number of solution corners. Instead, we turn to the so-called *simplex* method (Dantzig, 1963), which is the method of choice for solving constrained linear optimization problems (Gyr, 1978). This method is applied as follows: First locate a corner of the feasible set and then proceed from corner to corner along the edges of the feasible solutions. At a given corner there are several edges to chose from, some leading away from the optimal solution and others leading gradually toward it. In the simplex method one moves along the edge that is guaranteed to decrease (or increase) the objective function. That edge leads to a new corner with a lower (or higher) value, with no possibility of returning to any corner that yields a higher (or lower) value for the objective function. Eventually, a special corner is reached, from which all edges lead to higher objective function values, and this corner represents the optimal solution. Despite its simple description, the simplex method is quite complex to implement, but robust algorithms have been developed [see, for example, Gyr (1978)] and are available in several software packages, such as Mathematica and Matlab.

EXAMPLE 8.8

Stoichiometric Interpretation of *Escherichia coli* Metabolism

Optimal flux distributions have been obtained by linear programming for several metabolic models. In this example, we will consider the analysis of *E. coli* metabolism as described by Varma *et al.* (1993a,b). Based on a detailed stoichiometric model for *E. coli* metabolism, they analyzed glucose catabolism and the biochemical production capabilities of this organism. Their model consists of 107 metabolites (including substrates and metabolic products) and 95 reversible reactions, and because each reaction has to be included in both a forward and backward direction this gives a stoichiometric matrix of 107 x 190, *i.e.*, there are 83 degrees of freedom. None of the fluxes were measured, but one flux (the specific glucose uptake rate) was used to scale the values of the fluxes such that they all have units millimoles per gram dry weight per hour. The only experimental data used for the flux calculations were the following:

- Metabolic demands for growth were used to calculate the drain of metabolites for biomass synthesis. In practice, the drain of metabolites is included as stoichiometric coefficients in a reaction leading to biomass synthesis.
- ATP requirements for maintenance. By fitting the specific glucose uptake rate at different specific growth rates to experimental data, a requirement of 23 mmol of ATP $(g \, DW)^{-1}$ for growth-associated maintenance and 5.87 mmol of ATP (g DW h) for non-growth-associated

maintenance was estimated, and these values are used in the calculations.

- A maximum specific oxygen uptake rate of 20 mmol (g DW h)$^{-1}$.

The objective function was the maximization of the specific growth rate. Thus, if a specific glucose uptake rate is given, the model determines the corresponding maximum specific growth rate along with all the fluxes in the network. Selected network fluxes can be plotted against the corresponding specific growth rate, as illustrated in Fig. 8.7. It is observed that at low specific glucose uptake rates (corresponding to low specific growth rates), no byproducts are formed, i.e., there is complete respiration. As a result, in this regime the specific oxygen uptake rate increases linearly with the specific growth rate (see also Section 3.4). When the specific glucose uptake rate approaches about 8 mmol (g DW h)$^{-1}$ (corresponding to a specific growth rate of about 0.9 h^{-1}) the oxygen requirements for complete oxidation of glucose to carbon dioxide exceed the specified maximum of 20 mmol (g DW h)$^{-1}$. The cells therefore shift to a mixed metabolism where there is both respiration and fermentation. The first fermentative metabolite that is excreted is acetate, and at higher glycolytic fluxes (or higher specific growth rates) formate is also excreted. Finally, at very high glycolytic fluxes ethanol is excreted. It is interesting that the model predicts the right sequence of excretion of metabolites, i.e., first acetate, then formate, and finally ethanol, which is consistent with experimental findings. Varma *et al.*

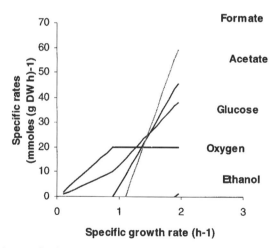

FIGURE 8.7   The specific glucose and oxygen uptake rates and the specific rates of formation of acetate, formate, and ethanol as function of the specific growth rate calculated by maximizing the specific growth rate of the *E. coli* metabolic model using linear programming.

(1993a) explain this as a consequence of the low energy content of acetate (its ATP yield upon oxidation is lower than that for formate and ethanol).

The critical glycolytic fluxes (or specific growth rates) were also calculated for values of the maximum specific oxygen uptake rate lower than 20 mmol (g DW h)$^{-1}$ (which could be a consequence of limitations in oxygen supply to the bioreactor). It was found that there is a linear relationship between these flux values and the oxygen supply to the cell.

An advantage of linear programming is that additional information can be obtained by calculating the sensitivity of the objective function z with respect to the system variables. These sensitivities are called *shadow prices*, which are given for the E. coli metabolic model by

$$\lambda_i = \frac{\partial \mu}{\partial r_i} \tag{1}$$

where r_i is the net specific rate of consumption of the ith compound, *e.g.*, oxygen, acetate, or an amino acid. Thus, the shadow price reflects the effects that changes in metabolic rates have on the specific growth rate. Table 8.2 gives some of the shadow prices calculated by the metabolic model for different specific oxygen uptake rates (r_o). At low r_o the shadow price for oxygen is positive, *i.e.*, by increasing r_o the specific growth rate can be increased. When r_o is at its maximum, the specific growth rate cannot be increased further by increasing r_o and the shadow price is therefore zero. At anaerobic conditions ($r_o = 0$), the shadow price for NADH is negative due to the inability to oxidize this cofactor, *i.e.*, if NADH production increases the specific growth rate will decrease. This indicates that cofactor regeneration can limit growth at anaerobic conditions. For ATP it is observed that the

TABLE 8.2   Shadow Prices at Different Specific Oxygen Uptake Rates [in mmol (g DW h)$^{-1}$] Calculated Using the Metabolic Model of Varma *et al.* (1993a)[a]

Metabolite	Shadow price [g DW (mmol metabolite)$^{-1}$]		
	0	12	20
Oxygen	0.0399	0.0282	0
ATP	0.0109	0.0106	0.0049
NADH	-0.0054	0	0.0065
Acetate	0	0	0.0242
Ethanol	0	0.0106	0.0422
Lactate	0.0054	0.0106	0.0422
Succinate	0.0109	0.0177	0.0504

[a] The specific glucose uptake rate is 10 mmol (g DW h)$^{-1}$.

shadow price is high at anaerobic conditions and decreases with increasing r_o. Thus, supply of ATP for growth becomes less limiting for cells growing aerobically compared with anaerobic growth. The shadow prices for all the metabolites increase with r_o, because at aerobic conditions these compounds can be oxidized by the cell and give additional ATP required for growth. It is interesting to note that at anaerobic conditions the shadow price for succinate is higher than that for the other metabolites. Oxidation of this compound will give rise to increasing NADH production (which the cell has difficulty reoxidizing at these conditions), but it is likely that this compound may be valuable for the supply of precursor metabolites.

The model also allowed calculation of the fluxes through the different pathways. Here it was found that at anaerobic conditions there is no flux through the oxidative branch of the PP pathway, i.e., glucose-6-P dehydrogenase is not active, and the NADPH required for biosynthesis is supplied by a transhydrogenase reaction. For high specific oxygen uptake rates, approximately 60% of the glucose-6-P is metabolized via the PP -route and the transhydrogenase is not active.

The model was also applied to investigate the potential of E. coli to produce additional metabolites, e. g., amino acids (Varma et al., 1993b). This potential can be determined from the magnitude of the shadow prices, as the latter provide a measure of the trade off in biomass growth for the production of a specific metabolite, e. g., amino acids. It was found that the shadow prices are low for amino acids like glycine, alanine, and aspartate, which have simple biosynthetic routes, whereas the shadow prices are high for the aromatic amino acids phenylalanine, tryptophan, and tyrosine, which have complex biosynthetic routes. Thus, if the cell is going to produce a product by a long and complex biosynthetic route, the requirement of cell metabolism is large and the effect on biomass growth is therefore severe. On the contrary, a compound with a simple metabolic route can be produced by the cells at little cost, and the effect on biomass growth is therefore small.

8.4. SENSITIVITY ANALYSIS

In the previous sections, we derived solutions for the mass balance equations of determined, overdetermined, and underdetermined systems. Although in each case a unique solution was obtained, it is important to measure the sensitivity, or "vulnerability", of the solution with respect to small perturbations in the measurements. The first step in this sensitivity analysis is to examine whether the system is well-posed, i.e., check whether the stoichiometric matrix is well-conditioned. With an ill-conditioned stoichiometric matrix, even small variations in the measurements may have a very large impact on the calculated fluxes. To illustrate this, we consider the two equation systems:

$$\begin{pmatrix} 1 & 1 \\ 1 & 1.0001 \end{pmatrix} \begin{pmatrix} v_1 \\ v_2 \end{pmatrix} = \begin{pmatrix} 2 \\ 2 \end{pmatrix} \quad \text{and} \quad \begin{pmatrix} 1 & 1 \\ 1 & 1.0001 \end{pmatrix} \begin{pmatrix} v_1 \\ v_2 \end{pmatrix} = \begin{pmatrix} 2 \\ 2.0001 \end{pmatrix}$$

(8.27)

Even though the right-hand side differs by a meager 0.0001, the solution changes from $v_1 = 2$ and $v_2 = 0$ to $v_1 = v_2 = 1$. Because solutions are generally derived using numerical techniques, errors in flux estimates caused by small computational roundoff errors may be amplified if the stoichiometric matrix is ill-conditioned. A measure of the sensitivity of a matrix is the so-called *condition number*, which is given by

$$C(G^T) = \|G^T\| \|(G^T)^\#\|$$

(8.28)

where $\| \ \|$ indicates any matrix norm and $(G^T)^\#$ is the pseudo-inverse of the stoichiometric matrix (see Section 8.2). Calculation of the condition number is quite complex (for details, see Box 8.1), but in practice it is found using commercial software packages like Matlab, Mathematica, or Maple. For a symmetric matrix with positive eigenvalues, the condition number is given as the ratio of the largest to the smallest eigenvalue. The condition number therefore is always greater than 1 (this is also true for asymmetric matrices), and a large condition number indicates an ill-conditioned matrix.[2] Thus, for the matrix of eq. (8.27), which is symmetric and has the eigenvalues 2 and $10^{-4}/2$, the condition number is 10^4. The magnitude of the condition number provides important information about requirements for the accuracy of the measured fluxes, as the measurements have to be carried out with a precision that carries the same number of digits as there are in the condition number. Thus, in the preceding example where the condition number has five digits, it is necessary to measure fluxes precisely at the fifth digit. Because fermentation rates rarely can be quantified with precision greater than two digits (e.g., the specific growth rate of microorganisms can be specified as 0.37 h^{-1} with good precision), a requirement for a well-condi-

---

[2] With this definition of the condition number, there is a direct analogy between an ill-conditioned matrix in a set of linear, algebraic equations and a stiff system of coupled, first-order differential equations, which is well-known to many chemical engineers.

BOX 8.1

Calculation of the Condition Number

The general definition of the condition number is given in eq. (8.28). In order to use this equation it is, however, necessary to calculate the norm of a matrix, and on this note we give a short description of how this is done. For more details we refer to textbooks on linear algebra.

The norm of a matrix measures the largest amount by which any vector x is amplified by matrix multiplication:

$$\|G^T\| = \max \frac{\|G^T x\|}{\|x\|} \tag{1}$$

and it can be shown that it is equal to the square root of the largest eigenvalue of GG^T:

$$\|G^T\| = \sqrt{\lambda_{\max}(GG^T)} \tag{2}$$

Thus, finding the norm is equivalent to an eigenvalue problem, and from the eigenvalues of GG^T and the eigenvalues of the similar matrix for the pseudo-inverse (G^T) [given by $(GG^T)^{-1}G$] the condition number can be calculated using eq. (8.28). However, because it is generally time-consuming to calculate the eigenvalues, a different approach is used in computer algorithms. The squares of the eigenvalues of the matrix GG^T are equal to the so-called singular values of G^T, which are easily found by singular value decomposition, a procedure rapidly performed by most mathematical software packages. Thus, the largest singular values are found for both G^T and its pseudo-inverse (G^T) , and the two values are multiplied.

tioned stoichiometric matrix is that the condition number be between 1 and 100. This is fulfilled for most metabolic models. Thus, the condition number for the model of Vallino and Stephanopoulos (1993) is 62 and that of the model of Nissen *et al.* (1997) is 22. If the condition number is greater than 100 the stoichiometric matrix is ill-conditioned, and it may be necessary to modify the biochemistry used in the development of the model. It should be noted that no measurements are needed for the calculation of the condition number, and the metabolic model therefore can be tested before any calculations are carried out.

As discussed earlier calculation of the condition number should always be the first step in sensitivity analysis of a metabolic model. This, however, does not provide any information about the *sensitivity* of the calculations with respect to variations in the measured fluxes. This information is supplied by the elements of the solution matrix, and for a determined system we have

$$\frac{\partial \mathbf{v}_c}{\partial \mathbf{v}_m} = -\left(\mathbf{G}_c^T\right)^{-1}\mathbf{G}_m^T \tag{8.29}$$

In eq. (8.29) the element in the *j*th row and *i*th column specifies the sensitivity of the *j*th flux (which is calculated) with respect to variations in the measurement of the *i*th flux. Thus, the elements of the solution matrix provide information about the sensitivity of the system.

REFERENCES

Aiba, S. & Matsuoka, M. (1979). Identification of metabolic model: Citrate production from glucose by *Candida lipolytica*. *Biotechnology and Bioengineering*. **21**, 1373-1386.

Dantzig, G. B. (1963). Linear programming and extensions. Princeton: Princeton University Press.

Gyr, M. (1978). Linear optimization using the simplex algorithm (simple). CERN Computer Center Program Library, CERN, Geneva, Switzerland.

Madron, F., Veverka, V. & Vanecek, V. (1977). Statistical analysis of material balance of a chemical reactor. *AIChE Journal*. **23**, 482-486.

Nielsen, J. (1997). *Physiological Engineering Aspects of Penicillium chrysogenum*. Singapore: World Scientific.

Nielsen, J. & Villadsen, J. (1994). *Bioreaction Engineering Principles*. New York: Plenum Press.

Nissen, T. L., Schulze, U., Nielsen, J. & Villadsen, J. (1997). Flux distributions in anaerobic, glucose-limited continuous cultures of *Saccharomyces cerevisiae*. *Microbiology* **143**, 203-218.

Reardon, K. F., Scheper, T. H. & Bailey, J. E. (1987). Metabolic pathway rates and culture fluorescence in batch fermentations of *Clostridium acetobutylicum*. *Biotechnology Progress* **3**, 153-167.

Stephanopoulos, G. & Vallino, J. J. (1991). Network rigidity and metabolic engineering in metabolite overproduction. *Science* **252**, 1675-1681.

Theobald, U., Mailinger, W., Baltes, M., Rizzi, M. & Reuss, M. (1997). In vivo analysis of metabolic dynamics in *Saccharomyces cerevisiae*. *Biotechnology and Bioengineering*. **55**, 303-316.

Tsai, S. P. & Lee, Y. H. (1988). Application of metabolic pathway stoichiometry to statistical analysis of bioreactor measurement data. *Biotechnology and Bioengineering* **32**, 713-715.

Vallino, J. J. & Stephanopoulos, G. (1993). Metabolic flux distributions in *Corynebacterium glutamicum* during growth and lysine overproduction. *Biotechnology and Bioengineering* **41**, 633-646.

Varma, A., Boesch, B. W. & Palsson, B. O. (1993a). Stoichiometric interpretation of *Escherichia coli* glucose catabolism under various oxygenation rates. *Applied and Environmental Microbiology* **59**, 2465-2473.

Varma, A., Boesch, B. W. & Palsson, B. O. (1993b). Biochemical production capabilities of *Escherichia coli. Biotechnology and Bioengineering* **42**, 59-73.

CHAPTER 9

Methods for the Experimental Determination of Metabolic Fluxes by Isotope Labeling

In the previous chapter we presented the general methodology for the determination of intracellular metabolic fluxes from the measurement of extracellular metabolite concentrations. Several reasons were provided to justify this exercise. It was noted that intracellular fluxes are very valuable in establishing the physiological state of a biological system. As we will show in subsequent chapters, metabolic fluxes play another important role, namely, they provide a comprehensive picture of the control architecture of metabolic networks. We will show that metabolic control coefficients, which have been proposed as measures of the degree of control exercised in metabolic networks, can be determined directly from the magnitude of metabolic fluxes and flux changes in response to the introduction of metabolic perturbations. In this regard, methods for the accurate determination of intracellular fluxes become extremely important as they can lead directly to the elucidation of the control of flux in metabolic networks.

The primary inputs required for the determination of intracellular fluxes, as discussed in the previous chapter, are fluxes of extracellular metabolites, namely, rates of substrate uptake and rates of product secretion. These rates can best be obtained from steady state chemostat experiments, in which concentrations of all metabolites are measured at steady state. The corresponding rates of uptake/secretion can then be calculated by multiplying these concentrations by the chemostat dilution rate. Because chemostat experiments are very long in duration, often batch experiments are employed for the determination of fluxes. Batch cultivation of microorganisms can generate large amounts of data in a relatively short period of time; however, the determination of extracellular metabolite rates is significantly more involved as it requires calculation of the derivative of the corresponding concentrations with respect to time, an operation that usually introduces large errors. Furthermore, the lack of a physiological steady state during a batch culture experiment can complicate the interpretation of batch culture data.

Whether a chemostat or a batch culture experiment is used, one should bear in mind that the extent to which metabolic networks can be elucidated by using concentration measurements of extracellular metabolites alone is limited. At most, such measurements provide information about the distribution of carbon flux at a small number of branch points in the metabolic network. This point may be obscured by complexities in formulating the equations of the network, and several examples are provided in this book (see Chapter 12), illustrating the reduction of rather complex sets of biochemical reactions to their simple network equivalents. In particular, extracellular measurements have limited potential in elucidating aspects of metabolic networks, such as (a) flux distribution at split points that converge at another point of the network, (b) metabolic cycles, and (c) unraveling network structure to a finer biochemical resolution. These points are illustrated in the schematic of Fig. 9.1, which shows the split of a primary compound A between two competing pathways proceeding at net rates v_1 and v_2. Measurement of the consumption rate of A and the secretion rates of

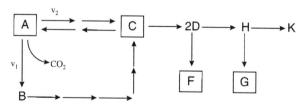

FIGURE 9.1   Schematic of flux split in a metabolic network that is only partially resolved by the extracellular metabolite (in boxes) balancing methods presented in Chapter 8.

F and G are sufficient to determine the distribution of fluxes at branch points D and H and the independent calculation of another secreted metabolite, K. However, the flux split ratio at metabolite A between the two pathways 1 and 2 or, equivalently, the flux through the cycle A → B → C → A is indeterminable. Similarly, if more reactions are involved in the conversion of C to 2D, they must all be lumped into a single reaction. If more information about the split ratio between pathways 1 and 2 or a finer biochemical resolution of the depicted steps is desirable, different methods must be applied.

One class of such methods that will be described in this chapter is the use of compounds labeled with ^{13}C or ^{14}C at specific carbon locations. Through the use of such compounds, pathways that introduce asymmetries in the distribution of carbon atoms of intermediate metabolites may be distinguished from one another, even though they lead to the formation of the same final product. This asymmetry leads to fundamentally different patterns of label enrichment in one or more metabolites, the measurement of which provides the necessary information for the determination of the rates of the competing reactions. ^{13}C- or ^{14}C-labeled compounds are used in connection with the measurement of label enrichment in extracellular or intracellular metabolites. Furthermore, the type of information that one can obtain from the application of such methods greatly depends on the type of labeled substrate used, the particular carbon atoms that are labeled, and the particular metabolite (intracellular or extracellular) whose degree of enrichment is measured. This is a very important point, as it is often the case that decisions about the labeled carbon substrate(s) to be used and the particular compounds to be measured are based solely on familiarity and prior experience, with little consideration given to the type of information that can be obtained from a particular combination of labeled substrate and specific metabolite measurement.

In this chapter, we present methods for the determination of intracellular metabolic fluxes through the use of labeled compounds. Although the basic idea remains the same throughout the chapter, we have divided the possible methods into three categories, primarily depending on the size and complexity of the network that is being analyzed. In the first case we present methods applicable to situations with directly measurable metabolites or simple networks that are usually amenable to analytical solutions (Section 9.1). For metabolic networks involving cycles, straightforward accounting of label transfer is inadequate. An alternative method, based on the enumeration of all possible metabolite isotopomers, is the method of choice in such cases (Section 9.2). In addition to facilitating the exact determination of the degree of label enrichment of specific metabolites, isotopomer enumeration also allows the estimation of molecular weight distributions of specific metabolites, measurable by gas chromatography-mass spectrometry instruments (GC-MS). Because molecular weight distributions depend on metabolic fluxes, GC-MS combined with ^{13}C-NMR spectrometry is another source of valuable information regarding intracellular metabolic fluxes. A final point regarding metabolite isotopomers is the role such isotopomers play in determining the fine structure of NMR spectra of the corresponding metabolites. It will be shown that the fine structure of NMR spectra depends on the relative amounts of the corresponding isotopomers, and, as such, it is directly related to the rates of intracellular reactions leading to the formation of such isotopomers. Finally, for complex situations of large networks, numerical solutions are required. For such cases, the concept of atom mapping matrices introduced in Section 9.3 provides a convenient framework for carrying out efficiently the label distribution calculations, especially when iterative procedures are needed. The analysis of such networks involves a fair amount of trial and error in order to ensure consistency of the large number of measurements possible through the application of NMR spectrometry. These calculations are facilitated with the use of atom mapping matrices.

9.1. DIRECT FLUX DETERMINATION FROM FRACTIONAL LABEL ENRICHMENT

This class of methods relies on the measurement of the intensity (or degree of enrichment) of a label in a carefully selected metabolite for the direct determination of metabolic fluxes. Two basic approaches are presented, requiring either transient or steady state intensity measurements.

9.1.1. FLUX DETERMINATION FROM TRANSIENT INTENSITY MEASUREMENTS

Transient intensity measurements of radiolabeled compounds can be used to probe directly a specific metabolic pathway. It is required that the labeled compound be transported through the cell membrane and that its intensity be measured as a function of time. Due to its specificity and the high accuracy of radioactivity measurements, this method can be applied to the determination of metabolic fluxes of small magnitude that cannot be observed through material balancing techniques. On the other hand, it requires dedicated equipment for the separation and measurement of compounds contaminated with radioactivity.

This method is illustrated in Fig. 9.2, depicting a schematic pathway involving metabolite M that is synthesized and consumed at a rate equal to

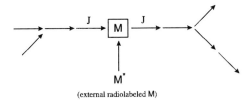

FIGURE 9.2 Illustration of radiolabeled metabolite use for the direct determination of metabolic fluxes.

the pathway flux, J. The flux J can be determined by introducing a pulse of radiolabeled M (M^*) and measuring the radioactivity in samples of purified M^*. By assuming metabolic steady state for the duration of the experiment, the pool of intracellular radiolabeled M will change according to the following equation:

$$\frac{dM^*}{dt} = -\left(\frac{M^*}{M}\right)J \qquad (9.1)$$

which, upon integration for constant J and M (due to the metabolic steady state), gives

$$\ln\left(\frac{M^*(t)}{M^*(0)}\right) = -\frac{J}{M}t \qquad (9.2)$$

Equation (9.2) indicates that the unknown flux J can be determined from a semilog plot of radioactivity counts and the total intracellular concentration of M. It should be noted that metabolic steady state is required, but only for the duration of the labeling experiment. Due to the short duration of such experiments, they can be repeated on small samples withdrawn from a large vessel to determine metabolic fluxes during a transient. Of course, this will be possible only if the time scales of the labeling measurement and actual experiment are well-separated, and if sufficient care is exercised to ensure minimal perturbation of the actual environmental conditions during the labeling measurement. To this end, the magnitude of the pulse as a function of the intracellular metabolite pool should be selected carefully.

9.1.2. METABOLIC AND ISOTOPIC STEADY STATE EXPERIMENTS

Another way of using labeled compounds for the determination of intracellular fluxes is depicted in Fig. 9.3. This figure shows a schematic of label distribution corresponding to the sample network of Fig. 9.1. If metabolite A

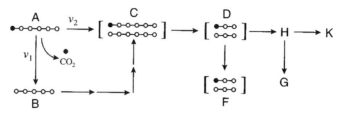

FIGURE 9.3 Schematic of flux split ratio determination of diverging metabolic pathways by means of isotope-labeled compounds (labeled carbon atoms marked by a solid circle).

is a six-carbon compound whose first carbon is labeled, then, assuming that the reaction leading to metabolite B is a decarboxylation reaction removing the labeled carbon, the pathway through metabolite B will produce an unlabeled molecule of metabolite C while the molecule of C produced via the direct pathway (of rate v_2) will retain the carbon atom at position 1 as indicated in the figure. The enrichment of metabolite C in labeled molecules will be directly proportional to the rate of reaction 2 relative to the total rate of consumption of metabolite A. This is also true for metabolite D and, in particular, the secreted metabolite F, which can be measured through extracellular medium analysis. The degree of ^{13}C enrichment on the first position of metabolite F thus is found to provide information about the relative rate of reaction 2, which can be used in conjunction with total flux measurements to differentiate between rates v_1 and v_2.

We will illustrate the details of these methods by means of two examples. Before we do so, however, two important points should be noted. The first is that this approach is valid only at metabolic and isotopic steady state. A metabolic steady state is the state in which the concentrations of all intracellular metabolites participating in the metabolic network are constant. At an isotopic steady state, the relative populations of metabolite isotopomers are constant as well. During a metabolic or isotopic transient, intracellular flux determination is significantly more involved, requiring extensive *in vivo* kinetic information. As such information is unavailable, it is imperative that experiments are designed in such way as to ensure that steady state is reached, and this is the second point that should be noted. Clearly, a steady state chemostat is the preferred system in which to conduct experiments; however, there are also constraints on the manner in which the labeled substrate is introduced to the system. One way is by a finite pulse, whose concentration is such that the substrate concentration in the feed is not altered and whose duration is of sufficient length to allow isotopic steady state to be reached. The examples provide certain check points that can help in assessing whether a satisfactory steady state has been reached. We also note that the simple approach illustrated in the schematic of Fig. 9.3 can have a number of variations, depending on the specific mechanisms by which the asymmetry between the competing pathways is generated. Besides decarboxylation of labeled CO_2, epimerase, isomerase, and other reactions can generate the required asymmetry.

EXAMPLE 9.1

Determination of Bypass Pathway Flux in Lysine Biosynthesis

This example illustrates the use of labeled pyruvate for the differentiation between two competing pathways in the final branch of the lysine production pathway. As shown in Fig. 9.4, the final steps in lysine biosynthesis involve

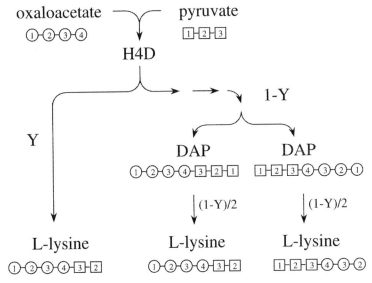

FIGURE 9.4 Template for lysine labeling from pyruvate and oxaloacetate through the dual pathways. The four-step succinylase pathway of flux $(1-y)v_{lysine}$ contains the enzymatic sequence N-succinyl-2,6-ketopimelate synthase, N-succinylaminoketopimelate:glutamate aminotransferase, N-succinyldiaminopimelate desuccinylase, and diaminopimelate epimerase. The one-step pathway (of flux yv_{lysine}) involves the action of *meso*-DAP dehydrogenase. Subscripts of O and P indicate the original carbon positions.

the transformation of tetrahydrodipicolinate (H4D) to *meso-α,ε*-diamino-pimelate (*meso*-DAP), which can proceed either through a four-step pathway or through the single-step reaction catalyzed by *meso*-DAP dehydrogenase (DDH). Lysine is formed from meso-DAP following a decarboxylation reaction. The rate of lysine production alone cannot differentiate between these two pathways, the combined flux of which equals the total rate of lysine secretion. Through the use, however, of labeled pyruvate, it is possible to obtain estimates of the fraction of flux that is converted to lysine through each pathway.

H4D is formed from oxaloacetate (OAA) and pyruvate (Pyr) through a condensation reaction. In the conversion of H4D to lysine through the single-step pathway (left branch of Fig. 9.4), the arrangement of carbon atoms remains unaltered. On the other hand, the last reaction of the four-step pathway is an epimerase reaction leading to the formation of two types of *meso*-DAP, and by extension lysine, molecules that are mirror images of one another and can be differentiated through the use of labeled carbon atoms. It is this last step that introduces the asymmetry in the transformation of H4D to lysine and, as such, provides the basis for the differentiation between the four-step and the single-step, or bypass, lysine formation pathway.

The use of pyruvate labeled at position 3 ([3-^{13}C]pyruvate) as carbon source will give rise to OAA molecules that are also labeled with ^{13}C as some of its carbon atoms. In this analysis, we focus on the label carried by the third and fifth carbon atoms of the compounds appearing in Fig. 9.4, and we let [%OAA$_3$] and [%Pyr$_3$] represent the fraction of OAA and Pyr pools, respectively, that are labeled at the third carbon position. We note that these fractions may include OAA and Pyr molecules that are labeled at other carbon positions as well, as long as they carry a ^{13}C label at their third carbon atom. At a metabolic steady state, there is no accumulation of intracellular metabolites, yielding the following equation for the pathway flux of Fig. 9.4:

$$v_{condensation} = v_{single\ step} + v_{four\ step} = v_{lysine} \tag{1}$$

We now derive equations for the pools of the different isotopomers of H4D. Abbreviations are as follows: the isotopomer that is labeled at carbon 3 only, H4D$_{3,only}$; that labeled at carbon 5 only; H4D$_{5,only}$, that labeled at both carbons 3 and 5, H4D$_{35}$; the *total* pool of the H4D isotopomer labeled at carbon 3, (H4D$_3$ = H4D$_{3,only}$ + H4D$_{35}$); and the *total* pool of H4D labeled at carbon 5, (H4D$_5$ = H4D$_{5,only}$ + H4D$_{35}$). The balance for the pool of one of these isotopomers, for example, $H4D_3$, can be written as

$$\frac{dc_{H4D,3}}{dt} = v_{\text{condensation}}[\%OAA_3](1 - [\%Pyr_3])$$

$$+ v_{\text{condensation}}[\%OAA_3][\%Pyr_3]$$

$$- (v_{\text{single step}} + v_{\text{four step}})[\%H4D_3]$$

$$= v_{\text{lysine}}[\%OAA_3] - v_{\text{lysine}}[\%H4D_3] \tag{2}$$

where the first two terms in eq. (2) represent the rate of synthesis of the $H4D_{3,\,only}$ and $H4D_{35}$ isotopomers, respectively, and the third term is the rate of $H4D_3$ depletion by the two pathways of Fig. 9.4. At an isotopic steady state, the derivative of the preceding equation is set equal to zero, yielding an expression for the pool of $H4D_3$:

$$[\%H4D_3] = [\%OAA_3] \tag{3}$$

The pools of the remaining isotopomers of H4D can be obtained similarly and are given by the following equations:

$$[\%H4D_5] = [\%Pyr_3] \tag{4}$$

$$[\%H4D_{35}] = [\%OAA_3][\%Pyr_3] \tag{5}$$

$$[\%H4D_{3,\,only}] = [\%OAA_3](1 - [\%Pyr_3]) \tag{6}$$

$$[\%H4D_{5,\,only}] = [\%Pyr_3](1 - [\%OAA_3]) \tag{7}$$

$$[\%H4D_{\text{unlabeled}}] = (1 - [\%Pyr_3])(1 - [\%OAA_3]) \tag{8}$$

The preceding H4D isotopomers will generate lysine isotopomers labeled at carbon atoms 3 and/or 5. By assuming that the fraction of the lysine flux synthesized via the single-step pathway is y, i.e., $v_{\text{single step}} = yv_{\text{lysine}}$, the rates at which the various lysine isotopomers are formed from the indicated H4D isotopomer precursors are as follows:

$$L_{35} \text{ from } H4D_{35}: \quad v_{\text{lysine}}[\%OAA_3][\%Pyr_3] \tag{9}$$

$$L_{3,\,only} \text{ from } H4D_{3,\,only}: \quad v_{\text{single step}}[\%H4D_{3,\,only}]$$

$$+ (v_{\text{four step}}/2)[\%H4D_{3,\,only}]$$

$$= v_{\text{lysine}}[(1 + y)/2][\%OAA_3](1 - [\%Pyr_3]) \tag{10}$$

$L_{3,\text{only}}$ from H4D$_{5,\text{only}}$: $\left(v_{\text{four step}}/2\right)\left[\%\text{H4D}_{5,\text{only}}\right]$

$$= v_{\text{lysine}}\left[(1-y)/2\right]\left[\%\text{Pyr}_3\right](1-\left[\%\text{OAA}_3\right]) \qquad (11)$$

$L_{5,\text{only}}$ from H4D$_{3,\text{only}}$: $\left(v_{\text{four step}}/2\right)\left[\%\text{H4D}_{3,\text{only}}\right]$

$$= v_{\text{lysine}}\left[(1-y)/2\right]\left[\%\text{OAA}_3\right](1-\left[\%\text{Pyr}_3\right]) \qquad (12)$$

$L_{5,\text{only}}$ from H4D$_{5,\text{only}}$: $v_{\text{single step}}\left[\%\text{H4D}_{5,\text{only}}\right]$

$$+\left(v_{\text{four step}}/2\right)\left[\%\text{H4D}_{5,\text{only}}\right]$$

$$= v_{\text{lysine}}\left[(1+y)/2\right]\left[\%\text{Pyr}_3\right](1-\left[\%\text{OAA}_3\right]) \qquad (13)$$

By using eqs. (9)-(13), a balance for the sum of all lysine isotopomers carrying ^{13}C at position 3, L_3, can be written for a steady state chemostat as follows:

$$\frac{dc_{L,3}}{dt} = v_{\text{lysine}}\{\left[\%\text{OAA}_3\right]\left[\%\text{PYR}_3\right]$$

$$+\left[(1+y)/2\right]\left[\%\text{OAA}_3\right](1-\left[\%\text{PYR}_3\right])$$

$$+\left[(1-y)/2\right]\left[\%\text{PYR}_3\right](1-\left[\%\text{OAA}_3\right])\}$$

$$-Dc_{L,3} = 0 \qquad (14)$$

and, similarly, for the sum of all lysine isotopomers carrying ^{13}C on position 5, L_5:

$$\frac{dc_{L,5}}{dt} = v_{\text{lysine}}\{\left[\%\text{OAA}_3\right]\left[\%\text{PYR}_3\right]$$

$$+\left[(1-y)/2\right]\left[\%\text{OAA}_3\right](1-\left[\%\text{PYR}_3\right])$$

$$+\left[(1+y)/2\right]\left[\%\text{PYR}_3\right](1-\left[\%\text{OAA}_3\right])\}$$

$$-Dc_{L,5} = 0 \qquad (15)$$

where D is the chemostat dilution rate. Equations (14) and (15) can be combined with the steady state balance for total lysine,

$$\frac{dc_L}{dt} = v_{\text{lysine}} - Dc_L = 0 \qquad (16)$$

to eliminate the difficult to measure fraction of OAA enrichment ($[\%OAA_3]$) and solve for the flux split ratio y to yield:

$$y = \{[c_{L,5} - c_{L,3}]/c_L\}/\{2[\%Pyr_3] - [c_{L,3} + c_{L,5}]/c_L\} \quad (17)$$

Equation (17) shows how the flux split ratio for the two lysine-producing pathways can be determined from the relative enrichments of lysine at positions 3 and 5 upon administration of pyruvate labeled at carbon position 3. The calculation requires the enrichment of intracellular pyruvate, which is best determined by direct measurement. An alternative approach can be based on the measurement of the label enrichment of secreted valine and/or alanine, two amino acids synthesized directly from pyruvate. It is not recommended that $[\%Pyr_3]$ be set equal to the enrichment of extracellular pyruvate because of possible dilution of the intracellular pyruvate pool, the extent of which is, in general, not known.

The method described is valid only at metabolic and isotopic steady state. Such a steady state can be reached in a chemostat, however, under conditions of steady lysine production, a batch reactor could also be employed experimentally. The preceding formulas are still valid for a batch reactor, provided that the fractions of labeled lysine, $c_{L,3}/c_L$ or $c_{L,5}/c_L$, in eq. (17) are replaced by the ratios $(dc_{L,3}/dt)/(dc_L/dt)$ and $(dc_{L,5}/dt)/(dc_L/dt)$, respectively. Of course, in the case of a batch reactor the validity of the metabolic and isotopic steady state assumption is questionable. It is noted, however, that intracellular metabolites can be at steady state for a period of time, even in an inherently transient system. This would be, for example, the case of balanced growth or steady production under constant biomass concentration, conditions that are frequently encountered in batch operations. It is, nevertheless, important to seek internal checks that will provide evidence for the validity of the steady state assumption, as inappropriate application of these methods will produce incorrect results. One way to test the validity of the steady state assumption in the example considered here is by plotting the fractions of [3-^{13}C]lysine and [5-^{13}C]lysine, along with total lysine concentration, as the product accumulates in the reactor. A linear graph would provide support for a metabolic and isotopic steady state for the corresponding time period.

The preceding approach was tested experimentally in shake flask as well as batch experiments. The former yielded estimates in the range of 30-50% total lysine flux contributed by the single-step pathway. More careful experiments, conducted in a controlled batch experiment over a short period of time, yielded surprising information. The flux through the single-step pathway or, equivalently, the flux split ratio at the H4D branch point is not constant with time, but varies in the course of the fermentation. In particular, the flux through the single-step pathway increased significantly from almost zero to close to 100% in the short period following the introduction of labeled glucose pulse and then declined over a period of 5-7 h to a level similar to that before the introduction of the pulse. This result is a demonstration of the control of flux exercised by intracellular metabolites and *in vivo* regulation of the enzymes catalyzing the corresponding reactions, both important factors in determining flux distributions in metabolic networks. It should be noted, however, that these findings should be regarded with caution, as no adequate measures were taken to ensure or check for a steady state environment. The results possibly are correct in providing a sense of the general trends; however, carefully designed experiments must be conducted for quantitative assessment of the *in vivo* fluxes of the two pathways of Fig. 9.4.

EXAMPLE 9.2

Analysis of Label Distribution in Acetyl-CoA Metabolism

A second example illustrating isotope label distribution and metabolic flux estimation is presented for the simple network of Fig. 9.5, representing a two-pool model of AcCoA metabolism [adapted from Blum and Stein (1982)]. Such a model could be used to describe the distribution of AcCoA between mitochondria and the cytosol. In this system, pyruvate and hexanoate are substrates that can be labeled potentially with ^{13}C or ^{14}C. The distribution of isotope label in the atoms of AcCoA in pools I and II will be a function of the carbon fluxes v_1 through v_6 and the level and pattern of substrate labeling. Fluxes v_1 through v_6 have units of millimoles per cell per hour. Here, because AcCoA is an intracellular metabolite, we assume an isotopic steady state, implying that the flow of label into a specific carbon of an intracellular

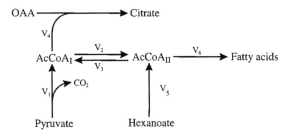

FIGURE 9.5 Simple metabolic network for modeling AcCoA metabolism. Two pools of AcCoA are considered, modeling mitochondrial and cytosolic compartments. v_1-v_6 represent fluxes.

intermediate is exactly balanced by the flow of labeled carbons out. We derive the mathematical relationships for the fractional enrichment of AcCoA carbons from which the intracellular fluxes can be determined. The fractional enrichment of a particular carbon atom is represented by the metabolite name followed by the carbon atom in parentheses.

Let us construct the steady state balance for the first carbon of AcCoA in pool I. We designate, by $AcCoA_I(1)$ and $AcCoA_I(2)$, the label enrichment in the first and second carbon atoms, respectively, of AcCoA in pool 1. There are two reactions generating $AcCoA_I$: (1) decarboxylation of pyruvate with a total flux of v_1 and (2) transfer of $AcCoA_{II}$ from pool II to pool I with a flux of v_3.

Carbon 1 of pyruvate is lost to CO_2 whereas carbons 2 and 3 of pyruvate become carbons 1 and 2, respectively, of $AcCoA_I$. Therefore, the flux of label from pyruvate to carbon 1 of $AcCoA_I$ is given by the product $v_1 Pyr(2)$. The conversion of $AcCoA_{II}$ to $AcCoA_I$ does not involve a chemical reaction, so that carbons 1 and 2 of $AcCoA_{II}$ become carbons 1 and 2, respectively, of $AcCoA_I$. Thus, the flux of label from $AcCoA_{II}$ to carbon 1 of $AcCoA_I$ is given by the product $v_3 AcCoA_{II}(1)$. The total flux of label into $AcCoA_I(1)$ is the sum of the two individual fluxes:

$$\text{flux into } AcCoA_I(1) = v_1 Pyr(2) + v_3 AcCoA_{II}(1) \qquad (1)$$

The two reactions that remove $AcCoA_I$ are the condensation with oxaloacetate (OAA) to form citrate, which has a flux of v_4, and the transfer of AcCoA from pool I to pool II, which has a flux of v_2. The flux of label out of carbon 1 of $AcCoA_I$ is simply $AcCoA_I(1)$ multiplied by the sum of the two fluxes, v_2 and v_4:

$$\text{flux out of } AcCoA_I(1) = (v_2 + v_4)AcCoA_I(1) \qquad (2)$$

At steady state the flux label into $AcCoA_I(1)$ will be exactly balanced by the flux out. Equating eqs. (1) and (2) yields the steady state balance of label for carbon 1 of $AcCoA_I$:

$$(v_2 + v_4)AcCoA_I(1) = v_1 Pyr(2) + v_3 AcCoA_{II}(1) \qquad (3)$$

Similarly, the steady state isotope balance for carbon 2 of $AcCoA_I$ is given by

$$(v_2 + v_4)AcCoA_I(2) = v_1 Pyr(3) + v_3 AcCoA_{II}(2) \qquad (4)$$

The derivation of the steady state equations can be repeated for AcCoA in pool II. There are two reactions that generate $AcCoA_{II}$: (1) oxidation of hexanoate, which has a total flux of v_5, and (2) transfer of AcCoA from pool I to pool II, which has a flux of v_2. The six carbons of hexanoate are removed in pairs via β-oxidation, so that the odd and even carbons each contribute one-third of the specific activity to carbons 1 and 2 of $AcCoA_{II}$, respectively. Therefore, the flux of label from hexanoate to carbon 1 of $AcCoA_{II}$ is given by the product $(v_5/3)[\text{Hex}(1) + \text{Hex}(3) + \text{Hex}(5)]$. The conversion of $AcCoA_{II}$ to $AcCoA_I$ does not involve a chemical reaction, so that the flux of label from $AcCoA_I$ to carbon 1 of $AcCoA_{II}$ is given by the product $v_2 AcCoA_I(1)$. The total flux of label into $AcCoA_{II}(1)$ is the sum of the two individual fluxes:

$$\text{flux into } AcCoA_{II}(1) = (v_5/3)[\text{Hex}(1) + \text{Hex}(3) + \text{Hex}(5)]$$
$$+ v_2 AcCoA_I(1) \qquad (5)$$

The flux of label out of carbon 1 of $AcCoA_{II}$ is the specific activity $AcCoA_{II}(1)$ multiplied by the sum of the two fluxes, v_3 and v_6:

$$\text{flux out of } AcCoA_{II}(1) = (v_3 + v_6)AcCoA_{II}(1) \qquad (6)$$

Equating eqs. (5) and (6) yields the steady state balance of label for carbon 1 of $AcCoA_{II}$:

$$(v_3 + v_6)AcCoA_{II}(1) = (v_5/3)[\text{Hex}(1) + \text{Hex}(3) + \text{Hex}(5)]$$
$$+ v_2 AcCoA_I(1) \qquad (7)$$

Similarly, the steady state isotope balance for carbon 2 of $AcCoA_{II}$ is given by

$$(v_3 + v_6)AcCoA_{II}(2) = (v_5/3)[\text{Hex}(2) + \text{Hex}(4) + \text{Hex}(6)]$$
$$+ v_2 AcCoA_I(2) \qquad (8)$$

Equations (3), (4), (7), and (8) can be represented in matrix form as follows:

$$
\begin{pmatrix}
\dfrac{v_2 + v_4}{v_1} & 0 & -\dfrac{v_3}{v_1} & 0 \\[2mm]
0 & \dfrac{v_2 + v_4}{v_1} & 0 & -\dfrac{v_3}{v_1} \\[2mm]
-\dfrac{3v_2}{v_5} & 0 & \dfrac{3(v_3 + v_6)}{v_5} & 0 \\[2mm]
0 & -\dfrac{3v_2}{v_5} & 0 & \dfrac{3(v_3 + v_6)}{v_5}
\end{pmatrix}
\begin{pmatrix}
AcCoA_I(1) \\
AcCoA_I(2) \\
AcCoA_{II}(1) \\
AcCoA_{II}(2)
\end{pmatrix}
$$

$$
= \begin{pmatrix}
\text{Pyr}(2) \\
\text{Pyr}(3) \\
\text{Hex}(1) + \text{Hex}(3) + \text{Hex}(5) \\
\text{Hex}(2) + \text{Hex}(4) + \text{Hex}(6)
\end{pmatrix} \qquad (9)
$$

Equation (9) can be solved analytically for the fractional enrichment of the two pools of AcCoA. The solution for $AcCoA_{II}(1)$, for example, is given by

$$AcCoA_{II}(1) = \frac{v_1 v_2}{v_2 v_6 + v_3 v_4 + v_4 v_6} Pyr(2)$$

$$+ \frac{v_5(v_2 + v_4)}{3(v_2 v_6 + v_3 v_4 + v_4 v_6)} (Hex(1) + Hex(3) + Hex(5))$$

(10)

We note that eq. (9) provides for relationships among the fractional enrichments of the indicated metabolites and the six metabolic reaction rates v_1 through v_6. In an experiment that employs a predetermined label for pyruvate and hexanoate the right-hand side of eq. (9) is known. Furthermore measurement of the total pyruvate and hexanoate uptake and fatty acid accumulation rates would yield estimates for reaction rates v_1, v_5, and v_6, respectively. Therefore, the measurement of the fractional enrichment at the two carbon atoms of the two AcCoA pools would yield estimates for reaction rates v_2, v_3, and v_4, as well as provide an extra degree of redundancy to test the consistency of the overall scheme. Solution of eq. (9) for the measurement of metabolite uptake and fractional enrichments would yield valuable estimates for the interconversion of the two pools of AcCoA, as well as the rate of carbon incorporation into the citric acid cycle for energy generation.

9.2. APPLICATIONS INVOLVING COMPLETE ENUMERATION OF METABOLITE ISOTOPOMERS

As mentioned earlier, the previous methods of direct accounting for the fate of isotope labels introduced through labeled substrates can be applied to relatively simple networks that do not involve metabolic cycles. More complicated networks, especially those involving cycles, are very difficult to handle with such methods. In these cases, one needs to account for the generation of all possible metabolite isotopomers and determine, through them, the predicted degree of label enrichment at various metabolite carbon positions.

To illustrate the additional complication introduced with the inclusion of metabolic cycles, we follow the fate of ^{13}C label present in the third carbon of pyruvate ([3-^{13}C]pyruvate) after one turn through the TCA cycle (see Fig. 9.6). Starting with 100% [3-^{13}C]pyruvate, all acetyl-CoA will be labeled at the C2 position via the pyruvate dehydrogenase reaction. OAA derived from the anaplerotic reaction via pyruvate carboxylase will be of the form O_{34} if the bicarbonate fixed is labeled with $^{13}CO_2$ and O_3 if the bicarbonate fixed is unlabeled. Furthermore, it is assumed that the reverse reactions from OAA to fumarate are very rapid compared to the reactions of the citrate

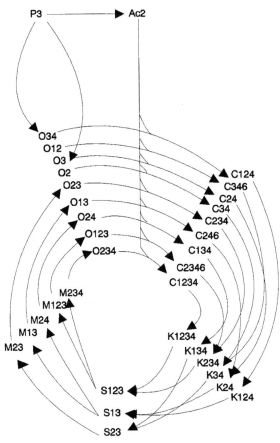

FIGURE 9.6 Schematic diagram of the sequential labeling of the carbons of intermediates via multiple turns of the TCA cycle using 100% enriched [3-^{13}C]pyruvate. Abbreviations: Ac, O, C, K, S and M for acetyl-CoA, oxaloacetate, citrate, a-ketoglutarate, succinate, and malate, respectively. The numbers next to the metabolite symbol indicate the carbon position labeled by ^{13}C.

synthase and PEP carboxykinase and, thus, results in full symmetrical equilibration of carbon label between C1 and C4 and between C2 and C3 in OAA, succinate, malate, and fumarate. This assumption results in equal concentrations of O_{34} and O_{12} and of O_3 and O_2. Following the TCA cycle, O_{34} condensing with Ac_2 will yield sequentially C_{124}, K_{124}, S_{13}, $0.5F_{13} + 0.5F_{24}$, $0.5M_{13} + 0.5M_{24}$, and $0.5O_{13} + 0.5O_{24}$. It is thus seen that the two isotopomers, O_{34} and O_3, have now produced two additional isotopomers for oxaloacetate, namely, O_{13} and O_{24}. The latter will be modified further through additional turns of the TCA cycle, leading to the generation of more isotopomers for the TCA cycle intermediates. It is clear that the previous approach of accounting for the fate of ^{13}C label is inadequate in handling the additional complexity introduced by multiple turns of the metabolic cycles. An alternative method for handling these situations makes use of balances of metabolite isotopomers. In this section, we first provide a general description of the overall method, followed by a complete analysis of the TCA cycle metabolite isotopomers. The section is completed with specific examples illustrating various modes of application in elucidating cellular metabolism.

To apply this method, one begins with the complete enumeration of all possible metabolite isotopomers. When applied to the TCA cycle with 100% [3-^{13}C]pyruvate as labeling substrate, the possible isotopomers of oxaloacetate, citrate, α-ketoglutarate, succinate, and malate are shown in Fig. 9.6. No other species are formed in this case. It should be noted that not all combinatorially possible isotopomer species are present. In this case, for example, there are $2^5 = 32$ α-ketoglutarate species, but only 6 species are present on the basis of the assumed biochemistry.

The next step is to write balance equations for all metabolites and their isotopomers appearing in the network. These equations involve the concentrations of the metabolites and isotopomer species, as well as the rates of the intracellular conversions that one wants to determine (fluxes). *Under the assumption of metabolic and isotopic steady state*, these balances are linear equations with respect to the unknown metabolic fluxes and can be solved explicitly for the relative concentrations of metabolite isotopomers as a function of the fluxes. The final step is an iterative trial and error process by which the unknown fluxes are determined, such as to best describe experimental data. Starting with guessed values for the unknown metabolic fluxes, relative isotopomer populations are determined first and then used to calculate one or more of the following: (a) the label enrichment of various metabolite carbon atoms, (b) metabolite molecular weight distributions measurable by GC-MS, or (c) the fine structure of metabolite NMR spectra. Comparison of model predictions with the corresponding experimental values of label enrichment, molecular weight distributions, and line intensities of NMR spectra produces new estimates for the unknown fluxes. This iterative process is repeated until satisfactory agreement between model predictions and measured values is obtained, eventually, converging, to the determination of the otherwise unobservable metabolic fluxes. Figure 9.7 provides a schematic of the trial and error process.

In the following two sections, we derive expressions for the distribution of all TCA cycle isotopomers generated upon administration of labeled pyruvate or acetate substrates. These derivations serve to illustrate the method of isotopomer enumeration in a very common metabolic cycle. Furthermore, the reported TCA results are used in several examples to show how additional insights on cell physiology can be obtained from experiments with labeled substrates. Finally, in Section 9.2.3 we indicate how isotope label enrichment, molecular weight distribution, and fine structure of NMR spectra can be calculated from the relative isotopomer populations obtained in the previous sections.

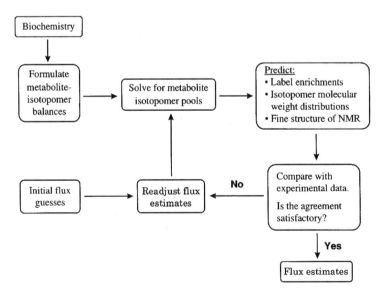

FIGURE 9.7   Schematic diagram of the iterative process leading to the determination of *in vivo* intracellular metabolic fluxes from measurements of isotope enrichment, GC-MS, and fine structure NMR spectra.

9.2.1. DISTRIBUTION OF TCA CYCLE METABOLITE ISOTOPOMERS FROM LABELED PYRUVATE

Figure 9.8 depicts the reactions of pyruvate utilization via the TCA cycle and gluconeogenic pathways commonly found in eukaryotic systems. We will assume that phosphoenolpyruvate (PEP) is not formed directly from pyruvate

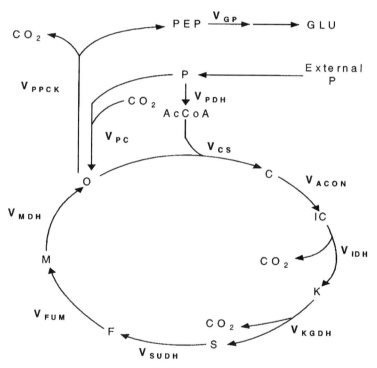

FIGURE 9.8   Pyruvate utilization via the citric acid cycle and the gluconeogenic pathway. The abbreviations used in the subscripts of the fluxes are as follows: GP, gluconeogenic pathway; PDH, pyruvate dehydrogenase; CS, citrate synthase; ACON, aconitase; KGDH, α-ketoglutarate dehydrogenase; SUDH, succinate dehydrogenase; FUM, fumarase; MDH, malate dehydrogenase; PPCK, phosphoenolpyruvate carboxykinase; PC, pyruvate carboxylase. The abbreviations for metabolites are as follows: PEP, phosphoenolpyruvate; P, pyruvate; AcCoA, acetyl-CoA; C, citrate; IC, isocitrate; K, α-ketoglutarate; S, succinate; F, fumarate; M, malate; O, oxaloacetate; CO_2, carbon dioxide; GLU, glucose. The operation of futile cycles such as PEP $\rightarrow$ P $\rightarrow$ O $\rightarrow$ PEP, is assumed inactive. Other reactions such as glutamate-oxaloacetate transaminase and glutamate dehydrogenase for biosynthesis are ignored.

via PEP synthetase, but rather is formed from oxaloacetate (OAA) via PEP carboxykinase. Pyruvate is converted to OAA by direct carboxylation, or it can be converted to acetyl-CoA, which then condenses with OAA to form citrate that traverses the TCA cycle and eventually regenerates OAA. Finally, glucose can be formed from PEP via the reverse Embden-Meyerhof scheme of glycolysis or gluconeogenic pathway.

We begin by writing mass balances for the metabolites identified in the network of Fig. 9.8.

$$\frac{dc_{AcCoA}}{dt} = v_{PDH} - v_{CS} \tag{9.3}$$

$$\frac{dc_C}{dt} = v_{cs} - v_{ACON} \tag{9.4}$$

$$\frac{dc_{IC}}{dt} = v_{ACON} - v_{IDH} \tag{9.5}$$

$$\frac{dc_K}{dt} = v_{IDH} - v_{KGDH} \tag{9.6}$$

$$\frac{dc_S}{dt} = v_{KGDH} - v_{SUDH} \tag{9.7}$$

$$\frac{dc_F}{dt} = v_{SUDH} - v_{FUM} \tag{9.8}$$

$$\frac{dc_M}{dt} = v_{FUM} - v_{MDH} \tag{9.9}$$

$$\frac{dc_o}{dt} = v_{MDH} + v_{PC} - v_{CS} - v_{PPCK} \tag{9.10}$$

$$r_{CO_2} = v_{IDH} + v_{KGDH} - v_{PC} + v_{PPCK} \tag{9.11}$$

$$\frac{dc_{PEP}}{dt} = v_{PPCK} - v_{GP} \tag{9.12}$$

$$\frac{dc_P}{dt} = v_{IMPORT} - v_{PC} - v_{PDH} \tag{9.13}$$

At metabolite steady state there is no accumulation of intracellular metabolites and eqs. (9.3)-(9.13) yield:

$$v_{CS} = v_{PDH} = v_{ACON} = v_{IDH} = v_{KGDH} = v_{SUDH} = v_{FUM} = v_{MDH} \quad (9.14)$$

$$v_{PC} = v_{PPCK} \quad (9.15)$$

$$r_{CO_2} = 2v_{CS} \quad (9.16)$$

$$v_{PC} = v_{GP} \quad (9.17)$$

$$v_{IMPORT} = v_{PC} + v_{CS} \quad (9.18)$$

Similar to the total metabolite concentrations, balances can be written for the concentration of each metabolite isotopomer shown in Fig. 9.6 upon administration of 100% [3-^{13}C]pyruvate:

$$\frac{dc_o}{dt} = \begin{bmatrix} O_3 \\ O_2 \\ O_{34} \\ O_{12} \\ O_{13} \\ O_{24} \\ O_{123} \\ O_{234} \\ O_{23} \end{bmatrix} = v_{MDH} \begin{bmatrix} 0 \\ 0 \\ 0 \\ 0 \\ M_{13} \\ M_{24} \\ M_{123} \\ M_{234} \\ M_{23} \end{bmatrix} + v_{PC} \begin{bmatrix} \left(\dfrac{1 - CO_2^*}{2}\right) \\ \left(\dfrac{1 - CO_2^*}{2}\right) \\ \left(\dfrac{CO_2^*}{2}\right) \\ \left(\dfrac{CO_2^*}{2}\right) \\ 0 \\ 0 \\ 0 \\ 0 \\ 0 \end{bmatrix}$$

$$- v_{CS} \begin{bmatrix} O_3 \\ O_2 \\ O_{34} \\ O_{12} \\ O_{13} \\ O_{24} \\ O_{123} \\ O_{234} \\ O_{23} \end{bmatrix} - v_{PPCK} \begin{bmatrix} O_3 \\ O_2 \\ O_{34} \\ O_{12} \\ O_{13} \\ O_{24} \\ O_{123} \\ O_{234} \\ O_{23} \end{bmatrix} \quad (9.19)$$

In the preceding equation, each isotopomer species is represented in relative population, *i.e.*, its concentration is normalized by the total concentration of the same metabolite (for example, $O_3 = c_{O,3}/c_O$), and CO_2^* is the fraction of CO_2 labeled with ^{13}C.

Similar to oxaloacetate, isotopomer balances can be written for citrate/isocitrate (indicated by C), α-ketoglutarate, succinate, malate, and the fraction of CO_2 labeled with ^{13}C, CO_2^* [eqs. (9.20)-(9.24), respectively]:

$$
\frac{dc_C}{dt}
\begin{bmatrix}
C_{24} \\
C_{34} \\
C_{124} \\
C_{346} \\
C_{246} \\
C_{134} \\
C_{2346} \\
C_{1234} \\
C_{234}
\end{bmatrix}
= v_{CS}
\begin{bmatrix}
O_3 \\
O_2 \\
O_{34} \\
O_{12} \\
O_{13} \\
O_{24} \\
O_{123} \\
O_{234} \\
O_{23}
\end{bmatrix}
- V_{IDH}
\begin{bmatrix}
C_{24} \\
C_{34} \\
C_{124} \\
C_{346} \\
C_{246} \\
C_{134} \\
C_{2346} \\
C_{1234} \\
C_{234}
\end{bmatrix}
\tag{9.20}
$$

$$
\frac{dc_K}{dt}
\begin{bmatrix}
K_{24} \\
K_{34} \\
K_{124} \\
K_{134} \\
K_{234} \\
K_{1234}
\end{bmatrix}
= v_{IDH}
\begin{bmatrix}
C_{24} + C_{246} \\
C_{34} + C_{346} \\
C_{124} \\
C_{134} \\
C_{2346} + C_{234} \\
C_{1234}
\end{bmatrix}
- v_{KGDH}
\begin{bmatrix}
K_{24} \\
K_{34} \\
K_{124} \\
K_{134} \\
K_{234} \\
K_{1234}
\end{bmatrix}
\tag{9.21}
$$

$$
\frac{dc_S}{dt}
\begin{bmatrix}
S_{23} \\
S_{13} \\
S_{123}
\end{bmatrix}
= v_{KGDH}
\begin{bmatrix}
K_{134} + K_{34} \\
K_{24} + K_{124} \\
K_{1234} + K_{234}
\end{bmatrix}
- v_{SUDH}
\begin{bmatrix}
S_{23} \\
S_{13} \\
S_{123}
\end{bmatrix}
\tag{9.22}
$$

$$\frac{dc_M}{dt}\begin{bmatrix} M_{23} \\ M_{13} \\ M_{24} \\ M_{123} \\ M_{234} \end{bmatrix} = v_{SUDH}\begin{bmatrix} S_{23} \\ \frac{1}{2}S_{13} \\ \frac{1}{2}S_{13} \\ \frac{1}{2}S_{123} \\ \frac{1}{2}S_{123} \end{bmatrix} - v_{MDH}\begin{bmatrix} M_{23} \\ M_{13} \\ M_{24} \\ M_{123} \\ M_{234} \end{bmatrix} \qquad (9.23)$$

$$r_{CO_2}CO_2^* = v_{IDH}\left(C_{346} + C_{246} + C_{2346}\right) + v_{KGDH}\left(K_{124} + K_{134} + K_{1234}\right)$$

$$-v_{PC}\left(CO_2^*\right) + v_{PPCK}\left(O_{34} + O_{24} + O_{234}\right) \qquad (9.24)$$

At metabolite and isotope steady state, using eqs. (9.22), (9.23), and (9.14), we obtain:

$$\begin{bmatrix} M_{23} \\ M_{13} \\ M_{24} \\ M_{123} \\ M_{234} \end{bmatrix} = \begin{bmatrix} K_{134} + K_{34} \\ \frac{1}{2}(K_{24} + K_{124}) \\ \frac{1}{2}(K_{24} + K_{124}) \\ \frac{1}{2}(K_{1234} + K_{234}) \\ \frac{1}{2}(K_{1234} + K_{234}) \end{bmatrix} \qquad (9.25)$$

Similarly, from eqs. (9.20), (9.21), and (9.14):

$$\begin{bmatrix} K_{24} \\ K_{34} \\ K_{124} \\ K_{134} \\ K_{234} \\ K_{1234} \end{bmatrix} = \begin{bmatrix} O_3 + O_{13} \\ O_2 + O_{12} \\ O_{34} \\ O_{24} \\ O_{123} + O_{23} \\ O_{234} \end{bmatrix} \qquad (9.26)$$

By combining eqs. (9.25) and (9.26) and inserting the resulting expressions for malate isotopomers into the steady state version of eq. (9.19), the isotopomer balance of oxaloacetate at metabolic and isotopic steady state is obtained:

$$
v_{CS}\begin{bmatrix} 0 \\ 0 \\ 0 \\ 0 \\ \frac{1}{2}(O_3 + O_{13} + O_{34}) \\ \frac{1}{2}(O_3 + O_{13} + O_{34}) \\ \frac{1}{2}(O_{234} + O_{123} + O_{23}) \\ \frac{1}{2}(O_{234} + O_{123} + O_{23}) \\ (O_{24} + O_2 + O_{12}) \end{bmatrix} + v_{PC}\begin{bmatrix} \left(\dfrac{1 - CO_2^*}{2}\right) \\ \left(\dfrac{1 - CO_2^*}{2}\right) \\ \left(\dfrac{CO_2^*}{2}\right) \\ \left(\dfrac{CO_2^*}{2}\right) \\ 0 \\ 0 \\ 0 \\ 0 \\ 0 \end{bmatrix} = (v_{CS} + v_{PC})\begin{bmatrix} O_3 \\ O_2 \\ O_{34} \\ O_{12} \\ O_{13} \\ O_{24} \\ O_{123} \\ O_{234} \\ O_{23} \end{bmatrix}
$$

$$(9.27)$$

Equation (9.27) indicates that all oxaloacetate isotopomers can be obtained from the relative enrichment of carbon dioxide and the citrate synthase and pyruvate carboxylase fluxes only, v_{CS} and v_{PC}, respectively. The preceding analysis can also be applied in a straightforward manner to all other isotopomers to derive equations for their determination too in terms of v_{CS} and v_{PC}, under metabolic and isotopic steady state. Therefore, v_{CS} and v_{PC} are the only unknowns from which the relative populations of all isotopomers can be calculated and, through them, the degree of label enrichment of intermediate metabolites.

Solutions can be expressed conveniently in terms of x, defined as the probability that oxaloacetate exits the TCA cycle via the PEP carboxykinase (PPCK) reaction [then $(1 - x)$ will denote the probability that oxaloacetate will re-enter the TCA cycle via citrate synthase (CS)], and y, defined as the probability of bicarbonate fixed into pyruvate to be labeled with ^{13}C, $^*CO_2$. In terms of x, one can write:

$$\frac{x}{1 - x} = \frac{v_{PPCK}}{v_{CS}} = \frac{v_{PC}}{v_{CS}} \tag{9.28}$$

Also, if there is no pathway that generates and consumes CO_2 other than the reactions indicated in this model, y can be expressed exclusively in terms of

x as well. Table 9.1 summarizes the results for the relative populations of oxaloacetate and glutamate isotopomers in terms of the two probabilities x and y when 100% enriched pyruvate is used as substrate. The legend of Table 9.1 provides expressions for the determination of y in terms of x. Table 9.1 also shows the relative enrichment of glucose and glutamate molecules synthesized from pyruvate labeled at three different carbon atoms, as indicated. The possible isotopomers generated when 100% enriched [2-^{13}C]pyruvate and 100% enriched [1-^{13}C]pyruvate, respectively, are used as labeling substrates are shown in Fig. 9.9. We note that the results of Table 9.1 allow the determination of any other isotopomer or metabolite enrichment for comparison with experimental data, when available. Also, G and K are used interchangeably as the isotopomer distributions of glutamate and α-ketoglutarate are identical.

We close this section with a summary of the assumptions invoked in derivation of the expressions of Table 9.1:

- There is no label recycling due to the operation of futile cycles such as pyruvate $\rightarrow$ OAA $\rightarrow$ PEP $\rightarrow$ pyruvate and malate $\rightarrow$ pyruvate $\rightarrow$ OAA $\rightarrow$ malate.
- There is no compartmentalization of metabolites, and homogeneous pools exist that are common to the TCA cycle and other pathways.
- There is no net flux to biosynthesis except in the formation of glucose.
- There is no label scrambling by the operation of the PPP that would redistribute the label of glucose synthesized via the gluconeogenic pathway.
- Metabolic and isotopic steady states are reached.
- Input substrates are isotopically pure. They are 100% enriched at the designated position and none at the undesignated positions.
- Isotopic dilution due to the presence of unlabeled endogenous pools is negligible.

9.2.2. DISTRIBUTIONS OF TCA CYCLE METABOLITE ISOTOPOMERS FROM LABELED ACETATE

Figure 9.10 summarizes three possible models for acetate utilization. Model I depicts acetate utilization via the glyoxylate shunt (GS) pathway, active in

TABLE 9.1 Steady State Distribution of Oxaloacetate (O) and Glutamate (G) Isotopomers and Relative Carbon Enrichments in Glucose and Glutamate Following the Utilization of 100% [3-^{13}C]Pyruvate, 100% [2-^{13}C]Pyruvate, and 100% [1-^{13}C]Pyruvate through the Major Pathways Described in Fig. 9.8[a]

	[3-^{13}C]Pyruvate	[2-^{13}C]Pyruvate	[1-^{13}C]Pyruvate
y	$y = \dfrac{1-x}{2+x-x^2}$	$y = \dfrac{1+x-2x^2}{2+x-x^2}$	$y = \dfrac{x}{2-x}$
Oxaloacetate isotopomer distribution	$O_2 = O_3 = \dfrac{x(1-y)}{2}$ $O_{12} = O_{34} = \dfrac{xy}{2}$ $O_{23} = \dfrac{(1-x)x}{(1+x)}$ $O_{13} = O_{24} = \dfrac{(1-x)x}{2(1+x)}$ $O_{123} = O_{234} = \dfrac{(1-x)^2}{2(1+x)}$	$O_2 = O_3 = \dfrac{x(1-y)}{2}$ $O_{13} = O_{24} = \dfrac{x(1-x+y+xy)}{2(1+x)}$ $O_{14} = \dfrac{x(1-x)}{(1+x)}$ $O_1 = O_4 = \dfrac{(1-x)^2}{2(1+x)}$	$O_1 = O_4 = \dfrac{x(1-y)}{2}$ $O_{14} = xy$ $O = 1-x$
Glucose enrichment pattern	$C-1 = \dfrac{1}{(1+x)}$ $C-2 = \dfrac{1}{(1+x)}$ $C-3 = \dfrac{1-x+xy+x^2y}{2(1+x)}$	$C-1 = \dfrac{x}{(1+x)}$ $C-2 = \dfrac{x}{(1+x)}$ $C-3 = \dfrac{1+x-2x^2+xy+x^2y}{2(1+x)}$	$C-1 = 0$ $C-2 = 0$ $C-3 = \dfrac{x(1+y)}{2}$

Glutamate isotopomer distribution			
$K_{24} = \dfrac{x(2-y-xy)}{2(1+x)}$	$K_{35} = \dfrac{x(1-y)}{2}$	$K = \dfrac{2-x-xy}{2}$	
$K_{34} = \dfrac{x}{2}$	$K_{135} = \dfrac{x(1-x+y+xy)}{2(1+x)}$	$K_1 = \dfrac{x(1+y)}{2}$	
$K_{124} = \dfrac{xy}{2}$	$K_{25} = \dfrac{x}{1+x}$		
$K_{134} = \dfrac{x(1-x)}{2(1+x)}$	$K_{15} = \dfrac{1-x}{2}$		
$K_{234} = \dfrac{1-x}{2}$	$K_5 = \dfrac{(1-x)^2}{2(1+x)}$		
$K_{1234} = \dfrac{(1-x)^2}{2(1+x)}$			

Glutamate enrichment pattern		
$C-1 = \dfrac{1-x+xy+x^2y}{2(1+x)}$	$C-1 = \dfrac{1+x-2x^2+xy+x^2y}{2(1+x)}$	$C-1 = \dfrac{x(1+y)}{2}$
$C-2 = \dfrac{1}{1+x}$	$C-2 = \dfrac{x}{1+x}$	$C-2 = 0$
$C-3 = \dfrac{1}{1+x}$	$C-3 = \dfrac{x}{1+x}$	$C-3 = 0$
$C-4 = 1$	$C-4 = 0$	$C-4 = 0$
$C-5 = 0$	$C-5 = 1$	$C-5 = 0$

[a] y can be obtained exclusively as a function of x, as indicated in the first row. Glutamate (G) and α-ketoglutarate (K) are used interchangeably as the labeling patterns of the two are identical. Glucose synthesized via the gluconeogenic pathway with these substrates has the following enrichment pattern: $C-4 = C-3$; $C-5 = C-2$; $C-6 = C-1$.

379

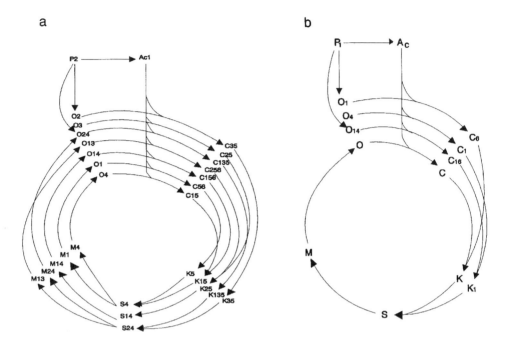

FIGURE 9.9 Schematic diagrams of the sequential labeling of metabolite intermediates via multiple turns of the TCA cycle using (a) 100% enriched [2-^{13}C]pyruvate and (b) 100% enriched [1-^{13}C]pyruvate.

many bacterial systems. The operation of this pathway is essential for growth on acetate as the sole carbon source because it replenishes TCA cycle intermediates drained off for biosynthesis. Bacteria maintain a precise balance of carbon utilization for energetic and biosynthetic requirements by controlling the carbon flow at the isocitrate branch point. Isocitrate, when catalyzed by the action of isocitrate dehydrogenase and subsequently by the enzymes of the TCA cycle, will lead to the production of NADH, which is used for energy generation. Alternatively, isocitrate, by the action of isocitrate lyase, can be converted to succinate and glyoxylate, and glyoxylate condenses with acetyl-CoA to produce malate. Thus, via the GS pathway, one molecule of oxaloacetate is formed from two molecules of acetate.

It has been established that mammals utilize acetate solely via conversion to acetyl-CoA, which subsequently is catabolized in the TCA cycle (Fig. 9.10, Model II). However, in order to maintain a balanced steady state flow, a separate anaplerotic source for the synthesis of OAA must exist as some of

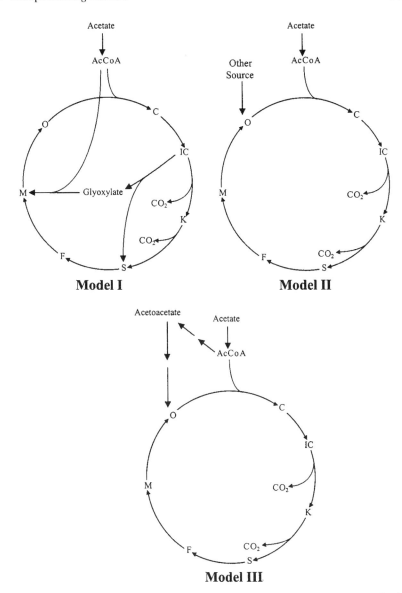

FIGURE 9.10   Three proposed models for acetate utilization. In Model I, acetate is utilized via the dual pathways of the TCA cycle and the glyoxylate shunt pathway. In Model II, acetate is utilized solely via the conversion to acetyl-CoA, which traverses the TCA cycle. Oxaloacetate is replenished from the other metabolites present endogenously. In Model III, acetate is converted to acetyl-CoA and also becomes the source for oxaloacetate synthesis via the route of acetoacetate to lactate to pyruvate pathway. The operation of the Model III pathway is not clearly established yet.

the intermediates will be drained off the cycle to supply biosynthetic precursors, such as glutamate and glutamine. We will consider two scenarios for the utilization of acetate: The first, depicted as Model II, is a generally accepted pathway in which acetate is utilized only via acetyl-CoA conversion by acetyl-CoA synthase. In this case, alternative metabolite sources exist that provide for oxaloacetate formation. The second scenario of oxaloacetate replenishment (Model III) is a hypothetical pathway in which oxaloacetate is derived directly from acetate. This can be accomplished by the condensation of two acetyl-CoA molecules to form acetoacetate, which is, in turn, decarboxylated to form lactate or pyruvate to eventually become the source of oxaloacetate. The pathway leading to the synthesis of C_3 metabolites such as lactate or pyruvate from acetoacetate is not clearly established. However, this hypothetical pathway is not entirely infeasible because there is evidence for the formation of 3-hydroxybutyrate from acetate in sheep (Annison *et al.*, 1963) and the formation of labeled acetoacetate from labeled acetate in rat liver (Desmoulin *et al.*, 1985). The purpose of including Model III is to test whether any of the labeling studies are consistent with it, as in this case, oxaloacetate entering the TCA cycle will always be labeled if the input acetate is labeled. The GS pathway is not found in mammalian systems.

To analyze this case, we proceed in exactly the same way as described in Section 9.2.1 by enumerating first the possible isotopomers as shown in Figs. 9.11 and 9.12 for the cases of 100% [2-^{13}C]- and 100% [1-^{13}C]acetate, respectively. Then metabolite and isotopic balances are written, and the steady state assumption is applied to obtain solutions for the relative populations of oxaloacetate isotopomers and, through them, the glucose and glutamate enrichment patterns. The results are summarized in Tables 9.2 and 9.3 for [2-^{13}C]- and [1-^{13}C]acetate, respectively. In the case of acetate utilization, we introduce yet another variable, z to denote the probability of isocitrate to be utilized via the GS pathway. Then $(1 - z)$ is the fraction of isocitrate utilized via the TCA cycle. Therefore, the ratio of flux via the GS pathway to flux via the TCA cycle can be expressed as

$$\frac{z}{1 - z} = \frac{v_{GS}}{v_{TCA}} \qquad (9.29)$$

It is noted that the results of Tables 9.1-9.3 are valid for 100% enriched pyruvate and acetate.

9.2.3. INTERPRETATION OF EXPERIMENTAL DATA

The focus of most metabolic NMR experiments is primarily on ^{13}C label enrichment of specific metabolite carbon atoms. The results of Tables 9.1-9.3

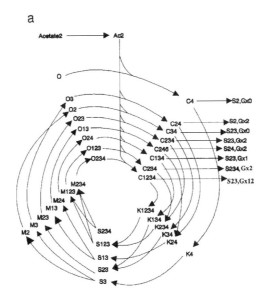

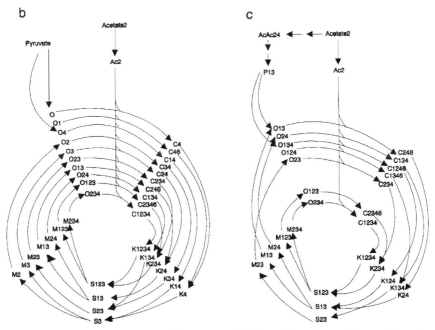

FIGURE 9.11   Schematic diagram of the sequential labeling of the carbons of intermediates via multiple turns of the TCA cycle for 100% [2-^{13}C]acetate. Models I-III of Fig. 9.10 correspond to a-c, respectively.

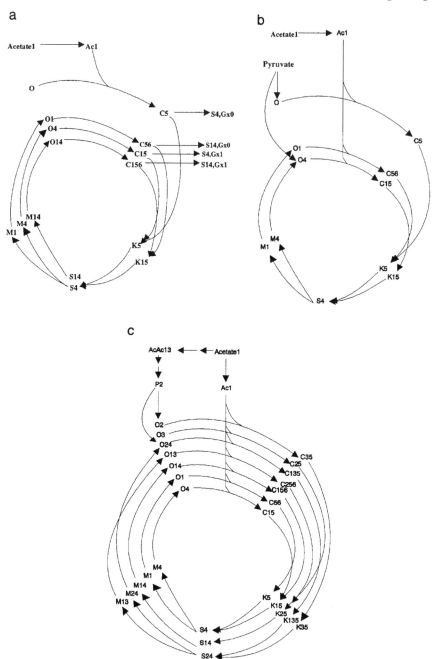

FIGURE 9.12 Schematic diagram of the sequential labeling of the carbons of intermediates via multiple turns of the TCA cycle for 100% [1-^{13}C]acetate. Models I-III of Fig. 9.10 correspond to a-c, respectively.

not only allow the correct interpretation of such label enrichment data but also extend the amount of information that can be derived from NMR experiments through the analysis of the fine structure of NMR spectra as well as, the measurement of the molecular weight distribution of labeled metabolites.

With respect to ^{13}C enrichment data, the degree of enrichment of various carbon atoms can be obtained as the sum of the relative populations of the corresponding isotopomers using the expressions reported in Tables 9.1-9.3. Referring, for example, to the use of 100% [3-^{13}C]pyruvate, the relative degree of enrichment of each carbon position of α-ketoglutarate can be calculated as the sum of all isotopomer species that contain labeled carbon at that particular position. This is reflected in the following equations:

$$C - 1 = K_{124} + K_{134} + K_{1234}$$

$$C - 2 = K_{24} + K_{124} + K_{234} + K_{1234}$$

$$C - 3 = K_{34} + K_{134} + K_{234} + K_{1234} \tag{9.30}$$

$$C - 4 = K_{24} + K_{34} + K_{124} + K_{134} + K_{234} + K_{1234} = 1$$

$$C - 5 = 0$$

Because the relative populations of α-ketoglutarate isotopomers are functions of the metabolic fluxes or the metabolic flux ratios, as indicated in the expressions of Tables 9.1-9.3, label enrichment data provide information about the magnitude of *in vivo* metabolic fluxes.

From the isotopomer distributions of Tables 9.1-9.3, the fine structure of NMR peaks at specific carbon atom positions can be analyzed further. Because the relative amounts of isotopomers depend on relative metabolic flux distributions (as represented by the parameters x and z of the equations of Tables 9.1-9.3), it is clear that the NMR spectra fine structure contains useful information about the overall distribution of metabolic fluxes. The same expressions for the relative amounts of isotopomers can be used further for the determination of molecular weight distributions of different labeled metabolites. For example, the relative amount of glutamate molecules with normal molecular weight M (corresponding to the absence of ^{13}C label from all carbon atoms) can be found from Table 9.1 to be equal to zero, as all glutamate isotopomers are expected to carry at least one ^{13}C carbon. Similarly, the relative amount of isotopomers with molecular weight $(M + 1)$ is also equal to zero. However, the relative amount of glutamate with molecular

TABLE 9.2 Steady State Distribution of Oxaloacetate and Glutamate Isotopomers and Relative Carbon Enrichment in Glucose and Glutamate Following 100% [2-¹³C]Acetate Administration through the Three Different Model Pathways Described in Fig. 9.10[a]

	[2-¹³C]Acetate via Model I	[2-¹³C]Acetate via Model II	[2-¹³C]Acetate via Model III
y		$y = \dfrac{(1-x)^2}{2 + x - x^2}$	$y = \dfrac{1 + x^2}{2 + x - x^2}$
Oxaloacetate isotopomer distribution	$O_{23} = z$ $O_{123} = O_{234} = \dfrac{1-z}{2}$	$O = x - xy$ $O_1 = O_4 = \dfrac{xy}{2}$ $O_2 = O_3 = \dfrac{x(1-x)}{2}$ $O_{23} = \dfrac{(1-x)^2 x}{(1+x)}$ $O_{13} = O_{24} = \dfrac{(1-x)^2 x}{2(1+x)}$ $O_{123} = O_{234} = \dfrac{(1-x)^3}{2(1+x)}$	$O_{13} = O_{24} = \dfrac{x(2 - y - xy)}{2(1+x)}$ $O_{134} = O_{124} = \dfrac{xy}{2}$ $O_{23} = \dfrac{x(1-x)}{(1+x)}$ $O_{123} = O_{234} = \dfrac{(1-x)^2}{2(1+x)}$
Glucose enrichment pattern	$C-1 = 1$ $C-2 = 1$ $C-3 = \dfrac{1-z}{2}$	$C-1 = \dfrac{1-x}{(1+x)}$ $C-2 = \dfrac{1-x}{(1+x)}$ $C-3 = \dfrac{1 - 2x + x^2 + xy + x^2 y}{2(1+x)}$	$C-1 = \dfrac{1}{(1+x)}$ $C-2 = \dfrac{1}{(1+x)}$ $C-3 = \dfrac{1 + x^2 + xy + x^2 y}{2(1+x)}$

Glutamate isotopomer distribution	$K_{234} = \dfrac{1+z}{2}$ $K_{1234} = \dfrac{1-z}{2}$	$K_4 = x - \dfrac{xy}{2}$ $K_{14} = \dfrac{xy}{2}$ $K_{24} = \dfrac{x(1-x)}{(1+x)}$ $K_{34} = \dfrac{(1-x)x}{2}$ $K_{134} = \dfrac{x(1-x)^2}{2(1+x)}$ $K_{234} = \dfrac{(1-x)^2}{2}$ $K_{1234} = \dfrac{(1-x)^3}{2(1+x)}$	$K_{24} = \dfrac{x(2-y-xy)}{2(1+x)}$ $K_{134} = \dfrac{x}{(1+x)}$ $K_{124} = \dfrac{xy}{2}$ $K_{234} = \dfrac{1-x}{2}$ $K_{1234} = \dfrac{(1-x)^2}{2(1+x)}$
Glutamate enrichment pattern	$C-1 = \dfrac{1-z}{2}$ $C-2 = 1$ $C-3 = 1$ $C-4 = 1$ $C-5 = 0$	$C-1 = \dfrac{1-2x+x^2+xy+x^2y}{2(1+x)}$ $C-2 = \dfrac{1-x}{1+x}$ $C-3 = \dfrac{1-x}{1+x}$ $C-4 = 1$ $C-5 = 0$	$C-1 = \dfrac{xy+x^2y+1+x^2}{2(1+x)}$ $C-2 = \dfrac{1}{(1+x)}$ $C-3 = \dfrac{1}{1+x}$ $C-4 = 1$ $C-5 = 0$

[a] In Models II and III, x specifies the relative flux into the oxaloacetate pool via pyruvate carboxylase as described in Section 9.2.1. Again, y can be given as function of x, as indicated in the first row. This table corresponds to the pathways depicted in Fig. 9.11.

TABLE 9.3 Steady State Distribution of Oxaloacetate and Glutamate Isotopomers and Relative Carbon Enrichment in Glucose and Glutamate following 100% [1-^{13}C]Acetate Administration through the Three Different Model Pathways Described in Fig. 9.10[a]

	[1-^{13}C]Acetate via Model I	[1-^{13}C]Acetate via Model II	[1-^{13}C]Acetate via Model III
y		$y = \dfrac{1-x}{2-x}$	$y = \dfrac{1+x-2x^2}{2+x-x^2}$
Oxaloacetate isotopomer distribution	$O_{14} = z$ $O_1 = O_4 = \dfrac{1-z}{2}$	$O = x - xy$ $O_1 = O_4 = \dfrac{1-x+xy}{2}$	$O_2 = O_3 = \dfrac{x(1-y)}{2}$ $O_{13} = O_{24} = \dfrac{x(1-x+y+xy)}{2(1+x)}$ $O_{14} = \dfrac{x(1-x)}{(1+x)}$ $O_1 = O_4 = \dfrac{(1-x)^2}{2(1+x)}$
Glucose enrichment pattern	$C-1 = 0$	$C-1 = 0$	$C-1 = \dfrac{x}{(1+x)}$
	$C-2 = 0$	$C-2 = 0$	$C-2 = \dfrac{x}{(1+x)}$
	$C-3 = \dfrac{1+z}{2}$	$C-3 = \dfrac{1-x+xy}{2}$	$C-3 = \dfrac{1+x-2x^2+xy+x^2y}{2(1+x)}$
Glutamate isotopomer distribution	$K_{15} = \dfrac{1-z}{2}$ $K_5 = \dfrac{1+z}{2}$	$K_{15} = \dfrac{1-x+xy}{2}$ $K_5 = \dfrac{1+x-xy}{2}$	$K_{35} = \dfrac{x(1-y)}{2}$ $K_{135} = \dfrac{x(1-x+y+xy)}{2(1+x)}$ $K_{25} = \dfrac{x}{1+x}$ $K_{15} = \dfrac{1-x}{2}$ $K_5 = \dfrac{(1-x)^2}{2(1+x)}$
Glutamate enrichment pattern	$C-1 = \dfrac{1-z}{2}$ $C-2 = 0$ $C-3 = 0$ $C-4 = 0$ $C-5 = 1$	$C-1 = \dfrac{1-x+xy}{2}$ $C-2 = 0$ $C-3 = 0$ $C-4 = 0$ $C-5 = 1$	$C-1 = \dfrac{1+x-2x^2+xy+x^2y}{2(1+x)}$ $C-2 = \dfrac{x}{1+x}$ $C-3 = \dfrac{x}{1+x}$ $C-4 = 0$ $C-5 = 1$

[a] This table corresponds to the pathways depicted in Fig. 9.12.

weight $(M + 2)$ will be equal to the sum of K_{24} and K_{34}, the relative amount of glutamate with molecular weight of $(M + 3)$ will be equal to the sum of K_{124}, K_{134} and K_{234}, and, finally, the relative amount of glutamate with molecular weight $(M + 4)$ will be equal to K_{1234}. Such molecular weight distributions can be obtained by gas chromatography-mass spectrometry and can provide further information about the origin of the isotopomers and the corresponding metabolic fluxes.

It is noted that a great deal of redundancy is built into the calculations. For example, isotopomer distributions, molecular weight distributions, and fine structure of NMR spectra depend on the two parameters x and z in the cases considered in Tables 9.1-9.3. Any two of these measurements could be used for the determination of the unknown parameters x and z, and the rest of the data could then be employed to test the consistency of the obtained values of x and z. In practice, a better approach may be to fit all measurements to the values of the two unknown parameters x and z and use the residual mean square as a measure of the accuracy of the fit. Low values of the residual mean square will indicate a good fit, whereas large values will be an indication of the presence of errors either in the measurements or in the assumed biochemical pathways. In the event that inconsistencies are identified, one can try to eliminate various measurements one at a time and test the accuracy of the fit obtained with the remaining measurements. If a significant improvement is obtained upon eliminating one measurement, this would make the eliminated measurement suspected of containing gross errors, suggesting that it should not be considered in the fit. If none of the measurements are identified as containing gross errors, then the process can be repeated by altering the biochemical pathways included in the derivation of the corresponding equations until a satisfactory fit is obtained. The procedures of Chapter 4 for the analysis of data consistency can be applied profitably in the reconciliation of spectroscopic and metabolic data and accurate flux determination. Finally, it should be noted that the method relies on measurements of secreted metabolites or intracellular metabolites in cell lysates. There is no need for *in situ* NMR measurements and the elaborate arrangements required for this purpose.

Clearly, a great deal of information can be extracted from conventional NMR data in light of the analysis presented. The availability of the expressions of Tables 9.1-9.3 facilitates the analysis of such measurements. However, analytical solutions for isotopomer distributions cannot be expected to be available in general for any configuration of biochemical pathways analyzed. It is possible that certain network structures will not be amenable to exact analytical solutions of the type reported. In such cases, the isotopomer distributions will have to be obtained numerically. Suitable computer programs that facilitate the analysis of label transfer and calculation of isotopomer distribution will be invaluable in maximizing the information obtainable from these experiments and providing the best experimental design [see, for example, Schmidt et al., 1997a]. This software will also facilitate the trial and error investigations into metabolic pathway structures that best fit a set of metabolic, spectroscopic, and other data (Schmidt et al., 1997b).

EXAMPLE 9.3

Pyruvate Utilization in Mammals

The validity of the preceding modeling concept can be tested by comparing the model predictions with experimental data. Chance et al. (1983) perfused rat heart tissue with 90% enriched [3-^{13}C]pyruvate and obtained a ^{13}C-NMR spectrum that shows resonances and line splittings at the C-2, C-3, and C-4 carbons of glutamate (Fig. 9.13). The glutamate C-2 resonance is split into nine lines (centered at 55.5 ppm), C-4 is split into three lines (centered at 34.2 ppm), and C-3 is split into five lines (centered at 27.8 ppm). The line splittings are caused by ^{13}C-^{13}C spin coupling among carbon atoms.

With 90% enriched [3-^{13}C]pyruvate as substrate, one needs to derive the new profile of possible isotopomers as well as expressions analogous to those of Tables 9.1-9.3 for the relative populations of the isotopomers. It should be emphasized that these differ from the case of 100% enrichment, and serious errors can result if the correct pyruvate labeling is not taken into consideration.

Pyruvate and acetyl-CoA may be labeled at C-3 and C-2, respectively, or unlabeled. Oxaloacetate produced from labeled pyruvate via the anaplerotic reaction will be labeled at carbon 3, if the bicarbonate is unlabeled, or at carbons 3 and 4, if the bicarbonate is labeled. However, also present in the oxaloacetate pool will be unlabeled oxaloacetate formed from unlabeled pyruvate, if the bicarbonate is unlabeled, and isotopomer O_4, if the bicarbonate is labeled. The seven isotopomers of oxaloacetate, O, O_3, O_{34}, O_4, O_2, O_{12}, and O_1 [the last three formed because of the equilibrium between carbons (2,3) and (1,4)], can condense with labeled (at carbon 2) and unlabeled AcCoA to produce citrate in the TCA cycle. At steady state, there are 12 isotopomers of oxaloacetate in the oxaloacetate pool: O_4, O_2, O_3, O_{34}, O_{12}, O_{23}, O_{13}, O_{24}, O_{123}, and O_{234}. Accordingly, there are 16 isotopomers in the glutamate pool: G, G_4, G_{14}, G_1, G_{124}, G_{12}, G_{34}, G_3, G_{24}, G_2, G_{234}, G_{23}, G_{134}, G_{13}, G_{1234} and G_{123}.

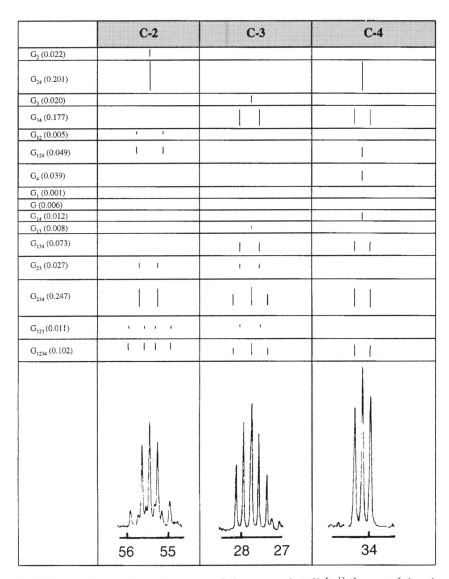

FIGURE 9.13 Observed multiplet pattern of glutamate with 90% [3-^{13}C]pyruvate [adapted from Chance *et al.* (1983)]. The value of x is estimated to be 0.35. The numbers in parentheses are the relative populations of the corresponding isotopomers. The intensity of each line within the multiplet should be the relative population normalized by the number of lines.

The occurrence of multiple patterns in the NMR spectrum, presented in Fig. 9.13, is consistent with the prediction of different isotopomer species. The nine lines splitting at C-2 is the sum of a singlet (superposition of two different singlets) due to G_{24} and G_2, a doublet with coupling constant J_{12} due to G_{124} and G_{12}, a doublet with coupling constant J_{23} due to G_{234} and G_{23}, and a quartet due to G_{1234} and G_{123}. Two different doublets occur as a result of the difference in the coupling constants between J_{12} (53.5 Hz) and J_{23} (34.6 Hz). Because of this difference, the line splitting at C-2 due to G_{1234} and G_{123} is a quartet, whereas the line splitting at C-3 due to G_{1234} and G_{234} is a triplet with an intensity ratio 1:2:1 due to the similar values of J_{23} and J_{34}. For the same reason ($J_{23} = J_{34}$), the isotopomers G_{123}, G_{23}, G_{134}, and G_{34} provide the same doublet at C-3. Thus, the five lines splitting at C-3 is the sum of a singlet due to G_3 and G_{13}, a doublet due to G_{123}, G_{23}, G_{134} and G_{34}, and a triplet due to G_{1234} and G_{234}. Similarly, the line splitting should be three lines at C-4, three lines at C-1, and none at C-5. Thus, the fine structure of NMR spectra at all glutamate resonances is consistent with the prediction of the different glutamate isotopomer species.

The intensity of each line is proportional to the sum of the isotopomer concentrations contributing to the corresponding line splitting (doublet, triplet, or quartet), normalized by the number of lines. For example, the intensity of one of the two lines of the doublet at C-2 due to G_{124} and G_{12} is proportional to $([G_{124}] + [G_{12}])/2$, where $[G_i]$ is the concentration of G_i. By using our model, the relative distribution of glutamate isotopomer species that determine the line intensity of the NMR spectra can be calculated uniquely from the values of x (the fraction of oxaloacetate exiting the TCA cycle) and y (the fraction of bicarbonate fixed in pyruvate, which is labeled with ^{13}C). Assuming that there are no reactions that generate CO_2 other than those depicted in Fig. 9.8, the value of y can be determined uniquely from x. Predictions of the relative populations of the 16 glutamate isotopomers for different values of x are shown in Fig. 9.14.

Because the line intensities of the NMR spectra of Fig. 9.13 can be predicted uniquely from the relative populations of glutamate isotopomers, which, in turn, are functions of the fraction x, flux information can be extracted from the fine structure of NMR spectra. Any of the three resonances at the C-2, C-3, or C-4 carbons of glutamate can be used for the determination of x, whereas the other two can be used to validate the estimate of x thus obtained. Table 9.4 shows the line intensity ratios as estimated from the NMR spectra of Fig. 9.13. The values of x obtained when only one carbon resonance at a time is used is approximately the same as the least square estimate obtained when all three resonances are used in the calculation and equal to 0.35. The corresponding value of y is 0.26. Table 9.4 shows the line intensity ratios calculated for these values of x and

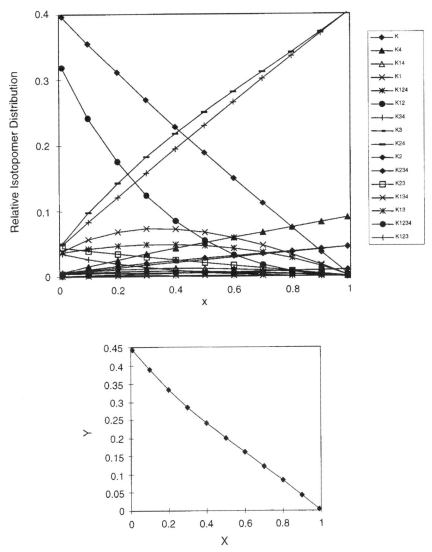

FIGURE 9.14   Distribution of α-ketoglutarate isotopomers as a function of x for the case of 90% [3-^{13}C]pyruvate. Sixteen species are present, and y is obtained exclusively as a function of x by assuming that there are no reactions that generate CO_2 other than those depicted in Fig. 9.8.

TABLE 9.4 Determination of the Fraction x from the Line Intensities of the NMR Spectra of Fig. 9.13[a]

C-2			
	S/D_{12}	S/D_{23}	S/Q
NMR spectrum	4.20 ± 0.3	0.73 ± 0.1	1.95 ± 0.3
$x = 0.35$	*4.13*	*0.81*	*1.97*

C-3	
	T/D
NMR spectrum	1.3 ± 0.1
$x = 0.35$	*1.3*

C-4	
	S/D
NMR spectrum	0.5 ± 0.2
$x = 0.35$	*0.5*

[a] The best estimate for x is 0.35. The values in italics represent the line intensity ratios calculated for the estimated value of x. (S, D, T: the intensities of the singlet, doublet (both lines), and triplet (all three lines), respectively. D_{ij}: the intensity of the doublet with coupling constant J_{ij}).

y. The satisfactory agreement with the experimental data supports the validity of our modeling methodology. Differences observed between the experimental data and the theoretical predictions at C-2 can be explained by the uncertainty in the estimation of the line intensity ratios from the published NMR spectra of Fig. 9.13. It is noted that line intensity ratios calculated for $x = 0.35$ fall within the range of experimental error. Similar multiplet patterns in glutamate and glutamine were also obtained in isolated livers of mice perfused with [3-^{13}C]alanine (Hall *et al.*, 1988).

EXAMPLE 9.4

Acetate Utilization in *Escherichia coli*

Walsh and Koshland (1984) obtained a ^{13}-C NMR spectrum of purified glutamate from *E. coli* grown with [2-^{13}C]acetate as the sole carbon source (Fig. 9.15). It shows a doublet at C-1, six lines at C-2, a triplet at C-3, and a doublet at C-4. This pattern is consistent with the operation of the glyoxylate shunt pathway (Model I) and exactly matches the predictions of Table 9.2.

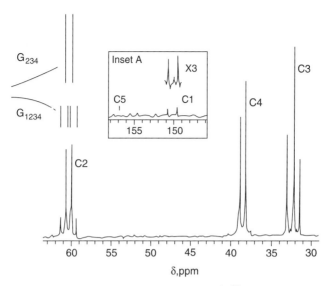

FIGURE 9.15 Observed glutamate multiplet patterns with [2-^{13}C]acetate [adapted from Walsh and Koshland (1984)]. Peak at C-2 is six lines consisting of a doublet due to G_{234} with each line intensity 0.15 and a quartet due to G_{1234} with each line intensity 0.05. Therefore, $Q/D = 2/3$.

There is only one variable in this case, z, which essentially determines the relative intensity of each peak and each line within each peak. z is defined as the probability of isocitrate utilized via the glyoxylate shunt pathway (or the fraction of isocitrate utilized in the glyoxylate pathway, the balance being converted in the TCA cycle). According to Walsh and Koshland (1984), the relative enrichment at the five carbons of intracellular glutamate was 0.4:1.0:0.9:1.0:0 for C-1 through C-5, respectively. Our model predicts C-2 = C-3 = C-4 = 1 and C-5 = 0, which is in total agreement with the experimental results, whereas C-1 = $(1 - z)/2$. By inserting the experimental value of 0.4 for C-1, z is calculated equal to 0.2. This is identical to the value estimated by Walsh and Koshland (1984) using an involved method based on input-output equations for intermediary metabolite carbon.

The model results can also explain exactly the fine structure of the obtained NMR spectra, because we consider all the isotopomers of the intermediate metabolites and not just isotope enrichment at specific metabolite carbon positions. Even though the labeling diagram (Fig. 9.11) shows the presence of up to six α-ketoglutarate isotopomers, the results of Table 9.2 show that, under steady state, only two α-ketoglutarate species should be present: K_{1234} (or G_{1234}) with a relative population of $(1 - z)/2$ and K_{234} (or G_{234}) with a relative population $(1 + z)/2$. The multiple pattern can be analyzed in a manner similar to the case of [3-^{13}C]pyruvate in Section 9.2.1. Of particular interest is the pattern at C-2. Here G_{1234} should give rise to a quartet at C-2 with each line intensity $(1 - z)/8 = 0.1$, and G_{234} should give a doublet at C-2 with each line intensity $(1 + z)/4 = 0.3$. These predictions cannot be compared directly with the results of Fig. 9.15 due to a lack of intensity data for the individual peaks of the C-2 resonance. However, comparison of the heights of the individual peaks shows general agreement between experimental and predicted values. The intensities at the C-3 and C-4 resonances can be analyzed similarly. They contain no information for the value of the flux ratio z; however, they are very consistent with the preceding patterns.

Another experiment for verifying the operation of the glyoxylate shunt pathway in bacteria was performed with 99% [1-^{13}C]acetate. In this case too the glutamate isotopomer species can be obtained in terms of z. Only two species should be present (Table 9.3): G_5 with relative population $(1 + z)/2$ and G_{15} with relative population $(1 - z)/2$. Therefore, enrichment of glutamate should be C-1 with $(1 - z)/2$, C-5 with 1, and none at C-2, C-3, and C-4. This is exactly what is observed in proline synthesis by E. coli (Crawford et al., 1987) and glutamate synthesis by Brevibacterium flavum (Walker and London, 1987). Because the system of Crawford et al. (1987) was similar to that of Walsh and Koshland (1985), one can assume the same value $z = 0.2$ that yields an estimate of 0.40 for glutamate C-1 enrichment. From the spectra reported in Crawford et al. (1987), the C-1 glutamate enrichment can be estimated as 30%, in fair agreement with the model estimates.

EXAMPLE 9.5

Glyoxylate Shunt Pathway vs TCA Cycle in Bacteria

The value of z, indicating the fraction of isocitrate converted via the glyoxylate shunt pathway, can be determined by three independent methods involving the use of GC-MS and ^{13}C NMR. First, as done by Walsh and Koshland (1984), the relative enrichment at glutamate C-1 can be used. This approach requires additional measurement of known standards in order to establish the calibration required for converting the absolute peak intensity into relative enrichment. Secondly, within the C-2 peak, the ratio of quartet (Q) to doublet (D) can be used to find the value of z from Table 9.3:

$$\frac{Q}{D} = \frac{1 - z}{1 + z} \tag{1}$$

From Fig. 9.15, this ratio is estimated to be equal to $2/3$, giving a value of z equal to 0.2, which is the same result obtained with the first method. Thirdly, GC-MS can be used to measure the fractions of the same metabolite with different molecular weights differentiated by one atomic mass unit. In the example of Walsh and Koshland (1984), GC-MS should produce two peaks for glutamate, one corresponding to molecular weight $(M + 3)$ and another corresponding to $(M + 4)$, where M represents the molecular weight of glutamate with all its carbons being ^{12}C. The glutamate species, G_{234}, has a molecular weight of $(M + 3)$, and G_{1234} has a molecular weight of $(M + 4)$. Therefore, the ratio of the $(M + 3)$ to the $(M + 4)$ species is

$$\frac{(M + 3)}{(M + 4)} = \frac{1 + z}{1 - z} \tag{2}$$

Similarly, in the case of $[1\text{-}^{13}C]$acetate, the value of z can be obtained in two different ways: from the ^{13}C NMR measurement of relative enrichment at each carbon and from the GC-MS measurement of the ratio of $(M + 2)$ to $(M + 1)$.

9.3. CARBON BALANCES

Detailed enumeration of all metabolite isotopomers may be difficult in large metabolic networks, or simply unnecessary in the absence of information about isotopomer molecular weights or fine structure NMR data. Often, only total isotope enrichments (or intensities) are measured in conjunction with isotope labeling experiments, and these data can be analyzed without the additional complication of full isotopomer enumeration and balancing. In this section, we illustrate two methods for the analysis of isotope intensity data using direct metabolite carbon balances and atom mapping matrices.

9.3.1. DIRECT CARBON BALANCES

This method is fairly straightforward as it consists of overall metabolite balances as well as balances for each carbon atom of the network metabolites. In this way, the fate of each carbon is accounted for and the enrichment of a particular carbon atom of a specific metabolite can be determined. We illustrate the method through application to the pentosephosphate pathway (PPP).

Figure 9.16 depicts a schematic of the pentosephosphate pathway. As mentioned in Chapter 2, PPP has an oxidative branch that begins with the

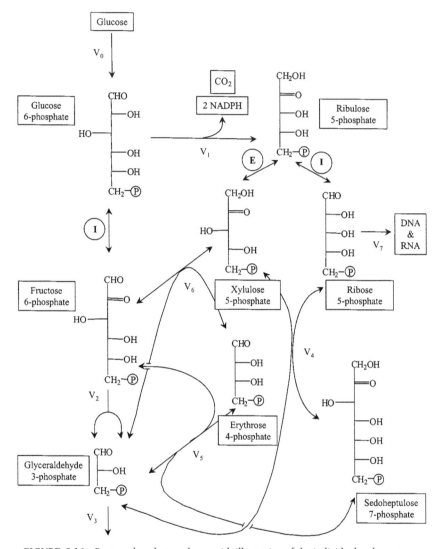

FIGURE 9.16   Pentosephosphate pathway with illustration of the individual carbon atoms.

dehydrogenation of Glc6P to form, first, 6-phosphogluconate and, after two more reactions, ribose-6-phosphate, which is used for RNA and DNA nucleotide sugar biosynthesis:

$$-Glc6P - 2NADP^+ - H_2O + ribose\text{-}5\text{-}phosphate + 2NADPH$$
$$+ 2H^+ + CO_2 = 0$$

Because most cells require more NADPH than ribose-5-phosphate, excess ribose can be converted back to glycolytic intermediates through the transfer of ketose and aldose units using transketolase and transaldolase enzymes. The net sum of this nonoxidative branch of PPP is

$$-3\text{ribose-5-phosphate} + 2\text{fructose-6-phosphate}$$

$$+ \text{glyceraldehyde-3-phosphate} = 0$$

The stoichiometry of the PPP reactions is also summarized in Table 10.1. Flux analysis through central carbon metabolism requires exact accounting of the amount of carbon oxidized in the PPP and the reducing power generated there for cell biosynthesis. Overall material balances can provide estimates of the *net* flux through this pathway and the flux distribution between PPP and glycolysis, as demonstrated in Section 10.1. However, it is very desirable to obtain independent confirmation of flux split ratios determined by material balancing, as well as estimates of the extent of reversibility of the key PPP reactions, namely, the transketolase and transaldolase reactions. This can be achieved through the introduction of labeled glucose and measurement of the isotope intensities of selected PPP metabolite intermediates, as shown in the following.

We begin with the balances for the metabolite pools of Fig. 9.16. It is noted that the isomerase reaction between Glc6P and Fru6P, as well as the isomerase and epimerase reactions among the pentosephosphate pools, is fast and assumed to be at equilibrium. Therefore, a single pool is considered for the hexoses (H6P) and pentoses (R5P) in the network:

$$\frac{dc_{H6P}}{dt} = v_0 - v_1 - v_2 + v_5 + v_6 \tag{9.31a}$$

$$\frac{dc_{R5P}}{dt} = v_1 - 2v_4 - v_6 - v_7 \tag{9.31b}$$

$$\frac{dc_{G3P}}{dt} = 2v_2 - v_3 + v_4 - v_5 + v_6 \tag{9.31c}$$

$$\frac{dc_{E4P}}{dt} = v_5 - v_6 \tag{9.31d}$$

$$\frac{dc_{S7P}}{dt} = v_4 - v_5 \tag{9.31e}$$

At steady state, the preceding equations are reduced to the following relationships, assuming negligible consumption of the pentosephosphate pool for nucleotide synthesis.

$$v_1 = 3xv_0 \tag{9.32a}$$

$$v_2 = (1 - x)v_0 \tag{9.32b}$$

$$v_3 = (2 - x)v_0 \tag{9.32c}$$

$$v_4 = v_5 = v_6 = xv_0 \tag{9.32d}$$

The value x in the preceding equations represents the fraction of the hexose flux that is directed through the PPP. The metabolite balances refer only to the overall flux through the pathway, without yet accounting for any measure of reversibility. It is seen that the net fluxes through the nonoxidative pathway reactions are equal, and when x equals unity all of the hexose input flows through the PPP. Thus according to eq. (9.32c), $v_3 = v_0$ and therefore 1 mol of pyruvate is formed from each mole of hexose consumed. As a result, half of the hexose is released as CO_2.

The preceding model can, of course, be tested by introducing labeled glucose as substrate and measuring the label distribution in one or more of the PPP metabolite intermediates. To facilitate the interpretation of such labeling data, metabolite carbon balance equations must be written to describe the overall balance on each carbon atom considered in the network. These balances are shown in Table 9.5, corresponding to the carbon atom transfer scheme of Fig. 9.17 for the nonoxidative reactions of the PPP. It is further noted that full reversibility has been assumed for each of the nonoxidative PPP reactions, so that each net reaction is the composite of a forward (v_i^+) and a reverse (v_i^-) reaction. Our convention is to denote both forward and reverse reactions as positive. The sign of the net reaction (determined by the relative magnitudes of the forward and reverse reactions) indicates whether the net flux is in the direction of the forward (positive) or the reverse (negative) reaction. This reversibility can lead to further redistribution of the label due to carbon atom exchange with glycolytic intermediates and must be considered for full accounting of the isotopic labeling pattern. An account of the reversibility of the three nonoxidative reactions of the PPP requires three additional unknowns describing the extent of such reversibility, or exchange rates ζ_i, defined, for a *positive* net forward reaction rate, as the difference between the forward rate and the net flux: $\zeta_i = v_i^+ - v_i = v_i^-$. Exchange rates are important as they affect directly the labeling

TABLE 9.5 Metabolite Carbon Balances for the Pentosephosphate Pathway

$$\frac{d[H6P]}{dt} = v_0 \begin{pmatrix} Hexose(1) \\ Hexose(2) \\ Hexose(3) \\ Hexose(4) \\ Hexose(5) \\ Hexose(6) \end{pmatrix} - (v_1 + v_2) \begin{pmatrix} H6P(1) \\ H6P(2) \\ H6P(3) \\ H6P(4) \\ H6P(5) \\ H6P(6) \end{pmatrix} + v_5^+ \begin{pmatrix} S7P(1) \\ S7P(2) \\ S7P(3) \\ G3P(1) \\ G3P(2) \\ G3P(3) \end{pmatrix} - (v_5^- + v_6^-) \begin{pmatrix} H6P(1) \\ H6P(2) \\ H6P(3) \\ H6P(4) \\ H6P(5) \\ H6P(6) \end{pmatrix} + v_6^+ \begin{pmatrix} R5P(1) \\ R5P(2) \\ E4P(1) \\ E4P(2) \\ E4P(3) \\ E4P(4) \end{pmatrix}$$

$$\frac{d[R5P]}{dt} = v_1 \begin{pmatrix} H6P(2) \\ H6P(3) \\ H6P(4) \\ H6P(5) \\ H6P(6) \end{pmatrix} - 2v_4^+ \begin{pmatrix} R5P(1) \\ R5P(2) \\ R5P(3) \\ R5P(4) \\ R5P(5) \end{pmatrix} + v_4^- \begin{pmatrix} S7P(1) \\ S7P(2) \\ G3P(1) \\ G3P(2) \\ G3P(3) \end{pmatrix} - v_6^+ \begin{pmatrix} R5P(1) \\ R5P(2) \\ R5P(3) \\ R5P(4) \\ R5P(5) \end{pmatrix} + v_6^- \begin{pmatrix} H6P(1) \\ H6P(2) \\ G3P(1) \\ G3P(2) \\ G3P(3) \end{pmatrix}$$

$$\frac{d[G3P]}{dt} = v_2 \begin{pmatrix} H6P(3) \\ H6P(2) \\ H6P(1) \end{pmatrix} + v_2 \begin{pmatrix} H6P(4) \\ H6P(5) \\ H6P(6) \end{pmatrix} + v_4^+ \begin{pmatrix} R5P(3) \\ R5P(4) \\ R5P(5) \end{pmatrix} - (v_4^- + v_5^- + v_6^- + v_3) \begin{pmatrix} G3P(1) \\ G3P(2) \\ G3P(3) \end{pmatrix} + v_6^- \begin{pmatrix} H6P(4) \\ H6P(5) \\ H6P(6) \end{pmatrix}$$

$$\frac{d[E4P]}{dt} = v_5^+ \begin{pmatrix} S7P(4) \\ S7P(5) \\ S7P(6) \\ S7P(7) \end{pmatrix} - (v_5^- + v_6^+) \begin{pmatrix} E4P(1) \\ E4P(2) \\ E4P(3) \\ E4P(4) \end{pmatrix} + v_6^- \begin{pmatrix} H6P(3) \\ H6P(4) \\ H6P(5) \\ H6P(6) \end{pmatrix}$$

$$\frac{d[S7P]}{dt} = v_4^+ \begin{pmatrix} R5P(1) \\ R5P(2) \\ R5P(1) \\ R5P(2) \\ R5P(3) \\ R5P(4) \\ R5P(5) \end{pmatrix} - (v_4^- + v_5^+) \begin{pmatrix} S7P(1) \\ S7P(2) \\ S7P(3) \\ S7P(4) \\ S7P(5) \\ S7P(6) \\ S7P(7) \end{pmatrix} + v_5^- \begin{pmatrix} H6P(1) \\ H6P(2) \\ H6P(3) \\ E4P(1) \\ E4P(2) \\ E4P(3) \\ E4P(4) \end{pmatrix}$$

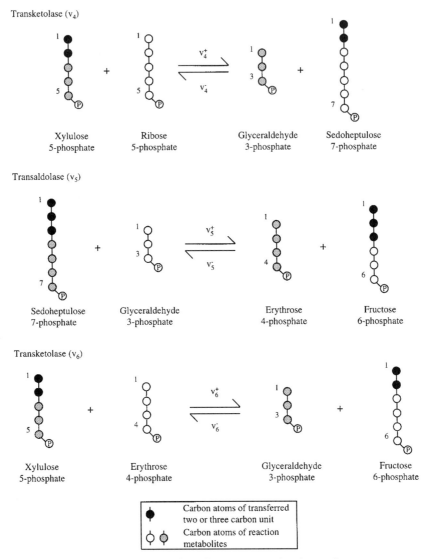

FIGURE 9.17   Transfer scheme for carbons in pentosephosphate pathway reactions.

pattern of the intermediate metabolites. Furthermore, they may vary significantly while yielding the same net flux determined from overall metabolite balances. The introduction of the exchange rates for the preceding three reactions allows all reaction rates of the equations of Table 9.5 to be expressed as functions of the net PPP flux (equal to xv_0) and the three exchange rates by making use of

$$v_4^+ = (xv_0 + \zeta_4) \tag{9.33a}$$

$$v_4^- = \zeta_4 \tag{9.33b}$$

$$v_5^+ = (xv_0 + \zeta_5) \tag{9.33c}$$

$$v_5^- = -\zeta_5 \tag{9.33d}$$

$$v_6^+ = (xv_0 + \zeta_6) \tag{9.33e}$$

$$v_6^- = -\zeta_6 \tag{9.33f}$$

   Clearly, the determination of the net PPP flux and exchange rates require label distribution measurements. There are many identification methods that can be employed for this determination. The simplest is the trial and error method of Fig. 9.7, whereby by starting with some initial guesses for the PPP flux and exchange rates the distribution of label is determined for a given labeled substrate (hexose) by solving the linear system of equations of Table 9.5. Upon comparison with isotope intensity measurements, the initial guesses are readjusted and the procedure repeated until convergence is achieved. Standard routines found in Mathematica or Maple can be employed for the solution. Two points should be kept in mind in applying this approach. First, the problem is, in essence, one of solving a system of nonlinear algebraic equations, and as such it may be sensitive to the initial guesses of the solution. Second, the choice of measured intensities is critical for the quality of the final solution or, even, for obtaining a solution at all. This point will be elaborated further.

   The approach outlined in this section was applied to the analysis of the ^{14}C label distribution measured in glycogen from rat epidymal adipose tissue incubated with [2-^{14}C]glucose either with or without insulin (Landau and Katz, 1964). Table 9.6 shows the measured specific activities of glycogen (H6P) carbon atoms relative to C-2, along with those calculated using the preceding model and solution procedure. Two entries are provided for the model predictions: one assuming no reversibility and the other allowing full reversibility of the three PPP reactions. It can be seen that the agreement between model predictions and measured intensities is superior for the latter case, which also provides estimates for the exchange rates of the reversible reactions. Table 9.7 summarizes the results of a similar study on the PPP for the lysine-producing bacterium *C. glutamicum* (Marx *et al.*, 1996). Again, the assumption of no reversibility leads to gross errors for the predicted carbon enrichments of erythrose-4-phosphate, glyceraldehyde phosphate, and oth-

TABLE 9.6　^{14}C Label Data for H6P from [2-^{14}C]Glucose[a]

	H6P	C-1	C-2	C-3	C-4	C-5	C-6
Experimental specific	Ins$^-$	15.2	100	12.8	1.9	13.9	2.9
activity ^{14}C	Ins$^+$	30.7	100	17.9	3.5	14.8	5.3
Model: no reversibility	Ins$^-$	20.6	100	11.5	0.6	5.4	1.1
	Ins$^+$	31.5	100	18.7	1.5	8.1	2.5
Model: reversibility	Ins$^-$	18.8	100	15.8	2.1	13.4	2.5
	Ins$^+$	30.5	100	20.9	2.7	13.1	4.0

[a] Experimental values compared with model predictions assuming no reversibility and reversibility for transketolase and transaldolase reactions (Landau and Katz, 1964). The calculations used $x_{Ins-} = 0.13$ and $x_{Ins+} = 0.23$. For the model with reversibility, the following parameters were used: $\zeta_4 = 0.5$, $\zeta_5 = 0.2$, and $\zeta_6 = 0$ and $\zeta_4 = 0.8$, $\zeta_5 = 0.08$, and $\zeta_6 = 0$ without and with insulin, respectively.

ers. This is corrected through the inclusion of the exchange rates and results in good agreement between predicted and measured label enrichment values. It should be noted here that the use of the E4P enrichment data alone was insufficient to allow robust determination of the reversible reaction exchange rates. This was corrected through the inclusion of ribose-5-phosphate enrichment data (particularly C-1 enrichment, measured by isolating guanosine derivatives from cell material), which apparently converted a previously undetermined system to a determined one. Glyceraldehyde phosphate enrichment can be used instead of R5P, indicating that there are several

TABLE 9.7　Measured and Estimated ^{13}C Enrichments in Erythrose 4-Phosphate (Ery4P) and Glyceraldehyde 3-Phosphate (G3P) in *Corynebacterium glutamicum* Growing in a Chemostat Culture with [1-^{13}C]Glucose as Substrate[a]

Method	Metabolite	^{13}C Enrichment at			
		C1	C2	C3	C4
Experimental					
	Ery4P	2.5%	2.0%	1.9%	15.3%
	G3P	2.7%	2.6%	26.3%	-
Model 1					
$\zeta_4 = \zeta_5 = \zeta_6 = 0$	Ery4P	0.0%	0.0%	0.0%	5.2%
	G3P	0.0%	0.0%	32.4%	-
Model 2					
$\zeta_4 = 0.5$, $\zeta_5 = 0.1$,	Ery4P	3.7%	0.9%	0.3%	15.6%
$\zeta_6 = 0.2$	G3P	1.9%	0.6%	31.0%	-

[a] Experimental values are compared with model predictions assuming no reversibility (model 1) and reversibility for transketolase and transaldolase reactions (model 2) [Marx et al. (1996), Follstad and Stephanopoulos (1998)].

combinations of enrichment measurements that can result in an observable system. Similarly, for the glycogen results of Table 9.6, if glycogen were to be replaced by G3P (measured through lactate), an undetermined system would result.

9.3.2. USE OF ATOM MAPPING MATRICES

An alternative method of modeling isotope distributions avoids several of the shortcomings of conventional analyses and is especially suited for application to large metabolic networks. In this method, atom mapping matrices (AMMs) are introduced to describe the transfer of carbon atoms from reactants to products (Zupke and Stephanopoulos, 1994). The AMMs decouple the details of the biochemical network from the formulation of the steady state equations governing isotope distribution. The resulting equations are easy to derive, modify, and solve. The method of AMMs is particularly suited for database development and application to large metabolic networks, as well as to trial and error situations where the structure of the network is modified after each iteration. The method is introduced using the simple biochemical network of Fig. 9.5.

The first step is to represent the specific activities (*i.e.*, degrees of enrichment) of the carbon atoms of each metabolite in vector form. The nth element of a metabolite vector contains the specific activity of the nth carbon. For the reaction network in Fig. 9.5, the following metabolite vectors are formed:

$$\text{Pyr} = \begin{pmatrix} \text{Pyr}(1) \\ \text{Pyr}(2) \\ \text{Pyr}(3) \end{pmatrix}; \text{AcCoA}_I = \begin{pmatrix} \text{AcCoA}_I(1) \\ \text{AcCoA}_I(2) \end{pmatrix} \tag{9.34}$$

$$\text{Hex} = \begin{pmatrix} \text{Hex}(1) \\ \text{Hex}(2) \\ \text{Hex}(3) \\ \text{Hex}(4) \\ \text{Hex}(5) \\ \text{Hex}(6) \end{pmatrix}; \text{AcCoA}_{II} = \begin{pmatrix} \text{AcCoA}_{II}(1) \\ \text{AcCoA}_{II}(2) \end{pmatrix} \tag{9.35}$$

With vector notation, all of the carbon atoms of a metabolite are represented in a clear and concise form.

The next step is to construct atom mapping matrices for the reactions in the metabolic network. These matrices describe the transfer of atoms from reactants to products. For each reaction, there will be a mapping matrix for every reactant-product pair. For example, consider a general reaction involving two reactants A and B, and two products C and D, catalyzed by enzyme E:

$$A + B \overset{E}{\leftrightarrow} C + D \tag{9.36}$$

This reaction will have four mapping matrices describing the transfer of carbons from A to C, A to D, B to C, and B to D. We designate the mapping matrices for each reaction by square brackets followed by the subscripted enzyme (or reaction) name. Inside the square brackets the particular reactant-product pair is indicated, separated by a $>$ (to indicate direction). Thus, the four mapping matrices for the sample reaction are as follows:

$[A > C]_E$ describes transfer of carbon from A to C

$[A > D]_E$ describes transfer of carbon from A to D

$[B > C]_E$ describes transfer of carbon from B to C

$[B > D]_E$ describes transfer of carbon from B to D

The atom mapping matrices are constructed such that multiplication of a reactant's specific activity vector by the AMM specifies the contribution of that reactant to the product specific activity vector. The specific activity of the resulting products will be the sum of the contribution from each reactant:

$$[A > C]_E A + [B > C]_E B = C \tag{9.37}$$

$$[A > D]_E A + [B > D]_E B = D \tag{9.38}$$

The dimensions of the mapping matrices are determined by the number of carbons in the reactant-product pair. The number of columns equals the number of atoms in the reactant, whereas the number of rows equals the number of carbons in the product. The element in the ith row and the jth column of the mapping matrix specifies the amount of the ith carbon of the product that is derived from the jth carbon of the reactant. Typically, there is a definite and unique mapping of reactant carbons to product carbons, so that the elements of the mapping matrix are usually 0 or 1. However, fractional elements are possible.

To return to the sample metabolic network, let us construct the mapping matrices for that system. Reaction 1 is catalyzed by pyruvate dehydrogenase (PDH) with pyruvate as the only reactant and $AcCoA_1$ and CO_2 as the two products. In this analysis we have not included CO_2 as a measurable metabolite, so that only one mapping matrix is needed: $[Pyr > AcCoA_1]_{PDH}$. We know that carbon 1 of pyruvate is lost to CO_2, and carbons 2 and 3 go to carbons 1 and 2, respectively, of $AcCoA_1$. This results in the following atom mapping matrix:

$$[Pyr > AcCoA]_{PDH} = \begin{pmatrix} 0 & 1 & 0 \\ 0 & 0 & 1 \end{pmatrix} \tag{9.39}$$

If $AcCoA_1$ were derived only from pyruvate, then its specific activity would be given by

$$[Pyr > AcCoA_I]_{PDH}Pyr = \begin{pmatrix} 0 & 1 & 0 \\ 0 & 0 & 1 \end{pmatrix} \begin{pmatrix} Pyr(1) \\ Pyr(2) \\ Pyr(3) \end{pmatrix} = \begin{pmatrix} Pyr(2) \\ Pyr(3) \end{pmatrix}$$

$$= \begin{pmatrix} AcCoA_I(1) \\ AcCoA_I(2) \end{pmatrix} = AcCoA_I \tag{9.40}$$

However, there is also transfer of AcCoA between pools I and II, so that Eq. 9.40 does not completely determine the specific activity of $AcCoA_1$.

Reaction 2 is the transport of AcCoA from pool I to pool II and is designated by the name transI, whereas reaction 3 is the transport of AcCoA from pool II to pool I and is named transII. Because the carbons in $AcCoA_1$ and $AcCoA_{II}$ map directly to each other, $[AcCoA_1 > AcCoA_{II}]_{transI}$ and $[AcCoA_{II} > AcCoA_1]_{transII}$ are both 2 x 2 identity matrices:

$$[AcCoA_I > AcCoA_{II}]_{transI} = [AcCoA_{II} > AcCoA_I]_{transII} = \begin{pmatrix} 1 & 0 \\ 0 & 1 \end{pmatrix}$$

$$\tag{9.41}$$

The final reaction for which there is an atom mapping matrix is reaction 5, the oxidation of hexanoate to $AcCoA_{II}$, designated βox. The β-oxidation of hexanoate removes carbons in pairs to form three AcCoA molecules. The odd carbons of hexanoate are indistinguishable upon conversion to $AcCoA_{II}$, so that their coefficients in the matrix $[Hex > AcCoA_1]_{\beta ox}$ are one-third, and the same holds true for the even carbons. The odd and even carbons are mapped to carbons 1 and 2, respectively, of $AcCoA_{II}$, giving

$$[\text{Hex} > AcCoA_{II}]_{\beta ox} = \begin{pmatrix} \frac{1}{3} & 0 & \frac{1}{3} & 0 & \frac{1}{3} & 0 \\ 0 & \frac{1}{3} & 0 & \frac{1}{3} & 0 & \frac{1}{3} \end{pmatrix} \qquad (9.42)$$

All of the necessary atom mapping matrices have been constructed and can be used to formulate the steady state isotope balance equations. The flux of label into a metabolite is simply the sum of the products of the mapping matrices and the reactant specific activity vectors, weighted by the corresponding reaction flux. For $AcCoA_1$, the two contributing reactions are PDH with flux v_1 and transII with flux rate v_2:

$$\text{flux into } AcCoA_1: v_1 [Pyr > AcCoA_I]_{\text{PDH}} \, Pyr$$
$$+ v_3 [AcCoA_{II} > AcCoA_I]_{\text{transII}} \, AcCoA_{II} \qquad (9.43)$$

The flux of label out of $AcCoA_1$ is

$$(v_2 + v_4) AcCoA_1 \qquad (9.44)$$

Equating eqs. (9.43) and (9.44) gives the steady state isotope balance for $AcCoA_1$:

$$(v_2 + v_4) AcCoA_I = v_1 [Pyr > AcCoA_I]_{\text{PDH}} Pyr$$
$$+ v_3 [AcCoA_{II} > AcCoA_I]_{\text{transII}} AcCoA_{II} \qquad (9.45)$$

Similarly, the steady state isotope balance for $AcCoA_{II}$ is

$$(v_3 + v_6) AcCoA_{II} = v_2 [AcCoA_I > AcCoA_{II}]_{\text{transI}} AcCoA_I$$
$$+ v_5 [\text{Hex} > AcCoA_{II}]_{\beta ox} \text{Hex} \qquad (9.46)$$

Equations (9.45) and (9.46) are equivalent to eqs. (3), (4), (7), and (8) of Example 9.2, obtained by treating each atom separately. In a small network like this example, the level of complexity of the two methods is similar. However, for larger networks, especially those with many metabolites possessing more than three carbons, construction of atom mapping matrices and formulation of the steady state balances in matrix form result in a more representative and compact system of equations. In addition, if new information about the way the carbons are transferred from reactant to product becomes available, only the atom mapping matrices describing the affected reaction need to be altered. This is rather straightforward and requires no new algebra.

The set of equations describing isotope distributions in a metabolic network using atom mapping matrices can be solved iteratively via computer. This requires that the specific activities of the substrate carbon atoms be initialized (to values between 0 and 1 for fractional enrichment) and a consistent set of flux values be provided. Then, each of the steady state equations is solved sequentially for the activities of the output metabolites, and the process is repeated until convergence is achieved. This is essentially equivalent to solving $\mathbf{As} = \mathbf{b}$ using the Gauss-Seidel method. All of the matrices are small and no matrix inversions are necessary, so that this method is not computationally demanding even for very large biochemical reactions.

Similar to the atom mapping matrices, one can construct isotopomer mapping matrices that describe the conversion of isotopomers in biochemical reactions. Together with isotopomer distribution vectors, these matrices are very useful when complex metabolic networks are to be analyzed, but we will not treat these and will only refer to Schmidt *et al.* (1997a,b), who give a detailed description of how these matrices are constructed and applied for analysis of complex metabolic networks.

REFERENCES

Annison, E. F., Leng, R. A., Lindsay, D. B., & White, R. R. (1963). The metabolism of acetic acid, propionic acid and butyric acid in sheep. *Biochem. J.* **88**: 248-252.

Blum, J. J. & Stein, R. B. (1982). On the analysis of metabolic networks. *In Biological Regulation and Development* pp. 99-125. Edited by R. F. Goldenberger & K. R. Yamamoto. New York: Plenum Press.

Chance, E. M., Seeholzer, S. H., Kobayashi, K. & Williamson, J. R. (1983). Mathematical analysis of isotope labeling in the citric acid cycle with applications to ^{13}C NMR studies in perfused rat hearts. *Journal of Biological Chemistry* **258**, 13785-13794.

Crawford, A., Hunter, B. K. & Wood, J. M. (1987). Nuclear Magnetic Resonance spectroscopy reveals the metabolic origins of proline excreted by an *Escherichia coli* derivative during growth on [13C]acetate. *Applied and Environmental Microbiology* **53**: 2445-2451.

Desmoulin, F., Canioni, P. & Cozzone, P. J. (1985). Glutamate-glutamine metabolism in the perfused rat liver: 13C-NMR study using (2-13C)- enriched acetate. *FEBS. Letters* **185**: 29-32.

Follstad, B. D. & Stephanopoulos, G. (1998). Effect of reversible reactions on isotope label distributions. Analysis of the pentose phosphate pathway. *European Journal of Biochemistry* **252**(3): 360-372.

Hall, J. D., Mackenzie, N. E., Mansfield, J. M., McCloskey, D. E. & Scotts, A. I. (1988). 13C-NMR analysis of alanine metabolism by isolated perfused rat livers from C3HeB/FeJ mice infected with African Trypanosomes. *Comp. Biochemical Physiology* **89B**: 679-685.

Marx, A., de Graaf, A. A., Wiechert, W., Eggeling, L. & Sahm, H. (1996). Determination of the fluxes in the central metabolism of *Corynebacterium glutamicum* by nuclear magnetic resonance spectroscopy combined with metabolite balancing. *Biotechnology and Bioengineering* 49, 111-129.

Schmidt, K., Carlsen, M., Nielsen, J. & Villadsen, J. (1997a). Modeling isotopomer distributions in biochemical networks using isotopomer mapping matrices. *Biotechnology and Bioengineering* 55, 831-840.

Schmidt, K., Nielsen, J. & Villadsen, J. (1997b). Quantitative analysis of metabolic fluxes in *E. coli* using 2 dimensional NMR spectroscopy and complete isotopomer models. *Journal of Biotechnology*, in press.

Walker, T. E. and London, R. E. (1987). Biosynthetic preparation of L-[13C]-and [15N]glutamate by Brevibacterium flavum. *Applied and Environmental Microbiology* 53:92-98.

Walsh, K. & Koshland, D. E. (1984). Determination of flux through the branch point of two metabolic cycles-The tricarboxylic acid cycle and the glyoxylate shunt. *Journal of Biological Chemistry* 259, 9646-9654.

Zupke, G. & Stephanopoulos, G. (1994). Modeling of isotope distributions and intracellular fluxes in metabolic networks using atom mapping matrices. *Biotechnology Progress* 10, 489-498.

CHAPTER 10

Applications of Metabolic Flux Analysis

In the previous two chapters, we reviewed methods for intracellular metabolic flux determination, along with their importance in providing a comprehensive picture of the metabolic state of cells. It should be emphasized that metabolic fluxes are the most fundamental measure of cell physiology. Together with intracellular metabolite concentrations, they provide the necessary information for deciphering the complex mechanisms of *metabolic flux control*, a central element of metabolic engineering. As such, metabolic flux analysis is not simply a mathematical exercise in matrix inversion. It is rather an attempt to obtain the most complete picture of the cellular state, consistent with the measured extracellular fluxes, isotope label distribution, GC-MS, and other specific reaction-probing data. It should be further noted that the extent to which intracellular flux estimates are accepted as reliable measures of actual *in vivo* metabolic fluxes depends on the degree of redundancy built into the calculations. Intracellular fluxes simply calculated from an equal number of algebraic metabolite balances should be regarded with caution as they depend entirely on the assumed biochemistry and accuracy of extracellular metabolite measurements. On the other hand, calculated metabolic fluxes that are also consistent with additional measurements of quantities directly dependent upon the fluxes (such as the degree of enrichment in ^{13}C label at specific carbon positions of intracellular and/or extracellular metabolites, the fine structure of metabolite NMR spectra, the molecular weight distribution of metabolites produced upon labeling with carbon isotopes, etc.) are more reliable measures of the corresponding *in vivo* intracellular fluxes.

In this chapter, we provide two extended case studies of metabolic flux analysis. The purpose of these examples is threefold. First, they can be used to test the level of understanding, by the reader, of the presented concepts and calculation procedures. To facilitate this exercise, in most cases we have provided sufficient information to allow the independent determination of fluxes and other quantities, which can be compared with results reported in the chapter. In those cases where the volume of data needed was simply too large to be included in this book, adequate references are provided for the missing information. The second objective of the case studies reviewed in this chapter is to show how experimental information can be upgraded, through metabolic flux analysis, to provide additional insights about the metabolic state of cells followed by useful suggestions for further experimentation. In many instances, the additional information extracted from the original measurements, along with the results of derivative experiments, more than compensated for the marginal effort required for carrying out metabolic flux analysis of the original fermentation data. Finally, the steps of the outlined methodology can serve as a *blueprint* for similar research programs aiming at the identification of critical branch points in metabolic networks and the reaction(s) that most likely limit product yield and productivity.

In the literature one may find many other case studies of metabolic flux analysis, for example, for the following systems:

- *Penicillin production by the filamentous fungus Penicillium chrysogenum.* This system has been analyzed by Jørgensen *et al.* (1995) and Henriksen *et al.* (1996), who used metabolic flux analysis (MFA) to calculate metabolic flux distributions during fed-batch and continuous cultures. Furthermore, they used MFA to calculate the maximum theoretical yields for different biosynthetic pathways leading to cysteine (which is a precursor for penicillin biosynthesis). In their analysis, they found a correlation between the flux through the pentosephosphate pathway and penicillin production. Their model is the first example of a consideration of intracellular compartmentation, *i.e.*, it distinguishes between cytosolic and mitochondrial reactions.

- *Anaerobic growth of Saccharomyces cerevisiae.* This system has been analyzed by Nissen *et al.* (1997), and one of the flux schemes of their analysis is shown in Fig. 8.1. In addition to calculation of the metabolic flux distribution, they used MFA to analyze the possible role of various isoenzymes in *S. cerevisiae.* They also considered compartmentation in their analysis and demonstrated that this may actually explain the role of isoenzymes of alcohol dehydrogenase.

- *Growth of S. cerevisiae on glucose/ethanol mixtures.* This system has been studied by Van Gulik and Heijnen (1995), who used linear programming to estimate the fluxes during growth at different glucose/ethanol mixtures. They demonstrated that the gluconeogenic flux increased when the ethanol fraction increased.

- *Growth of Escherichia coli.* This system has been studied extensively by the group of Palsson, and some of their results are discussed in Example 8.8.

All of these case studies form an excellent basis for exercises for students and for more elaborate discussions in analogy with those we carry out in the two case studies treated here.

10.1. AMINO ACID PRODUCTION BY GLUTAMIC ACID BACTERIA

Commercially the most important amino acid of the aspartate family is lysine. It is found in limiting amounts in most animal feed grains, such as corn, rice, and wheat; consequently, the nutritional value of these grains can be improved significantly by supplementation with external lysine. Initially isolated from protein hydrolysates, lysine is presently produced in large scale by microbial fermentation from inexpensive carbon sources. Numerous organisms have been isolated that can excrete lysine and glutamic acid. As a group, these bacteria are referred to as glutamic acid bacteria. Although this group appears to span several different genera, this classification has been found to be unwarranted and the majority of the glutamic acid bacteria can be classified under the genus *Corynebacterium sensu stricto.* Studies have also shown that *Brevibacterium flavum* should be classified as *Corynebacterium glutamicum.* We will use the latter name in this chapter to describe all microorganisms employed in the production of aspartate family amino acids.

We focus our analysis on lysine overproduction, with particular emphasis on issues of yield and productivity. Even small improvements in these two figures of merit can be critical for the economics of a large volume, low added value product like lysine. With reported industrial molar yields on glucose in the range of 30-40%, there is room for substantial improvement considering that the theoretical yield is estimated to be upward of 75% (see Section 10.1.2). The following steps are presented as part of a systematic effort to identify factors of importance in improving lysine fermentation performance: (a) Determine the theoretical yield for the conversion of a carbohydrate source, such as glucose, to lysine. (b) Perform flux analysis of standard as well as selectively perturbed fermentations in order to prioritize the importance of metabolic branch points as determining factors of product yield. (c) Demonstrate the use of labeled compounds together with mutants of special genetic background to differentiate between two possible anaplerotic pathways. (d) Critically analyze chemostat and other data to evaluate the outcome of competition for limited carbon precursors between anabolic and catabolic reactions. (e) Illustrate the use of flux analysis in identifying enzymatic transhydrogenase activity and its role in balancing reducing biosynthetic equivalents.

10.1.1. BIOCHEMISTRY AND REGULATION OF GLUTAMIC ACID BACTERIA

Table 10.1 provides a comprehensive list of all reactions considered in describing the biochemistry of glutamic acid bacteria. The Ph.D. thesis of J. J. Vallino and references therein can be consulted for evidence of the detected enzymatic activities that support the biochemistry depicted in Table 10.1. One can locate the main glucose-processing, ammonia uptake, and product-forming pathways. Table 10.1 thus constitutes a base case that will be analyzed first, with variations (such as, for example, the operation of the glyoxylate shunt in the TCA cycle) to be investigated later.

A number of important points should be noted for the biochemistry depicted in Table 10.1:

(a) On several occasions, sequential reactions have been lumped into a single reaction step by eliminating intermediate metabolites. This is the case, for example, of reaction 4 of the EMP pathway, which lumps together phosphofructokinase, Fru16dP aldolase, and triosephosphate isomerase. This lumping reduces the number of reaction steps without affecting the obtained flux results. Lumped reactions are assumed to proceed at the same (steady state) rate, and intermediate metabolites are assumed to be at steady state.

(b) PEP carboxylase (reaction 9) is shown as a representative *overall* anaplerotic reaction, not as the *only* anaplerotic reaction. In fact, pyruvate carboxylase, the combination of isocitrate lyase and malate synthase, the malic enzyme, OAA decarboxylase, and PEP carboxykinase have been proposed, among other possibilities, as anaplerotic routes for these bacteria. The exact reaction is still a matter of debate, and attempts to identify its nature will occupy part of this section.

(c) Ammonium uptake is fulfilled predominantly by glutamate dehydrogenase and glutamine synthetase, because aspartase exhibits little or no activity and alanine and leucine dehydrogenases have not been detected. Additionally, the GS/GOGAT ammonium assimilation route is not believed to operate at the high ammonium ion concentrations employed in the fermentation experiments. Of the five amino acid transferases detected in these organisms, only aspartate aminotransferase is included at it accounts for more than 90% of the total aminotransferase activity.

(d) The lysine formation pathway is shown in more detail in Fig. 9.4. In addition to the four-step *meso*-diaminopimelate (*meso*-DAP) pathway, an alternate pathway has also been identified for the direct conversion of tetrahydrodipicolinate (H4D) to *meso*-DAP, and both pathways appear to support significant carbon flux.

(e) A lumped equation has also been included in Table 10.1 to represent biomass synthesis, with metabolite yield coefficients determined to match the measured elemental composition of *C. glutamicum*: C, 47.6%; O, 31.0%; N, 11.8%; ash, 3.02%.

(f) Although three energy coupling sites are possible in the respiratory chain of *B. flavum*, only two sites appear to translocate protons in *C. glutamicum*, so that the P/O ratio was set equal to 2. To account for maintenance requirements and futile cycles, a reaction (34) has been included for the dissipation of excess ATP. It is noted, however, that the energy balance is not used for flux determination due to these uncertainties. It is simply included to provide an estimate of excess energy availability and possible energy limitations during the fermentation process.

(g) Glucose transport is by the phosphotransferase system with the simultaneous conversion of PEP to pyruvate. This can have profound implications in the absence of pyruvate-carboxylating activity for the theoretical yields, especially for the production of threonine.

(h) No transhydrogenase (THD) reaction is included in the set of equations, as no such activity initially had been detected. This omission led to violation of the steady state assumption for NADPH (detected by

TABLE 10.1 Biochemical Reactions Included in the Metabolic Model for *Corynebacterium glutamicum*

C. glutamicum biochemistry	Metabolite accumulation rate vector
PEP: glucose transferase system	(1) AC — Acetate
(1) GLC + PEP = GLC6P + PYR	(2) ACCOA — Acetyl coenzyme A
	(3) AKG — α-Ketoglutarate
Storage compound: trehalose	(4) ALA — Alanine
(2) GLC6P + 0.5ATP = 0.5TREHAL + 0.5ADP	(5) ASP — Aspartate
	(6) ATP — Adenosine-5'-triphosphate
Embden-Meyerhof-Parnas pathway	(7) BIOMAS — Biomass
(3) GLC6P = FRU6P	(8) CO_2 — Carbon dioxide
(4) FRU6P + ATP = 2GAP + ADP	(9) E4P — Erythrose-4-phosphate
(5) GAP + ADP + NAD = NADH + G3P + ATP	(10) FADH — Flavine adenine dinucleotide, reduced
(6) G3P = PEP + H_2O	(11) FRU6P — Fructose-6-phosphate
(7) PEP + ADP = ATP + PYR	(12) G3P — 3-Phosphoglycerate
(8) PYR + NADH = LAC + NAD	(13) GAP — Glyceraldehyde-3-phosphate
	(14) GLC — Glucose
Anaplerotic reaction: PEP carboxylase	(15) GLC6P — Glucose-6-phosphate
(9) PEP + CO_2 = OAA	(16) GLUM — Glutamine
	(17) GLUT — Glutamate
Tricarboxylic acid cycle	(18) ISOCIT — Isocitrate
(10) PYR + COA + NAD = ACCOA + CO_2 + NADH	(19) LAC — Lactate
(11) ACCOA + OAA + H_2O = ISOCIT + COA	(20) LYSE — Lysine, extracellular
(12) ISOCIT + NADP = AKG + NADPH + CO_2	(21) LYSI — Lysine, intracellular
(13) AKG + COA + NAD = SUCCOA + CO_2 + NADH	(22) MAL — Malate
(14) SUCCOA + ADP = SUC + COA + ATP	(23) NADH — Nicotinamide adenine dinucleotide, reduced
(15) SUC + H_2O + FAD = MAL + FADH	(24) NADPH — Nicotinamide adenine dinucleotide phosphate, reduced
(16) MAL + NAD = OAA + NADH	
	(25) NH_3 — Ammonium
Acetate production or consumption	(26) O_2 — Oxygen
(17) ACCOA + ADP = AC + COA + ATP	

416

(27) OAA	Oxaloacetate
(28) PEP	Phosphoenolpyruvate
(29) PYR	Pyruvate
(30) RIB5P	Ribose-5-phosphate
(31) RIBU5P	Ribulose-5-phosphate
(32) SED7P	Sedoheptulose-7-phosphate
(33) SUC	Succinate
(34) SUCCOA	Succinate coenzyme A
(35) TREHAL	Trehalose
(36) VAL	Valine
(37) XYL5P	Xylulose-5-phosphate

Glutamate, glutamine, alanine, and valine production

(18) NH_3 + AKG + NADPH = GLUT + H_2O + NADP

(19) GLUT + NH_3 + ATP = GLUM + ADP

(20) PYR + GLUT = ALA + AKG

(21) 2 PYR + NADPH + GLUT = VAL + CO_2 + H_2O + NADP + AKG

Pentosephosphate pathway

(22) GLC6P + H_2O + 2NADP = RIBU5P + CO_2 + 2 NADPH

(23) RIBU5P = RIB5P

(24) RIBU5P = XYL5P

(25) XYL5P + RIB5P = SED7P + GAP

(26) SED7P + GAP = FRU6P + E4P

(27) XYL5P + E4P = FRU6P + GAP

Oxidative phosphorylation: P/O = 2

(28) 2NADH + O_2 + 4ADP = $2H_2O$ + 4ATP + 2NAD

(29) 2FADH + O_2 + 2ADP = $2H_2O$ + 2ATP + 2FAD

Aspartate amino acid family

(30) OAA + GLUT = ASP + AKG

(31) ASP + PYR + 2 NADPH + SUCCOA + GLUT + ATP = SUC + AKG + CO_2 + LYSI + 2 NADP + COA + ADP

(32) LYSI = LYSE

Biomass synthesis: $C_{1.97}$, $H_{6.46}$, $O_{1.94}$, $N_{0.345}$, 3.02 % ash

(33) 0.021GLC6P + 0.007FRU6P + 0.09RIB5P + 0.036E4P + 0.013GAP + 0.15G3P + 0.052PEP + 0.03PYR + 0.332ACCOA + 0.08ASP + 0.033LYSI + 0.446GLUT + 0.025GLUM + 0.054ALA + 0.04VAL + 0.052THR + 0.015MET + 0.043LEU + 3.82ATP + 0.476NADPH + 0.312NAD = BIOMAS + 3.82ADP + 0.364AKG + 0.476NADP + 0.312NADH + 0.143CO_2

(34) ATP = ADP + P_i

consistency analysis) in certain mutants and the eventual experimental verification of THD in these cases.

Figure 10.1 summarizes the regulation of the lysine-producing pathway. A key enzyme in lysine synthesis is aspartate kinase (AK), which is subject to concerted-multivalent-feedback inhibition by threonine plus lysine but is not significantly inhibited by either amino acid separately. The first lysine-producing strains were deficient in homoserine dehydrogenase (HDH) activity and, as such, were unable to synthesize threonine (such as ATCC strain 21253). They were thus able to accumulate large concentrations of lysine without feedback inhibition, as long as the medium was sufficiently supplemented with homoserine, or threonine plus methionine, that the organism was unable to synthesize on its own. This supplementation, of course, had to be very carefully balanced as any excess threonine would lead to the interruption of lysine production and resumption of cell growth. More recent lysine fermentation processes utilize *C. glutamicum* strains with AK insensitive to feedback inhibition, which can accumulate large concentrations of lysine. Such strains, for example ATCC strain 21799, are called AEC-resistant strains due to their resistance to feedback inhibition by the nonmetabolizable lysine analogue S-(2-aminoethyl)-L-cysteine (AEC). Another point of regulation is the first enzyme after the ASA branch point, homoserine dehydrogenase, which is strongly inhibited by threonine and weakly by isoleucine and is also repressed by methionine. It has been documented as an

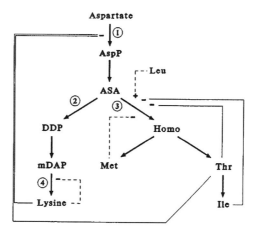

FIGURE 10.1 Regulation of the aspartate amino acid family in *Corynebacterium glutamicum*. The regulated enzymes are (1) aspartate kinase; (2) dihydrodipicolinate synthase; (3) homoserine dehydrogenase; and (4) diaminopimelate decarboxylase. Solid lines illustrate inhibition (-) or activation (+), and dashed lines illustrate repression (-) or induction (+).

allosteric enzyme in *B. flavum*. In even more recent mutants, homoserine dehydrogenase activity has been attenuated to a level sufficient to endogenously supply the downstream amino acids of threonine and methionine, yet low enough to prevent their accumulation to inhibitory levels. Hence, in such strains, bioreactor feed control of a threonine-containing medium has been substituted by an equivalent genetic regulation of the total supply of threonine and methionine. As the supply of these amino acids limits growth in these strains, the latter have been termed *bradytrophs*, i.e., slow- growing organisms.

10.1.2. CALCULATION OF THEORETICAL YIELDS

Theoretical yields can be calculated from an overall reaction for substrate conversion to product, from more detailed balances that take cofactor requirements into consideration, or, finally, from theoretical flux analysis of a bioreaction network. We demonstrate these methods for the case of lysine biosynthesis. Two points should be made at the outset: First, all three approaches should yield the same result, and, second, the maximum theoretical yield *is not an inherent property of the particular product-substrate pair* (as is frequently implied in publications). Rather, it critically depends on the particular metabolic pathway catalyzing the overall conversion.

The conversion of glucose to lysine can be expressed by the following overall reaction:

$$- aC_6H_{12}O_6 - bO_2 - c\,NH_3 + C_6H_{14}N_2O_2 + dCO_2 + eH_2O = 0 \quad (10.1)$$

Four of the five stoichiometric coefficients of eq. (10.1) can be determined from the balances for carbon, nitrogen, and hydrogen and degree of reductance, yielding:

$$- [(4 + e)/6]C_6H_{12}O_6 - (e - 3)O_2 - 2NH_3 + C_6H_{14}N_2O_2$$
$$+ (e - 2)CO_2 + eH_2O = 0 \quad (10.2)$$

The molar yield of lysine thus is seen to be equal to $Y = 6/(4 + e)$. The stoichiometric coefficient for oxygen is indeterminable from first principles. As there can be no net production of oxygen, $e \geq 3$, and the maximum stoichiometric yield of lysine is obtained for $e = 3$: $Y = 0.857$ (6/7) mol of lysine/mol of glucose.

The preceding calculation, of course, does not take any cofactor requirements into consideration. In other words, the fact that this equation provides sufficient carbon, nitrogen, hydrogen, and oxygen for the synthesis of lysine does not necessarily imply that energy currency metabolites and reducing equivalents are also available in the required amounts and that the balances for all intermediate metabolites are satisfied. Theoretical yield calculations should take such constraints into consideration.

We begin with the formulation of the overall stoichiometries for the major metabolic pathways of Table 10.1:

Glycolysis: $- GLC + PEP + Pyr + 2NADH + ATP = 0$ (10.3)

PEP carboxylase: $- PEP - CO_2 + OAA = 0$ (10.4)

Transaminase: $- OAA - GLUT + ASP + AKG = 0$ (10.5)

Lysine pathway: $- ASP - Pyr - 2NADPH - GLUT - 2ATP$

$$+ LYS + AKG + CO_2 = 0$$ (10.6)

Glutamate synthesis: $- NH_3 - AKG - NADPH + GLUT = 0$ (10.7)

Summing of reactions (10.3)-(10.7) yields the following, more detailed equation for the biosynthesis of lysine:

$$- GLC - 4NADPH - 2NH_3 - ATP + LYS + 2NADH = 0$$ (10.8)

It is thus seen that the synthesis of 1 mol of lysine requires an additional 2 or 4 mol of NADPH, depending on whether a transhydrogenase (THD) activity, reversibly converting NADH to NADPH, is present. This is a reflection of the fact that lysine is more reduced than glucose (with a degree of reductance of 4.67 compared to 4 for glucose). The required NADPH is supplied primarily from the pentosephosphate pathway, whose overall stoichiometry under conditions of complete carbon oxidation is given by

PPP (*complete oxidative pathway*): $- GLC6P + 6CO_2 + 12NADPH = 0$

(10.9)

In the simple case where glucose would be transported and phosphorylated directly by a kinase, the preceding equation indicates that an additional 1/6 or 1/3 glucose mol would be required to provide the NADPH reducing equivalents for the lysine synthesis reaction [e.g. (10.8)], depending, again, on the presence or not, respectively, of THD activity. The resulting molar lysine yield would be $6/7$ ($= 0.857$) or $6/8$ ($= 0.75$), respectively, corresponding to $e = 3$ or $e = 4$ in the overall conversion reaction. An additional consideration arises, however, in connection with the glucose phosphotransferase system (PTS), requiring the conversion of 1 mol of PEP to pyruvate per mole of glucose transported into the cell:

$$- GLC - PEP + GLC6P + Pyr = 0$$ (10.10)

Equations (10.3), (10.9), and (10.10) can be multiplied by 1/6 and added to the lysine eq. (10.8), assuming NADPH and NADH equivalence by THD activity, to yield:

$$- (8/6)GLC - 2NH_3 - (5/6)ATP + LYS + (1/3)Pyr + CO_2$$

$$+ (1/3)NADH = 0 \tag{10.11}$$

As long as a pyruvate-carboxylating reaction does not occur to convert pyruvate to OAA or PEP synthetase activity is not present for the recycle of the pyruvate formed from glucose transport, the preceding equation shows a theoretical yield of $6/8 = 0.75$ mol of lysine per mole of glucose. This is so because any extra pyruvate formed beyond what is needed for lysine synthesis is oxidized in the TCA cycle and does not contribute to product yield. In the opposite case, however, the so-formed pyruvate could be carboxylated further in the anaplerotic pathway, increasing the amount of lysine produced but also changing the amount of NADPH required for this purpose. This in turn would change the amount of glucose that would have to be oxidized in the PPP for the production of the extra NADPH, and so forth. It is easily seen that, although this direct method of theoretical yield determination is attractive in certain simpler cases, it can lead to rather involved calculations when metabolic products are recycled in the pathways considered.

A more general approach begins to make use of metabolite balances: For every mole of glucose transported into the cell, 1 mol of PEP is consumed and 1 mol each of pyruvate and GLC6P are produced per eq. (10.10). Letting x be the fraction of GLC6P oxidized completely in the PP pathway (to produce $12x$ mol of NADPH), $(1 - x)$ will be the fraction catabolized in glycolysis, producing $2(1 - x)$ mol of PEP for a total of $(1 - 2x)$. In the presence of THD activity but without a pyruvate carboxylase, the fraction x can be determined from the following balance on NADPH:

$$\text{NADPH produced: } 12x = 2(1 - 2x) \text{:NADPH consumed} \tag{10.12}$$

Equation (10.12) yields $x = 1/8$, from which the theoretical yield is calculated as $1 - 2x = 0.75$. When pyruvate carboxylase is included as an alternative anaplerotic pathway, or a PEP synthetase activity can recycle excess pyruvate formed by the PTS, an additional variable y is introduced for the amount of pyruvate converted to PEP. This conversion should produce equal amounts of PEP and Pyr to ensure full carbon utilization (see also reaction 31, Table 10.1), thus yielding, together with the NADPH balance, the following equations for the determination of x and y:

$$\text{PEP formed: } 1 - 2x + y = 1 - y \text{ :Pyr formed} \qquad (10.13)$$

$$\text{NADPH balance: } 2(1 - 2x + y) = 12x \qquad (10.14)$$

The solution of eqs. (10.13) and (10.14) yields $y = x = 1/7$ for a theoretical lysine yield of 0.857 (6/7). This is obtained by utilizing completely all available carbon, with only the necessary amount of glucose consumed for NADPH production. Similarly, in the absence of THD, the theoretical yield is found to be equal to 0.75 and 0.60, depending on whether or not, respectively, a pyruvate carboxylase activity complements the PEP anaplerotic pathway. It is noted that the ATP requirements are minimal and easily satisfied in this pathway.

Although the preceding approaches certainly are correct, they are prone to errors as the example with the PTS complication amply demonstrates. The main sources of error are in the formulation of the overall stoichiometry of partial pathways and in the accounting of *all* sources and sinks of intermediate and currency metabolites. The procedure outlined here suggests that a more structured approach for metabolite balancing could avoid these problems and yield a formal method that is generally applicable. Metabolic flux analysis is well-suited for this purpose. In the yield calculation mode, the aim of MFA is no longer the determination of internal metabolic fluxes, as most of them are *set to values that ensure maximum product yield*. The goal is rather to determine a few flux split ratios so that all metabolite balances are satisfied in a network that yields maximum product. In the lysine pathway, for example, such a network is achieved by setting the rate of biomass formation equal to zero, the rates of all secreted products other than lysine (acetate, lactate, and trehalose) also equal to zero, the glucose uptake rate equal to -100, and the lysine production rate equal to the yield Y. For a given value of Y there is a total of 34 metabolite balances that can be solved to determine the fluxes of the 34 reactions of the network of Table 10.1.

The matrix equations of Chapter 8 can be employed to formulate and conveniently solve the set of 34 equations of the lysine metabolic network of Table 10.1. Theoretical fluxes thus are determined for ever increasing yields of lysine until an infeasible flux distribution results. For a network without THD and pyruvate carboxylase activity, this happens when the lysine yield approaches 60%, at which point the pyruvate kinase (PK) flux reaches zero (Fig. 10.2). The zero PK flux is the direct result of the PTS discussed earlier. If PEP synthetase or another outlet for the utilization of the pyruvate formed (like pyruvate carboxylase) is added to the reaction network, then the next infeasibility arises when the lysine yield reaches 75%, at which point the flux supported by the TCA cycle drops to zero (Fig. 10.3). Any further increase in

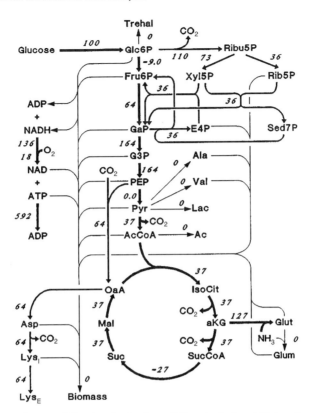

FIGURE 10.2   Theoretical flux distribution for a 64% molar lysine yield, based on constraints dictated by the model of Table 10.1. Limitation is due to irreversibility of pyruvate kinase. The 64% yield exceeds the 60% figure determined in the text due to additional NADPH produced in the TCA cycle. From Vallino (1991). © 1991 MIT.

the yield results in a negative TCA cycle flux. Therefore, under these conditions, the maximum theoretical yield of lysine is 75%. If, further, transhydrogenase activity is added to the network for the interconversion of NADPH and NADH, then similar calculations yield a maximum theoretical yield for lysine equal to 75% or 85.7%, depending on the presence of a pyruvate carboxylase, and this is consistent with the findings of the cofactor balancing method discussed earlier in this section. It is noted that the yield is not constrained by ATP availability, as the ATP dissipation reaction (reaction 34 in Table 10.1) is nonzero.

Figure 10.4a illustrates the theoretical fluxes obtained for the preceding network under the assumption of no THD and pyruvate carboxylase activity and a lysine yield Y equal to 35%. In Fig. 10.4b we provide the theoretical

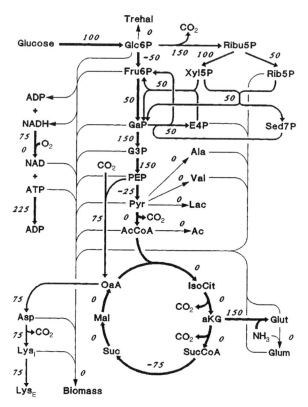

FIGURE 10.3 Theoretical flux distribution at 75% lysine yield. Limitation is due to constraints of TCA cycle flux (pyruvate kinase irreversibility constraint relaxed). From Vallino (1991). © 1991 MIT.

flux map, also for 35% lysine yield, but for a modified bioreaction network in which the anaplerotic reaction of PEP carboxylase has been replaced by the glyoxylate shunt and OAADC with the simultaneous removal of α-ketoglutarate dehydrogenase (αKGDH). The purpose of this calculation is to draw attention to the significant differences that are to be expected for the flux distributions of even slightly altered bioreaction networks.

We close this section with four additional points:

- After a theoretical flux calculation has been completed, the CO_2 evolution rate, as well as the O_2 uptake rate, can be calculated to determine the theoretical value of the respiratory quotient, RQ. This turns out to

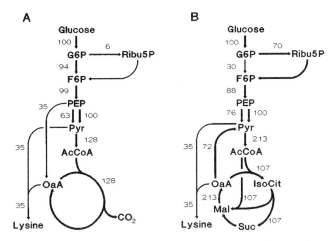

FIGURE 10.4   Theoretical flux distributions necessary to support a lysine molar yield of 35% in the *C. glutamicum* network (illustrated in condensed form) based on (a) the TCA cycle or (b) the glyoxylate shunt. The two fluxes between PEP and pyruvate account for reactions catalyzed by pyruvate kinase (left) and the glucose PTS (right). From Vallino and Stephanopoulos, 1993. © 1993 John Wiley & Sons, Inc. Reprinted by permission of John Wiley & Sons, Inc.

be equal to 2.0 at the maximum theoretical yield of 75%. This number can be used in benchmarking fermentation processes in terms of their closeness to the theoretical maximum and, also, in the design of bioreactor feeding strategies seeking an optimal point of operation.

- One can repeat the previous calculations by setting the rate of lysine production equal to zero and varying the rate of biomass production instead. This produces a maximum yield of biomass of 73%, which is certainly an overestimate as no maintenance requirements and futile cycles have been taken into consideration.

- The removal of the balances for CO_2, NH_3, and O_2 radically increases the condition number of the resulting stoichiometric matrix, as shown in Table 10.2. However, the value is still acceptable, especially when one considers the fact that there is no uncertainty in the value of the measurement vector in this case. In general, a condition number of that magnitude could be a cause for concern if the balance equations were to be solved with actual experimental data.

- The final point is that there should be sufficient equations to determine all theoretical internal fluxes in a theoretical yield calculation. One simply needs to be careful with the extra constraints imposed on the network so that a maximum product yield is obtained.

TABLE 10.2 Condition Numbers (See Section 8.4) of the Lysine Network When Selected Reactions (Indicated in Parentheses) Leading to Extracellular Metabolites Are Deleted from the Network

Metabolites deleted from the network			Condition number
None			59
Biomass (7)			140
CO_2 (8)			59
Glucose (14)			61
Lysine (20)			60
NH_3 (25)			61
O_2 (26)			132
Glucose	O_2		136
NH_3	O_2		138
Biomass	CO_2		143
Lysine	O_2		144
Biomass	Glucose		155
Biomass	NH_3		174
Biomass	O_2		207
Biomass	Lysine		445
CO_2	O_2		762
Biomass	Lysine	NH_3	462
Biomass	Glucose	NH_3	519
Biomass	Glucose	Lysine	726
Biomass	CO_2	O_2	845
CO_2	NH_3	O_2	881

10.1.3. METABOLIC FLUX ANALYSIS OF LYSINE BIOSYNTHETIC NETWORK IN C. GLUTAMICUM

In this section, we demonstrate how metabolic flux analysis can provide additional insights about the control architecture of selected branch points of a metabolic network. The focus is on enhancing lysine yield, which, as mentioned earlier, can be increased from current industrial levels significantly. Usually the enzymatic activities of reactions in the product pathway are amplified in order to increase product yield. However, product yield ultimately is controlled by the flux split ratios of critical branch points. For example, in the network A → B, B → C, and B → D, the yield of D on A strictly depends on the split ratio at the B node. Certainly the amplification of limiting enzymes in the product pathway may affect *indirectly* nodal split ratios; however, product yield ultimately depends on the flexibility of these nodes to flux perturbations. In a rigid node, the flux split ratio is insensitive to product branch activity, and the yield of the product will not be enhanced by mere amplification of the product pathway. Furthermore, if the product is synthesized at adequate rates under normal conditions, then it may be more beneficial to affect split ratios by attenuating byproduct branches. One focus of metabolic engineering research should be on altering nodal split ratios as the main mechanism of improving product yields. A critical issue in this regard is the identification of the branch points that are crucial to product synthesis or lead to byproduct formation. These branch points constitute the *principle nodes* of the network and must be identified first.

Once the critical branch points in a network have been identified, specific perturbations must be carried out in order to characterize the control of these branch points. In the course of a regular fermentation, there is variation of metabolic fluxes and metabolic flux split ratios that could provide some insight on nodal flexibility or rigidity. However, because metabolite effector concentrations (which also vary in such diverse experiments) strongly affect nodal rigidity, the information provided from global variations induced by shifts in biomass and product synthesis rates must be interpreted with caution. It is possible that such global perturbations involve many additional effects that cannot be anticipated and accounted for accurately in the metabolic balances. Consequently, more local perturbations are needed to elucidate the specific aspects of nodal control. This will be demonstrated in the present chapter by examining the metabolic control at two specific nodes, glucose-6-phosphate and the PEP/pyruvate composite branch point.

It should be noted that our discussion in this section makes use of the notions of nodal flexibility and rigidity introduced in Section 5.4 for the control of flux. That section should be reviewed to allow full appreciation of the interpretation of the results presented herein. Furthermore, the concepts of rigidity and flexibility that were introduced in qualitative terms in Section 5.4 will be expanded further and described quantitatively in Chapter 12 in the context of metabolic control analysis (MCA). MCA usually describes the flexibility of individual reaction segments in linear or branched metabolic pathways. The same concepts have been extended in Chapter 12 to characterize the rigidity of *branch points* of metabolic networks.

In accordance with the general objectives of this chapter, sufficient information is supplied to allow the independent calculation of fluxes, flux split ratios, and other results that can be obtained through the application of metabolic flux analysis (MFA). To demonstrate the general philosophy of MFA, we also provide a review of the reasons for and the strategy used in designing the presented experiments. *Bionet* can be used to reproduce the results presented here.

Identification of Principal Nodes

Although metabolic networks contain a large number of nodes, it is usually the case that flux split ratios at only a few nodes actually change as the yield of the product changes. These nodes are referred to as *principal* nodes as they directly affect product yield. The split ratios of the remaining nodes are relatively unaffected and do not merit further investigation. To locate the principal nodes in a network, the product, byproducts, and substrates are first identified. This is then followed by a *theoretical* flux analysis similar to that presented in the previous section for the calculation of the theoretical product yield. Principal nodes are identified by systematic variation of the product yield and observation of the flux split ratios of various nodes.

In applying the method to the lysine biosynthetic network, as the lysine yield increases, the split ratios at five branch points are significantly affected, namely, Glc6P, Fru6P, PEP, pyruvate, and OAA. Of these branch points, only two, Glc6P and the PEP/Pyr group, are considered as principal nodes of interest for the following reasons. Fru6P turns out to be a condensation point when the yield of lysine is less than 60% and a branch point for lysine yields greater than 60%. In the latter case the isomerase reaction reverses and Glc6P becomes a condensation point. Therefore, between Glc6P and Fru6P, one is always a branch point and the other a condensation point, and because typical yields are below 60%, we consider only Glc6P in our analysis as a principal branch point. With regard to OAA, close inspection of this node reveals that it is basically a trivial one, as all OAA consumed by citrate synthase must be returned by malate dehydrogenase (last enzyme in the TCA cycle). In essence, the TCA cycle flux simply passes through the OAA node, and, as a result, all OAA consumed for lysine synthesis is produced by the anaplerotic pathway(s). This underlines the importance of understanding mass balance constraints in the operation of the TCA cycle and associated anaplerotic reactions. It should be noted that all carbon entering the TCA cycle via AcCoA must be oxidized to CO_2, whereas carbon that enters the TCA cycle by other reactions cannot be oxidized and eventually must leave the cycle to supply precursors for biomass and product synthesis. Due to misunderstanding of these concepts, it has been suggested that aspartase would improve aspartate availability by tapping directly into the fumarate pool. Fumarate, however, is no more available than oxaloacetate as the *net production* of either is governed by the flux of the anaplerotic pathways. This point was confirmed by the failure of the introduction of aspartase activity to obtain increased lysine yield.

The preceding principal nodes, Glc6P and the PEP/Pyr group, simply reflect the fact that, of the four precursors necessary for lysine synthesis (carbon, NADPH, ATP, and ammonia), two (carbon and NADPH) are dependent upon the glucose supply and, hence, the metabolic flux distribution at the preceding branch points. The remaining two (ATP and ammonia) apparently present no limitation as (a) adequate ATP is generated by reactions accompanying glucose catabolism for the production of carbon precursors and (b) the reactions responsible for nitrogen uptake for amino acid and lysine biosynthesis are amply supplied through the glutamine synthetase and glutamate dehydrogenase assimilation systems. If the latter were not true, then reactions supplying ATP and replenishing glutamate used in the ammonia assimilation pathway should also be included in the list of principal branch points of the lysine biosynthetic network.

To summarize the dependence of lysine yield on the split ratios at the two principal nodes, it is noted that, as the lysine yield increases, a higher fraction of glucose must be diverted into the pentosephosphate pathway (PPP) to meet the increasing NADPH requirements. If the actual enzymatic kinetics at this node are such that glycolysis outcompetes the PPP for Glc6P, then lysine will be NADPH-limited and the excess carbon entering glycolysis ultimately must lead to byproduct formation. However, if the Glc6P node split ratio is flexible to vary so as to exactly meet the NADPH demand, then lysine yield limitations are caused by suboptimal split at the PEP/Pyr principal node. If the anaplerotic pathway split ratio at the PEP/Pyr branch is less than 50%, then insufficient OAA will be synthesized and excess pyruvate will be formed that will most likely be oxidized in the TCA cycle. If the anaplerotic branch split ratio is greater than 50%, then excessive amounts of OAA will be synthesized, which could yield to secretion of aspartate or glutamate, although this has never been observed. Optimal splits at the Glc6P and PEP/Pyr nodes yield optimal lysine synthesis and theoretical product yields. Whether this is feasible depends on the kinetics and regulation of the enzymatic reactions affecting the split ratios at the preceding branch points. The extent to which they limit the production of lysine can be determined by introducing specific local perturbations, as shown next.

Analysis of Flux Perturbations at the Glc6P Branchpoint

To investigate the flexibility of the Glc6P branch point, two specific experimental perturbations were conducted. The first perturbation was the attenuation of Glc6P isomerase (GPI), the first enzyme of glycolysis, followed by fermentation and flux analysis of the resulting *C. glutamicum* mutant of ATCC strain 21253, NFG068. This mutant was isolated in an attempt to redirect metabolic flux into the pentosephosphate pathway. The second perturbation was the use of gluconate as the sole carbon energy source in the fermentation of *C. glutamicum* strain ATCC 21253. By supplying carbon directly into the pentosephosphate pathway, gluconate effectively bypasses the Glc6P branch point and provides additional information about the control exerted by the latter on lysine synthesis.

The original reference of Vallino and Stephanopoulos (1994a) should be consulted for details of the fermentation protocol and obtained results. Compared to a control fermentation with ATCC 21253, the GPI mutant exhibited attenuated specific production and consumption rates of all extracellular metabolites during growth. It also grew at a lower specific growth rate and respired at 50% of the original strain. A lower final biomass concentration was obtained. The final lysine titer was 25% greater than that for the control fermentation, and the instantaneous yield at the start of lysine production was 34% molar, compared to the 30% standard molar yield. The preceding changes could easily be interpreted as a departure from the standard fermentation. However, this would lead to erroneous conclusions, because upon closer examination of the metabolic flux map constructed from the data of this fermentation, it can be seen that differences from the base fermentation are minimal. Furthermore, the rise in product yield was not sustained, and the higher final product titer was the result of a longer run rather than a prolonged higher yield.

Table 10.3 shows the measured and estimated metabolite accumulation rates along with a calculated consistency index reflecting the degree of

TABLE 10.3 Measured and Estimated Metabolite Accumulation Rates and Standard Deviations (σ) for the NFG068 Fermentation at 34.3 h from Measurements Taken at 31 and 37.5 h[a]

	Accumulation rates [mmol $(L\ h)^{-1}$]	
Metabolites	Measured	Estimated[b]
Acetate	0 ± 1	0.02
Alanine	0 ± 1	0
Biomass	-0.17 ± 0.9	-0.15
CO_2	26.5 ± 2.7	26.2
Glucose	-9.0 ± 2.5	-8.7
Lactate	0 ± 1	0.03
Lysine	3.16 ± 0.2	3.16
NH_3	-5.5 ± 5.8	-6.3
O_2	-22.7 ± 2.3	-23.0
Pyruvate	0 ± 1	0
Trehalose	0.46 ± 1	0.57
Valine	0 ± 1	0.08

[a] Consistency index $h = 0.09$. Data are taken from Vallino and Stephanopoulos (1994a).
[b] Estimated rates are those that exactly satisfy mass balance constraints and are derived from the estimated fluxes.

closure of mass balance constraints. Compared with the control fermentation, metabolite accumulation rates for the NFG068 fermentation were
approximately 2 - 3 times smaller, largely a result of the reduced biomass
growth. The flux distribution map, Fig. 10.5, is remarkably similar to that
observed during the control fermentation under similar conditions. Consequently, no significant alteration of flux partitioning at the Glc6P branch
point resulted from a 90% attenuation of GPI activity in the NFG068
mutant, *even though the mutation produced significant alterations in overall
growth and production kinetics.* The mutation did not alter the instantaneous
lysine yield significantly, but reduced the specific flux through the network
during growth and the initial production phase. These results are consistent
with the notion of a dependent network that harbors a rigid branch point(s)
(see the following and Section 5.4).

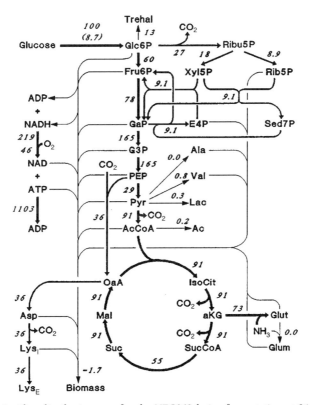

FIGURE 10.5   Flux distribution map for the NFG068 lysine fermentation at 34.3 h. Fluxes
were estimated from measurements taken at 31 and 37.5 h and normalized by the glucose
uptake rate [shown in parentheses, mmol $(L\ h)^{-1}$]. Adapted from Vallino and Stephanopoulos
(1994a). © 1994 American Chemical Society. Reprinted with permission.

To classify the Glc6P branch point as either flexible, weakly rigid, or strongly rigid (see Section 5.4), it is instructive to compare the actual results with the expected consequences of the perturbation for each of the preceding cases. The Glc6P branch point is weakly rigid if carbon flux is directed preferentially to the glycolytic pathway due to the higher affinity of GPI for Glc6P than that of Glc6P dehydrogenase (first enzyme in the PPP). If the lysine yield were constrained by a weakly rigid Glc6P branch point, then the GPI perturbation should have improved the yield, as the attenuation of GPI should have increased Glc6P and allowed more Glc6P to enter the PPP. This was not observed. To investigate the possibility that Glc6P might be strongly rigid (i.e., flux split ratios are not amenable to change due to strong regulation of both enzymes in the branch point), the second perturbation experiment was conducted using gluconate as carbon source. Gluconate essentially bypasses the Glc6P branch point and directly enters the pentosephosphate pathway. Consequently, if the Glc6P branch point is strongly rigid and limits the yield of lysine, cultivation of C. glutamicum ATCC 21253 on gluconate should improve lysine yield. In the absence of any improvements, the implication should be that Glc6P is indeed a flexible branch point and that it can respond to changing demands of the metabolic network by adapting its flux split ratio accordingly. Table 10.4 shows measured and estimated metabolite accumulation rates with gluconate as the carbon-energy source. It is noted that C. glutamicum ATCC 21253 did not grow well on gluconate alone; however, upon addition of glucose supplement, its growth rate was restored to normal (approximately 0.3 h^{-1}). After lysine production commenced, the data of Table 10.4 were obtained once all glucose had been depleted and gluconate was the only carbon source in the medium. Results showed that the instantaneous molar yield did not exceed 34%. Other notable features of the fermentation included a higher than normal respiratory quotient (1.35) and a complete lack of the typically observed fermentation byproducts. The lack of byproduct formation yielded a very satisfactory consistency index. Figure 10.6 shows the flux map obtained for these data. The following modifications were made to the metabolic network of Table 10.1 to represent the additional chemistry of gluconate: The glucokinase reaction was represented by

$$- \text{Glcn} - \text{ATP} + \text{Glcn6P} + \text{ADP} = 0 \qquad (10.15)$$

and the oxidative branch of the PPP was broken into two reactions,

$$- \text{Glc6P} - \text{H}_2\text{O} - \text{NADP} + \text{Glcn6P} + \text{NADPH} = 0 \qquad (10.16)$$

$$- \text{Glcn6P} - \text{NADP} + \text{Ribu5P} + \text{CO}_2 + \text{NADPH} = 0 \qquad (10.17)$$

TABLE 10.4 Metabolite Accumulation Rates in a Lysine
Fermentation with Gluconate as Carbon Source[a, b]

| Metabolites | Accumulation rates [mmol (L h)⁻¹] | |
	Measured	Estimated[b]
Acetate	0 ± 2	0
Alanine	0 ± 2	0.02
Biomass	1.67 ± 3.2	1.72
CO_2	62.4 ± 6.2	62.4
Glucose	-17.0 ± 3.9	-17.0
Lactate	0 ± 2	0
Lysine	5.48 ± 0.4	5.48
NH_3	-12.8 ± 1.1	-12.3
O_2	-48.0 ± 4.8	-48.0
Pyruvate	0 ± 2	0
Trehalose	0 ± 2	0.02
Valine	0 ± 2	0.03

[a] Consistency index $h = 0.003$. Data are for samples taken at
18 and 21 h into the fermentation. Data are taken from
Vallino and Stephanopoulos (1994a).
[b] Estimated rates are those that exactly satisfy mass balance
constraints and derived from the estimated fluxes.

where 6-phosphogluconolactonase has been lumped with G6PDH in reaction
(10.16).

When the flux distribution map was calculated from the rates of Table
10.4, reaction (10.16) exhibited a negative flux. This is caused by a significantly higher rate of NADPH production than needed for biomass and lysine
synthesis. However, reaction (10.16) cannot proceed in the reverse direction
due to a very large and positive standard free energy change that renders this
reaction essentially irreversible. In order to allow NADPH to reach a
pseudo-steady state, reaction (10.16) was deleted from the network and
replaced with a direct NADPH oxidation:

$$- 2NADPH - O_2 + 2H_2O + 2NADP = 0 \qquad (10.18)$$

It is noted that the preceding reaction simply represents NADPH depletion
by a simple oxidation. An alternative could be a transhydrogenase reaction
by which NADPH is oxidized to NADP with a simultaneous reduction of
NAD to $NADH^+$. At the time the preceding research was conducted, the
presence of transhydrogenase activity had not been confirmed, so that the
inclusion of reaction (10.18) allowed the NADPH pseudo-steady state and
metabolite material balances to be satisfied. With this modification, the flux

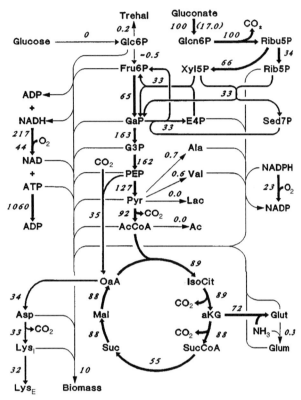

FIGURE 10.6 Metabolic flux distribution for the gluconate fermentation at 19 h. Fluxes were estimated from measurements taken at 18 and 21 h and normalized by the gluconate uptake rate [shown in parentheses, mmol (L h)$^{-1}$]. Note that the first enzyme of the PPP has been removed, and reactions for gluconokinase and direct oxidation of NADPH have been added to eliminate inconsistencies, per discussion in the text. Adapted from Vallino and Stephanopoulos (1994a). © 1994 American Chemical Society. Reprinted with permission.

distribution map was obtained as shown in Fig. 10.6. With the exception of the PPP, the flux distribution map is fairly similar to that of the control fermentation. The excess NADPH obtained from the use of gluconate did not increase the lysine yield. This, as suggested earlier, is a strong indication that the lysine yield is not limited by NADPH availability and, furthermore, Glc6P is not a rigid branch point. Combined with the results of the GPI flux analysis, it is concluded that the Glc6P branch point is indeed a flexible one, readily responding to meet the changing demand of NADPH for product and biomass production.

Although the discussion in this section has focused on the derivation of flux maps characteristic of constant physiological conditions over extended periods of time, it should be pointed out that intracellular flux maps can also be obtained during transients, so long as the steady state hypothesis for intracellular metabolites is not violated and on-line measurements are available for the rates of extracellular metabolites. Under these conditions, *on-line metabolic flux analysis* can be carried out, often yielding results of great value and significance.

Such an on-line metabolic flux analysis was carried out for a batch lysine fermentation by *C. glutamicum* (Takiguchi *et al.*, 1997), and Fig. 10.7 depicts the observed relationship between the calculated flux through the pentosephosphate pathway (r_7) and the lysine flux (r_8). Three important points should be noted:

- First, one can see two distinct physiological states, one of cell growth (state 1) and one of lysine production (state 2), along with the transient points obtained during the transition from state 1 to state 2.

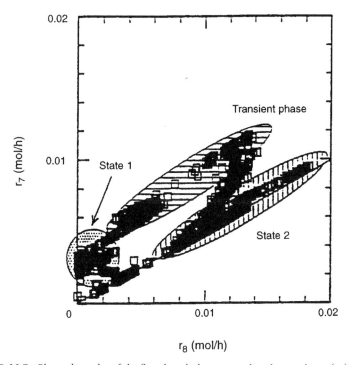

FIGURE 10.7   Phase plane plot of the flux though the pentosephosphate pathway (r_7) and the flux toward lysine (r_8). The data are from different phases of a batch fermentation. Adapted from Takiguchi *et al.* (1997). © 1997 John Wiley & Sons, Inc. Reprinted by permission of John Wiley & Sons, Inc.

- Second, the slope of the line in state 2 is equal to $1/2$, reflecting the stoichiometric need for NADPH for lysine synthesis. Referring to eq. (10.8), it can be seen that 2 mol of NADPH is needed for each mole of lysine, *under the assumption that a transhydrogenase activity is present*, in essence equating reducing equivalents in the form of NADH and NADPH. Absent THD activity, the slope of the line representing state 2 in Fig. 10.7 should have been equal to $1/4$ to reflect the new stoichiometry of lysine synthesis.
- Third, the linear correlation between the flux through the pentosephosphate pathway and lysine production is evidence of the ability of the PPP flux to self-adjust in order to meet the biosynthetic demand of lysine. This is an additional confirmation of the flexibility of the Glc6P branch point discussed in this section. It should be noted that none of this information can be obtained from assays of enzymatic activities, which remained rather flat during the experiment of Fig. 10.7.

Analysis of Flux Perturbations at the PEP/Pyr Branchpoint

We group PEP and pyruvate together for the simple reason that the contribution of the two anaplerotic reactions, PEP carboxylase and Pyr carboxylase, to the production of lysine has not been delineated yet. In the absence of such information, it is not possible to differentiate between the two carboxylation reactions, which are therefore grouped together as one. Pyruvate represents an important branch point because, in addition to the anaplerotic reaction, it also condenses in the dihydrodipicolinate synthase reaction with aspartic semialdehyde in the lysine pathway. Furthermore, it can form acetyl-CoA in the pyruvate dehydrogenase complex (PDC) reaction, which is further oxidized in the TCA cycle. If the PEP/Pyr branch point is weakly rigid, the flux of carbon preferentially is directed to the TCA cycle due to the much higher affinity of the PDC enzyme for pyruvate in comparison to the affinity for pyruvate of the PEP and Pyr carboxylating enzymes. In this case the TCA flux is relatively unaffected by flux changes in the lysine pathway, and product yield is limited by pyruvate availability. Therefore, if the yield of lysine suffers from a weakly rigid pyruvate branch point, then attenuation of PDC should increase the availability of pyruvate and improve lysine yield. If, on the other hand, poor lysine yield is due to strong rigidity of the PEP/Pyr branch point or very low anaplerotic activity, then PDC attenuation should result in either the excretion of some intermediate metabolite or overall network flux attenuation.

To examine the possibility that rigid flux partitioning at the PEP/Pyr branch point limits lysine yield, two perturbation experiments were conducted. The first experiment involved the isolation, fermentation, and flux analysis of a PDC-attenuated mutant of *C. glutamicum* ATCC 21253. The second experiment involved monitoring flux alterations at the PEP/Pyr branch point following inhibition of PDC activity by the addition of the specific inhibitor fluoropyruvate (FP) after the start of lysine overproduction.

The PDC-attenuated mutant exhibited difficulty growing in regular glucose medium, so potassium acetate was added to the culture. Growth stimulation by acetate addition is consistent with PDC attenuation for this strain. The addition of acetate did not complicate flux analysis because (a) all flux calculations were carried out after the acetate had been depleted and (b) the glyoxylate shunt is repressed in the presence of glucose. Growth and accumulation-depletion rates for all extracellular metabolites were severely attenuated compared to the control fermentation, but their time profiles were similar to those observed in the control. Specific rates were reduced to approximately one-half to one-third of those of the control fermentation. However, the resulting flux distribution map [provided in Vallino and Stephanopoulos (1994b)] is very similar to that of the control fermentation during the lysine production phase. Furthermore, the flux split ratio at the PEP/Pyr branch point is similar to that of the control. Therefore, a 98% attenuation of PDC activity did not change flux distributions at the principal branch points nor did it result in lysine yield improvement, although it did cause a significant flux attenuation of the overall network.

If lysine yield were limited solely by a weakly rigid PEP/Pyr branch point (*i.e.*, if carbon preferentially enters the TCA cycle so that lysine synthesis is, in essence, pyruvate-limited), then PDC attenuation should have increased pyruvate availability and lysine yield. As this was not observed, it can be concluded that lysine yield is not limited by a weakly rigid PEP/Pyr branch point.

As mentioned earlier, lysine biosynthesis occurs in a *dependent* network. In such a network, flux must be partitioned in unison at all principal branch points to obtain high yield. If one or more of the principal branch points in a dependent network are rigid or comprise reactions with very low activity, then flux partitioning cannot vary in unison to meet the stoichiometric needs of the product. Furthermore, if metabolic controls prevent the excretion or accumulation of intermediate metabolites, then the attenuation of any branch in a rigid node of a dependent network will result in overall flux attenuation. Hence, the observed results from the attenuation of PDC are consistent with a dependent network that harbors a branch point that is rigid or limited by the activity of one branch. Because the Glc6P branch point is flexible, the rigidity in the network must be due to the rigidity or limitation by one of the reactions of the PEP/Pyr branch point. This conclusion was confirmed further by an inhibition experiment using the fluoropyruvate (FB) inhibitor of PDC activity. Immediately following the addition of FP, a dramatic drop in respiration occurred accompanied by a decrease in the specific growth rate and an accumulation of extracellular pyruvate. Lysine synthesis and glucose consumption rates were unaffected. A few hours after FP addition, respiration resumed, excreted pyruvate was reconsumed, and glucose uptake exhibited a transient decrease probably due to pyruvate reconsumption. The temporary nature of the perturbation undoubtedly is due to the breakdown of FP, which can be metabolized after a short period of time. The remainder of the fermentation exhibited characteristics similar to those of the control fermentation. Fluxes were calculated in this case, and the resulting flux map illustrated the diversion of pyruvate from the TCA cycle to pyruvate excretion. The flux distribution in the remaining pathways was relatively unaffected compared to the control. These effects are illustrated in Fig. 10.8, which shows the flux distributions around the PEP/Pyr branch point for the control fermentation and the perturbed fermentation, respectively. Fluxes have been normalized with respect to the PEP synthesis rate. The inhibition of PDC by FP caused an approximate 50% reduction in the flux supported by PDC; however, flux distributions at the branch point remain relatively unaffected.

The fluoropyruvate inhibition experiment can be viewed as the transient analogue of the more permanent disruption of the TCA flux exhibited in the PDC-attenuated mutant. In the latter, glucose uptake, glycolytic and TCA fluxes were reduced to one-third to one-half of the wild-type magnitude, and no accummulation or secretion of intermediates was observed. Under fluoropyruvate inhibition, the normally high PDC flux was drastically reduced, forcing the secretion of pyruvate as an outlet for the glycolytic flux, which remained high following the addition of fluoropyruvate. *Changes in flux partitioning to favor the anaplerotic pathway were not observed in either perturbation.* As a result, the lysine yield was unaffected despite the significantly higher availability of pyruvate for carboxylation and condensation with ASA in the dihydrodipicolinate synthase reaction. It can be concluded that lysine yield is not limited by the preferential consumption of pyruvate in the TCA cycle, and, therefore, the PEP/Pyr branch point is not weakly rigid. Lysine production therefore is limited by either a strongly rigid PEP/Pyr branch point or low activity of the anaplerotic reaction(s). To increase lysine yield, one should focus on deregulating some of the reactions of the PEP/Pyr branch point (PDC, Pyr carboxylase, PEP carboxylase) or attempt to increase the flux of the anaplerotic reactions, most likely the one catalyzing pyruvate carboxylation.

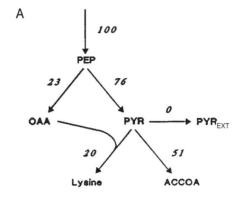

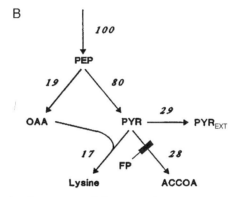

FIGURE 10.8   Flux distributions around the PEP and pyruvate principal branch points normalized by the PEP synthesis rate for (a) the control fermentation and (b) the fluoropyruvate (FP) perturbed fermentation at 13.5 h. The solid rectangle notes the site of FP inhibition and Pyr_{ext} represents extracellular pyruvate. Not all fluxes involving PEP, oxaloacetate, and/or pyruvate are illustrated. From Vallino and Stephanopoulos (1994b). © 1994 American Chemical Society. Reprinted with permission.

10.1.4. Metabolic Flux Analysis of Specific Deletion Mutants of *C. glutamicum*

Two issues of fundamental physiological and biotechnological significance were identified in the previous sections, namely, elucidation of the means of dissipating excess reducing equivalents generated by a hyperactive pentosephosphate pathway and determination of the exact anaplerotic route by which carbon is supplied for biomass and product synthesis. As these questions cannot be resolved with the usual wild-type strains, mutants with simplified genetic backgrounds were sought to facilitate the research. Specifically, PEP carboxylase (PPC) and pyruvate kinase (PK) were targeted, as these enzymes play key roles in the supply routes of carbon metabolites either for energy generation via the TCA cycle or for biosynthesis through the synthesis of the OAA precursor. Disruption mutants of combinations of the two enzymes were constructed via transconjugation and cultured in a standard batch fermentation, and results were evaluated in the framework of metabolic flux analysis (Park *et al.*, 1997a). The bioreaction network of Table 10.1 was augmented with the following five reactions to account for the additional possibility of pyruvate carboxylation (10.19), transhydrogenase activity (10.20), and three additional extracellular metabolites [propionate (PROP, 10.21), glyceraldehyde (GLY, 10.22), and dihydroxyacetone (DHA, 10.23)] that were found to accumulate in the fermentation medium of the mutants described:

$$- PYR - CO_2 - ATP + OAA + ADP = 0 \qquad (10.19)$$

$$- NADPH - NAD + NADP + NADH = 0 \qquad (10.20)$$

$$- SUC + PROP + CO_2 = 0 \qquad (10.21)$$

$$- G3P - ADP + GLY + ATP = 0 \qquad (10.22)$$

$$- G3P - ADP + DHA + ATP = 0 \qquad (10.23)$$

Experimental results are summarized in Table 10.5, showing the yields of cell mass and lysine on glucose and the specific rates of growth, glucose

TABLE 10.5   Comparison of Fermentation Results of *C. glutamicum* ATCC21253[a] and Deletion Mutants Δppc, Δpyk, and $\Delta ppc \Delta pyk$ in Two Distinct Phases[b]

	Phase I				Phase II			
Parameter	wt	Δppc	Δpyk	$\Delta ppc \Delta pyk$	wt	Δppc	Δpyr	$\Delta ppc \Delta pyr$
μ	0.35	0.26	0.27	0.12				
Y_{sx}	0.54	0.59	0.50	0.36				
r_s	0.59	0.59	0.53	0.33	0.35	0.29	0.25	0.11
r_p	0	0	0	0	0.08	0.11	0.04	0.02
Y_{sp}	0	0	0	0	0.25	0.27	0.17	0.21

[a] Wild-type strain designated wt.
[b] Specific growth rate, specific rates, and yields are given in h^{-1}, g(g DW h)$^{-1}$, and g (g glucose consumed)$^{-1}$, respectively.

uptake, and lysine production for the four strains during the phases of pure growth (I), and no growth and high lysine production (II).

Compensation of PPC by Pyruvate Carboxylation in *C. glutamicum*

The results of Table 10.5 should be evaluated in the light of Fig. 10.9, summarizing the possible anaplerotic routes leading to the formation of OAA. The fact that the disruption of the *ppc* gene leaves growth and lysine production essentially unaffected indicates that an additional anaplerotic pathway(s) exists in *C. glutamicum*. Potentially, there may be two alternative

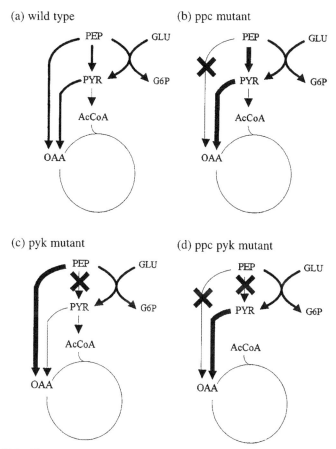

FIGURE 10.9 Flux maps around the PEP and pyruvate nodes in different mutants. The thickness of the arrows indicates the magnitude of the fluxes. From Park *et al.* (1997a). © 1997 John Wiley & Sons, Inc. Reprinted by permission of John Wiley & Sons, Inc.

pathways for the formation of OAA: (1) one via carboxylation of PEP by the action of enzymes other than PPC, such as PEP carboxykinase, PEP carboxy-transphosphorylase, or transcarboxylase, and (2) via carboxylation of pyruvate.

The presence of an alternative PEP carboxylating pathway is not supported by the results obtained with the *pyk* and *ppc pyk* mutants. If such an alternative PEP carboxylating pathway truly existed, then blockage of PEP conversion to pyruvate in the *pyk* mutant should result in PEP accumulation and increased OAA (hence lysine) synthesis (note that there is sufficient pyruvate from the PTS transport of glucose for condensation with aspartyl semialdehyde). On the contrary, lysine productivity and yield are reduced by approximately 50% and 30%, respectively, consistent with the viewpoint that pyruvate carboxylation is the main anaplerotic pathway: As pyruvate availability is reduced in the *pyk* mutant, anaplerotic flux and lysine productivity decline.

Similarly, if PEP carboxylation was the main anaplerotic pathway, a decline in lysine production should be observed in the case of the *ppc* mutant. This, however, is not supported by the data of Table 10.5. Finally, the hypothesis that pyruvate carboxylation is the main anaplerotic route in this *C. glutamicum* strain is corroborated by the attenuation of both growth and lysine production observed with the *ppc pyk* double mutant. In this case the only route of pyruvate formation is via the PEP:glucose PTS, which produces one molecule of pyruvate for every molecule of glucose imported. Here there is, obviously, less pyruvate available for OAA synthesis via pyruvate carboxylation, as well as for acetyl-CoA synthesis via pyruvate dehydrogenase. This limits the generation of both energy and biosynthetic precursors. This effect is more severe during the lysine production phase, as some pyruvate must also be drained off in the condensation step with aspartate semialdehyde to synthesize lysine. Consequently, these data provide strong support for the presence of a pyruvate carboxylating activity as the main anaplerotic pathway in these strains.

Discovery of Transhydrogenase Activity

There are 40 metabolites participating in a total of 39 reactions in this network. This means that there are 39 unknown fluxes and 40 equations, which is sufficient, in principle, to provide a unique solution to the system along with one redundancy for testing the overall consistency of the biochemical assumptions and fermentation measurements. Before, however, one attempts to obtain a solution, it must first be established that the stoichiometric matrix is nonsingular and that a unique solution exists. If, for example, both a PEP and a pyruvate carboxylating reaction are present in the network, matrix G is singular, implying that the PEP carboxylation flux cannot be determined independently from the pyruvate carboxylation flux via extracellular measurements alone. This is true when at least one of the THD or PYK activities is present. However, in the case of a strain deficient in both THD and PYK activities, matrix G becomes nonsingular and PEP and pyruvate carboxylation fluxes *can be determined independently!*

There are, obviously, many possibilities to be considered, depending on the genetic background of a strain. Table 10.6 summarizes the outcome regarding the nature of matrix G for the 16 cases corresponding to all possible permutations of the four key enzymes, PPC, PC, THD, and PYK. Along with matrix singularity, a row for the biochemical feasibility of each combination is provided: Because no strain can grow in the absence of both anaplerotic reactions, strains deficient in both PPC and PC activities are marked infeasible.

For each mutant strain, the stoichiometric matrix G must be reconfigured to reflect the corresponding genetic background and enzymatic data. The wild-type strain reflecting the presence of PYK activity, while allowing flexibility with respect to the exact nature of the anaplerotic reaction, can be described by any of G_1, G_2, G_5, G_6, G_9, or G_{10}. G_1, G_2, and G_5 are singular, and therefore only cases G_6, G_9, and G_{10} can be solved for the intracellular fluxes. Results from matrix G_9 suggest a very high THD flux similar in magnitude to that of glycolysis. As such a high flux was not supported by the measured THD activity, this case was rejected. Discrimination between the possibilities of PEP (G_6) and pyruvate (G_{10}) carboxylation

TABLE 10.6 Properties of Stoichiometric Matrix G for Various Pathway Configurations[a]

Reaction	1	2	3	4	5	6	7	8	9	10	11	12	13	14	15	16
36 (PPC)	+	+	+	+	+	+	+	+	-	-	-	-	-	-	-	-
37 (PC)	+	+	+	+	-	-	-	-	+	+	+	+	-	-	-	-
38 (PYK)	+	+	-	-	+	+	-	-	+	+	-	-	+	+	-	-
39 (THD)	+	-	+	-	+	-	+	-	+	-	+	-	+	-	+	-
Matrix singularity	S	S	S	N	S	N	N	N	N	N	N	N	N	N	N	N
Biochemical feasibility	F	F	F	F	F	F	F	F	F	F	F	F	I	I	I	I

[a] Sixteen cases are considered. Each case consists of a basic reaction set [comprising reactions (10.21)-(10.23) and all reactions of Table 10.1 except PYK and PPC] and additional reactions indicated in each column. Reaction 36 represents the enzymatic step for PEP carboxylase, reaction 37 that for pyruvate carboxylase, reaction 38 that for pyruvate kinase, and reaction 39 that for transhydrogenase. Abbreviations: +, presence of reaction; -, absence of reaction; S, matrix G is singular; N, matrix G is nonsingular; F, pathway is biochemically feasible; I, pathway is biochemically infeasible.

pathways is not possible based on the flux analysis because both lead to the same error estimation. In fact, both backgrounds give identical flux solution except that the anaplerotic flux is through PPC, or PYK and PC, respectively. This means that any linear combination of solutions based on G_6 and G_{10} is likely, namely, carbon flow through both PEP carboxylation and pyruvate carboxylation is possible.

For the ppc mutant, we focus on the cases where a pyruvate carboxylating activity is present (G_9-G_{12}), based on the preceding results and further support provided by ^{13}C- labeling studies that indicate that this pathway is operative in $vivo$ (Park et $al.$, 1997b). Of the four possibilities represented by matrices G_9-G_{12}, G_{11} and G_{12} are rejected as they correspond to no PYK activity, in contrast to what was measured. Between the remaining two possibilities, G_9 and G_{10}, G_{10} is most likely on the basis of its significantly higher value of the overall consistency index. This conclusion, namely, small, if any, THD activity for the ppc mutant, is supported by the in $vitro$ activity measurements (Table 10.7; Park et $al.$, 1997a).

In the case of the pyk mutant, the following stoichiometries were analyzed (note that G_3 is singular and G_{15} and G_{16} are not feasible): G_4, G_7, G_8, G_{11}, and G_{12}. Cases G_8 and G_{12} were rejected for not satisfying the redundant equations and yielding very high values for the consistency index. The solution of G_{11} was also rejected as it resulted in negative TCA cycle fluxes. Likewise, the solution based on G_4 was unlikely as it produced a negative flux through the pyruvate carboxylation pathway. Consequently, G_7 is the most likely stoichiometry in this mutant. It should be noted that a high flux via the THD step is also supported by the enzymatic data (Table 10.7). The fermentation data of the ppc pyk mutant were analyzed in a manner similar to those of the pyk mutant. Similarly, the estimation based on G_{11} is the most likely solution for this mutant.

In order to experimentally corroborate the functioning of THD as suggested by the metabolic flux analysis, the activity of THD was measured in crude extracts for the four strains, which were grown with PMB medium in

TABLE 10.7   Activities [nmol (g protein min)$^{-1}$] of Selected Enzymes in C. $glutamicum$ ATCC 21253 (Wild Type) and Three Deletion Mutants

Strain	Pyruvate kinase	PEP carboxylase	Transhydrogenase
Wild type	822	344	220
Δppc	677	n.d.	150
Δpyk	n.d.[a]	279	310
$\Delta ppc \Delta pyk$	n.d.	n.d.	490

[a] n.d., not detected.

1-L shake flasks and harvested at midexponential stage. The data are reported in column 4 of Table 10.7. Significantly higher activities were detected for the *pyk* and *ppc pyk* mutants than for the other two strains, as also suggested by the intracellular flux analysis, which showed increased PPP flux and NADPH synthesis rate.

It should be noted that the presence of THD was suggested by the failure of flux consistency tests that also pointed to NADPH accumulation as the most likely cause of the observed inconsistency. Additional observations resulting from flux analysis are the significant reduction in glycolytic fluxes and the concomitant increase in the PPP flux in the *pyk* mutants and, in particular, the *ppc pyk* mutant. Fluxes through glucose-6-phosphate isomerase are in the reverse direction for the preceding mutants, which also exhibit reduced anaplerotic fluxes. Finally, specific glucose transport is reduced in the course of the batch, following a similar trend in ATP requirements. The latter, however, were found to be quite similar for the four strains, pointing to relatively *invariant specific metabolic energy requirements for the four genetic backgrounds*. Figure 10.9 summarizes the results of the analysis in this section by listing the most likely pathways for the three mutants.

10.2. METABOLIC FLUXES IN MAMMALIAN CELL CULTURES

Our second example applies metabolic flux analysis to the pathways of energy metabolism in mammalian cell cultures. Cell culture has emerged as an important technology for the manufacturing of complex pharmaceutical proteins. After the first wave of recombinant products which were simple, nonglycosylated proteins expressed intracellularly in prokaryotic cells (human growth hormone, insulin, α-interferon), the second generation of biotechnology products has focused predominantly on the production of active, glycosylated, complex glycoproteins secreted in the medium of mammalian cells. Chinese hamster ovary cells (CHO), BHK cells, and hybridoma cells for the production of monoclonal antibodies are the most common systems. Presently, a number of biopharmaceuticals produced by mammalian cells are on the market, such as erythropoietin, Factor VIII, tissue plasminogen activator, and granulocyte-colony stimulating factor (G-CSF). Many more are at various stages of the regulatory approval process and are expected to be introduced into the market in the near future. Mammalian cells are the means for manufacturing these molecules, and, as a result, their importance has increased dramatically during the past decade.

As soon as it became apparent that mammalian cells would be employed in the production of a broad spectrum of complex glycoproteins, intense efforts were undertaken to improve the density and productivity of these systems. Mammalian cells are much more fragile than microbial cells and more demanding in their needs for growth factors and nutrients. A main obstacle in achieving high productivities in mammalian cell cultures has been the low cell densities attainable in these systems. The primary reason for that, especially in batch culture operations, has been the accumulation of inhibitory byproducts of glucose and glutamine catabolism, such as lactate and ammonia, during cell growth. Reduction of the amounts of these byproducts in the medium has contributed to higher cell densities, with commensurate increases in product accumulation rates. As the preceding byproducts are directly related to central carbon metabolism, it is natural to carry out metabolic flux analysis and determine the distributions of primary carbohydrate and amino acid fluxes leading to byproduct formation.

Frequently, an additional problem in cell culture is low cell viability, especially at high cell densities. This is due to the induction of programmed cell death (apoptosis), a process that has been correlated extensively in industrial and laboratory observations with the uptake of lactate and the secretion of alanine and other metabolic products directly related to energy metabolism. Finally, numerous observations point to the importance of glucose availability as a determining factor of the quality of secreted products, particularly, as quality relates to glycosylation site occupancy. The availability of sugars for the glycosylation reactions is an issue that can be elucidated by analyzing carbon flux distribution at the glucose-6-phosphate branch point. Glc6P is the main entry point into the pentosephosphate pathway that supplies the carbon precursors for the production of fully glycosylated proteins. Determination of intracellular fluxes, especially when combined with intracellular metabolite concentrations and enzymatic activities, can improve our understanding of protein glycosylation in these cells.

10.2.1. DETERMINATION OF INTRACELLULAR FLUXES

In this section, we summarize work carried out on the analysis of metabolic fluxes in the hybridoma cell line, ATCC CRL 1606, producing an immunoglobulin G (IgG) against human fibronectin. We will show how these fluxes are estimated from cell culture data and how they can be validated further by comparing predicted isotope enrichment with the measured isotope distributions in secreted lactic acid following the introduction of

^{13}C-labeled glucose in the cell culture medium. The cells were grown in a 2-L well-stirred Celligen bioreactor using standard Dulbecco's Modified Eagle's Medium (DMEM). The original papers (Zupke and Stephanopoulos, 1994; Zupke et al., 1995) can be consulted for further details on the experimental protocols. The metabolites measured in the cell culture experiments are listed in Table 10.8 for two different experiments. In experiment 1, labeled glucose was added at a cell density of 4.8×10^5 cells mL^{-1}, and lactate was allowed to accumulate for 20 h. In experiment 2, glucose was added at a cell density 6.8×10^5 cells mL^{-1}, and lactate was allowed to accumulate for 10 h. The two experiments are represented in the two columns of Table 10.8.

The specific rate of oxygen uptake, r_{O2} [mmol (cell h)$^{-1}$], was measured dynamically by flushing the head space with nitrogen and monitoring the decline of dissolved oxygen concentration (c_o) from 60% to approximately 20%. The following equation was used for the determination of the specific oxygen uptake rate using independently measured values for the mass transfer coefficient, $k_L a$:

$$\frac{dc_o}{dt} = k_L a(c_o^* - c_o) - r_o x \tag{10.24}$$

where c_o^* is the dissolved oxygen concentration at equilibrium with the gas phase and x is the cell concentration in the bioreactor.

Measurement of the total CO_2 production rate in mammalian cell cultures is complicated by the presence of bicarbonate in the medium for pH control. Due to the CO_2/bicarbonate interaction, there may be loss or accumulation of CO_2 in the aqueous phase, which is not detected by the measurement of

TABLE 10.8   Measured Rates in Hybridoma Cell Culture Used for Flux Estimations[a]

Metabolite	Experiment 1	Experiment 2
Glucose	-2.92 ± 0.16	-3.9 ± 0.19
Alanine	0.31 ± 0.03	0.31 ± 0.03
Lactate	6.2 ± 0.32	7.03 ± 0.35
Glutamine	-0.86 ± 0.09	-1.1 ± 0.1
NH$_3$	0.73 ± 0.07	-0.70 ± 0.07
Biomass	3.71 ± 0.19	3.97 ± 0.20
O$_2$	-2.0 ± 0.2	-2.0 ± 0.2
CO$_2$	2.6 ± 0.3	2.6 ± 0.3
IgG	0.96 ± 0.1	0.99 ± 0.1

[a] All rates are in 10^{-10} mmol (cell h)$^{-1}$. Negative rates signify consumption. The units for biomass and IgG are in C- and N-molar, respectively.

the exit gas CO_2 concentration. Under conditions typically found in cell culture media, bicarbonate formation is favored; however, there may be stripping of CO_2 under high aeration rates. To properly account for the total CO_2 production rate (CPR), the aqueous CO_2 accumulation rate (CAR) must be added to the gaseous CO_2 evolution rate (CER). The CO_2 evolution rate was measured as $F(c_{out} - c_{in})$, where F is the gas flow rate and c_{in} and c_{out} are the concentrations of CO_2 in the inlet and outlet streams of the bioreactor gases, respectively. By using a bottled gas without CO_2 (i.e., $c_{in} = 0$), the outlet concentration was measured by bubbling the exhaust gas through a saturated $CaCO_3$ solution and measuring the pH of the solution. At equilibrium, the pH of the $CaCO_3$ solution is a linear correlation of the CO_2 concentration in the gas phase. Liquid phase CO_2 contents were measured by storing cell-free samples at 4°C and measuring the total CO_2/bicarbonate content enzymatically using assay kits. It was found that the aqueous CO_2 content over the course of the hybridoma batch culture (approximately 3.5 days) varied significantly, first decreasing due to the stripping effect by the aeration gases and then increasing due to the continuous accumulation of CO_2 from the respiration of the cells. The specific rates of production of extracellular metabolites were determined by measuring extracellular metabolite concentrations over relatively brief periods (10-15 h), so that culture conditions and cellular metabolism did not change significantly, and by fitting a least squares equation to the concentration data according to the following equation:

$$c_s = c_{s,0} + \frac{r_s x_0}{\mu(e^{\mu t} - 1)} \tag{10.25}$$

where c_s is the measured concentration, $c_{s,0}$ is the initial concentration of the metabolite (i.e., the concentration at the beginning of the 10-15 h measurement period), r_s is the specific uptake or consumption rate of the metabolite, x_0 is the initial cell concentration, and μ is the specific cell growth rate. The preceding equation was modified for the case of glutamine to account for the spontaneous decomposition of glutamine to form ammonia.

The primary carbon and energy sources for mammalian cells in submerged cultures are glucose and glutamine. Glutamine also serves as the main nitrogen source. Other amino acids besides glutamine are consumed, and they can approach 50% of glutamine consumption in hybridomas. However, more than 80% of that consumption is for protein synthesis; therefore, amino acids other than glutamine contribute very little to energy production. Additional components such as vitamins, trace minerals, and growth factors are significant for cell growth, but contribute very little to energy metabolism. The major products of glucose and glutamine metabolism are biomass, secreted protein in the form of antibody product, energy, reducing power for biosynthesis, carbon dioxide, and the waste products lactic acid and ammonia. Additionally, aspartic acid, alanine, and glutamic acid may be secreted depending on the cell line and the growth conditions. Contrary to normal tissue, cells in culture tend to exhibit high rates of aerobic glycolysis. This is the result of glucose metabolism via glycolysis to pyruvate. Glucose can also be metabolized in the pentosephosphate pathway to produce NADPH and biosynthetic intermediates. Pyruvate can be converted to lactic acid that is secreted, oxidized in the TCA cycle, or transaminated to form alanine. Additionally, pyruvate can be carboxylated through the anaplerotic pathway or back synthesized by the malic enzyme reaction. In many transformed and tumor cells, the production of cholesterol is elevated and the alterations in membrane composition may contribute to the altered metabolism they usually exhibit. However, CRL 1606 typically converts over 90% of consumed glucose to lactate, indicating that there is no significant diversion of carbon from AcCoA to cholesterol synthesis. The synthesis of lipids is taken into account in the biomass synthesis equation, but not in the calculation of ^{13}C distribution in lactate. Glutamine can be incorporated into biomass or oxidized via the TCA cycle. Glutamine has been found to be the major source of energy in many cell lines. For CRL 1606 hybridomas, glucose and glutamine are the major energy sources, with ammonia, lactic acid, and alanine as the major products.

The biochemistry of energy metabolism in mammalian cells is depicted in Fig. 10.10 and summarized in Table 10.9. By lumping intermediates appearing in reaction chains, the network has been simplified. The pathways provide the energy (ATP), reducing power (NADPH), and precursors for the synthesis of biomass and antibody. A lumped equation is also provided for the synthesis of biomass and antibody following the procedure illustrated in Section 2.5, *i.e.*, these equations were derived by using reactions for macromolecule synthesis from precursors present in the biochemical network, combined with an equation specifying the macromolecular composition of the cells. A typical protein composition was assumed for the antibody secreted by these cells, as indicated in Table 2.8.

By using rate data from two different experiments (described in more detail in the next section and shown in Table 10.8), the intracellular fluxes of Table 10.10 were obtained. Random measurement errors were sufficient to explain discrepancies in the data with a 95% confidence level. For the three redundant measurements present in this study, a consistency index (CI) of less than 7.81 was calculated. It is emphasized again that the consistency index is very useful for identifying erroneous measurements. In the present study, the CI identified the problem with the measurement of CPR. It was

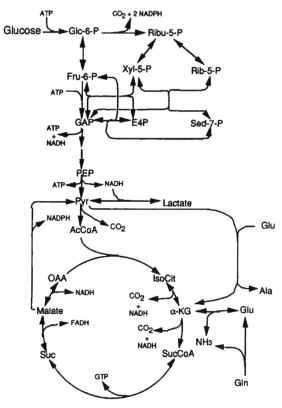

FIGURE 10.10 The biochemistry of energy metabolism in hybridomas. The pathway consists of glycolysis, glutaminolysis, the pentosephosphate pathway, and the TCA cycle. The malic enzyme can operate in either direction as an anaplerotic reaction or for pyruvate synthesis.

found that the use of CPR values obtained directly from gaseous CO_2 measurements of CER yielded significant inconsistencies, which were eliminated by removing CER from the list of measured rates. This pointed to CER as a source of gross measurement errors, which was indeed found to be the case due to the interaction between aqueous and gas phase CO_2. The error was corrected by including the CO_2 accumulation rate in the determination of the true CPR, as pointed out earlier.

Intracellular fluxes were determined at different levels of dissolved oxygen and were used to evaluate oxygen effects on cellular metabolism. Table 10.11 summarizes the most important fluxes and their uncertainties. Pyruvate flux to the TCA cycle is significant at normal DO, but only slightly positive at 1% DO and essentially zero under oxygen limitation. The flux through alanine

TABLE 10.9   Set of Reactions Describing Energy Metabolism in Hybridoma Cells[a]

---

1. Glc + ATP → Glc-6-P + ADP
2. Glc-P → Fru-6-P
3. Fru-6-P + ATP → 2GAP + ADP
4. GAP + 2ADP + NAD → Pyruvate + NADH + 2ATP
5. Pyruvate + NADH → Lactate + NAD
6. Glc-6-P + 2NADP → Ribulose-5-P + 2NADPH + CO_2
7. Ribulose-5-P → Xylulose-5-P
8. Ribulose-5-P + Xylulose-5-P → Sed-7-P + GAP
9. Sed-7-P + GAP → Fru-6-P + Erythrose-4-P
10. Xylulose-5-P + Erythrose-4-P → Fru-6-P + GAP
11. Pyruvate + Malate + 3NAD → α-KG + 3NADH + $2CO_2$
12. α-KG + 2NAD + ADP → 2NADH + CO_2 + ATP + Malate
13. Glutamine → Glutamate + NH_3
14. Pyruvate + Glutamate → Alanine + α-KG
15. Glutamate + NAD → α-KG + NADH + NH_3
16. 2NADH + 6ADP + O_2 → 2NAD + 6ATP
17. Malate + NADP → Pyruvate + CO_2 + NADPH
18. ATP → ADP
19. 0.036Glutamine + 0.062Glutamate + 0.0031Glc-6-P + 0.013Ribulose-5-P + 0.042GAP
    + 0.123Pyruvate + 0.019Malate + 0.74ATP + 0.188NADPH + 0.23NAD → Biomass +
    0.073α-KG + 0.23NADH + $0.12CO_2$ + 0.74ADP + 0.188NADP
20. 0.012Glutamine + 0.07Glutamate + 0.04GAP + 0.011NADPH + 0.012Malate +
    0.082NAD → Antibody + 0.082NADH + 0.057α-KG + 0.011NADP

---

[a] In reaction 12, part of the TCA cycle, FAD has been replaced with NAD and the GTP produced has been omitted. Because one NADH can generate three ATP via the electron transport chain and $FADH_2$ can generate two, the correct number of ATP equivalents is produced. Abbreviations: FAD, flavine adenine dinucleotide; NAD, nicotinamide adenine dinucleotide; GTP, guanosine 5'-triphosphate; ADP, adenosine 5'-diphosphate; GAP, glyceraldehyde-3-phosphate; α-KG, α-ketoglutarate.

aminotransferase is roughly constant at all DO levels. The flux through glutamate dehydrogenase is positive at DO of 60% and negative at low DO. A number of interesting quantities can also be calculated from the fluxes, as shown in Table 10.11. Total ATP produced decreases but not significantly, from 1.95 to 1.58 pmol/cell h, when DO is reduced. This is reminiscent of the result obtained with microbial cells, according to which specific ATP production seems to be fairly independent of culture conditions. This amount of ATP increasingly is derived from glucose as dissolved oxygen is reduced and parallels the fraction of ATP derived from glycolysis, which increased from 0.34 to 0.69. The results obtained also suggest that dissolved oxygen has a significant effect on nitrogen metabolism and the flux through glutamate dehydrogenase, which reverses at low DO to favor glutamate formation. This was something that was made clear through flux analysis and points to

TABLE 10.10   Flux Estimates in Hybridoma Cultures from Measured Rates
$[10^{-10}$ mmol (cell h)$^{-1}]^a$

Reaction	Experiment 1	Experiment 2	Reaction	Experiment 1	Experiment 2
1	3.18 ± 0.11	3.91 ± 0.13	11	0.211 ± 0.060	0.203 ± 0.060
2	2.85 ± 0.11	3.54 ± 0.13	12	0.922 ± 0.063	0.925 ± 0.063
3	3.03 ± 0.11	3.74 ± 0.13	13	0.683 ± 0.036	0.691 ± 0.036
4	5.96 ± 0.22	7.38 ± 0.26	14	0.314 ± 0.031	0.318 ± 0.031
5	5.61 ± 0.22	7.00 ± 0.26	15	0.073 ± 0.039	0.054 ± 0.038
6	0.315 ± 0.028	0.355 ± 0.029	16	1.88 ± 0.17	1.92 ± 0.17
7	0.178 ± 0.018	0.202 ± 0.019	17	0.629 ± 0.036	0.634 ± 0.037
8	0.089 ± 0.009	0.101 ± 0.010	18	15.2 ± 1.1	16.6 ± 1.2
9	0.089 ± 0.010	0.101 ± 0.010	19	3.69 ± 0.18	4.01 ± 0.20
10	0.090 ± 0.010	0.101 ± 0.010	20	0.961 ± 0.096	0.999 ± 0.099

^{a} Reaction numbers refer to the biochemical reactions of Table 10.9. The data are taken from
Zupke and Stephanopoulos (1995).

TABLE 10.11   Selected Flux Estimates [mol (cell h)$^{-1}$] Calculated from Measured
Rates of Change of Extracellular Metabolites in Hybridoma Cultures^a

	Flux		
Reaction/quantity	DO = 0%	DO = 1%	DO = 60%
Pyruvate into TCA cycle	0.0 ± 0.3	0.28 ± 0.3	0.94 ± 0.28
Pyruvate to alanine	0.33 ± 0.3	0.26 ± 0.2	0.33 ± 0.3
Glu to α-ketoglutarate via GDH	-0.29 ± 0.3	-0.14 ± 0.2	0.4 ± 0.3
Malate to pyruvate	-0.24 ± 0.3	0.41 ± 0.3	0.59 ± 0.4
Total ATP [pmol (cell h)$^{-1}$]	1.58	1.67	1.95
Fraction of ATP from glycolysis	0.69	0.5	0.34
Fraction of ATP from glucose	0.85	0.75	0.67
Fraction of lactate from glucose	0.99	0.98	0.96
Fraction of ATP for maintenance	0.29	0.46	0.62

^{a} Total ATP production is also shown along with the contribution of various pathways and
unaccounted ATP amount dissipated by maintenance reactions. The data are taken from Zupke
et al (1995).

the fundamental role of intracellular oxidation/reduction in cell physiology
and the possibility of controlling physiological processes through DO modulation.

10.2.2. VALIDATION OF FLUX ESTIMATES BY ^{13}C LABELING STUDIES

In order to validate the flux estimates obtained by material balancing,
^{13}C-labeled glucose was used as a special probe in the cell culture medium.
Similar to the approach employed by Walsh and Koshland (1984) in investi-

gating the glyoxylate shunt in *E. coli*, the measured isotope distribution in lactate was compared to that predicted from the intracellular flux estimates. It is noted again that for a given set of intracellular fluxes (determined, for example, by material balancing), and a pathway for carbon atom mapping from reactants to products, steady state isotope distributions can be determined for all the metabolites in the metabolic network. It is important to select labeling substrates that can discriminate among metabolic fluxes. It was found that if glucose is used as the labeling compound, the degree of ^{13}C enrichment at the three carbon positions of secreted lactic acid exhibits significant dependence on the intracellular fluxes. Specifically, the isotope enrichment distribution in secreted lactic acid is a function of two flux ratios: f_1, the fraction of pyruvate produced from glycolysis, and f_2, the fraction of α-ketoglutarate derived from glutamate.

The labeling experiment was conducted by growing cells on unlabeled glucose for a brief period of time and then, once exponential growth was established, adding 1 g of 99% $[1-^{13}C]$ glucose to the bioreactor. The medium was collected after 10 or 20 h, and cells were removed by centrifugation. The supernatant was concentrated by lyophilization and analyzed via NMR. NMR data were obtained for the degree of enrichment of the three carbon atoms of lactate. They are expressed as the ratios of C1:C2 and C3:C1 enrichment and summarized in Table 10.12. Isotope enrichment ratios were used in the comparison with predicted enrichment data in order to minimize errors due to baseline variation.

TABLE 10.12   Isotope Enrichment Data for Two Labeling Experiments Conducted To Validate the Hybridoma Metabolic Fluxes Determined by Material Balancing Techniques[a]

Quantity	Experiment 1	Experiment 2
^{13}C ratio, C1:C2 of lactate[b]	0.98 ± 0.01	0.98 ± 0.01
^{13}C ratio, C3:C1 of lactate[c]	3.5 ± 0.11	3.0 ± 0.09
Fraction of unlabeled lactate	0.39 ± 0.02	0.55 ± 0.03
Fraction of labeled glucose	0.115 ± 0.006	0.117 ± 0.006
Fraction of pyruvate from GAP (f_1)	0.905 ± 0.006	0.921 ± 0.005
Fraction of α-KG from glutamine (f_2)	0.77 ± 0.05	0.78 ± 0.05

[a] The areas of the three lactate ^{13}C peaks were quantified and used to calculate the enrichment ratios. The enrichment levels of glucose and lactate were determined by measurement of concentrations before and after the addition of labeled glucose. The flux ratios f_1 and f_2 were estimated from the network stoichiometry and measurements of metabolite rates of change. The data are taken from Zupke *et al.* (1995).

[b] The probability of error in the estimation of the C1:C2 ratio was 16% and 0.5% for the two experiments, respectively.

[c] The probability of error in the estimation of the C3:C1 ratio was 5% and 1% for the two experiments, respectively.

The experimental isotope enrichments can be compared to those calculated from the intracellular fluxes by material balancing on the secreted metabolites. In general, the two flux ratios f_1 and f_2 are sufficient to determine the ^{13}C enrichment of lactate. The dependence of ^{13}C lactate enrichment ratios on the two flux ratios is shown in Figs. 10.11 and 10.12. These figures show calculated contours of constant enrichment ratio as a function of the flux ratios f_1 and f_2. A single isotope ratio restricts the flux ratio to a single contour line, but does not uniquely determine f_1 and f_2. Determination of the two fluxes requires the measurement of two independent isotope ratios, and then f_1 and f_2 can be determined at the intersection of two such contours of the families shown in Figs. 10.11 and 10.12. In reverse order, knowledge of the flux ratios f_1 and f_2 allows the determination of the isotopic ratios C1:C2 and C3:C1, through the use of the contour plots of Figs. 10.11 and 10.12 and their direct comparison to experimentally determined values of the isotope ratios. Therefore, measurements of isotope enrichment in secreted lactate can be used either to validate intracellular flux estimates obtained by other means or to directly determine intracellular flux ratios at the intersection of the contours of Fig. 10.11 with the contours of Fig. 10.12. It is noted that, although intracellular flux validation can be

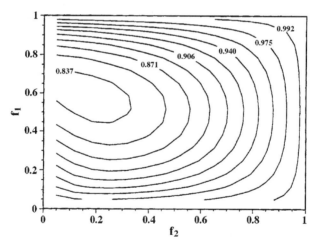

FIGURE 10.11 Contours of constant C1:C2 ratio in lactate as function of f_1 and f_2 for hybridoma culture experiment 1, where C1 and C2 are the amounts of ^{13}C present in carbons 1 and 2, respectively, of lactate. The flux ratio f_1 is the fraction of pyruvate that is produced from glycolysis, and f_2 is the fraction of α-ketoglutarate derived from glutamate. The contours were calculated by computer simulation of the biochemical network using the experimental conditions of 11.5% of glucose labeled with ^{13}C in the 1 position and 39% of the accumulated lactate not enriched. Adapted from Zupke and Stephanopoulos (1994). © John Wiley and Sons, Inc. Reprinted by permission of John Wiley & Sons, Inc.

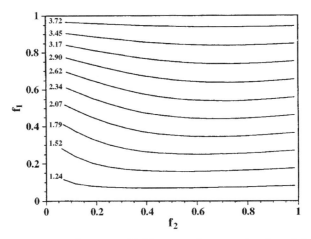

FIGURE 10.12   Contours of constant C3:C1 ratio in lactate as function of f_1 and f_2 for hybridoma culture experiment 1, where C3 and C1 are the amounts of ^{13}C present in carbons 3 and 1, respectively, of lactate. The flux ratio f_1 is the fraction of pyruvate that is produced from glycolysis, and f_2 is the fraction of α-ketoglutarate derived from glutamate. The contours were calculated by computer simulation of the biochemical network using the experimental conditions of 11.5% of glucose labeled with ^{13}C in the 1 position and 39% of the accumulated lactate not enriched. Adapted from Zupke and Stephanopoulos (1994). © 1994 John Wiley & Sons, Inc. Reprinted by permission of John Wiley & Sons, Inc.

carried out with a single isotopic ratio, intracellular flux determination usually requires that more than one isotopic ratio be measured. Moreover, considering the shape of the contours of Figs. 10.11 and 10.12, the fidelity of flux validation is significantly enhanced through the combined use of more than one isotopic ratio measurement.

Results from experiments 1 and 2 comparing the measured isotope ratios with the isotope ratios predicted from the metabolic flux estimates are shown in Fig. 10.13. The figure presents the intersection of the contour lines that correspond to the two measured isotope ratios along with their estimated uncertainty (± 1 standard deviation). Also on the same graph are shown the flux ratios computed by material balancing, along with their uncertainties. There is good agreement between the two methods, although the estimates of the flux ratios from NMR have large uncertainties. To check the statistical significance of the agreement, two tests were performed. First, the hypothesis that the computed and measured ratios are the same was tested and found to be true within the 90% confidence interval. Second, the hypothesis that the ratios were different by more than a specified delta was tested. The delta was chosen to correspond to 20% difference in f_1 or f_2. For the ratios C1:C2 and C3:C1, the delta equaled 0.04 and 0.4, respectively. The hypothesis that

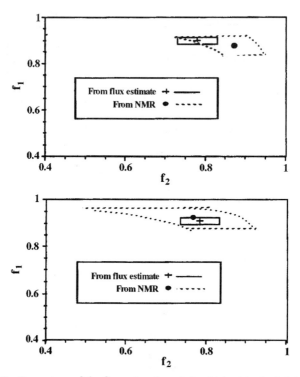

FIGURE 10.13 Comparison of the flux ratios f_1 and f_2, obtained in the hybridoma culture from the lactate ^{13}C isotope ratios and from the flux estimates made using material balances and extracellular measurements. The intersection of the two contours of constant C1:C2 and C3:C1 determine f_1 and f_2 from the *in vitro* NMR measurements of ^{13}C in secreted lactate. The curved dotted lines represent contours corresponding to plus or minus the standard deviation of the NMR measurements. The solid boxes represent the uncertainties in the flux ratios calculated from the estimated intracellular fluxes. (Top) Experiment 1; (bottom) Experiment 2. Adapted from Zupke and Stephanopoulos (1994). © 1994 John Wiley & Sons, Inc. Reprinted by permission of John Wiley & Sons, Inc.

the measured and computed ratios were different by more than delta can be rejected with the probabilities of error indicated in Table 10.12. From this, one can conclude that there is good agreement between the NMR measurements and the ratios predicted from the fluxes.

This example shows how labeling experiments can be employed profitably to validate the flux maps obtained by material balances. It should be noted that the choice of labeled substrates has a significant effect on the ability to quantify intracellular fluxes from isotope enrichments and, in particular, to differentiate between the contribution of different substrates. In this example, two lactate ratios were used. As one can see in Fig. 10.12, measurement of the C3:C1 ratio is sensitive to the value of f_1 but rather insensitive to the value of f_2. This is corrected by combining the C3:C1 ratio measurement with that of C1:C2 ratio. A similar analysis would indicate that a better determination of f_2 is possible by using labeled glutamine because of its more direct involvement in the reaction split, leading to lactate labeling. The use of labeled glutamine, together with labeled glucose, could produce a better system for discriminating between the intracellular fluxes and increasing the accuracy of the agreement between expected and measured values of label ratios.

10.2.3. APPLICATION OF FLUX ANALYSIS TO THE DESIGN OF CELL CULTURE MEDIA

In the previous section, a lumped equation was provided for the synthesis of biomass and antibody. It should be noted that, in addition to the metabolites identified in that equation, other nutrients are also needed for biomass synthesis that cannot be derived from the basic carbohydrate and glutamine sources. Such nutrients are, for example, essential amino acids that are provided as additional components in the medium. However, some of the components that can be synthesized from glucose and glutamine could be included in the medium to minimize the amount of glucose and glutamine consumed for this purpose. Such is the case, for example, of known nonessential amino acids, which, in typical cell culture media, are derived from the glutamine in the medium. This has the undesirable effect of ammonia production, which accumulates in the medium and, eventually, causes the end of the cell culture process. By supplementing the medium with those nonessential amino acids that can be derived from glutamine, the amount of the latter in the medium can be reduced and, consequently, the amount of accumulated ammonia similarly is minimized. This approach was applied in the development of new medium formulation for mammalian cells (Xie and Wang, 1996). The formulation reflects an exact stoichiometric balance based on a general equation for biomass and antibody production. The criteria in the formulation of the medium are to (a) supply all components required for the synthesis of biomass and antibody, (b) minimize the amount of glutamine used for nonessential amino acids by supplementing the medium with stoichiometrically balanced amounts of the latter, and (c) feed the stoichiometrically balanced medium so that the concentration of glucose in the medium is maintained at a limiting level and the osmolality of the medium is not altered significantly in the course of the feeding.

As a result of the combined use of the stoichiometrically balanced medium along with the preceding feeding strategy, the cell densities and product titers that could be obtained in CRL 1606 cultures were improved dramatically. Cells grew to densities of several million per milliliter, and antibody accumulated to the extraordinary level of 2.4 g per L. The production of this antibody was found to be directly correlated to the total number of viable cells multiplied by the duration of time that they were present in the bioreactor medium. This, of course, implies a constant specific antibody production rate at low as well as very high cell densities, a very important finding in the overall development effort of cell culture technology.

REFERENCES

Henriksen, C. M., Christensen, L. H., Nielsen, J. & Villadsen, J. (1996). Growth energetics and metabolic fluxes in continuous cultures of *Penicillium chrysogenum*. *Journal of Biotechnology*. 45, 149-164.

Jørgensen, H., Nielsen, J. Villadsen, J. & Møllgaard, H. (1995). Metabolic flux distributions in *Penicillium chrysogenum* during fed-batch cultivations. *Biotechnology and Bioengineering* 46, 117-131.

Nissen, T. L., Schulze, U., Nielsen, J. & Villadsen, J. (1997). Flux distributions in anaerobic, glucose-limited continuous cultures of *Saccharomyces cerevisiae*. *Microbiology* 143, 203-218.

Park, S. M., Sinskey, A. J. and Stephanopoulos, G. (1997a). Metabolic and physiological studies of *Corynebacterium glutamicum* mutants. *Biotechnology and Bioengineering* 55, 864-879.

Park, S. M., Shaw-Reid, C., Sinskey, A. J. & Stephanopoulos, G. (1997b). Elucidation of anaplerotic pathways in *Corynebacterium glutamicum* via ^{13}C-NMR spectroscopy and GC-MS. *Applied Microbiology and Biotechnology* 47, 430-440.

Takiguchi, N., Shimizu, H. & Shioya, S. (1997). An online physiological-state recognition system for the lysine fermentation process-based on a metabolic reaction model. *Biotechnology and Bioengineering* 55:170-181.

Vallino, J. J. (1991). Identification of branch-point restrictions in microbial metabolism through metabolic flux analysis and local network pertubations. Ph.D. thesis, MIT, Cambridge, MA.

Vallino, J. J. & Stephanopoulos, G. (1993). Metabolic flux distributions in *Corynebacterium glutamicum* during growth and lysine overproduction. *Biotechnology and Bioengineering* 41, 633-646.

Vallino, J. J. & Stephanopoulos, G. (1994a). Carbon flux distribution at the glucose 6- phosphate branch point in *Corynebacterium glutamicum* during lysine overproduction. *Biotechnology Progress* 10, 327-334.

Vallino, J. J. & Stephanopoulos, G. (1994b). Carbon flux distribution at the pyruvate branch point in *Corynebacterium glutamicum* during lysine overproduction. *Biotechnology Progress* 10, 320-326.

van Gulik, W. M. & Heijnen, J. J. (1995). A metabolic network stoichiometry analysis of microbial growth and product formation. *Biotechnology and Bioengineering*. 48, 681-698.

Walsh, K. & Koshland, D. E. (1984). Determination of flux through the branch point of two metabolic cycles - The tricarboxylic acid cycle and the glyoxylate shunt. *Journal of Biological Chemistry* 259, 9646-9654.

Xie, L. & Wang, D. I. C. (1996). Material balance studies on animal cell metabolism using a stoichiometrically based reaction network. *Biotechnology and Bioengineering* 52, 579-590.

Zupke, G. & Stephanopoulos, G. (1994). Modeling of isotope distributions and intracellular fluxes in metabolic networks using atom mapping matrices. *Biotechnology Progress* **10**, 489-498.

Zupke, G., Sinskey, A. J. & Stephanopoulos, G. (1995). Intracellular flux analysis applied to the effect of dissolved oxygen on hybridomas. *Applied Microbiology Biotechnol.* **44**, 27- 36.

CHAPTER 11

Metabolic Control Analysis

One of the most important goals of metabolic engineering is to elucidate the parameters responsible for the *control of flux*. In the preceding chapters, we described how metabolic fluxes can be determined from mass balances around intracellular metabolites and more specialized methods involving isotopic labels. The concept of metabolic flux analysis (MFA) is useful for studying interactions between different pathways and quantifying flux distributions around metabolic branch points. However, MFA by itself does not provide any quantitative measures of the control of flux. Flux control is important for keeping the rates of synthesis and conversion of metabolites closely balanced over a very wide range of external conditions without the catastrophic rise or fall of intracellular metabolite concentrations. Additionally, understanding flux control is important for the rational modification of metabolic fluxes, which is a central goal of metabolic engineering. The discoveries in the 1950s of feedback inhibition, cooperativity and covalent enzyme modifications, and control of enzyme synthesis revealed a number of molecular mechanisms that can play a role in flux control. With such a variety of flux control mechanisms, it is not surprising that disputes often arose over how the flux through a given pathway is controlled. Because most reports on enzyme control typically are qualitative (*e.g.*, "Phosphofructokinase is the major flux-controlling enzyme of glycolysis in muscle." (Voet and Voet, 1990), it is difficult to assess the importance of various research findings in flux control and, in many cases, such findings may seem to conflict. Furthermore, statements and references to *rate-limiting steps* and *bottleneck enzymes* abound when control of flux is discussed. These may be interpreted in many different ways, often leading to misconceptions regarding the actual role of these reaction steps in controlling metabolism. For example, the early finding that the first enzyme in a pathway, or the first enzyme after a branch point, is under some type of regulation (such as by feedback inhibition) often resulted in statements of the type, "the first step in a pathway is the rate-limiting step," which is not necessarily correct.

The introduction of *Metabolic control analysis* (MCA) helped to establish some rigor in the mostly qualitative treatment of enzyme kinetics interactions and flux control. Furthermore, it has provided a useful framework for the systematic evaluation and description of metabolic parameters important in the control of flux. MCA is, in essence, a *linear perturbation theory* of the inherently nonlinear problem of enzymatic kinetics of metabolic networks. As such, MCA predictions are local in nature and any extrapolations should be tentative at best. Despite these limitations, MCA has proven very useful for providing measures of metabolic flux control by individual reactions, elucidating the concept of a rate-controlling step in enzymatic reaction networks, describing the effects of enzymatic activity on intracellular metabolite concentrations, and coupling local enzymatic kinetics with systemic metabolic behavior, among others.

The concept of MCA was developed from the landmark papers of Kacser and Burns (1973) and Heinrich and Rapoport (1974), which were both building on ideas initially developed by Higgins (1963, 1965). It has many similarities with two other frameworks developed to quantify metabolic control: Biochemical Systems Theory (see Box 11.1) and Crabtree and Newsholme's flux-oriented theory (Crabtree and Newsholme, 1987a, b). Although initial attention focused on the differences among these three approaches, it was later realized that they indeed converge on the same basic concepts despite differences in formalism. In this chapter, we will present the basic ideas and results of MCA, due to its more specific orientation toward the description of flux control in metabolic networks.

BOX 11.1

Biochemical Systems Theory

Biochemical Systems Theory, which was introduced by Savageau in the late 1960s (Savageau, 1969a, b, 1970), deals with the *modeling of complex systems* where interactions may be present at many different levels. Its starting point is that reaction rates can be described by general power-law expressions of the type

$$v_i = \alpha_i \prod_j X_j^{g_{ij}} \tag{1}$$

where X_j are system variables (metabolite concentrations, enzyme activities, concentrations of effectors, etc.). The parameters α_i and g_{ij} are called apparent rate constants and apparent kinetic orders, respectively. Equation (1) can be interpreted as a linearization of nonlinear kinetics in logarithmic coordinates, and as a consequence of the nonlinearity of most biochemical reactions, it is a better approximation of reaction kinetics than linear expressions. According to the theory, all reactions that generate a metabolite X_j are combined into a single reaction with net rate v_i, and all reactions that consume the same metabolite similarly are combined into another reaction with net rate v_{-i}. The rate of each of these combined reactions is approximated by power-law expressions and from mass balances around all the metabolites, a system of differential equations can be written that can be studied in detail for its control characteristics. Such a system is called a synergistic system or an *S-system* (Savageau, 1985).

Biochemical Systems Theory (BST) is an elegant approach to modeling complex systems, and with model parameters determined from fitting the model to experimental data, it can also be applied to extract information on flux control. On this account BST can be regarded as a general theory, with MCA representing a special case applicable to linearized systems (Savageau, *et al.*, 1987a, b). This has been an oft-debated issue (Discussion Forum, 1987; Savageau, *et al.*, 1987a, b; Kacser, 1991), which most likely reflects a misconception of the objectives and assumptions of MCA (Cornish-Bowden, 1989). Where the objective of BST is to set up kinetic models that may quantify fluxes through pathways, not only at steady state but also at transient conditions, the primary objective of MCA is to assign clear meaning by quantitative parameters to concepts used in the discussion of metabolic control. To support their claim, Savageau, *et al.*, (1987a, b) have shown that the fundamental equations of MCA (the summation and connectivity theorems, discussed later in this chapter) can be derived from the equations of BST and that the parameters of MCA can also be obtained from the kinetic parameters in the power-law expressions. This comparison is, however, based on another misconception, namely, that the parameters of MCA (the control coefficients and elasticity coefficients) are constants, which is by no means required (Cornish-Bowden, 1989).

11.1. FUNDAMENTALS OF METABOLIC CONTROL ANALYSIS

Metabolic control analysis strictly applies only to steady-state (or pseudo-steady-state) conditions, and a basic assumption is that a stable steady state is uniquely defined by the activities of the enzymes catalyzing the individual steps in a pathway. Enzyme activities therefore are considered to be *system parameters*, along with the concentrations of the substrate for the first reaction and the product of the last reaction in the metabolic pathway. The concentrations of the first substrate and final product of the pathway may be kept constant either by controlling the environmental conditions, *e.g.*, in a steady-state chemostat, or as a result of intracellular regulation. System parameters can, in principle, be changed at will, and as such they completely define the system. Properties that are determined by the values of the parameters, *e.g.*, the flux through the pathway or intermediary metabolite concentrations, are considered to be *system variables*.

To illustrate this definition of parameters and variables in the context of enzyme kinetics, we consider the simple two-step pathway where the substrate S is converted to the product P via an intermediate X:

$$S \overset{E_1}{\leftrightarrow} X \overset{E_2}{\leftrightarrow} P \qquad (11.1)$$

The net rate (or flux) of conversion of S to P at steady state is given by J.[1]

---

[1] In metabolic flux analysis (Chapter 8) we used the term v for fluxes, because the basic assumption of MFA, namely, steady state (or pseudo-steady state) in the pathway intermediates, implies that the flux through a linear pathway is equal to the reaction rate (or velocity) of the individual reactions in this pathway. In MCA the aim is to study the influence of the parameters, *i.e.*, enzyme activities, on the steady-state flux, and we therefore want to distinguish between the rate of the individual reactions v and the overall steady-state flux, which is therefore termed J.

Clearly, a steady state is uniquely determined by the parameters of the system, *i.e.*, the levels of the enzyme activities E_1 and E_2 and the concentrations of the substrate S and product P. The definition of the steady state entails determination of the intermediate metabolite concentration c_X, along with the pathway flux J and other derivative quantities. If the parameters are changed, *e.g.*, if E_1 is increased, a new steady state will emerge that is characterized by other values for the variables, *i.e.*, the concentration of the intermediate X and the net flux *J* through the pathway. Because metabolite concentrations are considered to be unique variables, it is assumed that they are distributed homogeneously over the enzymes that act on them. Subcellular compartments are not a problem in MCA as transport processes can be included in the analysis as additional processing steps with their own rates. However, spatial distribution of metabolite concentrations within a compartment cannot be considered as this would require a complex model providing positional information of metabolite concentrations.

11.1.1. CONTROL COEFFICIENTS AND THE SUMMATION THEOREMS

One objective of Metabolic control analysis is to relate the variables of a metabolic system to its parameters. Once this is done, the sensitivity of a system variable, such as the flux, with respect to the system parameters, namely, the enzyme activities, can be determined. These sensitivities represent fundamental aspects of flux control as they summarize the extent of systemic flux control exercised by the activity of a single enzyme in the pathway. Similarly, one can solve for the concentrations of intracellular metabolites and also determine their sensitivity with respect to enzyme activities, effector concentrations, and other system parameters. These sensitivities are summarily described by a set of coefficients, of which the most prominent are the *control coefficients*. They describe how a parameter, *e.g.*, the activity of an enzyme in the pathway, affects the variables of the system, *e.g.*, the flux through the pathway. The control coefficients only apply to the steady-state conditions studied, and this explains why the component parts, *e.g.*, enzyme activities, are described as parameters, whereas the properties of the system that may change as a result of a change in the parameters, *e.g.*, fluxes and metabolite concentrations, are referred to as variables.

The most important control coefficients are the so-called *flux control coefficients*, often abbreviated as FCCs. They are defined by the *relative* change in the steady-state flux resulting from an infinitesimal change in the activity of an enzyme of the pathway divided by the *relative* change of the enzymatic activity. It is noted that, because the activity is an independent system parameter, its change will affect the flux both directly as well as indirectly through changes imparted on other system variables. The full derivative symbol used in Eq. (11.2) is intended to indicate this point. In a plot of steady-state flux vs. enzyme activity (Fig. 11.1), this is equivalent to the slope of the tangent at the particular enzymatic activity normalized by the corresponding steady-state flux and enzyme activity:

$$C^J = \frac{E}{J}\frac{dJ}{dE} = \frac{d \ln J}{d \ln E} \qquad (11.2)$$

Because the FCCs are defined in terms of *relative* flux and activity values, they are dimensionless and their magnitude is independent of flux and activity units used. For a linear pathway they have values between 0 and 1. In the general case of branched pathways, FCCs are introduced to describe the effect of each of the L enzyme activities on each of the L fluxes through the different reactions:

$$C_i^{J_k} = \frac{E_i}{J_k}\frac{dJ_k}{dE_i} = \frac{d \ln J_k}{d \ln E_i} \qquad i, k \in \{1, 2, \ldots, L\} \qquad (11.3)$$

where J_k is the steady-state flux through the kth reaction in the pathway, and E_i is the activity of ith enzyme. For such a general system, FCCs may attain any value (negative or positive).

The definition in Eq. (11.3) is the original one proposed by Kacser and Burns (1973). A more general definition is based on the *rate* of the ith

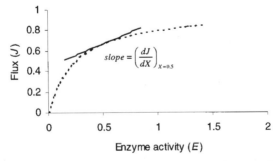

$$slope = \left(\frac{dJ}{dX}\right)_{X=0.5}$$

FIGURE 11.1 Schematic of the steady-state pathway flux J as a function of the activity of an intermediate enzyme in the pathway. The flux control coefficient for the enzyme at a particular enzyme activity is found by multiplying the slope of the tangent at that enzymatic activity by the enzyme activity and normalizing with respect to the steady-state flux.

reaction rather than the enzyme activity:

$$C_i^{J_k} = \frac{v_i}{J_k} \frac{dJ_k}{dv_i} = \frac{d \ln J_k}{d \ln v_i} \qquad i, k \in \{1, 2, \ldots, L\} \qquad (11.4)$$

which Heinrich, *et al.*, (1977) presented as

$$C_i^{J_k} = \frac{v_i}{J_k} \frac{dJ_k}{dp} \left(\frac{\partial v_k}{\partial p} \right)^{-1} \qquad i, k \in \{1, 2, \ldots, L\} \qquad (11.5)$$

where p can be any parameter that acts exclusively on the ith reaction rate. If enzyme activity is chosen as the parameter p, Eq. (11.5) reduces to the original formulation of the FCCs given by Eq. (11.3), provided that the reaction rate is directly proportional to the enzyme activity, as is often the case. The general definition of the FCCs offered by Eq. (11.4) or (11.5), however, allows for the handling of special situations with enzyme-enzyme interactions, as in metabolite channeling where metabolites are transported directly from one enzyme to the next enzyme in the sequence.

From the definition of the FCCs it is obvious that the enzyme with the largest flux control coefficient exerts the largest control of flux at the particular steady state, as an increase in the activity of this enzyme results in the largest overall flux increase. An important consequence of the normalization of the FCCs is that, with respect to each flux, they all must sum to unity. This is known as the *flux-control summation theorem*:

$$\sum_{i=1}^{L} C_i^{J_k} = 1 \qquad k \in \{1, 2, \ldots, L\} \qquad (11.6)$$

From this equation, it is clear that FCCs are completely dependent upon the structure of the system and that nothing general can be said about their individual values. For very long pathways, most FCCs may have small values; however, one step may still exist that exerts significant control over flux if the magnitude of its FCC is significantly larger than that of the other FCCs. For short pathways the FCCs may have much greater magnitudes, even for the case where flux control is distributed among more than one step. FCCs therefore should only be compared with each other within the same pathway and never with FCCs of other pathways. The small value of flux control coefficients in long pathways explains why so many successive rounds of mutation and selection steps usually are needed in order to improve strains for the production of metabolites such as amino acids and antibiotics.

It should be noted that there have been objections both to the name and concept of FCCs. Some of the principal grounds for these objections have been reviewed by Fell (1992) and are summarized briefly:

1. In the original formulation, enzyme concentration[2] was used as the parameter descriptive of the enzymatic reaction rate. Enzyme concentration, however, is not particularly relevant to enzymatic activity and flux control, considering the fact that enzymatic activity principally is affected by the action of effectors binding to allosteric enzymes. To correct for this, Kacser and Burns (1973) introduced another set of control coefficients, termed *response coefficients*, that allow for quantification of the flux sensitivities with respect to such effectors:

$$R_{X_i}^{J_k} = \frac{e_i}{J_k} \frac{dJ_k}{de_i} \qquad i, k \in \{1, 2, \ldots, L\} \qquad (11.7)$$

The preceding definition of the response coefficients is analogous to that of the FCCs and in effect may be regarded as flux control coefficients with respect to external effectors. Furthermore, it is easily shown [see Eq. (11.12)] that the response coefficients represent net sensitivities with respect to external effectors, and, as such, they can be partitioned into an FCC component and a local kinetic component (described by an elasticity, as discussed below).

2. The sensitivity of a flux to an enzyme activity is not a measure of whether that enzyme is a control or regulatory enzyme, and some, therefore, find the term control coefficient misleading (Crabtree and Newsholme, 1987; Savageau, *et al.*, 1987a, b; Atkinson, 1990). In many pathways the synthesis rate of the end metabolite is regulated via feedback inhibition of the first step. In such cases the first enzyme is a regulatory enzyme in the classical sense, but it will normally have a small flux control coefficient. The apparent paradox that a regulatory enzyme has a low FCC was specifically discussed in the original paper of Kacser and Burns (1973). Before the introduction of the FCCs, it was often implied that a regulatory enzyme was a potential candidate as the rate-limiting step, and the introduction of the FCCs revealed that this is not necessarily the case. Furthermore, as mentioned in the

---

[2] In the present text, we prefer to use the term enzyme activity rather than enzyme concentration, as the term activity is related only to the active form of the enzyme, which may be quite different from the enzyme concentration determined, *e.g.*, by Western blotting. Thus, we interpret enzyme activity as being equal to the v_{max} of the enzyme, whereas we do not imply anything about the *in-vivo* activity of the enzyme, which, because of allosteric regulation, may be completely different from its v_{max}.

introduction to this chapter, MCA tells us that the concept of a rate-limiting step is nonsense, and instead the FCCs quantify the degree of rate limitation.

3. The FCCs have little predictive value as they only apply to the steady state under study, and, as the state of the system changes, the values of the FCCs may change. This cannot be disputed, but MCA has not been promoted as a tool for system modeling like Biochemical Systems Theory, but rather as a tool for describing metabolic control. It obviously would be desirable to have a theory that would allow one to make predictions of how large perturbations in enzyme activity would affect the flux, but this is a difficult task due to the nonlinearity of cellular reaction kinetics and the ill-described nature of *in-vivo* enzymatic kinetics.

Similar to the FCCs one can define sensitivities for the effect of system parameters on intracellular metabolite concentrations. Such sensitivities are termed *concentration control coefficients*, abbreviated here as CCCs, where the variable affected by the enzyme activity E_i is a metabolite concentration c_j:

$$C_i^{X_j} = \frac{E_i}{c_j} \frac{dc_j}{dE_i} = \frac{d \ln c_j}{d \ln E_i} \qquad i \in \{1, 2, \ldots, L\}, j \in \{1, 2, \ldots, K\} \quad (11.8)$$

or, more generally,

$$C_i^{X_j} = \frac{v_i}{c_j} \frac{dc_j}{dv_i} = \frac{d \ln c_j}{d \ln v_i} \qquad i \in \{1, 2, \ldots, L\}, j \in \{1, 2, \ldots, K\} \quad (11.9)$$

These coefficients specify the relative change in the level of the jth intermediate X_j when the activity of the ith enzyme is changed. Because the level of any intermediate remains unchanged when all enzyme activities are changed by the same factor, it follows that, for each of the K metabolites, the sum of all the concentration control coefficients must equal zero:

$$\sum_{i=1}^{L} C_i^{X_j} = 0 \qquad j \in \{1, 2, \ldots, K\} \tag{11.10}$$

Equation (11.10) implies that for each metabolite at least one enzyme must exert negative control, *i.e.*, when the level of that enzyme increases, the metabolite concentration decreases. Thus, in the simple two-step pathway of Eq. (11.1) the concentration control coefficient C_2^X normally will be negative because the metabolite concentration c_X will decrease when the activity of the second enzyme is increased.

11.1.2. ELASTICITY COEFFICIENTS AND THE CONNECTIVITY THEOREMS

Another important concept in MCA is the *elasticity coefficient*. Unlike control coefficients, which are *systemic* properties of the overall metabolic system, elasticities are *local* properties of the individual enzymes in the metabolic network. The most common elasticity coefficients are the sensitivities (or elasticities) of the reaction rates with respect to metabolite concentrations. As with control coefficients, elasticity coefficients too are normalized with respect to the reaction rate and metabolite concentration. Thus, the elasticity of the ith reaction rate with respect to the concentration of metabolite X_j is defined as the ratio of the relative change in the reaction rate brought about by an infinitesimal change in the metabolite concentration, assuming that all other system variables did not change from their steady-state values:

$$\varepsilon_{X_j}^i = \frac{c_j}{v_i}\frac{\partial v_i}{\partial X_j} = \frac{\partial \ln v_i}{\partial \ln c_j} \qquad i \in \{1, 2, \ldots, L\}, j \in \{1, 2, \ldots, K\} \quad (11.11)$$

Because the elasticity coefficients represent derivatives with respect to variables, partial derivatives are used to indicate that all other variables must be kept constant. The elasticity coefficients can be regarded as apparent reaction rate orders with respect to the particular metabolite, and therefore they are equal to the power-law exponents used in Biochemical Systems Theory (see Box 11.1). They have positive values for metabolites that stimulate a reaction, *e.g.*, a substrate and an activator, whereas they have negative values for metabolites that slow the reaction, *e.g.*, a product and an inhibitor. The elasticity coefficients are the quantitative equivalent of the vague concept of enzyme responsiveness with respect to a metabolite concentration (Fell, 1992). They also allow quantification of the influence of several substrates on the reaction rate (most cellular reactions use two or more substrates), as well as the influence of product concentrations, which often have a significant effect due to the reversibility of cellular reactions.

In addition elasticities with respect to metabolites, elasticity coefficients may also be introduced for effectors that are not pathway intermediates, *i.e.*, X_j may be interpreted more broadly as any compound that influences the ith reaction rate. With this broader definition of elasticity coefficients, it can be shown that the response coefficient of an effector, introduced in Eq. (11.7), is the product of the elasticity coefficient of the affected enzyme and the flux control coefficient with respect to the same enzyme:

$$R_{X_j}^{J_i} = C_k^{J_i}\varepsilon_{X_j}^k \qquad i, k \in \{1, 2, \ldots, L\}, j \in \{1, 2, \ldots, K\} \quad (11.12)$$

If a metabolite or an effector acts on more than one enzyme, the total response coefficient will be given by the sum of the responses from each enzyme:

$$R^{J_i}_{X_j} = \sum_{k=1}^{L} C^{J_i}_k \varepsilon^k_{X_j} \qquad i \in \{1, 2, \ldots, L\}, j \in \{1, 2, \ldots, K\} \quad (11.13)$$

Equation (11.12) provides the valuable information that even if the elasticity of an enzyme with respect to a metabolite or an effector is large, the response of the flux to changes in this compound is only significant if the corresponding flux control coefficient is nonzero.

The FCCs of enzymes that are highly responsive to changes in the concentration of their substrates or products, i.e., enzymes that have high elasticity coefficients with respect to the pathway intermediates, tend to be low (Westerhoff, et al., 1984). This can be visualized as follows using the simple two-reaction pathway of Eq. (11.1): A reduction in the activity of the second enzyme E_2 (brought about, for example, by a specific inhibitor) will result in an increase in the concentration of its substrate S and its product P. If enzyme E_2 has a large elasticity coefficient with respect to S and P, the preceding changes in metabolite concentrations will compensate for the decreasing enzyme activity. Consequently, the reduction in enzyme activity will have a small effect on the overall steady-state flux, meaning that the flux control coefficient of the perturbed enzyme is small. Similarly, perturbation of an enzyme with small elasticity with respect to its intermediate metabolite would cause a significant flux change. This relationship between flux control coefficients and elasticities is expressed in the following *flux-control connectivity theorem* derived by Kacser and Burns (1973):

$$\sum_{i=1}^{L} C^{J_k}_i \varepsilon^i_{X_j} = 0 \qquad i \in \{1, 2, \ldots, L\}, j \in \{1, 2, \ldots, K\} \quad (11.14)$$

The connectivity theorem generally is considered to be the most important of the MCA theorems because it provides the means to understand how local enzyme kinetics affect flux control. As a further illustration, consider the simple two-step pathway of Eq. (11.1). For this pathway, the connectivity theorem gives

$$C^J_1 \varepsilon^1_X + C^J_2 \varepsilon^2_X = 0 \qquad (11.15)$$

or

$$\frac{C_1^J}{C_2^J} = -\frac{\varepsilon_X^2}{\varepsilon_X^1} \tag{11.16}$$

From Eq. (11.16) it is quite clear that large elasticities are associated with small FCCs and vice versa. Reactions operating close to thermodynamic equilibrium normally are very sensitive to variations in the metabolite concentrations, i.e., their elasticities are large, also indicating that the flux control of such reactions generally will be small.

As with FCCs, connectivity theorems have also been derived for concentration control coefficients, although they are slightly more complex. When the control coefficient and the elasticity refer to different metabolites, the connectivity theorem takes the form:

$$\sum_{i=1}^{L} C_i^{X_l} \varepsilon_{X_j}^i = 0 \qquad j, l \in \{1, 2, \ldots, K\}, j \neq l \tag{11.17}$$

whereas if the metabolites are the same it takes the form:

$$\sum_{i=1}^{L} C_i^{X_j} \varepsilon_{X_j}^i = -1 \qquad j \in \{1, 2, \ldots, K\} \tag{11.18}$$

Generally, less attention is paid to the CCCs, but as will be demonstrated in Chapter 13, the latter provide important information when the results of MCA are to be used in the design of enzymatic amplifications. Furthermore, in the original derivation by Heinrich and Rapoport (1974), they were used to derive expressions for the FCCs [see Eqs. (11.24) and (11.25)].

The elasticity coefficients with respect to metabolite concentrations are the most commonly used, but Kacser, et al., (1990) introduced the so-called *parameter elasticity coefficients* as measures of enzymatic sensitivity to changes in other parameters, e.g., enzyme activity or an inhibitor that is not a metabolite in the pathway under study. For a general parameter p_j, they are defined as

$$\pi_{p_l}^i = \frac{p_l}{v_i} \frac{\partial v_i}{\partial p_l} = \frac{\partial \ln(v_i)}{\partial \ln(p_i)} \qquad i \in \{1, 2, \ldots, L\} \tag{11.19}$$

If the enzyme activity is selected as the parameter, the parameter elasticity coefficient is normally 1 when $i = l$ and 0 when $i \neq l$, i.e., there is direct proportionality between the rate of the i'th reaction and the activity of the ith enzyme whereas there is no influence of the ith enzyme, on the rate of the other reactions in the pathway. When the enzyme activity is the parameter, an important exception to these rules is the case of enzyme-enzyme interactions.

11.1.3. GENERALIZATION OF MCA THEOREMS

The concepts and theorems presented in the previous two sections are the foundation of MCA. They will be expanded here to the case of a general multireaction network. Before we do that, however, it is interesting to look further into how the systemic response of the flux to enzymatic modulation can be partitioned into its source components. Thus, a change in an enzymatic activity will give rise to both a *direct* effect on the pathway flux as well as an *indirect* effect through changes in the metabolite concentrations. With the introduction of the parameter elasticity coefficient in Eq. (11.19), there is, furthermore, the possible effect on the flux from other parameters. Because the steady-state flux is a function of the system parameters (some of which are the enzyme activities) and the metabolite concentrations, *i.e.*, $J_i(p_1, c_j)$, the FCCs can be defined as follows by expanding the derivative in the original definition of Eq. (11.3):

$$C_i^{J_k} = \frac{E_i}{J_k} \left(\sum_l \frac{\partial J_k}{\partial p_l} \frac{dp_l}{dE_i} + \sum_{j=1}^{K} \frac{\partial J_k}{\partial c_j} \frac{dc_j}{dE_i} \right) \qquad i, k \in \{1, 2, \ldots, L\} \quad (11.20)$$

In most cases the parameters are not coupled, so a change in one parameter does not change the others, and consequently:

$$\frac{dp_l}{dE_i} = \begin{cases} 0; & p_l \neq E_i \\ 1; & p_l = E_i \end{cases} \qquad (11.21)$$

Equation (11.20) therefore can be rearranged to yield:

$$C_i^{J_k} = \frac{E_i}{J_k} \frac{\partial J_k}{\partial E_i} + \sum_{j=1}^{K} \frac{\partial J_k}{\partial c_j} \frac{c_j}{J_k} \frac{dc_j}{dE_i} \frac{E_i}{c_j} \qquad i, k \in \{1, 2, \ldots, L\} \quad (11.22)$$

Because the rate of the kth reaction (v_k) at steady state is equal to the steady-state flux J_k, the elasticity coefficient and the parameter elasticity coefficients can be recognized in Eq. (11.22), and the equation can be rewritten as

$$C_i^{J_k} = \pi_i^j + \sum_{j=1}^{K} \varepsilon_{X_j}^k C_i^{X_j} \qquad i, k \in \{1, 2, \ldots, L\} \qquad (11.23)$$

In most cases the parameter elasticity coefficients reduce to 1 when $i = k$ (i.e., when the ith enzyme modulates the flux response through its own (ith) reaction rate and the change in reaction rate is proportional to the activity change) and 0 when $i \neq k$ (i.e., when there is no *direct* effect of a change in the activity of the ith enzyme on the flux through the kth reaction). Equation (11.23) thus is further reduced to:

$$C_i^{J_i} = 1 + \sum_{j=1}^{K} \varepsilon_{X_j}^i C_i^{X_j} \qquad i \in \{1, 2, \ldots, L\} \qquad (11.24)$$

$$C_i^{J_k} = \sum_{j=1}^{K} \varepsilon_{X_j}^k C_i^{X_j} \qquad i, k \in \{1, 2, \ldots, L\}, i \neq k \qquad (11.25)$$

These equations, which were first derived in the original paper of Heinrich and Rapoport (1974), illustrate how the FCCs depend on the elasticity coefficients (i.e., the kinetics of the individual enzymes) and the concentration control coefficients (i.e., enzyme interactions through changes in the metabolite levels). Thus, Eq. (11.24) shows that, when an enzyme activity changes, the *direct* effect of modulation of the enzyme activity is combined with the effects from changes in the metabolite concentrations to produce the overall flux change. This overall change of flux is represented as the sum of the effect of each metabolite, with the effect from each metabolite given by the product of the change in the metabolite concentration (the CCC) and the effect of this change on the flux (the elasticity coefficient). Equation (11.25) indicates that modulation of the ith enzyme affects the steady-state flux only through changes in the metabolite concentrations brought about by indirect effects on the other enzymatic reactions.

There are a total of L^2 equations in Eq. (11.23), which can be written in a condensed form through the use of matrix notation (Ehlde and Zacchi,

1997):

$$
\begin{pmatrix}
C_1^{J_1} & C_2^{J_1} & \cdots & C_L^{J_1} \\
C_1^{J_2} & C_2^{J_2} & \cdots & C_L^{J_2} \\
\cdot & \cdot & \cdots & \cdot \\
C_1^{J_L} & C_2^{J_L} & \cdots & C_L^{J_L}
\end{pmatrix}
=
\begin{pmatrix}
\pi_{E_1}^1 & \pi_{E_2}^1 & \cdots & \pi_{E_L}^1 \\
\pi_{E_1}^2 & \pi_{E_2}^2 & \cdots & \pi_{E_L}^2 \\
\cdot & \cdot & \cdots & \cdot \\
\pi_{E_1}^L & \pi_{E_2}^L & \cdots & \pi_{E_L}^L
\end{pmatrix}
$$

$$
+
\begin{pmatrix}
\varepsilon_{X_1}^1 & \varepsilon_{X_2}^1 & \cdots & \varepsilon_{X_K}^1 \\
\varepsilon_{X_1}^2 & \varepsilon_{X_2}^2 & \cdots & \varepsilon_{X_K}^2 \\
\cdot & \cdot & \cdots & \cdot \\
\varepsilon_{X_1}^L & \varepsilon_{X_2}^L & \cdots & \varepsilon_{X_K}^L
\end{pmatrix}
$$

$$
\times
\begin{pmatrix}
C_1^{X_1} & C_2^{X_1} & \cdots & C_L^{X_1} \\
C_1^{X_2} & C_2^{X_2} & \cdots & C_L^{X_L} \\
\cdot & \cdot & \cdots & \cdot \\
C_1^{X_K} & C_2^{X_K} & \cdots & C_L^{X_K}
\end{pmatrix}
\tag{11.26}
$$

or

$$
\mathbf{C}^J = \mathbf{P} + \mathbf{E} \cdot \mathbf{C}^X \tag{11.27}
$$

With the assumptions applied in the derivation of Eqs. (11.24) and (11.25) the parameter elasticity coefficient matrix $\mathbf{P}$ becomes an identity matrix, *i.e.*, a diagonal matrix with all the diagonal elements being 1. Equation (11.27) is a complete and general formulation of all the theorems of MCA. It is a compact form that includes both the summation and the connectivity theorems, and is therefore useful in later analysis, especially for branched pathways. However, a disadvantage of the equation is that the illustrative theorems of MCA are implicit and not easily recognized.

11.2. DETERMINATION OF FLUX CONTROL COEFFICIENTS

The magnitude of a flux control coefficient provides a relative measure of the flux increase that is expected from the amplification of a particular enzymatic activity. This, of course, does not mean that the increase in flux will be proportional to the FCC for *any* activity amplification. However, the measure provided by FCCs can be a reasonable approximation, especially for small changes of enzymatic activity. In particular, enzymatic steps with FCCs close to unity qualify for rate-limiting steps, and should be addressed first in a flux amplification program. On the other hand, small magnitude FCCs indicate dispersion of flux control among many steps, with none playing a determining role in the control of flux. It follows that FCCs are an important measure of flux control and their determination is an important milestone in the effort to unveil the architecture of flux control in metabolic networks.

A number of experimental methods have been proposed to determine both the control coefficients and the elasticity coefficients. All of these methods involve, in one way or another, the determination of derivatives of nonlinear functions, i.e., the slopes of tangents to nonlinear functions at specific points. As discussed in Box 11.2, this implies a fundamental prob-

---

BOX 11.2

Derivatives of Nonlinear Functions

Ehlde (1995) discussed the central problem in determining derivatives of nonlinear functions, which is the basis of the experimental determination of the parameters of MCA. The only way to measure the derivative of an *unknown* function is to approximate it by a finite difference. This requires that measurements be taken in the vicinity of the point of interest, followed by determination of the slope of a line drawn through the measurements. The accuracy of the approximation of the derivative at this point decreases as the magnitude of the finite difference increases. Also, if the finite difference is not centered around the point of interest, the estimate of the derivative will be biased. Random measurement errors will also affect the result, and the sensitivity to measurement errors increases as the size of the finite approximation decreases. These points are illustrated in Fig. 11.2. Here we wish to determine the slope of the tangent at the position of the filled circle. If the finite difference is based on the measurements shown as filled triangles, the finite difference is small, and a good estimate of the actual slope may be obtained. However, in this case the measurement errors can be large due to the small magnitude of the perturbation, so that the sensitivity of the estimate will also be very large. If, instead, the finite difference is based on the measurements shown as filled squares, the finite difference will be larger and a systematic error in the estimate may appear. However, in this case the estimate is much less sensitive to measurement errors.

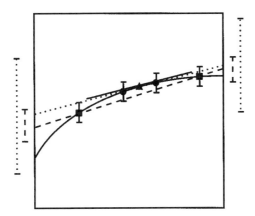

FIGURE 11.2 Two approximations of the derivative based on finite differences. The error bars on the data points indicate the random measurement errors. The error bars on the abscissa indicates the sensitivity in the estimate of the derivative from each of the two sets of data.

lem, as infinitesimal differences are mathematical abstractions and the only way to measure the derivative of an unknown function is to approximate it with a finite difference.

Various methods have been proposed in the literature for the determination of FCCs. These methods can be classified in three groups as follows:

- Direct methods, where the control coefficients are determined directly from flux and activity measurements following small but finite activity changes.
- Indirect methods, where the elasticity coefficients are determined first and the control coefficients subsequently are calculated from the theorems of MCA. It is noted that the latter produces a closed system whereby FCCs and elasticities can be determined from one another through the use of theorems.
- Determination of FCCs from transient metabolite concentration measurements.

A fourth approach also has been proposed that makes use of flux and activity measurements resulting from large magnitude activity perturbations. As this approach holds significant promise due to its simplicity and practical implementation, it will be examined in more detail in Section 11.5.

Recently, it has been argued that no experimental method can give satisfactory results unless the experimental data are evaluated first by hyperbolic regression analysis and the actual flux vs enzyme activity profile must closely conform to a hyperbolic relationship similar to that depicted in

Fig. 11.1 (Pettersson, 1996). This argument implies that, if this condition is not satisfied, MCA should be based on kinetic models that enable direct calculation of the FCCs, the concentration control coefficients, and the elasticity coefficients. Although there can be no argument about the accuracy of FCC determination from kinetic models, it should be noted that this approach is severely limited by model availability and validation. Therefore, experimental procedures for FCC determination are very valuable, and more effort should be directed toward their improvement. Furthermore, the results of Metabolic control analysis of a given pathway should not be intended as a final answer, but only as an indication of how flux control is distributed in a pathway.

11.2.1. DIRECT METHODS FOR FCC DETERMINATION

The most direct approach for the determination of FCCs is to observe the flux consequences of enzymatic activity manipulation in a metabolic pathway while maintaining all other metabolic parameters at constant values. In light of the inherent problems associated with the determination of derivatives from finite changes in enzymes (see Box 11.3), one should perform sufficient experiments to establish first the basic relationship between enzyme activity and flux, i.e., by generating a plot similar to that of Fig. 11.1. Once such a plot is available, FCCs can be determined from the slope of the tangent at the activity of interest. This leads to a very accurate determination of FCCs, the major drawback being the requirement of a large number of experiments to establish the flux vs enzyme activity curve.

Genetic Alteration of Expressed Enzyme Activity

With the rapid development of molecular biological techniques, the most direct method for altering enzyme activity is by controlling expression at the genetic level. This approach allows one to study the effects of activity changes *in vivo*; however, realistic activity changes yield large perturbations that may not be suitable for a slope determination of the type suggested by an infinitesimal analysis. The large perturbation approach of Section 11.5 is better suited to utilize the results of this method. Alternatively, one needs to perform sufficient experiments to generate a plot analogous to Fig. 11.1, as discussed earlier.

Changes in enzyme activity by genetic means may be introduced either by changing the gene dosage for the enzyme(s) of interest or by inserting

BOX 11.3

Derivation of Equation (11.39)

Equation (11.39) was derived in the original paper of Delgado and Liao (1991) by assuming linearized kinetics with respect to metabolite concentrations:

$$v_i = \sum_{j=1}^{K} k_{ij}c_j + b_i \qquad i \in \{1, 2, \ldots, L\} \qquad (1)$$

where k_{ij} are the kinetic constants and b_i are constants due to the linearization of the reaction rates. From the definition of the elasticity coefficients, one can write

$$\varepsilon_{X_j}^i = \frac{c_j}{J_i} k_{ij} \qquad i \in \{1, 2, \ldots, L\}, j \in \{1, 2, \ldots, K\} \qquad (2)$$

where J_i is the steady-state flux through the ith reaction. Insertion of Eq. (2) into the connectivity theorem gives

$$\sum_{i=1}^{L} C_i^{J_k} k_{ij} \frac{c_j}{J_i} = 0 \qquad i \in \{1, 2, \ldots, L\}, j \in \{1, 2, \ldots, K\} \qquad (3)$$

or

$$\sum_{i=1}^{L} C_i^{J_k} \frac{k_{ij}}{J_i} = 0 \qquad i \in \{1, 2, \ldots, L\}, j \in \{1, 2, \ldots, K\} \qquad (4)$$

because $c_j \neq 0$. Upon multiplication by Δc_j and summation over all the metabolites, one obtains:

$$\sum_{j=1}^{K} \sum_{i=1}^{L} C_i^{J_k} \frac{k_{ij}}{J_i} \Delta c_j = 0 \qquad k \in \{1, 2, \ldots, L\} \qquad (5)$$

which yields

$$\sum_{i=1}^{L} C_i^{J_k} \frac{\Delta v_i}{J_i} = 0 \qquad k \in \{1, 2, \ldots, L\} \qquad (6)$$

Because $\Delta v_i = v_i(t) - J_i$, the difference between the transient reaction rate and the steady-state flux, Eq. (11.39) is easily derived using the summation theorem.

regulatory promoters to modulate gene expression at will. Gene dosage may be changed by classical methods of breeding homozygotes and heterozygotes for different alleles of the gene, as illustrated by Flint, *et al.*, (1981), who studied the arginine biosynthetic pathway in the filamentous fungus *Neurospora crassa*. In addition to changes in the effective gene dosage, this approach also enables alteration in the catalytic activity by using polymorphic alleles of different specific activity. On the negative side, the classical approach does not allow fine adjustments in enzyme activity. Furthermore, the genetic work required for the generation of the many different mutants is substantial.

A more targeted approach to increasing gene dosage is by inserting multicopy plasmids carrying the gene of interest or by performing multiple gene integration into the chromosome. This approach does not provide the flexibility for decreasing gene dosage, but with multiploid strains it is possible to down-modulate the enzyme activity by constructing deletion mutants lacking the gene of interest (or in which a point deletion has been introduced to disrupt the gene coding for the enzyme of interest) on one or several of the chromosomes. This has been illustrated by Niederberger, *et al.*, (1992) who studied the tryptophan biosynthetic pathway in a tetraploid strain of *Saccharomyces cerevisiae*. Five of the enzymes in the pathway were down-modulated by varying the gene dosage. The flux through the pathway was not determined directly, but the effect of the enzyme activity on the specific growth rate of the yeast was studied and the corresponding FCCs were determined. All the FCCs were found to be small[3] and this was confirmed by up-modulation experiments, where an increase in the enzyme activity (obtained by the use of multicopy plasmids) had only a small effect on the flux. Whereas down-modulation is relatively easy in well-characterized species such as *S. cerevisiae* (where the complete genome has been sequenced) and *Escherichia coli* (where deletion mutants often are available or easily constructed), it is more difficult to carry out with less well-characterized species as a substantial amount of work is involved in generating the necessary mutants.

The most elegant approach in genetically modulating enzyme activity is by inserting regulatory promoters, *e.g.*, the *tac* promoter in *E. coli* (a hybrid between the *lac* and *trp* promoters), in front of the genes coding for the enzymes of the studied pathway. This has the dual benefit of allowing the

---

[3] It is quite obvious that when the flux through the tryptophan pathway is measured through ie specific growth rate, small FCCs are obtained as what is actually determined is the influence of the individual enzymes on the overall flux to biomass synthesis. In fact it is surprising that the authors find FCCs on the order of 0.05 and 0.17, as the flux control on biomass synthesis is expected to be distributed over a large number of enzymes in all the biosynthetic pathways.

introduction of changes of different magnitude in the enzyme activity, as well as enabling both up- and down-modulation of the enzyme activity. The use of regulatory promoters has been illustrated by Ruyter, et al., (1991), who studied the genetic control (via the tac promoter) of enzyme II^{Glc} (one of the proteins in glucose-PTS see Section 2.2.3) and its effect on the glucose metabolism in E. coli. The plasmid also contained the gene for the lac^{Iq} repressor which allowed control of expression from the tac promoter through variations in the amount of the inducer isopropyl β-D-thiogalactopyranoside (IPTG) in the medium. The plasmid was inserted into a strain that did not contain any chromosomal encoded II^{Glc}, and it was therefore possible to vary the activity of this protein between 20 and 600% of the wild-type chromosomal level. It was concluded from this study that this PTS protein had no (or only very little) effect on the specific growth rate or on the rate of glucose oxidation in a glucose-limited medium. It was, however, found that II^{Glc} exerted substantial control on the uptake and phosphorylation of the glucose analog methyl α-glucopyranoside (with an FCC of approximately 0.6).

Titration with Purified Enzyme

Cell-free extracts can be titrated with purified enzyme, and if the enzyme activity is measured as a parameter, the flux control coefficients can be determined directly. A drawback of this method is the large sensitivity to experimental errors, especially if the FCCs are small. Torres, et al., (1986, 1991) proposed a procedure to decrease this sensitivity to experimental errors. The procedure is based on pathway shortening, whereby all other enzymes (so-called auxiliary enzymes) of the pathway are supplied in excess such that their contribution to flux control is reduced. Under these conditions, flux control is distributed among the enzymes studied, and the relative distribution of flux control between these enzymes can be quantified by enzyme titration. The thus determined in vitro FCCs are proportional to the in-vivo FCCs but are greater in magnitude. Important assumptions in this approach are that all feedback and feed-forward regulation loops are confined within the pathway segment studied and that the degree of saturation of the enzymes does not change during titration. The method of enzyme titration has been applied by Torres, et al., (1986) to study the first enzymes of glycolysis in cell extracts of rat liver. The enzymes aldolase, triosephosphate isomerase, and glycerol-3-phosphate dehydrogenase were added in excess, and enzyme titration was performed with hexokinase (HK), glucose-6-phosphate isomerase (GPI), and phosphofructokinase (PFK). From their analysis it was found that GPI exerts practically no control on glycolytic flux (which is consistent with the general finding that the hexosephosphates constitute a single metabolite pool; see Section 2.3.1), whereas the flux control coefficients for HK and PFK were determined to be 0.77 and 0.24, respectively (notice that the summation theorem is satisfied). The procedure of enzyme titration is elegant, but it is obviously applicable only for systems where the purified enzymes are available. Furthermore, the requirement of decoupling the pathway segment of interest from the rest of the metabolism imposes a limitation, especially for systems where all regulatory loops have not been identified.

Titration with Specific Inhibitors

For many enzymes, specific inhibitors exist that can be used to titrate the *in vivo* enzyme activity. When the resulting flux is also measured, an estimate of the response coefficient [see Eq. (11.7)] with respect to the inhibitor can be obtained. If the enzyme response (*i.e.*, the elasticity) with respect to the inhibitor is known, then the flux control coefficient can be calculated using Eq. (11.12). Because the FCC is to be determined for the situation where no inhibitor is present, the response coefficient has to be evaluated for the case where the inhibitor concentration is zero:

$$\left. \frac{c_1}{J} \frac{dJ}{dc_I} \right|_{c_I=0} = R_I^J = C_i^J \varepsilon_I^i = C_i^J \frac{c_I}{v_i} \left. \frac{\partial v_i}{\partial c_I} \right|_{c_I=0} \qquad (11.28)$$

or

$$C_i^J = \left. \frac{1}{J} \frac{dJ}{dc_I} \right|_{c_I=0} \left(\left. \frac{1}{v_i} \frac{\partial v_i}{\partial c_I} \right|_{c_I=0} \right)^{-1} \qquad (11.29)$$

For an irreversible inhibitor, the enzyme activity decreases *linearly* with the inhibitor concentration. In such a case, the quantity in the brackets of Eq. (11.29) is equal to the negative of the inhibitor concentration at which there is complete inhibition of the enzyme activity ($-c_{I,max}$), so that the flux control coefficient can be obtained as (Groen, *et al.*, 1982):

$$C_i^J = -\frac{c_{I,max}}{J} \left. \frac{dJ}{dc_I} \right|_{c_I=0} \qquad (11.30)$$

In Eq. (11.30) the last term can be evaluated from the initial slope of the inhibition curve. For reversible competitive and noncompetitive inhibitors other expressions hold (Groen, *et al.*, 1982), but if the influence of the inhibitor on the reaction rate is known, the last term in Eq. (11.29) is easily evaluated.

Inhibitor titration is perhaps the most widely-used approach to determine flux control coefficients, especially in studies of respiration in isolated mitochondria and cells. In their study of rat liver mitochondria, Groen, *et al.*, (1982) determined the FCCs of the individual steps in the oxidative phosphorylation with succinate as the substrate using specific inhibitors. Their results are summarized in Table 11.1. Fell (1992) reviewed a number of other applications of inhibitor titration for FCC determination. Among these can be mentioned a study by Walter, *et al.*, (1987) of the role of the glucose PTS system in the regulation of glycolysis in the bacterium *Clostridium pasteurianum*. In this study, xylitol was used as specific inhibitor of the glucose PTS, and it was found that the FCC of the transport system on the glycolytic flux is equal to 0.14.

A major problem with the inhibitor titration method is that the initial slope of the flux vs inhibitor curve has to be extrapolated to zero inhibitor concentration in order to determine the last term of Eq. (11.30). Because the effect of inhibitors on flux is often highly nonlinear this gives rise to the problems discussed in Box 11.3. Fitting a line to the quasilinear initial part of the curve is unreliable, whereas fitting a nonlinear function, *e.g.*, polynomial functions, does give some improvement in the estimate (Small, 1993). Another limitation of the method is that the inhibitor has to be completely specific with absolutely no effect on other reactions in the system.

TABLE 11.1   Flux Control Coefficients of Different Steps in the Oxidative Phosphorylation in Rat Liver Mitochondria

Enzyme	Step	Inhibitor	FCC
Adenine nucleotide translocator	Transfer of ATP from mitochondria to cytosol	Carboxyatractyloside	0.29 ± 0.05
Proton leak[a]	Leak of proton gradient	Proton uncoupler	0.04 ± 0.01
Dicarboxylate carrier	Transport of succinate into mitochondria	Phenylsuccinate	0.33 ± 0.04
Cytochrome *c* oxidase	Transfer of electrons to oxygen	Azide	0.17 ± 0.01
bc_1 complex	Succinate dehydrogenase	Hydroxyquinoline-N-oxide	0.03 ± 0.005
Hexokinase[b]	ATP drain	—	0

[a] In the estimation of the FCC for the proton leak, the oxidative phosphorylation was blocked by the addition of oligomycin and a proton uncoupler [glycol bis(β-aminoethyl ether)-N,N,N',N'-tetraacetic acid] was added in varying amounts. This procedure implicitly assumes that the FCC for the proton leak is 1 when there is no ADP regeneration, which has been criticized by Brand, *et al.*, (1988) (see also Example 12.3). The data are from Groen, *et al.*, (1982b).

[b] The FCC for hexokinase was determined by enzyme titration.

11.2.2. INDIRECT METHODS FOR FCC DETERMINATION

The methods described in this section are characterized as indirect because the elasticities are found first experimentally, and flux control coefficients are then determined by making use of the theorems of MCA. Two major assumptions underlying this approach are (1) that the metabolic system is adequately described, *i.e.*, all reactions and relevant regulatory interactions have been included in the system description, and (2) that the system is at steady state and the initial substrate and final product concentrations are fixed at a constant value. To ensure that these assumptions are satisfied it is advisable to check the results by direct determination of at least one of the control coefficients of the system. In their analysis of gluconeogenesis in rat liver cells Groen, *et al.*, (1986) determined all the elasticity coefficients (see also Example 11.2) and used the theorems of MCA to calculate the FCCs of this pathway. To check the results, they also determined the response coefficient of the pathway with respect to pyruvate, the substrate. Because the elasticity of the first reaction (pyruvate carboxylase) with respect to pyruvate was also determined, the FCC of the first step could be calculated using the definition of the response coefficient [Eq. (11.12)].

A number of different approaches can be applied to determine the elasticity coefficients. A brief description of the most common are provided next.

Double Modulation

To illustrate this method, consider the hexose isomerase reaction of the EMP pathway:

$$\cdots \to \text{glucose - 6 - phosphate} \overset{\text{GPI}}{\to} \text{fructose - 6- phasphate} \to \cdots \quad (11.31)$$

The rate of the isomerase reaction depends only on the concentrations of the compounds shown, *i.e.*, $v_{\text{GPI}} = f(c_{\text{G6P}}, c_{\text{F6P}})$. At steady state the rate of this reaction equals the flux J through the EMP pathway and consequently,

$$dJ = \frac{\partial v_{\text{GPI}}}{\partial c_{\text{G6P}}} dc_{\text{G6P}} + \frac{\partial v_{\text{F6P}}}{\partial c_{\text{F6P}}} dc_{\text{F6P}} \quad (11.32)$$

Scaling of this equation with the steady-state flux and rearrangement of the result yields:

$$d \ln J = \varepsilon_{\text{G6P}}^{\text{GPI}} d \ln c_{\text{G6P}} + \varepsilon_{\text{F6P}}^{\text{GPI}} d \ln c_{\text{F6P}} \quad (11.33)$$

In a control experiment, it is possible to measure the concentrations of the two metabolites and the steady-state flux. A perturbation can then be introduced, e.g., by changing the extracellular glucose concentration, to yield a new set of steady-state measurements for the flux and metabolite concentrations.[4] By approximating the differentials in Eq. (11.33) by differences, this perturbation experiment yields an equation relating flux and concentration measurements to the two elasticity coefficients. If a second perturbation is introduced, an additional set of data for the steady-state flux and metabolite concentrations can produce a second equation, from which the two elasticity coefficients can be calculated.

This approach, termed double modulation, was first described by Kacser and Burns (1979). It is obvious that the two perturbations must be so designed as to produce two linearly independent equations, i.e., the following inequality must be satisfied:

$$\frac{d \ln c_{G6P}^1}{d \ln c_{F6P}^1} \neq \frac{d \ln c_{G6P}^2}{d \ln c_{F6P}^2} \qquad (11.34)$$

The superscripts in Eq. (11.34) refer to measurements obtained following the corresponding perturbation. If the difference between the two terms of Eq. (11.34) is small, the two equations are ill-conditioned for the calculation of elasticities and the solution becomes very sensitive to experimental error. Actually, in many pathways it may be experimentally difficult to obtain a linearly independent set of measurements that satisfies Eq. (11.34). To increase the probability that such measurements are indeed independent, it is suggested that modulation be performed both upstream and downstream of the pathway under investigation (Fell, 1992). Another disadvantage of this approach is that it requires small changes in order to satisfactorily approximate the differentials by finite differences, and such changes are likely to be dominated by disproportionate experimental errors. It is noted that for most enzymatic reactions there are other effectors besides the reactant and reaction product, e.g., cofactors and inhibitors. This means that the reaction velocity is a function of more than the two variables used in the preceding example. As a different elasticity is needed to describe the effect of each of these variables, it therefore becomes necessary to perform more than two perturbation experiments that yield linearly-independent equations.

---

[4] It is quite obvious that chemostat experiments are excellent for this type of study as a steady state can be obtained for each extracellular glucose concentration. Furthermore, the flux through the EMP pathway can be estimated with high precision in a chemostat, and sufficiently large samples can be withdrawn to obtain good data for the intracellular metabolite concentrations.

Single Modulation

If one of the elasticity coefficients is known for a reaction sequence like that of Eq. (11.31), the other elasticity coefficient can be determined from a single modulation experiment. The advantage of this approach is that several modulations of the same type but different magnitude can be applied and a graphical method can be used to determine the differential terms in Eq. (11.33). Groen, et al., (1986) used this method to determine the elasticity coefficient of pyruvate kinase with respect to phosphoenolpyruvate and the elasticity coefficient of pyruvate carboxylase with respect to cytoplasmic oxaloacetate (see also Example 11.2). The approach obviously is more robust than that of double modulation, but it still relies on a good estimation of the differential terms in Eq. (11.33) and, of course, the knowledge of one elasticity coefficient.

Top-Down Approach

In many situations the detailed information supplied by *all* the FCCs is not really necessary, whereas it is more important to identify a group of reactions in which the majority of flux control is located. This can help one focus on relevant parts of metabolism. Also, by repeating this procedure one can localize flux control to progressively smaller groups of reactions. Reaction grouping is a powerful concept in the analysis of metabolic pathway flux control and is discussed in detail in Chapter 13. The basic idea derives from the *top-down* approach to MCA introduced by Brand, Brown, and co-workers whereby the pathway under study is divided into segments (or groups) with only one common metabolite (Brown, et al., 1990a; Hafner, et al., 1990):

$$S \xrightarrow{\text{Group 1}} X \xrightarrow{\text{Group 2}} P \qquad (11.35)$$

In a branched pathway the common intermediate X is obviously the branch point metabolite, whereas for a linear pathway X will be one of the K metabolites in the linear pathway (see the following). Group flux control coefficients can be introduced to provide a measure of the extent of flux control exercised by the reactions in the group. It is easily shown that the flux control coefficients for each of the two groups in Eq. (11.35) equal the sum of FCCs for the individual steps in the groups (Brown, et al., 1990a) and that the group flux control coefficients obey the summation theorem,

$$C^J_{\text{Group 1}} + C^J_{\text{Group 2}} = 1 \qquad (11.36)$$

In addition to the group FCCs, a set of elasticity coefficients with respect to the intermediate metabolite can be introduced to describe the effect of that metabolite on the rate of the reactions in the group (see also Chapter 13):

$$\varepsilon_X^{\text{Group } i} = \frac{c_X}{v_{\text{Group } i}} \frac{\partial v_{\text{Group } i}}{\partial c_X} \qquad i \in \{1, 2\} \tag{11.37}$$

In analogy with the connectivity theorem, the group elasticity coefficients and the group flux control coefficients are related according to

$$C_{\text{Group } 1}^J \varepsilon_X^{\text{Group } 1} + C_{\text{Group } 2}^J \varepsilon_X^{\text{Group } 2} = 0 \tag{11.38}$$

Similar theorems hold for the group concentration control coefficients (Brown, et al., 1990a).

When the pathway has been segmented, the overall elasticity coefficients can be determined, e.g., by double-modulation experiments, or if one of the elasticity coefficients is known, the other can be determined by single modulation experiments. From Eqs. (11.36) and (11.38), the group FCCs can then be calculated, to provide a measure of the flux control residing in each of the two pathway segments. Because the pathway, in principle, can be segmented at any arbitrary position, it is possible to progressively narrow the group of individual reactions with the largest group FCC. A prerequisite for the application of the double (or single) modulation approach is, however, that there are no effectors other than the pathway intermediate X, in other words, no cross-talk between the reaction groups is allowed other than that occurring through the intermediate metabolite X. This can be a major drawback as there are many situations where interactions between the two groups can occur other than through the common metabolite X. A method to evaluate the strength of such interactions and the extent to which it affects FCC determination from flux measurements is presented in Chapter 13. The top-down approach has been applied by Brand, Brown, and co-workers in the analysis of oxidative phosphorylation (see Example 11.3), and it is also useful for the analysis of complex reaction networks, as illustrated in Chapter 13, where the approach is described in more detail.

Calculation of Elasticity Coefficients from Kinetic Models

If a valid mathematical model is available for the rate of an enzymatic reaction, its elasticity with respect to effectors and substrates can easily be calculated using the definition of the elasticity coefficient. For the purpose of control analysis, it is not required that the kinetic expression be mechanistically based, as long as it correctly describes the kinetic effect of the various effectors. This approach is straightforward and very robust, but it raises a fundamental question: whether the *in vitro* enzyme kinetics correctly describe the *in vivo* enzyme function. Fortunately, v_{max} values cancel out in the calculation of the elasticity coefficients, and the problem is reduced to one of affinity (or K_m) determination. It is also important to consider all possible effectors in these studies. Furthermore, because small elasticity coefficients generally are associated with large control coefficients, the effectors with low elasticity coefficients are the most important for flux control, whereas in most *in-vitro* studies of enzyme kinetics there is a tendency to focus on effectors with large elasticities. Another caveat in using *in vitro* kinetic data is that most studies of enzyme kinetics are based on initial velocity measurements where there is no reaction product present. Clearly, this experimental situation is not representative of *in-vivo* conditions.

There are many examples of elasticity determination from kinetic models. In their analysis of gluconeogenesis in rat liver cells, Groen, *et al.*, (1986) determined some of the elasticity coefficients using this approach (see Example 11.2), and in an analysis of glycolysis in *S. cerevisiae*, Galazzo and Bailey (1990) determined the group elasticity coefficients for all the key steps and used them to calculate the FCCs of the pathway from glucose to ethanol in suspended and immobilized cells. This topic is covered in greater detail in Chapters 12 and 13.

11.2.3. USE OF TRANSIENT METABOLITE MEASUREMENTS

Even though flux control coefficients are defined for systems at steady state, Delgado and Liao (1992a, b) proposed an approach to determine FCCs directly from transient measurements of metabolite concentrations. The method makes four assumptions:

1. The external pool of metabolites does not affect the pathway kinetics, or their concentration is controlled at a steady level. This is a general MCA assumption.

2. A linear approximation of enzymatic kinetics around the steady state is valid over a broad region of metabolite concentrations. This assumption can be relaxed as discussed later.

3. It must be theoretically possible to determine transient fluxes through each reaction in the pathway from metabolite concentration measurements.

4. Metabolites are homogeneously distributed in the system (another general MCA assumption).

On the basis of the preceding assumptions, it is possible to derive the following relationship between FCCs and the transient flux through each enzyme (see Box 11.3):

$$\sum_{i=1}^{L} C_i^J \left(\frac{v_i(t)}{J_i} \right) = 1 \tag{11.39}$$

where $v_i(t)$ is the transient flux through the ith reaction and J_i is the steady-state flux. Transient fluxes can be evaluated from the measurement of the metabolite concentrations, and Eq. (11.39) can be used to calculate the FCCs of the pathway. However, precise estimation of transient fluxes from metabolite measurements involves the determination of derivatives, a procedure that is difficult to implement and prone to errors. For this reason, Delgado and Liao (1992a) introduced an alternative approach based on the integral version of Eq. (11.39). According to this approach, a set of coefficients α_j is determined first from the transient metabolite measurements by making use of the following equation, which is the integral equivalent of Eq. (11.39):

$$\sum_{j=1}^{K} \alpha_j \big(c_j(t) - c_j(0) \big) = t \tag{11.40}$$

The FCCs of the pathway are related to the coefficients α_j of Eq. (11.40). This relationship is as follows for a linear pathway:

$$\left(C_1^J \quad C_2^J \quad \cdots \quad C_L^J \right) = \left(\alpha_1 \quad \alpha_2 \quad \cdots \quad \alpha_K \right) G^* J \tag{11.41}$$

where J is the steady-state flux through the pathway and G^* is the stoichiometric matrix of the pathway. The stoichiometric matrix in Eq. (11.41) is not quite identical to the stoichiometric matrix G for the intracellular metabolites introduced in Chapter 3, as it needs to be augmented to include the stoichiometric coefficients for the substrate and the metabolic product of the pathway (with the stoichiometric coefficients for the substrate positioned in the first column and the stoichiometric coefficients for the metabolic product positioned in the last column). For a branched pathway, the equivalent of Eq. (11.41) for the determination of the FCC is

$$C^J = \left(\alpha_1 \quad \alpha_2 \quad \cdots \quad \alpha_K \right) G^* J \tag{11.42}$$

where C^J is the matrix containing all the FCCs [see Eq. (11.26)] and J is a diagonal matrix containing the flux through the individual branches of the pathway in the diagonal.

The coefficients α_j can be determined by regression fitting measurements of the metabolite concentrations during a transient, $e.g.$, after a pulse addition to a steady-state chemostat, to Eq. (11.40). These coefficients are then used to calculate the FCCs from either Eq. (11.41) or Eq. (11.42). Least-squares fitting to Eq. (11.40) is not possible in the presence of linear constraints among metabolite concentrations, *i.e.*, if one concentration is a linear combination of other concentrations. An example of such a constraint is the mass balance equation.

$$\sum_{i=1}^{K+2} [c_i(t) - c_i(0)] = 0 \qquad (11.43)$$

Stoichiometric constraints may also exist for conserved moieties such as NAD^+ and NADH.[5] Even though these constraints may be filtered out by Eq. (11.40), it is advisable to eliminate them by deleting the metabolite measurement and its corresponding row in the stoichiometric matrix. The obvious choice of measurement to be deleted is the one with the largest uncertainty.

Even if there are no stoichiometric constraints, some of the metabolites may still be linearly correlated due to kinetic considerations. If, for example, one of the reactions is much faster than the others, this reaction may reach a quasi-equilibrium shortly after the introduction of the perturbation, with the result that the concentrations of the substrate and the product of this reaction will be related by the equilibrium constant. If equilibrium reactions are known *a priori*, Delgado and Liao (1992a) suggested that the corseting metabolites be lumped together into a common pool.

The approach based on transient metabolite measurement can also be used to determine concentration control coefficients (Delgado and Liao, 1992b). Here the analysis is based on an equation similar to Eq. (11.39):

$$\sum_{i=1}^{L} C_i^{X_j} \left(\frac{v_i(t)}{J_i} \right) = 1 - \frac{c_j(t)}{c_{j,\text{ss}}} \qquad (11.44)$$

where $c_{j,\text{ss}}$ is the steady-state metabolite concentration. Equations (11.39) and (11.44) clearly show that metabolite concentrations and fluxes through the individual reactions do not vary independently following a perturbation from a steady state because they are constrained by the control coefficients of the system at steady state. An integral version of Eq. (11.44) can be used for

---

[5] Notice that, in our definition of the stoichiometric matrix G for the intracellular metabolites, only one stoichiometric coefficient is included for all compounds in sets of conserved moieties, *e.g.*, stoichiometric coefficents for only one of the compounds NAD^+ or NADH are included. Thus, there are no linearly dependent metabolites in the stoichiometric matrix G, and therefore there will be no stoichiometric constraints with our definition of the matrix.

the determination of the concentration control coefficients, and this allows least squares regression of linear functions to the metabolite measurements, with subsequent calculation of the concentration control coefficients.

The procedure of transient metabolite measurement is quite elegant as it enables a relatively easy determination of the control coefficients. A major criticism of this procedure has been the assumption of linearized kinetics. It has been shown recently (Nielsen, 1997) however, that the procedure can also be applied when the reaction kinetics are described by linear functions with respect to the logarithm of the metabolite concentrations. This type of linearization generally gives a more accurate description of reaction kinetics. There has been no practical demonstration of the method in the analysis of a complete pathway. Delgado, *et al.*, (1993) applied it to determine the FCCs of hexokinase and phosphofructokinase in an *in vitro* reconstituted partial glycolytic pathway. The FCCs were also determined using enzyme titration, and a good agreement between the results of the two methods was found. Another problem of the approach is the high sensitivity of the determined control coefficients with respect to measurement errors (Ehlde and Zacchi, 1996), which is a consequence of a high degree of correlation between the individual metabolite concentrations. Based on a theoretical analysis of the procedure (using Monte Carlo simulations), Ehlde and Zacchi (1996) concluded that it is not possible to obtain a reasonable determination of the control coefficients from real experimental data, which are always corrupted with a certain degree of noise.

11.2.4. KINETIC MODELS

When a complete kinetic model has been set up for the pathway under study the concept of MCA is, in principle, not necessary. If the biochemical model is sufficiently robust, it can be used to predict the effect of both small perturbations and large changes in the enzyme activities. Furthermore, the structure of the biochemical model will normally reveal the effect of metabolite and effector levels on the reaction rates. Despite this fact it may still be valuable to apply the kinetic model to determine the MCA coefficients at different conditions, as they supply concise quantitative information about the flux control. A general criticism of the application of kinetic models for complete pathways is that, despite the level of detail included, they cannot include all possible interactions in the system, and therefore they represent only one model of the system. Particularly if the kinetic model is to be used for predictions, the robustness of the model is extremely important. Unfortunately, most biochemical models-even very detailed models-are only valid at operating conditions close to those where the parameters have been estimated, *i.e.*, the predictive strength is limited. For analysis of complex systems it is, however, not necessary that the model give a quantitatively correct description of all the variables, as even models that give a qualitatively correct description of the most important interactions in the system may be valuable in studies of flux control.

11.3. MCA OF LINEAR PATHWAYS

In this and the following section, we apply metabolic control analysis to linear and branched pathways. Special forms of the MCA equations are derived for these cases, and their application is demonstrated with illustrative case studies.

For a linear pathway the number of metabolites is $L - 1$, where L is the number of enzymatic reactions, *i.e.*, there is one fewer metabolite than there are enzymatic reactions. There is only one flux in this case that is equal to the rates of all reactions at steady state. Also, there are L unknown flux control coefficients, one for the effect of each reaction in the pathway on the overall pathway flux. The FCCs can be determined from the $L - 1$ connectivity theorems for each of the intermediate $L - 1$ metabolites, which, together with the summation theorem, provide exactly sufficient equations to calculate the L flux control coefficients from the elasticity coefficients. Similarly, the concentration control coefficients can be calculated from the corresponding summation and connectivity theorems. The equations for the (flux and concentration) control coefficients (a total of L^2 equations) can be conveniently summarized in matrix notation as proposed by Fell and Sauro (1985):

$$
\begin{pmatrix}
1 & 1 & \ldots & 1 \\
\varepsilon^1_{X_1} & \varepsilon^2_{X_1} & \ldots & \varepsilon^L_{X_1} \\
\cdot & \cdot & \ldots & \cdot \\
\varepsilon^1_{X_K} & \varepsilon^2_{X_K} & \ldots & \varepsilon^L_{X_K}
\end{pmatrix}
\begin{pmatrix}
C^J_1 & -C^{X_1}_1 & \ldots & -C^{X_{L-1}}_1 \\
C^J_2 & -C^{X_1}_2 & \ldots & -C^{X_{L-1}}_2 \\
\cdot & \cdot & \ldots & \cdot \\
C^J_L & C^{X_1}_L & \ldots & -C^{X_{L-1}}_L
\end{pmatrix}
$$

$$
=
\begin{pmatrix}
1 & 0 & \ldots & 0 \\
0 & 1 & \ldots & 0 \\
\cdot & \cdot & \ldots & \cdot \\
0 & 0 & \ldots & 1
\end{pmatrix}
\tag{11.45}
$$

This equation is easily shown to be identical to the general formulation in Eq. (11.26) for a linear pathway, where the parameter elasticity matrix **P** is equal to the identity matrix. If the elasticity coefficient matrix is nonsingular, the control coefficients can be calculated using matrix inversion. In this way, the systemic properties expressed by the control coefficients are related to the local enzymatic kinetics reflected in the elasticity coefficients.

To illustrate these concepts, we return to the simple two-step pathway of Eq. (11.1), for which Eq. (11.45) is reduced to the following:

$$
\begin{pmatrix} 1 & 1 \\ \varepsilon_X^1 & \varepsilon_X^2 \end{pmatrix} \begin{pmatrix} C_1^J & -C_1^X \\ C_2^J & -C_2^X \end{pmatrix} = \begin{pmatrix} 1 & 0 \\ 0 & 1 \end{pmatrix}
\tag{11.46}
$$

which has the solution:[6]

$$
\begin{pmatrix} C_1^J & C_1^X \\ C_2^J & C_2^X \end{pmatrix} = \begin{pmatrix} \dfrac{\varepsilon_X^2}{\varepsilon_X^2 - \varepsilon_X^1} & \dfrac{1}{\varepsilon_X^2 - \varepsilon_X^1} \\[2ex] -\dfrac{\varepsilon_X^1}{\varepsilon_X^2 - \varepsilon_X^1} & -\dfrac{1}{\varepsilon_X^2 - \varepsilon_X^1} \end{pmatrix}
\tag{11.47}
$$

Normally the elasticity of a reaction is negative with respect to its product and positive with respect to its substrate. Thus, ε_X^1 is negative and ε_X^2 is positive, and the denominator in the expressions for the control coefficient is positive. This makes both FCCs positive. The distribution of flux control depends on the value of the corresponding elasticity coefficients: a large elasticity is associated with a small FCC and vice versa. If the elasticity of the first reaction with respect to the metabolite is very low, as for a practically irreversible reaction, the FCC of the first reaction will approach 1, yielding a true *rate-limiting step* in this case. Only when the elasticity of the first reaction with respect to the metabolite is zero, *i.e.*, there is absolutely no influence of the reaction product on the reaction rate, will the FCCs of the two reactions exactly attain the values 1 and 0 independent of the metabolite concentration. However, even for reactions that are practically irreversible there is always some elasticity with respect to the reaction product, and the situation of a true bottleneck reaction is never attained in practice. It is interesting to note that if all reactions in a linear pathway are described by irreversible Michaelis-Menten type kinetics, the elasticities with respect to the reaction products become zero, and all FCCs are equal to zero except the first, which is exactly equal to unity. It is, however, quite obvious in practice

[6] Notice that the solution is consistent with the general formulation of MCA given in Eq. (11.26), which takes the form:

$$
\begin{pmatrix} C_1^J & C_2^J \\ C_1^J & C_2^J \end{pmatrix} = \begin{pmatrix} 1 & 0 \\ 0 & 1 \end{pmatrix} + \begin{pmatrix} \varepsilon_X^1 \\ \varepsilon_X^2 \end{pmatrix} \begin{pmatrix} C_1^X & C_2^X \end{pmatrix}
$$

that if the activity of the first enzyme is increased drastically, the intermediate metabolite concentration will increase and eventually the second reaction will become saturated in that metabolite. This yields a very low elasticity of this reaction with respect to the metabolite and, therefore, a very high flux control coefficient for the second reaction. Indeed, a small increase in the activity of the second reaction will have a large impact on the pathway flux as it readily catalyzes the depletion of the accumulated stores of the intermediate metabolite. It should noted also be that an ill-conditioned equation system can arise in this case, and the solution found from matrix inversion can be very sensitive to the available data for the metabolite concentration.

If we assume that the effect of metabolite X is negative for the first reaction and positive for the second, the concentration control coefficient of the first reaction is positive (meaning that the metabolite concentration increases as the rate of the first reaction increases), whereas the concentration control coefficient for the second reaction is negative (meaning that the concentration of the metabolite decreases when the rate of the second reaction increases). If the magnitude of at least one of the elasticity coefficients is large, the concentration control coefficients are small and vice versa. Thus, if the reactions are very elastic with respect to metabolite concentration changes, changing reaction rates will have a small effect on the metabolite concentration, as the reaction rates are adjusted to the new conditions.

EXAMPLE 11.1

MCA of the Penicillin Biosynthetic Pathway

The penicillin biosynthetic pathway is well-described [for a review see Nielsen (1996)]. The pathway consists of three enzymatic reactions (Fig. 11.3), of which the first two steps are identical in all biosynthetic pathways for β-lactam antibiotics (with a very high degree of homology among enzymes from different organisms). The first step in the pathway is the condensation of the three amino acids, L-α-aminoadipic acid, L-cysteine, and D-valine to form the tripeptide L-α-aminoadipyl-L-cysteinyl-D-valine (LLD-ACV) L-Cysteine and L-valine are well-known intermediates in the metabolism of all cells, whereas α-aminoadipic acid is an intermediate in the biosynthesis of lysine in fungi. The formation of the tripeptide LLD-ACV is catalyzed by a single multifunctional enzyme, the ACV synthetase (ACVS). The second step in the pathway is the oxidative ring closure of the LLD-ACV to form isopenicillin N. This is done by the enzyme isopenicillin N synthetase with free oxygen as electron acceptor. The final conversion of

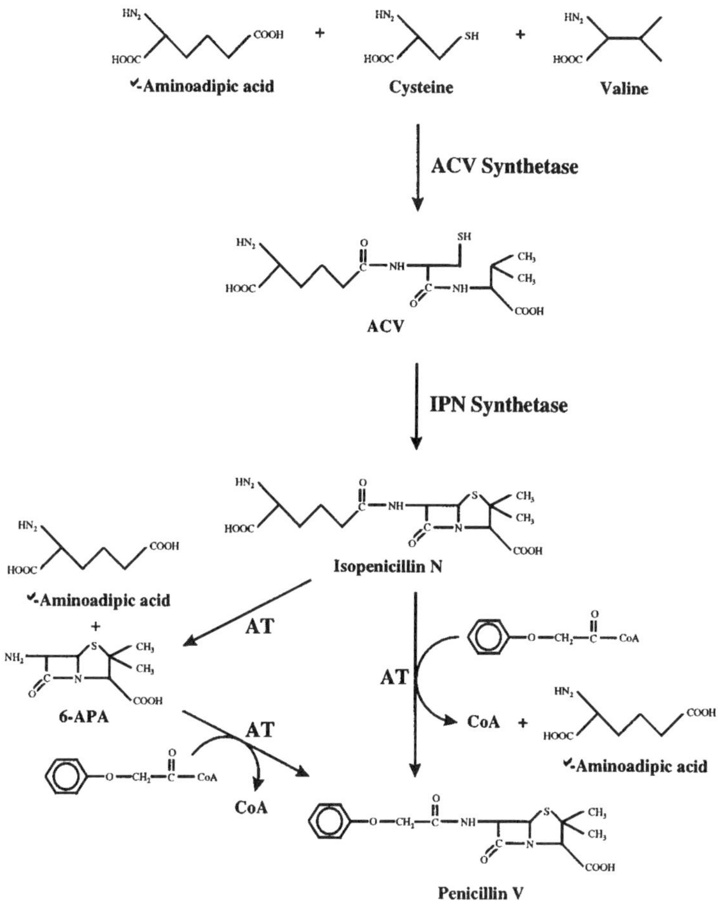

FIGURE 11.3   The penicillin biosynthetic pathway in *Penicillium chrysogenum.*

isopenicillin N into penicillin V is catalyzed by acyl-CoA:isopenicillin acyl-transferase and may occur by either a one- or two step mechanism. In the two-step reaction mechanism, the α-aminoadipic acid moiety is cleaved off and 6-APA is released. If an activated precursor, *i.e.*, phenoxyacetyl-CoA, is available, 6-APA may bind to the AT and subsequently be converted into penicillin V. In the one-step reaction mechanism the α-aminoadipic acid side chain of isopenicillin N is exchanged with the precursor without release of 6-APA from the enzyme.

Metabolic control analysis of the pathway has been carried out by Nielsen and Jørgensen (1995). In their analysis, kinetic expressions were proposed for the first two enzymes. From the kinetic expressions of the ACVS and the IPNS, the elasticity coefficients were determined, and the theorems of MCA were used to calculate the flux control coefficients for the first two steps. From activity measurements of the last enzyme (AT), it was concluded that the flux control exercised by AT was minimal. This finding was further confirmed by a more detailed analysis of all the reactions in the pathway carried out later by Pissara, *et al.*, (1996).

ACVS has been shown to exhibit Michaelis-Menten kinetics with respect to the individual amino acids participating in the synthetase reaction. In their analysis, Nielsen and Jørgensen (1995) proposed feedback inhibition by the reaction product LLD-ACV, and the following kinetics therefore was proposed for the ACVS-catalyzed reaction:

$$r_{ACV} = \frac{v_{max}}{\left(1 + K_{aaa}c_{aaa}^{-1} + K_{cys}c_{cys}^{-1} + K_{val}c_{val}^{-1}\right)\left(1 + c_{ACV}K_{ACV}^{-1}\right)} \tag{1}$$

Recently the *P. chrysogenum* ACVS was purified and the proposed feedback inhibition by LLD-ACV confirmed (Theilgaard, *et al.*, 1997). In Eq. (1) the maximum rate is a function of various effectors which may include ATP, AMP, pyrophosphate, phosphate, CoA, and Mg^{2+}, with ATP stimulating the reaction and AMP and pyrophosphate inhibiting the reaction. The three Michaelis-Menten parameters K_{aaa}, K_{cys} and K_{val} were taken to be equal to those reported for *S. clavuligerus*, *i.e.*, 0.63, 0.12, and 0.30 mM, respectively (the *P. chrysogenum* ACVS was not characterized at the time of the analysis and application of other K_m values does not affect the analysis significantly). The value of K_{ACV} was estimated from fitting r_{ACV} to the flux through the penicillin biosynthetic pathway.

Isopenicillin N synthetase (IPNS) is the best-characterized enzyme of the β-lactam biosynthetic pathway. This enzyme is an iron - dependent oxidase, which removes four hydrogen equivalents from LLD-ACV with the consumption of one mole of oxygen. IPNS has been purified from *P. chrysogenum* (Ramos *et al.*, 1985), and molecular oxygen, Fe^{2+}, dithiothreitol, and ascorbate are all required for *in-vitro* activity of the purified IPNS. The enzyme exhibits Michaelis-Menten kinetics with respect to the LLD-ACV concentration with a K_m of 0.13 mM. Furthermore, there seems to be competitive inhibition by glutathione with an apparent K_i of 8.9 mM. There are no reports on the influence of the oxygen concentration on the kinetics of the purified IPNS, but for a partly purified IPNS from *P. chrysogenum*, Bainbridge, *et al.*, (1992) found the kinetics to be first-order in oxygen concentration in the range 0.070-0.18 mM (corresponding to 25-70% of saturation with air). This implies that the conversion of LLD-ACV to isopenicillin N is very sensitive to variations in the dissolved oxygen concentration. This has also been observed experimentally for several β-lactam producers. On the basis of these results, the following rate expression was proposed for the kinetics of the conversion of LLD-ACV to isopenicillin N (IPN):

$$r_{IPN} = \frac{v_{max}\, c_{ACV}}{c_{ACV} + K_m\left(1 + c_{glut}\, K_i^{-1}\right)} \qquad (2)$$

Literature values were used for the two parameters K_m and K_i equal to 0.13 and 8.9 mM, respectively.

From measurements of the intracellular pools of the precursor amino acids and LLD-ACV and the activity of the ACVS, the rate of LLD-ACV formation (r_{ACV}) was calculated using the kinetic expression in Eq. (1). By fitting the calculated rate to measurements of total penicillin synthesis, K_{ACV} was estimated to be 12.5 mM, which is slightly higher than that found with the purified enzyme (Theilgaard, et al., 1997). Obviously, the inhibition parameter determined in vitro is likely to be the mechanistically correct value, whereas the value determined from the fit of the productivity may be overlaid by other effects present in vivo. The fitted value yields a very good agreement between the actual flux through the ACVS-catalyzed reaction and that predicted from the kinetic expression. Because a good agreement was also found for the IPNS reaction, it was concluded that the simple kinetic expressions of Eqs. (1) and (2) are adequate representations of the kinetics of the first two reactions to allow accurate determination of the elasticity coefficients and flux control coefficients.

The elasticity coefficients for the ACVS and IPNS reactions first were derived by taking the partial derivatives of the rate equations:

$$\varepsilon_{ACTs}^{ACVS} = -\frac{c_{ACV}\, K_{ACV}^{-1}}{1 + c_{ACV}\, K_{ACV}^{-1}} \qquad (3)$$

$$\varepsilon_{ACV}^{IPNS} = \frac{K_m\left(1 + c_{glut}\, K_i^{-1}\right)}{c_{ACV} + K_m\left(1 + c_{glut}\, K_i^{-1}\right)} \qquad (4)$$

Equations (3) and (4) show that the elasticity coefficients are functions of the concentrations of glutathione and LLD-ACV. The elasticity coefficient of the first reaction with respect to LLD-ACV is negative due to the feedback inhibition, whereas the second elasticity coefficient is positive. During a fed-batch cultivation changes occur in the intracellular concentrations of the pathway metabolites, so that the assumption of steady state is not strictly satisfied. However, from a time scale analysis it has been found that the pools of the pathway metabolites are at pseudo-steady state throughout the fed-batch cultivation, despite a significant accumulation of LLD-ACV (Pissara, et al., 1996). Thus, the theorems of MCA may still be applied, and from transient measurements of the glutathione and LLD-ACV concentrations, the elasticity coefficients were calculated at different times during a fed-batch cultivation, as shown in Fig. 11.4.

Note that the elasticity coefficient for ACVS is negative, i.e., an increasing concentration of LLD-ACV diminishes the rate of LLD-ACV synthesis by ACVS, and that the magnitude of this elasticity coefficient increases throughout the fed-batch cultivation due to the increasing concentration of LLD-ACV. The elasticity coefficient for IPNS, on the other hand, is positive, i.e., the effect of increasing concentrations of LLD-ACV is positive, but its magnitude decreases throughout the fed-batch cultivations as the IPNS becomes saturated with LLD-ACV.

The flux control coefficients were calculated from the elasticity coefficients using Eq. (11.47), and the results are shown in Fig. 11.5. Initially the flux control coefficient for the ACVS is high (close to 1) as there is very little inhibition from the intermediate LLD-ACV. However, when the LLD-ACV concentration increases, control of flux gradually is shifted from ACVS to IPNS, and after approximately 70 the flux is controlled primarily by IPNS. With the shift in flux control from the ACVS to IPNS, neither of the two enzymes can be identified as a rate-controlling step. This demonstrates that FCCs are not constant for a given pathway.

Another interesting observation is that the proposed kinetics for the IPN formation in Eq. (2) reveals that an increase in the dissolved oxygen concentration may result in a significant increase in the flux through the pathway, because only a small increase in the activity of the IPNS will prevent accumulation of LLD-ACV and, thereby, inhibition of the ACVS. The influence of dissolved oxygen on penicillin biosynthesis was examined by

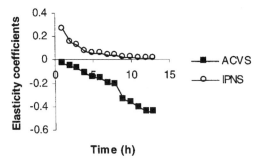

Time (h)

FIGURE 11.4   Elasticity coefficients for the two enzymes ACVS and IPNS with respect to LLD-ACV during fed-batch cultivation. Note that the elasticity coefficient for ACVS is negative.

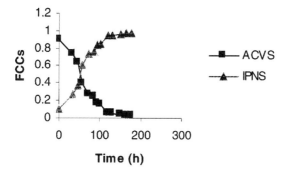

FIGURE 11.5   Flux control coefficients for the two enzymes ACVS and IPNS during fed-batch cultivation.

Pissara, *et al.*, (1996) using a complete kinetic model for the pathway, *i.e.*, with all reactions of Fig. 11.3 considered. They found that the penicillin flux is indeed sensitive to the concentration of oxygen and that at low dissolved oxygen tensions the FCC for the IPNS increases. This has recently been confirmed experimentally by Henriksen, *et al.*, (1997), who found increasing LLD-ACV concentration at low dissolved oxygen concentrations in a glucose-limited chemostat. From Eqs. (3) and (4) it is obvious that this will result in a numerically large elasticity coefficient for the ACVS-catalyzed reaction and a smaller elasticity coefficient for the IPNS-catalyzed reaction, which again means a higher FCC for IPNS.

From the elasticity coefficients, the concentration control coefficients can also be calculated using Eq. (11.47). For ACVS the concentration control coefficient is found to be about 5 in the beginning of the fed-batch cultivation, whereas it decreases to about 2 at the end of the cultivation. This shows that the metabolite pool becomes less sensitive to the activity of ACVS. This is also a consequence of the accumulation of LLD-ACV, as the feedback inhibition at high concentrations makes the reaction rate and, by extension, the metabolite concentration less sensitive to changes in enzyme activity.

11.4. MCA OF BRANCHED PATHWAYS

Application of MCA to branched pathways involves an additional degree of complexity. First, we are dealing with more than a single flux, as was the case for linear pathways. Each one of the fluxes of a branched pathway can be affected by any of the enzymatic reactions in the network, thus requiring a *matrix* of flux control coefficients for the complete description of flux control. Second, in branched pathways the number of metabolites is smaller than $L - 1$ (L being the number of reactions), and the connectivity and summation theorems in general cannot provide a sufficient number of equations to calculate the control coefficients from the elasticity coefficients. In such pathways, however, the fluxes are not independent, and mass conservation equations among the fluxes yield the additional constraints that allow, again, the determination of control coefficients from enzyme elasticities. These constraints have been referred to as structural relationships (Reder, 1988), and they are determined by the stoichiometry of the reactions in the network. It turns out that there are exactly sufficient structural relationships to supply the "missing" equations, such that the control coefficients always can be calculated from the elasticity coefficients. The general formulation of the structural relationships is relatively complex. To facilitate the presentation of the general case, we are going to discuss first the simple branched pathway depicted in Fig. 11.6.

The three fluxes J_1, J_2, and J_3 of Fig. 11.6 are not independent at steady state because they are related through a mass balance around the pathway metabolite X:

$$J_1 = J_2 + J_3 \tag{11.48}$$

or

$$1 = f_{12} + f_{13} \tag{11.49}$$

where f_{1k} are the fractional fluxes given by

$$f_{1k} = \frac{J_K}{J_1} \qquad k \in \{2, 3\} \tag{11.50}$$

If a perturbation is introduced such that the flux J_2 changes while J_1 is kept constant, the structural relationship of [Eq. (11.48)] directly shows that J_3

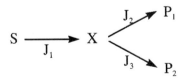

FIGURE 11.6 Simple branched pathway with one branch point metabolite X, one substrate S, and two metabolic products P_1 and P_2.

must also change. This imposes a constraint on the flux control coefficients which was first derived by Kacser (1983):

$$-f_{12}C_3^{J_1} + f_{13}C_2^{J_1} = 0 \tag{11.51}$$

Combined with the summation and connectivity theorems this equation allows the calculation of the FCCs from the elasticity coefficients and the fractional flux through one of the branches:

$$\begin{pmatrix} C_1^{J_1} \\ C_2^{J_1} \\ C_3^{J_1} \end{pmatrix} = \frac{1}{\varepsilon_X^1 - f_{12}\varepsilon_X^2 - (1 - f_{12})\varepsilon_X^3} \begin{pmatrix} -f_{12}\varepsilon_X^2 - (1 - f_{12})\varepsilon_X^3 \\ f_{12}\varepsilon_X^1 \\ (1 - f_{12})\varepsilon_X^1 \end{pmatrix} \tag{11.52}$$

In Eq. (11.51) the reference flux is taken to be J_1, but similar structural relationships hold for the FCCs of the other two fluxes in the pathway:

$$(1 - f_{12})C_1^{J_2} + C_3^{J_2} = 0 \tag{11.53}$$

$$f_{12}C_1^{J_3} + C_2^{J_3} = 0 \tag{11.54}$$

Combination of Eq. (11.53) [or (11.54)] with the summation and connectivity theorems yields similar expressions for the FCCs of the other two fluxes J_2 and J_3 [see Eqs. (11.68) and (11.69)].

In more complex pathways where several reactions may be occurring at each branch point, similar equations can be derived. Fell and Sauro (1985) derived Eq. (11.55) by relating the FCCs of the different branches leading to or emanating from a branch point metabolite X_j. The flux leading to the metabolite is taken to be J_i, and the fractional flux through the kth branch is f_{ik}, i.e., the flux through this branch is J_k.

$$-f_{1k} \underset{\text{Branch } k}{\sum} C_k^{J_i} + (1 - f_{ik}) \underset{\text{Branch } m}{\sum} C_m^{J_i} = 0 \tag{11.55}$$

By invoking the concepts of reaction grouping and top-down MCA, the sums in Eq. (11.55) are nothing but the *group* FCC of the reactions participating in the corresponding branch. As such, the preceding equation is the group analogue of Eq. (11.51). If the pathway branches further, the summation of Eq. (11.55) (or, equivalently, the reaction group) includes all the FCCs in all the subbranches of each branch from the metabolite X_j.

Notice that the signs of both terms in Eqs. (11.53) and (11.54) are positive, whereas the signs of the terms in Eq. (11.51) are opposite. If the flux J_2 is increased with J_1 held constant, J_3 must decrease and vice versa, and this explains the difference in signs in Eq. (11.51). On the contrary, if flux J_2 is increased while flux J_3 remains constant, flux J_1 must also increase and this explains the positive expressions in Eqs. (11.53) and (11.54). Obviously, stoichiometry also plays a role. This was demonstrated in the work of Reder (1988), who derived a set of general equations encompassing both the theorems of MCA and the structural relationships imposed by stoichiometric constraints. With these general equations it is possible to derive all independent relations among MCA coefficients even for very complex pathways.[7] The structural relationships derived by Reder (1988) are based on *nonscaled* control coefficients, but Ehlde and Zacchi (1997) derived similar structural relationships for scaled control coefficients as well. We provide the derivation of these general relationships next.

Consider a network of reactions comprising K internal metabolites interconnected by L enzymatic reactions. One can write K flux balances, one for each metabolite in the network. By introducing the $L \times K$ stoichiometric matrix G and the $L \times 1$ flux vector J, these flux balances can be summarized in the following matrix equation, which is identical to Eq. (8.3):[8]

$$G^T J = 0 \qquad (11.56)$$

This equation represents K linear equations relating the L steady-state fluxes, and, obviously, by rearranging the equation we can specify K of the fluxes (termed the *dependent* fluxes) as linear functions of the remaining $L - K$ fluxes (the *independent* fluxes). To do this we collect the rows of the stoichiometric matrix containing dependent fluxes in the sub-matrix G_c and the remaining rows in the sub-matrix G_0 to rewrite Eq. (11.56) as:

$$G_c^T J_{dep} + G_0^T J_{in} = 0 \qquad (11.57)$$

Submatrix G_c is square and nonsingular, and the independent fluxes therefore can be calculated from the dependent fluxes by

$$J_{dep} = -\left(G_c^T\right)^{-1} G_0^T J_{in} \qquad (11.58)$$

This equation is identical to Eq. (8.8), the basis for metabolic flux analysis of determined systems. By using Eq. (11.58), all fluxes can be specified in terms

---

[7] In the preceding analysis, we do not include metabolic cycles, for which another set of relationships can be specified (Fell and Sauro, 1985).

[8] Our definition of the stoichiometric matrix for the intracellular metabolites G does not allow linearly-dependent rows (or linearly-dependent metabolites), since it retains only one of the two compounds in cofactor couples such as $NAD^+/NADH$. In the analysis of Ehlde and Zacchi (1997) [as well as that of Reder (1988)] there is not such a requirement in the specification of the stoichiometric matrix, and their analysis therefore is more general (but also more complex).

of the dependent fluxes as follows:

$$J = L^J J_{in} \tag{11.59}$$

where the matrix L^J is given by

$$L^J = \begin{pmatrix} I_{L-K} \\ -(G_c^T)^{-1} G_0^T \end{pmatrix} \tag{11.60}$$

where I_{L-K} is an identity matrix of dimension $(L - K) \times (L - K)$. With Eq. (11.59) providing, in essence, the structural relationships among the fluxes, Ehlde and Zacchi (1997) derived a general equation for the determination of the FCC from the independent FCCs (see also Box 11.4):

$$C^J = L_F^J C_{in}^J \tag{11.61}$$

where the matrix L_F^J is generated by multiplying each element l_{jk} of L^J with the fractional flux f_{jk} given by Eq. (11.50). By substituting Eq. (11.61) into the general relation between the coefficients of MCA given by Eq. (11.27), we find:

$$L_F^J C_{in}^J = P + EC^X \tag{11.62}$$

which can be rearranged to

$$\begin{pmatrix} L_F^J & -E \end{pmatrix} \begin{pmatrix} C_{in}^J \\ C^X \end{pmatrix} = P \tag{11.63}$$

This is the general equation for the determination of the control coefficients from the elasticity coefficients, the parameter elasticity coefficients, and the fractional fluxes through the different branches of the pathway. Notice that Eq. (11.45), valid for linear pathways only, can be derived from Eq. (11.63), which now represents the general equation for all types of pathway structure. In their derivation of Eq. (11.65), Ehlde and Zacchi (1997) allowed the presence of dependent metabolites, $e.g.$, the presence of both compounds in cofactor couples. This necessitates the introduction of another matrix that specifies all the concentration control coefficients as functions of the independent concentration control coefficients. Earlier, we assumed that there are no dependent metabolites, $i.e.$, no linearly-dependent columns in the stoichiometric matrix G. Normally the stoichiometry can be arranged easily such that this requirement is satisfied, and if linearly-dependent columns are present, one or more of the linearly-dependent metabolites are removed from consideration without loss of generality.

BOX 11.4

Derivation of Equation (11.61)

The starting point for the derivation of Eq. (11.61) is the relationship between all pathway fluxes and the independent fluxes, as given by Eq. (11.59). Differentiation with respect to E_i yields:

$$\frac{d\mathbf{J}}{dE_i} = \mathbf{L}^{\mathbf{J}}\frac{d\mathbf{J}_{in}}{dE_i} \tag{1}$$

or

$$\frac{dJ_k}{dE_i} = \sum_{m=1}^{L-K} l_{km}\frac{dJ_{in,\,m}}{dE_i} \tag{2}$$

Equation (2) is multiplied by $E_i\mathbf{I}J_k$ to give:

$$C_i^{J_k} = \sum_{m=1}^{L-K} l_{km}\frac{E_i}{J_k}\frac{J_m}{J_m}\frac{dJ_{in,\,m}}{dE_i} \tag{3}$$

or

$$C_i^{J_k} = \sum_{m=1}^{L-K} l_{km}\frac{J_m}{J_k}C_{in,\,i}^{J_m} = \sum_{m=1}^{L-K} l_{km}f_{km}C_{in,\,i}^{J_m} \tag{4}$$

Since $l_{km}f_{km}$ is the element in the kth row and mth column of the matrix $\mathbf{L}_F^J$, Equation (4) is seen to be identical to Eq. (11.61).

To illustrate the application of Eq. (11.63), we return to the simple pathway of Fig. 11.6. For this pathway, the stoichiometric matrix $\mathbf{G}$ is

$$\mathbf{G} = \begin{pmatrix} 1 \\ -1 \\ -1 \end{pmatrix} \tag{11.64}$$

Because there are three fluxes ($L = 3$) and one metabolite ($K = 1$), there are two independent fluxes and one dependent flux. If J_3 is taken to be the dependent flux, L^J becomes

$$L^J = \begin{pmatrix} 1 & 0 \\ 0 & 1 \\ 1 & -1 \end{pmatrix} \tag{11.65}$$

The next step is to generate L_F^J by multiplication of the elements in L^J with f_{jk}:

$$L_F^J = \begin{pmatrix} 1 & 0 \\ 0 & 1 \\ f_{31} & -f_{32} \end{pmatrix} \tag{11.66}$$

and from Eq. (11.63):

$$\begin{pmatrix} 1 & 0 & -\varepsilon_X^1 \\ 0 & 1 & -\varepsilon_X^2 \\ f_{31} & -f_{32} & -\varepsilon_X^3 \end{pmatrix} \begin{pmatrix} C_1^{J_1} & C_2^{J_1} & C_3^{J_1} \\ C_1^{J_2} & C_2^{J_2} & C_3^{J_2} \\ C_1^X & C_2^X & C_3^X \end{pmatrix} = \begin{pmatrix} 1 & 0 & 0 \\ 0 & 1 & 0 \\ 0 & 0 & 1 \end{pmatrix} \tag{11.67}$$

When the control coefficients are calculated using Eq. (11.67), the FCCs of flux J_3 are obtained using Eq. (11.61). Solution of Eq. (11.67) may seem more tedious, involving a total of nine equations, but the result gives all the control coefficients immediately. Furthermore, there are several software packages that will easily solve the matrix equation numerically or even derive analytical expressions. Thus, by solving Eq. (11.67), we find

$$\begin{pmatrix} C_1^{J_1} & C_2^{J_1} & C_3^{J_1} \\ C_1^{J_2} & C_2^{J_2} & C_3^{J_2} \\ C_1^X & C_2^X & C_3^X \end{pmatrix} = \frac{1}{f_{31}\varepsilon_X^1 - f_{32}\varepsilon_X^2 - \varepsilon_X^3}$$

$$\times \begin{pmatrix} -\varepsilon_X^3 - f_{32}\varepsilon_X^2 & f_{32}\varepsilon_X^1 & \varepsilon_X^1 \\ -f_{31}\varepsilon_X^2 & f_{31}\varepsilon_X^1 - \varepsilon_X^3 & \varepsilon_X^2 \\ f_{31}\varepsilon_X^3 & f_{32}\varepsilon_X^3 & f_{31}\varepsilon_X^1 - f_{32}\varepsilon_X^2 \end{pmatrix}$$

$$\tag{11.68}$$

and by using Eq. (11.61) we find the FCCs with respect to flux J_3:

$$
\begin{pmatrix} C_1^{J_3} \\ C_2^{J_3} \\ C_3^{J_3} \end{pmatrix} = \frac{1}{f_{31}\varepsilon_X^1 - f_{32}\varepsilon_X^2 - \varepsilon_X^3} \begin{pmatrix} -f_{31} \\ f_{32} \\ 1 \end{pmatrix} \tag{11.69}
$$

The first row in Eq. (11.68) is identical to the solution found in Eq. (11.52) if multiplied by J_3 and divided by J_1 [whereby the fractional fluxes are converted to f_{12} and f_{13} and applied to Eq. (11.49)]. In addition to giving all the FCCs, the matrix solution also directly gives the concentration control coefficients.

EXAMPLE 11.2

Gluconeogenesis and Glycolysis in Rat Liver Cells

Very few studies are reported where the FCCs of a pathway have been determined by measuring all the relevant elasticity coefficients. In a detailed study of gluconeogenesis in rat liver cells, however, Groen, et al., (1983, 1986) determined all the FCCs of this pathway at high lactate and pyruvate concentrations. The pathway is depicted in Fig. 11.7. In their analysis, they determined the elasticity coefficients either from enzyme kinetics or from single or double modulation. In the determination of elasticity coefficients from enzyme kinetics, they applied Eqs. (2) and (3), which are based upon reversible Michaelis-Menten kinetics:

$$
v = v_f - v_r = \frac{v_{f,max}\, c_S / K_S - v_{r,max}\, c_P / K_P}{1 + c_S / K_S + c_P / K_P} \tag{1}
$$

where c_S and c_P are the concentrations of the substrate and product of the enzyme-catalyzed reaction, respectively. $v_{f,max}$ and $v_{r,max}$ are the maximum forward and reverse reaction rates, respectively. With these kinetics, the elasticity coefficients for the reaction with respect to the substrate and the product are given by

$$
\varepsilon_S^v = \frac{c_S}{v}\frac{\partial v}{\partial c_S} = \frac{1}{1 - \Gamma / K_{eq}} - \frac{v_f}{v_{f,max}} \tag{2}
$$

$$
\varepsilon_P^v = \frac{c_P}{v}\frac{\partial v}{\partial c_P} = \frac{1}{1 - \Gamma / K_{eq}} - \frac{v_r}{v_{r,max}} \tag{3}
$$

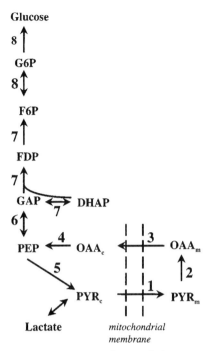

FIGURE 11.7   Overview of gluconeogenesis. The specified reactions are: (1) mitochondrial pyruvate translocator; (2) pyruvate carboxylase (PC); (3) transport of OAA; (4) PEP carboxykinase (PEPCK); (5) pyruvate kinase (PYK); (6) enolase, phosphoglycerate mutase (PGM), glyceraldehyde-3-phosphate dehydrogenase (GAPDH), and 3-phosphoglycerate kinase (3PG); (7) triosephosphate isomerase (TPI), aldolase, fructose-1,6-bisphosphatase; (8) phosphoglucoisomerase (PGI), glucose-6-phosphatase (G6PE).

where v_f and v_r are the rates of the forward and reverse reactions, respectively. Γ is the mass action ratio given by

$$\Gamma = \frac{c_P}{c_S} \tag{4}$$

and K_{eq} is the equilibrium constant given by:

$$K_{eq} = \frac{c_{P,eq}}{c_{S,eq}} = \frac{v_{f,max}}{v_{r,max}} \frac{K_P}{K_S} \tag{5}$$

If the enzyme operates far from equilibrium, i.e., $\Gamma/K_{eq} \ll 1$, the elasticity coefficients are almost exclusively determined by the last terms of Eqs. (2) and (3), i.e., by the degree of saturation of the enzyme by the substrate and product. On the other hand, if the enzyme operates close to equilibrium, *i.e.*, $\Gamma/K_{eq} \approx 1$, the elasticity coefficients are determined mainly by the first terms of Eqs. (2) and (3).

Table 11.2 summarizes the determined elasticity coefficients. For the pyruvate translocator (reaction 1), measurements of pyruvate in both the mitochondria and the cytosol (done at high saturating concentrations of lactate and pyruvate) showed that the ratio Γ/K_{eq} was 0.86 ± 0.11, and therefore it was concluded that this step operates close to equilibrium. Pyruvate carboxylase operates far from equilibrium, and the elasticity coefficients therefore were determined from the last terms of Eqs. (2) and (3). For calculation of the elasticity coefficient with respect to pyruvate the forward rate of the reaction was determined from the flux into glucose and the flux through the pyruvate kinase reaction. Similarly the maximum forward rate of the reaction was determined from the maximum flux into glucose. The elasticity coefficients of the oxaloacetate transport step were determined by double modulation. By varying the substrate concentration or by inhibiting PEP carboxykinase with 3-mercaptopicolinic acid the flux through this step could be varied. From measurements of the corresponding changes in the oxaloacetate concentrations the elasticity coefficients could be determined. For the PEP carboxykinase the elasticity coefficients were also calculated using Eqs. (2) and (3). The forward reaction rate was determined as the sum of the net flux and the reverse reaction rate, *i.e.*, $v_f = v + v_r$, and because $v_r/v_f = \Gamma/K_{eq}$, we have

$$v_f = \frac{v}{1 - \Gamma/K_{eq}} \qquad (6)$$

The net flux through the reaction was again determined from the rate of glucose formation and the flux through the pyruvate kinase reaction, and

TABLE 11.2   Elasticity Coefficients for the Pathway of Figure 11.7[a]

Reaction	PYR_c	PYR_m	OAA_m	OAA_c	PEP	GAP	G6P
1	7.1	-6.1					
2		0.05	-0.04				
3			0.86	-0.74			
4				0.35	-0.09		
5					3.5		
6					2.0	-1.0	
7						1.2	-0.08
8							1.0

[a] Cells without values specified are assumed to have elasticity coefficients of zero.

from measurements of the mass action ratio the forward reaction rate could be determined. The elasticity coefficient for the *pyruvate kinase* was determined by single modulation (note that this reaction is assumed to be practically irreversible with the elasticity coefficients with respect to pyruvate being zero). By varying the PEP concentration and measuring the corresponding flux, the elasticity coefficient could be determined directly. For the group of reactions enolase, phosphoglycerate mutase (PGM), glyceraldehyde-3-phosphate dehydrogenase (GAPDH), and 3-phosphoglycerate kinase (3PG), it was found that the overall reaction is close to equilibrium under all conditions, and Γ/K_{eq} was measured to be 0.51. The group of reactions triosephosphate isomerase (TPI), aldolase, and fructose-1,6-bisphosphatase was considered as one, and the elasticity coefficients were determined from Eq. (7):

$$
\frac{dJ_{glc}}{J_{glc}} = \varepsilon_{GAP}^{7}\frac{dc_{GAP}}{c_{GAP}} + \varepsilon_{F6P}^{7}\frac{dc_{F6P}}{c_{F6P}} + \varepsilon_{P_i}^{7}\frac{dc_{P}}{c_{P}}
\tag{7}
$$

The elasticity coefficient with respect to free phosphate was assumed to be very small, and the last term therefore was neglected. The elasticity coefficient with respect to fructose-6-phosphate (which was assumed to make up a common pool with glucose-6-phosphate) was determined to be -0.08 from knowledge of enzyme kinetics. The term dc_{F6P}/c_{F6P} is not very large, so the second term in Eq. (7) therefore could also be neglected. This allowed determination of the elasticity coefficients with respect to GAP from measurements of the flux at different concentrations of GAP. Finally, the glucose-6-phosphatase-catalyzed reaction is known to operate far from equilibrium. The glucose-6-phosphate concentration was found to be far below the K_m, and the elasticity coefficient therefore is about 1 at all conditions.

From the determined elasticity coefficients, the flux control coefficients were calculated and the results collected in Table 11.3. It is observed that for several of the steps that operate close to equilibrium the FCC is very low. However, the combined conversion of PEP to fructose-1,6-bisphosphate, which was assumed to be at equilibrium, has a relatively high FCC, and therefore it is not possible to draw the general conclusion that equilibrium reactions do not express flux control. The highest FCC is for the pyruvate carboxylase reaction (notice that this step also has the lowest elasticity coefficients), and this reaction certainly is a key reaction in gluconeogenesis. It is, however, interesting that the PEP carboxykinase has a very low FCC, but this must be due to the tight coupling of this reaction with the pyruvate carboxylase, which has a high flux control. Pyruvate kinase obviously has a negative FCC, and this demonstrates the appearance of negative FCCs in branched pathways.

TABLE 11.3   Flux Control Coefficients for the Pathway of Figure 11.7[a]

Reaction	FCCs
1. Pyruvate translocator	0.004
2. Pyruvate carboxylase	0.51
3. OAA transport	0.02
4. PEP carboxykinase	0.05
5. Pyruvate kinase	-0.17
6. Enolase/PGM/GAPDH/3PK	0.29
7. TIM/aldolase/fructose-1,6-bisphosphatase	0.27
8. PGI/glucose-6-phosphatase	0.02

[a] The FCCs are for the flux into glucose.

Groen, et al., (1983, 1986) also determined the FCCs at conditions other than high lactate concentrations, (corresponding to 5 mM). For decreasing lactate concentrations the FCC for pyruvate carboxylase increased further (up to 0.75 at 0.5 mM lactate), and when glucagon, a stimulator of gluconeogenesis, is added, there is a complete shift in flux control. Thus, pyruvate kinase expresses no flux control, and the FCC for pyruvate carboxylase increases to 0.83. Furthermore, the conversion of PEP to fructose-1,6-bisphosphate has a very low FCC (0.003), and the further conversion of fructose-1,6-bisphosphate to glucose also has a low FCC (0.03). The FCCs for the other steps are approximately the same. Thus, at these conditions, PEP carboxykinase is the step with the second largest FCC (0.08).

The analysis of Groen, et al., (1983, 1986) is quite thorough and is, as mentioned earlier, a good example of analysis of a complete pathway. The determined FCCs are, however, associated with quite high standard deviations, as demonstrated by Ehlde (1995). The standard deviation for the estimated FCCs range from about 50% up to 275%, and is especially high for small FCCs. But even with these high standard deviations, it is still clear that flux control is mainly at the pyruvate carboxylase.

EXAMPLE 11.3

Oxidative Phosphorylation

Brand and coworkers have analyzed the control of oxidative phosphorylation in isolated mitochondria (Brand, et al., 1988; Brown, et al., 1990b; Hafner, et al., 1990). The system was represented by three groups of reactions, which are linked through the proton-motive force Δp, as illus-

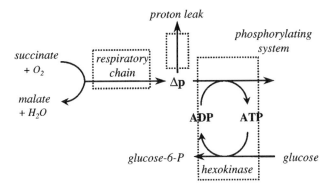

FIGURE 11.8 Schematic overview of the three groups of reactions in respiration. Succinate serves as substrate for respiration, and when electrons are pumped out of the mitochondria as a result of transfer of electrons through the electron transport chain, a proton-motive force Δp is generated (see also Fig. 2.11). The proton-motive force may drive ATP production by the F_0F_1-ATPase, or it may leak. To regenerate ADP, which is a substrate for the ATPase, hexokinase is added together with glucose. The phosphorylating system also includes the adenine-nucleotide translocator, which transports ATP and ADP through the mitochondrial membrane.

trated in Fig. 11.8. With this grouping of reactions they applied the top-down approach. Notice that this approach can be applied here as the proton-motive force is the only link between the three groups of reactions.

In the analysis the proton-motive force was perturbed in two different ways:

- By the addition of varying amounts of a proton uncoupler [carbonyl cyanide p-trifluoro-methoxyphenylhydrazone (FCCP)], which transports protons across the mitochondrial membrane.
- By the addition of malonate, which inhibits the respiratory chain.

Thereby the elasticity coefficients for each group reaction with respect to Δp could be determined from single modulation experiments. From linear regression to data of respiratory flux vs. Δp the elasticity coefficients for proton leak and respiration at *non-phosphorylating conditions* were estimated,[9]

$$\varepsilon_{\Delta p}^{\text{resp}} = -18.7; \quad \varepsilon_{\Delta p}^{\text{leak}} = 7.9 \tag{1}$$

[9] Ehlde (1995) analyzed the data of Brand, *et al.*, (1988) by using linear regression in a double logarithmic plot, which gave a much better fit to the experimental data. From this analysis he derived the elasticity coefficients to be $\varepsilon_{\Delta p}^{\text{resp}} = -12.9$; $\varepsilon_{\Delta p}^{\text{leak}} = 10.9$. This corresponds to the following FCCs: $C_{\text{leak}}^{J_s} = 0.54$; $C_{\text{resp}}^{J_{\text{resp}}} = 0.46$ which are quite different from those of Eq. (2).

which correspond to the following FCCs:

$$C_{\text{leak}}^{J_{\text{resp}}} = 0.70; \quad C_{\text{resp}}^{J_{\text{resp}}} = 0.30 \tag{2}$$

Thus, leak has a significant control of flux at these conditions, *i.e.*, removal of the proton motive force has a higher flux control than generation of the proton-motive force.

In analysis of the complete system, Brown, *et al.*, (1990b) used an inhibitor (oligomycin) of the ATPase together with FCCP, and from inhibitor experiments they determined the elasticity coefficients with respect to Δp and the FCCs in fed cells. Table 11.4 summarizes their results. For the respiratory flux all FCCs are positive, and for non-phosphorylating conditions, flux control is mainly in the processes removing the proton motive force, with the phosphorylation having the highest FCC. This is consistent with the findings of Groen, *et al.*, (1982), who found low FCCs for the step involved in respiration (see also Table 11.1). For the fluxes of proton leak and phosphorylation, there is highest flux control by their own steps, whereas the other step draining proton-motive force has a negative FCC. Brown, *et al.*, (1990b) also investigated starved cells and found very similar FCCs, which indicates that the preceding conclusions are generally valid for the system.

In another study, the flux control over respiration was determined at different respiration rates (Hafner, *et al.*, 1990). At low respiration rates, flux control was mainly by proton leak. For increasing respiration rates, flux control by the proton leak decreased to a very low value (almost zero), whereas flux control by phosphorylation increased to a value of about 0.8. For very high respiration rates the flux control by phosphorylation decreases again, and flux control is shared approximately equally between phosphorylation and respiration. Similarly, the flux control on the two other steps were determined at varying respiration rates. For the phosphorylating flux, phosphorylation itself had the most control except at very high respiration rates, where flux control is shared almost equally between phosphorylation and respiration. For the proton leak the picture is, however, more complicated.

TABLE 11.4 Elasticity Coefficients and FCCs for Oxidative Phosphorylation

Property	Respiration	Leak	Phosphorylation
$\varepsilon_{\Delta p}^{J_i}$	-7.56 ± 0.21	4.69 ± 0.78	$2.77 + 1.03$
$C_i^{J_{\text{resp}}}$	0.29 ± 0.05	0.22 ± 0.04	0.49 ± 0.04
$C_i^{J_{\text{leak}}}$	0.42 ± 0.03	0.87 ± 0.01	-0.29 ± 0.04
$C_i^{J_{\text{phos}}}$	0.23 ± 0.07	-0.07 ± 0.03	0.84 ± 0.05

At low respiration rates, flux control is mainly at the group of leak reactions itself. At high respiration rates there is still a high flux control by the leak reaction on its own flux, but now there is a large negative flux control by phosphorylation (a FCC of about -1.0) and a large flux control by respiration (a FCC of about 1.0). This demonstrates the more complex interpretation of FCCs in branched pathways, where an FCC close to 1.0 does not necessarily mean that the other steps have no flux control.

The analysis of Brand and co-workers is, as mentioned earlier, based on lumping of the individual steps in the oxidative phosphorylation into three groups of reactions (using the top-down approach). If one wants to obtain further information about the individual reactions in a group, a more detailed analysis is required. Often, however, it is possible to draw at least some qualitative conclusions on the contribution of the individual steps in the reaction groups, and the group FCCs therefore directly give information about the individual steps. In the phosphorylation reaction group of Fig. 11.8, the group FCC represents contributions from both the ATPase and the hexokinase. Since hexokinase and glucose are normally added in excess, the FCC for the hexokinase reaction is likely to be low, which is indeed what was found by Groen, *et al.*, (1983) in their analysis (see also Table 11.1). Thus the FCC for the phosphorylation group is likely to be identical to the FCC for the F_0F_1-ATPase.

11.5. THEORY OF LARGE DEVIATIONS

A central tenet of metabolic control analysis is the definition of the elasticities and control coefficients. In fact, the key relationships between these quantities (*i.e.*, the summation and connectivity theorems) are direct consequences of the definitions of these coefficients. It is important to realize, however, that the coefficients of MCA are defined only as full or partial derivatives and, as such, are directly measurable only through infinitesimal changes in systemic variables (*i.e.*, concentrations, activities, and fluxes). Alternatively, detailed kinetic models may be used for the mathematical calculation of the required derivatives, thereby allowing MCA theorems to be invoked for the calculation of the remaining quantities. We discussed earlier in Box 11.3 the dual difficulty of experimentally realizing a small perturbation that is also minimally corrupted by experimental error. Realistically, a change in a systemic variable can be quantified experimentally only if it is above a measurable threshold and, therefore, finite. Because the larger the perturbation, the smaller the experimental error, it is obviously desirable to

514

Metabolic Engineering

attempt to estimate control coefficients from large systemic perturbations, provided, of course, that the calculations provide accurate estimates of the control coefficients. Until recently, no method was available that could relate finite experimental results to the derivative coefficients of MCA. A new set of relationships, developed as an extension of MCA (Small and Kacser, 1993a, b), allows *large deviations* from a steady state to be used in the determination of elasticities and control coefficients. These relationships and their resulting implications are discussed in detail in this section. For simplicity, we start the discussion with a description of unbranched networks, followed by an extension of the theories to branched networks.

11.5.1. Unbranched Networks

As discussed in Section 11.1.1, the flux through an unbranched pathway increases as the activity of any particular enzyme is amplified. The degree of flux amplification achieved by activity enhancement depends on the degree of control exercised by the original activity, as defined by the FCC at the initial activity (see Fig. 11.1).

According to the definition of the flux control coefficient for a particular enzyme (Eq. 11.2), FCCs are obtained from an infinitesimal change in that enzyme's activity by determining the slope of the corresponding flux-activity curve (Fig. 11.1) at the steady-state condition. Clearly, at low enzyme activity an enzyme affects strongly the pathway flux; this is reflected by a large flux control coefficient. At higher activity, however, an enzyme has little effect on the flux, resulting in a small FCC. It should be noted that different numerical values for the FCC are obtained at different activities, as the slopes are measured at different sections of the flux vs. activity curve of Fig. 11.1.

A finite change in enzyme activity, on the other hand, will only approximate the slope of the flux-activity curve at the original steady state point, resulting in an inexact value of the flux control coefficient. Consider the effect of a large change in enzyme activity, as indicated in Fig. 11.9. Clearly, the slope of this large change, measured by $\Delta J / \Delta E$, does not closely approximate dJ/dE at the original point. Therefore, a method is needed to facilitate the determination of the true FCC from finite perturbation experiments. The solution to this seeming incongruity will become evident, through the use of Small and Kacser's theory of large deviations.

Basis for Large Deviation Theory

It is not possible to describe fully an arbitrary flux-enzyme relationship, as it is generally very complex and highly nonlinear. It is instructive, however,

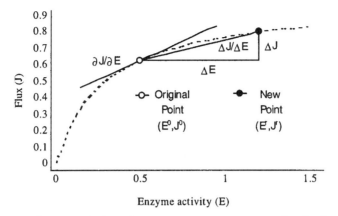

FIGURE 11.9   Illustration of a large deviation in enzyme activity. Note that the slope $\Delta J/\Delta E$ differs from dJ/dE, the intercept slope at the original point. Consequently, direct replacement of dJ/dE with $\Delta J/\Delta E$ in the calculation of the flux control coefficient will lead to an incorrect result.

to instead define the simplest system that yields the expected flux-enzyme relationship depicted by the curve of Fig. 11.1. Hence, Small and Kacser (1993a) simplified the standard formulation for the rate v of a reversible unimolecular enzymatic transformation of X_i to X_j:

$$v = \cfrac{\dfrac{E \cdot k_{\mathrm{cat}}}{K_{\mathrm{m}_i}}\left(c_i - \dfrac{c_j}{K}\right)}{1 + \dfrac{c_i}{K_{\mathrm{m}_i}} + \dfrac{c_i}{K_{\mathrm{m}_j}}} \qquad (11.70)$$

to an approximate form that is linear with respect to the enzyme activity:

$$v = e\left(c_i - \frac{c_j}{K}\right) \qquad (11.71)$$

in which

$$e = \frac{E \cdot k_{\mathrm{cat}}}{K_{\mathrm{m}_i} k_{\mathrm{s}}} = \frac{v_{\mathrm{max}}}{K_{\mathrm{m}_i} k_{\mathrm{s}}} \qquad (11.72)$$

where

$$k_{\mathrm{s}} = 1 + \frac{c_i}{K_{\mathrm{m}_i}} + \frac{c_j}{K_{\mathrm{m}_j}} \cong \text{constant} \qquad (11.73)$$

In the preceding equations k_{cat} is the reaction rate constant in the product-producing step of a Michaelis-Menten mechanism, and the subscribed K_m parameters are the equilibrium constants of the enzyme-substrate complex formation-dissociation reactions. The saturation function k_s is constant under the following conditions: (a) if $c_i \ll K_{m_i}$ and $c_j \ll K_{m_j}$ when $k_s \cong 1$; or (b) if there is no significant change in enzyme saturation with respect to its substrates resulting from an observed change in enzyme activity. Through application of the large deviation theory to numerous systems, this simplification was shown a posteriori to be valid for a wide range of conditions (Small and Kacser, 1993a, b). If the constants in Eq. (11.71) are further combined, it is clear that this form is equivalent to a reversible, first-order reaction:

$$v = kc_i - k_{-1}c_j \qquad (11.74)$$

Consider now a pathway comprising a linear series of reactions converting X_0 to X_n:

$$X_0 \overset{e_1}{\leftrightarrow} X_1 \overset{e_2}{\leftrightarrow} X_2 \overset{e_3}{\leftrightarrow} X_3 \Lambda X_{n-1} \overset{e_n}{\leftrightarrow} X_n \qquad (11.75)$$

and assume that the reaction rate of each of the reaction steps in the series is given by an equation similar to Eq. (11.71), in which the activity and equilibrium constant for the reaction catalyzed by enzyme i are e_i and K_i, respectively. The flux J through such a pathway can be shown to be equal to (Kacser and Burns, 1973)

$$J = \frac{c_0 - \dfrac{c_n}{K_{1:n}}}{\dfrac{1}{e_1} + \dfrac{1}{e_2 K_1} + \dfrac{1}{e_3 K_{1:2}} + \Lambda + \dfrac{1}{e_n K_{1:n}}} \qquad (11.76)$$

in which

$$K_{1:n} = \prod_{i=1}^{n} K_i \qquad (11.77)$$

Note that this equation for the flux is nonlinear with respect to the activity of any particular enzyme. In fact, Eq. (11.76) can be rewritten as a function of the activity of any enzyme E_i, resulting in a general hyperbolic relationship:

$$J = \frac{E_i \cdot \chi_1}{1 + E_i \cdot \chi_2} \qquad (11.78)$$

in which χ_1 and χ_2 are different for each enzyme, but always contain only constants in the form of kinetic constants and fixed external concentrations. Note also that, if the activities of all other enzymes remain constant, the form of Eq. (11.78) not unexpectedly yields the shape of the flux-activity curve of Fig. 11.1, confirming that the simplified kinetics of Eq. (11.71) indeed fits that primary criterion.

Derivation of Expressions for the Estimation of Flux Control Coefficients

Recall that the goal of developing a large-deviation theory is to relate finite changes in concentrations and fluxes to the infinitesimal basis of MCA. With this in mind, we can apply the general flux-activity expression of Eq. (11.78) to the FCC definition of Eq. (11.2) to obtain the following equation for the control coefficient of flux J with respect to enzyme activity E_i:

$$C_i^J = \frac{1}{1 + E_i^0 \cdot \chi_2} \tag{11.79}$$

in which E_i^0 is the enzyme activity at the original steady state.

Now let E_i^r be the enzyme activity in the perturbed steady state following a change in the initial enzyme activity E_i^0 by a factor r, so that $E_i^r = r \cdot E_i^0$, and J^0 and J^r represent the original and perturbed fluxes, respectively. Because it is clear from Fig. 11.9 that the slope of a finite change cannot be substituted for the derivative in the calculation of the flux control coefficient from Eq. (11.2), it is necessary to find a new expression employing $\Delta J/\Delta E$. To calculate this ratio, we consider the shift from the original flux,

$$J^0 = \frac{E_i^0 \cdot \chi_1}{1 + E_i^0 \cdot \chi_2} \tag{11.80}$$

to its perturbed state, J^r,

$$J^r = \frac{r \cdot E_i^0 \cdot \chi_1}{1 + r \cdot E_i^0 \cdot \chi_2} \tag{11.81}$$

to obtain the slope of the large deviation in Fig. 11.9 as

$$\frac{\Delta J}{\Delta E} = \frac{J^r - J^0}{E_i^r - E^0} = \frac{\chi_1}{\left(1 + r \cdot E_i^0 \cdot \chi_2\right)\left(1 + E_i^0 \cdot \chi_2\right)} \tag{11.82}$$

By comparison with Eq. (11.79), it is clear that, if the preceding differences were normalized using the values of the *perturbed* flux and activity (as opposed to the original ones),

$$\frac{E_i^r}{J^r} = \frac{1 + r \cdot E_i^0 \cdot \chi_2}{\chi_1} \tag{11.83}$$

they would yield an expression for the normalized slope that, *although derived from finite differences, is identical to the actual flux control coefficient:*

$$D_i^J \equiv \frac{\Delta J}{\Delta E} \frac{E_i^r}{J^r} = \frac{1}{1 + E_i^0 \cdot \chi_2} = C_i^J \tag{11.84}$$

Small and Kacser (1993a) termed the new coefficient D_i^J a *deviation index*, in order to differentiate it from the flux control coefficient. Nevertheless, Eq. (11.84) shows that the deviation index provides a close estimate of its corresponding flux control coefficient for any unbranched pathway that satisfies the conditions discussed earlier. It is especially important to realize that the use of *perturbed* values in the normalization of the slope in Eq. (11.84) is entirely correct, even though the *original* values are used in the definition of the FCC in Eq. (11.2). This allows the accurate determination of flux control coefficients using *either one* of the two slopes shown on the flux-activity curve of Fig. 11.9.

Some important properties of the deviation index should be noted. First, the deviation index found through Eq. (11.84) is *independent* of r, the factor by which the enzyme activity was changed. As a result, any number of deviation indices, calculated from different magnitude changes from a steady state, should each be equal to the flux control coefficient at the original steady state. This provides a convenient internal check of the validity of FCC determination if more than one perturbation experiments are available. Second, because the hyperbolic flux-activity relationship shown in Eq. (11.78) is completely general for a pathway, the result of Eq. (11.84) is applicable to any enzyme in the pathway (each with its own parameter χ_2, of course). Third, the deviation indices of the enzymes in a linear pathway should sum to unity, just the same as the flux control coefficients. The partial sum of the deviation indices of a set of enzymes in a group similarly is equal to the group deviation index of the set of enzymes. Finally, as the finite change $\Delta J / \Delta E$ approaches an infinitesimal change, the definition of the deviation index converges to that of the flux control coefficient, as the initial and final values are then physically and mathematically indistinguishable, while the slope approaches dJ/dE.

EXAMPLE 11.4

Determination of Flux Control Coefficients from Large Deviations

Tryptophan metabolism in rat hepatocytes, studied by Salter, *et al.*, (1986), was one of the examples used by Small and Kacser (1993a) in introducing the theory of large deviations. This metabolism was considered to consist of two steps, tryptophan uptake by the cells and tryptophan degradation by tryptophan 2,3-dioxygenase. By using hormonal induction, the activity (v_{max}) of tryptophan 2,3-dioxygenase was increased to 101 μmol (g DW h)$^{-1}$ from its basal value of 13.7. The measured flux through this pathway was found to be 7.8 μmol (g DW h)$^{-1}$ under the induced conditions, compared to 2.6 in the base state. The researchers wished to know whether transport or degradation was the primary factor controlling the flux through the base state and to estimate the flux control coefficients of each step.

According to Eq. (11.84), the flux control coefficient of the pathway flux with respect to tryptophan 2,3-dioxygenase is equal to its deviation index. This quantity can be determined readily from the provided activity and flux measurements:

$$C_{TD}^J = D_{TD}^J \equiv \frac{\Delta J}{\Delta E} \frac{E_{TD}^r}{J^r} = \frac{(7.8 - 2.6)}{(101 - 13.7)} \frac{101}{7.8} = 0.77$$

By the summation theorem, the flux control coefficients for these two enzymes must total 1; therefore, the FCC of the transport step is 0.23. Clearly, the degradation enzyme is the primary controlling step in the base state. (Salter, *et al.*, also reached the same conclusion. Without the benefit of the large deviation theory, however, flux and activity measurements had to be plotted for seven different enzyme activities and the FCC estimated from the slope of the plot.)

Interestingly, these data allow another question to be answered: Is tryptophan 2,3-dioxygenase activity still the controlling factor in the induced state? By swapping the definitions of the base and perturbed states in the preceding calculation, it is possible to determine the flux control coefficients for the induced state. In doing so, it was determined that the flux control coefficient of the enzyme at the induced condition was, in fact, 0.31 and that of the transport step was 0.69. Thus, in the induced state, tryptophan uptake is the primary limitation of the pathway flux.

Predicting Changes in Flux

In the preceding section, it was demonstrated that measurements of the altered fluxes resulting from a change in enzyme activity can be used to estimate flux control coefficients. Conversely, the theory also provides the means for estimating the new fluxes that will result from a particular change in an enzyme activity, given the value of the appropriate FCCs. Such calculations are useful in determining the efficacy of potential genetic alterations before embarking on an elaborate program of genetic engineering. Simply stated, this estimate requires only a rearrangement of Eq. (11.84):

$$\frac{\Delta J}{J^r} = C_i^J \frac{\Delta E}{\Delta E_i^r} \tag{11.85}$$

A *dimensionless* formulation of Eq. (11.85) proves more convenient for flux estimation. This formulation requires the definition of two *amplification factors*, which characterize each flux and activity change. The *activity amplification factor* (*AAF*), r_i, describes the extent by which an enzyme activity E_i has been altered:

$$r_i = \frac{E_i^r}{E_i^0} \tag{11.86}$$

The *flux amplification factor* (*FAF*), f, is the ratio of the perturbed flux to the original:

$$f = \frac{J^r}{J^0} \tag{11.87}$$

By introducing these two dimensionless factors, Eq. (11.85) can be rewritten in a form that allows one to determine the projected flux at the new steady state from the value of the original flux and the flux amplification factor:

$$f = \frac{1}{1 - C_i^J \left(\dfrac{r_i - 1}{r_i} \right)} \tag{11.88}$$

The symmetry of this expression is revealed more easily by the following rearrangement:

$$\frac{f-1}{f} = C_i^J \left(\frac{r_i - 1}{r_i} \right) \tag{11.89}$$

Up to this point, we have considered only the amplification of a single enzyme. The large deviation theory can also be applied to situations involving the alteration of the activities of multiple enzymes (Small and Kacser, 1993a). In this case, a different activity amplification factor will be applied to each of the n enzymes in the pathway, and the flux change can be found from

$$\frac{f-1}{f} = \sum_{i=1}^{n} \left[C_i^J \left(\frac{r_i - 1}{r_i} \right) \right] \tag{11.90}$$

Keep in mind that Eq. (11.90) employs a different FCC and AAF for each of the n enzymes. Furthermore, even though this summation includes the unaltered enzymes, it should be noted that unaltered enzymes have r_i values of 1, resulting in the addition of a null term. Hence, the single-enzyme relationship of Eq. (11.89) is found to be a special case of the more general relationship in Eq. (11.90), as every term but one is 0.

EXAMPLE 11.5

Estimation of the Flux Change Resulting from a Change in Enzyme Activity

In their exploration of tryptophan metabolism in rat liver hepatocytes, Salter, et al., (1986) measured the effects of several levels of tryptophan 2,3-dioxygenase beyond those discussed in Example 11.4. Two such experiments measured enzyme activities (v_{max}) of 30 and 58 μmol (g DW h)$^{-1}$, respectively, with corresponding fluxes of 4.6 $\pm$ 0.35 and 6.7 $\pm$ 0.35 μmol (g DW h)$^{-1}$. Given that the pathway flux was 2.6 μmol (g DW h)$^{-1}$ under basal conditions, with a measured tryptophan 2,3-dioxygenase activity of 13.7 μmol (g DW h)$^{-1}$ and an FCC of 0.77, are the results of these two experiments consistent with those predicted by the theory of large deviations?

For the first of these two experiments, the activity amplification factor is 30/13.7 or 2.19. By using Eq. (11.88), the flux amplification factor for this experiment can be found:

$$f = \cfrac{1}{1 - C_{TD}^J \left(\cfrac{r_{TD} - 1}{r_{TD}} \right)} = \cfrac{1}{1 - 0.77 \left(\cfrac{2.19 - 1}{2.19} \right)} = 1.72$$

By multiplying the flux of the basal state by this result, the predicted flux for this experiment is 4.47 μmol (g DW h)$^{-1}$. The same steps for the second experiment yields a predicted flux of 6.31 μmol (g DW h)$^{-1}$. Because neither of these predictions lies far outside one standard deviation of the flux measurements, it must be concluded that these results are consistent.

11.5.2. BRANCHED NETWORKS

Metabolic networks generally include various branch points, at which a species S can be used by two different pathways, as shown in Fig. 11.10. In fact, nearly any network architecture can be constructed by connecting various unbranched pathways at particular branch points, often building a complex interweaving of branches. In addition, some networks may involve convergent branch points, in which S is produced by two different pathways and used by a single pathway. Because of the simplifying assumption of linearized kinetics made in Eq. (11.71), the mathematical treatment of a branch point by the theory of large perturbations is functionally equivalent for any of these cases.

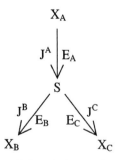

FIGURE 11.10 Illustration of a simple branched network. Each branch can consist of a single step, a series of unbranched steps, or a collection of sub-branches. Each branch is represented mathematically as a single *virtual reaction*, catalyzed by enzyme(s) of activities E_i, with linearized kinetics of the form shown in Eq. (11.71) and flux through the branch J_i.

In its formulation of branched networks, the large deviation theory assumes linearized kinetics in the form of Eq. (11.71), just as in the unbranched case. Consequently, each branch can be represented by a first-order, reversible kinetic expression:

$$v_A = E_A \cdot \left(c_A - \frac{c_S}{K_A} \right)$$

$$v_B = E_B \cdot \left(c_S - \frac{c_B}{K_B} \right) \tag{11.91}$$

$$v_C = E_C \cdot \left(c_S - \frac{c_C}{K_C} \right)$$

As a result of these linearized rate expressions, the kinetics of any branch of Fig. 11.10 will be of the form shown in Eq. (11.91), even if the branch consists of multiple steps and/or subbranches. Similarly, the flux through any subbranches must be of the same general form. Assume, for example, that branch A actually contains three reactions in a single, unbranched pathway. As shown earlier, the flux through that pathway is of the form shown in Eq. (11.76), which is of the same linearized form as Eq. (11.91). Because this flux is dependent upon the concentrations of an external species X_A and the branch point metabolite S, this does not represent a solution to the flux in the branched network, but instead defines the kinetics of the *virtual reaction* represented by branch A. In effect, this virtual reaction takes the place of the entire branch, combining the individual kinetic constants and concentrations into the constants of Eq. (11.91). By similar arguments, branches B and C can also be represented as virtual reactions of the form shown in Fig. 11.10, with kinetics equivalent to those of Eq. (11.91).

In order to determine the steady state fluxes from the kinetics of Eq. (11.91), it is necessary to invoke the balance around the branch point metabolite S:

$$J_A = J_B + J_C \tag{11.92}$$

If each rate v_i in Eq. (11.89) is replaced with its steady state flux counterpart J_i, Eq. (11.91) can be substituted into Eq. (11.92) to allow the determination of intermediate S and the fluxes through each of the three branches in terms of kinetic constants and external concentrations. The relationships between deviation indices and flux control coefficients subsequently were derived from these expressions by Small and Kacser (1993b), using methods analogous to those shown in the previous section. Because these involve rather lengthy derivations, only the results will be presented here.

In contrast to the results from the analysis of an unbranched pathway, this analysis revealed that the deviation indices in a branched system are *not necessarily* equal to the corresponding flux control coefficients. If an activity E_i is enhanced by a factor r_i, the general expression for the deviation index for the effect on flux J_k in terms of flux control coefficients is

$$D_i^{J_k} \equiv C_i^{J_k} \cdot F_i^k \qquad i, k \in \{A, B, C\} \qquad (11.93)$$

where

$$F_i^k = \cfrac{1}{1 - \left(C_i^{J_i} - C_i^{J_k}\right)\cfrac{r_i - 1}{r_i}} \qquad i, k \in \{A, B, C\} \qquad (11.94)$$

Thus, the deviation coefficient differs from its corresponding FCC by the factor F_i^k, which is dependent not only upon the flux amplification factor but also upon two flux control coefficients. Note, though, that if $i = k$ (i.e., when the effect of a change in the activity of a branch upon its own flux is measured), $F_i^k = 1$, and the deviation index *is equivalent* to the flux control coefficient. Hence,

$$D_i^{J_i} \equiv C_i^{J_i} \qquad i \in \{A, B, C\} \qquad (11.95)$$

This result should not be unexpected. Because branch i itself can be thought of as a single step or as an unbranched pathway, both the activity perturbation and the flux measurement in this case occur within a single unbranched pathway. Hence, the analysis of Section 11.5.1 would apply, as long as the branch point metabolite S does not intervene between the perturbed enzyme and the point at which the flux is measured.

Although Eq. (11.93) provides a general approach relating the deviation indices to FCCs, it is somewhat burdensome. Because the deviation index can be represented in terms of the flux and activity amplification factors introduced in Eqs. (11.86) and (11.87), respectively,

$$D_i^{J_k} \equiv \frac{\Delta J_k}{\Delta E_i} \frac{E_i^r}{J_k^r} = \frac{f_i^k - 1}{r_i - 1} \qquad (11.96)$$

the deviation index in Eq. (11.93) can be dispensed with and replaced by this expression. A subsequent rearrangement yields a simple, intuitive relationship between the three FCCs corresponding to an activity change in branch i (Stephanopoulos and Simpson, 1997):

$$K_i = \frac{C_i^{J_A}}{f_i^A - 1} = \frac{C_i^{J_B}}{f_i^B - 1} = \frac{C_i^{J_C}}{f_i^C - 1} \qquad i \in \{A, B, C\} \quad (11.97)$$

where

$$K_i = \frac{r_i}{f_i^i(r_i - 1)} \qquad i \in \{A, B, C\} \qquad (11.98)$$

Equation (11.97) reveals that the fractional change in the flux J_k resulting from a perturbation in the activity of branch i, $f_i^k - 1$, is proportional to the corresponding control coefficient, $C_i^{J_k}$; additionally, the same proportionality applies to the flux change in all three branches. This proportionality is characterized by the *perturbation constant*, K_i, which can be calculated from the AAF and FAF of the perturbed branch. Bear in mind, however, that this constant merely reflects the result of a particular perturbation in a branch and that different perturbations will yield different perturbation constants.

Predicting Changes in Flux

Once the flux control coefficients for one or more of the branches around the branch point metabolite are known, it is possible to predict the effects of certain perturbations on the system. For a change in a single branch i, characterized by an activity amplification factor r_i, the expected flux change for all three branches can be determined in the form of FAFs (Small and Kacser, 1993b):

$$f_i^k = \frac{1}{1 - C_i^{J_k}\left(\dfrac{r_i - 1}{r_i}\right)} \qquad k \in \{A, B, C\} \qquad (11.99)$$

Note that this result is functionally identical to Eq. (11.88), and thus can be rearranged into a more symmetrical form analogous to Eq. (11.89):

$$\frac{f_i^k - 1}{f_i^k} = C_i^{J_k}\left(\frac{r_i - 1}{r_i}\right) \qquad k \in \{A, B, C\} \qquad (11.100)$$

Finally, given the close similarities of these results to those of the un-branched system, it should come as no surprise that the expression for perturbations in multiple branches (characterized by r_A, r_B, and r_C, respectively) closely parallels Eq. (11.90):

$$
\frac{f_{ABC}^k - 1}{f_{ABC}^k} = \sum_i C_i^{J_k} \left(\frac{r_i - 1}{r_i} \right) \qquad i, k \in \{A, B, C\} \qquad (11.101)
$$

Concentration Control Coefficients and Concentration Changes

Because the concentration of the branch metabolite S was determined in the course of the preceding derivation, it is also possible to use large deviation theory to determine concentration control coefficient, and CCCs to predict the concentration change that will result from a particular perturbation. Furthermore, because of the parallels between branched and unbranched systems, CCCs can be determined in this manner not only for the branch metabolite S, but for any internal metabolite X_j. For a perturbation in enzyme (or branch) activity E_i from to E_i^0 to E_i^r, the CCC corresponding to the change in concentration c_j is found to be

$$
C_i^{X_j} = \frac{\left(c_j^r - c_j^0 \right)}{\left(E_i^r - E_i^0 \right)} \frac{E_i^r}{c_j^0} \frac{J_i^0}{J_i^r} = \frac{\Delta c_j}{\Delta E_i} \frac{E_i^r}{c_j^0} \frac{J_i^0}{J_i^r} \qquad (11.102)
$$

If this expression is compared to that of the flux control coefficient in Eq. (11.84), two interesting differences are immediately noticeable: not only is Eq. (11.102) normalized by the *perturbed* enzyme activity and the *original* metabolite concentration, but it is *further* normalized by the ratio of the perturbed and original fluxes through the altered enzymatic step. Hence, the determination of the concentration control coefficient requires measurements not only of concentration and activity changes but of flux changes as well. As expected, though, in the limit of an infinitesimal change, Eq. (11.102) tends to the conventional definition of the concentration control coefficient.

In order to consider the prediction of the change in metabolite concentration that will result from an activity change, it is useful to rearrange Eq. (11.102) into a dimensionless form. In doing so, we introduce a new quantity, the *metabolite amplification factor* (MAF), ϕ_i^j, which is completely analogous to the flux amplification factor:

$$
\phi_i^j \equiv \frac{c_j^r}{c_j^0} \qquad (11.103)
$$

For an activity change in a single step, characterized by an AAF of r_i, the MAF corresponding to a change in the concentration c_j can be predicted by

$$C_i^{X_j}\left(\frac{r_i - 1}{r_i}\right) = (\phi_i^j - 1)\left[1 - C_i^{J_i}\left(\frac{r_i - 1}{r_i}\right)\right] = \frac{\phi_i^j - 1}{f_i^i} \quad (11.104)$$

As we have seen before, the perturbation of a single enzyme activity is a special case. When large deviation theory is applied instead to the alteration of the activities of n enzymes (Small and Kacser, 1993b), the MAF can be found from

$$\sum_{i=1}^{n}\left[C_i^{X_j}\left(\frac{r_i - 1}{r_i}\right)\right] = (\phi_i^j - 1)\left\{1 - \sum_{i=1}^{n}\left[C_i^{J_i}\left(\frac{r_i - 1}{r_i}\right)\right]\right\} \quad (11.105)$$

Keep in mind that Eq. (11.105) employs different FCC, CCC, and AAF values for each of the n enzymes. For a branched network, the summation will be performed over the three branches (A, B, and C). Furthermore, even though this summation includes any enzymes whose activities have not been altered, all such enzymes have r_i values of 1, resulting in the addition of a null term.

11.5.3. RESPONSE TO CHANGES IN NUTRIENT CONCENTRATIONS AND EXTERNAL EFFECTORS

In addition to describing the effects of altered enzyme activities, large deviation theory also can be used to describe changes in flux resulting from changes in the concentrations of external nutrients and other effectors (Small and Kacser, 1993a). Any such effector is assumed to act by inducing a shift in the rate of one or more enzymes, which in turn produces changes in the steady state fluxes of the pathway. If the induced changes in the enzyme reaction rates are known, then the activity amplification factors (and, from these, the flux control coefficients) can be determined following the procedure of the preceding sections. If, on the other hand, the rate changes caused by the effector are not known, flux changes can instead be expressed in terms of the response coefficient given by Eq. (11.7). If the kinetic effect upon enzyme E_k is characterized by an effector elasticity, the response coefficient can be expressed through Eq. (11.12). If we assume that the effector, X_0 being perturbed is the external substrate of an unbranched pathway, the original flux can be represented by

$$J^0 = \frac{E_1^0 \cdot \left(c_0 - \dfrac{c_n}{K_{1:n}} \right)}{1 + E_1^0 \cdot \chi_2} \tag{11.106}$$

and the perturbed flux by

$$J^x = \frac{E_1^0 \cdot \left(x_0 \cdot c_0 - \dfrac{c_n}{K_{1:n}} \right)}{1 + E_1^0 \cdot \chi_2} \tag{11.107}$$

where x_0 is the factor by which c_0, the concentration of X_0, was changed (the *effector amplification factor*). Use of Eq. (11.106) in the definition of the response coefficient yields the following expression:

$$R_{X_0}^J = \frac{c_0}{c_0 - \dfrac{c_n}{K_{1:n}}} \tag{11.108}$$

In order to relate this result to large perturbations in c_0, the following ratio is determined from Eqs. (11.106) and (11.107) to be

$$\frac{\Delta J}{\Delta c_0} = \frac{J^x - J^0}{x_0 \cdot c_0 - c_0} = \frac{E_1^0}{1 + E_1^0 \cdot \chi_2} \tag{11.109}$$

From a comparison of the previous two equations, it is apparent that the response coefficient can be determined by normalizing Eq. (11.109) by c_0/J^0. Hence,

$$R_{X_0}^J = \frac{\Delta J}{\Delta c_0} \cdot \frac{c_0}{J^0} \tag{11.110}$$

Additionally, once both the response coefficient and the flux control coefficient for the affected enzyme are known, the effector elasticity can be found from Eq. (11.105). In addition, the flux change resulting from a change in the effector concentration can be predicted:

$$f_j^i - 1 = \frac{J_i^x - J_i^0}{J_i^0} = R_{X_j}^J \cdot (x_j - 1) \tag{11.111}$$

Although a change in the concentration of the initial substrate was used in this proof, Eq. (11.110) is general to the change in any external substrate or product concentration that does not affect enzyme kinetics except through its activity. If an effector directly affects the activity of an enzyme by changing its concentration, catalytic activity, or affinity for its substrate/product, the analysis in this section can instead be repeated using the appropriate kinetic effects in place of Eq. (11.107). Results for several such cases were derived by Small and Kacser (1993a).

11.5.4. DISCUSSION

The theory of large deviations is a powerful tool that extends the infinitesimal confines of metabolic control analysis into the measurable realm of experiment. Nevertheless, a few final notes are in order concerning the ramifications of this theory.

First, it should be emphasized that the preceding equations were derived using a simplified, first-order approximation of enzyme kinetics. Consequently, a system exhibiting highly nonlinear kinetics may give dubious results under analysis by these equations. In justification of their work, Small and Kacser (1993b) employed several nonlinear models in order to test the consistency of their analysis, with very good results. Nevertheless, the need for consistency tests cannot be overemphasized. This often can be done simply by repeating the calculation of an FCC using two or more different levels of amplification of the enzyme under study. In addition, an estimate of the error in FCC calculations arising from measurement uncertainties can easily be determined (Small and Kacser, 1993b).

Secondly, although the equations resulting from this theory are deceptively simple, they are not intuitive. There are some unexpected consequences, for example, in the effects of amplification or attenuation of an enzyme's activity. Because of the shape of the flux-activity curve of Fig. 11.9, a very small reduction in activity can have a significant effect, whereas a large increase in activity may be required to give measurable flux changes. Indeed, a decrease in activity to zero will nullify an unbranched pathway. Activity amplification, on the other hand, will only slowly reach its boundary of maximum flux.

Finally, it is critical to understand the limits of the applicability of this theory. The reduction of the activity of any enzyme to zero, for example, is predicted to reduce to zero any flux for which the corresponding FCC is positive; nonetheless, if the attenuated enzyme is in branch B of a branched pathway, the flux through branch A will not be reduced to zero, as long as some activity remains in branch C. These equations can also predict negative concentrations or negative fluxes through irreversible reactions.

It is, therefore, necessary to view any conclusions from large deviation theory, as with any theory, with a discriminating eye. Common sense should help to recognize when flux control coefficients are unreliable due to experimental error or nonlinearities. In addition, the next chapter will discuss some further issues of network stability, which will aid in the avoidance of gross inconsistencies in predictions. A method will also be introduced by which flux control coefficients can be determined in more general branched pathways, and even more complex networks, from a variety of perturbations.

REFERENCES

Atkinson, D. E. (1990). What should a theory of metabolic control offer to the experimenter? "Control of Metabolic Processes," NATO ASI Series A Vol. 190, pp. 3-27. Edited by A. Cornish-Bowden and M. L. Cardenas. New York: Plenum Press.

Bainbridge, Z. A., Scott, R. I., & Perry, D. (1992). Oxygen utilization by isopenicillin N synthase from Penicillium chrysogenum. Journal of Chemical Technology and Biotechnology 55; 233-238.

Brand, M. D., Hafner, R. P., & Brown, G. C. (1988). Control of respiration in non-phosphorylating mitochondria is shared between the proton leak and the respiratory chain. Biochemical Journal 255: 535-539.

Brown, C., Hafner, P. P., & Brand, M. D. (1990a). A "top-down" approach to determination of control coefficients in metabolic control theory. European Journal of Biochemistry 188; 321-325.

Brown, G. C., Lakin-Thomas, P. L., & Brand, M. D. (1990b). Control of respiration and oxidative phosphorylation in isolated rat liver cells. European Journal of Biochemistry 192; 355-362.

Cornish-Bowden, A. (1989). Metabolic control theory and biochemical systems theory: Different objectives, different assumptions, different results. Journal of Theoretical Biology 136; 365-377.

Crabtree, B., & Newsholme, E. A. (1987a). A systematic approach to describing and analyzing metabolic control systems. Trends in Biochemical Science 12; 4-12.

Crabtree, B., & Newsholme, E. A. (1987b). The derivation and interpretation of control coefficients. Biochemical Journal 247; 113-120.

Delgado, J., & Liao, J. C. (1991). Identifying rate-controlling enzymes in metabolic pathways without kinetic parameters. Biotechnology Progress 7; 15-20.

Delgado, J., & Liao, J. C. (1992a). Determination of flux control coefficients from transient metabolite concentrations. Biochemical Journal 282; 919-927.

Delgado, J., & Liao, J. C. (1992b). Metabolic control analysis using transient metabolite concentrations. Determination of metabolite concentration control coefficients. Biochemical Journal 285; 965-972.

Delgado, J., Meruane, J., & Liao, J. C. (1993). Experimental determination of flux control distribution in biochemical systems: In vitro model to analyze transient metabolite concentrations. Biotechnology and Bioengineering 41; 1121-1128.

Discussion Forum. (1987). Trends in Biochemical Science 12; 216-224.

Ehlde, M. (1995). Dynamic and steady state models of metabolic pathways. A theoretical evaluation. Ph.D. thesis, University of Lund, Sweden.

Ehlde, M., & Zacchi, G. (1997). A general formalism for Metabolic control analysis. *Chemical Engineering Science* 52; 2599-2606.

Ehlde, M., & Zacchi, G. (1996). Influence of experimental errors on the determination of flux control coefficients from transient metabolite measurements. *Biochemical Journal* 313; 721-727.

Fell, D. A. (1992). Metabolic control analysis: a survey of its theoretical and experimental development. *Biochemical Journal* 286; 313-330.

Fell, D. A. (1985). Metabolic control and its analysis. Additional relationships between elasticities and control coefficients. *Biochemical Journal* 269; 255-259.

Flint, H. J., Tateson, R. W., Barthelmess, I. B., Porteous, D. J., Donachie, W. D, & Kacser, H. (1981). Control of flux in the arginine pathway of *Neurospora crassa*. *Biochemical Journal* 200; 231-246.

Galazzo, J. L., & Bailey, J. E. (1990). Fermentation pathway kinetics and metabolic flux control in suspended and immobilized *Saccharomyces cerevisiae*. *Enzyme and Microbial Technology* 12; 162-172.

Groen, A. K., Wanders, R. J. A., Westerhoff, H. V., van der Meer, R., & Tager, J. M. (1982). Quantification of the contribution of various steps to the control of mitochondrial respiration. *Journal of Biological Chemistry* 257; 2754-2757.

Groen, A. K., Vervoorn, R. C., Van der Meer, R., & Tagger, J. M. (1983). Control of rat liver glyconeogenesis in rat liver cells. I. Kinetics of the individual enzymes and the effect of glucagon. *Journal of Biological Chemistry* 258; 14346-14353.

Groen, A. K., van Roermund, C. W. T., Vervoorn, R. C., & Tager, J. M. (1986). Control of gluconeogenesis in rat liver cells. Flux control coefficients of the enzymes in the gluconeogenic pathway in the absence and presence of glucagon. *Biochemical Journal* 237; 379-389.

Hafner, R. P., Brown, G. C., & Brand, M. D. (1990). Analysis of the control of respiration rate, phosphorylation rate, proton leak rate and protonmotive force in isolated mitochondria using the "top-down" approach of Metabolic control analysis. *European Journal of Biochemistry* 188; 313-319.

Heinrich, R., & Rapoport, T. A. (1974). A linear steady state treatment of enzymatic chains. General properties, control and effector strength. *European Journal of Biochemistry* 42; 89-95.

Heinrich, R., Rapoport, S. M., & Rapoport, T. A. (1977). Metabolic regulation and mathematical models. *Progress in Biophysics and Molecular Biology* 32; 1-82.

Henriksen, C. M., Nielsen, J., & Villadsen, J. (1997). Influence of dissolved oxygen concentration on the penicillin biosynthetic pathway in steady state cultures of *Penicillium chrysogenum*. *Biotechnology Progress* 13:776-782.

Higgins, J. (1963). Analysis of sequential reactions. *Annals of the New York Academy of Science* 108; 305-321.

Higgins, J. (1965). Dynamics and control in cellular reactions. "*Control of Energy Metabolism,*" pp. 13-36. Edited by B. Chance, R. W. Estabrook, and J. R. Williamson. New York: Academic Press.

Kacser, H. (1983). The control of enzyme systems *in vivo*: Elasticity analysis of the steady state. *Biochemical Society Transactions* 11; 35-43.

Kacser, H. (1991). A superior theory? *Journal of Theoretical Biology* 149; 141-144.

Kacser, H. & Burns, J. A. (1973). The control of flux. *Symposia of the Society for Experimental Biology* 27; 65-104.

Kacser, H. & Burns, J. A. (1979). Molecular democracy: Who shares the controls? *Biochemical Society Transactions* 7; 1149-1160.

Kacser, H., Sauro, H. M., & Acerenza, L. (1990). Enzyme-enzyme interactions and control analysis. 1. The case of non-additivity: monomer-oligomer associations. *European Journal of Biochemistry* 187; 481-491.

Liao, J. C., & Delgado, J. (1993). Advances in Metabolic control analysis. *Biotechnology Progress* 9; 221-233.

Niederberger, P., Prasad, R., Miozzari, G., & Kacser, H. (1992). A strategy for increasing an *in vivo* flux by genetic manipulations. The tryptophan system of yeast. *Biochemical Journal* 287; 473-479.

Nielsen, J. (1997). *Physiological engineering aspects of* Penicillium chrysogenum. Singapore: World Scientific.

Nielsen, J., & Jørgensen, H. S. (1995). Metabolic control analysis of the penicillin biosynthetic pathway in a high yielding strain of *Penicillium chrysogenum*. *Biotechnology Progress* 11; 299-305.

Pettersson, G. (1996). Errors associated with experimental determinations of enzyme flux control coefficients. *Journal of Theoretical Biology* 179; 191-197.

Pissarra, P. de N.; Nielsen, J.; & Bazin, M. J. (1996). Pathway kinetics and metabolic control analysis of a high-yielding strain of *Penicillium chrysogenum* during fed-batch cultivations. *Biotechnology and Bioengineering* 51; 168-176.

Ramos, F. R., López-Nieto, M. J., & Martín, J. F. (1985). Isopenicillin N synthetase of *Penicillium chrysogenum*, an enzyme that converts δ-(L-α-aminoadipyl)-L-cysteinyl-D-valine to isopenicillin N. *Antimicrobial Agents and Chemotherapy* 27; 380-387.

Reder, C. (1988). Metabolic control theory: a structural approach. *Journal of Theoretical Biology* 135; 175-201.

Ruyter, G. J. G., Postma, P. W., & van Dam, K. (1991). Control of glucose metabolism by enzyme II^{Glc} of the phosphoenolpyruvate-dependent phosphotransferase system in *Escherichia coli*. *Journal of Bacteriology* 173; 6184-6191.

Salter, M., Knowles, R. G., & Pogson, C. (1986). Quantification of the importance of individual steps in the control of aromatic amino acid metabolism. *Biochemical Journal* 234; 635-647.

Savageau, M. A. (1969a). Biochemical systems analysis. I. Some mathematical properties of the rate law for the component enzymatic reactions. *Journal of Theoretical Biology* 25; 365-369.

Savageau, M. A. (1969b). Biochemical systems analysis. II. The steady state solutions for an *n*-pool system using a power-law approximation. *Journal of Theoretical Biology* 25; 370-379.

Savageau, M. A. (1970). Biochemical systems analysis. III. Dynamic solutions using a power-law approximation. *Journal of Theoretical Biology* 26; 215-226.

Savageau, M. A., Voit, E. O., & Irvine, D. H. (1987a). Biochemical systems theory and metabolic control theory: 1. Fundamental similarities and differences. *Mathematical Biosciences* 86; 127-145.

Savageau, M. A., Voit, E. O., & Irvine, D. H. (1987b). Biochemical systems theory and metabolic control theory: 2. The role of summation and connectivity relationships. *Mathematical Biosciences* 86; 147-169.

Small, J. R. (1993). Flux control coefficients determined by inhibitor titration. The design and analysis of experiments to minimize errors. *Biochemical Journal* 296; 423-433.

Small, J. R. & Kacser, H. (1993a). Responses of metabolic systems to large changes in enzyme activities and effectors. I. The linear treatment of unbranched chains. *European Journal of Biochemistry* 213; 613-624.

Small, J. R. & Kacser, H. (1993b). Responses of metabolic systems to large changes in enzyme activities and effectors. II. The linear treatment of branched pathways and metabolite concentrations. Assessment of the general nonlinear case. *European Journal of Biochemistry* **213**; 625-640.

Stephanopoulos, G. & Simpson, T. W. (1997). Flux amplification in complex metabolic networks. *Chemical Engineering Science* **52**; 2607-2628.

Theilgaard, H. B. A., Henriksen, C. M., Kristiansen, K., & Nielsen, J. (1997). Purification and characterization of δ-(L-α-aminoadipyl)-L-cysteinyl-D-valine synthetase (ACVS) from *Penicillium chrysogenum*. *Biochemical Journal* **327**; 185-191.

Torres, N. V., Mateo, F., Melendez-Hevia, E., & Kacser, H. (1986). Kinetics of metabolic pathways. A system *in vitro* to study the control of flux. *Biochemical Journal* **234**; 169-174.

Torres, N. V., Sicilia, J., & Melendez-Hevia, E. (1991). Analysis and characterization of transition states in metabolic systems. Transition times and the passivity of the output flux. *Biochemical Journal* **276**; 231-236.

Voet, D. & Voet, J. G., editors (1995). Biochemistry 2. New York: John Wiley & Sons.

Walter, R., Morris, J., & Kell, D. (1987). The roles of osmotic stress and water activity in the inhibition of the growth, glycolysis and glucose phosphotransferase system of *Clostridium pasteuranum*. *Journal of General Microbiology* **133**; 259-266.

Westerhoff, H. V., Groen, A. K., & Wanders, R. J. A. (1984). Modern theories of metabolic control and their application. *Bioscience Reports* **4**; 1-22.

CHAPTER 12

Analysis of Structure of Metabolic Networks

In the previous chapter, we reviewed the basic concepts of Metabolic Control Analysis and methods for the determination of its key parameters. It was noted there that the framework of MCA provides the means of quantifying the degree of control exerted by each individual reaction step in a metabolic network upon each network flux and metabolite, in the form of control coefficients. In effect, control coefficients provide a characterization of the effects resulting from the perturbation of a particular reaction, so that the effects of other perturbations in that same step can be predicted. The methods presented in Chapter 11 permit the determination of control coefficients and other MCA parameters from *flux* estimates, even after large perturbations in the kinetics of the network. Chapters 8-10 provided a comprehensive approach to the estimation of fluxes required for these calculations, based upon metabolite balances, reaction stoichiometries, and isotopic labels.

In this chapter we turn to the challenging task of analyzing flux control in more complex networks than the simple unbranched or branched pathways discussed in Chapter 11. We define a complex metabolic network as a system consisting of many enzymatic reactions producing and depleting metabolites and interacting with one another through common precursors, products, or effectors. Such networks process external substrates through a series of reactions which ultimately produce one or more desirable products, often in competition with unwanted byproducts. An example, which will be used as a case study throughout this and the following chapter, is the aromatic amino acid biosynthetic pathway (Fig. 12.1). With the use of a model for the network reaction kinetics (Stephanopoulos and Simpson, 1997), we will illustrate the methods allowing experimental elucidation of the control structure of this network, or any similarly complex system.

The objectives of this and the following chapter are to provide general and systematic methods for

- defining the intrinsic structure of a network, in terms of identifiable branch points, based upon the system's stoichiometry;
- analyzing the kinetic behavior of complex metabolic networks in terms of MCA parameters;
- identifying critical reactions and branch points; and
- optimizing the amplification of a desired flux through enhancement/ attenuation of critical reactions.

We will provide first a general description of each method followed by illustrations using our case study, i.e., a kinetic model for the aromatic amino acid production pathways in baker's yeast. Keep in mind that this model, which is based upon the work of Galazzo and Bailey (1990, 1991), is employed only as a surrogate cell used for illustrative purposes. It is not intended to accurately portray the functioning of *Saccharomyces cerevisiae*; hence, the conclusions reached in this chapter may or may not be applicable to an actual production strain. The purpose here is not to simulate biological reality, but rather to take advantage of a system that exhibits many aspects of regulation, tight control, and feedback mechanisms that are typical of real biological systems in exploring various methods for analyzing such systems. Lessons learned from the analysis of surrogate cells can form the basis for the systematic experimental investigation of real biological systems.

Our discussion of complex networks begins with an investigation of branched pathways, as branching is the simplest trait common to complex systems. Since each branch point adds another level of complexity to a network, it is important to (a) locate the branch points of consequence in overall network flux control, and (b) understand the control of flux distribu-

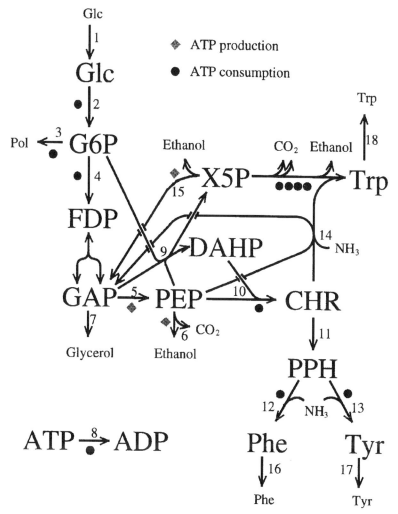

FIGURE 12.1   Aromatic amino acid biosynthetic network. Schematic diagram of key reactions participating in the production of aromatic amino acids in *Saccharomyces cerevisiae*, including reactions involved in free-energy production. Steady-state internal metabolites (see Section 12.2.2) are shown in large type. Reactions (numbered for reference) may comprise one or more coordinated enzymatic steps. The unnumbered reaction converting FDP into GAP is considered to be near equilibrium, with a net rate equal to that of reaction 4. Abbreviations: CHR, chorismate; DAHP, 3-deoxy-D-*arabino*-heptulosonate-7-phosphate; FDP, fructose-1,6-diphosphate; G6P, glucose-6-phosphate; GAP, glyceraldehyde-3-phosphate; Glc, glucose; PEP, phosphoenolpyruvate; Phe, phenylalanine; PPH, prephenate; Pol, polysaccharides; Trp, tryptophan; Tyr, tyrosine.

tion at branch points and provide comprehensive means of measuring such control.

Following this introductory material, we focus exclusively on methods for analyzing complex networks. It is important here to realize that determining the effect of each individual reaction on the flux is impractical for large biosynthetic networks. A more effective approach is to *group* reactions and investigate the role of each group in controlling overall pathway flux. This is the essence of the top-down approach mentioned earlier. This chapter presents a systematic procedure for implementing the top-down approach to realistic networks. Although the overall concept is rather simple, the implementation procedure can become involved. It is therefore important to maintain a clear vision of the goal of the process in relation to the individual steps, as outlined here:

1. The first step is to develop the rules for grouping reactions and pathways. This is straightforward for simple reaction systems; it becomes less clear, however, as the size and complexity of the network increases. In Section 12.2 we provide the methodology for systematically grouping reactions and identifying the critical branch points in a network. In Section 12.3, we demonstrate reaction grouping with the amino acid case study.

2. *Group flux control coefficients* (gFCCs) provide a measure of the effect of a group of reactions on a pathway flux. Methods for the determination of gFCCs based upon strategic perturbations and flux measurements within each group are presented in the next chapter, in Section 13.2. There are two ways one can make use of gFCCs: First, by comparing the magnitude of different gFCCs, it is possible to determine individual flux control coefficients, and, hence, to pinpoint reactions that are likely to impact strongly the fluxes of interest. Second, gFCCs allow flux control to be localized within a small number of reactions through consecutive grouping and analysis. These points are discussed in Section 13.4.

3. The optimization of a flux of interest, based upon the results of top-down MCA and reaction grouping, is detailed in Section 13.5. In the course of this optimization, it will become apparent that it is generally not possible to amplify at will the rate of any reaction in a biosynthetic network. There are constraints imposed by the allowable range of variation of intermediary metabolites. In connection with such constraints, group concentration control coefficients (gCCCs) emerge as an equally important parameter that provides a measure of the impact of kinetic changes on metabolite levels. Because extreme changes in metabolite levels can affect network stability, it is desirable to maintain them near their normal steady-state values. This implies that

the optimal flux amplification will be obtained by balancing the desire for significant flux increases against the constraint of limited metabolite changes allowed.

4. The last section of this analysis (Section 13.6) considers the effect of experimental error on the results and, more importantly, means of testing the validity of the assumptions inherent in the framework used. For this purpose, internal consistency tests are provided to ensure that the results obtained do not violate these key assumptions.

12.1. CONTROL OF FLUX DISTRIBUTION AT A SINGLE BRANCH POINT

Before discussing the properties of complex systems, it is beneficial to take an in-depth look at the functioning of a single branch point. Although this analysis involves the kinetic interactions of only three reactions, it nevertheless provides insights into the functioning of more complex systems. Important concepts which arise in this discussion include the *flexibility* of the flux distribution at the branch point (see also Chapters 5 and 10), the *distribution* of flux control within the three branches, and the possibility of *metabolic instabilities* resulting from extreme levels of the link metabolite concentration. We address these questions in this chapter by employing a detailed kinetic analysis. Similar conclusions can also be obtained by applying the concepts of MCA and investigating the nature of flux control coefficients as functions of the reaction elasticities, as noted in Eqs. (11.68) and (11.69) in Section 11.4. The analysis is applied in this section to the simple branch point of Fig. 12.2 involving the three reactions, A, B, and C. The *link*

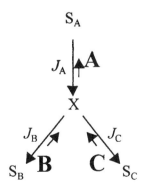

FIGURE 12.2   Representation of a single branch point. The branch point metabolite X produced by reaction A can be used by either reaction B or C. The external concentrations of S_A, S_B, and S_C are considered fixed, as are all kinetic parameters.

metabolite, X, produced by reaction A can be utilized by reaction B or reaction C, which can vary independently. At steady state, though, the flux through reaction A must equal the total flux through reactions B and C.

Although the kinetics of the three reactions in Fig. 12.2 can take any mathematical form, we here assume standard reversible Michaelis-Menten kinetics. Thus, for a reaction converting S into P, the rate is expressed as

$$v = v_{\max} \frac{c_S - c_P/K_{SP}}{K_{m,S} + c_S + c_P(K_{m,S}/K_{m,P})} \tag{12.1}$$

where c_S is the concentration of the substrate, c_P is the concentration of the product, and the remaining parameters are (nonnegative) kinetic constants of the particular reaction. Eq. (12.1) expresses the net reaction rate for each of the three reactions, with a positive value indicating a direction for the conversion from S_A to S_B and S_C. The flexibility of the branch point and the distribution of fluxes between the competing branches depend on the mutual disposition of the kinetic curves for the three reactions, as well as the concentration of the link metabolite X. Therefore, we will consider the effects of changes in the concentration of the branch point metabolite X, with all other concentrations and parameters fixed. In this case, the kinetics of the upstream reaction, A, takes the form

$$v_A = v_{\max, \mathrm{eff}_A} \left(\frac{1 - k_{1_A} c_X}{1 + k_{2_A} c_X} \right) \tag{12.2}$$

and each of the downstream reactions, B and C, take the form

$$v_i = v_{\max, \mathrm{eff}_i} \left(\frac{c_X - k_{1_i}}{c_X + k_{2_i}} \right) \qquad i \in \{B, C\} \tag{12.3}$$

where the three parameters k_1, k_2, and $v_{\max, \mathrm{eff}}$ for a particular reaction are each composites of the fixed kinetic parameters and concentrations of A, B, and C. Note that these values differ in the three preceding reaction rates. Furthermore, each of these parameters is nonnegative, as are the actual kinetic constants from which they arise. If each of these parameters is nonzero, Eqs. (12.2) and (12.3) take a hyperbolic form, as shown in Fig. 12.3. For each of these equations, the forward reaction rate is bounded by the effective maximum rate, $v_{\max, \mathrm{eff}}$. As the concentration of X increases (for the upstream branch), or decreases (for the downstream branches), the reaction rate decreases, finally reaching the maximum rate of the reverse reaction, $(k_1/k_2)v_{\max, \mathrm{eff}}$. If k_1 and/or k_2 are zero, these curves will take

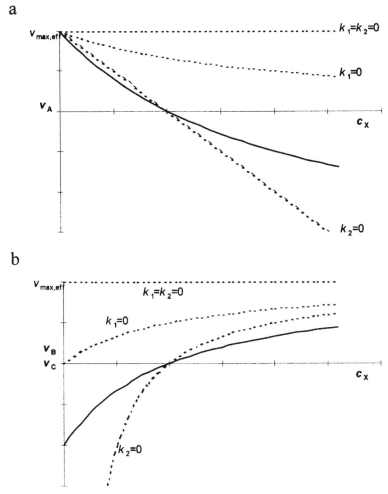

FIGURE 12.3 Schematics of reaction rate dependence with respect to the link metabolite concentration. (a) Rate of upstream branch (A). (b) Rate of a downstream branch (B or C). Profiles are derived from Eqs. (12.2) and (12.3). Dashed lines represent rates with indicated parameters fixed at zero.

different forms, which are also shown in Fig. 12.3. In the following discussion, each rate equation will be assumed to be of the full hyperbolic form.

At each branch point, there is a competition between the downstream branches for utilization of the link metabolite formed by the upstream branch. It is important to understand the ramifications of this competition upon the steady-state operation of a branch point, and thus the control distribution around the branch point, in order to be able to extend these concepts to the control of complex networks.

An important feature of a branch point is its *flux distribution*, namely, the flux ratio between the two downstream reactions. In general, the flux distribution is strongly dependent upon the concentration of the link metabolite. Consider, for example, the two downstream reactions (B and C) with typical rate curves as shown in Fig. 12.4. If the concentration of X is very high (c_1), both reactions will be at their effective maximum rates, and a change in c_X will have little effect upon either rate. In other words, the elasticity of each of these reactions with respect to c_X is near zero. This implies that the flux distribution between branches B and C is also fixed, which defines a *rigid* branch point. Another situation giving rise to a rigid branch point would be one in which a change in c_X would change the rates of both B and C by nearly the same proportion, so that the flux distribution would again remain fixed. In other words, the competition of each reaction for the increased amount of X would be roughly equal yielding a similar flux ratio. If, on the other hand, the concentration of X is c_2, a small change in c_X will produce a much larger change in the rate of reaction C than that of reaction B. Under these conditions, the elasticities of the two reactions differ with respect to c_X, thus defining a *flexible* branch point. In essence, reaction C competes more effectively for the increased amount of X than does reaction B.

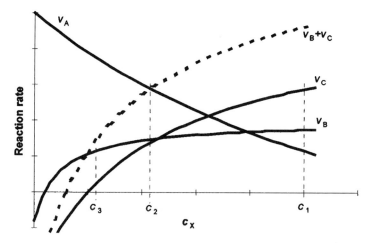

FIGURE 12.4 Rate balance for the three reactions around a branch point. The steady-state concentration of the link metabolite is found at the intersection of the rate curve for reaction A with the sum of the B and C curves, in this case at concentration c_2. The rigidity of a branch point depends upon the slope of each curve at the steady-state concentration.

Note that the *steady-state* concentration of the link metabolite actually depends upon the kinetics of all three reactions, since the sum of the rates of reactions B and C must equal the rate of reaction A at this steady-state concentration. Accordingly, if a change is induced in the effective rate of reaction A (*e.g.*, through a change in the concentration of S_A or amplification of the activity of enzyme A), c_X will shift to a different steady-state level, resulting in different fluxes for all three branches. The rigidity or flexibility of the branch point provides a qualitative interpretation of the tendency for the flux distribution to shift as a result. Furthermore, this resultant shift in the flux distribution can be quantified using the ratio of the elasticities of the downstream reactions with respect to the level of the link metabolite.

The rate curves of Fig. 12.4 also provide the basis for understanding flux control around the branch point. Any change in the intrinsic kinetics of one of the three reactions will be represented by a shift in its rate curve in Fig. 12.4. The magnitude of the shifts in the resultant fluxes indicates the degree of control the perturbed reaction exerts upon each branch. Consider, for example, a change in the rate of branch A, which would be transmitted to the other branches through a change in the steady-state concentration of X. For a rigid branch point, the flux of each branch would be changed by nearly the same proportion; therefore the flux control coefficients for such a change will be nearly equal for all three branches. Furthermore, if both reaction B and reaction C are near their effective maximum rates, the FCC for any change in branch A will be near zero for all three branches. However, the concentration control coefficient C_A^X may have a significant positive value. In the case of a flexible branch point, on the other hand, a change in reaction A will cause a much greater shift in one downstream branch than in the other. At initial concentration c_2 in Fig. 12.4, therefore, a small decrease in the activity of reaction A would lower the steady-state concentration of X, resulting in a much larger decrease in the flux of reaction C than that of reaction B. Indeed, if the steady-state concentration of X were decreased below c_3, reaction C could even reverse direction (assuming that there is a constant supply of S_C). In this case, the CCC C_A^X will be smaller than in the case of a rigid branch point, representing a smaller change in the level of the link metabolite.

The control exerted by the downstream branches can similarly be envisioned using Fig. 12.4. Consider first an increase in the activity of reaction B at an initial steady-state concentration c_1. This would effectively shift the rate curve of reaction B upward. In this case, because the rate curve of reaction C has already plateaued, the flux through reaction C will decrease only slightly because of the decreased steady-state concentration of X. The flux through reactions A and B, however, will both increase by approximately the same amount. Hence, the flux control coefficient for such a change upon reaction C, $C_B^{J_C}$, will have a very small negative value, while $C_B^{J_B}$ and $C_B^{J_A}$ will have significant positive values. Because the increase in the flux through reaction B represents a larger proportional change than that in reaction A, $C_B^{J_B}$ will be larger than $C_B^{J_A}$, but generally less than 1. The concentration control coefficient C_B^X will also be significantly negative.

On the other hand, if the branch point is flexible (as would be the case when the original steady-state level of X is c_2 in Fig. 12.4), the rate curve of the competing reaction will be relatively steep. In this case, an increase in the activity of reaction B (*i.e.*, an upward shift in the rate curve of reaction B), will increase the flux of reaction B, accompanied by a decrease of the concentration of X and a large drop in the flux of reaction C. In addition, the flux through reaction A will be increased moderately, whereas the flux through reaction B will be increased significantly. In effect, this activity change primarily causes a diversion of flux from reaction C to reaction B. Hence, $C_B^{J_C}$ will have a significant negative value, $C_B^{J_B}$ will have a large positive value, and $C_B^{J_A}$ will have a small positive value. Finally, because the concentration of X will change less in a flexible branch point than in a rigid branch point, the concentration control coefficient C_B^X will also be small and negative.

Although other situations may arise if the rate curves take different forms, these examples are representative of the typical branch point and provide a useful basis for the understanding of more complex network situations. In a complex network involving multiple branch points, however, a certain amount of feedback will exist between branch points. In this case, the effects of changes in the activity of a particular section of the network cannot be easily interpreted graphically. In the remainder of this and the following chapter, mathematical methods are employed for the analysis of the control in complex networks in analogy to the control around a single branch point.

Finally, it should be noted that certain conditions may exist around a branch point which constrain the allowable activity changes. In a rigid branch point, in particular, an increase in the activity of the upstream branch may lead to a precipitous increase of the concentration of the link metabolite without significantly changing the fluxes of the downstream branches. This would be true at a concentration c_1 in Fig. 12.4, since the curves for reactions B and C are both near their corresponding maxima. In addition, if the rate curve for reaction A does not intersect the summed rate curves of reactions B and C at all, there will be no real solution of the steady-state balance for the link metabolite concentration. This situation represents a type of metabolic instability, in which the actual metabolite level could either drop to zero, build up to toxic levels, or induce alternate metabolic routes for its use. This sort of metabolic instability is best avoided in practice, since the results are unpredictable and would likely be detrimental to the cell itself. In Section 13.5, we will consider the ramifications of such metabolic instabilities in greater detail.

12.2. GROUPING OF REACTIONS

As discussed in the previous chapter, Metabolic Control Analysis has classically been concerned with determining the effects of *individual* reactions upon network fluxes; consequently, the usual method of determining control coefficients is based on perturbing the kinetics of a *single* enzyme and measuring the impact of such kinetic perturbation on the flux. This *bottom-up* approach, in which each step is studied in turn, is very useful in the context of a simple pathway or a pathway in which a small number of steps are implicated in network control. In a complex pathway, however, which may involve literally hundreds of different enzymatic and transport steps, many of which exert little or no control on fluxes, this standard approach clearly becomes impractical. Top-down control analysis (TDCA; see Section 11.2.2) was developed as an alternative method. The main tenet of TDCA is to focus on *groups* of reactions instead of individual reaction steps and, following the evaluation of different reaction groups, to converge eventually to a single reaction or reaction group that exercises a significant fraction of the overall kinetic control of the network. Such a methodology is necessary considering the very large number of reactions typically present in metabolic networks.

One requirement of TDCA is that, strictly speaking, the reaction groups surrounding the link metabolite must be completely isolated from each other, *i.e.*, linked *only* through the link metabolite. In many cases, it may be impossible to find link metabolites fitting this criterion. The approach described here, however, relaxes the constraint of completely isolated groups, allowing analysis of networks exhibiting interaction effects such as feedback inhibition. Since there are limits to the extent that such interactions can be accommodated with the present method, this important issue will be revisited in Section 13.6. Additionally, this approach is able to effectively separate coupled biological processes, such as free energy production or expansion flux resulting from growth, from the main carbon-related biosynthetic fluxes. Because of the nature of these coupled processes, the reaction groupings defined through this analysis frequently differ from groupings suggested by biochemical maps.

12.2.1. GROUP FLUX CONTROL COEFFICIENTS

A group flux control coefficient (gFCC) is defined as the flux control coefficient which would exist, were the entire reaction group actually a single step. A simple example of a reaction group is provided by group A in Fig. 12.5a, which shows an unbranched reaction pathway. If the activity of each reaction step within box A were changed by the same relative magnitude, the gFCC of group A upon its own flux ($*C_A^{J_A}$) would be defined as the ratio of the resultant relative change in the pathway flux (dJ_A/J_A) to the relative change in the activity of each reaction in the group (also equal to the

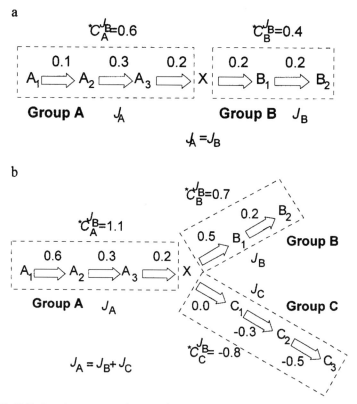

FIGURE 12.5 Reaction groups. The typical partitioning of a pathway into groups around a link metabolite is shown for (a) an unbranched pathway and (b) a branched pathway. Also shown are the required flux balances between groups, as well as the individual and group flux control coefficients affecting Group B in each case.

activity change of the overall group, $d^* v_A / {}^* v_A$):

$$^*C_A^{J_A} \equiv \frac{dJ_A}{d^* v_A} \frac{{}^* v_A}{J_A} \tag{12.4}$$

This is an extension of the classical definition of the FCC (Kacser and Burns, 1973) to TDCA (Brown, *et al.*, 1990). Note that a superscripted asterisk (*) is used on the FCC and reaction velocities to indicate that these quantities refer to groups of reactions rather than a single reaction.

Since it is practically impossible to implement changes of the same magnitude in all reactions in a group, the experimental determination of a gFCC must instead make use of measurements of fluxes and flux changes following perturbations in one or more reactions in the group. It must be emphasized that the grouping of reactions replaces the individual steps with a *virtual* overall reaction with a gFCC equal to the sum of the FCCs of the individual reactions in the group. (This is a consequence of the summation theorem of MCA.)

The concept of a gFCC can easily be extended to branched metabolic pathways such as the network depicted in Fig. 12.5b. An additional complication here is that whenever a branching of reactions occurs, one needs to establish the effect of a change in one of the branches on the fluxes through all three branches. The result is that, instead of a single flux control coefficient, there are now three gFCCs defined for each of the branches of the network, for a total of nine. In more complex networks, groups can be identified around several different metabolites. In such cases, the gFCCs for the groups surrounding each of these *link metabolites* are determined separately. From the magnitudes of the gFCCs obtained for different overlapping groupings, the magnitudes of the gFCCs of the common reactions between the overlapping groups can be obtained. These gFCCs in turn define the degree of control exercised by the smaller groups of the *intermetabolite linkages*, i.e., the short pathways intervening between two different link metabolites. As discussed in Section 13.4, these gFCCs can be used to narrow down the search for individual steps exhibiting large FCCs.

12.2.2. IDENTIFICATION OF INDEPENDENT PATHWAYS

An independent pathway is defined as the smallest set of reactions connecting a single network output with the necessary network inputs in a way that permits the levels of internal species to reach a steady state (Simpson, *et al.*, 1995). The requirement of invariant internal metabolite pools ensures that the system can attain a steady state under a constant input. For a network consisting of L reactions and K_0 independently-variable internal metabolites whose pools should be able to reach a steady state, the maximum number of independent pathways P can be shown to be

$$P = L - K_0 \tag{12.5}$$

For the simple examples of Figs. 12.6a and 12.6b, one can easily identify three independent pathways (P1, P2, and P3), as shown in the figures. For more complicated networks, such as that of Fig. 12.1, the independent

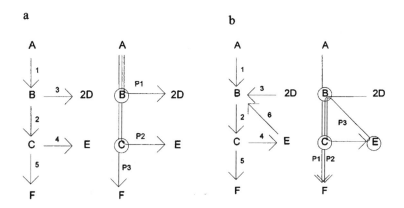

FIGURE 12.6 Identification of independent pathways. Two sample reaction networks are shown with their resulting independent pathways and their SIMS and kernel matrices. (a) Network with branches between multiple outputs. (b) Network with branches between multiple inputs and a recycle pathway.

pathways are not as obvious and a systematic approach is needed for their identification. In order to do so, one begins by first defining the *steady-state internal metabolite stoichiometry* (SIMS) matrix, which rigorously describes the stoichiometry of the network (Reder, 1988). The SIMS matrix N (otherwise referred to as the *scheme matrix*), is an $K \times L$ matrix, in which K is the number of explicit steady-state metabolites in the network, and L is the number of explicit reactions. Each element N_{ji} of this matrix is the stoichiometric coefficient g_{ji} of metabolite X_j participating in reaction i if each reaction is written as:

$$\sum_{\text{reactants}} \left(-g_{ji}X_j\right) + \sum_{\text{products}} \left(g_{ji}X_j\right) = 0 \qquad (12.6)$$

In order to ensure accurate grouping of reactions based upon the method described below, it is crucial that the direction of each reaction described by the above equation be the same as its net flux in the actual network. Thus, since the net flux through reaction 3 in Fig. 12.6a is opposite that in Fig. 12.6b, the signs in its columns of the corresponding SIMS matrices are reversed.

Strictly speaking, the SIMS matrix is a subset of the standard stoichiometry matrix G, introduced in Chapter 8 and also defined by the above equation but for *all* metabolites in the network. Therefore, any external metabolites which serve as feeds or sinks have corresponding rows in the stoichiometry matrix, but not in the SIMS matrix. Since B and C are the only metabolites in Fig. 12.6a that can be maintained at steady state, K for this system is 2 (as is K_0, since the concentrations of these two metabolites vary independently). The species A, D, E, and F are therefore not included in the resulting SIMS matrix.

Two important types of reactions are easily identified in the SIMS matrix. An *output step* produces only external species, and is represented in the SIMS matrix by a column that includes no positive values. An *input step* consumes only external species, and is represented by a column with no negative values. The identification of output steps, in particular, facilitates the enumeration of independent pathways as shown below.

Once the SIMS matrix has been satisfactorily constructed, the independent pathways are obtained as the vectors of the kernel matrix of N. By definition, a *kernel* matrix K is any nontrivial solution to Eq. (12.7):

$$N \cdot K = 0 \qquad (12.7)$$

Figures 12.6a and 12.6b illustrate these concepts with the indicated SIMS and kernel matrices for the corresponding networks. *Note that the entries of the kernel matrix vectors represent the relative rates of the reactions utilized exclusively in the corresponding independent pathway.* Thus, if K_{ij} are the elements of the kernel matrix **K**, K_{1j} represents the relative rate of depletion of metabolite A (*i.e.*, the relative rate of reaction 1), exclusively for pathway $j \in \{1, 2, 3\}$ in Fig. 12.6a. The *total* rate of reaction 1 (equal to the total rate of depletion of A) would be equal to $K_{11} + K_{12} + K_{13}$. The product of the kernel matrix and the SIMS matrix in Eq. (12.7) is nothing but the steady-state balances for the internal metabolites B and C for each of the three independent pathways (a total of 2×3 equations). Thus, the first column of the kernel matrix of Fig. 12.6a indicates that pathway P1 comprises reactions 1 and 3 and that with the indicated stoichiometry, A will be converted to 2D with no net production of either B or C through the series of reactions defining pathway P1:

$$
\begin{array}{c}
A \rightarrow B \\
\underline{B \rightarrow 2D} \\
A \rightarrow 2D
\end{array}
$$

If the stoichiometric ratio between steps is not 1:1, numbers corresponding to the appropriate ratio will instead appear in the entries of the kernel (see Section 12.3).

The exact composition of the kernel is admittedly arbitrary, since Eq. (12.7) will still be valid if any column is multiplied by a constant or added to any other column. In fact, the only formal constraints on the kernel are that it be of full rank with independent columns. For the purpose of determining reaction groups and link metabolites, though, it is sufficient to define the simplest kernel possible. For a network with K_0 independent steady-state metabolites, there are a total of K_0 steady-state relationships among the L reaction rates of *each pathway*. Consequently, there are $(L - K_0)$ degrees of freedom in the determination of the reaction rates of each pathway, or, equivalently, the columns of the kernel matrix. By choosing the elements of the kernel matrix so that each column corresponds to a fundamental vector of the base of the kernel, the structure of each independent pathway can be established. The base of the kernel usually consists of the $(L - K_0)$ rows of the kernel corresponding to steps which are each *unique to a single independent pathway.* We term these steps *eigenreactions* or *characteristic reactions*, because they uniquely define the stoichiometric composition and flux of each independent pathway, as discussed below. Such steps are generally identifiable as output steps leading to different products, but may also encompass steps required for recycles or alternate inputs. Most often, in fact, each independent pathway is associated with the production of a different product, in competition with the products of every other independent pathway.

From the illustration of Fig. 12.6a, it is apparent that the three degrees of freedom define a base of three reactions. Since the clear selections for the base are the three output steps (reactions 3, 4, and 5), each column of the kernel can be defined according to the fundamental vectors for these three steps. This is done by setting values for the three columns of the kernel matrix equal to 100, 010, and 001 for reactions 3, 4, and 5, respectively, and applying Eq. (12.7) to determine the remainder of the elements. If an acceptable base is used, this procedure will yield a complete kernel with no negative elements. Owing to the use of the fundamental vectors of the base of the kernel matrix, any steady state observed in a network will be a linear combination of the basic fluxes through the pathways established by the above method (i.e., the fluxes through the characteristic reactions). In addition, once the characteristic reaction for each pathway has been identified, the *actual fluxes* through each pathway can be found simply by multiplying each value within its column by the flux through the characteristic reaction. For the example of Fig. 12.6a, since the base consists of steps 3-5, the characteristic reactions for pathways P1, P2, and P3 are reactions 3, 4, and 5, respectively. Thus, the flux through each step can be expressed as

$$
\begin{bmatrix} J_1 \\ J_2 \\ J_3 \\ J_4 \\ J_5 \end{bmatrix} = \begin{bmatrix} K_{11} \\ K_{21} \\ 1 \\ 0 \\ 0 \end{bmatrix} J_3 + \begin{bmatrix} K_{12} \\ K_{22} \\ 0 \\ 1 \\ 0 \end{bmatrix} J_4 + \begin{bmatrix} K_{13} \\ K_{23} \\ 0 \\ 0 \\ 1 \end{bmatrix} J_5 \qquad (12.8)
$$

EXAMPLE 12.1

Determination of a Kernel Through Matrix Algebra

For the network of Fig. 12.6a and the indicated SIMS matrix, its kernel matrix can be obtained by applying Eq. (12.7). There are five reactions in this network and two independent steady-state metabolites, so, according to Eq. (12.5), there will be three independent pathways. Equation (12.7) for this system becomes

$$
\begin{bmatrix} 1 & -1 & -1 & 0 & 0 \\ 0 & 1 & 0 & -1 & -1 \end{bmatrix} \begin{bmatrix} K_{11} & K_{12} & K_{13} \\ K_{21} & K_{22} & K_{23} \\ K_{31} & K_{32} & K_{33} \\ K_{41} & K_{42} & K_{43} \\ K_{51} & K_{52} & K_{53} \end{bmatrix} = \begin{bmatrix} 0 & 0 & 0 \\ 0 & 0 & 0 \end{bmatrix} \qquad (1)
$$

The meaning of this equation can be better appreciated if the elements of the kernel matrix, K_{ij}, are interpreted as the rate of reaction i utilized exclusively by pathway Pj. In this case, the six equations summarized in this matrix expression simply reflect the steady-state balances for the two internal metabolites B and C for each of the three independent pathways.

Equation (1) represents 6 equations in 15 unknowns. Any 9 of the unknown K's can be chosen arbitrarily yielding a different kernel in each case. However, the columns of all such kernels are simply linear combinations of three basic columns that form the *base* of this system. Obviously, the simplest kernel will be the one that corresponds to this base. This is obtained by selecting the arbitrary K's to be equal to the identity matrix for the lower 3×3 section of the kernel. In this form, the base is defined by the three output reaction steps 3, 4, and 5 (which are the characteristic reactions for pathways P1, P2, and P3, respectively), since each of these steps occurs only in its corresponding independent pathway. Once the base has been defined, the remaining entries of the kernel matrix are found by solving the set of linear equations expressed in Eq. (2). Note also that the absolute values for the K's are not significant, since each column of the kernel can be scaled independently.

$$\begin{bmatrix} 1 & -1 & -1 & 0 & 0 \\ 0 & 1 & 0 & -1 & -1 \end{bmatrix} \begin{bmatrix} K_{11} & K_{12} & K_{13} \\ K_{21} & K_{22} & K_{23} \\ 1 & 0 & 0 \\ 0 & 1 & 0 \\ 0 & 0 & 1 \end{bmatrix} = \begin{bmatrix} 0 & 0 & 0 \\ 0 & 0 & 0 \end{bmatrix} \quad (2)$$

If the multiplication indicated in Eq. (2) is performed, a set of six linear equations results,

$$\begin{bmatrix} K_{11} - K_{21} - 1 & K_{12} - K_{22} & K_{13} - K_{23} \\ K_{21} & K_{22} - 1 & K_{23} - 1 \end{bmatrix} = \begin{bmatrix} 0 & 0 & 0 \\ 0 & 0 & 0 \end{bmatrix} \quad (3)$$

which can easily be solved:

$$K_{11} - K_{21} = 1$$
$$K_{12} - K_{22} = 0$$
$$K_{13} - K_{23} = 0$$
$$K_{21} = 0 \quad\quad (4)$$
$$K_{22} = 1$$
$$K_{23} = 1$$

By framing these equations in matrix form, the remaining elements of the kernel can be determined through matrix inversion:

$$
\begin{bmatrix} K_{11} \\ K_{12} \\ K_{13} \\ K_{21} \\ K_{22} \\ K_{23} \end{bmatrix}
=
\begin{bmatrix}
1 & 0 & 0 & -1 & 0 & 0 \\
0 & 1 & 0 & 0 & -1 & 0 \\
0 & 0 & 1 & 0 & 0 & -1 \\
0 & 0 & 0 & 1 & 0 & 0 \\
0 & 0 & 0 & 0 & 1 & 0 \\
0 & 0 & 0 & 0 & 0 & 1
\end{bmatrix}^{-1}
\begin{bmatrix} 1 \\ 0 \\ 0 \\ 0 \\ 1 \\ 1 \end{bmatrix}
=
\begin{bmatrix} 1 \\ 1 \\ 1 \\ 0 \\ 1 \\ 1 \end{bmatrix}
\tag{5}
$$

The resulting kernel is the same as that shown in Fig. 12.6a:

$$
\mathbf{K} =
\begin{bmatrix}
1 & 1 & 1 \\
0 & 1 & 1 \\
1 & 0 & 0 \\
0 & 1 & 0 \\
0 & 0 & 1
\end{bmatrix}
\tag{6}
$$

In a simple enough system, the kernel matrix can be constructed by simple inspection of the network schematic, according to the following procedure:

1. Starting with the first column of the kernel, put a 1 in the row of the first column corresponding to one of the output steps, and note the metabolite(s) consumed by that reaction.
2. Determine which step(s) produce(s) the metabolite consumed by the output step identified in step 1.
3. If only one reaction step produces the identified metabolite, place in the entry of the current column corresponding to that reaction the number of times that reaction must proceed in order to produce enough of the metabolite to balance the amount consumed by the previous step; this will usually be a 1. If more than one reaction can produce the metabolite, copy the entries of the entire current column into the next column(s) and continue constructing the current column using the first choice. Once this column is completed, construct the next column(s) using the alternate choices.
4. If the identified reaction is not an input step, repeat Steps 2-4 with that reaction. If it is an input step, the column is finished, and the values in its remaining rows are 0. If the identified reaction is one that was

already employed in this column, there will already be a number in the corresponding row. This indicates a cycle, in which case the entries downstream from the identified reaction should be replaced with 0.

5. Repeat Steps 1-4 beginning with any other output steps.

It is left as an exercise to reproduce the kernels shown in Fig. 12.6 via this method.

In complex networks, reactions are likely to have multiple reactants, complicating the choice of metabolites to balance. In such cases, it is often easiest to proceed first with a metabolite which can be produced by only one particular reaction. After the resolution of the production of this metabolite, any *remaining* unbalanced metabolites from the previous step(s) must subsequently be balanced. Additional network features which may confound the determination of a kernel by inspection include multiple inputs with multiple outputs, and the presence of coupled pathways. In such cases, it may be simpler to solve the matrix equations than to balance the reactions by hand through this procedure. These issues will be demonstrated more fully through the analysis of the case study of aromatic amino-acid production.

12.3. CASE STUDY: AROMATIC AMINO ACID BIOSYNTHETIC PATHWAY

The methodology of link metabolite and reaction group identification is demonstrated in this section with a case study of aromatic amino acid biosynthesis in *S. cerevisiae*. We first present a description of this system, followed by identification of the independent pathways and link metabolites.

12.3.1. MODEL OF AROMATIC AMINO ACID BIOSYNTHESIS IN *S. CEREVISIAE*

This case study is based on a published kinetic model of *S. cerevisiae* biochemistry (Galazzo and Bailey, 1990, 1991). Figure 12.1 shows a schematic of the biosynthetic network, which couples central carbon metabolism and free energy production with amino acid biosynthesis. This network depicts the production of the three aromatic amino acids tryptophan, phenylalanine, and tyrosine from simple precursors, with glucose as the sole carbon source. Other metabolic functions included in this model are the production of storage compounds in the form of polysaccharides and glycerol, the production and utilization of free energy, and the requirement of excess free energy production for maintenance.

In adapting this system for the purpose of illustrating our analytical methods, several specific adjustments to the original model were introduced either out of expediency or in an effort to better match the cell's known biochemistry. Most significantly, the effect of cell growth on metabolite levels and on amino acid balancing was neglected; this had the primary effect of simplifying the metabolite balances and pathway delineation. Inclusion of cell growth in a metabolite balance network generally is represented by a *flux to expansion*, J_j^E, which depletes each metabolite X_j by a factor proportional to its concentration:

$$J_j^E = \mu c_j \tag{12.9}$$

where μ is the specific growth rate and c_j is the concentration of metabolite X_j. This flux to expansion can, in essence, be treated as a single growth-related reaction that consumes every growth-related metabolite in the network at the preceding rate. These fluxes have not been accounted for explicitly in the analysis but have been lumped, instead, in a general first-order depletion reaction step because (a) their magnitude in general is small, (b) they are often inconsequential in the determination of the control structure and flux amplification of metabolic networks, and (c) they are not the only mechanism for metabolite depletion. The rate constants of these depletion reaction steps have been determined such as to yield the same overall amino acid concentrations and metabolic fluxes at steady state as the original model. These reaction steps (reactions 16, 17, and 18 in Fig. 12.1) can be viewed as the sum total of amino acid uptake for protein production or other cellular processes, as well as degradation and extracellular transport of these three amino acids. Additionally, the production and consumption of the pentose phosphates produced by the transcarboxylase step (reaction 9) have been explicitly taken into account in the stoichiometry, whereas they had originally been lumped with glucose-6-phosphate. This also prompted the introduction of a recycle reaction (15) for the return of xylulose-5-phosphate (X5P) into the glycolytic pathway. It should be noted that, despite these changes, the steady state rates and concentrations of this system very closely match those of the Galazzo and Bailey model.

The resulting metabolic network comprises 13 metabolites participating in 18 reactions, which form several reaction branches and recycles and exhibit significant feedback and allosteric inhibition. As such, this system is well-suited for the description and analysis of complex metabolic networks.

Table 12.1 provides a list of the reactions present in this network. Note that several of the reactions listed on Table 12.1 are the stoichiometric equivalent of a number of individual reactions lumped together to yield the overall reactions shown in the table. Thus, following transport into the cell

TABLE 12.1   Stoichiometry of Reactions in the Amino Acid
Biosynthetic Network

Reaction[a]	Reaction stoichiometry[b]
1	$Glc_{ext} \to Glc$
2	$Glc + ATP \to G6P + ADP$
3	$G6P + ATP \to ADP + Pol + 2P_i$
4	$G6P + ATP \to FDP + ADP$
FDP/GAP[c]	$FDP \leftrightarrow 2GAP$
5	$GAP + ADP + P_i \to PEP + ATP$
6	$PEP + ADP \to ATP + EtOH + CO_2$
7	$GAP \to Gol + P_i$
8	$ATP \to ADP + P_i$
9	$G6P + GAP + PEP \to X5P + DAHP + P_i$
10	$PEP + DAHP + ATP \to CHR + ADP + 3P_i$
11	$CHR \to PPH$
12	$PPH + ATP + NH_3 \to Phe + ADP + P_i$
13	$PPH + ATP + NH_3 \to Tyr + ADP + P_i$
14	$PEP + X5P + CHR + 4ATP + NH_3$
	$\quad \to GAP + Trp + 4ADP + EtOH + 2CO_2 + 3P_i + PP_i$
15	$X5P + ADP + P_i \to GAP + ATP + CO_2$
16	$Phe \to Phe_{ext}$
17	$Tyr \to Tyr_{ext}$
18	$Trp \to Trp_{ext}$
Adenosine[d]	$ATP + AMP \leftrightarrow 2ADP$

[a] Reaction numbers correspond to the numbering scheme of Fig. 12.1.

[b] Metabolites in bold-face type are maintained at steady state. Several of the steps shown are composites of coordinated sets of individual enzymatic reactions.

[c] This reaction represents the equilibrium maintained between FDP and triose sugars, which are henceforth considered to be a single metabolite pool.

[d] This reaction represents the equilibrium maintained between ATP, ADP, and AMP.

and phosphorylation (reactions 1 and 2), glucose-6-phosphate (G6P) can be converted to polysaccharides (Pol) by a series of reactions (lumped as reaction 3) or further phosphorylated in the glycolytic pathway to yield fructose-1,6-diphosphate (FDP) through reaction 4. FDP (or its equilibrium equivalent, glyceraldehyde-3-phosphate, GAP) is converted into phospho-enolpyruvate (PEP) and further to ethanol (via pyruvate) with the simultaneous release of free energy in the form of ATP (reactions 5 and 6). Alternatively, GAP can be converted to glycerol (Gol) through a different pathway (reaction 7). The initiation of the shikimate pathway, which ultimately leads to aromatic amino acids, occurs with the synthesis of the pentose phosphate pathway intermediate xylulose-5-phosphate (X5P) and erythrose-4-phosphate (not shown), which then condenses with PEP to form 3-deoxy-D-*arabino*-heptulosonate-7-phospate (DAHP). These two steps are combined in reaction 9. Chorismate (CHR) is formed from DAHP by a series of reactions depicted as reaction 10. Chorismate, in turn, can be converted to prephenate (PPH) and finally to the products phenylalanine (Phe) and tyrosine (Tyr) through a series of reaction steps lumped in reactions 11, 12, and 13, respectively. An alternative fate of chorismate lies in the production of tryptophan (Trp) by the tryptophan biosynthetic pathway represented by reaction 14; this pathway requires the consumption of X5P in addition to chorismate, as well as the recycle of PEP into GAP. The remaining X5P is recycled to GAP by reaction 15. Amino acid export, degradation, and utilization steps are depicted in reactions 16-18.

In addition to balancing the formation and depletion of the metabolites involved in the synthesis of the carbon skeletons, a steady state in this network must also balance the production and consumption of ATP, as identified in Table 12.1. These ATP-producing and -consuming reactions are also represented by circles and diamonds, respectively, in Fig. 12.1 to underline the strong coupling that exists between the various reactions and parts of the network beyond the obvious ones resulting from direct stoichiometric balances. Reaction 8 has been added for the conversion of ATP generated in excess of what is explicitly required by the other reaction steps. This reaction can be thought of as the total of all ATP consumed by other biosynthetic reactions, for cell maintenance, or in futile cycles.

Table 12.2 provides a summary of the kinetic expressions and material balances used to define the steady state conditions of the base state of this analysis. Solution of the metabolite balances of this model yields the steady state metabolite concentrations and fluxes shown in Table 12.3.

12.3.2. IDENTIFICATION OF INDEPENDENT PATHWAYS

The SIMS matrix of this network, constructed from the stoichiometry of Table 12.1, is given in Table 12.4. Note that only 12 of the 13 metabolites in this matrix are truly independent, as the rows corresponding to ATP and ADP mirror each other. Equation (12.5), therefore, shows that a total of 6 independent pathways can be constructed from these 18 reactions, each corresponding to 1 of the 6 output reactions. These reactions (3, 7, 8, 16, 17, and 18) are identified as the characteristic reactions of the network and, thus, form the base of the kernel. (Strictly speaking, reactions 3 and 8 are recycle steps of ATP to ADP, but this makes no difference in this analysis, as ADP is

TABLE 12.2   Kinetic Expressions and Material Balances of Reactions in the Amino Acid
Biosynthetic Network

Reaction[a]	Kinetic expression[b]

1   Glucose uptake:
$v_{Glc, in} = 200 - 132.5[G6P]$

2   Glucose phosphorylation:
$$v_{Glc} = 68.5 \left(\frac{0.00062}{[Glc][ATP]} + \frac{0.11}{[Glc]} + \frac{0.1}{[ATP]} + 1 \right)^{-1}$$

3   Polysaccharide formation:
$$v_{Pol} = 15.74 \left(\frac{[G6P]^{8.51}}{193 + [G6P]^{8.51}} \right) \left(\frac{2.558}{[G6P]^2} + \frac{2.326}{[G6P]} + 1 \right)^{-1}$$

4   Glucose-6-phosphate isomerase and phosphofructokinase:
$$v_{PFK} = \frac{3019[G6P][ATP]R}{R^2 + 6253L^2T^2}$$
$$R = 1 + 0.5714[G6P] + 16.67[ATP] + 95.24[G6P][ATP]$$
$$L = \left[1 + 0.76 \left(\frac{[ADP]^2}{[ATP]} \right) \right] \left[1 + 40 \left(\frac{[ADP]^2}{[ATP]} \right) \right]^{-1}$$
$$T = 1 + 0.0002857[G6P] + 16.67[ATP] + 0.004762[G6P][ATP]$$

5   Glyceraldehyde-3-phosphate dehydrogenase:
$$v_{GAPD} = 99.6 \left[1 + \frac{0.25}{[FDP]} + \left(0.09375 + \frac{6.273}{[FDP]} \right) A \right]^{-1}$$
$$A = 1 + \frac{\left(\frac{[ADP]^2}{[ATP]} \right)}{1.1} + \frac{[ADP]}{1.5} + \frac{[ATP]}{2.5}$$

6   Pyruvate kinase and reduction of pyruvate to CO_2 and ethanol:
$$v_{PK} = 9763[PEP][ADP] \frac{R + 0.3964L^2T}{R^2 + 311.2L^2T^2}$$
$$R = 1 + 157[PEP] + 0.2[ADP] + 3.14[PEP][ADP]$$
$$L = \frac{1 + 0.05[FDP]}{1 + 5[FDP]}$$
$$T = 1 + 0.02[PEP] + 0.2[ADP] + 0.004[PEP][ADP]$$

7   Glycerol production:
$v_{Gol} = 0.068 v_{GAPD}$

8   Unspecified ATP degradation and utilization:
$v_A = 12.1[ATP]$

(continues)

TABLE 12.2   (*continued*)

Reaction[a]	Kinetic expression[b]
9	Transcarboxylase reaction producing E4P and X5P, followed by production of DAHP from E4P and PEP:

$$v_T = 4568 \frac{\dfrac{0.79}{\left(1 + \dfrac{[\text{Phe}]}{53}\right)} + \dfrac{0.2}{\left(1 + \dfrac{[\text{Tyr}]}{40}\right)} + \dfrac{0.01}{\left(1 + \dfrac{[\text{Trp}]}{16}\right)}}{\left(\dfrac{0.0002}{[\text{PEP}][\text{G6P}]} + \dfrac{0.06}{[\text{PEP}]}\right)(2 + [\text{CHR}])(0.9281 + [\text{ATP}])\left(1 + \dfrac{[\text{Trp}]}{16}\right)}$$

| 10 | DAHP consumption through shikimate pathway, followed by conversion of shikimate into chorismate: |

$$v_D = \frac{116[\text{DAHP}][\text{PEP}][\text{ATP}]}{(2 + [\text{ATP}])(0.008665 + [\text{PEP}])(0.921 + [\text{ATP}])}$$

| 11 | Prephenate production: |

$$v_{PPH} = 475.4[\text{CHR}]\left[(2 + [\text{CHR}])\left(1 + \frac{[\text{Phe}]}{50}\right)\left(1 + \frac{[\text{Tyr}]}{40}\right)\right]^{-1}$$

| 12 | Phenylalanine production: |

$$v_{Phe} = 63.4[\text{PPH}]\left[(1 + [\text{PPH}])\left(1 + \frac{[\text{Phe}]}{50}\right)\right]^{-1}$$

| 13 | Tyrosine production: |

$$v_{Tyr} = \frac{10.48[\text{PPH}]}{1 + [\text{PPH}]}$$

| 14 | Tryptophan production: |

$$v_{Trp} = \frac{75.6[\text{X5P}][\text{CHR}][\text{ATP}]}{(1.269 + [\text{X5P}])(2 + [\text{CHR}])(0.9281 + [\text{ATP}])\left(1 + \frac{[\text{Trp}]}{16}\right)}$$

| 15 | X5P degradation into GAP and ethanol: |

$$v_X = 17[\text{ADP}]\frac{[\text{X5P}]}{0.8 + [\text{X5P}]}$$

| 16 | Phenylalanine export/degradation/utilization: |
| | $v_{Phe,out} = 0.013[\text{Phe}]$ |

| 17 | Tyrosine export/degradation/utilization: |
| | $v_{Tyr,out} = 0.013[\text{Tyr}]10$ |

(*continues*)

TABLE 12.2 (*continued*)

Reaction[a]	Kinetic expression[b]
18	Tryptophan export/degradation/utilization: $v_{\text{Trp,out}} = 0.013[\text{Trp}]$

Adenosine[c] Equilibrium between and total amount of adenosine-phosphate species:

$$[\text{ATP}] + [\text{ADP}] + \left(\frac{[\text{ADP}]^2}{[\text{ATP}]} \right) = 3$$

Metabolite	Material balance
Glc	$v_{\text{Glc,in}} = v_{\text{Glc}}$
G6P	$v_{\text{Glc}} = v_{\text{Pol}} + v_{\text{PFK}} + v_{\text{T}}$
FDP	$v_{\text{PFK}} + \frac{1}{2}v_{\text{Trp}} + \frac{1}{2}v_{\text{X}} = \frac{1}{2}v_{\text{GAPD}} + \frac{1}{2}v_{\text{Gol}} + \frac{1}{2}v_{\text{T}}$
PEP	$v_{\text{GAPD}} = v_{\text{PK}} + v_{\text{T}} + v_{\text{D}} + v_{\text{Trp}}$
ATP, ADP	$v_{\text{GAPD}} + v_{\text{PK}} + v_{\text{X}} = v_{\text{Glc}} + v_{\text{Pol}} + v_{\text{PFK}} + v_{\text{D}} + v_{\text{Phe}} + v_{\text{Tyr}} + 4v_{\text{Trp}}$
DAHP	$v_{\text{T}} = v_{\text{D}}$
CHR	$v_{\text{D}} = v_{\text{PPH}} + v_{\text{Trp}}$
PPH	$v_{\text{PPH}} = v_{\text{Phe}} + v_{\text{Tyr}}$
X5P	$v_{\text{T}} = v_{\text{Trp}} + v_{\text{X}}$
Phe	$v_{\text{Phe}} = v_{\text{Phe,out}}$
Tyr	$v_{\text{Tyr}} = v_{\text{Tyr,out}}$
Trp	$v_{\text{Trp}} = v_{\text{Trp,out}}$

[a] Reaction numbers correspond to the numbering scheme of Fig. 12.1.
[b] Molar reaction rates (mM/min) of the base state as functions of the concentrations (mM) of internal metabolites.
[c] The final expression represents a balance of the total adenosine concentration, as well as the equilibrium maintained between ATP, ADP, and AMP.

not an independent species.) At first glance, it would seem a straightforward exercise to derive the kernel matrix (Table 12.5) by working backward and balancing each metabolite according to the method described in Section 12.2.2. However, the interdependence of the energy-associated reactions introduces some additional complications.

Consider, for example, the pathway comprising reactions 1, 2, and 3, which results in the production of polysaccharides (Pol). From Fig. 12.1, it can be seen that reactions 1 and 2 should balance the glucose-6-phosphate (G6P) consumed by reaction 3. Hence, from the standpoint of the carbon conversion schematic of Fig. 12.1 alone, these three steps should complete the column for the polysaccharide pathway (pathway 4) in the kernel. However, inspection of the fourth column of the kernel shown in Table 12.5 reveals additional entries for this pathway. These entries correspond to the reactions that must also be included along with reactions 1, 2, and 3 to allow

TABLE 12.3 Steady-State Metabolite Concentrations
and Fluxes of Amino Acid Biosynthetic Network

Internal metabolite	Concentration (mM)
Glc	0.17
G6P	1.27
FDP	23.0
PEP	0.014
ATP	0.20
DAHP	3.74
CHR	7.87
PPH	0.55
X5P	1.48
Phe	271
Tyr	286
Trp	76.7
ADP	0.66

Reaction[a]	Molar flux (mM/min.)
1	31.67
2	31.67
3	0.14
4	23.30
5	43.63
6	26.15
7	2.97
8	2.44
9	8.24
10	8.24
11	7.24
12	3.52
13	3.72
14	1.00
15	7.24
16	3.52
17	3.72
18	1.00

[a] Reaction numbers correspond to the numbering scheme
of Fig. 12.1.

for the closure of the ATP (and, by extension, ADP) balance. As it happens, the only way this network can provide the ATP needed by any reactions is through the glycolytic pathway ending with reaction 6. Thus, the completion of the polysaccharide pathway requires the addition of reactions 1, 2, and 4, as well as 2 equivalents of reactions 5 and 6. If the same procedure is

TABLE 12.4   SIMS Matrix of Amino Acid Biosynthetic Network

Species	Reaction[a]																	
	1	2	3	4	5	6	7	8	9	10	11	12	13	14	15	16	17	18
Glc	1	-1	0	0	0	0	0	0	0	0	0	0	0	0	0	0	0	0
G6P	0	1	-1	-1	0	0	0	0	-1	0	0	0	0	0	0	0	0	0
FDP	0	0	0	1	-0.5	0	-0.5	0	-0.5	0	0	0	0	0.5	0.5	0	0	0
PEP	0	0	0	0	1	-1	0	0	-1	-1	0	0	0	-1	0	0	0	0
ATP	0	-1	-1	-1	1	1	0	-1	0	-1	0	-1	-1	-4	1	0	0	0
DAHP	0	0	0	0	0	0	0	0	1	-1	0	0	0	0	0	0	0	0
CHR	0	0	0	0	0	0	0	0	0	1	-1	0	0	-1	0	0	0	0
PPH	0	0	0	0	0	0	0	0	0	0	1	-1	-1	0	0	0	0	0
X5P	0	0	0	0	0	0	0	0	1	0	0	0	0	-1	-1	0	0	0
Phe	0	0	0	0	0	0	0	0	0	0	0	1	0	0	0	-1	0	0
Tyr	0	0	0	0	0	0	0	0	0	0	0	0	1	0	0	0	-1	0
Trp	0	0	0	0	0	0	0	0	0	0	0	0	0	1	0	0	0	-1
ADP	0	1	1	1	-1	-1	0	1	0	1	0	1	1	4	-1	0	0	0

[a] Reaction numbers correspond to the numbering scheme of Fig. 12.1.

TABLE 12.5   Kernel of the SIMS Matrix of Amino Acid Biosynthetic Network

Reaction[b]	Independent Pathway[a]					
	P1 Trp	P2 Phe	P3 Tyr	P4 Pol	P5 Gol	P6 ADP
1	5.5	3	3	2	1	0.5
2	5.5	3	3	2	1	0.5
3	0	0	0	**1**	0	0
4	4.5	2	2	1	1	0.5
5	9	4	4	2	1	1
6	6	2	2	2	1	1
7	0	0	0	0	**1**	0
8	0	0	0	0	0	**1**
9	1	1	1	0	0	0
10	1	1	1	0	0	0
11	0	1	1	0	0	0
12	0	1	0	0	0	0
13	0	0	1	0	0	0
14	1	0	0	0	0	0
15	0	1	1	0	0	0
16	0	**1**	0	0	0	0
17	0	0	**1**	0	0	0
18	**1**	0	0	0	0	0

[a] These independent pathways are shown schematically in Fig. 12.7. The characteristic reaction for each pathway is indicated by bold type and underline.
[b] Reaction numbers correspond to the numbering scheme of Fig. 12.1.

followed for each of the other output steps, the result will be the kernel matrix shown in Table 12.5. This kernel matrix can also be found through the solution of Eq. (12.8), using the six output steps identified previously as the base. Figure 12.7 provides a schematic representation of the balanced fluxes for each of the pathways identified in the kernel. As shown in Fig. 12.7, the glycolytic ATP-producing pathway is used by all six independent pathways.

12.3.3. IDENTIFICATION OF LINK METABOLITES AND GROUP FLUX DETERMINATION

We determined in the previous section that the aromatic amino acid biosynthetic network of Fig. 12.1 includes six independent pathways. This, of course, also implies that the network harbors several branch points at which the pathways diverge in order to form the final products. We demonstrate here a formal procedure for the location of branch points that can be generally applied to networks of any complexity. The procedure begins with the identification of the independent pathways (as shown in Fig. 12.7), as defined by the kernel, and proceeds through the following steps:

1. Identify the largest set of common reactions (*i.e.*, the longest common pathway) among the independent pathways defined by the kernel. This is shown for the six independent pathways of the amino acid network (Fig. 12.7) to be the glycolytic pathway. This common pathway produces the *link metabolite(s)* of the first branch point in the network. For the network used as our case study, this metabolite is ATP. The reactions participating in the common pathway form group A, the group upstream of this link metabolite.

2. Within the common pathway, identify the *production step*, which is the step responsible for the *net production* of the link metabolite. Scale each column of the kernel so that all entries for this step are equal. Balance any other reactions in the common pathway so as to have no net production/consumption of any intermediate metabolite within the pathway. The balanced stoichiometry of the reactions of the common pathway is shown in Fig. 12.7 for the network of our case study.

3. *Partition* the original kernel matrix into three submatrices (thereby creating three *group kernels*), each comprising the reactions in each group. Because the common pathway identified earlier constitutes the upstream group A, enter the stoichiometric coefficients of the balanced reactions of the common pathway into each column of its group

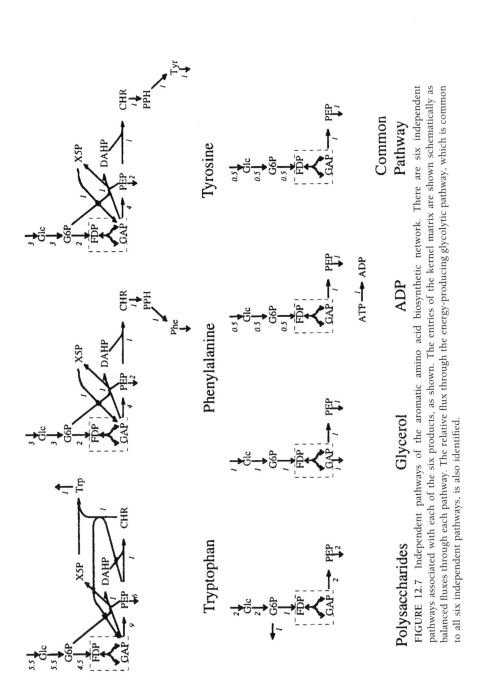

FIGURE 12.7 Independent pathways of the aromatic amino acid biosynthetic network. There are six independent pathways associated with each of the six products, as shown. The entries of the kernel matrix are shown schematically as balanced fluxes through each pathway. The relative flux through the energy-producing glycolytic pathway, which is common to all six independent pathways, is also identified.

kernel, leaving null entries for all other reactions. Subtract this group kernel from the original kernel to yield a reduced kernel comprising the reactions of the two downstream groups, B and C.

4. Find the largest set of reactions common to as many of the pathways of the reduced kernel as possible (*i.e.*, the new common pathway, which will be used to define the next branch point within group C). Partition into a new downstream group kernel B the entries of those kernel columns (*i.e.*, pathways) that *do not* contain the reactions of the new common pathway. The group kernel of group C is formed by subtracting the entries of group kernel B from those of kernel A. All pathways of group C include the new common pathway.

5. For any downstream group (B or C) with entries in two or more columns, repeat steps 1-4 in order to identify each link metabolite in turn. Note that the common pathway for group C has already been found in step 4. The group kernel corresponding to this common pathway defines group L for the new link metabolite, as it consists of the reactions linking it to the previous link metabolite. Nevertheless, the entire group A upstream of this link metabolite consists of group L, *in addition to* the group A upstream of the previous link metabolite and the group (B or C) that diverged from the new common pathway at the previous link metabolite (see Fig. 12.10). As a consequence, every step in the network will be accounted for in one of the three groups (A, B, and C) around any link metabolite. The separate accounting of groups A and L is necessary for the proper determination of group fluxes and group control coefficients, as discussed later.

6. Repeat the preceding procedure until no more common pathways can be identified. At that point, the final downstream groups will each consist of only a single column with nonzero entries. The number of branch points identified in this manner is equal to $P-1$, where P is the number of independent pathways.

It should be noted that the ultimate purpose of this procedure is to locate the branch points in a metabolic network and rationally define the reaction groups, and their corresponding fluxes, around each branch point. This is shown schematically in Fig. 12.8, depicting a branch point metabolite X_j along with the three reaction groups A, B, and C formed around it. As formulated in Chapter 13, group control coefficients will be determined for such reaction groups from the three fluxes J_A, J_B, and J_C around the branch point metabolite. Correct accounting of these fluxes will ensure the correct determination of the corresponding group control coefficients.

The application of this procedure to our case study is demonstrated next, applied iteratively to the kernel and group kernels. The first two of these

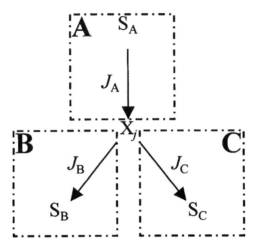

FIGURE 12.8 General representation of a branch point along with the reaction groups and metabolic fluxes defined around the branch point. Reaction groups A, B, and C include all network reactions upstream and downstream of the branch point metabolite.

iterations are shown schematically in Figs. 12.9 and 12.10. Note that, in order to facilitate the determination of the group fluxes, the original kernel (Table 12.5) has been supplemented with two new rows. The first of these is a row of ones which will later be scaled along with the other entries in each column. The second new row simply indicates the characteristic flux of each of the independent pathways and need never be scaled.

Branch Point 1: ATP

1. Inspection of the columns of the kernel of Table 12.5 reveals that the longest common pathway is that of glycolysis, comprising reactions 1, 2, 4, 5, and 6. The end metabolite of this pathway, ATP, therefore is identified as the first link metabolite in this network. (This can also be seen from the pathways in Fig. 12.7.)

2. The production step in the common pathway is reaction 6, which is responsible for the *net* formation of ATP, as the production and consumption of ATP by reactions 2, 4, and 5 cancel each other out. Hence, the columns of the kernel must be scaled with respect to reaction 6. By multiplying the entries in column 2 by 3, the entry for reaction 6 becomes 6. The entries in each of the other columns similarly are multiplied by the appropriate factor to make their entries for reaction 6 equal to 6. The resulting scaled matrix is shown in

Fig. 12.9, with the reactions common to all six pathways shown within boxes. To ensure that no intermediate metabolites are formed by this common pathway, its remaining reactions must be balanced as shown in the last frame of Fig. 12.7. This requirement is important in establishing the correct fluxes through each group of reactions, as will be discussed in more detail later.

3. The balanced stoichiometric reactions forming ATP are collected into group A upstream of ATP, resulting in the group kernel shown in Fig. 12.9. This group kernel (A) contains a copy of the balanced glycolytic pathway (P6) in each column. Note that the entries for reactions 1, 2, 4, and 5 differ in the six columns of the scaled matrix and that only a portion of these fluxes is included in the fluxes within group A. In fact, only the portion of the flux through these four steps that can be directly attributed to ATP formation is included in group A. The remainder of the flux through these steps is associated with carbon metabolites and will be distributed around other branch points.

4. The portion of the scaled kernel remaining after subtraction of the common pathway (*i.e.*, kernel matrix A) is a *reduced* kernel comprising only the reactions included in the downstream groups. The compositions of the two downstream groups B and C are found from this reduced kernel as follows: Inspection of the reduced kernel reveals that reactions 1 and 2 are common to the first five independent pathways, but not the sixth. As such, reactions 1 and 2 form the new common pathway, whereas reaction 8 is the only entry for pathway P6 in the reduced kernel matrix. Accordingly, this remaining reaction in path-way P6 (*i.e.*, reaction 8) forms group B, whereas the rest of the network reactions indicated by the first five pathways of the reduced kernel form group C. The compositions of all three groups are shown in the group kernels of Fig. 12.9. It is important to realize that, if these three group kernels are added, the result is the scaled kernel from which they were derived. At this point, the compositions of all three groups around the first link metabolite, ATP, have been determined.

5. The new common pathway (reactions 1 and 2) identified in 4 will be used in the following discussion to identify the remaining branch points within group C. Group A upstream of the next branch point will consist not only of this common pathway, but of groups A and B identified here, which have already been separated from group C.

It is critical to understand that the first link metabolite to be identified in this network is indeed ATP. This fact is not immediately apparent from the reaction schematic of Fig. 12.1, as it is drawn in order to depict the carbon metabolite conversions. Recall, however, that the glycolytic production of

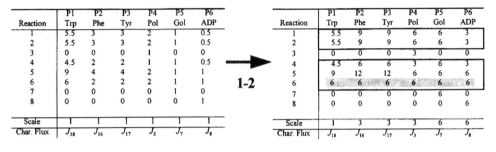

Original Kernel   **Scaled Kernel**

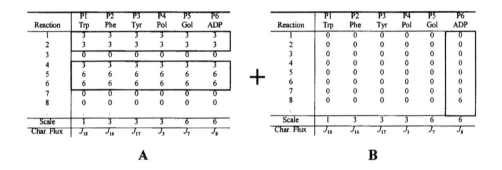

A   **B**

Group Kernels

FIGURE 12.9 Illustration of the scaling and partitioning of the kernel matrix at the ATP branch point. (1) The balanced common pathway (reactions 1-2 and 4-6) and the link metabolite (ATP) are identified. (2) The original kernel is scaled with respect to the production step of the common pathway (reaction 6, shown in shaded area) to produce the scaled kernel (only the first eight rows of which are shown).

ATP is a feature common to all independent pathways. In essence, the total flux through each of the independent pathways is the sum of the carbon flux required for the production of the corresponding end metabolite *and* the flux necessary to generate the ATP consumed by the same pathway. Because each independent pathway subsequently utilizes this ATP differently, the reaction grouping around the link metabolite ATP represents the net production of biochemical energy through glycolysis and its subsequent use by all energy-consuming reactions (group C). Group B represents the hydrolysis of any leftover ATP to ADP for maintenance or futile cycle ATP usage.

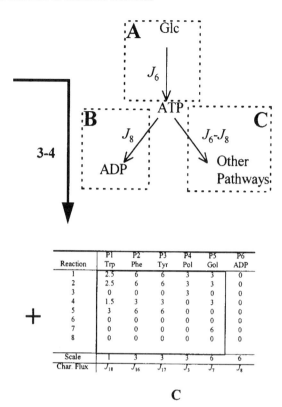

Reaction	P1 Trp	P2 Phe	P3 Tyr	P4 Pol	P5 Gol	P6 ADP
1	2.5	6	6	3	3	0
2	2.5	6	6	3	3	0
3	0	0	0	3	0	0
4	1.5	3	3	0	3	0
5	3	6	6	0	0	0
6	0	0	0	0	0	0
7	0	0	0	0	6	0
8	0	0	0	0	0	0
Scale	1	3	3	3	6	6
Char. Flux	J_{18}	J_{16}	J_{17}	J_3	J_7	J_8

C

(3) The group kernel of group A is formed by duplicating the entries of the balanced common pathway in all columns of kernel A. These entries are subtracted from the scaled kernel, resulting in a downstream reduced kernel equal to the sum of the group kernels of groups B and C. (4) Reactions 1 and 2 are identified as common to five downstream pathways. These pathways are placed into group C. The remaining downstream pathway is placed into group B. A schematic representation of this branch point and its group fluxes are shown for reference.

In fact, the fraction of glucose flux consumed for the generation of all ATP by the common pathway can be considered to be lumped into a single pool. The remaining carbon-associated glucose flux is distributed among pathways 1-5, whose excess energy needs are satisfied by the common ATP pool. In addition, it must be emphasized that the inappropriate grouping according to the carbon structure map, without the identification of ATP as a link metabolite, will in fact lead to gross inaccuracies in the estimation of control coefficients, if not the complete inability to obtain such estimates at all. This

Reaction	P1 Trp	P2 Phe	P3 Tyr	P4 Pol	P5 Gol
1	2.5	6	6	3	3
2	2.5	6	6	3	3
3	0	0	0	3	0
4	1.5	3	3	0	3
5	3	6	6	0	0
6	0	0	0	0	0
7	0	0	0	0	6
8	0	0	0	0	0
Scale	1	3	3	3	6
Char. Flux	J_{18}	J_{16}	J_{17}	J_3	J_7

Original Kernel

1-2

Reaction	P1 Trp	P2 Phe	P3 Tyr	P4 Pol	P5 Gol
1	6	6	6	6	6
2	6	6	6	6	6
3	0	0	0	6	0
4	3.6	3	3	0	6
5	7.2	6	6	0	0
6	0	0	0	0	0
7	0	0	0	0	12
8	0	0	0	0	0
Scale	2.4	3	3	6	12
Char. Flux	J_{18}	J_{16}	J_{17}	J_3	J_7

Scaled Kernel

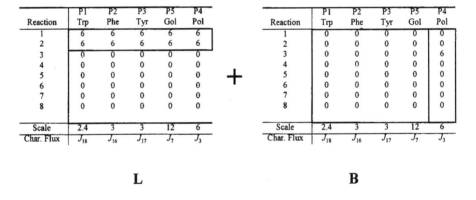

Reaction	P1 Trp	P2 Phe	P3 Tyr	P5 Gol	P4 Pol
1	6	6	6	6	6
2	6	6	6	6	6
3	0	0	0	0	0
4	0	0	0	0	0
5	0	0	0	0	0
6	0	0	0	0	0
7	0	0	0	0	0
8	0	0	0	0	0
Scale	2.4	3	3	12	6
Char. Flux	J_{18}	J_{16}	J_{17}	J_7	J_3

L

+

Reaction	P1 Trp	P2 Phe	P3 Tyr	P5 Gol	P4 Pol
1	0	0	0	0	0
2	0	0	0	0	0
3	0	0	0	0	6
4	0	0	0	0	0
5	0	0	0	0	0
6	0	0	0	0	0
7	0	0	0	0	0
8	0	0	0	0	0
Scale	2.4	3	3	12	6
Char. Flux	J_{18}	J_{16}	J_{17}	J_7	J_3

B

Group Kernels

FIGURE 12.10 Illustration of the scaling and partitioning of the kernel at the G6P branch point. (1) The balanced common pathway (reactions 1-2) and the link metabolite (G6P) are identified within the original kernel, which is the kernel of group C of the ATP branch point (FIGURE 12.9). (2) The kernel is scaled with respect to the production step of the common pathway (reaction 2) to produce the scaled kernel (only the first eight rows of which are shown).

same procedure of separating associated fluxes must be followed in handling any other link metabolites (*e.g.*, NADH) or growth-associated flux to expansion, which confound the standard reaction schematics.

After the identification of the first link metabolite and reaction groups, it remains only to define the actual flux of each group. Note that the units of these fluxes should be in moles of the *branch point metabolite* per unit time

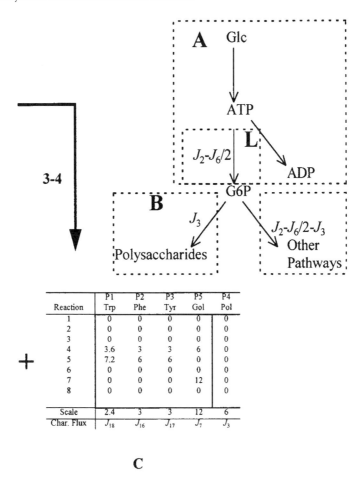

Reaction	P1 Trp	P2 Phe	P3 Tyr	P5 Gol	P4 Pol
1	0	0	0	0	0
2	0	0	0	0	0
3	0	0	0	0	0
4	3.6	3	3	6	0
5	7.2	6	6	0	0
6	0	0	0	0	0
7	0	0	0	12	0
8	0	0	0	0	0
Scale	2.4	3	3	12	6
Char. Flux	J_{18}	J_{16}	J_{17}	J_7	J_3

C

(3) The group kernel of group L is formed by duplicating the entries of the balanced common pathway in all columns of kernel A. These entries are subtracted from the scaled kernel, resulting in a downstream reduced kernel equal to the sums of the group kernels of groups B and C. (4) Reactions 1 and 2 are identified as common to four downstream pathways. They are placed into group C (which has been rearranged to keep these four together). The remaining downstream pathway is placed into group B. A schematic representation of this branch point, its group fluxes, and its relationship to the ATP branch point is shown for reference. Note that group L is a subset of the entire group A for this branch point, which also includes all the steps partitioned into groups A and B of the ATP branch point.

and unit of cells. In the case of group B, the group flux is clearly the flux through reaction 8. Similarly, the flux of group A equals the rate of reaction 6, which is the reaction step responsible for the net generation of ATP. Because ATP is consumed by numerous steps within group C, its group flux is determined by summing the contributions of all such steps. This flux is also equal to the difference between the fluxes through groups A and B.

The fluxes of these groups can also be found more rigorously through the direct use of the kernel matrix. Recall that the total flux through any reaction is equal to the sum of the fluxes through that reaction attributable to each independent pathway, according to Eq. (12.8). This equation also shows that the flux through any reaction can be found from the fluxes through the characteristic reactions defining the base of the kernel. This was the purpose of supplementing the preceding kernel with rows corresponding to the characteristic fluxes. By using this information, the group fluxes can be determined rigorously by the following procedure:

1. Identify the reaction i within group N (i.e., group A, B, or C) responsible for the net production or consumption of the link metabolite. This would be reaction 6 for group A and reaction 8 for group B.
2. For each pathway Pj included in group N, divide the group kernel entry K_{ij} by the scaling entry for that column and multiply by the flux through the characteristic reaction corresponding to pathway Pj. If the stoichiometry of reaction i in the SIMS matrix does not produce 1 equivalent of the link metabolite, multiply the resultant flux by the appropriate stoichiometric coefficient. The result will be the flux through group N attributable to pathway Pj in moles of link metabolite formed per unit time.
3. Sum the fluxes found in step 2 over all pathways in group N to obtain the total group flux.
4. Repeat the preceding steps for each remaining group. Alternatively, the flux through the final group can be found from the balance of the fluxes of the other groups.

Because all six columns of the kernel (i.e., all six independent pathways) contribute to group A, the flux of group A into ATP (i.e., the net flux of ATP formation) is the flux of reaction 6 attributable to all pathways. This is, of course, the total flux through reaction 6, and both results are shown in Eq. (12.10). The reaction consuming ATP in group B is reaction 8, the only reaction in the group. Hence, the flux of group B is equal to $(6/6)J_8$, or simply J_8. Because the reactions in group C corresponding to the net consumption of ATP are not readily apparent, its group flux is most easily found by the difference of the other two fluxes. In light of this discussion, the fluxes around the ATP branch point from the upstream reaction group (A) and into the downstream reaction groups (B and C) are as follows:

$$_{ATP}^{*}J_A = 6J_{18} + 2J_{16} + 2J_{17} + 2J_3 + J_7 + J_8 = J_6$$

$$_{ATP}^{*}J_B = J_8 \qquad\qquad (12.10)$$

$$_{ATP}^{*}J_C = 6J_{18} + 2J_{16} + 2J_{17} + 2J_3 + J_7 = J_6 - J_8$$

Branch Point 2: G6P

The second iteration of the branch point analysis uses, as a starting point, the kernel of group C. Note that, because pathway P6 has already been partitioned into a separate group and thus has only null entries in this kernel, it can be neglected in the analysis of this group. The initial kernel shown in Fig. 12.10, therefore, contains the first five columns of group C from Fig. 12.9. This kernel will be used to find groups L, B, and C around the new branch point. As shown in Fig. 12.10, however, the entire upstream group (A) for this branch point will comprise groups A and B of the ATP branch point *in addition to* group L identified here. Thus, the kernel of group A will be the sum of these three group kernels. This accounting is necessary in order to ensure that each reaction in the actual network is contained within one of the groups around this link metabolite.

The analysis of the second branch point proceeds as follows:

1. Inspection of this kernel reveals that reactions 1 and 2 constitute the longest pathway common to the remaining pathways P1-P5. As the end metabolite of this common pathway, G6P is identified as the next branch point.
2. Reaction 2 is identified as the production step of this branch metabolite. Each column of the matrix therefore is scaled with respect to reaction 2, as shown in Fig. 12.10. Reactions 1 and 2 are already balanced in this pathway in 1:1 proportion.
3. The kernel for group L is formed by replicating the 6:6 stoichiometry of the balanced common pathway into each column, resulting in pathways P1-P5 having entries equal to 6 for reactions 1 and 2. Subtraction of kernel L from the scaled kernel results in null values for the entries of reactions 1 and 2 in the downstream reduced kernel. Therefore, the remainder of the flux through these two steps is indeed considered to exist within group L. (Recall that group A for the G6P branch point consists of group L in addition to groups A and B of the ATP branch point.)

4. Inspection of the downstream reduced kernel (following subtraction of kernel L) reveals that reaction 4 is the only step common to the four pathways P1-P3 and P5. Reaction 4 therefore is the new common pathway, whereas reaction 3 is the sole entry for pathway 4 in the reduced kernel matrix. Accordingly, reaction 3 constitutes the new group B, whereas the remaining metabolic reactions downstream of G6P form group C. The resulting group kernels are shown in Fig. 12.10, with columns P4 and P5 rearranged for clarity.

5. The new common pathway (reaction 4) identified in step 4 will be used in the following discussion to locate the remaining branch points within group C. Group A, upstream of the next branch point, will consist not only of this common pathway, but of groups A and B of the G6P branch point, which have already been separated from group C.

The flux into the G6P branch point from its upstream branch (A) is equal to the *carbon-associated* flux through reaction 2. This flux can be determined either through the matrix inspection described earlier or by subtracting the energy-associated flux from the total flux through reaction 2. Stoichiometry indicates that the energy-associated flux through reaction 2 must equal half the total flux through reaction 6. Therefore,

$$_{G6P}^{*}J_A = \tfrac{5}{2}J_{18} + 2J_{16} + 2J_{17} + J_3 + \tfrac{1}{2}J_7 = J_2 - \tfrac{1}{2}J_6$$

$$_{G6P}^{*}J_B = J_3 \qquad\qquad\qquad\qquad\qquad\qquad (12.11)$$

$$_{G6P}^{*}J_C = \tfrac{5}{2}J_{18} + 2J_{16} + 2J_{17} + \tfrac{1}{2}J_7 = J_2 - \tfrac{1}{2}J_6 - J_3$$

Branch Point 3: FDP/GAP

1. The only step common to the pathways remaining within group C of the G6P branch point is reaction 4, which converts G6P into fructose-1,6-diphospate (FDP). FDP, therefore, can be identified as the third branch point. Note that, because FDP is considered to be at equilibrium with glyceraldehyde-3-phosphate (GAP) in this model, this branch point is actually the composite pool of FDP and GAP, and either of these can be considered to be the link metabolite.

2. Since reaction 4 is the only reaction of the common pathway, it is also clearly the production step. If the remaining four columns of this kernel are scaled with respect to this column, the following scaled

kernel results:

Reaction	P1 Trp	P2 Phe	P4 Tyr	P5 Gol
1	0	0	0	0
2	0	0	0	0
3	0	0	0	0
4	3	3	3	3
5	6	6	6	0
6	0	0	0	0
7	0	0	0	6
8	0	0	0	0
9	2	3	3	0
10	2	3	3	0
⋮	⋮	⋮	⋮	⋮
Scale	2	3	3	6
Char. Flux	J_{18}	J_{16}	J_{17}	J_7

3. Group L is formed by duplicating the entries in row 4 of the scaled matrix in each column with null entries everywhere else. Group A is, as before, formed from this group added to groups A and B of the previous (G6P) branch point. When group L is subtracted from the scaled matrix, the downstream reduced kernel results, comprising all the entries of the scaled matrix below row 4 and null entries everywhere else.

4. From the downstream reduced kernel, it is easily seen that reactions 5, 9, and 10 are common to pathways P1, P2, and P3. Because pathway P5 does not include any of these steps, the remaining entry in its column (*i.e.*, reaction 7) is placed into group B. Group C consists of the remaining reactions in the first three pathways, and its kernel consists of the entries in the first three columns of the scaled matrix, beginning with row 5, as shown by the box in the scaled matrix.

5. The common pathway (reactions 5, 9, and 10) within branch C will be used later to define the next branch point. Group A upstream of that downstream branch point will consist of this common pathway, in addition to groups A and B of this (FDP/GAP) branch point.

The fluxes of these branches are easily found from the scaled matrix and related to observable fluxes. Thus, the flux of branch A is the carbon-associated flux through reaction 4. The flux of branch B is only half that of reaction 7, as this step consumes only half an equivalent of FDP. Finally, the flux through branch C can be determined from the difference of the other two fluxes. Hence:

$$_{FDP}^{*}J_A = \tfrac{3}{2}J_{18} + J_{16} + J_{17} + \tfrac{1}{2}J_7 = J_4 - \tfrac{1}{2}J_6$$

$$_{FDP}^{*}J_B = \tfrac{1}{2}J_7 \qquad\qquad (12.12)$$

$$_{FDP}^{*}J_C = \tfrac{3}{2}J_{18} + J_{16} + J_{17} = J_4 - \tfrac{1}{2}J_6 - \tfrac{1}{2}J_7$$

Branch Point 4: X5P + CHR

1. Three reactions (5, 9, and 10) are common to pathways P1-P3 downstream of FDP/GAP. If these reactions are balanced in the stoichiometric proportion 2:1:1, both xylulose-5-phosphate (X5P) and chorismate (CHR) are formed with no net production of intermediate metabolites. Hence, this is an unusual case of a branch point with *dual* link metabolites. This, however, does not present a problem, as X5P and CHR *always* must be produced and consumed together at steady state.
2. Depending upon the choice of link metabolite, the production step of the common pathway is reaction 9 or 10. Either choice is acceptable, as the fluxes of the two steps are equivalent at steady state. The following scaled kernel results from either choice:

Reaction	P1 Trp	P2 Phe	P3 Tyr
5	3	2	2
6	0	0	0
7	0	0	0
8	0	0	0
9	1	1	1
10	1	1	1
11	0	1	1
12	0	1	0
13	0	0	1
14	1	0	0
15	0	1	1
16	0	1	0
17	0	0	1
18	1	0	0
Scale	1	1	1
Char. Flux	J_{18}	J_{16}	J_{17}

3. The three columns of group L upstream of this branch point are formed by placing entries equal to 2, 1, and 1 into rows 5, 9, and 10, respectively. When group L is subtracted from the scaled kernel, the

following downstream reduced kernel results:

Reaction	P1 Trp	P2 Phe	P3 Tyr
5	1	0	0
6	0	0	0
7	0	0	0
8	0	0	0
9	0	0	0
10	0	0	0
11	0	1	1
12	0	1	0
13	0	0	1
14	1	0	0
15	0	1	1
16	0	1	0
17	0	0	1
18	1	0	0
Scale	1	1	1
Char. Flux	J_{18}	J_{16}	J_{17}

4. One can see a nonzero entry in row 5, indicating that a remaining portion of the flux through this reaction will be included in the downstream branch corresponding to pathway P1. This portion of reaction 5 is required for the recycle of the GAP produced by reaction 14. It is clear from this downstream reduced kernel or from Fig. 12.7 that the tryptophan pathway consumes these two metabolites differently from the other two pathways. Therefore, group B consists of the remaining reactions in pathway P1, and group C consists of the remainder of pathways P2 and P3, which both employ reactions 11 and 15.

5. The common pathway (reactions 11 and 15) within branch C will be used later to define the next branch point. Group A upstream of that downstream branch point will consist of these two steps, in addition to groups A and B found earlier.

In terms of a single equivalent of chorismate, the flux of the upstream group into this branch point is equal to that of reaction 10. The flux into group B is equal to that of reaction 14 or, equivalently, reaction 18. The flux into group C clearly is the difference between the other two and is also equal to that of reaction 11. Therefore, the fluxes are defined by

$$_{X/C}^{*}J_A = J_{18} + J_{16} + J_{17} = J_{10}$$

$$_{X/C}^{*}J_B = J_{18} = J_{14}$$

$$_{X/C}^{*}J_C = J_{16} + J_{17} = J_{11}$$

(12.13)

Branch Point 5: PPH

1. The two remaining steps common to both the phenylalanine and tyrosine pathways (P1 and P2) are reactions 11 and 15. Reaction 15 merely recycles one unused equivalent of X5P back into the GAP branch point, whereas reaction 11 produces the final branch point metabolite, prephenate (PPH).

2. Reaction 11 clearly is the production step for this pathway. Because the entries for reactions 11 and 15 in group C are already balanced and equal, no further scaling is needed.

3. Group L consists of the common pathway reactions 11 and 15. Group A consists of these reactions as well as those included in groups A and B of the X5P/CHR branch point. When the entries in rows 11 and 15 are subtracted from the kernel, only the following entries remain in the downstream reduced kernel:

Reaction	P2 Phe	P3 Tyr
12	1	0
13	0	1
14	0	0
15	0	0
16	1	0
17	0	1
18	0	0
Scale	1	1
Char. Flux	J_{16}	J_{17}

4. Because no further commonalities exist within these two pathways, the remaining reactions within each pathway are placed into the two downstream groups. Branch B consists of reactions 12 and 16, which produce phenylalanine, and branch C consists of the dedicated tyrosine pathway, reactions 13 and 17. Reactions 12 and 13, respectively, are the consumption steps of these groups.

The determination of the fluxes around the prephenate branch point is certainly the simplest in this system. The flux from the upstream branch (A) is that of reaction 11, whereas the fluxes of the downstream branches are equal to the fluxes through either of their reactions. The fluxes around this branch point therefore are as follows:

$$\begin{aligned}
_{PPH}^{*}J_A &= J_{16} + J_{17} = J_{11} \\
_{PPH}^{*}J_B &= J_{16} = J_{12} \\
_{PPH}^{*}J_C &= J_{17} = J_{13}
\end{aligned} \qquad (12.14)$$

The branching structure identified by the preceding procedure is summarized in Fig. 12.11. From this diagram, showing both the carbon- and energy-associated fluxes, it should be clear that the metabolic network comprises a series of divergent branch points and that each independent pathway is associated with a secreted product formed by an output step. Measurement of the accumulation of these secreted metabolites, as well as other byproducts (in this case, CO_2 and ethanol), provides the means of estimating the flux through each independent pathway using the methods

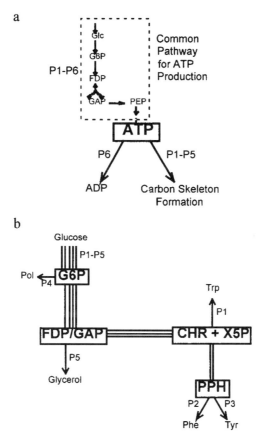

FIGURE 12.11   Branch points of the aromatic amino-acid biosynthetic network. The participation of each independent pathway in the system's five branch points is shown schematically. (a) The energy-related branch point meets the excess ATP requirements of products formed through pathways P1-P5, as well as the maintenance pathway P6. (b) Four downstream branch points separate the five product pathways P1-P5. Note that identification of the two distinct glucose fluxes that feed (a) and (b) is necessary for the analysis of reaction groups.

described in Chapters 8-10. Furthermore, changes in their accumulation during different phases of a fermentation process or in response to induced perturbations can be used for the quantitative evaluation of the kinetic control exercised by each reaction group, using the methods outlined in the beginning of Chapter 13. That section will begin with a discussion of the theoretical derivation of group control coefficients, as well as the methods for their calculation from flux estimates. This is followed by procedures for determining individual control coefficients from the analysis of overlapping groups. Finally, we will demonstrate the determination of the control structure of the *S. cerevisiae* case study using the branch points identified here, and their corresponding reaction groups will be used as a basis for the detailed examination of the network's control structure.

REFERENCES

Brown, C., Hafner, R. P., & Brand, M. D. (1990). A "top-down" approach to determination of control coefficients in metabolic control theory. *European Journal of Biochemistry* **188**; 321-325.

Galazzo, J. L. & Bailey, J. E. (1990). Fermentation pathway kinetics and metabolic flux control in suspended and immobilized *Saccharomyces cerevisiae*. *Enzyme and Microbial Technology* **12**; 162-172.

Galazzo, J. L. & Bailey, J. E. (1991). Errata. *Enzyme and Microbial Technology* **13**; 363.

Kacser, H. & Burns, J. A. (1973). The control of flux. *Symposia of the Society for Experimental Biology* **27**; 65-104.

Reder, C. (1988). Metabolic control theory: a structural approach. *Journal of Theoretical Biology* **135**; 175-201.

Simpson, T. W., Colón, G. E., & Stephanopoulos, G. (1995). Two paradigms of metabolic engineering applied to amino acid biosynthesis. *Biochemical Society Transactions* **23**; 381-387.

Stephanopoulos, G. & Simpson, T. W. (1997). Flux amplification in complex metabolic networks. *Chemical Engineering Science* **52**; 2607-2627.

CHAPTER 13

Flux Analysis of
Metabolic Networks

Control coefficients are additive by virtue of the summation theorem(s); therefore, it is possible to represent any number of reactions upstream or downstream of a metabolite as a single step. In fact, this reaction lumping is often necessary when the reactions within an unbranched pathway cannot easily be distinguished from each other experimentally. In our *Saccharomyces cerevisiae* case study, for example, this type of grouping of neighboring reactions was implemented extensively and helped reduce the number of discrete reactions considered by approximately an order of magnitude. The effect of grouping on the determination of the control structure of the network is minimal, as near-equilibrium reactions by definition have small control coefficients. Thus, their addition to the control coefficients of locally rate-controlling steps minimally affects the magnitude of the latter.

More importantly, it is possible to partition a reaction network into a small number of reaction groups, as illustrated in Chapter 12. Strictly speaking, this partitioning is applicable only if a system has an identifiable link metabolite, which is the sole conduit by which otherwise disparate pathways or processes are connected. With the formulation of the groups as described in the previous chapter, however, it is possible to form groups and define link metabolites, even if interactions exist between groups or single reactions participate in multiple groups. We will revisit this point later (Section 13.6). Here we note again that reaction grouping facilitates the identification of critical reaction steps by narrowing the search within groups with large group control coefficients.

In Sections 13.1 and 13.2, two different approaches for determining group control coefficients are presented. In Section 13.1, relationships between individual and group control coefficients are rigorously defined. These relationships allow the determination of gFCCs when individual FCCs are available. This procedure, termed the *bottom-up approach*, is illustrated with our case study in Section 13.3, where group control coefficients are determined from the kinetic model. Section 13.2 introduces a method by which group control coefficients may be determined using experimental flux measurements. By properly comparing these group control coefficients, individual control coefficients can be obtained (Section 13.4); this method therefore is termed the *top-down approach*. The results of these two approaches are compared in Section 13.3 for the case study. Figure 13.1 illustrates these two approaches schematically.

13.1. RELATIONSHIPS AMONG GROUP AND INDIVIDUAL CONTROL COEFFICIENTS (BOTTOM-UP APPROACH)

Mathematically, the gFCC for the effect of reaction group N, comprising several steps, on the flux J_k equals the sum of the individual FCCs of those steps upon the same flux J_k:

$$^*C_N^{J_k} = \sum_{i \in N} C_i^{J_k} \qquad k \in \{1, 2, \ldots, L\} \qquad (13.1)$$

In general, the gFCC subscript is a capital letter representing a clearly defined group, whereas that of an individual FCC is the number of an individual step. In addition, because groups can be described around different link metabolites, as described in the previous chapter, a preceding subscript can be used to indicate the link metabolite for which a group is defined.

Group concentration control coefficients (gCCCs) also satisfy a similar summation relationship. Thus, for the effect of reaction group N on metabo-

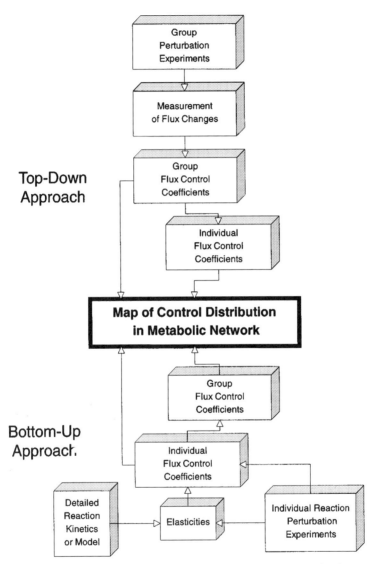

FIGURE 13.1   Schematic illustration of the flow of information involved in the determination of the control distribution using either the top-down or bottom-up approach.

lite X_j:

$$*C_N^{X_j} = \sum_{i \in N} C_i^{X_j} \qquad j \in \{1, 2, \ldots, K\} \tag{13.2}$$

Note that group control coefficients effectively obey the tenets of MCA as if they were individual control coefficients. In analogy to the significance of an individual FCC, a gFCC expresses the extent of control exerted *by a group of reactions* on a particular flux. For example, if it is determined that the group coefficient $*C_N^{X_j}$ is small, one can conclude that the *collective* effect of changes in the reactions within group N is small. By extension, the effects of the individual reactions within the group usually can be considered to be similarly small, although this may not strictly be the case if some of the reactions within group N exert strong *opposing* effects on a flux.

A special case of Eqs. (13.1) and (13.2) arises whenever the flux of a single reaction in a system is distributed into two (or more) different reaction groups formed around two (or more) branch points. Recall that this situation arose in our case study of amino acid production where, for example, the glycolytic reactions were part of both the up- and downstream groups around the ATP, G6P, and FDP branch points. In such a case, the portion of a reaction's control coefficient, which should be added to a particular group control coefficient, is equal to the fraction of the flux through that reaction attributed to that group. Thus, Eq. (13.1) can be rewritten more generally as

$$*C_N^{J_k} = \sum_{i \in N} \phi_i^N C_i^{J_k} \qquad k \in \{1, 2, \ldots, L\} \tag{13.3}$$

where ϕ_i^N is the fraction of the flux through reaction i which is considered to belong within group N (see the case study of Section 13.3). Additionally, several of the group fluxes in the case study of Section 12.3 were defined in terms of sums or differences of actual network fluxes as shown by Eqs. (12.10)-(12.14). In this case, the group flux control coefficient for the effect of a reaction or group of reactions *upon the combined group flux* is expressed as the weighted average of the gFCCs upon the individual fluxes. Thus, if the flux of group Z is the sum of fluxes J_1 and J_2, the gFCC describing the effect of a change in group N upon the flux of group Z is given by

$$*C_N^{J_Z} = \frac{*C_N^{J_1} J_1 + *C_N^{J_2} J_2}{J_1 + J_2} \tag{13.4}$$

More generally, if the flux into or out of any group Z is defined in terms of the network fluxes as

$$J_Z = \sum_{k=1}^{L} \sigma_k^Z J_k \tag{13.5}$$

then the resulting group flux control coefficient for the effect of any group N upon the flux through group Z is given by

$$^*C_N^{J_Z} = \frac{\sum_{k=1}^{L} \sigma_k^{Z*} C_N^{J_k} J_k}{J_Z} \tag{13.6}$$

where $^*C_N^k$ is defined by Eq. (13.3).

A similar set of equations can be written for the gCCC of a metabolite pool, which is the sum of the concentrations of two (or more) actual metabolites, as is the case with X5P and chorismate in the case study. For the case of a total metabolite pool equal to $c_1 + c_2$ (i.e., the sum of the concentrations of metabolites X_1 and X_2), for example, its group concentration control coefficient would be defined as

$$^*C_N^{X_1 + X_2} = \frac{^*C_N^{X_1} c_1 + {}^*C_N^{X_2} c_2}{c_1 + c_2} \tag{13.7}$$

Because group flux control coefficients emerge as important measures of network flux control, it is necessary to prescribe methods for their determination. Clearly, gFCCs must be derived from network fluxes and flux changes resulting from experimental perturbations to the network. In the next section, we show how flux change measurements around a link metabolite allow the determination of group control coefficients.

13.2. DETERMINATION OF GROUP CONTROL COEFFICIENTS FROM FLUX MEASUREMENTS (TOP-DOWN APPROACH)

In Chapter 11 we described various methods for the determination of individual flux control coefficients. These methods invariably require the measurement of flux changes *following a known change* in the enzyme activity or the rate of the reaction in question. Group flux control coefficients, on the other hand, cannot, in general, be determined by similar methods for the simple reason that usually it is not possible to control the change in the activity of an *entire reaction group*, unless the activity of every reaction within the group is changed equally. Thus, it is necessary instead to develop a method for the determination of group control coefficients from strategic perturbations within groups. Such a method is presented in this section, as adopted from the work of Stephanopoulos and Simpson (1997). This method is based on the large perturbation theory (Section 11.5) developed by Small and Kacser (1993a, b). For branched pathways and complex networks, we have extended Small and Kacser's theory to allow the determination of group control coefficients from perturbations originating in individual steps within reaction groups. In this analysis, the assumption of reversible first-order kinetics for each reaction naturally extends to the overall kinetics of each group, as shown in Section 11.5. Procedures for the determination of group and individual control coefficients in branched reaction networks are provided next. It should be noted that, in the case of an unbranched pathway, the procedures demonstrated in Section 11.5.1 are comprehensive.

The determination of group concentration control coefficients in a branched network requires the measurement of the concentration of the link metabolite, in addition to the fluxes leading into each group from that metabolite. Such measurements must be performed at the steady state condition under examination (i.e., the base state), as well as the steady states obtained following various induced perturbations from the base state. Furthermore, each of these perturbations must originate in one or more reactions *within a single group*. Examples of such perturbations include altered activity resulting from a change in the regulation of an enzyme, an increase in the copy number of the gene expressing an enzyme within the group, or enhancement of a transport step through a change in its external substrate concentration. Systemic perturbations that cannot be localized within a single group, such as changes in temperature or pH, generally would not be amenable to the analysis described next.

As a rule of thumb, a minimum of three perturbations, each originating within a different group, is required for the determination of all group coefficients in a branched network. Nevertheless, a well-designed perturbation may yield enough information that only one additional perturbation, rather than two, is required. Both cases are presented next.

13.2.1. DETERMINATION OF GFCCs FROM THREE PERTURBATIONS

We describe here the method for determining the group control coefficients around a divergent branch point from the fluxes at the base state and those resulting from perturbations within each of the three branches. In principle, measurements of the three group fluxes at the base state and at three perturbed states are necessary and sufficient for the determination of the complete set of gFCCs. Additionally, if the effects of each perturbation upon the concentration of metabolites are measured, group concentration control coefficients can also be determined.

The method is presented for a simple branch point depicted in Fig. 12.8. This schematic is applicable for all complex metabolic networks. Thus, each steady state around a properly defined branch point can be characterized by the level of the link metabolite X_j and the three group fluxes J_A, J_B, and J_C. Each of these fluxes represents the flux from the link metabolite into the corresponding group or vice versa (see Section 12.3). Because this analysis refers to fluxes in four separate states (the base state and those attained following the three perturbations), each state will be denoted by a subscript, so that $J_{k,0}$ refers to the flux through group k in the base state and $J_{k,A}$, $J_{k,B}$, and $J_{k,C}$ refer to the new steady state fluxes through group k resulting from perturbations *originating* in groups A, B, and C, respectively.

Recall from Section 11.5 that each perturbed state can be characterized by a single perturbation constant K_i, such that

$$K_i = \frac{{}^*C_i^{J_A}}{f_i^A - 1} = \frac{{}^*C_i^{J_B}}{f_i^B - 1} = \frac{{}^*C_i^{J_C}}{f_i^C - 1} \qquad i \in \{A, B, C\} \qquad (13.8)$$

where f_i^k is the flux amplification factor defined as the ratio of each perturbed flux to that of its base state:

$$f_i^k = \frac{J_{k,i}}{J_{k,0}} \qquad i, k \in \{A, B, C\} \qquad (13.9)$$

Equation (13.8) relates the gFCCs to the amplification factors and perturbation constants. The form of this equation reveals that the fractional change in each group flux resulting from a particular perturbation will be proportional to the gFCC of that group with respect to the group from which the perturbation originated. This proportionality is equivalent in all three branches and is defined by the perturbation constant K_i. The value of this constant is inverse to the magnitude of the corresponding perturbation. Furthermore, because the flux amplification factors are measured for each perturbation, gFCCs can be calculated from Eq. (13.8) if the perturbation constants for each perturbation are known. The determination of the perturbation constants is described next.

The summation theorem for this branch point is introduced to complement Eq. (13.8) and is expressed as follows for each of the three branches:

$$ {}^*C_A^{J_k} + {}^*C_B^{J_k} + {}^*C_C^{J_k} = 1 \qquad k \in \{A, B, C\} \qquad (13.10)$$

By using Eq. (13.8) to express the group control coefficients in Eq. (13.10) in terms of flux amplification factors and perturbation constants, the summation theorem yields:

$$K_A(f_A^k - 1) + K_B(f_B^k - 1) + K_C(f_C^k - 1) = 1 \qquad k \in \{A, B, C\} \quad (13.11)$$

This expression is simplified further by defining:

$$p_i^k = f_i^k - 1 = \frac{J_{k,i} - J_{k,0}}{J_{k,0}} \qquad i, k \in \{A, B, C\} \qquad (13.12)$$

which yields the following form of the three summation equations:

$$K_A\, p_A^A + K_B\, p_B^A + K_C\, p_C^A = 1$$
$$K_A\, p_A^B + K_B\, p_B^B + K_C\, p_C^B = 1 \qquad (13.13)$$
$$K_A\, p_A^C + K_B\, p_B^C + K_C\, p_C^C = 1$$

or in matrix notation:

$$\begin{bmatrix} p_A^A & p_B^A & p_C^A \\ p_A^B & p_B^B & p_C^B \\ p_A^C & p_B^C & p_C^C \end{bmatrix} \begin{bmatrix} K_A \\ K_B \\ K_C \end{bmatrix} = \begin{bmatrix} 1 \\ 1 \\ 1 \end{bmatrix} \qquad (13.14)$$

Although it appears that the three perturbation constants could be determined by matrix inversion of the preceding equation, this is not so because the square matrix in Eq. (13.14) is, in fact, singular by the nature of the material balance at the branch point. In essence, the flux changes in branch A are not independent, as its flux must always equal the sum of the other two branches' fluxes. Therefore, Eq. (13.14) provides only two of the equations necessary for the calculation of the perturbation constants. A third equation is provided by the branching theorem,

$$\frac{{}^*C_C^{J_A}}{{}^*C_B^{J_A}} = \frac{J_{C,0}}{J_{B,0}} \qquad (13.15)$$

which can also be expressed in terms of proportional flux changes and perturbation constants:

$$\frac{K_C}{K_B} = \frac{J_{C,0}\, p_B^A}{J_{B,0}\, p_C^A} = q \qquad (13.16)$$

When this expression is used in conjunction with Eq. (13.14), the following solution results for the first two perturbation constants:

$$
\begin{bmatrix} K_A \\ K_B \end{bmatrix} = \begin{bmatrix} p_A^B & p_B^B + p_C^B q \\ p_A^C & p_B^C + p_C^C q \end{bmatrix}^{-1} \begin{bmatrix} 1 \\ 1 \end{bmatrix} \tag{13.17}
$$

Once the third perturbation constant is determined from Eq. (13.16), all nine group control coefficients are calculated from:

$$
{}^*C_i^{J_k} = K_i\left(f_i^k - 1\right) \qquad i, k \in \{A, B, C\} \tag{13.18}
$$

EXAMPLE 13.1

Determination of gFCCs for the Lysine Biosynthetic Pathway

The preceding approach has been applied to the determination of the group flux control coefficients of the aspartate amino acid biosynthetic pathway (Simpson, *et al.,* 1998). This pathway, summarized in Table 10.1 and Fig. 10.2, was the subject of the flux determination case study of Chapter 10. It was determined there that a critical branch point for the production of lysine is pyruvate. Actually, due to the inability to differentiate between the anaplerotic pathways catalyzed by pyruvate carboxylase and PEP carboxylase, the two metabolites PYR and PEP were lumped together in defining the composite PYR/PEP branch point. By applying the preceding methodology to data obtained from a variety of perturbation experiments (including a glucose-6-phosphate dehydrogenase attenuated mutant, glucose and gluconate fermentations, addition of fluoropyruvate to specifically inhibit the activity of the pyruvate dehydrogenase complex, presence of threonine to inhibit the activity of aspartokinase, and a pyruvate kinase deficient mutant), group flux control coefficients were determined for the following three reaction groups: glycolytic reactions upstream of the branch point (group A), the respiratory cycle (group B), and the aspartate biosynthetic pathway (group C). The results are summarized in the following table in the form of gFCCs ($^*C_i^{J_k}$):

Perturbed group (i)		Affected flux (J_k)		
Group	Type of perturbation(s)	J_A	J_B	J_C
A	Decreased glucose feed Attenuated G6P isomerase Gluconate feed without glucose	0.08	0.50	0.42
B	Complex medium (industrial fermentation) Fluoropyruvate inhibition of pyruvate kinase	0.16	1.24	-0.40
C	Threonine inhibition of aspartate kinase	0.02	-0.22	1.20

By virtue of the nature of group flux control coefficients, this table summarizes the control exercised by each of the preceding reaction groups on the flux through every other group. It can be seen that most of the control on the flux to aspartate is from reactions in the same pathway, whereas the impact of the other two groups is rather marginal. This result is significant in light of the speculation about possible limiting reactions of lysine formation, which included limitation by NADPH availability, ATP, or pyruvate, among others.

We should also note that this table is the consensus outcome of a number (more than three) of different perturbation experiments essentially converging to the same final result. The good agreement among gFCC estimates obtained by entirely different and redundant perturbation experiments strengthens the confidence in the obtained gFCC values and supports the validity of the preceding approach.

13.2.2. DETERMINATION OF GFCCS FROM A CHARACTERIZED PERTURBATION

A perturbation is considered to be *characterized* if the fractional change in the overall activity of the originating reaction group is known, in which case gFCCs can be determined directly from their definition, by using relationships derived from the theory of large deviations (Section 11.5). Such perturbations are, in general, difficult to implement, as the overall activity of a reaction group depends upon the kinetics of all reactions in the group. As a result, characterized perturbations generally are possible only in the cases of groups comprising a single reaction, the near-complete shutdown of a group, or the simultaneous amplification of the entire group.

We introduce the activity amplification factor, r_i, to characterize the perturbation within group i:

$$r_i = \frac{{}^*v_{i,i}}{{}^*v_{i,0}} \qquad i \in \{A, B, C\} \qquad (13.19)$$

In this equation, ${}^*v_{i,0}$ and ${}^*v_{i,i}$ represent the overall activity of group i before and after the characterized perturbation, respectively. In the case of the reversible, first-order kinetics assumed by the large deviation equations, each overall activity in a branched or unbranched network is also reversible and first-order (see Section 11.5). These equations yield the following relationship between the amplification factor and the perturbation constant:

$$K_i = \frac{r_i}{f_i'(r_i - 1)} \qquad i \in \{A, B, C\} \tag{13.20}$$

If a characterized perturbation (i.e., with a known r_i) is introduced into group B of Fig. 12.8, for example, measurement of the fluxes at the base and perturbed states allows the immediate determination of K_B from Eq. (13.20). If K_B is so determined, K_C can be calculated from an uncharacterized perturbation in group C, using Eq. (13.16). From the two perturbation constants, six of the nine gFCCs can be determined through Eq. (13.18). The remaining three can be found through the summation theorem, Eq. (13.10).

A characterized perturbation in a single reaction also allows the individual flux control coefficients of that reaction to be determined, according to the individual FCC analogues of these expressions. Thus, for a characterized perturbation in reaction i, measurement of the change in flux through each reaction k allows an individual coefficient to be found (Small and Kacser, 1993b):

$$C_i^{J_k} = \frac{r_i(f_i^k - 1)}{f_i'(r_i - 1)} \qquad i, k \in \{1, 2, \ldots, L\} \tag{13.21}$$

In this equation, r_i is the activity amplification factor of the individual step.

13.2.3. DETERMINATION OF GCCCS

Group concentration control coefficients for the link metabolite can be determined once the group flux control coefficients are known. The only additional measurement required for this calculation is the concentration of the link metabolite X_j at the base state ($c_{j,0}$) and at each perturbed state ($c_{j,i}$). From these measurements, the gCCC can be found through the following relationship (Small and Kacser, 1993b):

$$^*C_i^{X_j} = K_i \left(\frac{c_{j,i}}{c_{j,0}} - 1 \right) \qquad i \in \{A, B, C\}, j \in \{1, 2, \ldots, K\} \tag{13.22}$$

The gCCC for any branch that was not perturbed can be found from the summation theorem:

$$C_A^{X_j} + C_B^{X_j} + C_C^{X_j} = 0 \qquad j \in \{1, 2, \ldots, K\} \tag{13.23}$$

Although strictly true only for the link metabolite, these equations can also be used to provide estimates of the group concentration control coefficients for other metabolites by substituting the appropriate concentrations. The importance of these gCCC estimates will be made clear in the discussion of flux amplification in Section 13.5.

13.2.4. OBSERVABILITY OF PERTURBATIONS

Even though the preceding equations were derived for the general case of any large kinetic perturbation, the actual permissible magnitude of each perturbation is functionally bounded. At one extreme of infinitesimally small changes in enzyme activity, each of these expressions ultimately reduces to the coefficient's definition. However, as pointed out in Chapter 11, perturbations of very small magnitude are subject to large measurement errors and are not particularly useful. A very large perturbation, at the other extreme, may upset the stability of the network, resulting in unpredictable metabolic changes or even failure of a system to reach a steady state (see Section 13.5). These two extremes should best be avoided by employing perturbations that are known to produce stable, measurable effects or by using a series of perturbations that are likely to be acceptable.

A final point is in order about the number of perturbations required. Although, in principle, group flux control coefficients can be determined from only two or three perturbations, it is wise to use more than the minimum required in order to test the consistency of the results (see Section 13.6). This point was demonstrated in Example 12.2. Nevertheless, the analysis of multiple branch points is not hindered significantly by the requirement of three or more perturbations around each branch point. In fact, if they are sufficiently localized (*i.e.*, within a single branch around each branch point), *the perturbations employed in the analysis of a branch point can also be used for any number of other branch points*, as long as the appropriate fluxes are measured. Thus, six strategically placed perturbations may permit the determination of the gFCCs around four different link metabolites while also providing a consistency check for each. The ramifications of this concept will be explored more fully in Section 13.4, where it will be shown that the analysis of neighboring branch points allows localization of the search for critical steps, while also permitting the determination of certain *individual* control coefficients.

13.3. CASE STUDY

In this section we illustrate the bottom-up and top-down approaches to flux control elucidation by applying them to the branch points of the *S. cerevisiae* case study.

13.3.1. ANALYTICAL DETERMINATION OF GROUP
CONTROL COEFFICIENTS (BOTTOM - UP APPROACH)

By using the kinetic expressions of Table 12.2 for the model network, all individual FCCs and CCCs were determined analytically. The calculation proceeded through the solution of the balances for the steady state metabolite concentrations and fluxes, followed by the determination of the control coefficients employing the methods of Chapter 11. The results are shown in Tables 13.1 and 13.2. An examination of the values in both tables reveals that reaction 4 exercises a significant degree of control on most network fluxes. This is the step catalyzed by phosphofructokinase (PFK), which exhibits strongly ATP-dependent kinetics (see Table 12.2). We will also show later how the large control coefficients associated with the PFK step are reflected in, and can indeed be deduced from, the magnitudes of the appropriate group control coefficients.

Branch Point 1: ATP

ATP is the uppermost link metabolite in this network, as demonstrated in Section 12.3 (to which the reader is encouraged to refer in the following discussion). The determination of the group control coefficients around this branch point must begin with a clear definition of the fluxes that should be counted within each reaction group. The flux to branch B, which has no reactions in common with either of the other branches, equals that of reaction 8. The reactions in branch A, on the other hand, overlap with those of branch C, because the energy-associated flux through the glycolytic pathway is only part of the flux through reactions 1, 2, 4, and 5. Thus, it is necessary to determine the exact fraction of each of these steps which should be counted to belong in branch A. As determined in Eq. (12.10), the true flux through branch A is defined by the flux through reaction 6, as this reaction belongs entirely in branch A. Thus, ϕ_1^A (the fraction of reaction 1 flux that belongs in branch A) is equal to $\frac{1}{2}J_6/J_1$, whereas ϕ_5^A is J_6/J_5, and ϕ_6^A is J_6/J_6 or 1. These fractions can also be rigorously determined by using the kernel of group A shown in Fig. 12.9. By applying Eq. (12.8) to reaction 1, for example, the total flux of reaction 1 within group A is found to be

$$J_1^A = 3J_{18} + J_{16} + J_{17} + J_3 + \tfrac{1}{2}J_7 + \tfrac{1}{2}J_8 = \tfrac{1}{2}J_6 \qquad (13.24)$$

[This result can be obtained by dividing the entries for reaction 1 in the group kernel A (Fig. 12.9) by the scale and multiplying by the characteristic flux shown there.] Clearly, ϕ_1^A is the ratio of this result to J_1, the total flux through reaction 1. The remaining flux through reactions 1, 2, 4, and 5 is placed within branch C, which also includes reactions 3, 7 and 9-18 (see

TABLE 13.1 Flux Control Coefficients of Amino Acid Biosynthetic Network[a]

							Perturbed reaction i[b]											
k	1	2	3	4	5	6	7	8	9	10	11	12	13	14	15	16	17	18
1	0.07	0	0	0.94	0.03	-0.07	-0.03	-0.01	0.01	0.04	0.01	0	0	-0.01	0.01	0.01	0.01	-0.01
2	0.07	0	0	0.94	0.03	-0.07	-0.03	-0.01	0.01	0.04	0.01	0	0	-0.01	0.01	0.01	0.01	-0.01
3	1.64	0	0.99	-1.64	-0.05	0.13	0.04	0.01	-0.02	-0.06	-0.01	0	0	0.01	-0.02	-0.02	-0.01	0.01
4	0.06	0	0	0.95	0	-0.04	0.01	0	0.01	0.02	0	0	0	0	0	0	0	0
5	0.06	0	0	0.95	0.06	-0.04	-0.06	0	0.01	0.02	0	0	0	0	0	0	0	0
6	0.07	0	0	0.92	0.01	0.04	-0.02	0.04	-0.01	-0.02	-0.02	0	0	0.02	-0.02	-0.02	-0.02	0.01
7	0.06	0	0	0.95	-0.94	-0.04	0.94	0	0.01	0.02	0	0	0	0	0	0	0	0
8	-0.04	0	-0.01	-0.25	0.07	1.16	-0.1	0.9	-0.2	-0.57	0.07	0.04	-0.04	-0.07	0.1	-0.04	0.04	-0.06
9	0.05	0	0	0.95	0.12	-0.17	-0.11	-0.04	0.03	0.08	0.04	-0.01	0.01	-0.04	0.05	0.05	0.04	-0.03
10	0.05	0	0	0.95	0.12	-0.17	-0.11	-0.04	0.03	0.08	0.04	-0.01	0.01	-0.04	0.05	0.05	0.04	-0.03
11	0.05	0	0	0.84	0.1	-0.23	-0.09	-0.03	0.04	0.11	0.09	-0.01	0.01	-0.09	0.11	0.09	0.08	-0.07
12	0.03	0	0	0.59	0.07	-0.16	-0.06	-0.02	0.03	0.08	0.06	0.35	-0.35	-0.06	0.08	0.37	0.06	-0.05
13	0.06	0	0	1.08	0.13	-0.3	-0.12	-0.04	0.05	0.15	0.11	-0.35	0.35	-0.11	0.14	-0.17	0.1	-0.09
14	0.08	0	-0.02	1.7	0.26	0.26	-0.27	-0.13	-0.05	-0.13	-0.31	0.01	-0.01	0.31	-0.41	-0.28	-0.28	0.26
15	0.05	0	0	0.84	0.1	-0.23	-0.09	-0.03	0.04	0.11	0.09	-0.01	0.01	-0.09	0.11	0.09	0.08	-0.07
16	0.03	0	0	0.59	0.07	-0.16	-0.06	-0.02	0.03	0.08	0.06	0.35	-0.35	-0.06	0.08	0.37	0.06	-0.05
17	0.06	0	0	1.08	0.13	-0.3	-0.12	-0.04	0.05	0.15	0.11	-0.35	0.35	-0.11	0.14	-0.17	0.1	-0.09
18	0.08	0	-0.02	1.7	0.26	0.26	-0.27	-0.13	-0.05	-0.13	-0.31	0.01	-0.01	0.31	-0.41	-0.28	-0.28	0.26

[a] Flux control coefficients corresponding to the network's base state were determined analytically using the structural method (Reder, 1988). Each coefficient $C_i^{J_k}$ describes the effect of a perturbation of reaction i upon the flux through reaction k.

[b] Reaction numbers correspond to the numbering scheme of Fig. 12.1.

TABLE 13.2 Concentration Control Coefficients of Amino Acid Biosynthetic Network[a]

X_j	Perturbed reaction i[b]																	
	1	2	3	4	5	6	7	8	9	10	11	12	13	14	15	16	17	18
Glc	0.24	-3.24	0.02	3.23	0.03	-1.13	0	0.05	0.2	0.55	-0.03	-0.04	0.04	0.03	-0.04	0.08	0	0.03
G6P	0.18	0	0	-0.18	-0.01	0.01	0	0	0	-0.01	0	0	0	0	0	0	0	0
FDP	0.19	0	0.01	2.56	-2.33	-1.52	-0.01	0.13	0.26	0.74	-0.1	-0.05	0.05	0.1	-0.13	0.06	-0.05	0.08
PEP	0.15	0	0	2.28	0.17	-1.69	-0.13	0.02	-0.19	-0.53	0.01	0.04	-0.04	-0.01	0.02	-0.09	-0.01	-0.01
ATP	-0.04	0	-0.01	-0.25	0.07	1.16	-0.1	-0.1	-0.2	-0.57	0.07	0.04	-0.04	-0.07	0.1	-0.04	0.04	-0.06
DAHP	0.07	0	0.01	0.74	-0.02	-1.36	0.05	0.09	0.77	-0.7	-0.08	-0.16	0.16	0.08	-0.1	0.34	0.04	0.07
CHR	0.64	0	-0.05	11.3	1.32	-3.14	-1.23	-0.4	0.54	1.53	-3.79	-0.09	0.09	-1.15	1.51	-2.9	-3.25	-0.94
PPH	0.1	0	-0.01	1.68	0.2	-0.47	-0.18	-0.06	0.08	0.23	0.17	-0.54	-1.01	-0.17	0.22	-0.26	0.16	-0.14
X5P	0.1	0	-0.02	2.18	0.34	0.35	-0.35	-0.17	-0.06	-0.17	0.31	0.01	-0.01	-0.31	-2.44	0.23	0.26	-0.25
Phe	0.03	0	0	0.59	0.07	-0.16	-0.06	-0.02	0.03	0.08	0.06	0.35	-0.35	-0.06	0.08	-0.63	0.06	-0.05
Tyr	0.06	0	0	1.08	0.13	-0.3	-0.12	-0.04	0.05	0.15	0.11	-0.35	0.35	-0.11	0.14	-0.17	-0.9	-0.09
Trp	0.08	0	-0.02	1.7	0.26	0.26	-0.27	-0.13	-0.05	-0.13	-0.31	0.01	-0.01	0.31	-0.41	-0.28	-0.28	-0.74
ADP	0.01	0	0	0.08	-0.02	-0.36	0.03	0.03	0.06	0.17	-0.02	-0.01	0.01	0.02	-0.03	0.01	-0.01	0.02

[a] Concentration control coefficients corresponding to the network's base state were determined analytically using the structural method (Reder, 1988). Each coefficient $C_i^{X_j}$ describes the effect of a perturbation of reaction i upon the concentration of metabolite X_j.
[b] Reaction numbers correspond to the numbering scheme of Fig. 12.1.

595

group kernel shown in Fig. 12.9). Fluxes belonging to each group are determined similarly.

Once the flux fractions associated with each group have been clarified, gFCCs are calculated from the individual flux control coefficients of Table 13.1 using the equations of Section 13.1. For simplicity, we begin with the gFCCs with respect to a perturbation in branch B, because this branch comprises only reaction 8. Since the flux through branch A is J_6, $^*C_B^{J_A} = C_8^{J_6} = 0.04$. Similarly, $^*C_B^{J_B} = C_8^{J_8} = 0.9$. Because the flux of branch C is the difference in two individual reaction fluxes, $^*C_B^{J_C}$ must be found by applying Eq. (13.6), using the fluxes shown in Table 12.4. Thus,

$$^*C_B^{J_C} = \frac{J_6 C_8^{J_6} - J_8 C_8^{J_8}}{J_6 - J_8} = -0.05 \tag{13.25}$$

In fact, it is a general result of Eq. (13.6) that the third gFCC with respect to a particular perturbed branch can be determined from the first two:

$$^*C_i^{J_C} = \frac{J_A{}^*C_i^{J_A} - J_B{}^*C_i^{J_B}}{J_A - J_B} \qquad i \in \{A, B, C\} \tag{13.26}$$

For the effect of a perturbation in branch A upon itself, it is necessary to employ Eq. (13.3):

$$^*C_A^{J_A} = \left(\tfrac{1}{2}J_6/J_1\right)C_1^{J_6} + \left(\tfrac{1}{2}J_6/J_2\right)C_2^{J_6} + \left(\tfrac{1}{2}J_6/J_4\right)C_4^{J_6}$$
$$+ \left(J_6/J_5\right)C_5^{J_6} + C_6^{J_6} = 0.59 \tag{13.27}$$

Similarly, $^*C_A^{J_B}$ can be determined by replacing each $C_i^{J_6}$ in Eq. (13.27) with $C_i^{J_8}$ to yield 1.05. By using Eq. (13.26), $^*C_A^{J_C}$ is found from these two results to be 0.54. The remaining three gFCCs are determined from the summation theorem. These results are summarized in Table 13.3.

The resulting gFCCs indicate that branch A exerts significant control on all three branches. Branch B, on the other hand, only affects itself. Finally, branch C exerts significant positive control on both itself and branch A and strong negative control on branch B. Given the relatively small flux of branch B compared to the other two, however, this can be judged a noncritical branch point, as the flow from branch A into branch C clearly predominates and is virtually unaffected by branch B. Nevertheless, the distribution of control between branches A and C and the relatively significant magnitudes of the group concentration control coefficients may make this a branch point worthy of closer examination.

TABLE 13.3   Group Control Coefficients for ATP Branch Point[a]

Group	Constituent steps[b]	J_A	J_B	J_C	$*C_N^{ATP}$
	Perturbed branch N		gFCC of flux $*J_k$		gCCC
A	$1_{.41}\ 2_{.41}\ 4_{.56}\ 5_{.6}\ 6$	0.59	1.05	0.54	1.05
B	8	0.04	0.90	-0.05	-0.10
C	$1_{.59}\ 2_{.59}\ 3\ 4_{.44}\ 5_{.4}\ 7\ 9\text{-}18$	0.37	-0.95	0.51	-0.95

[a] gFCCs are shown in the form $*C_N^{J_k}$. gCCCs are of the form $*C_N^{ATP}$.
[b] Reaction numbers correspond to the numbering scheme of Fig. 12.1. Subscripted values represent ϕ_i^N, the fraction of the preceding reaction's flux that is included in branch N.

Branch Point 2: G6P

At the glucose-6-phosphate branch point, the pathway of polysaccharide formation diverges from the glycolytic pathway. Because the energy-associated glycolytic flux has already been partitioned upstream (at the ATP branch point), only the carbon-associated flux flows into the G6P branch point, as derived in Eq. (12.11). Recall from Fig. 12.10 that the entire energy-related fraction of the glycolytic pathway (group A of the ATP branch point or group A_{ATP}) is considered to lie within group A of the G6P branch point, *but that the flux into the G6P branch point from branch* A_{G6P} *is the carbon-associated flux only*. Branch B_{G6P} consists entirely of reaction 3, and branch C consists of the remaining steps downstream of G6P, as shown in the group kernels of Fig. 12.10.

Let us turn to the determination of the gFCCs for this branch point from the FCCs of Table 13.1. Because branch B is the simplest, consisting only of reaction 3, its corresponding gFCC is determined first. $*C_B^{J_B}$, of course, is equivalent to $C_3^{J_3}$ or 0.99. Because $C_3^{J_2}$ and $C_3^{J_6}$ are both zero, it follows that $*C_B^{J_A}$ is zero. From these, $*C_B^{J_C}$ is found to be very near zero through Eq. (13.25). The effect of a perturbation in branch A must be determined according to the equations in Section 13.1. Thus,

$$*C_A^{J_B} = *C_A^{J_3} = C_1^{J_3} + C_2^{J_3} + \left(\tfrac{1}{2}J_6/J_4\right)C_4^{J_3} + \left(J_6/J_5\right)C_5^{J_3} + C_6^{J_3} + C_8^{J_3} = 0.83 \tag{13.28}$$

Because both J_A and J_C are defined by Eq. (12.11) in terms of the fluxes of multiple reactions, Eq. (13.6) must be used to determine one of them. Let us consider the flux from branch A, which is equal to $J_2 - \tfrac{1}{2}J_6$. First, both $*C_A^{J_2}$ (0.53) and $*C_A^{J_6}$ (0.67) must be calculated in the same manner as Eq. (13.28), by replacing each superscripted 3 in that equation with a 2 and a 6, respectively. Once these have been found, the application of Eq. (13.6) to the

TABLE 13.4 Group Control Coefficients for G6P Branch Point[a]

Perturbed branch N		gFCC of flux $^*J_k$			gCCC
Group	Constituent steps[b]	J_A	J_B	J_C	$^*C_N^{G6P}$
A	$1\ 2\ 4_{.56}\ 5_{.6}\ 6\ 8$	0.43	0.83	0.43	0.08
B	3	0	0.99	0	0
C	$4_{.44}\ 5_{.4}\ 7\ 9\text{-}18$	0.57	-0.82	0.57	-0.08

[a] gFCCs are shown in the form $^*C_N^{J_k}$. gCCCs are of the form $^*C_N^{G6P}$.
[b] Reaction numbers correspond to the numbering scheme of Fig. 12.1. Subscripted values represent ϕ_i^N, the fraction of the preceding reaction's flux that is included in branch N.

flux of branch A yields

$$^*C_A^{J_A} = \frac{J_2\,^*C_A^{J_2} - \frac{1}{2}J_6\,^*C_A^{J_6}}{J_2 - \frac{1}{2}J_6} = 0.43 \tag{13.29}$$

The remaining gFCCs can be determined through the summation theorem and Eq. (13.26). The results are shown in Table 13.4. This branch point can effectively be treated as an unbranched pathway, as the flux of branch B is insignificant and branch B has little effect on the other two branches. With the additional divulgence of the extremely small group concentration control coefficients, it is clear that this is not a critical branch point.

Branch Point 3: FDP / GAP

At the FDP/GAP branch point, the glycerol-producing pathway diverges from the pathways of amino acid biosynthesis. Hence, one downstream group (B) exclusively consists of reaction 7. The upstream group (A) is made up of reactions 1-4 and 8, as well as the energy-associated pathway through reactions 5 and 6, which was placed upstream of the first branch point. Branch C includes the remainder of the flux of reaction 5 and the fluxes of reactions 9-18. Because the procedure for calculating this branch point's group flux control coefficients is essentially identical to that of the previous branch point, we will proceed straight to the results (Table 13.5). Yet again, because the flux through branch B is relatively small and branch B exerts little control over the other two branches, this branch point can be ruled noncritical. In addition, the large gFCCs with respect to branch A indicate that the primary controlling reaction lies upstream of this branch point.

TABLE 13.5   Group Control Coefficients for FDP / GAP Branch Point[a]

Perturbed branch N		gFCC of flux $^*J_k$			gCCC
Group	Constituent steps[b]	J_A	J_B	J_C	$^*C_N^{FDP}$
A	1-4 $5_{.6}$ 6 8	0.83	0.41	0.90	-0.02
B	7	0	0.94	-0.16	-0.01
C	$5_{.4}$ 9-18	0.17	-0.35	0.26	0.03

[a] gFCCs are shown in the form $^*C_N^{J_k}$. gCCCs are of the form $^*C_N^{FDP}$.
[b] Reaction numbers correspond to the numbering scheme of Fig. 12.1. Subscripted values represent ϕ_i^N, the fraction of the preceding reaction's flux that is included in branch N.

Branch Point 4: X5P + CHR

The fluxes through the fourth branch point of this network are much simpler than those of the previous three, as the coupled energy-associated pathway is entirely upstream of this node. Only one step, reaction 5, is separated into different groups around this branch point. This is because a small amount of the flux through reaction 5 is required to counter the recycle of phosphoenolpyruvate (PEP) into GAP by reaction 18. This flux is governed by reaction 18; hence, ϕ_5^B, the amount of reaction 5 that is counted within branch B, is equal to J_{18}/J_5. Because each of the group fluxes around this branch point is equivalent to individual reaction fluxes, the calculation of gFCCs is straightforward. The results are given in Table 13.6. The determination of the group concentration control coefficient for the link metabolite(s), however, is not as simple, as this branch point is marked by the production and consumption of both X5P and CHR in fixed proportion.

TABLE 13.6   Group Control Coefficients for X5P + CHR Branch Point[a]

Perturbed branch N		gFCC of flux $^*J_k$			gCCC
Group	Constituent steps[b]	J_A	J_B	J_C	$^*C_N^{X/C}$
A	1-4 $5_{.98}$ 6-10	0.89	1.30	0.79	9.18
B	$5_{.02}$ 14 18	-0.07	0.58	-0.16	-1.82
C	11-13 15-17	0.18	-1.28	0.37	-7.36

[a] gFCCs are shown in the form $^*C_N^{J_k}$. gCCCs are of the form $^*C_N^{X/C}$.
[b] Reaction numbers correspond to the numbering scheme of Fig. 12.1. Subscripted values represent ϕ_i^N, the fraction of the preceding reaction's flux that is included in branch N.

First, beginning with branch B, the gCCCs for both X5P and CHR can be determined from the CCC analogue of Eq. (13.3):

$$*C_B^{X5P} = (J_{18}/J_5)C_5^{X5P} + C_{14}^{X5P} + C_{18}^{X5P} = -0.55 \qquad (13.30)$$

$$*C_B^{CHR} = (J_{18}/J_5)C_5^{CHR} + C_{14}^{CHR} + C_{18}^{CHR} = -2.06 \qquad (13.31)$$

Then, by making use of Eq. (13.7):

$$*C_B^{X/C} = \frac{*C_B^{X5P}c_{X5P} + *C_B^{CHR}c_{CHR}}{c_{X5P} + c_{CHR}} = -1.82 \qquad (13.32)$$

Similar calculations can be performed for the other two branches. The resulting gCCCs (Table 13.6) are extremely large, indicating a probable critical branch point. In fact, the large concentration control coefficients suggest that it will be far too easy to raise or lower these metabolites to unstable levels. In addition, the group flux control coefficients are distributed across all three branches, more than in any of the previous branch points. In conclusion, this branch point appears to be a critical one.

Branch Point 5: PPH

The last branch point in this network differentiates the tryptophan and tyrosine pathways. This is by far the simplest branch point to analyze, as all the group fluxes are actual reaction fluxes and each step fits squarely within a single branch. The results are shown in Table 13.7.

Notice that, in this case, neither of the downstream branches exerts a significant effect upon the upstream branch. Yet both downstream branches strongly control themselves and each other. This is clearly the mark of a flexible branch point, as the branch point metabolite can be diverted significantly toward either product. These data, as well as the significant values of

TABLE 13.7 Group Control Coefficients for PPH Branch Point[a]

Perturbed branch N		gFCC of flux $*J_k$			gCCC
Group	Constituent steps[b]	J_A	J_B	J_C	$*C_N^{PPH}$
A	1-11 14 18	0.83	0.57	0.07	1.75
B	12 16	0.08	0.72	-0.52	-0.80
C	13 17	0.09	-0.29	0.45	-0.85

[a] gFCCs are shown in the form $*C_N^{J_k}$. gCCCs are of the form $*C_N^{PPH}$.
[b] Reaction numbers correspond to the numbering scheme of Fig. 12.1.

the gCCCs, combine to indicate that this branch point could also be a critical one.

The magnitudes of the group control coefficients for each of these five branch points provide significant insight into the potential role that they may play in controlling flux through the network. In Section 13.4, we illustrate additional results that can be obtained through comparisons of the gFCCs of neighboring branch points. Keep in mind, however, that group control coefficients usually cannot be calculated analytically through the bottom-up approach, as we have done here, due to the lack of information about individual flux control coefficients. In the next section we show how group control coefficients can be obtained experimentally from flux measurements using the top-down approach to yield conclusions similar to those reached here.

13.3.2. SIMULATION OF EXPERIMENTAL DETERMINATION OF gFCCs (TOP - DOWN APPROACH)

Section 13.2 outlined a method for the calculation of gFCCs from experimental flux measurements made after strategic perturbation experiments. In this section, we demonstrate this method using *simulated* perturbation experiments with the *S. cerevisiae* kinetic model. Each of these simulated perturbations was effected by altering a kinetic parameter of a particular reaction, running the perturbed simulation, and calculating the resulting fluxes and steady state concentrations. By comparing these conditions with those of the base state (characterized in Table 12.3), flux amplification factors (FAF) were determined for each of the simulated perturbations. These FAFs were subsequently applied to the equations of Section 13.2 to provide estimates of the group control coefficients at each branch point.

Table 13.8 shows the results obtained for the group FCCs of each branch point. As discussed earlier, these techniques require that separate network perturbations be introduced within each of the three branches identified around a link metabolite or in two downstream branches, provided that one of these is a characterized perturbation. The step(s) to be perturbed in each branch can be selected arbitrarily, as long as the perturbation is localized within a single branch. At the ATP branch point, for example, because branch B consists of only a single step (reaction 8), the perturbation in this branch must be made to that step. Although branch A comprises five reactions, only one of these (reaction 6) is contained completely within this branch. Therefore, in order to ensure that the perturbation of branch A is

TABLE 13.8   Group Control Coefficients for Each Branch Point[a]

	Perturbed branch N			gFCC of flux $*J_k$			gCCC
N_{ATP}	Perturbed step[b] i	r_i	K_N	J_A	J_B	J_C	$*C_N^{ATP}$
A	-	-	-	1.18	2.27	1.07	2.3
B	8[c]	0.5	-1.4	-0.02	0.39	-0.06	-0.61
C	14	2	6.2	-0.16	-1.66	-0.01	-1.7
N_{G6P}	Perturbed step i	r_i	K_N	J_A	J_B	J_C	$*C_N^{G6P}$
A	-	-	-	0.89	0.18	0.90	0.02
B	3[c]	0.5	-2.0	0.00	1.00	-0.01	0.00
C	14	2	-3.3	0.11	-0.18	0.11	-0.02
N_{FDP}	Perturbed step i	r_i	K_N	J_A	J_B	J_C	$*C_N^{FDP}$
A	8	0.5	28.4	0.94	0.6	1.00	0.68
B	7	0.5	-0.74	0.01	0.35	-0.05	-0.08
C	14	2	-1.9	0.05	0.05	0.05	0.09
$N_{X/C}$	Perturbed step i	r_i	K_N	J_A	J_B	J_C	$*C_N^{X/C}$
A	5	0.75	-19.0	0.83	1.93	0.68	5.2
B	14	2	-0.5	0.02	-0.09	0.04	0.21
C	17	0.75	-11.3	0.15	-0.85	0.28	-48
N_{PPH}	Perturbed step i	r_i	K_N	J_A	J_B	J_C	$*C_N^{PPH}$
A	14	2	-14.4	1.02	0.72	1.31	1.9
B	12	0.5	-1.3	-0.01	0.29	-0.3	-0.53
C	17	0.75	0.27	-0.01	0	-0.01	-0.01

[a] gFCCs are shown in the form $*C_N^{J_k}$. gCCCs are of the form $*C_N^{X_j}$. Activity amplification factors r_i and perturbation constants K_N are also shown.
[b] Reaction numbers correspond to the numbering scheme of Fig. 12.1.
[c] A characterized perturbation was used in this branch.

completely localized *within* branch A, reaction 6 must be the only reaction affected. The selection of steps to perturb within branch C is much wider, consisting of reactions 3, 7, and 9-18. In an actual experiment, this decision would be based upon convenience or experimental expediency. For reasons that will be made apparent in Section 13.4, our choice here was reaction 14. In addition, the use of a characterized perturbation to branch B rendered a perturbation in branch A unnecessary in this case; the resulting group control coefficients are shown in the first subsection of Table 13.8.

Although the perturbations to the network kinetics employed in these simulations could have taken any mathematical form, we chose to employ the simplest and most intuitive: proportional scaling. This is convenient because proportional changes are physically meaningful and easier to interpret. Thus, the characterized perturbation in branch B was introduced by multiplying the kinetic expression of reaction 8 by a factor of 0.5. Branch C was perturbed by doubling the rate of reaction 14. By using each of these perturbed kinetic expressions, new steady states were calculated. The fluxes at these steady states were compared to their base values to determine the flux amplification factors, according to Eq. (13.9). Equations (13.12)-(13.20) were used to determine the group flux control coefficients from these three FAFs at this branch point. The results of these calculations are shown in Table 13.8. This table also includes the group concentration control coefficients for the effect of each of these groups upon the link metabolite, ATP. The remaining four sections of Table 13.8 illustrate the analysis of similar perturbations to each of the other four branch points of the network.

Upon comparison of the results of Table 13.8 with the corresponding gFCCs calculated by the bottom-up approach (shown in Tables 13.3-13.7, which are considered to be the correct ones for the network of our study), several observations are in order. First, although the quantitative agreement is not particularly good, the results are qualitatively comparable in that they identify the same steps with the largest magnitude gFCCs and gCCCs, whereas the others exhibit similar trends. For the ATP branch, for example, they show that branches A and, to a lesser extent, C are strongly controlling. Similarly, the significant control of branches A_{G6P} and A_{FDP} is also indicated. This tendency of a branch to influence one other branch more strongly than another is also reliably estimated. Second, discrepancies are the result of either significant deviations of the actual kinetics from the linearized kinetics assumed in the large perturbation theory or a near-singularity of the matrix of Eq. (13.17). This invariably is associated with exaggerated values for one of the K_i parameters. In such cases, the gFCCs calculated from this particular K_i generally are unreliable. For example, the fact that the perturbation constant of group C_{PPH} is much smaller than those of groups A or B indicates that the gFCCs of branch C_{PPH} are underestimated, as confirmed by Table 13.7.

Third, another reason for poor gFCC estimates is strong interbranch interaction despite efforts to localize such effects within each branch. Because it generally is very difficult *a priori* to predict and avoid such interactions due to broad ranging effects of certain metabolites, a method is presented in Section 13.6 that allows the detection of such interactions as well as the design of subsequent perturbations to minimize them. The method is based on the particular structure of the equations used for gFCC determination and resulting redundancies. These redundancies basically allow one to carry out self-consistency tests and select those perturbations that maximize the closure of the redundant equations. The summation theorem for group concentration control coefficients, requiring that all the gCCCs sum to zero, is a case in point. It can be seen that this criterion is not satisfied by some of the sets of gCCCs in sections of Table 13.8, a sure sign that some of the estimated control coefficients need to be corrected. This correction is brought about by either considering the outcome of some additional perturbations or including more than the necessary minimum number of perturbations in the calculations and applying a regression method to determine the control coefficients.

Finally, it is notable that the same perturbation in reaction 14 was used in each set of calculations shown in Table 13.8. The repeated use of a single perturbation for several branch points results from the natural overlap of groups around different branch points. As explored further in the next section, this overlap not only reduces the number of experimental perturbations needed to elucidate large networks, but allows one to focus the search of critical reactions within a small portion of the overall network.

13.4. EXTENSION OF CONTROL ANALYSIS TO INTERMETABOLITE REACTION GROUPS

In this section, we extend the application of the top-down approach by illustrating how the gFCCs of intermetabolite reaction groups or, in certain cases, individual reactions can be obtained from the gFCCs of different link metabolites.

13.4.1. THE PERTURBATION CONSTANT

The *perturbation constant* K_i, as introduced in Section 13.2, defines a special relationship between group flux control coefficients and the fractional flux changes in each branch around a link metabolite. Specifically, Eq. (13.8) shows that the perturbation constant K_i characterizes a particular experimental perturbation in group i and is equal to the ratio of the gFCC, $^*C_i^{J_k}$, for the effect of group i on the flux of any other group k, to the fractional change in the flux of group k, $f_i^k - 1$. In a different form, the same relationship can be used to estimate the gFCCs from flux changes and the perturbation constants:

$$^*C_i^{J_k} = K_i(f_i^k - 1) \qquad i, k \in \{A, B, C\} \qquad (13.33)$$

The importance of the perturbation constant lies in the fact that it allows an essential *decoupling* between the group (i) where the perturbation originates and the group (k) that is affected by it. Due to this decoupling, the gFCC for the effect of group i on the flux of group k is expressed as the product of a generic perturbation constant for group i and the fractional flux change in group k brought about by that perturbation. This decoupling is a direct consequence of the fundamental assumption in the derivation of Small and Kacser's large deviation theory, which was discussed in Section 11.5 (Small and Kacser, 1993a, b). Namely, any (individual or group) reaction in a branched network has linear kinetics of the form of Eq. (11.69), leading to fluxes of the form of Eq. (11.76). Because the grouping methods of the previous chapter have demonstrated that a complex network can be treated as a series of branches, it follows that the same forms can be used to describe a complex network as well. Therefore, *the perturbation constant calculated for a particular group (i) around a link metabolite can also be used to determine the (g)FCC of the same group on any other arbitrary group or individual reaction (k), so long as the fractional flux change of the latter is measured.* Furthermore, if group i consists of a single step, the individual flux control coefficient for the effect of that step upon any other can be found using this equation.

More importantly, information about the control exercised by groups of reactions between the link metabolites can be obtained at no extra cost. By comparing the gFCCs of these overlapping groups, the (g)FCCs of short pathways and individual reactions between link metabolites upon the flux of interest can be found. Once this detailed mapping of the control structure has been achieved, the search for critical reactions and pathways becomes trivial.

13.4.2. ANALYSIS OF OVERLAPPING REACTION GROUPS IN MULTIPLE BRANCH POINTS

Note that the sum total of all reactions in the groups formed around a branch point must include all the reactions in a network. The groups identified around different branch points are, therefore, formed simply by allocating the network reactions differently. Comparison of neighboring branch points can yield information that is not readily available from the analysis of single branch points, such as determining the control coefficients of the short linear pathways or even individual steps between link metabolites.

This point is illustrated in Fig. 13.2, which shows the groups formed around each of the two branch points of the network introduced in Fig. 12.6a. The individual FCCs of four of the reactions in this network can be determined directly from the gFCCs, as these steps (1, 3, 4, and 5) each constitute an entire group around one link metabolite or the other. Reaction 2, in contrast, shares a group downstream of B in Fig. 13.2a with reactions 4

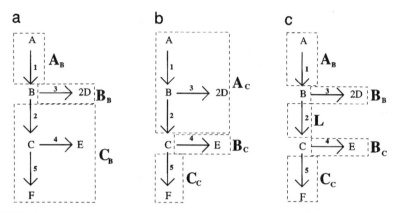

FIGURE 13.2 Comparison of reaction groups around two link metabolites. (a, b) Groups formed around metabolites B and C. (c) Groups identified from comparison of the overlapping groups around B and C.

and 5, whereas it also shares the group upstream of C in Fig. 13.2b with reactions 1 and 3. Thus, the flux control coefficients corresponding to reaction 2 cannot be determined directly from the analysis of either branch point. The simultaneous use of the results from both branch points, however, allows the determination of the FCC for reaction 2 through the use of the summation theorem of MCA. Reaction 2 can, in effect, be placed within its own group (L), as shown in Fig. 13.2c. Because the gFCCs for the four other groups in Fig. 13.2c are available from the analysis of both branch points, a gFCC for group L is simply the difference between 1 and the sum of the gFCCs of the four other groups. In general, if group L lies within group C in the grouping around the upstream link metabolite and within group A in the grouping around the downstream link metabolite, (see Fig. 13.2), the gFCC for group L upon any flux J_k is

$$*C_L^{J_k} = 1 - \left(_{\text{upstream}}*C_A^{J_k} + _{\text{upstream}} *C_B^{J_k} + _{\text{downstream}} *C_B^{J_k} + _{\text{upstream}} *C_C^{J_k} \right)$$

$$(13.34)$$

By applying Eq. (13.34), it is possible to determine gFCCs corresponding to any *intermetabolite linkage* in a network, *i.e.*, any pathway that converts one link metabolite into another. When combined with the group control coefficients already determined using the methods of Sections 13.2 and 13.4.1, the group FCCs for the effect of any input or output pathway or intermetabolite linkage upon any measured flux can be determined. Finally, using this information, it is possible to narrow the search for critical reaction steps (as measured by the FCC of such a step upon a flux of interest) to only those groups with sufficiently large group FCCs. In addition, it should be noted that a similar formulation holds for group concentration control coefficients:

$$*C_L^{X_k} = -\left(_{\text{upstream}}*C_A^{X_k} + _{\text{upstream}}*C_B^{X_k} + _{\text{downstream}}*C_B^{X_k} + _{\text{upstream}}*C_C^{X_k}\right)$$

$$(13.35)$$

Because a single reaction can participate in several reaction groups, a single perturbation in such a reaction can be utilized in the analysis of any number of branch points, as long as that perturbation is localized within a single group around each link metabolite and yields measurable flux changes on other reaction groups. Recall from Section 13.2 that the analysis of a single branch point (*e.g.*, metabolite B of Fig. 13.2) generally requires three perturbations, each of which should be in a different branch. In this case, perturbations are required in steps 1 and 3, but the third perturbation can be made to any other step. Three perturbations also are needed for the analysis of a second branch point downstream of the first (*e.g.*, metabolite C), two of which must be in reactions 4 and 5. Hence, in order to properly analyze both branch points, perturbations must be introduced in steps 1, 3, 4, and 5 (the input and output steps). The groups around metabolite B can, therefore, be analyzed using either reaction 4 or reaction 5 as the perturbation to group C. Because the data from both perturbations would be available, however, a comparison of the two results provides a consistency check for the results (see Section 13.6). The same logic holds for the second branch point. Thus, it is evident that the addition of a single well-placed perturbation to the three already required from the analysis of the first branch point not only allows the analysis of a second branch point, but also provides a consistency check for both. In fact, the number of strategically placed perturbations generally necessary for the complete analysis of a complex network is only two more than the number of branch points.

13.4.3. Case Study

In this section, we use the gFCC estimates obtained through the simulated experimental perturbations (shown in Table 13.8) to calculate the gFCCs of intermetabolite linkages. The resulting gFCCs summarized in Table 13.9 can be used to guide our search for the particular reaction step(s) exerting the largest control on a flux of interest. The flux chosen for this purpose is that of tryptophan release (reaction 18).

TABLE 13.9 Estimated Group Control Coefficients for Tryptophan Flux[a]

Group N	Perturbed step[b] i	f_N^{18}	K_N	$*C_N^{J_{18}}$
A_{ATP}	-	-	-	0.05
B_{ATP}	8	1.109	-1.4	-0.15
C_{ATP}	14	1.177	6.2	1.10
$L_{ATP/G6P}$	-	-	-	1.70
A_{G6P}	-	-	-	1.60
B_{G6P}	3	1.010	-2.0	-0.02
C_{G6P}	14	1.177	-3.3	-0.58
$L_{G6P/FDP}$	-	-	-	0.75[c]
A_{FDP}	8	1.109	28.4	3.10
B_{FDP}	7	1.160	-0.74	-0.12
C_{FDP}	14	1.177	-1.9	-0.34
$L_{FDP/X/C}$	-	-	-	-0.27[c]
$A_{X/C}$	5	0.898	-19.0	1.94
$B_{X/C}$	14	1.177	-0.5	-0.09
$C_{X/C}$	17	1.075	-11.3	-0.85
$L_{X/C/PPH}$	-	-	-	-0.92
A_{PPH}	14	1.177	-14.4	-2.55
B_{PPH}	12	0.962	-1.3	0.05
C_{PPH}	17	1.075	0.27	0.02

[a] gFCCs are shown in the form $*C_N^{J_{18}}$. Flux amplification factors f_N^{18} and perturbation constants K_N are also shown.
[b] Reaction numbers correspond to the numbering scheme of Fig. 12.1.
[c] Because summations were not satisfied for the FDP branch point, these values are the averaged results using the gFCCs calculated for the perturbation in either A_{FDP} or C_{FDP}.

From these results, it is clear that the linkages on either side of G6P are important in the control of the tryptophan flux. The G6P/FDP linkage, corresponding to reaction 4, exhibits a significant gFCC, which is borne out by the actual FCC value of reaction 4 (Table 13.1). In addition, the linkage that diverts chorismate into the prephenate pathway strongly competes with the tryptophan flux. Thus, these data lead to the conclusion that tryptophan production can best be improved either by amplifying the glycolytic pathway (i.e., the phosphofructokinase reaction) or by attenuating the competing pathways into phenylalanine and tyrosine.

Note that the exaggerated perturbation constants identified in Section 13.3 also distort some of the gFCCs in these tables, as evidenced by the fact that these coefficients do not correctly sum to unity for certain sets of groups. By comparing gFCCs validated through the more accurate methods of Section 13.6, however, very good consensus estimates of the gFCCs of the intermetabolite linkages can indeed be obtained. The value of these data will be accentuated in the next section, which describes a general procedure to optimize the amplification of a particular flux, based upon known control coefficients.

13.5. OPTIMIZATION OF FLUX AMPLIFICATION

The main motivation for determining control coefficients is to characterize the control architecture of a metabolic network in order to provide a sound basis for efficient and methodical flux amplification and redistribution. Nevertheless, it is not immediately obvious how to best utilize these coefficients for optimal flux enhancement. The large magnitudes of the phosphofructokinase (reaction 4) FCCs found for the case study of the previous sections would indicate, for example, that this particular step exerts the greatest amount of control on the tryptophan and phenylalanine fluxes. Following accepted wisdom, PFK would be the primary target of activity amplification in order to bring about the desired effect on amino acid overproduction. However, simulations with gradually increased kinetic rates for this reaction revealed that the structure of the network prevents stable PFK amplification beyond an 11% increase in activity. At higher levels, the overall system becomes unstable, *i.e.*, unable to converge to a steady state condition.

Mathematically, the reason for this particular instability is a bifurcation at the noted PFK amplification value into a space where a real solution for the steady state of the metabolite chorismate does not exist. In other words, there are no acceptable intracellular metabolite concentrations that can balance the rates of chorismate production and depletion. The reason for this phenomenon, as discussed in Section 12.1, is that the activities of the reactions that consume chorismate are not sufficient to balance the increased production rate of chorismate resulting from the amplification of the phosphofructokinase step. Although it cannot be claimed that this would indeed happen in a similarly modified strain of *S. cerevisiae*, this result is typical of cellular responses to the introduction of a catastrophic metabolic disturbance. In such cases, secretion of metabolites, induction of degradation pathways, and drastic changes in secreted product profiles commonly are observed. Because the actual cellular response to such an instability is highly unpredictable, it is best to avoid introducing such disturbances in the first place. Hence, a method of optimizing flux amplification while avoiding metabolic instabilities is desired.

One way to limit network instabilities would be to introduce additional pathways serving as sinks or shunts for offending metabolites. This approach, however, is burdensome and tends to remove one potential instability while contributing to others, which will subsequently have to be dealt with in the same manner. Another possibility is to minimize the extent of kinetic enhancement, which is ultimately self-defeating. A bolder approach is to design modifications that will increase the flux through the network *while maintaining intracellular metabolite levels near those of the base state.* This strategy extends a suggestion originally proposed by Henrik Kacser (Kacser and Acerenza, 1993) that flux amplification should only be attempted in a manner that disallows *any* change in intracellular metabolite concentrations. The corollary of this postulate is that, in order to achieve an overall flux amplification in any particular pathway, *all* steps involved in its corresponding independent pathway must be amplified, according to a certain formula. Whereas this requirement of full pathway amplification is entirely general, it is rarely feasible in practice. Our strategy here instead depends upon relaxing the rigid constraint that prevents any change in the intracellular metabolite concentrations. By allowing modest changes in metabolite levels, it can be shown that significant increases in the overall network flux can be obtained from the modification of *only a small number* of carefully selected enzymatic steps.

The optimization algorithm described in Section 13.5.1 can make use of either group or individual flux control coefficients, as determined according to the methods of the previous sections. In addition, group or individual concentration control coefficients emerge as key parameters in the optimization algorithm, as they provide a measure of metabolite sensitivity to reaction rate modifications. These concepts and the optimization procedure are illustrated using the amino acid biosynthetic pathway. For simplicity, this procedure was derived using individual control coefficients; nevertheless, this method should work equally well with accurately determined gFCCs and gCCCs.

13.5.1. DERIVATION OF OPTIMIZATION ALGORITHM

The problem addressed here is the determination of the best single reaction step, or perhaps the best two or three steps, that should be amplified in order to effect the largest possible increase in the flux of the network, subject to the constraint that *all* intracellular metabolite levels remain within a reasonable range of their original steady state values. The following optimization procedure not only locates the best reaction(s) to change, but also suggests an upper (or lower) limit to the activity change as well. In the case of changing the activities of two or more steps simultaneously, these limits become critical. Furthermore, the results of the optimization are rather insensitive to the allowed metabolite range as long as the metabolite concentration remains away from the neighborhood of the point of instability, due to the fact that, once the bifurcation borders of a metabolite are approached, progression toward network instability occurs rather precipitously. These simplifications allow one to reduce the optimization to a simple, linear form.

It has been shown through the theory of large deviations (Small and Kacser, 1993b) that $J_{k,L}$ the change in the flux J_k resulting from changes in the activity of any or all of the L reactions in the network, is given by

$$f_L^k = \frac{J_{k,L}}{J_{k,0}} = \frac{1}{1 - \sum_{i=1}^{L} C_i^{J_k}\left(\dfrac{r_i - 1}{r_i}\right)} \qquad k \in \{1, 2, \ldots, L\} \quad (13.36)$$

The maximum flux change is obtained when the activity amplification factors (AAF), r_i, as defined by Eq. (13.19), are optimized. The objective function of the optimization problem is obtained by recasting the preceding equation in the following form:

$$1 - \frac{1}{f_L^k} = \frac{J_{k,L} - J_{k,0}}{J_{k,L}} = \sum_{i=1}^{L} C_i^{J_k}\left(\frac{r_i - 1}{r_i}\right) \qquad k \in \{1, 2, \ldots, L\} \quad (13.37)$$

The extent of a flux increase depends upon the ability of the metabolite network to maintain metabolite levels at tolerable levels. According to Small and Kacser (1993b), the change in any metabolite level c_j resulting from altered activities of any number of reactions $(c_{j,L})$ can be found through

$$\phi_L^j = \frac{c_{j,L}}{c_{j,0}} = 1 + \frac{\sum_{i=1}^{L} C_i^{X_j}\left(\dfrac{r_i - 1}{r_i}\right)}{1 - \sum_{i=1}^{L} C_i^{J_i}\left(\dfrac{r_i - 1}{r_i}\right)} \qquad j \in \{1, 2, \ldots, K\} \quad (13.38)$$

Clearly, the lower limit of Eq. (13.38) is for ϕ_L^j to be equal to 0, as a metabolite level physically cannot be negative. Although the upper boundary is somewhat arbitrary, it was determined that limiting the value of ϕ_L^j at 2 provides a practical heuristic constraint that can also be relaxed as needed. Hence, the constraints upon the concentration become:

$$-1 \leq \phi_L^j - 1 = \frac{\sum_{i=1}^{L} C_i^{X_j}\left(\dfrac{r_i - 1}{r_i}\right)}{1 - \sum_{i=1}^{L} C_i^{J_i}\left(\dfrac{r_i - 1}{r_i}\right)} \leq 1 \qquad j \in \{1, 2, \ldots, K\}$$

$$(13.39)$$

It can be seen from Eqs. (13.37) and (13.39) that the problem of flux maximization subject to the constraint of minimal metabolite change can be defined as a linear constrained optimization problem with respect to the *activity amplification parameter* (AAP), defined as

$$R_i = \begin{cases} \dfrac{r_i - 1}{r_i}, & r_i \geq 1 \\ r_i - 1, & r_i < 1 \end{cases} \qquad i \in \{1, 2, \ldots, L\} \qquad (13.40)$$

The AAP thus is identical to the actual hyperbolic function of r_i invoked by Eqs. (13.36)–(13.39), when the activity of reaction i is amplified ($r_i \geq 1$), and it approximates this function by a linear function when the activity of reaction i is attenuated ($r_i < 1$). This approximation is good except when r_i becomes very close to zero. Extensive simulations using the *S. cerevisiae* kinetic model of the case study showed that the use of this AAP provides reasonably accurate predictions of the effects of attenuation of competing reactions. In addition, the AAP is a very convenient optimization parameter because physically it is bounded between -1 and 1, due to the fact that r_i must be positive.

The introduction of the AAP eliminates the exaggerated flux increases predicted by Eq. (13.36) from attenuating competing reactions (*i.e.*, when r_i is very small and $C_i^{J_k}$ is negative). At the same time it allows both the objective function and constraints to be cast in a convenient linear form as follows: Determine R_i so as to maximize

$$\sum_{i=1}^{L} C_i^{J_k} R_i \qquad k \in \{1, 2, \ldots, L\} \qquad (13.41)$$

subject to the constraints of limited metabolite variation expressed as:

$$\sum_{i=1}^{L} \left(C_i^{J_i} \pm C_i^{X_j} \right) R_i \leq 1 \qquad \forall j \in \{1, 2, \ldots, K\}$$
$$-1 \leq R_i \leq 1 \qquad \forall i \in \{1, 2, \ldots, L\} \qquad (13.42)$$

It must be emphasized that the summation in this inequality invokes *both* flux and concentration control coefficients. Note that the summations of Eqs. (13.41) and (13.42) *extend over all steps in the network, but contain nonzero terms only for the steps that are to be adjusted.* Thus, for a single step, the sum contains a single nonzero term, for two steps it contains two nonzero terms, and so on.

The preceding construct is a linear optimization problem that can be solved with available techniques to yield the optimal activity amplification for each bioreaction step or combination of steps considered simultaneously.

By comparing the optimal fluxes obtained for *all combinations* of reaction steps in sets of one, two, and so on, the optimal step or combination of steps is determined along with the corresponding optimal activity amplification(s). In addition, the optimized AAP indicates the extent by which an activity can be changed without causing network instabilities from perturbed metabolite levels.

When a single step is to be modified, the change in activity should be such that R_i has the same sign as $C_i^{J_k}$ and lies at the boundary of the most stringent constraint of Eq. (13.42). This is a result of the linear nature of the optimization problem. In general, this means that the maximal change in the activity of step i is defined by

$$R_i = \frac{1}{C_i^{J_i} \pm C_i^{X_j}} \quad i \in \{1, 2, \ldots, L\}, j \in \{1, 2, \ldots, K\} \quad (13.43)$$

Because R_i must satisfy the constraints of Eq. (13.42) for all K metabolites in the system, j refers to the metabolite exhibiting the largest magnitude denominator in Eq. (13.43). If, however, the absolute value of this denominator is smaller than 1 for every metabolite, R_i instead can be taken to its upper or lower bound (*i.e.*, 1 or -1); these represent, respectively, infinite amplification or complete removal of step i. If the activities of n steps can be varied simultaneously, the optimization becomes an n-dimensional linear problem, the solution of which lies at the intersection of n boundary conditions. In practice, n should be limited to about 3 in order to ensure reliable predictions of the network's response.

Certain consequences arising from this formulation of the optimization problem are discussed in Section 13.5.2. Specifically, the values of R_i required to enact sufficient change in the flux are also those that are likely to exceed the boundaries of Eq. (13.42). Hence, the optimal alteration will be a change in the kinetics of one or two reactions such that the flux of interest is strongly affected, while metabolite concentrations are not.

13.5.2. CASE STUDY

The optimization procedure described in the previous section was applied to the *S. cerevisiae* system, using the control coefficients in Tables 13.1 and 13.2. The tryptophan release step (reaction 18) was chosen as the flux to be optimized. Table 13.10 summarizes the results for single- and dual-step optimizations of this flux. The values in the right-hand column are the predicted flux changes found through the optimization procedure of Eqs. (13.41) and (13.42). These predictions are, in fact, quite close to the actual maximal stable fluxes that can be obtained from the complete network model through proportional changes in the step(s) indicated.

TABLE 13.10   Optimal Steps for the Amplification of the Tryptophan Flux

Adjusted step(s)[a] i		Activity amplification r_i		Tryptophan flux amplification $f_k^{J_{18}}$
7		0		1.37
14		3.19		1.27
18		5.99		1.27
15		0.57		1.21
4		1.09		1.16
16		0.60		1.13
5		1.72		1.12
4	14	1.17	50.4	2.25
14	16	∞	0.27	2.07
4	18	1.16	∞	1.96
15	16	0.37	0.2	1.95
14	17	∞	0.42	1.89
11	14	0.5	∞	1.86
16	18	0.33	∞	1.80

[a] Reaction numbers correspond to the numbering scheme of Fig. 12.1.

It can be seen from Table 13.10 and the control coefficients in Tables 13.1 and 13.2 that, although PFK (reaction 4) clearly is a limiting reaction, if a single step is to be amplified, it is instead most profitable to completely remove step 7, which is a competing branch producing glycerol, to triple the activity of step 14, which produces tryptophan, or to sextuple the activity of step 18, which releases tryptophan. The large FCC for PFK is effectively nullified by a much larger CCC for the metabolite chorismate that restricts the allowable kinetic amplification of PFK. Reactions 7, 14, and 18, in fact, emerge as the optimal single steps for tryptophan production precisely because they offer the best balance between the magnitude of the *permissible* amplification (measured by the CCC) and the impact on the network flux per unit of amplification (measured by the FCC). Moreover, according to the constraints of Eq. (13.42), only steps exhibiting moderate concentration control coefficients can be amplified or attenuated to any level arbitrarily. Nevertheless, the identification of four different steps that yield a 20-30% increase in the tryptophan flux provides further evidence that there can be no single "rate-limiting" step.

In this network, the concentration control coefficients for chorismate are so large that nearly any change in the network kinetics produces a sizable change in the level of chorismate. Thus, the chorismate branch point forms a *constriction point* and is considered a critical branch point. Near any such constriction point, it is necessary to limit the size of activity changes in order to maintain the metabolite at an acceptable level. If the goal is to increase a flux that bifurcates at such a branch point, three possible single-step activity perturbations may be effective: (a) increasing the rate of production of the metabolite by the upstream branch; (b) increasing the rate of consumption of the metabolite by the downstream branch of interest; or (c) decreasing the rate of consumption of the metabolite by the competing downstream branch. These three cases are illustrated schematically in Fig. 13.3a.

When the activity of two steps can be changed simultaneously, the optimization procedure predicts that the flux into tryptophan can be approximately doubled. The choice of steps that can form these pairs of reactions is not arbitrary, however, but rather follows a set pattern around critical branch points in order to modulate the level of the link metabolite(s). In this case, the reaction pairs indicated nearly always involve amplifying the downstream pathway of interest, while either (a) amplifying the upstream branch or (b) attenuating the competing downstream branch, thus diverting the metabolite into the branch of interest. In general, adjusting two reactions *within the same branch* around a constriction point will not greatly enhance the flux of interest, nor will amplifying the upstream branch while attenuating the

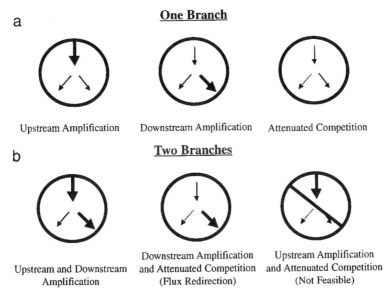

FIGURE 13.3   Amplification of downstream flux at a critical branch point. Acceptable methods of amplifying a flux based upon amplification (thickened arrow) or attenuation (dashed arrow) of (a) a single branch or (b) two branches around a critical branch point. Note that two of the three combinations in (b) are acceptable, but that the third combination cannot adequately regulate the level of the link metabolite and thus is an unacceptable option.

competing branch, as this will only add to the buildup of the link metabolite. The acceptable types of simultaneous adjustments of reaction pairs are shown in Fig. 13.3b.

The modulation of the chorismate level by two altered steps is illustrated by the reaction pair 4/14, which is one of the optimal selections for tryptophan production. In essence, the effect of this pairing is to moderate the level of chorismate by *pulling* away from chorismate (through reaction 14) part of the overwhelming carbon flux that is *pushed* into chorismate by the amplification of PFK. Thus, adjustment of one reaction serves to alleviate the metabolic instabilities otherwise caused by amplification of the second. Furthermore, the pairing of reactions permits each reaction to be amplified more fully than it could be alone. With the exception of 15/16, each of the other pairs of reactions in Table 13.10 fits the patterns described by Fig. 13.3.

A final point is in order concerning the possibility of amplification of more than two steps simultaneously. In general, if perturbing a single step within a branch is sufficient to reach the concentration constraints set out in Eq. (13.42), no further activity changes will be allowable in that branch. In the unlikely case that several steps can be amplified without reaching these constraints, other activity changes should be made elsewhere in the network. With a single constriction point (usually identified through the magnitude of the CCCs), it is unlikely that activity changes at more than two branches will be necessary. If two different constriction points have been identified, two or three branches may have to be amplified in order to amplify the network flux while modulating the levels of the constriction metabolites. In general, due to the nonlinear nature of metabolic networks, it is best to limit simultaneous adjustments to two or three steps at a time in order to avoid unanticipated instabilities. Subsequently, once a stable mutation has been developed that exhibits the desired flux amplification, a second round of flux analysis and optimization can be considered. Through the coordinated adjustment of a few key steps, followed by repeated rounds of analysis and optimization, *rational* process improvements can be achieved.

13.6. CONSISTENCY TESTS AND EXPERIMENTAL VALIDATION

In Section 13.2, we described a method for the calculation of group flux control coefficients from flux measurements and also provided an illustration of the gFCC calculation for the aromatic amino acid case study. The resulting gFCCs, as shown in Table 13.8, exhibited qualitative agreement with the analytically derived gFCCs of Tables 13.3-13.7; however, general quantitative agreement was lacking. The discrepancies between the (correct) gFCC values calculated from the individual FCCs and those determined from the simulated experimental perturbations using the entire bioreaction network are due to two primary reasons. First, the kinetics of the various reaction steps with respect to either product or substrate concentration may deviate from the linear form assumed in the theory of large perturbations. Additionally, if the kinetics of a reaction is dependent upon the levels of effectors other than its substrate(s) and product(s), its rate curve also will not exhibit the assumed linear form. Second, direct interaction between branches beyond that allowed through the link metabolite (interbranch cross-talk) may invalidate the basis of the top-down approach inherent in the formulation of reaction groups and the derivation of accompanying expressions for the calculation of group coefficients from fluxes alone. In fact, the presence of metabolite effectors or inhibition can effectively bypass the link metabolite and introduce interbranch effects not accounted for in the equations presented in this chapter. In the model used by the case study, for example, the kinetics of reaction 9 (Table 12.2) includes feedback inhibition by phenylalanine, tyrosine, and tryptophan. Thus, in the analysis of the branch points intervening between reaction 9 and the three amino acid products (*i.e.*, the CHR/X5P and PPH nodes), direct interactions between the upstream and downstream branches will likely produce nonlinear kinetics, resulting in miscalculated group flux control coefficients.

As discussed by Small and Kacser (1993b), kinetic nonlinearities do not result in significantly skewed estimates of *individual* flux control coefficients found through the theory of large deviations (Section 11.5). The reason is that this method is based on *characterized* perturbations, *i.e.*, perturbations where the enzyme activity amplification is known. In this case, errors in the estimate of an FCC can only result from inappropriate use of fluxes because of strong inter-branch interactions. Such interactions, and related errors, are reduced when small enough perturbations are used, therefore improving the accuracy of the FCCs thus obtained. In the case of reaction groups, on the other hand, the enzyme activity changes of each perturbation generally will not be known. Consequently, the gFCCs given by Eqs. (13.17) and (13.18) in essence are determined by applying the summation and branching theorems of MCA to the experimental data. In the process, a perturbation constant (K_i) characterizing the magnitude of the originating experimental perturbation is calculated. gFCCs are determined in Eq. (13.18) by multiplying the perturbation constant of the originating perturbation by the flux change of the reaction group affected by such perturbation. Clearly, kinetic nonlinearities and cross-interaction will distort these gFCC estimates.

We address these issues in this section. As such effects generally are not easily observable, our emphasis is on the development of self-consistency tests that can be applied to gauge the accuracy of the gFCC estimates and help to improve it by designing perturbations where such interactions are reduced.

13.6.1. DEVELOPMENT OF CONSISTENCY TESTS USING MULTIPLE PERTURBATIONS

As mentioned earlier, kinetic nonlinearities do not significantly affect the calculation of FCCs from characterized perturbations using the large deviation theory. Similarly, if the activity change in a branch is known, the values of the gFCCs determined by Eq. (13.21) are fairly reliable. Thus, a comparison of the gFCCs for the G6P branch point found from a characterized perturbation in reaction 3 (Table 12.13) with the true gFCCs in Table 12.9 shows excellent agreement. In the case of the ATP branch point, a similar comparison of the results of the perturbation in reaction 8 (Table 13.8) with the actual gFCCs (Table 13.3) is less favorable, with the largest coefficient underestimated by over 50%. These errors further affect the calculation of the gFCCs of the remaining branch points.

Fortunately, the presence of such errors can be detected if more than one perturbation is introduced in the same branch. Two different perturbations in branch i (indicated by subscripts 1 and 2) would yield two different sets of flux changes (as measured by p_i^k) and, consequently, two different values for the perturbation constant; nevertheless, their product ideally should yield identical estimates of the corresponding gFCCs. As can be seen from Eq. (13.18), the two sets of gFCCs calculated from two distinct perturbations in branch i will coincide if and only if the resultant fractional flux changes are proportional to each other:

$$\begin{bmatrix} *C_i^{J_A} \\ *C_i^{J_B} \\ *C_i^{J_C} \end{bmatrix} = \left(K_i \begin{bmatrix} p_i^A \\ p_i^B \\ p_i^C \end{bmatrix} \right)_1 = \left(K_i \begin{bmatrix} p_i^A \\ p_i^B \\ p_i^C \end{bmatrix} \right)_2 \qquad i \in \{A, B, C\} \quad (13.44)$$

As a result of Eq. (13.44), a clear test of the consistency of the two perturbations exists, such that

$$\frac{\left(p_i^A \right)_1}{\left(p_i^A \right)_2} \approx \frac{\left(p_i^B \right)_1}{\left(p_i^B \right)_2} \approx \frac{\left(p_i^C \right)_1}{\left(p_i^C \right)_2} \qquad i \in \{A, B, C\} \qquad (13.45)$$

Note that the ratios of Eq. (13.45) are directly observable through the flux changes. Large differences in these ratios would indicate inconsistent responses of the network to perturbations that were assumed to be directly comparable. Because the mass balance at the branch point constrains the flux change of one branch, only two fluxes are needed to express Eq. (13.44), which is then easily represented graphically in a two-dimensional plane. Thus, for each perturbation, a single *characteristic angle* θ_i can be defined (Simpson *et al.*, 1998):

$$\theta_i = \tan^{-1}\left(\frac{p_i^C}{p_i^B}\right) \qquad i \in \{A, B, C\} \qquad (13.46)$$

and the angles of different perturbations can be compared:

$$(\theta_i)_1 \approx (\theta_i)_2 \approx (\theta_i)_{\text{actual}} \qquad i \in \{A, B, C\} \qquad (13.47)$$

BOX 13.1

Graphical Representation of Flux Control Coefficients

For any particular branch point, we can define nine group flux control coefficients measuring the effect of a perturbation in each of the three branches upon each of the three group fluxes. The three gFCCs corresponding to a perturbation in any branch i can be represented by a vector in three dimensions:

$$*\vec{C}_i \equiv \left(*C_i^{J_A}, *C_i^{J_B}, *C_i^{J_C}\right) \qquad i \in \{A, B, C\} \qquad (1)$$

In fact, only two dimensions are needed to represent the flux control coefficients of a branch point. This is a result of the steady state mass balance of the link metabolite, which allows the gFCC of group A to be related to those of the other two branches by

$$*C_i^{J_A} = \frac{J_{B,0}*C_i^{J_B} + J_{C,0}*C_i^{J_C}}{J_{B,0} + J_{C,0}} \qquad i \in \{A, B, C\} \qquad (2)$$

Consequently, the group flux control coefficients can be adequately represented in the two-dimensional plane $(*C_i^{J_B}, *C_i^{J_C})$. If the gFCC vectors corresponding to three perturbations originating in each of the three branches are laid end to end, the summation theorem,

$$*\vec{C}_A + *\vec{C}_B + *\vec{C}_C = (1, 1, 1) \qquad (3)$$

(continues)

(continued)

indicates that the end point in the plane will be the point $(1, 1)$. Furthermore, if the individual flux control coefficients for each reaction within a group similarly are plotted, it is clear that their vector sums must yield the corresponding gFCCs. This is illustrated in Fig. 13.4, which shows the individual and group flux control coefficients corresponding to the prephenate branch point of the case study:

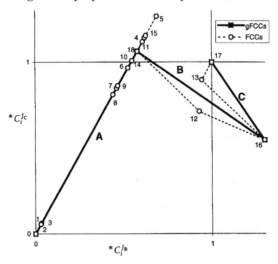

Figure 13.4 Graphical representation of the control coefficients at the prephenate branch point of the amino acid biosynthetic network. The magnitude of each vector indicates the degree of control that the corresponding reaction or group has on a particular group flux. Solid lines are the vectors $^*\vec{C_i}$ corresponding to the three group flux control coefficients. Dashed lines are the vectors of the individual FCCs constituting each gFCC. The number beside each open circle indicates the reaction corresponding to the most recently added vector. Thus, the vector corresponding to group B $(^*\vec{C_B})$ is the sum of the contributions of reactions 12 and 16 $(\vec{C_{12}}$ and $\vec{C_{16}}$, respectively). Reaction numbers correspond to the numbering scheme of Fig. 12.1.

The angle of each vector in Fig. 13.4 indicates how responsive a particular branch is to a perturbation in branch i. If a vector is very flat (i.e., with a characteristic angle near 0°), for example, it denotes that branch i has much greater control over the flux of branch B than that of branch C. A vector with a characteristic angle between 0° and 90° indicates that increased activity in branch i will increase the fluxes of both branches B and C. A vector between 90° and 180° or between -90° and 0°, on the other hand, suggests that an increased flux in one branch will occur concomitantly with a decreased flux in the other.

Equation (13.44) indicates that the unknown gFCCs are proportional to the flux changes that are obtained from introduced perturbations. Therefore, it is crucial that the experimental flux changes are parallel to the characteristic direction of the actual gFCCs. For the branch point illustrated in the preceding figure, it is clear that this requirement should be satisfied for any perturbation within branch A, as each individual FCC projects into nearly the same direction. The FCCs of the two reactions comprising group C, on the other hand, have radically different directions. As a result, if either of these directions is used to estimate the characteristic angle, it will yield incorrect estimates of the characteristic angle of a perturbation in group C and, therefore, incorrect gFCCs. In such cases more perturbations are needed, including some characterized ones.

Because there are only three degrees of freedom in the gFCC calculation, knowledge of the three actual characteristic angles would be sufficient for the determination of all nine group flux control coefficients. The characteristic angle determined from each perturbation, however, only approximates the actual characteristic angle of the corresponding branch. It is, therefore, necessary either to find *consensus* angles from multiple perturbations or to determine the characteristic angle of one or more branches from *characterized* perturbations. A consensus angle can be the average of the angles from two or more perturbations that do not differ significantly from one another. Once certain characteristic angles have been independently confirmed through one of these methods, the remaining unknown quantities can be found from these more reliable values, rather than from the raw perturbation data. Thus, if this angle is assumed to be φ_i, a *representative* perturbation can be defined that can replace the *actual* perturbations in the gFCC calculation according to

$$
\begin{bmatrix} p_i^A \\ p_i^B \\ p_i^C \end{bmatrix}_r = \begin{bmatrix} \dfrac{J_0^B + J_0^C \tan \varphi_i}{J_0^B + J_0^C} \\ 1 \\ \tan \varphi_i \end{bmatrix} \qquad i \in \{A, B, C\} \qquad (13.48)
$$

In determining such a consensus or average angle φ_i, the characteristic angles obtained from the different perturbation experiments cannot differ significantly. We limit, somewhat arbitrarily, the degree of disagreement to about $5°$. If reliable estimates of any of the gFCCs are available (*e.g.*, from a characterized perturbation), the summation theorems of Eq. (13.13) can be reformulated to include these values, and the remaining coefficients can be determined from the resulting equations. Because one known coefficient reduces the set of equations by one degree of freedom, however, the remaining coefficients must be calculated either from two of the three original balances or from a least squares regression of all three.

Two perturbations within a particular branch may exhibit markedly different characteristic angles that cannot be reconciled; this is evidence of nonlinearity in the kinetics of that branch and/or strong interactions between this branch and other branches. In such a case, it may be useful to use a representative perturbation that is an aggregate of the two. By using an averaging form of two sets of experimental perturbations in branch i, $(p_i^k)_1$ and $(p_i^k)_2$, the assumed characteristic angle φ_i in Eq. (13.48) could be as follows:

$$\varphi_i = \tan^{-1} \frac{1}{2} \left(\frac{(p_i^C)_1}{(p_i^B)_1} + \frac{(p_i^C)_2}{(p_i^B)_2} \right) \qquad i \in \{A, B, C\} \qquad (13.49)$$

An alternative to such an average is to use one or more characterized perturbations within the branch to find the FCCs of the individual steps in the branch. The gFCCs of the branch can then be determined from the appropriate summations. Furthermore, even when all of the individual FCCs of a branch cannot be determined, an FCC found from a characterized perturbation in any reaction i that lies within the branch generally provides a lower limit for the gFCC of the branch, as long as other steps in the branch do not compete significantly with reaction i (i.e., the characteristic angle of the other steps is similar to that of reaction i). Finally, if two different perturbations in a single reaction i result in flux changes exhibiting radically different characteristic angles that cannot be reconciled, it may be beneficial simply to use the results of other perturbations and determine the FCCs of reaction i from other known FCCs through the summation theorems.

A further difficulty in the determination of gFCCs through the methods of Section 13.2 arises from the matrix inversion in Eq. (13.17). In particular, if this matrix is nearly singular, the resulting perturbation constants can be greatly exaggerated; this yields gFCCs of very large magnitude. In Table 13.8, as you recall, excessively large perturbation constants and gFCC estimates resulted from several sets of calculations (in particular, for the last three branch points). Near-singularity of the inverted matrix generally arises from the use of small perturbations to the branch point. In fact, whenever K_i values greater than 5-10 result, it is best to attempt further experiments or to use representative perturbations as described earlier. If necessary, further control coefficient balances can be used in order to constrain the values of the perturbation constants. In the formulation of Eq. (13.17), three such balances were used: the summation theorems for two independent fluxes around the branch point and the branching theorem at the branch point. Two other sets of summations can be added to these to yield an overdetermined system, from which the perturbation constants K_i are determined by regression. These equations are the FCC and CCC summation theorems for any fluxes independent of those at the branch point or any metabolite concentration, including that of the corresponding link metabolite. For any independent flux J_k or concentration c_j, these take the following forms:

$$K_A\, p_A^k + K_B\, p_B^k + K_C\, p_C^k = 1 \qquad k \in \{1, 2, \ldots, L\} \qquad (13.50)$$

$$K_A\, \pi_A^j + K_B\, \pi_B^j + K_C\, \pi_C^j = 0 \qquad j \in \{1, 2, \ldots, K\} \qquad (13.51)$$

In these equations, p_i^k is the proportional change in flux J_k arising from a perturbation in branch i, as defined by Eq. (13.12), and π_i^j is similarly defined for a concentration change as

$$\pi_i^j = \frac{c_{j,i}}{c_{j,i}} - 1 = \phi_i^j - 1 \qquad i \in \{A, B, C\}, j \in \{1, 2, \ldots, K\} \quad (13.52)$$

These additional balances can also be useful if a perturbation has no discernible effect upon the flux of the upstream branch (A) into the node, as the use of the branching theorem in Eq. (13.16) will be stymied. In this case, the use of the summation theorem for the link metabolite of the branch point should yield the best result, as nonidealities in the system kinetics become more likely farther away from the branch point.

13.6.2. APPLICATION TO THE PREPHENATE BRANCH POINT

The preceding ideas are demonstrated with multiple perturbations around the prephenate branch point of the *S. cerevisiae* case study. Table 13.11 shows six perturbations around this branch point effected through changes in the activities of six different kinetic steps in the simulation.

From the data in Table 13.11, it is clear that the two perturbations within group A yield flux changes with very similar characteristic angles. Thus, a consensus angle of 62° seems appropriate to this group. The angles of the two perturbations in group B show some disagreement, but could be averaged according to Eq. (13.49) if necessary. As the perturbations in group C

TABLE 13.11 Effects of Multiple Perturbations Around the Prephenate Branch Point[a]

Perturbation			Change in flux J_k			Characteristic
Group	Reaction i[b]	r_i	J_A	J_B	J_C	angle θ_t
A	8	2.0	-0.038	-0.026	-0.048	61.6°
A	18	0.5	0.050	0.034	0.0645	62.2°
B	12	1.5	0.00	0.15	-0.14	-43.0°
B	16	1.5	0.0345	0.15	-0.075	-26.6°
C	13	1.5	0.003	-0.139	0.137	135.4°
C	17	1.5	0.033	0.023	0.043	61.8°

[a] Proportional change p_i^k in flux J_k resulting from perturbation in reaction i.
[b] Reaction numbers correspond to the numbering scheme of Fig. 12.1.

exhibit profoundly different behavior (as also shown in Box 13.1), however, it is prudent to investigate further the reactions of that branch. Because the activity amplification factors of the perturbations in reactions 13 and 17 are known, Eq. (13.20) can be used to find the FCCs corresponding to the effects of each of these reactions upon the group fluxes. The resulting FCCs for perturbations of reaction 13 upon branches A, B, and C are 0.01, -0.37, and 0.36, respectively, and those of reaction 17 are 0.09, 0.07, and 0.12 respectively. The group flux control coefficients for perturbations within branch C can be found by summing the preceding FCCs of reactions 13 and 17, yielding the following results:

$$*C_C^{J_A} = 0.10$$

$$*C_C^{J_B} = -0.30 \qquad (13.53)$$

$$*C_C^{J_C} = 0.48$$

The branching theorem of MCA can be used to directly determine from $*C_B^{J_A}$ and $*C_C^{J_A}$ the known fluxes:

$$*C_B^{J_A} = *C_C^{J_A}\left(\frac{J_{B,0}}{J_{C,0}}\right) = 0.10\left(\frac{3.52}{3.72}\right) = 0.09 \qquad (13.54)$$

By placing these results into the summation theorem formulation of Eq. (13.13) and making use of the consensus angle of branch A, we obtain the following relationships:

$$1.45K_A + 0.09 + 0.10 = 1$$

$$K_A + *C_B^{J_B} - 0.30 = 1 \qquad (13.55)$$

$$1.88K_A + *C_B^{J_C} + 0.48 = 1$$

The three unknowns in these three equations can easily be found, yielding the remaining gFCCs, as shown in Table 13.12.

A comparison of Table 13.12 to the exact gFCCs shown in Table 13.7 shows excellent agreement. Note that the coefficients of the main diagonal are all positive and no coefficient is significantly larger than 1, meaning that these coefficients are all physically reasonable. Finally, using the gFCCs of branch B in Table 13.12, which were derived independently from the summation theorem, the characteristic angle of group B is found to be -35.6°, which is in the middle of the range of the perturbations in group B shown in Table 13.11. Hence, even if the actual coefficients in Table 13.7 were not available for comparison, these gFCCs would be considered quite reasonable.

13.6.3. EFFECTS OF MEASUREMENT ERROR

Accurate measurement of fluxes and flux changes is necessary for the successful implementation of these methods. In fact, random errors in measurements have an effect upon the gFCC calculations which mimic those of nonlinearities. Hence, it is crucial to repeat and validate measurements, in order to reduce measurement error and ascertain whether nonidealities are indeed present.

The effects of random measurement error are shown in Fig. 13.5. Each calculation depicted in this figure was carried out following the methods of Section 13.2.1, using perturbations mirroring the ideal characteristic angles of the gFCCs at the branch point, with a random statistical error introduced into each flux measurement. It is important to realize that even a 5-10% error level can result in significantly skewed results. It also should be noted that, because the characteristic angles of perturbations in branches B and C are similar (but opposite), the gFCCs corresponding to these perturbations tend to be under- or overestimated more often than those corresponding to

TABLE 13.12   Validated Group Control Coefficients for PPH Branch Point[a]

Perturbed branch N	gFCC of flux $*J^k$		
	J_A	J_B	J_C
A	0.81	0.56	1.05
B	0.09	0.74	-0.53
C	0.10	-0.30	0.48

[a] gFCCs are shown in the form $*C_N^{J_k}$.

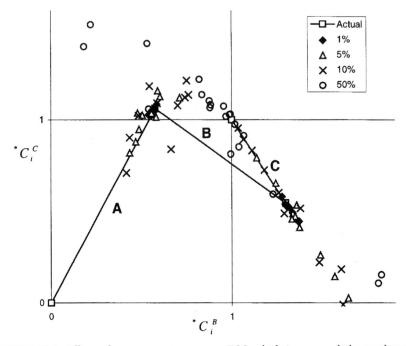

FIGURE 13.5 Effects of measurement error on gFCC calculations around the prephenate branch point. The lines represent the true group flux control coefficients. The scattered data points represent the endpoints of these lines calculated by assuming random statistical errors of 1, 5, 10, or 50% in the measurement of each flux change.

the A branch. This is a result of the near-singularity of the matrix inverted in Eq. (13.17). A comparison of the results of different levels of measurement error reveals that a 10% error allows good qualitative estimation of the gFCCs, but that an error level under 5% is necessary for good quantitative assessments. Because the accuracy of flux measurements may be beyond the control of the experimenter, however, the preferred method of improving the accuracy of control coefficient estimates is through regression analysis of the results of multiple perturbation experiments.

REFERENCES

Kacser, H. & Acerenza, L. (1993). A universal method for achieving increases in metabolite production. *European Journal of Biochemistry* **216**; 361-367.

Reder, C. (1988). Metabolic control theory: a structural approach. *Journal of Theoretical Biology* **135**; 175-201.

Small, J. R. & Kacser, H. (1993a) Responses of metabolic systems to large changes in enzyme activities and effectors. I. The linear treatment of unbranched chains. *European Journal of Biochemistry* **213**; 613-624.

Small, J. R. & Kacser, H. (1993b) Responses of metabolic systems to large changes in enzyme activities and effectors. II. The linear treatment of branched pathways and metabolite concentrations. Assessment of the general nonlinear case. *European Journal of Biochemistry* **213**; 625-640.

Simpson, T. W., Shimizu, H., & Stephanopoulos, G. (1998). Experimental determination of group flux control coefficients in metabolic networks. *Biotechnology & Bioengineering* **58**; 149–153.

Stephanopoulos, G. & Simpson, T. W. (1997). Flux amplification in complex metabolic networks. *Chemical Engineering Science* **52**; 2607-2627.

CHAPTER 14

Thermodynamics of Cellular Processes

Our analysis of cellular reactions in the preceding chapters has relied solely on stoichiometry and kinetics. Thus, in Chapter 7 we illustrated how all possible pathways for an overall conversion process can be generated from stoichiometric constraints around intracellular metabolites. In Chapters 8-10 we showed methods for the calculation of intracellular metabolic fluxes through the different cellular pathways from a stoichiometric model and measurements of extracellular metabolites and isotope label distributions. Finally, in Chapter 11, the issue of kinetic control was addressed and measures of the degree of control exerted by individual reactions, or groups of reactions (Chapters 12 and 13), were obtained in the form of control coefficients. We have not considered so far one important aspect of cellular reactions, namely, thermodynamic considerations, in the feasibility of the individual reactions and overall pathways' underlying metabolism. This is the subject of this chapter. We focus, in particular, on two important questions: First, the issue of thermodynamic feasibility is discussed in the context of single reactions and integrated pathways. The latter is important for the assessment of the numerous stoichiometrically possible pathways generated by the methods of Chapter 7. Second, although thermodynamics deals primarily with equilibrium processes, its extension into nonequilibrium processes can provide useful information about the kinetics of biological processes viewed as open systems. This enables a description of biological reaction kinetics, from which rate information may be obtained for the determination of the various control coefficients through the metabolic control analyses (MCA) theorems. We illustrate these applications of thermodynamics in this chapter, after we present, first, a short review of thermodynamic principles.

14.1. THERMODYNAMIC PRINCIPLES: A REVIEW

Thermodynamics is divided into *equilibrium* (or classic) *thermodynamics* and *nonequilibrium thermodynamics*. Classic thermodynamics considers only states at equilibrium, and therefore it provides little insight into the characteristics of transformations occuring in cellular pathways. Thus, the second law of thermodynamics, together with the Gibbs phase equilibrium principle, gives information on whether a reaction or conversion may proceed in a certain direction, *i.e.*, whether it is feasible (see Section 14.2), but it gives no information about its rate. It applies mainly to reversible reactions in closed systems, where equilibrium is reached almost invariably. Living systems are, however, open systems and, therefore, they can never be at equilibrium. They are constantly acquiring free energy by converting high-enthalpy, low-entropy substrates to low-enthalpy, high-entropy metabolic products. The conversion of substrates to metabolic products occurs via several individual reaction steps, some of which are close to equilibrium whereas others are far from equilibrium. Thus, in the EMP pathway several of the reaction steps are close to equilibrium, *e.g.*, reactions catalyzed by phosphoglucose isomerase, phosphoglycerate mutase, and enolase, whereas others operate far from equilibrium, *e.g.*, hexokinase, phosphofructokinase, and pyruvate kinase. In this review of thermodynamic principles applied to cellular processes, we consider only equilibrium thermodynamics, whereas nonequilibrium thermodynamics is treated in Section 14.3.

In thermodynamics, a *system* is defined as that part of the universe that is of interest, such as a bioreactor or a cell, whereas the rest of the universe is referred to as its surroundings. A system is said to be *open* or *closed* according to whether it can exchange matter and energy with its surroundings. Because living cells take up nutrients, release metabolites, and generate work and heat, they are open systems. The state of a system is defined by a set of state functions. These state functions include the internal energy (U, a measure of the total intrinsic energy of the system), enthalpy (H, equal to the heat absorbed under constant pressure when the only type of work is due to volume change), and entropy (S, a measure of the degree of order in the system). These state functions can be used in the formulation and interpretation of the two laws of thermodynamics, upon which classic thermodynamics is founded. The two laws postulate the following:

- **The First Law of Thermodynamics.** *Energy can be neither created nor destroyed* or, mathematically, $\Delta U = 0$.
- **The Second Law of Thermodynamics.** *Spontaneous processes occur in directions that increase the overall disorder (or entropy) of the <u>universe</u>* or, mathematically, $\Delta S > 0$.

Whereas the first law of thermodynamics specifies that energy is conserved in any process, *i.e.*, energy generated by a system must be taken up by its surroundings, the second law states that spontaneity of a process is determined from the *overall* change in entropy. In the study of cellular processes spontaneity is indeed an important issue, but disordering of the universe by spontaneous processes is an impractical criterion to use for assessing spontaneity as it is impossible to determine changes in the entropy of the entire universe. Furthermore, the spontaneity of a process cannot be decided from the entropy change of the system in question alone, because an exothermic process ($\Delta H_{system} < 0$, *i.e.*, heat is evolved from the system) may be spontaneous even if it is accompanied by a decrease in the entropy of the system, $\Delta S_{system} < 0$ (of course, the *total* entropy change is positive in this case too, due to an increase in the entropy of the environment that more than counterbalances the earlier decrease in the entropy of the system). An example is the spontaneous folding of denatured proteins to their highly ordered (*i.e.*, $\Delta S_{system} < 0$) native conformation.

Due to these difficulties with the use of entropy, spontaneity is determined by using another state function, the *Gibbs free energy*:

$$G = H - TS \qquad (14.1)$$

which was introduced by J. Willard Gibbs in 1878. The meaning of the free energy is that, for a process at constant temperature and pressure, the maximum work that can be done by the system (but not including the work of displacement) is equal to the decrease in the free energy of the system. For constant temperature and pressure processes, which describe the vast majority of biological systems, the criterion for spontaneity is $\Delta G \leq 0$.

Spontaneous processes, *i.e.*, those with $\Delta G \leq 0$, are said to be *exogenic*, and they can be utilized to do work. Processes that are not spontaneous, *i.e.*, those that have positive ΔG values, are termed *endergonic*, and they must be driven by the input of free energy. Processes at equilibrium, *i.e.*, the forward and backward processes are exactly balanced, are characterized by $\Delta G = 0$. Notice that the Gibbs free energy varies with temperature, which therefore must always be specified. This dependence of temperature explains the spontaneous denaturation of proteins above a certain temperature. As mentioned earlier, formation of a native protein from its denatured form has both a negative ΔH and a negative ΔS. However, for this process ΔH and ΔS change at different rates with temperature. Below the temperature where ΔH equals $T\Delta S$, the Gibbs free energy of the denaturation process is positive and the reverse reaction (*i.e.*, folding) is a spontaneous process. For temperatures above this temperature, ΔG_{denat} is negative and the native protein will tend to denature.

It is important to note that a large negative value of ΔG does not necessarily imply that a chemical reaction will proceed at a measurable rate. Thus, the free energy change of the phosphorylation of glucose to glucose-6-phosphate by ATP is large and negative, but this reaction does not run when glucose and ATP simply are mixed together. Only when the enzyme hexokinase is added does the reaction run. Similarly, most biological molecules, including proteins, nucleic acids, carbohydrates, and lipids, are thermodynamically unstable to hydrolysis, but they spontaneously hydrolyze at insignificant rates. Only when hydrolytic enzymes are added do the hydrolysis reactions proceed at a reasonable rate. Despite their importance in accelerating a reaction, enzymes do not change the ΔG for the reaction. As catalysts they can only accelerate the attainment of thermodynamic equilibrium, but do not allow a reaction with a positive ΔG to proceed.

The preceding definitions of the basic thermodynamic state functions are valid for systems of *fixed composition* only. For systems that may undergo changes in composition, this form of the state functions is insufficient to describe changes that may occur due to chemical reactions or diffusive processes with substances that initially were separated. To correct for this deficiency, it is recognized that U, H, S, and G are also functions of the total amount of each substance in the solution, in addition to the two intrinsic thermodynamic variables (such as, for example, temperature and pressure) used for systems of fixed composition. Thus, for the free energy:

$$G = G(T, p, n_1, n_2, \ldots, n_N) \qquad (14.2)$$

where $n_1, n_2, \ldots, n_N$ are the total number of moles of species 1 to N, respectively. From the preceding equation, we obtain

$$dG = -S\,dT + V\,dp + \Sigma\mu_i\,dn_i \qquad (14.3)$$

where

$$S = \frac{\partial G}{\partial T}\bigg|_{p,n_i}; \quad V = \frac{\partial G}{\partial p}\bigg|_{T,n_i}; \quad \mu_i = \frac{\partial G}{\partial n_i}\bigg|_{T,p,n_{j\neq i}} \quad (14.4)$$

In the preceding equations, μ_i is called the *chemical potential*, or *Gibbs free energy of compound i*, also introduced by Gibbs to simplify the analysis of open systems as well as closed ones undergoing changes in composition. It was defined as the increase in the free energy of the mass of a system due to the addition of an infinitesimal amount of any substance under conditions of constant temperature and pressure, divided by the quantity of the substance added. Just as a temperature difference determines the tendency of heat to flow and a pressure difference determines the tendency toward bodily movement, a difference in the chemical potential can be interpreted as a measure of the tendency for a chemical reaction to occur or a substance to diffuse down a gradient of the chemical potential. In this sense, the chemical potential may be regarded as a kind of chemical pressure, and it is an intensive property like the temperature and pressure themselves.

The chemical potential may also be defined similarly in terms of the other thermodynamic state functions U or H from which one can obtain:

$$dU = T\,dS - p\,dV + \Sigma \mu_i\,dn_i \quad (14.5)$$

$$dH = T\,dS + V\,dp + \Sigma \mu_i\,dn_i \quad (14.6)$$

with the chemical potential given as

$$\mu_i = \frac{\partial G}{\partial n_i}\bigg|_{T,p,n_{j\neq i}} = \frac{\partial U}{\partial n_i}\bigg|_{S,V,n_{j\neq i}} = \frac{\partial H}{\partial n_i}\bigg|_{S,p,n_{j\neq i}} \quad (14.7)$$

Considering the conditions under which the state thermodynamic functions are defined, it can be shown that:

$$U = TS - pV + \Sigma \mu_i n_i \quad (14.8a)$$

$$H = TS + \Sigma \mu_i n_i \quad (14.8b)$$

$$G = \Sigma \mu_i n_i \quad (14.8c)$$

These expressions, especially eq. (14.8c), are important in defining the criteria for the spontaneity of chemical reactions.

The Gibbs free energy of a compound is related to its concentration through the well-known equation for chemical potentials:

$$\mu_i = \mu_i^0(p,T) + RT \ln\left(\frac{f_i c_i}{f_{i,\text{ref}} c_{i,\text{ref}}}\right) \quad (14.9)$$

where f_i is the activity coefficient of compound i, and μ_i^0 is the chemical potential of compound i at a reference state represented by the activity coefficient $f_{i,\,ref}$ and concentration $c_{i,\,ref}$. The reference chemical potential is a function of temperature and pressure. Normally, the reference state is taken to be a concentration $c_{i,\,ref} = 1$ M and an activity coefficient $f_{i,\,ref} = 1$. Furthermore, biological systems normally are assumed to be dilute solutions (which is a reasonable assumption for many - but not all - intracellular compounds). This means that $f_i = 1$ and eq. (14.9) therefore reduces to

$$\mu_i = \mu_i^0(p, T) + RT \ln(c_i) \qquad (14.10)$$

where c_i is the molar concentration of the ith compound.

From the chemical potential of the reactants and products in a chemical reaction (or any type of process), we can calculate the Gibbs free energy change of the reaction (or process). Thus, for the general reaction:

$$cC + dD - aA - bB = 0 \qquad (14.11)$$

the free energy change, in light of eq. (14.8c), can be shown to be given by

$$\Delta G = c\mu_C + d\mu_D - a\mu_A - b\mu_B \qquad (14.12)$$

and, after inserting eq. (14.9) into eq. (14.12), we obtain

$$\Delta G = \Delta G^{0\prime} + RT \ln\left(\frac{c_C^c c_D^d}{c_A^a c_B^b}\right) \qquad (14.13)$$

where $\Delta G^{0\prime}$ is the free energy change of the reaction when all of its reactants and products are at their standard states (see Box 14.1). Thus, the free energy change consists of two terms: (1) a constant term that depends only on the type of reaction taking place (as well as temperature and pressure) and (2) a variable term that depends on the temperature, the concentrations of the reactants and products, and the stoichiometric coefficients. For a reaction at equilibrium, the Gibbs free energy is zero and eq. (14.13) becomes:

$$\Delta G^{0\prime} = -RT \ln(K_{eq}) \qquad (14.14)$$

where K_{eq} is the equilibrium constant of the reaction:

$$K_{eq} = \frac{c_{C,\,eq}^c c_{D,\,eq}^d}{c_{A,\,eq}^a c_{B,\,eq}^b} = e^{-\Delta G^0/RT} \qquad (14.15)$$

BOX 14.1

Standard State Conventions

The usual convention used in physical chemistry defines the standard state of a solute as that with unit concentration (or activity) at 25°C and 1 atm. Because biochemical reactions usually occur in dilute aqueous solutions near neutral pH, a somewhat different standard state normally is used for biological systems:

- The standard state of water is defined as that of the pure liquid. Thus, the concentration (or activity) of water is taken to be 1 despite the fact that its concentration is 55.5 M.
- The hydrogen ion activity is defined as unity at the physiologically relevant pH of 7 rather than at the chemical standard state of pH 0.
- The standard state of a compound that can undergo an acid-base reaction is defined in terms of the *total* concentration of its naturally occuring ion mixture at pH 7. The advantage is that normally it is easier to measure the total concentration of a compound than the concentration of one of its ionic species. Because the ionic composition of an acid or base varies with pH, the standard free energies calculated using total concentrations are, however, valid only at pH 7.

The standard free energy changes with the preceding references normally are symbolized by $\Delta G^{0\prime}$, with the prime inserted to distinguish it from free energy changes at the normal standard state. Because the free energy *change* is independent of the reference state, the choice of reference does not matter. In the formulation of the influence of water and protons on the free energy change it is, however, important to specify the reference state. Thus, if we use the biological standard state as reference, the concentration of water can be left out of eq. (14.13). Similarly, if the reaction is carried out at pH 7, the proton concentration does not have to be included. If the reaction is carried out at a pH different from pH 7, the proton concentration should be taken relative to 10^{-7}, *i.e.*, it should appear as $[H^+]/10^{-7}$.

From eq. (14.15) it is obvious that the equilibrium constant may be calculated from standard free energy data and vice versa. As a consequence of the exponential dependence, even relatively small negative standard free energy changes can result in equilibrium constants much greater than 1. Thus, an equilibrium constant of 100 is associated with a $\Delta G^{0'}$ of only -11.4 kJ/mol. According to eq. (14.13), it is seen that if the reactants are in excess of their equilibrium concentrations, the net reaction will proceed in the forward direction until the excess reactants have been converted to products and equilibrium is attained. Similarly, if the products are in excess of the equilibrium concentrations, the net reaction will proceed in the backward direction until the excess products have been converted to reactants in order to establish equilibrium. Thus, as Le Chatelier's principle states, *any deviation from equilibrium stimulates a process that tends to restore the equilibrium, and all closed systems therefore must inevitably reach equilibrium.* Living cells escape this thermodynamic *cul-de-sac* by being open systems.

The standard free energy change for a reaction can easily be determined from the free energies of *formation* of the reactants and products, and Table 14.1 summarizes the free energies of formation of some compounds of biochemical interest. Furthermore, through the use of group contribution methods, the free energy of formation may be calculated for compounds where there is no *a priori* knowledge of this value (see Section 14.2.2). From inspection of some of the $\Delta G^{0'}$ values for the reactions in the EMP pathway

TABLE 14.1 Standard Free Energies of Formation of Various Compounds

Compound	$\Delta G_f^{0'}$ (kJ mol^{-1})	Compound	$\Delta G_f^{0'}$ (kJ mol^{-1})
Acetaldehyde	139.7	Glyceraldehyde-3-phosphate^{2-}	1285.6
Acetate$^-$	369.2	H$_2$O	237.2
Acetyl-CoA	374.1^a	Isocitrate^{3-}	1160.0
cis-Aconitate^{3-}	920.9	α-Ketoglutarate^{2-}	798.0
CO$_2$ (aq)	386.2	Lactate$^-$	516.6
Citrate^{3-}	1166.6	Malate^{2-}	845.1
Dihydroxyacetonephosphate^{2-}	1293.2	OH$^-$	157.3
Ethanol	181.5	Oxaloacetate	797.2
Fructose	915.4	Phosphoenolpyruvate	1269.5
Fructose-6-P^{2-}	1758.3	2-Phosphoglycerate^{3-}	1285.6
Fructose-1,6-bisP^{4-}	2600.8	3-Phosphoglycerate^{3-}	1515.7
Fumarate^{2-}	604.2	Pyruvate$^-$	474.5
Glucose	917.2	Succinate^{2-}	690.2
Glucose-6-P^{2-}	1760.2	Succinyl-CoA	686.7^a

^{a} For formation from free elements plus free CoA.

(see Example 14.1) it is observed that some have positive values whereas others have negative values. Because there normally is a net flux through this pathway, the concentrations of the metabolites and cofactors adjust themselves such that the Gibbs free energy change of each reaction is negative (see Example 14.1). Thus, the high level of $NAD^+/NADH$, which is controlled by many reactions inside the cell, ensures that the conversion of glyceraldehyde-3-phosphate to 3-phosphoglycerol phosphate is thermodynamically favorable, even though the $\Delta G^{0'}$ for this reaction is 6.28 kJ mol^{-1}. Furthermore, because the subsequent reaction in the EMP pathway has a negative standard free energy change of -18.83 kJ mol^{-1}, it ensures spontaneity of this reaction even for a low concentration of 3-phosphoglycerol phosphate and thereby may drive the reaction preceding it. Consequently, reactions, which are not thermodynamically favoured to run in a forward direction may be driven within a pathway by other reactions which have large negative free energy changes. This is the basis for the thermodynamic feasibility analysis of Section 14.2.

It is interesting to note that in most pathways the first step and the last step are associated with large negative standard Gibbs free energies, e.g., the phosphorylation of glucose to glucose-6-phosphate has a $\Delta G^{0'}$ of -16.7 kJ mol^{-1} and the conversion of pyruvate to lactate has a $\Delta G^{0'}$ of -25.1 kJ mol^{-1}. This may play a role in ensuring that these reactions are thermodynamically favored in the forward direction even for low substrate concentrations and high product concentrations (the conversion of pyruvate to lactate is further thermodynamically favored in the forward direction due to the high level of $NAD^+/NADH$).

EXAMPLE 14.1

Free Energy Changes of Glycolytic Reactions

To determine the free energy changes of cellular reactions, it is necessary to know the concentration of all metabolites and cofactors participating in these reactions. Such data are available only for a few pathways, and thermodynamic considerations therefore are often based on evaluation of standard free energy changes. This may, however, lead to erroneous conclusions as the use of standard free energy changes assumes certain fixed concentrations for reactants and products (those of the standard state) that may be different from the actual intracellular metabolite concentrations. Furthermore, the latter may have a significant effect on the actual free energy change. To illustrate this point, we calculate the free energy change for some

of the reactions in the EMP pathway. Table 14.2 lists the intracellular
concentrations of some of the intermediates, ATP, ADP, and orthophosphate
in the human erythrocyte, and Table 14.3 lists the calculated free energy
changes.

From the calculated free energy changes, it is observed that, except for the
hexokinase-phosphofructokinase-, pyruvate kinase-, and, possibly, the triose-
P-isomerase-catalyzed reactions, all the reactions of the EMP pathway are
close to equilibrium (the free energy change for the reaction series 3-P-
glyceraldehyde dehydrogenase and 3-phosphoglycerate kinase is also close to
zero). Thus, the *in vivo* activity of several of the enzymes is sufficiently high
to equilibrate the conversions, or, in other words, the forward and backward
reactions of these conversions are much faster than the net flux through the
pathway. Obviously, these equilibrium reactions are very sensitive to changes
in the concentration of pathway intermediates, *i.e,.* they have large elasticity
coefficients, and therefore they rapidly communicate changes in flux generated by one of the reactions with a high negative free energy change
throughout the rest of the pathway.

The three reactions with large negative free energies are *thermodynamically irreversible* and are often considered as key control points in the
pathway (see also the discussion on equilibrium enzymes in Chapter 5).
Obviously, the *in vivo* activity of the three enzymes hexokinase, phosphofructokinase, and pyruvate kinase is too low to equilibrate the reactions they
catalyze. This may be the result of either too low gene expression, *i.e.*, the
in vivo v_{max} of the enzymes is too low, or regulation at the enzyme level, *e.g.*,
allosteric regulation or covalent enzyme modifications. For example, allosteric regulators for the phosphofructokinase enzyme have been discussed
extensively in Section 2.3.1.

TABLE 14.2   Concentrations of Intermediates and Cofactors of the EMP Pathway
in the Human Erythrocyte[a]

Metabolite/cofactor	Concentration (μM)	Metabolite/cofactor	Concentration (μM)
Glucose (GLC)	5000	2-Phosphoglycerate (2PG)	29.5
Glucose-6-P (G6P)	83	Phosphoenolpyruvate (PEP)	23
Fructose-6-P (F6P)	14	Pyruvate (PYR)	51
Fructose-1,6-bisP (FDP)	31	ATP	1850
Dihydroxyacetone P (DHAP)	138	ADP	138
Glyceraldehyde-3-P (GAP)	18.5	P_i	1000
3-Phosphoglycerate (3PG)	118		

[a] The data are taken from Lehninger (1975).

TABLE 14.3   Free Energy Changes over Reactions of the EMP Pathway in the Human Erythrocyte

Reaction	$\Delta G^{0\prime}$ (kJ mol^{-1})	ΔG (expression)	ΔG (kJ mol^{-1})
Hexokinase	-16.74	$\Delta G^{0\prime} + RT \ln \dfrac{[G6P][ADP]}{[GLC][ATP]}$	-33.3
Glucose-6-P isomerase	1.67	$\Delta G^{0\prime} + RT \ln \dfrac{[F6P]}{[G6P]}$	-2.7
Phosphofructokinase	-14.22	$\Delta G^{0\prime} + RT \ln \dfrac{[FDP][ADP]}{[F6P][ATP]}$	-18.7
Aldolase	23.97	$\Delta G^{0\prime} + RT \ln \dfrac{[DHAP][GAP]}{[FDP]}$	0.7
Triose-P-isomerase	7.66	$\Delta G^{0\prime} + RT \ln \dfrac{[GAP]}{[DHAP]}$	2.7
Phosphoglycerate mutase	4.44	$\Delta G^{0\prime} + RT \ln \dfrac{[2PG]}{[3PG]}$	1.0
Enolase	1.84	$\Delta G^{0\prime} + RT \ln \dfrac{[PEP]}{[2PG]}$	1.2
Pyruvate kinase	-31.38	$\Delta G^{0\prime} + RT \ln \dfrac{[PYR][ATP]}{[PEP][ADP]}$	-23.0

14.2. THERMODYNAMIC FEASIBILITY

As discussed in the previous section, a chemical reaction or transport process is feasible if ΔG is negative. Furthermore, for a system of biochemical reactions and transport processes to be feasible, the preceding criterion must be satisfied by *all participating steps, i.e.,* the ΔG of all participating reactions must be negative. If the ΔG for at least one reaction of the pathway is greater than zero, then the pathway is not feasible. If only one reaction step is not feasible, we refer to it as a *localized thermodynamic bottleneck*, whereas if several reactions violate the criterion of $\Delta G < 0$, we refer to them as a *distributed* thermodynamic bottleneck.

Consider now the two-reaction sequence:

$$A \rightarrow B \rightarrow C \qquad (14.16)$$

The two reactions will proceed from left to right if the following inequalities are satisfied at the same time:

$$\Delta G_1 = \Delta G_1^{0'} + RT \ln \frac{c_B}{c_A} < 0 \qquad (14.17)$$

$$\Delta G_2 = \Delta G_2^{0'} + RT \ln \frac{c_C}{c_B} < 0 \qquad (14.18)$$

If both $\Delta G^{0'}$ values are less than zero, the preceding system of reactions can proceed from left to right with a rather flat concentration profile for metabolites A, B, and C. If, however, $\Delta G_1^{0'}$ is less than zero but $\Delta G_2^{0'}$ is greater than zero, then a larger gradient will be required between the concentrations of metabolites B and C in order to overcome the positive standard Gibbs free energy changes of the second reaction. This is depicted graphically in Fig. 14.1, which also shows the situations for the cases of $\Delta G_1^{0'} > 0$ but $\Delta G_2^{0'} < 0$ and $\Delta G_1^{0'}, \Delta G_2^{0'} > 0$. In the latter case a rather large concentration gradient is required between the reactant metabolite concentration c_A and final product concentration c_C in order to overcome the positive $\Delta G^{0'}$ of both reactions. Clearly, positive values of $\Delta G^{0'}$ can be overcome by metabolite concentration differences; however, there is a limit to the extent to which metabolite concentrations can overcome unfavorable thermodynamic conditions.

When the preceding system is extended to include several reactions in a sequence, it becomes clear that in order to overcome several positive $\Delta G^{0'}$ values, a very low final metabolite concentration may be required that may exceed a reasonable physiological concentration range. Considering that

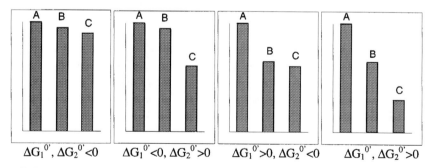

FIGURE 14.1   Concentration gradients required for making the reactions in the system e.g. (14.16) thermodynamically feasible.

metabolites can participate in several bioreactions simultaneously, imposing different constraints on them, one can see that their concentration becomes the key determinant of the feasibility of the reactions and, through them, the thermodynamic feasibility of the overall network. A systematic method clearly is required to determine the extent to which all thermodynamic constraints can be satisfied within an allowed range of metabolite concentrations and, therefore, establish the thermodynamic feasibility of a metabolic network (Mavrovouniotis, 1993). Such a method is presented in the sequel following some convenient transformations of the original variables that facilitate the analysis.

14.2.1. THE ALGORITHM

We now consider a general set of reactions with the following stoichiometry:

$$\sum_{i=1}^{K} g_{ji} X_i = 0; \qquad j = 1, \dots, J \tag{14.19}$$

The Gibbs free energy change for each reaction can be written as

$$\Delta G_j = \Delta G_j^{0\prime} + \sum_{i=1}^{K} g_{ji} RT \ln c_i; \qquad j = 1, \dots, J \tag{14.20}$$

where c_i is the concentration of metabolite X_i of reaction (14.19). We further assume that some of the metabolites have constant concentrations, while others are allowed to vary within a range. Typically, the concentrations of ATP, NADH, and other cofactors (referred to as currency metabolites) are controlled within a narrow range such that they can be assumed to be practically constant. With this grouping of metabolites, we rewrite eq. (14.20) as

$$\Delta G_i = \Delta G_i^{0\prime} + \sum_{\substack{currency \\ metabolites}} g_{ji} RT \ln c_i + \sum_{i=1}^{K'} g_{ji} RT \ln c_i \tag{14.21}$$

where K' represents the number of metabolites that are allowed to vary within a concentration range. If the first two terms are grouped together and renamed as $\Delta G^{0\prime\prime}$, eq. (14.21) can be written as:

$$\Delta G_i = \Delta G_i^{0\prime\prime} + \sum_{i=1}^{K'} g_{ji} RT \ln c_i \tag{14.22}$$

The requirements for the metabolic pathway to be thermodynamically feasible can then be specified by the following inequality:

$$\frac{\Delta G_j^{0\prime\prime}}{RT} + \sum_{i=1}^{K'} g_{ji} \ln c_i < 0; \qquad j = 1, \ldots, J \tag{14.23}$$

It is convenient to scale the concentration of each metabolite with respect to the range within which each such metabolite is allowed to vary. This is accomplished by first rewriting eq. (14.23) in the following form:

$$\frac{\Delta G_j^{0\prime\prime}}{RT} + \sum_{i=1}^{K'} g_{ji} \ln c_i^{\min} + \sum_{i=1}^{K'} g_{ji} \ln \frac{c_i}{c_i^{\min}} < 0 \tag{14.24}$$

or

$$\frac{\Delta G_j^{0\prime\prime}}{RT} + \sum_{i=1}^{K'} g_{ji} \ln c_i^{\min} + \sum_{i=1}^{K'} g_{ji} \ln \frac{c_i^{\max}}{c_i^{\min}} \frac{\ln \dfrac{c_i}{c_i^{\min}}}{\ln \dfrac{c_i^{\max}}{c_i^{\min}}} < 0 \tag{14.25}$$

and redefining a new dimensionless concentration f_i in terms of the relative location of the actual metabolite concentration in the allowed concentration range. Thus, by letting

$$h_j = \frac{\Delta G_j^{0\prime\prime}}{RT} + \sum_{i=1}^{K'} g_{ji} \ln c_i^{\min}; \qquad j = 1, \ldots, J \tag{14.26}$$

$$w_{ji} = g_{ji} \ln \frac{c_i^{\max}}{c_i^{\min}}; \qquad j = 1, \ldots, J \tag{14.27}$$

$$f_i = \frac{\ln \dfrac{c_i}{c_i^{\min}}}{\ln \dfrac{c_i^{\max}}{c_i^{\min}}} \tag{14.28}$$

eq. (14.25) can be rewritten as

$$H_j = h_j + \sum_{i=1}^{K'} w_{ji} f_i < 0; \qquad j = 1, \dots, J \qquad (14.29)$$

A different thermodynamic/stoichiometric meaning can be attributed to each of the preceding variables. For example, h_j can be interpreted as the Gibbs free energy change at a different state where each metabolite has its minimum permissible concentration rather than the predetermined value of 1 M (which is the state for which the Gibbs free energy is equal to $\Delta G^{0\prime}$). Similarly w_{ji} is a redefined stoichiometric coefficient that preserves the sign convention of g_{ji}. Furthermore, reactions are still additive in terms of w_{ji} and they can be combined in just the same way as in terms of g_{ji}. Finally, f_i is a scaled representation of the concentration, which allows the specification of the feasibility criteria in a linear form, something that simplifies the calculations a great deal.

In the framework of the preceding transformation, the problem of thermo-dynamic feasibility is reduced to one of identifying a set of reduced metabo-lite concentrations f_i for which all inequalities $H_j < 0$ are satisfied simulta-neously. If no such set can be found to satisfy these constraints, the set of reactions eq. (14.19) is thermodynamically infeasible. With the scaled con-centrations f_i lying between 0 and 1, we can introduce two sets of values for H_j, namely, a maximum and a minimum. The maximum value is obtained when the concentration of the reactants (which have negative stoichiometric coefficients, $w_{ji} < 0$) attains the minimum concentration (corresponding to $f_i = 0$) and the products (which have stoichiometric coefficients character-ized by $w_{ji} > 0$) attain the maximum concentration (corresponding to $f_i = 1$). At such conditions, the thermodynamic driving force of the reaction obvi-ously is minimal. Thus,

$$H_{j,\,max} = h_j + \sum_{i=1}^{K'} w_{ji}; \qquad w_{ji} > 0 \qquad (14.30)$$

Similarly, we can find a minimum value for H_j given by

$$H_{j,\,min} = h_j + \sum_{i=1}^{K'} w_{ji}; \qquad w_{ji} < 0 \qquad (14.31)$$

It can be seen in the preceding variables that, $H_{j,\,min}$ represents the scaled Gibbs free energy when all metabolites assume concentrations that are most favorable from a thermodynamic point of view. The new scaling permits $H_{j,\,min}$ to be interpreted as the most favorable distance of the reaction from equilibrium. Therefore, $H_{j,\,\min} < 0$ is a necessary condition for reaction j to be thermodynamically feasible, and consequently $H_{j,\,\min} > 0$ is a sufficient condition for reaction j to be thermodynamically infeasible. Similarly, $H_{j,\,\max}$ represents the least favorable distance from equilibrium. Therefore, $H_{j,\,\max} < 0$ is a sufficient condition for reaction j to be thermodynamically feasible and $H_{j,\,\max} > 0$ is a necessary condition for reaction j to be thermodynamically infeasible. Furthermore, it can be shown that any linear combination of thermodynamically feasible reactions also will produce a thermodynamically feasible reaction, and any linear combination of thermodynamically infeasible reactions also will produce a thermodynamically infeasible reaction. Therefore, if $H_{j,\,\max} < 0$, then reaction j can be eliminated from the set of reactions that are thermodynamically infeasible.

The linearity of the previous transformation offers some additional advantages. If a reaction j is a linear combination of two other reactions 1 and 2, then the scaled Gibbs parameter h_j for the composite reaction will be the same linear combination of the Gibbs parameters h_1 and h_2. Similarly, H_j will be the same linear combination of H_1 and H_2. Similar equations can be written for $H_{\min}$ and $H_{\max}$, provided that the constituent reactions do not share a common metabolite. In this case, i.e., when a metabolite is the product of one reaction and the reactant of another, it can be shown that the $H_{\min}$ of the resulting reaction will be greater than the linear combination of the constituent reactions, and $H_{\max}$ of the resulting reaction will be lower than the linear combination of the constituent reactions.

In the implementation of an algorithm for testing the thermodynamic feasibility of metabolic pathways, one should examine the signs of $H_{j,\,\min}$ and $H_{j,\,\max}$:

- If $H_{j,\,\min} > 0$, then reaction j is always thermodynamically infeasible.
- If $H_{j,\,\min} < 0$ and $H_{j,\,\max} > 0$, then no conclusion can be drawn regarding the feasibility or infeasibility of reaction j. In this case, reaction j should be combined with a reaction sharing a common metabolite and $H_{\min}$ and $H_{\max}$ should be recalculated for the resulting reaction, which can then be reassessed with respect to its feasibility. Because, as was pointed out before, the resulting reaction will have a greater $H_{\min}$ and a lower $H_{\max}$ than the constituent reactions, successive combinations of reactions eventually will lead to a definite determination of their thermodynamic feasibility or infeasibility.
- If $H_{j,\,\max} < 0$, then reaction j is always feasible.

The preceding ideas can be implemented in a simple algorithm testing the thermodynamic feasibility of biochemical pathways (see Box 14.2).

BOX 14.2

Algorithm for Determining Thermodynamic Pathway Feasibility

Mavrovouniotis (1993) described an algorithm based on the transformation of the feasibility criteria described in the text. The algorithm has the following steps:

1. Postulate upper and lower bounds for the concentrations of each metabolite.
2. Identify currency metabolites (those that assume constant concentrations).
3. Compute, in sequence, $\Delta G_j^{0"}$, $\Delta G_j^{0"}/RT$, h_j, w_{ji}, $H_{j,\,min}$, and $H_{j,\,max}$.
4. Reject from the set of bottleneck reactions all reactions with $H_{j,\,max} < 0$. These reactions are always feasible with the specified concentration range of the metabolites.
5. Retain all reactions with $H_{j,\,min} > 0$. These reactions are always infeasible with the specified concentration range of the metabolites.
6. For the remaining reactions, construct combinations of two reactions such that an intermediate metabolite is always eliminated. In those cases where no such intermediate metabolite exists, reactions cannot be combined further, and cannot be infeasible. If no undecided subpathways exist or no further combinations are possible, end the search. Otherwise go back to step 3.

Thermodynamic bottlenecks are identified from step 6. It is important to emphasize that the bottlenecks exist only for the specified range of metabolite concentrations. If it is known that there is flow through the pathway, the algorithm can be used to specify the likely *in vivo* range of metabolite concentrations.

EXAMPLE 14.2

Thermodynamic Feasibility Analysis of the EMP Pathway

In this example, we illustrate the thermodynamic feasibility algorithm with an analysis of the EMP pathway, which also was considered in Example 14.1. The example is based on the work of Mavrovouniotis (1993), where the preceding algorithm was presented for the first time. The EMP pathway is depicted in Fig. 2.6. Abbreviations for the metabolites used in this example are given in Table 14.2, whereas Table 14.3 gives the enzyme names and the standard Gibbs free energies for the enzymatic reactions of the pathway. As discussed in Example 14.1, it is obvious that reaction and pathway feasibility cannot be assessed from the $\Delta G^{0'}$ values directly as several reactions have positive standard Gibbs free energies, yet it is well-known that this is a feasible pathway. It is assumed that pH is 7, so that water, H^+, and OH^- concentrations are at their standard state, meaning that they can be omitted from the calculations. Constant concentrations are assumed for the energy currency metabolites ADP, ATP, and P_i. This assumption is consistent with numerous observations suggesting that global controls maintain the concentration of such metabolites within narrow predefined ranges. It is further assumed that all activity coefficients are constant and equal to 1. For this example, the following values for the currency metabolites are used, [AMP] = 0.82 mM, [ADP] = 1.04 mM, [ATP] = 7.9 mM, and $[P_i]$ = 7.9 mM, which correspond to an energy charge of 0.87. Furthermore, $[NAD^+]$ = 4 mM and [NADH] = 0.2 mM, which corresponds to a catabolic reduction charge of 0.05. In the following, we investigate several different possibilities regarding the allowed concentration range for the remaining metabolites of the EMP pathway.

First, we examine the concentration range 0.1-1.0 mM, and Table 14.4 shows the computations outlined in the algorithm for the 10 reaction steps.

TABLE 14.4   Computation of Individual Reaction Parameters for the First Interval (0.1-1.0 mM)[a]

Index	$\Delta G_j^{0'}$ (kJ mol^{-1})	$\Delta G_j^{0'}/RT$	$\Delta G_j^{0''}/RT$	h_j	H_{min}	H_{max}
1	-16.74	-6.753	-8.781	-8.781	-11.083	-6.478
2	1.67	0.675	0.675	0.675	-1.627	2.978
3	-14.22	-5.740	-7.768	-7.768	-10.070	-5.465
4	23.97	9.674	9.674	0.464	-1.839	5.069
5	7.66	3.090	3.090	3.090	0.787	5.392
6	6.28	2.532	12.895	3.867	1.564	7.017
7	-18.83	-7.597	-5.570	-5.570	-7.872	-3.267
8	4.44	1.790	1.790	1.790	-0.513	4.092
9	1.84	0.743	0.743	0.743	-1.560	3.045
10	-31.38	-12.662	-10.635	-10.635	-12.937	-8.332

[a] Only the last three columns actually depend on c^{min} and c^{max}. The index for the reactions refers to (1) hexokinase; (2) glucose-6-P-isomerase; (3) phosphofructokinase; (4) aldolase; (5) triose-P-isomerase; (6) 3-P-glyceraldehyde dehydrogenase; (7) 3-phosphoglycerate kinase; (8) phosphoglycerate mutase; (9) enolase; (10) pyruvate kinase.

On the basis of the signs calculated for H_{min} and H_{max}, it can be concluded that reactions 1, 3, 7, and 10 are always feasible where as reactions 5 and 6 are identified as two localized bottlenecks. No conclusion can be drawn for reactions 2, 4, 8, and 9 based on the calculated values of H_{min} and H_{max}. Because reactions 1 and 3 are determined to always be feasible within the examined concentration range, they cannot be a bottleneck by themselves, but they also can never be part of a distributed bottleneck either. Therefore, reactions 2 and 4 can be eliminated from the list of bottleneck reactions as they cannot be part of a distributed bottleneck in combination with reaction 1 or 3. With respect to reactions 5 and 6, they are both localized bottlenecks and they are infeasible for all concentration values in the examined concentration range.

Reactions 8 and 9 need to be examined further. For this purpose, we construct all combinations of two-reaction subpathways, such that one intermediate is eliminated. Only one such combination is possible, namely, reaction 8 plus 9, leading to the elimination of 2-phosphoglycerate. In the next iteration the scaled Gibbs parameter h is calculated for the combination of reactions 8 and 9. This is found to be equal to 2.533, so that $H_{min}(8 + 9)$ is equal to 0.23 > 0. It thus is seen that the combination of reaction, 8 and 9 is a distributed bottleneck. Considered in isolation from each other, reactions 8 and 9 are feasible, *i.e.*, for each reaction there is a region in the permissible concentration range that leads to negative Gibbs free energies for either reaction and makes the forward direction thermodynamically feasible. However, their combination is not thermodynamically feasible as there is no overlap between the portions of the concentration ranges that makes them feasible. Thus, the final result for the assumed metabolite concentration range is that reactions 5 and 6 are localized bottlenecks whereas the combination of reactions 8 and 9 constitutes a distributed bottleneck. It should be noted that reaction 4, which has the largest positive standard Gibbs free energy, actually is not a bottleneck reaction at all. This clearly illustrates, as discussed in Example 14.1, that it is not possible to draw any conclusions based on the standard Gibbs free energy data alone.

We now consider three other concentration ranges:

- $c_{min} = 0.1$ mM and $c_{max} = 2$ mM
- $c_{min} = 0.02$ mM and $c_{max} = 4$ mM
- $c_{min} = 0.004$ mM and $c_{max} = 5$ mM

The results of algorithm application in these concentration ranges are given in Table 14.5. For the first concentration range examined, the upper bound is twice the value of the previous one, *i.e.*, 2 mM instead of 1 mM, and the only difference from the previous analysis is that, when we compute the scaled Gibbs free energy h for the combination of reactions 8 and 9,

TABLE 14.5 Computations of Individual Reaction Parameters for the Three
Concentration Ranges Given in the Text

	c_{min} = 0.1 mM c_{max} = 2 mM			c_{min} = 0.02 mM c_{max} = 4 mM			c_{min} = 0.004 mM c_{max} = 5 mM		
Index	h_j	H_{min}	H_{max}	h_j	H_{min}	H_{max}	h_j	H_{min}	H_{max}
1	-8.781	-11.77	-5.785	-8.781	-14.08	-3.483	-8.781	-15.91	-1.650
2	0.675	-2.320	3.671	0.675	-4.623	5.974	0.675	-6.456	7.806
3	-7.768	-10.76	-4.772	-7.768	-13.07	-2.470	-7.768	-14.90	-0.637
4	0.464	-2.532	6.455	-1.146	-6.444	9.451	-2.755	-9.886	11.51
5	3.090	0.094	6.082	3.090	-2.209	8.388	3.090	-4.041	10.22
6	3.867	0.871	7.710	3.867	-1.432	10.01	3.867	-3.264	11.85
7	-5.570	-8.565	-2.574	-5.570	-10.87	-0.271	-5.570	-12.70	1.561
8	1.790	-1.206		1.790	-3.509	7.088	1.790	-5.341	8.921
9	0.743	-2.253		0.743	-4.555	6.041	0.743	-6.388	7.874
10	-10.64	-13.63	-7.639	-10.64	-15.93	-5.336	-10.64	-17.77	0.508

the result is 2.533 and $H_{min}(8 + 9)$ = -0.463 < 0. Therefore, contrary to the previous case, the subpathway of reactions 8 and 9 can no longer be classified as a bottleneck, but neither can it be rejected because $H_{max}(8 + 9)$ = 5.529 > 0. As no further combinations are allowed between reactions 8 and 9 and any of the other reactions sharing common metabolites with them, the algorithm ends at this point with the final result that there are two localized bottlenecks, namely, reactions 5 and 6. In the second case, both the lower and the upper bounds for the concentrations of the metabolites have been relaxed. By examining the results of Table 14.5, we see that there are no longer individual reactions with H_{min} greater than zero. Therefore, no localized bottlenecks are uncovered in the first iteration. Reactions 2, 4, 5, 6, 8, and 9 remain possible targets for distributed bottlenecks, and the combinations that allow the elimination of common intermediate metabolites are 8 + 9, 4 + 5, 4 + 6, and 5 + 6. For these possibilities, one can compute:

$$h_8 + h_9 = 2.533, H_{min}(8 + 9) = -2.765 < 0, H_{max}(8 + 9) = 7.831 > 0$$

$$h_4 + h_5 = 1.944, H_{min}(4 + 5) = -3.354 < 0, H_{max}(4 + 5) = 12.54 > 0$$

$$h_4 + h_6 = 2.721, H_{min}(4 + 6) = -2.577 < 0, H_{max}(4 + 6) = 14.164 > 0$$

$$h_5 + h_6 = 6.957, H_{min}(5 + 6) = 1.659 > 0$$

It thus can be seen that the combination of reactions 5 and 6 is a distributed bottleneck, whereas combinations of reactions 8 + 9, 4 + 5, and 4 + 6 must be examined further. Because, however, no new combinations can be created from 8 + 9 and 4 + 5, they are rejected. It can be concluded that, by relaxing the concentration interval, reactions 5 and 6 were eliminated as localized bottlenecks; however, their combination arose as a distributed bottleneck. If the concentration range is relaxed further, an even greater number of candidates for the formation of distributed bottlenecks appear. The combinations are from among the reactions 2, 4, 5, 6, 7, 8, and 9 and they are formed in a manner similar to the previous case. The second iteration, however, is not sufficient because the combination 5 + 6 remains a candidate bottleneck, along with 4 + 6 and others. This requires a third iteration in which the three-reaction combination 4, 5, and two times 6 is found to be a distributed bottleneck with:

$$h_4 + h_5 + 2h_6 = 8.069, \ H_{min} = 0.938 > 0$$

If all distributed bottlenecks are to be eliminated, the lower bound can be determined to be 0.0025 mM (with the upper bound being 5 mM). In this case it is found that H_{min} for the preceding combination of reactions 4, 5, and two times 6 is -0.003 < 0. This suggests, that the composite transformation of these reactions, which corresponds to

- *fructose-2,6-bisphosphate* - 2 NAD$^+$ - 2 P$_i$ + 2 *3-phosphoglycerol phosphate*

    + 2 *NADH*

thermodynamically is very difficult as a whole. In order to take place, it requires fructose-2,6-bisphosphate to assume the maximum and 3-phosphoglycerol phosphate the minimum possible concentrations. Furthermore, it points to the crucial role of the ratio between NADH and NAD$^+$. If this ratio is too high, thermodynamically it shuts down the pathway at this precise location.

14.2.2. DETERMINATION OF $\Delta G^{0\prime}$ FROM GROUP CONTRIBUTIONS

The importance of the standard Gibbs free energy of reaction, $\Delta G^{0\prime}$, in determining the thermodynamic feasibility of reactions is apparent from the previous sections. $\Delta G^{0\prime}$ is available for several reactions (see Table 14.1), but in many instances there are no experimental data on standard Gibbs free energies of reaction (or equilibrium constants). In such cases, $\Delta G^{0\prime}$ can be determined from the corresponding standard chemical potentials, $\mu_i^0(p, T)$, or standard Gibbs free energies of formation, $\Delta G_{f,i}^{0\,\prime}$, of the reactants and products of the reaction by replacing the chemical potentials in eq. (14.12) with these quantities:

$$\Delta G^{0'} = \Sigma g_i \, \Delta G^0_{f,i} \qquad (14.32)$$

A difficulty arises in that the standard Gibbs free energies of formation are not *a priori* available for different biochemical compounds. Therefore, a method is needed to determine these Gibbs energies of formation from available experimental data. It is noted that once such parameters are determined, they can be used to estimate the $\Delta G^{0'}$ of any other biochemical reaction in which the corresponding metabolites participate.

A method that has been applied extensively for the estimation of standard Gibbs free energies, as well as other thermodynamic properties, is by group contributions. In order to estimate the properties of a particular compound, one decomposes the compound into groups whose contributions are added to yield the value of the property for the compound of interest. If a certain functional group appears more than once within the structure of the compound, then its contribution must be multiplied by its number of occurrences. In certain instances, additional operations must be performed to account for corrections for special structural features of the compound. Frequently, the contributions of the constituent groups must be added to an origin, which is a starting value used in the estimation of the property and constant for all compounds for the corresponding property. The preceding approach is expressed by eq. (14.33), which states that the Gibbs free energy of formation of a compound is the sum of contributions from an origin P_0 and the various functional groups. In eq. (14.33) P_j represents the contribution from group j and n_j represents the number of occurrences of group j in the compound:

$$\Delta G^{0'} = P_0 + \sum_j n_j P_j \qquad (14.33)$$

As indicated in eq. (14.33), the value of $\Delta G^{0'}$ can be estimated once the contributions of the various groups are known. To obtain values for the different groups, one can fit the unknown contributions P_j to experimental data for the $\Delta G^{0'}$ of a large number of compounds. As the contributions of each group to $\Delta G^{0'}$ are assumed to be linear, this operation is, in essence, a straightforward linear regression that can be carried out by available numerical algorithms. The data employed in the multilinear regression must, however, be carefully selected. First, they must be relevant for the type of compounds involved in the biotransformation of interest. Second, they must come from a variety of sources to minimize systematic errors. Third, they must be screened to ensure that they represent the contributions of all possible functional groups in the molecular structure of the compounds. Finally, corrections will have to be introduced to account for group interactions, such as interaction between a nitrogen and an adjacent carbonyl group. In deciding about the exact types of groups that should be represented, one should attempt to include as many such groups as possible in order to represent the effect of all possible biochemical components on the property under consideration. There is a limit, of course, to the extent such groups can be further decomposed as some of them simply are not decomposable, whereas some others could be decomposed, but this would result in large errors. In such cases, the smallest compounds will be represented as consisting of a special single group.

Frequently in biological transformations, certain sets of complex biochemical compounds with important metabolic roles appear together, $e.g.$, cofactor pairs like NAD^+/NADH or ATP/ADP. In such cases, the pairs can be represented by a single group that stands for the transformation of NAD^+ to NADH or ATP to ADP, respectively. By considering such pairs as single groups, the complex calculations involved in the determination of the Gibbs free energy of formation of the corresponding compounds is carried out once, and the resulting difference is incorporated as a single group in the calculation of the $\Delta G^{0\prime}$ of the reaction in which the groups participate. In this way, calculations are greatly simplified, and, furthermore, errors that could result from the complex structure of the corresponding compounds are minimized. In the development and application of the group contribution method, all compounds are assumed to be in their common state in aqueous solution. Thus, amine groups normally have a protein attached ($R\text{-}NH_3^+$ rather than $R\text{-}NH_2$), carboxylic acids are in their anion form ($R\text{-}CO\text{-}O^-$ rather than $R\text{-}CO\text{-}OH$), amino acids are in their zwitterion forms, etc.

Contributions of various groups have been determined for a large number of relevant biochemical compounds. These contributions are listed in Tables 14.6-14.12, obtained from the work of Mavrovouniotis (1990, 1991), from which most of the discussion in this section has been adopted. Additionally, because the decomposition into groups and the application of various corrections are especially cumbersome for many biochemical compounds, the Gibbs free energies of formation of a large number of cyclic compounds have been precomputed and are listed in Table 14.13. These numbers can be used directly in the determination of $\Delta G^{0\prime}$ of any reaction in which the corresponding compound participates.

TABLE 14.6   Contributions of Groups with One Free Single Bond

Group	Contribution (kJ mol^{-1})
-CH$_3$	35.6
-NH$_2$	46.0
-OH (attached to benzene aromatic ring)	-130.5
-OH (primary)	-119.7
-OH (secondary)	-131.4
-OH (tertiary)	-125.1
-NH$_3^+$	20.1
-CH = O	-72.4
-SH	58.6
-COO$^-$	-298.7
-PO$_3^{2-}$	42.7
-SO$_3^-$	-439.7
-CO-OPO$_3^{2-}$	-301.7
-OPO$_3^{2-}$ (primary)	-120.5
-OPO$_3^{2-}$ (secondary)	-123.0
-OPO$_3^{2-}$ (tertiary)	-104.2

TABLE 14.7   Contributions of Groups with Two or More Bonds, Not Participating in a Ring

Open-chain group	Contribution (kJ mol^{-1})
$\equiv$ C-	100.4
$\equiv$ N	64.9
$\equiv$ CH	151.9
= C <	18.4
-CH =	46.0
= CH$_2$	79.5
-NH$_2^+$	3.3
= NH	59.8
> C <	-58.6
-CH <	-22.6
> N-	24.3
-NH$^+$ <	33.5
-CH$_2$-	7.1
> CO	-114.2
-O-PO$_2$-	-21.8
-S-S-	24.7
-CO-O$^-$	-307.9
-S-	39.7
> NH$_2^+$	28.9
-O- (attached to -O-PO$_2$- group)	-102.9
-O-	-94.1
-NH-	32.6

TABLE 14.8   Contributions of Groups Participating in One Ring[a]

Ring group	Contribution (kJ mol^{-1})
$> N =^+$ (a single and a double bond in a nonaromatic ring)	-0.4
$> C$- (in one benzene ring)	4.6
$> C =$ (two single bonds in a nonbenzene ring)	95.4
$> C =$ (a single bond and a double bond in a nonbenzene ring)	33.1
$> C <$	-57.3
$= N$-	43.9
$> CH$ (in one benzene ring)	36.0
$-CH =$	40.2
$-N <$	31.0
$-CH <$	-10.8
$-O$-CO-	-227.2
$-CO$-	-114.6
$-O$-	-100.8
$-NH$-	41.4
$-CH_2$-	26.4
$-O$-PO_2^{1-}	65.3

[a] The rings in which the compounds participate are **not** aromatic, unless it is otherwise indicated.

TABLE 14.9   Contributions of Groups Participating in Two Fused Rings

Two-ring group	Contribution (kJ mol^{-1})
$> C$- (in two fused benzene rings)	9.6
$-N <$ (in two fused nonbenzene rings)	79.9
$-CH <$ (in two fused nonbenzene rings)	-5.4
$> C =$ (in two fused nonbenzene rings)	70.7

TABLE 14.10   The Origin and Additional Corrections to the Gibbs Energy of Formation in Aqueous Solution[a]

Correction	Contribution (kJ mol^{-1})
Origin	-103.3
Hydrocarbon molecule (*i.e.*, containing only C and H)	16.7
Each aromatic ring	-25.1
Each three-carbon ring	123.4
Each amide (*i.e.*, each nitrogen attached to a carbonyl group)	-43.9
Each heteroaromatic ring (containing nitrogen, sulfur, or oxygen)	-24.7

[a] Note that the Gibbs energy of formation has been precomputed for many ring compounds (see Table 14.13).

TABLE 14.11   Gibbs Energy of Small Compounds That Were Not Broken into Groups

Formula	Name	Gibbs energy (kJ mol^{-1})
$^-$OCO-COO$^-$	Oxalate	-668.6
CH_4	Methane	-34.3
HCOO$^-$	Formate	-355.2
$CH_2 = O$	Formaldehyde	-151.0
H_2CO_3	Hydrated carbon dioxide	-623.0
HCO_3^-	Bicarbonate	-586.6
$HP_2O_7^{3-}$	Pyrophosphate	-240.2
HPO_4^{2-}	Phosphate	-249.4
SO_3^{2-}	Sulfite	-495.4
NO_3^-	Nitrate	-115.1
NO_2^-	Nitrite	-29.3
NH_4^+	Ammonium	-75.7
H_2O	Water	-236.8
OH$^-$	Hydroxide anion	-197.1
H$^+$	Hydrogen cation	-39.7

TABLE 14.12   Gibbs Energy Differences for Special Pairs of Compounds

Group	Contribution (kJ mol^{-1})
NADH (reduced) *minus* NAD$^+$ (oxidized)	19.8
NADPH (reduced) *minus* NADP$^+$ (oxidized)	19.8
Substituted coenzyme-A *minus* SH-coenzyme-A (to calculate the Gibbs energy difference, the contributions of the substituent's groups must be added to the correction)	-55.2
Pyocyamine reduced *minus* pyocyamine oxidized	-74.5

EXAMPLE 14.3

Gibbs Free Energy for the Formation of Glutamate

The syntactic formula of glutamate is shown in Fig. 14.2a. In order to calculate the Gibbs free energy of formation, this formula is broken down into groups as shown in Fig. 14.2b, and Table 14.14 summarizes the calculation details for the contribution of the various groups multiplied by the number of occurrences of each group. An origin taken from Table 14.10 has been used, and no special corrections have been incorporated. The result

TABLE 14.13 Estimated Standard Gibbs Free Energies of Formation of Selected Classes of Cyclic Compounds[a]

Compound	Standard Gibbs energy of formation in aqueous solution (kJ mol^{-1})
Derivatives of coenzyme A (relative Gibbs energies)[b]	
Acetyl coenzyme A	-138.5
Oxalyl coenzyme A	-472.8
Acrylyl coenzyme A	-48.5
Propionyl-coenzyme A	-131.4
Malonyl-coenzyme A	-465.7
Crotonoyl-coenzyme A	-46.0
Acetoacetyl-coenzyme A	-245.2
Ketobutyryl-coenzyme A	-245.2
Butyryl-coenzyme A	-124.3
Methylmalonyl-coenzyme A	459.4
Succinyl-coenzyme A	-458.6
Alanyl-coenzyme A	-139.7
3-Hydroxybutyryl-coenzyme A	-284.9
Ketohexanoyl-coenzyme A	-231.0
Hexanoyl-coenzyme A	-110.0
3-Hydroxyhexanoyl-coenzyme A	-270.7
Hydroxycrotonyl-coenzyme A	-299.6
Palmitoyl-coenzyme A	-38.5
Nucleosides and bases	
Uracil	-282.0
4,5-Dihydrouracil	-284.9
Hypoxanthine	81.2
Inosine	81.2
Thymine	-253.6
Adenine	321.3
Xanthine	-120.1
Guanine	120.1
Uric acid	-365.3
Orotate	-587.9
Cytosine	-41.8
Thymidine	-614.6
2'-Deoxyadenosine	-39.7
Adenosine	-207.9
Guanosine	-409.2
Nucleotides	
2'-Deoxycytidine monophosphate (dCMP)	-403.3
Uridine monophosphate (UMP)	-812.1
3',5'-Cyclic adenosine monophosphate (cAMP)	37.7
Cytidine monophosphate (CMP)	-572.0
3',5'-Cyclic guanoside monophosphate (cGMP)	-163.6

(*continues*)

TABLE 14.13 (*continued*)

Compound	Standard Gibbs energy of formation in aqueous solution (kJ mol⁻¹)
Nucleotides	
2'-Deoxythymidine monophosphate (dTMP)	-615.0
Inosine monophosphate (IMP)	-448.9
Thymidine monophosphate (TMP)	-783.7
Adenosine monophosphate (AMP)	-208.8
Guanosine monophosphate (GMP)	-409.6
2'-Deoxycytidine diphosphate (dCDP)	-425.5
Uridine diphosphate (UDP)	-834.3
Cytidine diphosphate (CDP)	-593.7
2'-Deoxythymidine diphosphate (dTDP)	-637.2
Inosine diphosphate (IDP)	-471.1
Thymidine diphosphate (TDP)	-805.4
Adenosine diphosphate (ADP)	-230.5
Guanosine diphosphate (GDP)	-431.8
Uridine triphosphate (UTP)	-856.0
Cytidine triphosphate (CTP)	-615.9
Inosine triphosphate (ITP)	-492.9
Thymidine triphosphate (TTP)	-827.6
Adenosine triphosphate (ATP)	-252.7
Guanosine triphosphate (GTP)	-453.5
Derivatives of nucleosides and nucleotides	
UDP-glucuronate	-1685.7
UDP-glucose	-1499.5
CDP-glucose	-1259.4
GDP-galactose	-1097.5
GDP-glucose	-1097.5
GDP-mannose	-1097.5
Adenosylhomocysteine	-335.1
Adenylosuccinate	-835.1
Adenylyl-α-aminoadipate	-697.9
Tetrahydrofolate and derivatives	
7,8-Dihydrofolate	-310.0
Tetrahydrofolate	-356.5
N^5,N^{10}-Methenyltetrahydrofolate	-286.6
N^5,N^{10}-Methylenetetrahydrofolate	-268.6
N^{10}-Formyltetrahydrofolate	-480.3
N^5-Formyltetrahydrofolate	-482.8
N^5-Methyltetrahydrofolate	-331.0
N^5-Formiminotetrahydrofolate	-317.6

TABLE 14.13

Compound	Standard Gibbs energy of formation in aqueous solution (kJ mol^{-1})
Cyclic sugars	
Arabinose	-746.0
Lyxose	-746.0
Ribose	-753.5
Ribulose	-746.8
Xylose	-746.0
Xylulose	-746.8
6-Deoxygalactose	-747.3
Fuculose	-758.6
Rhamnose	-747.3
Rhamnulose	-758.6
Allose	-895.8
Altrose	-895.8
Fructose	-906.7
Galactose	-895.8
Glucose	-895.8
Gulose	-895.8
Idose	-895.8
Inositol	-955.6
Mannose	-895.8
Psicose	-906.7
Sorbose	-906.7
Tagatose	-906.7
Talose	-895.8
Cellobiose	-1519.6
α-Lactose	-1519.6
β-Maltose	-1519.6
Sucrose	-1536.4
Trehalose	-1519.6
Cellotriose	-2143.0
Sugar phosphates	
2'-Deoxyribose-5'-phosphate	-586.2
Arabinose-5-phosphate	-754.4
Ribose-1-phosphate	-745.6
Ribose-5-phosphate	-754.4
Fuculose-1-phosphate	-761.5
Rhamnulose-1-phosphate	-761.5
Glucuronate-1-phosphate	-1073.6
Fructose-1-phosphate	-907.5
Fructose-6-phosphate	-907.5
Galactose-1-phosphate	-887.4

(*continues*)

TABLE 14.13 (*continued*)

Compound	Standard Gibbs energy of formation in aqueous solution (kJ mol^{-1})
Sugar phosphates	
Glucose-1-phosphate	-887.4
Glucose-6-phosphate	-896.6
Mannose-1-phosphate	-887.4
Mannose-6-phosphate	-896.6
Fructose-1,6-diphosphate	-908.3
Glucose-1,6-diphosphate	-888.3
Sucrose-6-diphosphate	-1537.2
Hydrocarbons	
Cyclopropane	115.5
Benzene	130.1
Cyclo-hexane	70.7
Naphthalene	221.3
Simple alcohols, aldehydes, and ethers	
Furan	-68.6
Tetrahydrofuran	-99.2
Phenol	-48.5
1,4-Dioxane	-200.0
Benzaldehyde	9.6
Methyl phenyl ketone	3.8
Cyclohexanol	114.2
Other compounds	
Pyridine	116.3
Niacine	-189.1
δ-Pyrroline-5-carboxylate	-276.1
Nicotinamide	-2.5
Pyrrolidone carboxylate	-477.4
Urocanate	-136.0
Creatinine	-28.9
Dihydroorotate	-620.4
2,3-Dihydrodipicolinate	527.6
5-Dehydroshikimate	-701.2
1-Piperidine-2,6-dicarboxylate	-555.6
Histidine	-223.4
Shikimate	-728.4
Gluconic acid γ-lactone	-892.0
Gluconic acid δ-lactone	-880.3
Histidinol	-37.7
5-Dehydroquinate	-930.9
Fructuronate	-1092.9
Glucuronate	-1081.6

TABLE 14.13

Compound	Standard Gibbs energy of formation in aqueous solution (kJ mol⁻¹)
Other compounds	
Imidazole acetol phosphate	-149.8
Phenylalanine	-212.1
Quinate	-958.1
Tyrosine	-374.0
δ-Gluconolactone-6-phosphate	-880.7
Histidinol phosphate	-38.1
7,8-Dihydrobiopterin	-107.1
5,6,7,8-Tetrahydrobiopterin	-153.6
Pteroate	113.8
Pteroyl glutamate	-318.4

[a] Within each class, the components are in increasing order of total number of atoms.
[b] The energies listed for CoA derivatives use coenzyme A as a reference point. In effect, the energy shown is equal to the Gibbs energy of each derivative minus the Gibbs energy of coenzyme A.

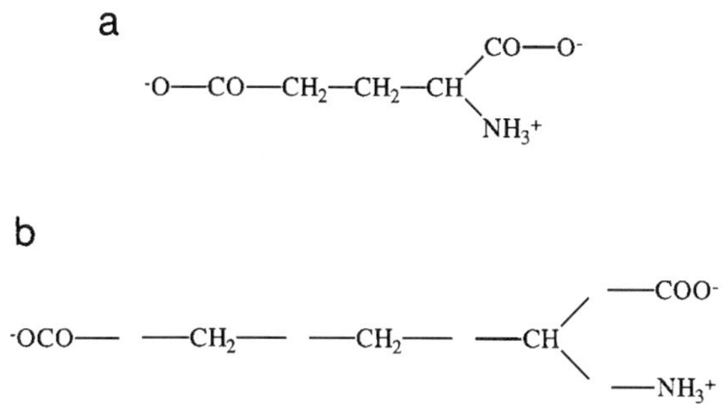

FIGURE 14.2   The structure of glutamate (a) and the decomposition of the structure into groups (b).

is a value of -689 kJ mol⁻¹ which deviates by 10 kJ mol⁻¹ from the literature value of 699 kJ mol⁻¹.

EXAMPLE 14.4

Gibbs Free Energy for the Formation of ATP

ATP is a complex cyclic compound for which the structure is given in Fig. 14.3. The structure is broken into groups as shown in Fig. 14.4. Note

TABLE 14.14   Calculation of Gibbs Free Energy of Formation of Glutamate from
Contributions of Groups

Group or correction	Number of occurrences	Contribution (kJ mol^{-1})	Source	Total contribution
Origin	1	-103.3	Table 14.10	-103.3
-NH$_3^+$	1	20.1	Table 14.6	20.1
-COO$^-$	2	-298.7	Table 14.6	-597.4
-CH$_2$-	2	7.1	Table 14.7	14.2
-CH <	1	-22.6	Table 14.7	-22.6
Total				-689.0

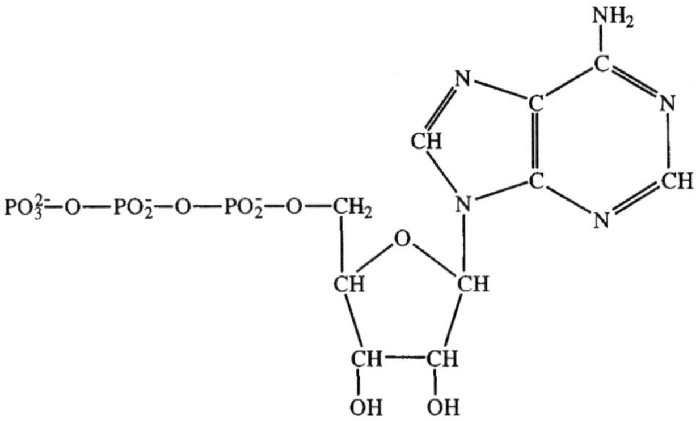

$$PO_3^{2-}-O-PO_2^--O-PO_2^--O-CH_2$$

FIGURE 14.3   The structure of ATP.

the classification of hydroxyl and phosphate groups as primary, secondary, or tertiary, the participation of bonds in rings, and the correction for two heteroaromatic rings identified in Fig. 14.4. Table 14.15 provides a summary of the calculation of the Gibbs free energy from these contributions. The result, 253.5 kJ mol^{-1}, is close to the precomputed value of -252.7 kJ mol^{-1} given in Table 14.13 (the small deviation is due to round-off error).

It should be clear that the most demanding task in the application of the group contribution method is the determination of the constituent groups of a specific compound. This becomes particularly challenging for the case of complex molecules such as ATP. In this regard, the energies of Table 14.13 are very useful as they provide a reference point to which a compound similar to the one given in the table can be related. To use this table, one

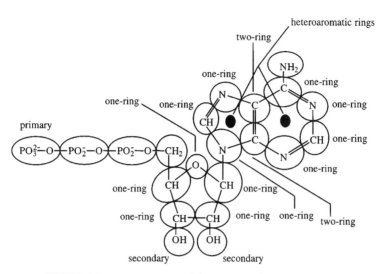

FIGURE 14.4   Decomposition of the structure of ATP into groups.

TABLE 14.15   Calculation of the Gibbs Free Energy of Formation of ATP from
Contributions of Groups

Group or correction	Number of occurrences	Contribution (kJ mol⁻¹)	Source	Total contribution
Origin	1	-103.3	Table 14.10	-103.3
-NH$_2$	1	46.0	Table 14.6	46.0
-OPO$_3^-$	1	-120.5	Table 14.6	-120.5
-OH secondary	2	-131.4	Table 14.6	-262.8
-CH$_2$-	1	7.1	Table 14.7	7.1
-OPO$_2$-	2	-21.8	Table 14.7	-43.6
Ring-O-	1	-100.8	Table 14.8	-100.8
Ring-CH <	4	-10.8	Table 14.8	-43.6
Ring-N <	1	31.0	Table 14.8	31.0
Ring-CH =	2	40.2	Table 14.8	80.4
Ring = N-	3	43.9	Table 14.8	131.7
Ring > C =	1	33.1	Table 14.8	33.1
Two-ring > C =	2	70.7	Table 14.9	141.4
Heteroaromatic ring	2	-24.7	Table 14.10	-49.4
Total				-253.5

needs to identify only the difference between the compound at hand and a similar one listed in Table 14.13 and then add or subtract the appropriate contribution corresponding to such differences.

EXAMPLE 14.5

Calculation of $\Delta G^{0\prime}$ for the Reaction Catalyzed by Alcohol Dehydrogenase

To illustrate how one can calculate the Gibbs free energy of a reaction directly from changes in group contributions, consider the reaction catalyzed by alcohol dehydrogenase. In this reaction, ethanol is converted to acetaldehyde with concurrent conversion of NAD^+ to NADH. As illustrated in Fig. 14.5, the reaction can be decomposed into various groups. The pair $NADH/NAD^+$ is considered as a single group. The calculation is shown in Table 14.16, which ignores the contributions of the origin and the group CH_3 because both are the same in ethanol and acetaldehyde. Their net result is, therefore, equal to zero. The value for the $\Delta G^{0\prime}$ of this reaction is calculated to be 20.2 kJ mol^{-1}.

The group contribution method presented has broad applicability because it provides the contribution of a comprehensive set of groups. For the data

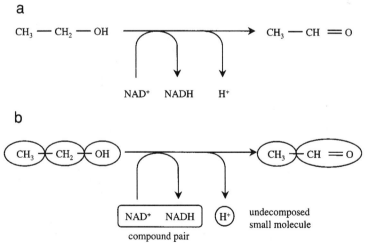

FIGURE 14.5 The reaction catalyzed by alcohol dehydrogenase (a) and the reaction decomposed into groups, so that its Gibbs energy can be estimated (b).

TABLE 14.16   Calculation of Gibbs Free Energy of the Reaction Catalyzed by
Alcohol Dehydrogenase from Contributions of Groups

Group or correction	Number of occurrences	Contribution (kJ mol^{-1})	Source	Total contribution
Origin	0			
H$^+$	1	-39.7	Table 14.11	-39.7
NADH-NAD$^+$	1	19.8	Table 14.12	19.8
-CH$_3$	0			
-CH$_2$-	-1	7.1	Table 14.7	-7.1
-OH primary	-1	-119.7	Table 14.6	119.7
-CH = O	1	-72.4	Table 14.7	-72.4
Total				20.2

points used in the regression, the error is estimated to be less than 8 kJ mol^{-1} for 85% of the data points and 21 kJ mol^{-1} for 95% of the data points. This is an acceptable first approximation to Gibbs energies and equilibrium constants in biochemical systems, and the approach therefore is very useful in thermodynamic analysis of biochemical pathways.

14.3. NONEQUILIBRIUM THERMODYNAMICS

As mentioned previously, cellular systems are open systems where many processes operate far from equilibrium. If they were at equilibrium, there would be no flow through the cellular pathways and the cells would stop functioning. This can be illustrated with the process of passive (free diffusive) transport of species across cellular membranes, which is the mechanism by which many species are transported to and from the surroundings of the cells. At equilibrium, the concentrations of species on either side of the membrane are the same and there is no flow. With a concentration gradient, there is a force that drives the flow of the compound down the concentration gradient. In the case of passive transport, we have a thermodynamic driving force, i.e., the concentration gradient, and there is a direct relationship between the driving force and the resulting flow, which is normally referred to as *conjugate flow*. In most cellular processes flows are conjugate, but on several occasions a thermodynamic force may promote a *nonconjugate flow*. An example of this is the phosphorylation of ADP to ATP in oxidative phosphorylation, which is driven by the proton gradient across the inner mitochondrial membrane of eukaryotes (or the cytoplasmic membrane of prokaryotes). This thermodynamic stimulation of a nonconjugate flow is often referred to as *energy transduction*.

Nonequilibrium thermodynamics is an extension of classic thermodynamics into nonequilibrium states. Its main concern is the relationship between flows (such as rates of conversion or reaction rates) and thermodynamic driving forces, for both conjugate and nonconjugate flows. Additionally, nonequilibrium thermodynamics concerns itself with the *interactions* between different cellular processes, *e.g.*, the individual processes of the oxidative phosphorylation where energy transduction plays an important role. The foundation of nonequilibrium thermodynamics is flow-force relationships, where a flow is obtained in response to and, furthermore, is specified as a function of the thermodynamic driving force. The driving force normally is quantified by the affinity A_i (in kilojoules per mole), which, for a chemical reaction, is equal to minus the free energy change of the reaction. Thus, for the ith reaction:

$$A_i = -\Delta G_i = -\sum_{j=1}^{C} g_{ij} \mu_j \qquad (14.34)$$

where g_{ij} and μ_j are, respectively, the stoichiometric coefficient and chemical potential of the jth compound. Any other thermodynamic force, *e.g.*, the proton-motive force of the oxidative phosphorylation (see Example 14.6), may play the role of the driving force, and in the following discussion we will use the variable A_i to express any type of such driving force. As discussed in the previous section, a spontaneous reaction requires a negative free energy change, which is associated with $A_i \geq 0$. However, due to thermodynamic coupling, not all processes require a nonnegative affinity for the process, as they may be driven by other processes that have a positive affinity. An example is the triosephosphate isomerase-catalyzed reaction in the EMP pathway, which has a negative affinity (or a positive free energy change, see Table 14.3). However, the conversion of dihydroxyacetone phosphate to glyceraldehyde-3-phosphate does take place because it is driven by other reactions downstream in the pathway. Thus, the specific requirement of $A_i \geq 0$ for the spontaneity of a bioprocess i should be replaced with the requirement that the *overall affinity*, resulting from all individual processes driving the overall bioprocess i, must be nonnegative. In evaluating the overall affinity of process i, A_i, also called *the dissipation function* Φ_i (kilojoules per hour), it is important to take into account the relative rates of the individual processes driving process i:

$$A_i = \Phi_i = \sum_{j=1}^{J} v_j A_j \geq 0 \qquad (14.35)$$

where v_j is the rate of the jth reaction or process (in moles per hour), whose affinity A_{ij} contributes to the overall affinity (or driving force) of process i. Notice that the dissipation function is a function of the reaction rates and therefore introduces time. Furthermore, it represents the overall dissipation from many different processes and, therefore, allows some processes with individual negative affinities (or positive free energy change) to operate, as long as the dissipation function remains nonnegative. These two aspects represent a major departure from equilibrium thermodynamics, where the feasibility of each reaction is evaluated separately on the basis of its own free energy change.

To illustrate the thermodynamic coupling expressed in the dissipation function, let us consider a reaction consisting of two partial reactions, such as the hexokinase reaction of the EMP pathway. The first partial reaction is phosphorylation of glucose to glucose-6-phosphate:

$$glucose\text{-}6\text{-}phosphate \text{ - } glucose \text{ - } P_i = 0 \qquad (14.36)$$

and the second partial reaction is the hydrolysis of ATP to ADP and orthophosphate:

$$ADP + P_i \text{ - } ATP = 0 \qquad (14.37)$$

The first reaction has a standard free energy change of 14.8 kJ mol^{-1} whereas the second reaction has a standard free energy change of -30.5 kJ mol^{-1}. Even at physiological conditions the free energy change (ΔG) for the first reaction is positive (corresponding to a negative affinity), and therefore it cannot run spontaneously in the forward direction. However, through coupling with the second reaction, there may be an overall dissipation of free energy. The total dissipation function is

$$\Phi = v_{ATP} A_{ATP} + v_{glc} A_{glc} \qquad (14.38)$$

or, with the introduction of the free energy eqs. (14.12) for the affinities:

$$\Phi = v_{ATP}(\mu_{ATP} - \mu_{ATP} - \mu_{P_i}) + v_{glc}(\mu_{glc} + \mu_{P_i} - \mu_{glc6P}) \qquad (14.39)$$

The rates of the two reactions are not independent as hexokinase catalyzes the phosphorylation of glucose to glucose-6-phosphate only when the other reaction runs simultaneously. Thus, $v_{glc} = v_{ATP}$ and eq. (14.39) reduces to

$$\Phi = v_{glc}(\mu_{glc} + \mu_{ATP} - \mu_{ADP} - \mu_{glc6P}) \qquad (14.40)$$

which is positive at all physiological conditions.

The preceding example of the hexokinase reaction illustrates thermodynamic coupling of two reactions that normally are considered as one overall reaction. There are many similar examples, however, it is more interesting to consider thermodynamic coupling of processes (reactions) that are independent, i.e., catalyzed by different enzymes. Examples are the oxidative phosphorylation, where the individual processes are coupled via thermodynamic constraints (see Example 14.6), or intracellular reactions coupled with cofactor conversion, e.g., NAD^+ to NADH, the latter providing the means for thermodynamic coupling between many different pathways. Furthermore, two consecutive reactions in a pathway also are thermodynamically coupled via the intermediate, which affects, through its concentration, the affinity of both the reaction forming it and the reaction utilizing it. These different thermodynamic couplings are quantified by the so-called *phenomenological equations*, which relate the forces with the driving forces as follows:

$$v_i = \sum_{j=1}^{J} L_{ij} A_j \qquad (14.41)$$

where L_{ij} are phenomenological coefficients. It is observed that each flow v_i (or reaction rate) is related to its conjugate driving force A_i through a direct coefficient L_{ii} and to any other driving force A_j through a cross-coefficient L_{ij}.

The phenomenological equations are analogous to other linear relationships between flow and forces, e.g., Ohm's law for flow of electric current, Fick's law of diffusion, and Poisseseuile's law for fluid flow, etc. Furthermore, the phenomenological equations are extended to cover all possible cross-effects and interferences. They are strictly valid only for slow processes, i.e., for processes operating sufficiently close to equilibrium (see Box 14.3). Faster reactions require the addition of higher order terms [see Westerhoff and van Dam (1987) for details], which limits the practical application of the equations. However, for many cellular processes, especially transport processes, linearity often is observed over a surprisingly wide range of magnitudes of the driving forces (Caplan, 1971). For chemical reactions the criterion usually is more restrictive, but, as we will demonstrate later, the linearity often is a good approximation.

The phenomenological coefficients are functions of the (kinetic, transport, and other) parameters of the system and, as such, are not constant (see Box 14.3 and later). They cannot attain arbitrary values, however, as they are subject to important thermodynamic constraints as illustrated in the classical work of Onsager (1931). First, the matrix of coefficients is symmetrical, so that

$$L_{ij} = L_{ji}; \quad i \neq j \tag{14.42}$$

These are commonly referred to as *Onsager's reciprocal relations* (or simply Onsager's law), which state that the first-order relative effect of any force A_j on any flow v_i is identical to the first-order relative effect of the inverse force-flow pair. The reciprocal relations strictly hold only close to equilibrium (see Box 14.3). Second, the direct coefficients can never be negative, i.e., $L_{ii} \geq 0$, and the cross-coefficient must satisfy the following condition:

$$L_{ij}^2 \leq L_{ii} L_{jj}; \quad i \neq j \tag{14.43}$$

These two constraints on the phenomenological coefficients can be derived from the requirement that the dissipation function be nonnegative at all conditions [see Westerhoff and van Dam (1987) for details].

Phenomenological equations may be employed to describe interactions between many different processes, but often only a two flow coupled system is considered:

$$v_1 = L_{11} A_1 + L_{12} A_2 \tag{14.44a}$$

$$v_2 = L_{21} A_1 + L_{22} A_2 \tag{14.44b}$$

where the reciprocity implies that $L_{12} = L_{21}$. Typically, one of the flows is driven against a back pressure by the free energy supplied by the other flow. In this case the flow that supplies the free energy is considered as an *input flow*, and the flow that is driven by the input flows through expenditure of some free energy supplied by these flows is the *output flow*. For such a system, the cross-coefficient L_{12} quantifies the extent of coupling between the input and the output processes. However, the magnitude of L_{12} *per se* is not a good measure of coupling. The latter is better defined by the magnitude of the cross-coefficient relative to the direct coefficients. The so-called *degree of coupling*, q, has been defined to express this as follows:

$$q = \frac{L_{12}}{\sqrt{L_{11} L_{22}}} \tag{14.45}$$

As a consequence of eq. (14.43), the absolute magnitude of the degree of coupling is always less than or equal to 1. When $q = 1$ the system is completely coupled, i.e., the two flows are proportional to one another, and the system is best expressed as a single stoichiometric process. When $q = 0$ the two processes are completely uncoupled and therefore are of no interest as far as energy conversion is concerned.

BOX 14.3

Flow-Force Relations and Reciprocity

In this note we elaborate further on the validity of the linear flow-force relationships of eq. (14.42) and Onsager's reciprocal relations of eq. (14.43).

Flow-force relations

To illustrate the derivation and validity of the linear flow-force relationships, we consider a simple chemical reaction for the conversion of substrate S to product P:

$$P - S = 0 \tag{1}$$

for which the net rate of forward reaction is given by

$$v = k_1 c_S - k_{-1} c_P \tag{2}$$

From the definition of the chemical potential [eq. (14.10)], we can specify the concentrations of S and P as

$$c_i = c_{i,ref} \exp\left(\frac{\mu_i - \mu_i^{ref}}{RT}\right), \qquad I = S, P \tag{3}$$

which, when inserted in eq. (2), yield:

$$v = k_1 c_{S,ref} \exp\left(\frac{\mu_S - \mu_S^{ref}}{RT}\right) - k_{-1} c_{P,ref} \exp\left(\frac{\mu_P - \mu_P^{ref}}{RT}\right) \tag{4}$$

This is a complete nonequilibrium thermodynamic description of the chemical reaction, which gives the reaction rate as a function of the chemical potentials of the substrate and product along with a set of kinetic parameters. If we consider concentrations in regions close to the reference state, where

$$|\mu_S - \mu_S^{ref}| \ll RT \gg |\mu_P - \mu_P^{ref}| \tag{5}$$

we can use the approximation $e^x \approx 1 + x$ for $x \ll 1$, whereby eq. (4) can be approximated by

$$v = \frac{k_1 c_{S,ref}}{RT}\left(\mu_S - \mu_S^{ref}\right) - \frac{k_{-1} c_{P,ref}}{RT}\left(\mu_P - \mu_P^{ref}\right) + k_1 c_{S,ref} - k_1 c_{P,ref} \tag{6}$$

If the reference state is taken to be the equilibrium, then $k_1 c_{S,ref} = k_{-1} c_{P,ref}$. Also, the chemical potentials of the substrate and product at equilibrium are the same. Equation (6) therefore reduces to

$$v = \frac{k_1 c_{S,ref}}{RT}(\mu_S - \mu_P) = L(\mu_S - \mu_P) \qquad (7)$$

which demonstrates the validity of the flow-force relation for the simple chemical reaction eq. (1) operating close to equilibrium. It is seen that the phenomenological coefficient is a function of the kinetic parameters of the reaction, and therefore they cannot be considered as constant, especially for biological systems where the kinetic parameters depend on the *in vivo* enzyme activity. Westerhoff and van Dam (1987) demonstrated that the flow-force relations could be derived for many other systems as well, e.g., coupled reactions and transport processes. In the derivation of eq. (7), the near-equilibrium requirement can be quantified, namely, through eq. (5), which has to be satisfied. Because RT is 2.48 kJ mol^{-1} at 25°C, it is seen from Table 14.3 that several of the reactions in the EMP pathway have free energy changes smaller than RT (or quite close to it), and the linear flow-force relations therefore may be applied to them. However, for reactions where the free energy change is large (the thermodynamically irreversible reactions), the near-equilibrium assumption obviously is a poor one, and caution is advised in the application of the linear flow-force relations (see also discussion later).

Reciprocity

Cellular growth is the outcome of a large number of processes, including transport of species and chemical reactions. Many of these processes are coupled whereas others are independent, and the system therefore can be considered as a set of independent subprocesses, where each subprocess may consist either of a single process or of two or more coupled processes. In analyzing the overall growth process, one often uses a black box approach where steady state is assumed for all the individual subprocesses. In this type of analysis, it is interesting to know whether Onsager's reciprocal relations apply to the overall conversion of substrate to metabolic products and biomass (or any other process that represents the sum of several subprocesses). Westerhoff and van Dam (1987) considered this aspect and showed that *if reciprocity applies to each set of coupled processes, then it also applies to the overall process.*

(continues)

(continued)

To illustrate the strict requirement of near-equilibrium conditions for the validity of reciprocity for flow-force relations, we again consider the simple chemical reaction (1) with the net forward reaction rate given by eq. (2). With this kinetics the rate of formation of P, v_P, is equal to v, whereas the rate of formation of S is $v_S = -v$. We can then derive so-called generalized phenomenological coefficients given by

$$L_{SP}^{gen} = \frac{\partial v_S}{\partial \mu_P} = \frac{k_{-1} c_P}{RT} \tag{8}$$

$$L_{PS}^{gen} = \frac{\partial v_P}{\partial \mu_S} = \frac{k_1 c_s}{RT} \tag{9}$$

(Hint: Use the chemical potentials, eq. (14.10), to derive μ_i and then determine the derivatives). Obviously, the two generalized phenomenological coefficients are identical only at equilibrium where the net rate of reaction is zero (hence, $k_1 c_S = k_{-1} c_P$). Away from equilibrium, the two generalized phenomenological coefficients do not obey the reciprocal relation, and their difference is equal to the net rate of reaction divided by RT. Westerhoff and van Dam (1987) derived generalized phenomenological coefficients for chemical reactions with more complex kinetics and for two partially coupled processes, and, in both cases, the preceding conclusion is valid: the reciprocal relations only hold at (or very close to) equilibrium. Despite this fact, reciprocal relations are, however, found empirically for some systems operating far from equilibrium (Rottenberg, 1979), which indicates a more general applicability of Onsager's reciprocal relations.

An interesting variable for the two-flow system is the so-called *flow ratio*, which is the ratio of the output flow to the input flow:

$$j = \frac{v_1}{v_2} = \frac{L_{11} A_1 + L_{12} A_2}{L_{12} A_1 + L_{22} A_2} \tag{14.46}$$

where v_1 is taken to be the output flow (with A_1 being the corresponding output force). Rearrangement of eq. (14.46) gives

$$j = Z\frac{(Z\chi) + q}{q(Z\chi) + 1} \tag{14.47}$$

where Z is the *phenomenological stoichiometry* (Westerhoff and van Dam, 1987) given by

$$Z = \sqrt{\frac{L_{11}}{L_{22}}} \tag{14.48}$$

and χ is the *force ratio* given by the ratio of the output force to the input force:

$$\chi = \frac{A_1}{A_2} \tag{14.49}$$

Because the forces normally have opposite signs (a process with a positive force drives a process with a negative force), the force ratio generally is negative. For $q = \pm 1$, *i.e.*, complete coupling, the flow ratio becomes equal to Z, and the phenomenological stoichiometry becomes identical to the mechanistic stoichiometry. For $q \neq \pm 1$, this is not generally true, even though Z often is associated with the mechanistic stoichiometry in oxidative phosphorylation (see Example 14.6). Equation (14.47) has been the basis for quantification of the *operational stoichiometry* (equal to the flow ratio) for a number of coupled processes, and Fig. 14.6 shows the flow ratio (relative to Z) as a function of the force ratio for various degrees of coupling. It is seen that the nearer q is to 1 (and the same holds for q close to -1), the weaker

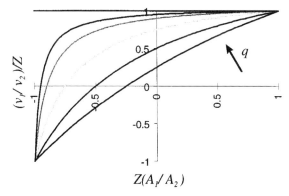

FIGURE 14.6 The flow ratio relative to the phenomenological stoichiometry $[j/Z = (v_1/v_2)/Z]$ for the free energy transducer described by eq. (14.47), as a function of the normalized force ratio $[Z\chi = Z(A_1/A_2)]$ at different degrees of coupling (q).

the dependence of the flow ratio on the force ratio. When q is zero, the flow ratio is proportional to the force ratio, i.e., each flow is proportional to its conjugate force without any influence from the other force. Through measurements of the force ratio and the degree of coupling, the operational stoichiometry can be specified as a function of Z. Furthermore, from conditions where there is no output flow ($v_1 = 0$, the flow ratio is zero), a condition referred to as *static head*, the phenomenological stoichiometry can be determined from measurements of the force ratio. This allows the determination of the operational stoichiometry, as illustrated in Example 14.6. Equation (14.47) also has been applied for the determination of the stoichiometry of various transport processes, e.g., lactose uptake by proton symport in *E. coli* [see Rottenberg (1979) for a review]. The phenomenological stoichiometry can also be determined from measurements of the flow ratio at conditions where the force ratio is zero, i.e., $A_1 = 0$, a condition referred to as *level flow*.

A quantity of interest in the general process of energy transduction is the *thermodynamic efficiency*. This is defined as the rate of production of output free energy divided by the rate at which input free energy is consumed:

$$\eta_{th} = -\frac{v_1 A_1}{v_2 A_2} = -j\chi = -(Z\chi)\frac{(Z\chi) + q}{q(Z\chi) + 1} \qquad (14.50)$$

The thermodynamic efficiency is seen to be a function of three variables: the degree of coupling (q), the phenomenological stoichiometry (Z), and the force ratio (χ). However, because Z and χ appear together, the thermodynamic efficiency actually is a function of two variables only, i.e., q and $Z\chi$, and Fig. 14.7 shows the thermodynamic efficiency as function of $Z\chi$ for different degrees of coupling. It is observed that for $q < 1$ the thermodynamic efficiency passes through a maximum and that this maximal value depends on the degree of coupling. From eq. (14.50) it is easily shown that the maximal thermodynamic efficiency is obtained for a force ratio of

$$\chi_{opt} = -\frac{q}{Z\left(1 + \sqrt{1 - q^2}\right)} \qquad (14.51)$$

which corresponds to an optimal thermodynamic efficiency of

$$\eta_{th, opt} = \left(\chi_{opt}Z\right)^2 = \frac{q^2}{\left(1 + \sqrt{1 - q^2}\right)^2} \qquad (14.52)$$

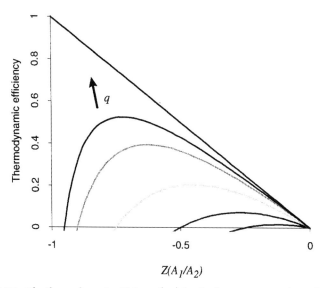

FIGURE 14.7 The thermodynamic efficiency (η_{th}) for the free energy transducer described by eq. (14.44), as a function of the normalized force ratio [$Z\chi = Z(A_1/A_2)$] at different degrees of coupling (q).

Although it may seem attractive to try maximize the thermodynamic efficiency of a system, it is not clear whether such a maximum would correspond to a true optimum for the overall system. Consider, for example, the case of complete coupling, i.e., $q = 1$. In this case the maximal thermodynamic efficiency is 1 at a force ratio of $-1/Z$. However, this situation corresponds to equilibrium, and therefore, there is no net flow. Thus, although the process is very efficient, nonetheless it is useless. To make the process run, the force ratio has to attain a value larger than $-1/Z$, i.e., $Z\chi$ has to be larger than -1, which implies that the thermodynamic efficiency decreases (see Fig. 14.7). Thus, to make the free energy transduction run, part of its efficiency must be sacrificed (Westerhoff and van Dam, 1987). It is not known which force ratio represents the optimal compromise between rate and efficiency, and, probably, the compromise is likely to vary from system to system. Kedem and Caplan (1965) explored one possible compromise, the so-called state of maximal output, where $v_1 A_1$ is optimized. It can be shown that, for any degree of coupling, the force ratio that gives the maximal output is always lower than the force ratio corresponding to the optimal thermodynamic efficiency (Westerhoff and van Dam, 1987). These considerations clearly indicate that maximal efficiency does not correspond, in general, to the optimal state of the system, as the rate of the output process at this state may be too low.

EXAMPLE 14.6

Oxidative Phosphorylation

The mechanism of oxidative phosphorylation was reviewed recently by Senior (1988) and also was described in Section 2.2.3. We recommend that the reader review that section before reading the nonequilibrium thermodynamic treatment in this example. Here we use nonequilibrium thermodynamics to perform a quantitative analysis of the process and derive an estimate for the operational P/O ratio, an energetically important parameter for aerobic processes (see also Example 3.3).

Basically, the huge amount of free energy made available by the oxidation of NADH to NAD^+ by oxygen is used to drive the phosphorylation process. The two processes can be specified as

$$NAD^+ + H_2O - NADH - 0.5O_2 - H^+ = 0 \tag{1}$$

$$ATP + H_2O - ADP - P_i = 0 \tag{2}$$

The standard free energy change associated with the oxidation of NADH can be obtained by considering the two single-electrode processes:

$$H_2O - 0.5O_2 - 2H^+ - 2e^- = 0; \qquad \varepsilon^{0'} = -0.815 \text{ V} \tag{3}$$

$$-NADH + NAD^+ + H^+ + 2e^- = 0; \qquad \varepsilon^{0'} = -0.315 \text{ V} \tag{4}$$

Equation (1) is obtained by adding eqs. (3) and (4), yielding the following estimate for the standard free energy change of the former:

$$\Delta G^{0'} = nF\Delta\varepsilon^{0'} = (2)(96.494 \text{ kJ V}^{-1} \text{ mol}^{-1})(-1.13 \text{ V}) = -218 \text{ kJ mol}^{-1} \tag{5}$$

There is plenty of free energy in this process to drive the phosphorylation of ADP to ATP (the standard free energy change of reaction (2) is 30.5 kJ mol⁻¹), which can allow more than 1 mol of ATP to be formed for each mole of NADH oxidized. According to the chemiosmotic theory (see Fig. 2.11), the two processes are coupled by the generation of a proton [eq. (3)] and electrical potential gradient by reactions (3) and (4) and by utilization of the proton gradient to drive the phosphorylation reaction (2). Synthesis of ATP is mediated by the membrane-bound F_0F_1-ATPase complex. Reaction between ADP and orthophosphate on the knoblike F_1 part of the protein, which sticks into the cytoplasm (for prokaryotes) or into the mitochondrial matrix (for eukaryotes), proceeds readily enough, but unless a steady stream of protons flows through the stemlike F_0 part of the protein, which transcends the membrane (cytoplasmic or inner mitochondrial), the synthesized ATP does not detach from F_1 and ATP production stops. Thus, the flow of protons to F_0 is necessary to accomplish the release of ATP from F_1, and the overall effect of the proton flow is to release a considerable amount of free energy, enough to drive the phosphorylation reaction (2) against a negative affinity. The driving force for the phosphorylation reaction, called the *proton-motive force*, Δp, by Mitchell (1961), who introduced the chemiosmotic theory, is given by

$$\Delta p = \frac{\Delta \mu_{H^+}}{F} = \Delta \Psi - 2.303 \frac{RT}{F} \Delta pH = \Delta \Psi - 0.059 \Delta pH \qquad (6)$$

$\Delta \mu_{H+}$ is determined as the sum of a contribution from the membrane potential (equal to $\Delta \Psi / F$) and a contribution from the proton gradient. Δp has the unit volts mole^{-1}. $\Delta \psi$ is the membrane potential in volts per mole of protons transported (the membrane is more negative on the inside; hence, $\Delta \psi$ is positive and it has been found to be approximately 0.15 V). ΔpH is the pH difference across the membrane (normally about 0.05 and negative). The proton-motive force therefore is about 0.153 V per mol of H^+, corresponding to a free energy of 14.75 kJ per mol of protons transported. Notice that the majority of the proton-motive force is due to the electrical potential difference. The contribution by the pH difference is minimal, which explains why the oxidative phosphorylation in prokaryotes may operate even at a very high extracellular pH.

According to the preceding description, oxidative phosphorylation can be conceived as a cyclic process with four steps: two scalar chemical reactions and two vectorial transport processes. The affinities of one transport process (outward flow of protons) and one chemical reaction (phosphorylation) are negative, whereas the affinities for the two other processes (inward flow of protons and oxidation) are sufficiently positive to ensure that the overall dissipation function [eq. (14.35)] for the system is positive. Because the oxidation reaction (1) and outward flow of protons, on one hand, and the phosphorylation reaction and the downhill flow of protons, on the other, are each tightly coupled, we can combine the four processes into two pairs: (1) oxidation with outward flow of protons and (2) phosphorylation with inward flow of protons. The overall affinities for these two processes are given by the following equations per mole of NADH oxidized:

$$A_o = A_o^0 - n_o F \Delta p \qquad (7)$$

$$A_p = A_p^0 + n_p F \Delta p \qquad (8)$$

A_p is often referred to as the phosphorylation potential. A_o^0 equals minus the free energy change of the oxidation reaction; it is large and positive. A_p^0 is negative and equal to minus the free energy change of the phosphorylation reaction. n_0 is the number of protons pumped out in association with the oxidation reaction, and n_p is the number of protons flowing inward in association with phosphorylation. Their values are not known exactly, but in eukaryotes n_0 is often taken to be 12 (corresponding to 4 protons at each of the complexes I, II, and IV, see Fig. 2.11), whereas n_p is 2 or 3 (the inward flow of one additional proton is associated with the exchange of ATP with ADP and P_i between the cytoplasm and the mitochondrial matrix).

In line with the phenomenological equations, we now write the rate of the two processes as linear functions of their affinities:

$$v_o = L_{oo} A_o + L_{op} A_p \tag{9}$$

$$v_p = L_{po} A_o + L_{pp} A_p \tag{10}$$

where Onsager's reciprocal relation implies that $L_{op} = L_{po}$. With the preceding model of oxidative phosphorylation, the flow ratio (v_p/v_o) is equal to the operational P/O ratio, and it is, therefore, interesting to quantify this ratio at physiological conditions. The force ratio χ can be measured using isolated mitochondria, and, if the degree of coupling and phenomenological stoichiometry are known, the operational stoichiometry can be calculated using eq. (14.47).

The degree of coupling can be determined from measurements of the rate of respiration (the rate of input) at static head (no output flow, i.e., $v_p = 0$) and level flow (the output force being zero, i.e., $A_p = 0$). Application of these conditions to eqs. (9) and (10) and elimination of the phenomenological coefficients yields:

$$\frac{(v_p)_{v_o = 0}}{(v_p)_{A_o = 0}} = 1 - q^2 \tag{11}$$

In most energy-conserving membranes, static head ($v_p = 0$) is obtained quickly as the output force is driven to its maximal potential, and the numerator of eq. (11) therefore is easily determined. However, it is more difficult to obtain the situation of level flow, as it generally is impossible to eliminate completely the potential of the output force (Rottenberg, 1979). Nevertheless, it has been shown that q can be calculated from the rates of the input reaction (in this case the rate of respiration) at static head and any other reference state of lower potential. In this case, however, the values of the output potential at static head (A_p^{static}) and at the reference state (A_p^{ref}) also must be known. The degree of coupling then can be calculated from

(Rottenberg, 1979):

$$q^2 = \frac{A_p^{\text{static}}(1 - R_c)}{A_p^{\text{ref}} - A_p^{\text{static}} R_c} \tag{12}$$

where R_c is the ratio of the rate of oxidation at the reference state to its value at static head:

$$R_c = \frac{v_o^{\text{ref}}}{v_o^{\text{static}}} \tag{13}$$

For quantification of the degree of coupling, the reference state is taken to be that of phosphorylation, and the ratio of the rate of oxidation during phosphorylation to the rate of oxidation after ADP is depleted is about 6 ($= R_c$). Furthermore, the phosphorylation potential at static head (A_p^{static}) is about 63 kJ mol^{-1}, and with the addition of ADP it is about 42 kJ mol^{-1} (Rottenberg, 1979). Thus, by using eq. (12) we find a degree of coupling of about 0.97.

The phenomenological stoichiometry Z can be determined from measurements of the force ratio χ at static head, where the flow ratio j is zero, and therefore we find from eq. (14.47):

$$Z\chi^{\text{static}} + q = 0 \tag{14}$$

The force ratio at static head, χ^{static}, is about -0.31 for the oxidation of NADH and about -0.46 for the oxidation of succinate (Rottenberg, 1979). This gives a phenomenological stoichiometry of about 3.0 for the oxidation of NADH and 2.0 for the oxidation of succinate. This corresponds very well with the often reported mechanistic stoichiometries for the oxidative phosphorylation, but this is a coincidence, as discussed later.

By using the value of $Z = 3$ (oxidation of NADH), we can now specify the operational P/O ratio as a function of the force ratio by using eq. (14.47):

$$\text{P} / \text{O} = \frac{v_p}{v_o} = 3.0 \frac{3.0\chi + 0.94}{2.82\chi + 1} \tag{15}$$

and Fig. 14.8 summarizes the results. For force ratios below -0.313, the P/O ratio becomes negative. This corresponds to the situation where the ATPase starts to work in the opposite direction, i.e., it consumes ATP by pumping protons outward. For force ratios above this value the P/O ratio increases quite rapidly, and for force ratios above -0.15 the P/O ratio has attained 85% of its maximum value, which is 2.82. With the above-determined parameters, the force ratio that corresponds to maximal thermodynamic

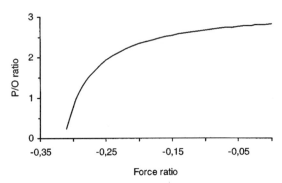

FIGURE 14.8   The P/O ratio as a function of the force ratio in the oxidative phosphorylation.

efficiency is found from eq. (14.51) to be equal to -0.23, and this corresponds to a P/O ratio of 2.1. At this force ratio the thermodynamic efficiency is 0.49 [found from eq. (14.52)]. Notice the dramatic decrease in the optimal thermodynamic efficiency when the degree of coupling decreases from 1 to 0.94.

By using isolated liver mitochondria, Stucki (1980) investigated the process of oxidative phosphorylation. By adding varying activities of hexokinase to the isolated mitochondria, the phosphate potential could be varied. From measurements of the rate of phosphorylation and respiration at different phosphorylation potentials, the phenomenological coefficients L_{op} and L_{pp} of eqs. (9) and (10) were determined from the slopes of the rates with respect to the phosphorylation potential. From calculation of the input force, the other two phenomenological coefficients could be determined from the intercepts of the same plots. By using the phenomenological coefficients, the degree of coupling was calculated to be 0.95 and the phenomenological stoichiometry to be 2.84. These parameters are quite close to the values found earlier. In these experiments the force ratio varied between -0.31 and -0.25 [similar figures are given in Westerhoff and van Dam (1987)]. By using the values for q and Z found by Stucki (1980), this corresponds to a range for the operational P/O ratio between 1.2 and 2.1. It is not known whether the force ratio changes this much at *in vivo* conditions, but it is interesting to note that the force ratio seems to be in the range where it has a large influence on the operational P/O ratio (see Fig. 14.8). When the force ratio is in the preceding interval, the thermodynamic efficiency is between 0.37 and 0.52.

The preceding procedure for the quantification of the operational P/O ratio has been seriously questioned by Westerhoff and van Dam (1987). First of all, the linear relations (9) and (10), which are the basis for the analysis, only hold at conditions close to equilibrium. However, both Rottenberg (1979) and Stucki (1980) (and several others) report experimental data that indicate that the linearity is quite good. The experimental determination of the phenomenological coefficients by Stucki (1980) even indicates that the reciprocal relation is a reasonable assumption. Secondly, the analysis extrapolates from special conditions to *in vivo* conditions: Both the degree of coupling and the phenomenological stoichiometry are determined from measurements at extreme conditions (static head). Both q and Z are functions of the phenomenological coefficients, which again depend on the particular physiological conditions, *e.g.*, the activity of the ATPase. Therefore it is likely that they attain different values at different physiological conditions. The extrapolation of Z is particularly critical, as it may deviate significantly from the mechanistic stoichiometry at physiological conditions. In accepting the limits of the preceding nonequilibrium thermodynamic model for oxidative phosphorylation, Lemasters and co-workers (Lemasters and Billica, 1981; Lemasters *et al.*, 1984), however, stressed that the values of q and Z still provide both an upper and a lower limit to the mechanistic stoichiometry, namely, that the mechanistic stoichiometry is bounded by -qZ and -Z/q. With the value of q normally being close to 1, the error of taking Z as the mechanistic stoichiometry (and therefore to be constant) therefore is quite small.

The preceding nonequilibrium thermodynamic model for oxidative phosphorylation is useful to illustrate the application of nonequilibrium thermodynamics in the analysis of complex biological systems. One should remember, however, that the black box model represents a significant simplification. In their extensive monograph on nonequilibrium thermodynamics of biological systems, Westerhoff and van Dam (1987) present a more detailed model for oxidative phosphorylation. In that model, a net flux of protons across the membrane is also included (see also Example 11.3). By using that model, the stoichiometry of the ATPase and the respiratory chain proton pump were determined separately, from which a mechanistic stoichiometry was determined. Furthermore, by using the more detailed model, they showed that the phenomenological stoichiometry is a function of the kinetic parameters of the system, such as the ATPase activity.

EXAMPLE 14.7

Description of Cellular Growth by Nonequilibrium Thermodynamics

In a black box description, cellular growth is represented as an overall process with an input of free energy in the form of substrates and output of free energy in the form of biomass. Input and output flows are coupled through many intracellular reactions and, also, through the generation and utilization of ATP. Cellular growth thus can be described as a free energy converter with input flow r_c, i.e., rate of catabolism [C-moles of substrate (C-mole of biomass)$^{-1}$ h^{-1}], and output flow r_a, i.e., rate of anabolism [C-moles of biomass (C-mole of biomass)$^{-1}$ h^{-1}]:

$$r_c = L_{cc} A_c + L_{ac} A_a \tag{1}$$

$$r_a = L_{ac} A_c + L_{aa} A_a \tag{2}$$

where Onsager's reciprocal relationship has been applied. For aerobic growth on glucose, the driving force for the input flow (A_c) is the affinity (or minus the free energy change) for the oxidation of glucose to carbon dioxide and water:

$$A_c = -\Delta G_c^{0\prime} + RT \ln\left(\frac{c_{glc}\, c_{O_2}^6}{c_{CO_2}^6}\right) \tag{3}$$

where the standard free energy change for glucose oxidation is about 2.9 MJ mol^{-1}. Thus, the driving force for the input flow depends on the glucose concentration. Similarly, for the output flow, the driving force is given by

$$A_a = -\Delta G_a^{0\prime} - RT \ln\left(\frac{c_x}{\prod_i c_{i,\,anab}}\right) \tag{4}$$

where $c_{i,\,anab}$ is the concentration of the anabolic substrates. In the preceding black box model, the catabolic flow is taken to be given by the glucose uptake rate, and glucose therefore should not be included as an anabolic substrate. Instead, the carbon source of the anabolic reactions should be taken to be carbon dioxide. The standard free energy for biomass synthesis from CO_2, N_2, and water is about 536 kJ C-mol^{-1}, whereas it is about 490 kJ C-mol^{-1} for synthesis from CO_2, NH_4^+, and water (Roels, 1983).

If we consider growth at steady state in a glucose-limited chemostat (where concentrations are constant), A_a normally will be constant. If we then eliminate the varying force A_c from eqs. (1) and (2), we obtain

$$r_c = \frac{1}{qZ} r_a + L_{ac}\left(1 - \frac{1}{q^2}\right) A_a \tag{5}$$

which is identical with the classical linear relationship between specific substrate consumption rate and specific growth rate given by eq. (3.26), where

$$Y_{sx}^{true} = \frac{1}{Y_{xs}^{true}} = qZ \tag{6}$$

$$m_s = L_{ac}\left(1 - \frac{1}{q^2}\right)A_a \tag{7}$$

From eq. (6) it can be seen that only when $q = 1$ does the phenomenological stoichiometry Z equal the mechanistic stoichiometry Y_{xs}^{true} of the system. Obviously, in this case there is perfect coupling and no free energy is wasted in futile cycles or for the support of maintenance processes. Equation (7) shows that the phenomenological coefficient L_{ac} is not constant, but a function of the maintenance coefficient, specific growth rate, and coupling parameter.

Combination of eq. (2) and (3) gives a relationship between the specific growth rate and glucose concentration. If we also have glucose limitation, i.e., A_a is constant, and dissolved oxygen and carbon dioxide are kept constant at the same time, we obtain:

$$r_a = L_{ac} RT \ln(c_{glc}) + a \tag{8}$$

where a is a constant. Equation (8) represents a logarithmic relationship between the specific growth rate and the limiting substrate concentration that is different from the classical saturation type (Monod) kinetics. It can, however, describe most experimental data as well as the Monod type kinetics. Figure 14.9 shows a comparison of the saturation type kinetics and the logarithmic function. Only at very high substrate concentrations does the logarithmic expression deviate from the saturation type kinetics. From the comparison between the two types of kinetics, we find that the parameters in the logarithmic function are related to the Monod parameters as

$$L_{ac} RT \approx 0.23\mu_{max}; \qquad a \approx 0.5 - 0.23\mu_{max} \ln(K_m) \tag{9}$$

If the maximum specific growth rate is known, eq. (9) can be used to determine the phenomenological coefficient L_{ac}.

The nonequilibrium thermodynamic black box model presented in this example is well-suited to illustrate the concept of coupled processes. In the case of growth there is a close coupling between catabolism and anabolism, but the coupling is not perfect due to maintenance processes, which can be compared with the leak of protons in oxidative phosphorylation. Westerhoff

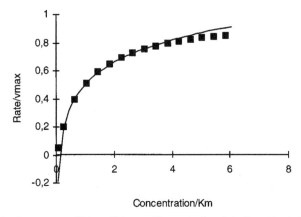

FIGURE 14.9 Comparison of Monod kinetics (data points) and the logarithmic function given by eq. (8). The parameters in the logarithmic function are $L_{ac} RT = 0.23$ and $a = 0.5$. The axes are scaled with the parameters of the Monod model (v_{max} and K_m).

and van Dam (1987) describe a more detailed nonequilibrium model for growth, which includes a flow of ATP (they use the term "mosaic nonequilibrium thermodynamic model" to distinguish it from the simple black box model of our example). This more detailed model allows independent evaluation of growth parameters at a more fundamental level, namely, the Y_{xATP} and m_{ATP} (see Section 3.4).

The linear force relation of eq. (14.41) is the basis of nonequilibrium thermodynamics. As discussed in Box 14.3, these relations are valid only near equilibrium. Because many cellular processes operate far from equilibrium, e.g., some of the reactions in glycolysis (see Example 14.1), this limits the wider applicability of this approach. However, empirical analysis of various cellular processes has revealed that a linearity is often observed between the flow and driving force of a process, even when the process operates far from equilibrium. We provide a possible explanation of why this may be the case[1] from the analysis of a simple, reversible, enzyme-catalyzed

---

[1] The following analysis was inspired by Westerhoff and van Dam (1987), who give a far more indepth treatment of this subject.

reaction:

$$E + S \overset{k_1^S}{\underset{k_{-1}^S}{\leftrightarrow}} ES \overset{k_2^S}{\underset{k_2^P}{\leftrightarrow}} EP \overset{k_{-1}^P}{\underset{k_1^P}{\leftrightarrow}} E + P \tag{14.53}$$

In the preceding equation, E is the enzyme and ES and EP are the enzyme-substrate and enzyme-product complexes, respectively. The net forward rate v of this reaction is given by

$$v = \frac{v_{S,max} \dfrac{c_S}{K_S} - v_{P,max} \dfrac{c_P}{K_P}}{1 + \dfrac{c_S}{K_S} + \dfrac{c_P}{K_P}} \tag{14.54}$$

where $v_{S,max}$ is the maximum forward reaction rate (obtained at high substrate concentrations and $c_P = 0$), and $v_{P,max}$ is the maximum reverse reaction rate (obtained at high product concentrations and $c_S = 0$). K_S and K_P usually are called Michaelis-Menten constants and they are given by

$$K_S = \frac{k_{-1}^S + k_2^S}{k_1^S} ; \qquad K_P = \frac{k_{-1}^P + k_2^P}{k_{-1}^P} \tag{14.55}$$

We now rewrite the kinetics of eq. (14.54) as a function of the affinity of the reaction (equal to the change in chemical potential). From eqs. (14.13) and (14.14), we have:

$$e^{(\mu_S - \mu_P)/RT} = \frac{c_S}{c_P} \frac{c_{P,eq}}{c_{S,eq}} \tag{14.56}$$

Furthermore, at equilibrium the net reaction rate becomes zero and we find:

$$K_{eq} = \frac{c_{P,eq}}{c_{S,eq}} = \frac{K_P}{K_S} \frac{v_{S,max}}{v_{P,max}} \tag{14.57}$$

By using these two equations, the kinetics can be rewritten as

$$\frac{v}{v_{S,max}} = \frac{e^{(\mu_S - \mu_P)/RT} - 1}{\left(\dfrac{K_S}{c_S + c_P} + 1 \right) e^{(\mu_S - \mu_P)/RT} + \dfrac{v_{S,max}}{v_{P,max}} \left(\dfrac{K_P}{c_S + c_P} + 1 \right)} \tag{14.58}$$

which expresses the net forward reaction rate as a function of the affinity, $A = \mu_s - \mu_p$ (or the driving force), of the reaction.

Direct application of eq. (14.58) is limited by the fact that the net forward reaction rate is a function of the concentrations of both the substrate and the product (or the chemical potential of these two compounds). In practice, however, these two variables are often related by a constraint on their sum:

$$c_S + c_P = \text{constant} \qquad (14.59)$$

which allows one to express the net forward reaction rate as a function of the affinity of the reaction only. Figure 14.10 shows the result for two sets of kinetic parameters.

There is often a distinction between *reversible* and *irreversible* reactions, which is in contradiction with the principle of microscopic reversibility. In accordance with this principle, both curves in Fig. 14.10 should pass through the origin, and for negative reaction affinities the net forward reaction rate should also become negative. Thus, both reactions are, in principle, reversible, but when the maximum forward reaction rate ($v_{S,\,max}$) is much larger than the maximum reverse reaction rate ($v_{P,\,max}$), the reaction is effectively irreversible, i.e., only for numerically very large and negative reaction affinities does the reaction proceed in the backward direction. For the reversible reaction (solid line in Fig. 14.10), it can be seen that a linear relationship of the type

$$v = LA \qquad (14.60)$$

holds for a wide range of net forward reaction rates. On the other hand, this proportionality between v and reaction affinity only holds for very small net,

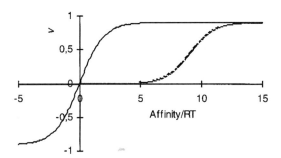

FIGURE 14.10 The dependence of an enzyme-catalyzed reaction rate on its free energy difference (or reaction affinity). The sum of substrate and product concentrations is assumed to be constant (and equal to 1). The function is calculated using eq. (14.58). The solid line depicts a thermodynamically reversible reaction with $v_{S,\,max} = v_{P,\,max} = K_S = K_P = 10$ (corresponding to an equilibrium constant K_{eq} of 1). The dashed line depicts a thermodynamically irreversible reaction with $v_{S,\,max} = K_S = 10$, $v_{P,\,max} = 0.1$, and $K_P = 100$ (corresponding to an equilibrium constant K_{eq} of 1). The affinity is equal to $\mu_S - \mu_P$.

forward reaction rates for the irreversible reaction (dashed line in Fig. 14.10). For irreversible reactions, eq. (14.60) would be a very poor approximation of the actual relationship between v and reaction affinity in the relevant range of reaction rates (between 0.1 and 0.9). Yet, around $v = 0.5$, a linear approximation of the relationship seems possible:

$$v = L^{\#}(A - A^{\#})$$

(14.61)

where $A^{\#}$ is the reaction affinity at the inflection point of the curve and $L^{\#}$ is the slope of the tangent at the inflection point. These two parameters are functions of the kinetic parameters, and Westerhoff and van Dam (1987) showed that if they are given by

$$L^{\#} = \frac{v_S + v_P}{4RT}$$

(14.62)

$$\frac{A^{\#}}{RT} = \ln\left(\frac{v_S}{v_P}\right) - 2\frac{v_S - v_P}{v_S + v_P}$$

(14.63)

the net forward reaction rate calculated using eq. (14.61) deviates by less than 15% from the true value when v is in the range between 0.18 and 0.93. v_S and v_P are given by

$$v_S = \frac{v_{S,\max}}{1 + K_S/(c_S + c_P)} \; ; \quad v_P = \frac{v_{P,\max}}{1 + K_P/(c_S + c_P)}$$

(14.64)

Linear relationships between reaction rates or transmembrane fluxes and free energy differences have been observed experimentally for many different systems, even for reactions or processes operating far from equilibrium [see Westerhoff and van Dam (1987) for a review on this topic]. Westerhoff and van Dam (1987) also analyzed a number of other kinetic expressions, and in all cases found that in a certain range of free energy differences a linear relationship could be assumed.

With a linear flow-force relationship holding more generally for processes operating far from equilibrium, it is of interest to know whether reciprocity in the phenomenological coefficients also is true in these cases. It was mentioned in Example 14.6 that Stucki (1980) experimentally observed reciprocity for the oxidative phosphorylation operating far from equilibrium. Furthermore, Caplan (1981) showed that, for highly coupled enzymatic reactions, operation around a multidimensional inflection point may lead not only to linearity in the flow-force relations but also to reciprocity [see also Pietrobon and Caplan (1985)]. However, it remains to be shown that this is a general situation. One therefore should only assume reciprocity for those systems where there is sufficient experimental evidence.

One assumption in the preceding analysis is that the sum of concentrations for the substrate and the product is constant. For some pathway reactions this assumption is not reasonable, but in many cases there are physical constraints on the concentrations. Oftentimes, one of the concentrations is kept constant by external controls or tight intracellular regulation, or its concentration vastly exceeds its Michaelis-Menten constant of the enzyme acting upon it. For these cases, Rottenberg (1973) also showed that the relationship between reaction rate and driving force becomes nearly linear. This is illustrated clearly for the simple situation where the product concentration is kept constant. In this case, the driving force can be written as a linear function of $\ln(c_S)$. The reaction rate therefore can also be written as a linear function of the logarithm of the substrate concentration, and, as illustrated in Fig. 14.9, such an expression corresponds very well with the classical saturation type kinetics often found for enzymatic reactions.

14.4. APPLICATION OF THERMOKINETICS TO MCA

The subject of metabolic control was reviewed extensively in Chapter 11. Among others, the concept of a flux control coefficient (FCC) was introduced to provide a quantitative measure of the degree of control exerted by individual reactions or groups of reactions (Chapter 12) on the overall pathway flux. It was also shown that FCCs can be derived analytically from kinetic models of the particular reaction rates through the application of the summation and connectivity theorems of MCA [eqs. (11.6), (11.14), and (11.27)]. Of particular importance in this analysis are the elasticities [eq. (11.11)], as the parameters embodying the actual enzyme kinetics. It was also pointed out that a lack of reliable *in vivo* kinetic models hinders the direct derivation of FCCs and pathway control analysis.

One of the goals of nonequilibrium thermodynamics is the derivation of expressions for the rate of a process in terms of the applicable driving force(s) [eq. (14.41)]. These rate expressions could, in principle, be used for the calculation of the elasticities and, through them, flux control coefficients. In this section, we demonstrate the general procedure along with an example from penicillin biosynthesis.[2]

[2] The method for extracting MCA parameters from a thermokinetic description of reaction rates was derived by Nielsen (1997), and the present description is adapted from this paper.

Consider the simple two-step pathway converting X_1 to X_3 via X_2 with the following stoichiometry:

$$g_{11} X_1 + g_{12} X_2 = 0$$
$$g_{22} X_2 + g_{23} X_3 = 0$$

$$(14.65)$$

This pathway may be for two actual enzymatic reactions or represent several enzymatic reactions lumped into the two overall reactions. The affinity of the ith reaction is given by

$$A_i = -\Delta G_i^{0\prime} - RT \ln \left(\prod_{k=1}^{3} X_k^{g_{ik}} \right); \qquad i = 1, 2 \qquad (14.66)$$

where $\Delta G_i^{0\prime}$ is the standard free energy change of the ith reaction. We further assume that the rate of each of the two reactions can be described in terms of the corresponding reaction affinity according to eq. (14.61). With this *thermokinetic* description of the reaction kinetics, the elasticity coefficients for the two reactions with respect to the intermediate X_2 are given by

$$\varepsilon_{2,i} = \frac{1}{v_i} \frac{\partial v_i}{\partial \ln(X_2)} = \frac{1}{v_i} \frac{\partial v_i}{\partial A_i} \frac{\partial A_i}{\partial \ln(X_2)}; \qquad i = 1, 2 \quad (14.67)$$

where the two partial derivatives can be evaluated using eqs. (14.61) and (14.66):

$$\varepsilon_{2,i} = -\frac{g_{i2} RT}{A_i - A_i^{\#}}; \qquad i = 1, 2 \qquad (14.68)$$

Thus, if $A_1^{\#}$ and $A_2^{\#}$ are known, the elasticity coefficients for the two reactions can be calculated. For reversible reactions, $A_i^{\#}$ is zero and eq. (14.68) allows the calculation of the elasticity coefficients directly from the affinity, which can be determined from measurements of the metabolite levels at steady state using eq. (14.66). Westerhoff *et al.* (1984) derived a similar relation for the elasticity coefficient for enzyme reactions at equilibrium, but used the mass action ratio rather than the reaction affinity for the rate. To facilitate flux control through the pathway, at least one step normally is irreversible, *i.e.*, it operates far from equilibrium. The parameter $A_i^{\#}$ is not zero for such a step, but may be determined by plotting the reaction rate as a function of the reaction affinity. Because $A_i^{\#}$ only depends on the Michaelis-Menten constants, the ratio of the maximal forward and maximal reverse reaction rates, and $c_S + c_P$ [*i.e.*, is independent of the *in vivo* activity of the enzyme, see eqs. (14.63) and (14.64)] it is, in principle, possible to determine $A_i^{\#}$ from *in vitro* experiments. It is, however, important to ensure that the enzyme activity is constant, otherwise $L^{\#}$ may vary [see eq. (14.62)] and linearity between v and $A^{\#}$ may not hold. For the case of a variable enzyme activity, one could normalize the reaction rate with the enzyme activity as $L^{\#}$ is proportional to the enzyme activity (or v_{max}) (see Example 14.8).

With the elasticity coefficients calculated from eq. (14.68), the FCCs can be determined from the summation and connectivity theorems as described in Sections 11.1-11.2. Thus, the thermokinetic description of reaction rates allows a complete evaluation of the MCA coefficients from measurements of the metabolite pool levels at steady state.

Equation (14.68) can also be applied to pathways with more than two reactions, under the strict requirement that there be no regulatory loops that span more than one reaction, $e.g.$, when the last metabolite feedback inhibits the first reaction in a pathway with more than two steps. This is due to the fact that the thermokinetic description of the rates allows quantification of the influence only of the substrate and product on the reaction rate and does not include any influence of effectors. Another point regarding the outcome of reaction lumping is in order. Many pathways include only one or two irreversible reactions, with the other reactions being close to equilibrium. In this case reactions may be lumped into two overall steps to yield a pathway structure like eq. (14.65), for which reaction affinities are determined for the overall conversions. Because $A_i^{\#} = 0$ for all reactions operating close to equilibrium, the value of $A_i^{\#}$ for the lumped set of reactions will equal the corresponding value of the irreversible reaction, which is not zero. Similarly, for the FCCs, when equilibrium reactions are lumped together with one irreversible reaction, the value of the overall FCC for the lumped set of reactions is approximately equal to that of the irreversible reaction.

In cases where it is not possible to lump the individual pathway reactions into two overall reactions and where regulatory loops extend beyond a single reaction, the preceding approach is not directly applicable. It is possible, however, to still apply the thermokinetic expressions to modify the Delgado-Liao approach for the direct determination of the FCCs (see Section 11.2.3 and Box 11.4). For this purpose, we rewrite eq. (14.61) as follows after introducing eq. (14.66) for the affinity:

$$v_i = a_i \sum_{j=1}^{L+1} k_{ij} \ln(X_j) + b_i \qquad (14.69)$$

As mentioned in Section 11.2.3, this approach enables a direct determination of the FCCs from measurements of the metabolite pools during a transient. In eq. (14.69) a_i, b_i, and k_{ij} are kinetic parameters. For substrates

BOX 14.4

Extension of the Delgado and Liao (1992) Approach

The starting point of the approach of Delgado and Liao (1992) is eq. (1):

$$\sum_{i=1}^{L} C_i \Delta v_i = 0 \tag{1}$$

where $\Delta v_i = v_i - v_{ss}$ is the change in ith reaction rate during the transient compared with the steady state reaction rate v_{ss}. This equation is derived from the connectivity theorem, assuming linear kinetics. In the following we show that eq. (1) also holds when the kinetics is given by eq. (14.69). First, we find the elasticity coefficient:

$$\varepsilon_{j,i} = \frac{b_i k_{ij}}{v_{ss}} \tag{2}$$

where v_{ss} is the steady state flux through the pathway. When eq. (2) is inserted in the connectivity theorem, we obtain:

$$\sum_{i=1}^{L} C_i b_i k_{ij} = 0 \tag{3}$$

Upon multiplication of eq. (3) by $\Delta \ln(X_j) = \ln(X_j(t_2)) - \ln(X_j(t_1))$, we obtain:

$$\sum_{i=1}^{L} C_i b_i k_{ij} \Delta \ln(X_j) = 0 \tag{4}$$

By summing the preceding equation for all j, we obtain:

$$\sum_{j=2}^{L} \sum_{i=1}^{L} C_i b_i k_{ij} \Delta \ln(X_j) = \sum_{i=1}^{L} C_i b_i \sum_{j=2}^{L} k_{ij} \Delta \ln(X_j) = 0 \tag{5}$$

which is seen to reduce to eq. (1) because from eq. (14.61) we have:

$$\Delta v_i = b_i \sum_{j=2}^{L} k_{ij} \Delta \ln(X_j) \tag{6}$$

and products the k_{ij}'s are identical with the stoichiometric coefficients, whereas for effectors they are empirical parameters. For compounds (substrates, products, or effectors) that do not influence the kinetics, k_{ij} is zero. For the simple case where there are no effectors, a_i becomes equal to $L_i^\#$ and b_i becomes equal to $-L_i^\# A_i^\#$, whereas in the general case they should be considered as empirical parameters. Based on the discussion in Section 14.3, eq. (14.69) will always be a better approximation of reaction kinetics than the linearized kinetics assumed by Delgado and Liao (1992), ensuring a wider applicability of their approach to FCC determination.

EXAMPLE 14.8

MCA of the Penicillin Biosynthetic Pathway Using the Thermokinetic Approach

The preceding methodology is illustrated with the calculation of the elasticity coefficients and FCCs of the penicillin biosynthetic pathway, which also was analyzed using a kinetic model in Example 11.1. It was mentioned there that flux control is exerted mainly by the first two steps of the pathway, and in the following we only consider these steps. The free energy change for these reactions has been calculated by Pissarra and Nielsen (1997). Both reactions were found to be very exogenic, with free energy changes on the order of -130 kJ mol^{-1} for the ACVS-catalyzed reaction and -480 kJ mol^{-1} for the IPNS-catalyzed reaction. Thus, both reactions are operating far from equilibrium.

By using fed-batch cultivation data, the free energy change and affinity for the first reaction were calculated. Plotting of the rate of the ACVS-catalyzed reaction (which varied during the fed-batch cultivation) as a function of the calculated reaction affinity yields a linear relationship (Fig. 14.11a). Furthermore, the *activity* of ACVS was found to be approximately constant (Nielsen and Jørgensen, 1995), thus allowing determination of the parameters in the thermokinetic description of eq. (14.61). $L^\#$ was estimated to be 0.82 x 10^{-6} mol^2 (g DW)$^{-1}$ h^{-1} kJ^{-1} and $A^\# = 115$ kJ mol^{-1}.

For the IPNS-catalyzed reaction, not only is a linear relationship between the reaction rate and reaction affinity not observed (Fig. 14.11b), but, in fact, the reaction rate decreases as reaction affinity increases. This is explained by a decrease in the activity of the corresponding enzyme throughout the fed-batch cultivation (Nielsen and Jørgensen, 1995). The thermokinetics of eq. (14.61) cannot be applied directly in this case. However, if, as discussed previously, the reaction rate is normalized with respect to the measured

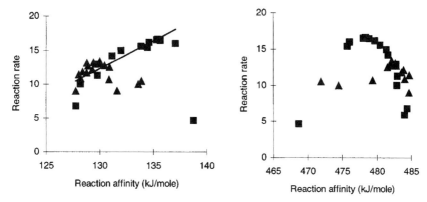

FIGURE 14.11   Reaction rate [μmol (g DW)$^{-1}$ h^{-1}] vs reaction affinity (kJ mol^{-1}) for the first two steps of the penicillin biosynthetic pathway. The data were obtained from two different fed-batch cultivations during which the reaction rate of the two reactions varied slowly. ($\blacktriangle$) Data for the fed-batch cultivation FB023. ($\blacksquare$) Data for the fed-batch cultivation FB028. The reaction affinities were calculated as described in Pissarra and Nielsen (1997) using data from Jørgensen *et al.* (1995a,b). (a) Reaction rate vs reaction affinity for the ACVS-catalyzed reaction. (b) Reaction rate vs reaction affinity for the IPNS-catalyzed reaction.

enzyme activity, a linear relation between relative rate and reaction affinity is obtained (Fig. 14.12) and $A^{\#}$ is determined to be 465 kJ mol^{-1}.

With the parameters of the thermokinetic description determined, we can calculate the elasticity coefficients and the FCCs at different times of the fed-batch cultivation. Strictly speaking, the summation and connectivity

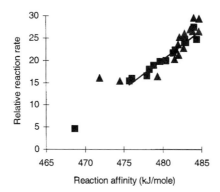

FIGURE 14.12   Relative reaction rate (reaction rate over enzyme activity) vs reaction affinity (kJ mol^{-1}) for the IPNS-catalyzed reaction. The reaction rate was normalized with respect to the measured enzyme activity reported by Nielsen and Jørgensen (1995).

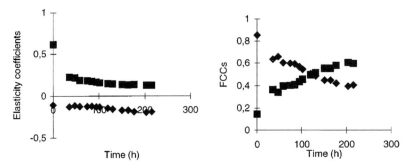

FIGURE 14.13 MCA based on the thermokinetic description of the two first steps in the penicillin biosynthetic pathway. The elasticity coefficients and FCCs were calculated at different times of cultivation during a fed-batch cultivation [FB028 of Jørgensen *et al.* (1995b)]. (a) The elasticity coefficients for ACVS (♦) and IPNS (■). The elasticity coefficients were calculated from the free energy change at the given time and by using eq. (14) with $A^{\#}_{ACVS} = 115$ kJ mol^{-1} and $A^{\#}_{IPNS} = 465$ kJ mol^{-1}. (b) The FCCs for ACVS (♦) and IPNS (■). The FCCs were calculated from the elasticity coefficients using the summation and connectivity theorems.

theorems are applicable at steady state and not during the preceding fed-batch experiments, but, as discussed in Pissarra *et al.* (1996), pseudo-steady state can be assumed for the biosynthetic pathway. The results of the calculations for one fed-batch cultivation are shown in Fig. 14.13. It is observed that there is a shift in flux control from the first reaction to the second reaction during the cultivation, consistent with the results of the kinetic analysis of the pathway presented in Example 11.1.

REFERENCES

Caplan, S. R. (1971). Nonequilibrium thermodynamics and its application to bioenergetics. *Current Topics in Bioenergetics* 4, 1-79.

Caplan, S. R. (1981). Reciprocity or near-reciprocity of highly coupled enzymatic processes at the multidimensional inflection point. *Proceedings of the National Academy of Science. USA* 78, 4314-4318.

Christensen, L. H., Henriksen, C. M., Nielsen, J., Villadsen, J. & Egel-Mitani, M. (1995). Continuous cultivation of *Penicillium chrysogenum*. Growth on glucose and penicillin production. *Journal of Biotechnology* 42, 95-107.

Delgado, J. & Liao, J. C. (1991). Identifying rate-controlling enzymes in metabolite pathways without kinetic parameters. *Biotechnology Progress* 7, 15-20.

Delgado, J. & Liao, J. C. (1992). Metabolic Control analysis from transient metabolite concentrations. *Biochemical Journal* 282, 919-927.

Groen, A. K., Wanders, R. J. A., Westerhoff, H. V., van der Meer, R. & Tager, J. M. (1982). Quantification of the contribution of various steps to the control of mitochondrial respiration. *Journal of Biological Chemistry* **257**, 2754-2757.

Jørgensen, H. S., Nielsen, J., Villadsen, J. & Møllgaard, H. (1995a). Metabolic flux distributions in *Penicillium chrysogenum* during fed-batch cultivations. *Biotechnology and Bioengineering.* **46**, 117-131.

Jørgensen, H. S., Nielsen, J., Villadsen, J. & Møllgaard, H. (1995b). Analysis of the penicillin V biosynthesis during fed-batch cultivations with a high yielding strain. *Applied Microbiology Biotechnology* **43**, 123-130.

Kedem, O. & Caplan, S. R. (1965). Degree of coupling and its relation to efficiency of energy conversion. *Transactions of the Faraday Society* **6**, 1897-1911.

Lehninger, A. E. (1975). *Biochemistry*, 2nd ed. New York: Worth.

Lemasters, J. J. & Billica, W. H. (1981). Non-equilibrium thermodynamics of oxidative phosphorylation by inverted inner membrane vesicles of rat liver mitochondria. *Journal of Biological Chemistry*. **256**, 12949-12957.

Lemasters, J. J., Grunwald, R. & Emaus, R. K. (1984). Thermodynamic limits to the ATP/site stoichiometries of oxidative phosphorylation by rat liver mitochondria. *Journal of Biological Chemistry* **259**, 3058-3063.

Mavrovouniotis, M. L. (1990). Group contributions for estimating standard Gibbs energies of formation of biochemical compounds in aqueous solutions. *Biotechnology and Bioengineering* **36**, 1070-1082.

Mavrovouniotis, M. L. (1991). Estimation of standard Gibbs energy changes of biotransformations. *Journal Biological Chemistry* **266**, 1440-1445.

Mavrovouniotis, M. L. (1993). Identification of localised and distributed bottlenecks in metabolic pathways. International Conference on Intelligent Systems for Molecular Biology, Washington DC.

Mitchell, P. (1961). Coupling of phosphorylation to electron and hydrogen transfer by a chemiosomotic type of mechanism. *Nature* **191**, 144-148.

Nielsen, J. (1997). Metabolic control analysis of biochemical pathways based on a thermokinetic description of reaction rates. *Biochemical Journal* **321**, 133-138.

Nielsen, J. & Jørgensen, H. S. 1995. Metabolic control analysis of the penicillin biosynthetic pathway in a high yielding strain of *Penicillium chrysogenum. Biotechnology Progress* **11**, 299-305.

Onsager, L. (1931). Reciprocal relations in irreversible processes. *Physical Reviews* **37**, 405- 426.

Pietrobon, D. & Caplan, S. R. (1985). Flow-force relationships for a six-state proton pump model: Intrinsic uncoupling, kinetic equivalance of input and output forces, and domain of approximate linearity. *Biochemistry* **24**, 5764-5776.

Pissarra, P. N. & Nielsen, J. (1997). Thermodynamics of metabolic pathways for penicillin production: Analysis of thermodynamic feasibility and free energy changes during fed- batch cultivation. *Biotechnology Progress*. **13**, 156-165.

Pissarra, P. N., Nielsen, J. & Bazin, M. J. (1996). Pathway kinetics and metabolic control analysis of a high-yielding strain of *Penicillium chrysogenum* during fed-batch cultivations. *Biotechnology and Bioengineering* **51**, 168-176.

Roels, J. A. (1983). Energetics and Kinetics in Biotechnology. Amsterdam: Elsevier Biomedical Press.

Rottenberg, H. (1973). The thermodynamic description of enzyme-catalyzed reactions. The linear relation between the reaction rate and the affinity. *Biophysical Journal* **13**, 503-511.

Rottenberg, H. (1979). Non-equilibrium thermodynamics of energy conversion in bioenergetics. *Biochimica et Biophysica Acta* **549**, 225-253.

Senior, A. E. (1988). ATP synthesis by oxidative phosphorylation. *Phys. Rev.* **68**, 177-231.

Stucki, J. W. (1980). The optimal efficiency and the economic degrees of coupling of oxidative phosphorylation. *European Journal of Biochemistry* **109**, 269-283.

Westerhoff, H. V. & van Dam, K. (1987). Thermodynamics and Control of Biological Freeenergy Transduction. Amsterdam: Elsevier

Westerhoff, H. V., Groen, A. K. & Wanders, R. J. A. (1984). Modern theories of metabolic control and their applications. *Bioscience Reports* **4**, 1-22

GLOSSARY

Adenine (A) A nitrogenous base, one member of the base pair A-T (adenine-thymine).

Alleles Alternative forms of a genetic locus; a single allele for each locus is inherited separately from each parent (*e.g.*, at a locus for eye color the allele might result in blue or brown eyes).

Allosteric A term that describes a protein, especially enzymes, in which a compound combines with a site on the protein other than the active site. This may result in a conformational change at the active site so that the normal substrate cannot bind to it. The allosteric property is useful in the regulation of enzyme activity.

Amino acid Any of a class of 20 molecules that are combined to form proteins in living systems. The sequence of amino acids in a protein and, hence, protein function are determined by the genetic code.

Amplification An increase in the number of copies of a specific DNA fragment; can be *in vivo* or *in vitro*. See cloning, polymerase chain reaction.

Anabolism Refers to those metabolic processes involved in the synthesis of cell constituents from simpler molecules, such as organic and/or inorganic precursors. An anabolic process usually requires energy.

Anaerobic respiration Respiration under anaerobic conditions. The terminal electron acceptor, instead of oxygen in the case of regular respiration, can be CO_2, Fe^{2+}, fumarate, nitrate, nitrite, nitrous oxide, sulfur, sulfate, etc. Note that anaerobic respiration still uses an electron transport chain to dump the electron whereas fermentation does not.

Anticodon A sequence of three bases in tRNA that base pairs with a codon in mRNA.

Antigen A substance, usually macromolecular, that induces a specific immune response.

Base pair (bp) Two nitrogenous bases (adenine and thymine, or guanine and cytosine) held together by weak bonds. Two strands of DNA are held together in the shape of a double helix by the bonds between base pairs.

Base sequence   The order of nucleotide bases in a DNA molecule.

Biotechnology   A set of biological techniques developed through basic research and now applied to research and product development. In particular, the use by industry of recombinant DNA, cell fusion, and new bioprocessing techniques.

Capsid   The protein coat of a virus.

Catabolism   The biochemical processes involved in the breakdown of organic compounds, usually leading to the production of energy.

Cellular microbiology   A new discipline emerging at the interface between cell biology and microbiology. One major focus of this new field is on the interference of pathogenic bacteria with many eukaryotic cell functions, such as maturation of intracellular compartments, internal cellular communication, or even cell division and differentiation. The study of cellular microbiology in this respect is providing a sophisticated tool kit for mammalian cell biologists [(1996). *Science* **271**, 315].

Chaperonin   A protein that aids in the correct folding of other proteins and the assembly of multi subunit structures.

Chemiosmosis   The use of iron gradients across membranes, especially proton gradients, to generate ATP. See proton-motive force.

Chromosome   A genetic element carrying genes essential to cellular metabolism. Prokayrotes typically have a single chromosome, consisting of a circular DNA molecule. Eukaryotic cells contain several chromosomes, each containing a linear DNA molecule complexed with specific proteins.

Clones   A group of cells derived from a single ancestor.

Cloning   The process of asexually producing a group of cells (clones), all genetically identical, from a single ancestor. In recombinant DNA technology, the use of DNA manipulation procedures to produce multiple copies of a single gene or segment of DNA is referred to as cloning DNA.

Cloning vector   DNA molecule originating from a virus, plasmid, or the cell of a higher organism into which another DNA fragment of appropriate size can be integrated without loss of the vector's capacity for self-replication; vectors introduce foreign DNA into host cells, where it can be reproduced in large quantities. Examples are plasmids, cosmids (large plasmids), and yeast artificial chromosomes; vectors often are recombinant molecules containing DNA sequences from several sources.

Coenzyme   A low-molecular-weight chemical that participates in an enzymatic reaction by accepting and donating electrons or functional groups. Examples: NAD^+, FAD.

Commodity chemical   Chemicals such as ethanol that have low monetary value thus are sold primarily in bulk.

Complementary DNA (cDNA)   DNA that is synthesized from a messenger RNA template; the single-stranded form often is used as a probe in physical mapping.

Conserved sequence A base sequence in a DNA molecule (or an amino acid sequence in a protein) that has remained essentially unchanged throughout evolution.

Cosmid Artificially constructed cloning vector containing the *cos* gene of phage lambda. Cosmids can be packaged in lambda phage particles for infection into *E. coli*; this permits the cloning of large DNA fragments (up to 45 kb) that can be introduced into bacterial hosts in plasmid vectors.

Cytosine (C) A nitrogenous base, one member of the base pair G-C (guanine and cytosine).

Diploid A full set of genetic material consisting of paired chromosomes, one chromosome from each parental set. Most animal cells except the gametes have a diploid set of chromosomes. The diploid human genome has 46 chromosomes. Compare with haploid.

DNA (deoxyribonucleic acid) The molecule that encodes genetic information. DNA is a double-stranded molecule held together by weak bonds between base pairs of nucleotides. The four nucleotides in DNA contain the bases adenine (A), guanine (G), cytosine (C), and thymine (T). In nature, base pairs form only between A and T and between G and C; thus, the base sequence of each single strand can be deduced from that of its partner.

DNA replication The use of existing DNA as a template for the synthesis of new DNA strands. In humans and other eukaryotes, replication occurs in the cell nucleus.

DNA sequence The relative order of base pairs, whether in a fragment of DNA, a gene, a chromosome, or an entire genome.

Electron acceptor A substance that accepts electrons during an oxidation-reduction reaction. An electron acceptor is an oxidant.

Electron transport phosphorylation Synthesis of ATP involving a membrane-associated electron transport chain and the creation of a proton-motive force. Also called oxidative phosphorylation.

Electrophoresis A method of separating large molecules (such as DNA fragments or proteins) from a mixture of similar molecules. An electric current is passed through a medium containing the mixture, and each kind of molecule travels through the medium at a different rate, depending on its electrical charge and size. Separation is based on these differences. Agarose and acrylamide gels are the media commonly used for electrophoresis of proteins and nucleic acids.

Embden-Meyerhof-Parnas pathway (Embden-Meyerhof pathway; EMP pathway) A pathway that degrades glucose to pyruvate; the six-carbon stage converts glucose to fructose-1,6-bisphosphate, and the three-carbon stage produces ATP while changing glyceraldehyde-3-phosphate to pyruvate. Compare with Entner-Doudoroff pathway.

Entner-Doudoroff pathway (ED pathway) A pathway that converts glucose to pyruvate and glyceraldehyde-3-phosphate by producing 6-phosphogluconate and then dehydrating it.

Enzyme A protein that acts as a catalyst, speeding the rate at which a biochemical reaction proceeds but not altering the direction or nature of the reaction.

Escherichia coli Common bacterium that has been studied intensively by geneticists because of lack of pathogenicity, and ease of growth in the laboratory.

Eukaryote Organism whose cells have (1) chromosomes with nucleosomal structure and separated from the cytoplasm by a two-membrane nuclear envelope and (2) compartmentalization of a function in distinct cytoplasmic organelles.

Exogenous DNA DNA originating outside an organism.

Exons The protein-coding DNA sequences of a gene. Compare with introns.

Expression The ability of a gene to function within a cell in such a way that the gene product is formed.

Expression vector A cloning vector that contains the necessary regulatory sequences to allow transcription and translation of a cloned gene or genes.

Feedback inhibition Inhibition by an end product of the biosynthetic pathway involved in its synthesis.

Fermentation 1. Catabolic reactions producing ATP in which organic compounds serve as both primary electron donor and ultimate electron acceptor. 2. A large-scale microbial process.

Fusion protein The result of translation of two or more genes joined such that they retain their correct reading frames but make a single protein.

Gel An inert polymer, usually made of agarose or polyacrylamide, used for separating macromolecules such as nucleic acids or proteins by electrophoresis.

Gene The fundamental physical and functional unit of heredity. A gene is an ordered sequence of nucleotides located in a particular position on a particular chromosome that encodes a specific functional product (*i.e.*, a protein or RNA molecule). See gene expression.

Gene disruption Use of both *in vitro* and *in vivo* recombination to substitute an easily selected mutant gene for a wild-type gene.

Gene expression The process by which a gene's coded information is converted into the structures present and operating in the cell. Expressed genes include those that are transcribed into mRNA and then translated into protein and those that are transcribed into RNA but not translated into protein (*e.g.*, transfer and ribosomal RNAs).

Gene mapping Determination of the relative positions of genes on a DNA molecule (chromosome or plasmid) and the distance, in linkage units or physical units, between them.

Gene product The biochemical material, either RNA or protein, resulting from expression of a gene. The amount of gene product is used to measure how active a gene is; abnormal amounts can be correlated with disease-causing alleles.

Genetic code The sequence of nucleotides, coded in triplets (codons) along the mRNA, that determines the sequence of amino acids in protein synthesis. The DNA sequence of a gene can be used to predict the mRNA sequence, and the genetic code in turn can be used to predict the amino acid sequence.

Genetic engineering The use of *in vitro* techniques in the isolation, manipulation, recombination, and expression of DNA.

Genetic map The physical arrangement and order of genes on the chromosome.

Genomic library A collection of clones made from a set of randomly generated overlapping DNA fragments representing the entire genome of an organism.

Growth factor Organic compounds that must be supplied in the diet for growth because they are essential cell components or precursors of such components and cannot be synthesized by the organisms themselves.

Guanine (G) A nitrogenous base, one member of the base pair G-C (guanine and cytosine).

Haploid A single set of chromosomes (half the full set of genetic material) present in the egg and sperm cells of animals and in the egg and pollen cells of plants. Human beings have 23 chromosomes in their reproductive cells.

Homologies Similarities in DNA or protein sequences between individuals of the same species or among different species.

Human gene therapy Insertion of normal DNA directly into cells to correct a genetic defect.

Human Genome Initiative Collective name for several projects begun in 1986 by DOE to (1) create an ordered set of DNA segments from known chromosomal locations, (2) develop new computational methods for analyzing genetic map and DNA sequence data, and (3) develop new techniques and instruments for detecting and analyzing DNA. This DOE initiative is now known as the Human Genome Program. The national effort, led by DOE and NIH, is known as the Human Genome Project.

Hybridization The process of joining two complementary strands of DNA or one each of DNA and RNA to form a double-stranded molecule.

Immune response The specific reactions induced in the human or animal body due to the contact with foreign material. The foreign material that induces the immune response is called an immunogen or antigen. The immune response may involve either antibody production, the activation of T-cells, or both.

In situ hybridization Use of a DNA or RNA probe to detect the presence of the complementary DNA sequence in cloned bacterial or cultured eukaryotic cells.

In vitro   Literally means "in glass," away from a living organism; it is used to describe whatever happens in a test tube or other receptacle, as opposed to *in vivo*. When a study or an experiment is done outside the living organism, *i.e.*, in a test tube, it is said to be done *in vitro*.

In vivo   In the body, in a living organism, as opposed to *in vitro*. When a study or an experiment is done in the living organism, it is said to be done *in vivo*.

Informatics   The study of the application of computer and statistical techniques to the management of information. In genome projects, informatics includes the development of methods to search databases quickly, to analyze DNA sequence information, and to predict protein sequence and structure from DNA sequence data.

Initiation factors   The set of catalytic proteins required, in addition to mRNA and ribosomes, for protein synthesis to begin. In bacteria, three distinct proteins have been identified: IF-1 (8 kDa), IF-2 (75 kDa), and IF-3 (30 kDa). At least 6-8 proteins have been identified in eukaryotes. IF-1 and -2 enhance the binding of initiator tRNA to the initiation complex.

Introns   The DNA base sequences interrupting the protein-coding sequences of a gene; these sequences are transcribed into RNA but are cut out of the message before it is translated into protein.

Kilobase (kb)   Unit of length for DNA fragments equal to 1000 nucleotides.

Lactose repressor   Protein (tetramer of 37 kDa subunits) that normally binds with very high affinity to the operator region of the lactose operon and inhibits transcription of the downstream genes by blocking access of the polymerase to the promoter region. When the lactose repressor binds allolactose, its binding to the operator is reduced, and the gene set is derepressed.

Library   An unordered collection of clones (*i.e.*, cloned DNA from a particular organism), whose relationship to each other can be established by physical mapping.

Locus (pl. loci)   The position on a chromosome of a gene or other chromosome marker; also, the DNA at that position. The use of *locus* is sometimes restricted to mean regions of DNA that are expressed.

Marker   An identifiable physical location on a chromosome (*e.g.*, restriction enzyme cutting site, gene) whose inheritance can be monitored. Markers can be expressed regions of DNA (genes) or some segment of DNA with no known coding function but whose pattern of inheritance can be determined.

Messenger RNA (mRNA)   RNA species that contains the information to specify the amino acid sequence of proteins and that is translated on the ribosome. In eukaryotes mRNA normally is formed by splicing a large primary transcript. In eukaryotes the mRNA acquires a GTP cap and ususlly a polyA tail.

Metabolism   All biochemical reactions in a cell, both anabolic and catabolic.

Microorganism   A living organism too small to be seen with the naked eye.

Includes bacteria, fungi, protozoans, and microscopic algae; also includes viruses.

Mitochondrion (pl. mitochondria) Eukaryotic organelle responsible for processes of respiration and electron transport phosphorylation.

Mutation An inheritable change in the base sequence of the DNA of an organism. *nonsense mutation*: A mutation that changes a sense codon into one that does not code for an amino acid. *amber mutation*: The mutation due to the introduction of a stop codon (UAG) within the coding sequence of a gene that results in premature termination of translation.

Nicotinamide adenine dinucleotide (phosphate) [NAD(P)] An important coenzyme, functioning as a hydrogen carrier in a wide range of redox reactions; the H is carried on the nicotinamide residue. The oxidized form of the coenzyme is written $NAD(P)^+$, the reduced form as NAD(P)H. Many oxidoreductases are specific for either NAD^+ or $NADP^+$, although some can function with either. As a broad generalization, NADP is more commonly associated with biosynthetic reactions and NAD^+ with catabolic and energy-yielding reactions.

Nucleotide A subunit of DNA or RNA consisting of a nitrogenous base (adenine, guanine, thymine, or cytosine in DNA; adenine, guanine, uracil, or cytosine in RNA), a phosphate molecule, and a sugar molecule (deoxyribose in DNA and ribose in RNA). Thousands of nucleotides are linked to form a DNA or RNA molecule.

Nucleus The cellular organelle in eukaryotes that contains the genetic material.

Oligonucleotide A short nucleic acid molecule, either obtained from an organism or synthesized chemically.

Oncogene A gene, one or more forms of which is associated with cancer. Many oncogenes are involved, directly or indirectly, in controlling the rate of cell growth.

Open reading frame (ORF) The entire length of a DNA molecule that starts with a start codon and ends with a stop codon.

Operator Site of repressor binding on a DNA molecule; part of an operon.

Operon A controllable unit of transcription consisting of a number of structural genes transcribed together. Contains at least two distinct regions: the operator and the promoter. The first described example was the *lac* operon.

Oxidation-reduction (redox) reaction A coupled pair of reactions, in which one compound becomes oxidized while another becomes reduced and takes up the electrons released in the oxidation reaction.

Periplasmic space The area between the cytoplasmic membrane and the cell wall in Gram-negative bacteria, containing certain enzymes involved in nutrition.

Phage A virus for which the natural host is a bacterial cell.

Physical map A map of the locations of identifiable landmarks on DNA (*e.g.*, restriction enzyme cutting sites, genes), regardless of inheritance. Distance is measured in base pairs. For the human genome, the lowest resolution physical map is the banding patterns on the 24 different chromosomes; the highest resolution map would be the complete nucleotide sequence of the chromosomes.

Plasmid Autonomously replicating, extrachromosomal circular DNA molecules, distinct from the normal bacterial genome and nonessential for cell survival under nonselective conditions. Some plasmids are capable of integrating into the host genome. A number of artificially constructed plasmids are used as cloning vectors.

Polycistronic mRNA A single mRNA molecule that is the product of the transcription of several tandemly arranged genes; typically the mRNA transcribed from an operon.

Polymerase chain reaction (PCR) A method for amplifying a DNA base sequence *in vitro* using a heat-stable polymerase and two 20-base primers, one complementary to the (+)-strand at one end of the sequence to be amplified and the other complementary to the (−)-strand at the other end. Because the newly synthesized DNA strands subsequently can serve as additional templates for the same primer sequences, successive rouonds of primer annealing, strand elongation, and dissociation produce rapid and highly specific amplification of the desired sequence. PCR also can be used to detect the existence of the defined sequence in a DNA sample.

Polypeptide Several amino acids linked together by peptide bonds.

Primer Short preexisting polynucleotide chain to which new deoxyribonucleotides can be added by DNA polymerase.

Probe Single-stranded DNA or RNA molecules of specific base sequence, labeled either radioactively of immunologically, that are used to detect the complementary base sequence by hybridization.

Prokaryotes Organisms, namely, bacteria and cyanobacteria (formerly known as blue-green algae), characterized by the possession of a simple naked DNA chromosome, or occasionally two such chromosomes, usually of circular structure, wihtout a nuclear membrane and possessing a very small range of organelles (generally only a plasma membrane and ribosomes).

Promoter A region of DNA to which RNA polymerase binds before initiating the transcription of DNA into RNA. The nucleotide at which transcription starts is designated +1 and nucleotides are numbered from this with negative numbers indicating upstream nucleotides and positive indicating downstream nucleotides. Most bacterial promoters contain two consenuss sequences that seem to be essential for the binding of the polymerase. The first, the Pribnow box, is at about −10 and has the consensus sequence 5′-TATAAT-3′. The second, the -35 sequence, is centered about -35 and has the consensus sequence 5′-TTGACA-3′. Most factors that regulate gene transcription do so by binding at or near the promoter and affecting the initiation of transcription. Much less is known about eukaryotic promoters; each of the three

RNA polymerases has a different promoter. RNA polymerase I recognizes a single promoter for the precursor of rRNA. RNA polymerase II, which transcribes all genes coding for polypeptides, recognizes many thousands of promoters. Most have the Goldberg-Hogness or TATA box that is centered around position -25 and has the consensus sequence 5'-TATAAAA-3'. Several promoters have a CAAT box around -90 with the consensus sequence 5'-GGC-CAATCT-3'. There is increasing evidence that all promoters for "housekeeping" genes contain multiple copies of a GC-rich element that includes the sequence 5'-GGGCGG-3'. Transcription by polymerase II also is affected by more distant elements known as enhancers. RNA polymerase III synthesizes 5S ribosomal RNA, all tRNAs, and a number of small RNAs. The promoter for RNA polymerase III is located within the gene either as a single sequence, as in the 5S RNA gene, or as two blocks, as in all tRNA genes.

Protein   A large molecule composed of one or more chains of amino acids in a specific order; which is determined by the base sequence of nucleotides in the gene coding for the protein. Proteins are required for the structure, function, and regulation of the body's cells, tissues, and organs, and each protein has unique functions. Examples are hormones, enzymes, and antibodies.

Proton-motive force (PMF)   An energized state of a membrane created by expulsion of protons through the action of an electron tranport chain. See also chemiosmosis.

Purine   A nitrogen-containing, single-ring, basic compound that occurs in nucleic acids. The purines in DNA and RNA are adenine and guanine.

Pyrimidine   A nitrogen-containing, double-ring, basic compound that occurs in nucleic acids. The pyrimidines in DNA are cytosine and thymine; in RNA, cytosine and uracil.

Recombinant clones   Clones containing recombinant DNA molecules. See recombinant DNA technologies.

Recombinant DNA molecules   A combination of DNA molecules of different origins that are joined using recombinant DNA technologies.

Recombinant DNA technologies   Procedures used to join together DNA segments in a cell-free system (an environment outside a cell or organism). Under appropriate conditions, a recombinant DNA molecule can enter a cell and replicate there, either autonomously or after it has become integrated into a cellular chromosome.

Recombination   The process by which progeny derive a combination of genes different from those of either parent. In higher organisms, this can occur by crossing over.

Regulation   Processes that control the rates of synthesis of proteins. Induction and repression are examples of regulation.

Regulatory regions or sequences   A DNA base sequence that controls gene expression.

Regulon   A situation in which two or more spatially-separated genes are regulated in a coordinated fashion by a common regulator molecule.

Repressor protein A protein that binds to an operator of a gene preventing the transcription of the gene. The binding affinity of repressors for the operator may be affected by other molecules. Inducers bind to repressors and decrease their binding to the operator, while corepressors increase the binding. The paradigm of repressor proteins is the lactose repressor protein, which acts on the *lac* operon and for which the inducers are β-galactosides such as lactose; it is a polypeptide of 360 amino acids that is active as a tetramer.

Restriction enzyme cutting site A specific nucleotide sequence of DNA at which a particular restriction enzyme cuts the DNA. Some sites occur frequently in DNA (*e.g.*, every several hundred base pairs), others much less frequently (rare cutter; *e.g.*, every 10,000 base pairs).

Restriction enzyme, endonuclease A protein that recognizes specific, short nucleotide sequences and cuts DNA at those sites. Bacteria contain over 400 such enzymes that recognize and cut over 100 different DNA sequences. See restriction enzyme cutting site.

Ribonucleic acid (RNA) A chemical found in the nucleus and cytoplasm of cells; it plays an important role in protein synthesis and other chemical activities of the cell. The structure of RNA is similar to that of DNA. There are several classes of RNA molecules, including messenger RNA, transfer RNA, ribosomal RNA, and other small RNAs, each serving a different purpose.

Ribosomal RNA (rRNA) A class of RNA found in the ribosomes of cells.

Ribosome A heterodimeric multisubunit enzyme composed of ribonucleoprotein and protein subunits. Interacts with mRNAs and aminoacylated tRNAs and translates protein coding sequences from messenger RNA. Similar ribosomes found in all living organisms, are all composed of large and small subunits, as well as chloroplasts and mitochondria. Differences are apparent between prokaryotic and eukaryotic ribosomes.

Ribosomes Small cellular components composed of specialized ribosomal RNA and protein; site of protein synthesis. See ribonucleic acid (RNA).

Shine-Dalgarno sequence A short stretch of nucleotides on a prokaryotic mRNA molecule upstream of the translational start site that serves to bind to ribosomal RNA and thereby bring the ribosome to the initiation codon on the mRNA.

Sigma factor Initiation factor (86 kDa) that binds to *E. coli* DNA dependent RNA polymerase and promotes attachment to specific initiation sites on DNA. Following attachment, the sigma factor is released.

Signal sequence (signal peptide) A short stretch of amino acids found at the beginning of proteins, typically rich in hydrophobic amino acids, which helps transport the entire polypeptide through the membrane.

SOS system The DNA repair system, also known as error-prone repair in which purinic DNA molecules are repaired by incorporation of a base that may be the wrong base but that permits replication. *RecA* protein is required for this type of repair. SOS genes function in control of the cell cycle in prokaryotes and eukaryotes.

Southern blotting Transfer by absorption of DNA fragments separated in electrophoretic gels to membrane filters for detection of specific base sequences by radiolabeled complementary probes.

Spore A general term for resistant resting structures formed by many prokaryotes and fungi.

Substrate-level phosphorylation Synthesis of high-energy phosphate bonds through reaction of inorganic phosphate with an activated (usually) organic substrate.

Termination 1. Stop of mRNA synthesis (*i.e.*, transcription) at the terminator site. 2. Stop of protein synthesis (*i.e.*, translation) at the stop codon.

Thymine (T) A nitrogenous base, one member of the base pair A-T (adenine-thymine).

Transcription Synthesis of RNA by RNA polymerases using a DNA template.

Transcriptional control Control of gene expression by controlling the number of RNA transcripts of a region of DNA. A major regulatory mechanism for differential control of protein synthesis in both prokaryotic and eukaryotic cells.

Transfer RNA (tRNA) A class of RNA having structures with triplet nucleotide sequences that are complementary to the triplet nucleotide coding sequences of mRNA. The role of tRNAs in protein synthesis is to bond with amino acids and transfer them to the ribosomes, where proteins are assembled according to the genetic code carried by mRNA.

Transformation A process by which the genetic material carried by an individual cell is altered by the incorporation of exogenous DNA into its genome.

Translation The process that occurs at the ribosome whereby the information in mRNA is used to specify the sequence of amino acids in a polypeptide chain.

Translational control Control of protein synthesis by regulation of the translation step, for example, by selective usage of preformed mRNA or instability of the mRNA.

Uracil (U) A nitrogenous base normally found in RNA but not DNA, uracil is capable of forming a base pair with adenine.

Virus A noncellular biological entity that can reproduce only within a host cell. Viruses consist of nucleic acid covered by protein; some animal viruses also are surrounded by membrane. Inside the infected cell, the virus uses the synthetic capability of the host to produce progeny virus.

INDEX